# MICROBIOLOGY

## Laboratory Theory & Application

BRIEF

THIRD EDITION

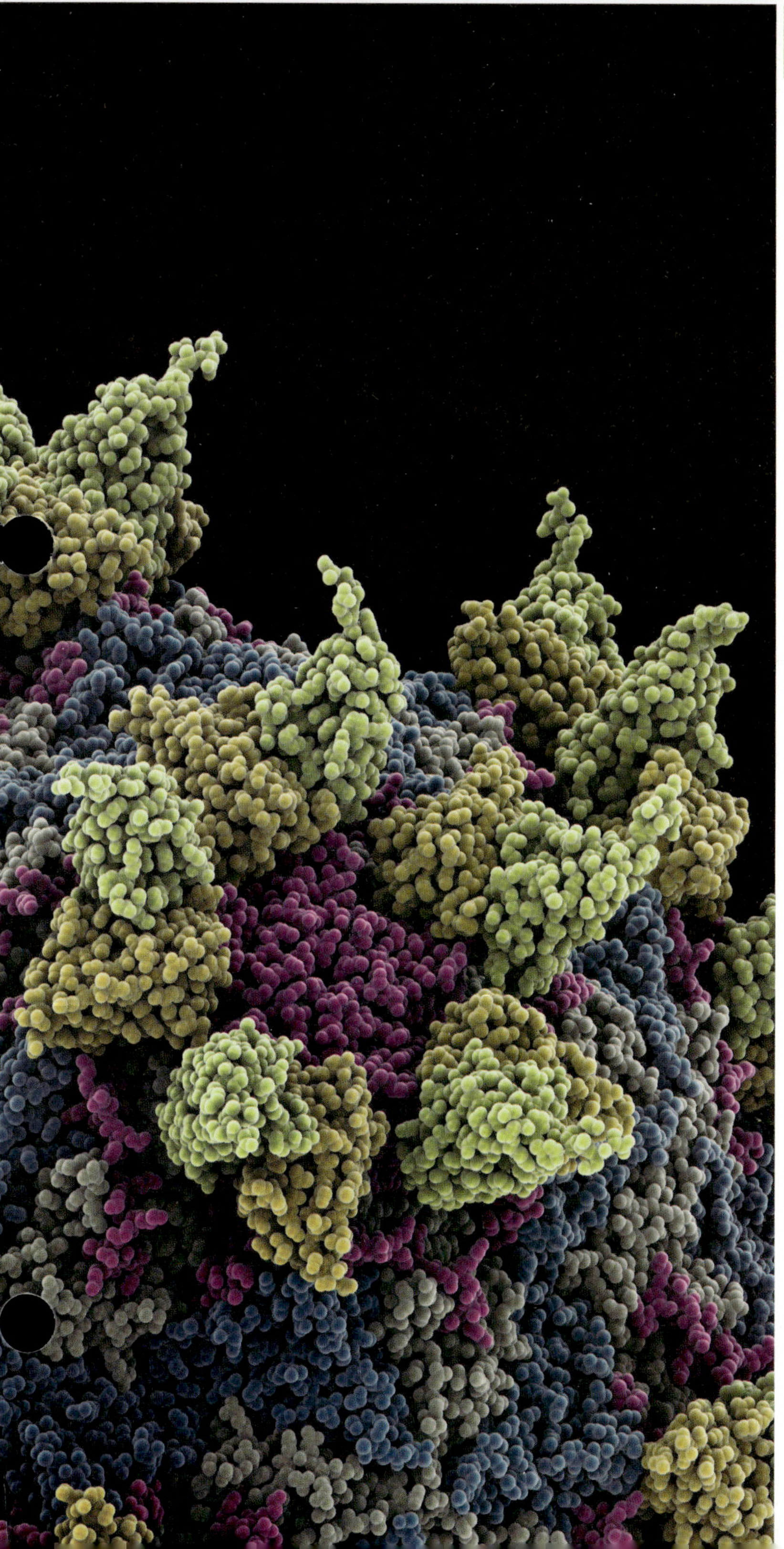

**Michael J. Leboffe**
San Diego City College

**Burton E. Pierce**

925 W. Kenyon Avenue, Unit 12
Englewood, CO 80110

www.morton-pub.com

## Book Team

| | |
|---|---|
| **Publisher** | Douglas N. Morton |
| **President** | David M. Ferguson |
| **Acquisitions Editor** | Marta R. Martins |
| **Project Editor** | Rayna S. Bailey |
| **Production Manager** | Joanne Saliger |
| **Production Assistant** | Will Kelley |
| **Illustrations** | Imagineering Media Services, Inc. |

*Cover image:*

Rhinovirus. Molecular model of the antigen-binding fragment (Fab) from a strongly neutralizing antibody bound to a human rhinovirus 14 (HRV-14) particle. This virus consists of a protein capsid enclosing an RNA (ribonucleic acid) genetic code (genome). The rhinovirus infects the upper respiratory tract and is the cause of the common cold. It is spread by coughs and sneezes.

Copyright © 2008, 2012, 2016 by Morton Publishing Company

All rights reserved. No part of this publication may be reproduced, stored in a retrieval system, or transmitted, in any form or by any means, electronic, mechanical, photocopying, recording, or otherwise, without the prior written permission of the publisher.

Printed in the United States of America

10 9 8 7 6 5 4

ISBN-10: 1-61731-477-3
ISBN-13: 978-1-61731-477-3

Library of Congress Control Number: 2015919196

# PREFACE

Back in college I remember balking at writing a term paper, or even an essay. I never dreamed I'd be writing the preface to a 600+ page, college-level, third edition microbiology laboratory manual. I imagine Dr. Kelly, my General Microbiology professor at San Diego State University, would have felt the same way had she given it any thought. I also figured that hers would be the last microbiology course I'd ever take or need. (I can state with pride, however, I *did* earn a "B.") Silly me, making plans.

Nevertheless, odds and odd career paths be damned: here I am and here it is. America, Land of Opportunity, indeed!

It must be said that "Here *I* am" is only a half-truth. My longtime friend and co-author, Burt Pierce, retired from writing in 2012. As a consequence, the books published since then have become my sole responsibility. But please notice in the descriptions of revisions, deletions, and additions that follow, I use the first person plural pronoun "we" because Burt's influence is still all over this book. Even though the discussions regarding what changes should be made were one-sided in my head, his commitment to quality and not underestimating our audience's abilities continued to influence me. *Microbiology Laboratory Theory & Application*, *Brief*, 3rd edition (hereafter referred to as *MLTAB* 3e), is a spinoff from the larger, more comprehensive *Microbiology Laboratory Theory & Application*, 4th edition (hereafter referred to as *MLTA* 4e).

Whereas *MLTA* 4e targets college-level microbiology courses for science majors, *MLTAB* 3e is designed for use in microbiology courses with more of a medical emphasis and whose students are pursuing an allied health or other health-related career path. Many of the exercises in biotechnology, food and environmental microbiology, and those with a heavy math emphasis have been omitted, but the rigor of those remaining is comparable.

## Global Changes

### Lab Safety

- From the first edition, we have emphasized laboratory safety and we continue to do so in this edition. Students are reminded to wear a lab coat, eye protection, and gloves in every exercise (when appropriate). They are also reminded to properly dispose of materials (often by telling them to follow the rules of their particular laboratory).
- We continue to emphasize techniques that minimize aerosol production during transfers.
- As in previous editions, BSL-2 organisms are identified in the Materials section of each exercise. Not all strains of a species are BSL-2 and we don't know what strains each college will be using, so we erred on the side of caution in identifying these. We checked ATCC strains and if any were identified as BSL-2 we listed them that way. The instructor may choose to announce that the strain(s) used in their lab are not BSL-2.
- We have continued to reduce the number of BSL-2 organisms used. Where possible, they have been replaced with suitable substitutes that give the same result. In other cases, they have been listed as "optional." In still others, where no replacements are available, they remain. Please pay attention to the organisms you are working with and use appropriate

caution. As always, it is ultimately up to the instructor to use his or her professional judgment and choose organisms that are suitable for the course level and lab facilities of their institution and to each student to follow standard safety guidelines.

- Because some professors may decide to use BSL-2 organisms, we have added a section in the Introduction with recommended BSL-2 procedures.

### Pedagogy: Theory, Application, and Instructions

Virtually all lab exercises were rewritten to a greater or lesser degree in an attempt to bring improved clarity to the theory, application, and instructions for each.

### Pedagogy: Photos, Micrographs, Artwork, and Tables

One feature of *MLTA* that has pleased adopters of previous editions has been the visual content—photos, micrographs, and art. Upholding our commitment to quality, more than 200 of these elements have been revised or replaced in this edition. Additionally, there are more than 60 brand new photographs, micrographs, and pieces of art.

### Pedagogy: Data Sheet Questions

From the first edition of *MLTA* in 2002, the data sheet questions have ranged from simple recall to explanation of procedures, interpretation of data, or extrapolation from what has been covered in the exercise. In this edition, they were examined carefully and in many exercises they have been re-sequenced, reworded, split into parts, or replaced. Some new questions have also been added.

Because there are more exercises in *MLTAB* 3e than can be done in any one-semester course, there is also repetition of certain questions from exercise to exercise. This was done because they address particularly important points and we wanted to ensure that students have to answer them regardless of which exercises are chosen for their course.

## Specific and Noteworthy Changes

Following are some highlights of specific topics added or revised in each section.

- **Introduction—Safety and Laboratory Guidelines** A section devoted exclusively to handling BSL-2 organisms has been added, as was a sample "safety contract" to be signed by the student agreeing to comply with safety regulations outlined in this book and as amended by their institution. The topic of controls was also revised.
- **Section 1—Fundamental Skills for the Microbiology Laboratory** In Exercise 1-5 (Streak Plate Methods of Isolation), the T-streak inoculation has been added as an option to the quadrant streak. As a safety precaution, Exercise 1-6 (Spread Plate Method of Isolation) now advises the use of a screw-cap jar to hold the alcohol for flaming the glass rod to minimize the chances of it catching fire.
- **Section 2—Microbial Growth** While the basic exercise protocols are unchanged, each has been rewritten for greater clarity and integration. The biggest change has been in Exercise 2-2 (Colony Morphology). The artwork illustrating colony features was revised and 15 new photos were added, bringing the total up to 44. Some of the new photos were shot through a stereo microscope and illustrate detail (and beauty!) not visible to the naked eye. Lastly, photos are now arranged by the colony features illustrated, such as margin, overall shape, color, etc., rather than randomly as in previous editions.

- **Section 3—Microscopy and Staining** The instructions on general microscope use in Exercise 3-1 (Introduction to the Light Microscope) were enhanced. Exercise 3-3 (Microscopic Examination of Eukaryotic Microbes) was totally reorganized to reflect a more current taxonomy of eukaryotes, and some new photos replaced older ones. Minor revisions were made to the bacterial stain exercises for clarity and safety. New artwork has been added to Exercise 3-6 (Gram Stain) illustrating the mechanism underlying the decolorization step.
- **Section 4—Selective Media** All photos in this section were replaced with better ones. A confusing point for some students in previous editions had been that the photos don't illustrate spot inoculations—and they still don't. The reason we made this choice is that spot inoculations don't illustrate the colors or growth patterns as clearly in photographs as short streaks do.
- **Section 5—Differential Tests** This section was the recipient of many replacement photos and new artwork. It also was treated to sequence changes, with related tests brought together. Additionally, test interpretation tables now address the possibility that a negative result is a false negative.
- **Section 6—Quantitative Techniques** Exercise 6-2 (Standard Plate Count) discusses other methods of counting colonies and explains the rationale for using "Original Sample Volume" rather than "Plate Dilution Factor" in calculations.
- **Section 7—Medical Microbiology** The Theory section of Exercise 7-1 (Snyder Test) was rewritten, and a Gram stain of a tooth scraping replaces a Gram stain of the gumline. Two optional antibiotics have been added to the materials list for the Kirby-Bauer test (Exercise 7-2), and the Theory was rewritten with more detail about the test and its history. There are also optional instructions for using a spectrophotometer in place of a MacFarland standard. Exercise 7-3 (Morbidity and Mortality Weekly Report [MMWR] Assignment) was updated to reflect changes in the CDC website and changes to the list of notifiable diseases. Instructions for Exercise 7-4 (Epidemic Simulation) were rewritten to emphasize the safe execution of the lab. Exercise 7-6 (Multiple Tube Fermentation Method for Total Coliform Determination) was rewritten.
- **Section 8—Microbial Genetics and Serology** Photos of the jellyfish from which the green fluorescent protein gene was obtained are included in Exercise 8-2 (Bacterial Transformation: The pGLO System). An extensive section introducing antigens and antibodies was written and precedes the serology exercises (Exercises 8-4 through 8-6). In addition, Exercise 8-5 (Blood Typing) now explains more thoroughly the genetic and molecular basis for blood types in the ABO system.
- **Section 9—Identification of Unknowns** Most BSL-2 organisms have been removed from the flowcharts in Exercises 9-1 through 9-3. In addition, results were verified and in some cases the flowcharts were modified. Exercise 9-4 (api® 20 E Identification System for *Enterobacteriaceae* and other Gram-Negative Rods) was updated based on the most recent instructions from bioMérieux, Inc., and screen shots of the analytical profile index from api*web* replace the images of its printed counterpart. In Exercise 9-5 (EnteroPluri-*Test*), the Enterotube has been retired and replaced with the EnteroPluri-*Test* system, which is comparable but slightly different than its predecessor. The Streptex Rapid Agglutination test has been removed because of cost and feedback from adopters (but still can be custom published, if desired).

- **Appendices** Artwork in Appendix A (Biochemical Pathways) has been modified greatly. Intersecting pathways have been removed from Figures A.2 (Glycolysis) and A.5 (Entry Step and Citric Acid Cycle), but a new diagram (Fig. A.1, Integrated Metabolism) shows integrated metabolism and references relevant lab exercises for the various pathways. It is designed to show how the biochemical tests (mostly in Section 5) fit into the bigger picture of metabolism. Figure A.6 (Sampling of Fermentation Pathways) has been redrawn and color-coded to indicate specific fermentations. Appendices B, C, and D (transfer methods) have been rewritten to include BSL-2 precautions, and the majority of photos in Appendices B and D have been replaced.

As you proceed through your microbiology lab, please step back and take a moment to marvel at how amazing the microbes you are studying really are and to cultivate an appreciation of them. Remember: you are outnumbered!

All the best,

Mike<br>
La Mesa, CA<br>
December 2015

## Acknowledgments

A favorite saying of mine that I picked up from some unknown or long-forgotten source is this: "Many hands make light work." I would like to consider it to be a universal truth, except that another saying, "Too many cooks spoil the broth" also has merit. So, granting that the world is not black and white, but full of nuance and shades of gray, it is still a favorite saying of mine and I try to apply it whenever I can.

When Burt (now retired) and I started writing for Morton Publishing in 1995, we did all the writing, photography, artwork, permissions, most of the proofreading, and a bunch of other tasks required to produce quality, college-level publications. In the last 20 years, Morton Publishing has grown, not only in the number and variety of books they publish, but in their number of employees. The addition of these many hands has made light(er) work for us, and we are very grateful for all they do and have done.

I must start with Doug Morton, who founded Morton Publishing in 1977 with the vision of producing high-quality textbooks at a reasonable price, a business model that has been wildly successful for his company and is still its guiding philosophy. Burt and I are grateful to Doug for seeing potential in our work and giving us a chance and a vehicle with which to present it to the greater college community. Thanks, Doug. You changed our lives.

I am extremely grateful for the support, encouragement, patience, and friendship of President David Ferguson, Vice President of Operations Chrissy DeMier, and Vice President of Sales and Marketing Carter Fenton. These people have been charged with the task of navigating Morton Publishing through the challenging publishing landscape of the early 21st century, and I feel very secure with them at the helm.

Thanks to Marta Martins, Senior Acquisitions Editor, who administered peer reviews of the previous edition and collated the responses into a manageable form that I could respond to easily. She also offered helpful advice/opinions at times when I was wrestling with alternative solutions to problems…er, "challenges." Honestly, I still have problems, but admitting that just dates me, so I'll be trendy and go with "challenges."

Special thanks go to Rayna Bailey, Project Editor. Rayna edited the manuscript, obtained permissions, coordinated the art program, and communicated with virtually everyone associated with this book so I could concentrate on writing. She has also meticulously gone over page proofs several times (which, to me, is akin to having one's fingernails pulled out with pliers) to ensure that the finished product is the absolute best it can be. Her positive attitude and gentle prodding kept me on task and moving forward. This is the fourth book Rayna and I have worked on since she started at Morton five years ago. If she had been working on our books from the beginning when Burt and I were performing the tasks she now performs, her contributions would have merited co-authorship!

The production team at Morton has once again done a masterful job in designing this book with their customary artistic flair. Art meets science at Morton Publishing! Thanks to Joanne Saliger, Production Manager, and Will Kelley, Production Assistant, for applying their talents to our book. Will not only did the layout and design, he also modified artwork and sometimes created it from scratch, a difficult task considering he was working with technical images represented by my freehand artistic style of scribbles, cross-outs, and arrows! Well done, Will!

Thanks to Scott Day, Sales Manager, and all the sales representatives who meet potential adopters and present my books in the best light possible. Without them, my garage would be full of unsold books! Their job is a difficult one and I appreciate their efforts more than they can know.

As I get older, I have found I'm more productive if I remove myself from my daily distractions, so I spent quite a bit of time working on this edition in the Morton offices. I extend my heartfelt gratitude to everyone who works there for embracing me as one of their own during my temporary occupancy of six different office spaces over the last year. Visit **www.morton-pub.com** and meet all of these people!

Other non-Morton contributors that rounded out the book team are Imagineeringart Inc., of Toronto, ON, who rendered much of the beautiful artwork; Carolyn Acheson, who compiled the index; and Trina Lambert, who applied a fresh pair of proofreader's eyes to the final product. Thanks to all of you.

Thanks are also due to my colleagues at San Diego City College—past and present. Current full-time faculty Jake Brashears, Jennifer Chambers, Roya Lahijani, Erin Rempala, Dave Singer, and Gary Wisehart tolerated my pinball-like presence in the hallways as I raced from classes, office hours, and meetings between writing obligations. "Hello!" became an in-depth conversation over the last year. So now, let's talk sometime. Soon.

City College microbiology instructors Tom Kaido, Sabine Kurz-Camcho, Martha Myers, and Brett Pickett used previous editions of this book in their classes, offered suggestions, and provided me with a sounding board for new ideas, as did my San Diego State University teaching interns Heather Heinz and Polly Parks. Dean (also friend and former office partner) of the School of Engineering & Technologies, Mathematics, Sciences, and Nursing, Minou Spradley, oversaw the process of my renting San Diego City College lab facilities. (These things go so smoothly when I let someone else handle them!)

Last but not least, biology lab technicians Ryan McWey, Deb Reed, Laura Steininger, and Muu Vu were always ready to cheerfully assist me in finding materials I needed that I should have learned the location of years ago. Thank you, one and all.

I engaged in many conversations about safety in the microbiology teaching laboratory with colleagues Marlene DeMers and Tom Gibson, both formerly of San Diego State University. They had a profound influence on my presentation of that topic in this edition. Thanks, you two. Enjoy your retirements!

I also want to recognize the contributions of hand models Kadija Amba, Diana Carrillo, Heather Heinz, Anita Hettena, Alicia Leboffe, Nathan Leboffe, Deb Reed, Burt Pierce, Carla Sweet, Rick Tenorio, and Gary Wisehart. I know which hands are yours and I get a warm, fuzzy feeling remembering each photograph and knowing which hands belong to whom even if the majority of readers probably don't notice the differences. Oh, and just to show that I'm keeping up with the times, there are also a couple of "selfies" in the book. See if you can find them…

Thanks to the reviewers of *Microbiology Laboratory Theory and Application,* 3e, who provided me with comprehensive feedback and many useful suggestions for revising that book into a fourth edition, from which much of this book is derived. They are: Patricia Clinard Alfing, Davidson County Community College; Richard Adler, University of Michigan–Dearborn; Amy Warenda Czura, Suffolk County Community College–Eastern Campus; Timothy Ladd, Millersville University; and Jeanette M. Loutsch, University of Science and Arts of Oklahoma.

Probably most of all, Burt Pierce, my retired former co-author, deserves recognition because without his contributions there wouldn't have been a first edition of any of our books, much less third and fourth editions. His contributions and influence are still vital parts of this book. Our creative strengths overlapped enough that we were able to see a common vision for our books, but more importantly our individual strengths complemented the other's, making our "whole" greater than the sum of our parts.

To all the others: There have been so many who have contributed in various ways over four editions of *Microbiology Laboratory Theory and Application (MLTA)*, two editions of *MLTA Brief*, four editions of the *Photographic Atlas for the Microbiology Laboratory*, and four editions of *Exercises for the Microbiology Laboratory* that it has become impractical to identify everyone individually, but your collective efforts have not been forgotten and continue to be appreciated. One big, hearty "Thanks" to you all.

Many hands…indeed.

## Dedication

There's no "I" in team, and the Leboffe and Pierce microbiology atlases and lab manuals published by Morton since 1995 have always been the product of teamwork, beginning with two members and expanding upward to its current level of a dozen or so. Although he is retired and no longer an active team member (instead working on his tan in Oregon), I dedicate this book to my longtime friend, San Diego City College colleague, and co-author Burton Pierce. Burt, your work ethic, attention to detail, dedication to doing things "the right way," and sense of humor continue to influence me. There may be no "I" in "team," but there is a "T" in Burt. Thank you for all you've done.

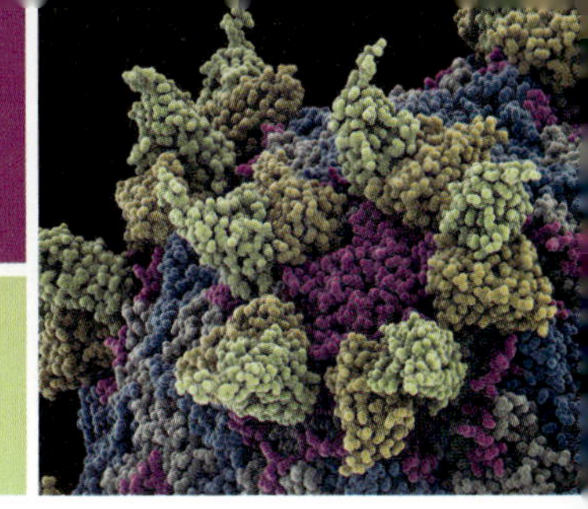

# CONTENTS

# Safety and Laboratory Guidelines

We hope that you find microbiology lab to be an interesting and exciting experience, but at the outset you must be made aware of some potential hazards. Improper handling of chemicals, equipment, and/or microbial cultures is dangerous and can result in injury or infection. Safety with lab equipment will be addressed when you first use that specific piece of equipment, as will specific examples of chemical safety. Our main concern here is to introduce you to safe handling and disposal of microbes.[1]

Because microorganisms present varying degrees of risk to laboratory personnel (oneself, other students, technicians, and faculty), people outside the laboratory, and the environment, microbial cultures must be handled safely. Classifying microbes into four biosafety levels (BSLs) provides a set of minimum standards for laboratory practices, facilities, and equipment to be used when handling organisms at each level. These biosafety levels, defined in the U.S. government publication, *Biosafety in Microbiological and Biomedical Laboratories,* 5th edition (2009), are summarized below and in Table I-1. For complete information, readers are referred to the original document.

**BSL-1:** Organisms do not typically cause disease in healthy individuals and present a minimal threat to the environment and lab personnel. Standard microbiological practices are adequate. These microbes may be handled in the open, and no special containment equipment is required. Examples include *Bacillus subtilis*, *Escherichia coli* (most strains), *Rhodospirillum rubrum*, and *Lactobacillus acidophilus*.

**BSL-2:** Organisms are commonly encountered in the community and present a moderate environmental and/or health hazard. These organisms are associated with a variety of human diseases, most of which can be successfully treated if identified in a timely manner. The infection routes of primary concern are ingestion, inhalation, or penetration of the skin (percutaneous). Individuals performing work prone to splashes or aerosol generation (even though these organisms are not generally known to be transmitted by aerosols) should work in a biological safety cabinet (BSC, Fig. I.1). Otherwise, laboratory work may be done using standard microbiological practices. Examples include *Salmonella*, *Staphylococcus aureus*, *Clostridium dificile*, and *Borrelia burgdorferi*.

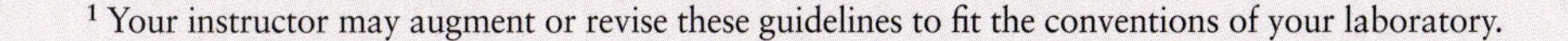

[1] Your instructor may augment or revise these guidelines to fit the conventions of your laboratory.

## Biosafety Cabinets

Biosafety cabinets are classified into three categories, all of which draw air into the cabinet to minimize microbial contamination back into the room.

**Class I BSCs** resemble chemical fume hoods with a protective glass in the front, but have a HEPA filter along the exhaust path to prevent microbes from entering the environment.

**Class II BSCs** use laminar airflow, which minimizes turbulence, and HEPA filters to protect the user and materials within the cabinet from contamination. The exhaust passes through a HEPA filter before its release into the environment.

**Class III BSCs** are completely sealed and gas tight with a fixed viewing window. Incoming air passes through a HEPA filter, whereas materials to be handled are passed through a double-door system where the intermediate compartment is an autoclave. (Other methods may also be used, but accomplish the same purpose.) Exhaust air passes through two HEPA filters or a HEPA filter and an air incinerator. Materials are handled with arm-length, heavy-duty gloves built into the wall of the cabinet. ■

**I.1 Biological Safety Cabinet (BSC) in a Teaching Laboratory** ■ In this Class II BSC, air is drawn in from the room and is passed through a HEPA filter prior to release into the environment. This airflow pattern is designed to keep aerosolized microbes from escaping from the cabinet. The microbiologist is pipetting a culture. When the BSC is not in use at the end of the day, an ultraviolet light is turned on to sterilize the air and the work surface.

(San Diego County Public Health Laboratory)

**BSL-3:** Organisms are of local or exotic origin and are associated with respiratory transmission and serious or lethal diseases where treatment and/or vaccines may or may not be available. Special ventilation systems are used to prevent aerosol transmission out of the laboratory, and access to the lab is restricted. Specially trained personnel handle microbes in a Class II or III BSC, not on the open bench. Examples include *Bacillus anthracis*, *Mycobacterium tuberculosis*, and West Nile virus.

**BSL-4:** Organisms have a great potential for lethal infection. Inhalation of infectious aerosols, exposure to infectious droplets, and autoinoculation are of primary concern. The lab is isolated from other facilities, and access is strictly controlled. Ventilation and waste management are under rigid control to prevent release of the microbial agents to the environment. Specially trained personnel perform transfers in Class III BSCs. Class II BSCs may be used as long as personnel wear positive pressure, one-piece body suits with a life-support system. Examples include agents causing hemorrhagic diseases, such as Ebola, Marburg, and Lassa fever viruses.

The microorganisms used in introductory microbiology courses depend on the institution, objectives of the course, and student preparation. Most introductory courses use organisms that may be handled at BSL-1 and BSL-2 levels so we have followed that practice in designing this set of exercises.

Following are general safety rules to reduce the chance of injury or infection to you and to others, both inside and outside the laboratory. Although both BSL-1 and BSL-2 guidelines are listed, we believe it is best to err on the side of caution and that students should learn and practice the safest level of standards (relative to the organisms they are likely to encounter) at all times. Please follow these and any other safety guidelines required by your college.

Chemical safety is also important in a microbiology laboratory. Be aware of the hazards presented by the chemicals you are handling. Containers should be labeled with a standard set of precautions as seen in Figure I.2. Numbers are assigned to the degree of health, fire, and reactivity hazard posed by the chemical. On stickers placed by the laboratory, there also is a space to enter specific hazards, such as acid, corrosive, and radioactivity.

TABLE **I-1** **Summary of Recommended Biosafety Levels for Infectious Agents**

| BSL | Agents | Practices | Safety Equipment (Primary Barriers) | Facilities (Secondary Barriers) |
|---|---|---|---|---|
| 1 | ■ Not known to consistently cause disease in healthy adults<br>■ e.g., *Lactobacillus casei*, *Bacillus subtilis*, *Rhizobium leguminosarum* | ■ Standard microbiological practices | ■ Primary barriers: none required<br>■ Personal Protective Equipment (PPE): laboratory coats and gloves; eye, face protection as needed | ■ Laboratory bench practices and sink required |
| 2 | ■ Agents associated with human disease; treatments and/or vaccines are usually available<br>■ Routes of transmission include percutaneous injury, ingestion, and mucous membrane exposure<br>■ e.g., *Staphylococcus aureus*, *Salmonella enterica*, measles virus | **BSL-1 practices plus:**<br>■ Limited lab access<br>■ Biohazard warning signs<br>■ "Sharps" precautions<br>■ Biosafety manual defining any needed waste decontamination | ■ Primary barriers: Class I or II BSCs or other physical containment devices used for all manipulations of agents that cause splashes or aerosols of infectious materials<br>■ PPEs: laboratory coats, gloves, face protection as needed | **BSL-1 plus:**<br>■ Autoclave available |
| 3 | ■ Indigenous or exotic agents with potential for aerosol transmission<br>■ Disease may have serious or lethal consequences; treatments and/or vaccines may be available<br>■ e.g., *Yersinia pestis*, *Mycobacterium tuberculosis*, rabies virus | **BSL-2 practices plus:**<br>■ Controlled lab access<br>■ Decontamination of all waste<br>■ Decontamination of all lab clothing before laundering<br>■ Baseline serum | ■ Primary barriers: Class I or II BSCs or other physical containment devices used for all manipulations of agents<br>■ PPEs: protective lab clothing, gloves, respiratory protection as needed | **BSL-2 plus:**<br>■ Physical separation from access corridors<br>■ Access to self-closing double door<br>■ Exhausted air not recirculated<br>■ Negative airflow into laboratory |
| 4 | ■ Dangerous/exotic agents that pose high risk of life-threatening disease, aerosol-transmitted lab infections, or related agents with unknown risk of transmission; treatment and vaccines are unavailable<br>■ e.g., Ebola virus, Lassa virus | **BSL-3 practices plus:**<br>■ Clothing change before entering<br>■ Shower on exit<br>■ All material decontaminated on exit from facility | ■ Primary barriers: all procedures conducted in Class III BSCs or Class I or II BSCs *in combination with* full-body, air-supplied, positive pressure personnel suit | **BSL-3 plus:**<br>■ Separate building or isolated zone<br>■ Dedicated supply and exhaust vacuum, and decon system<br>■ Other requirements outlined in the text |

**Source:** Adapted from *Biosafety in Microbiological and Biomedical Laboratories*, 5th edition (Washington, DC: U.S. Government Printing Office, 2007).

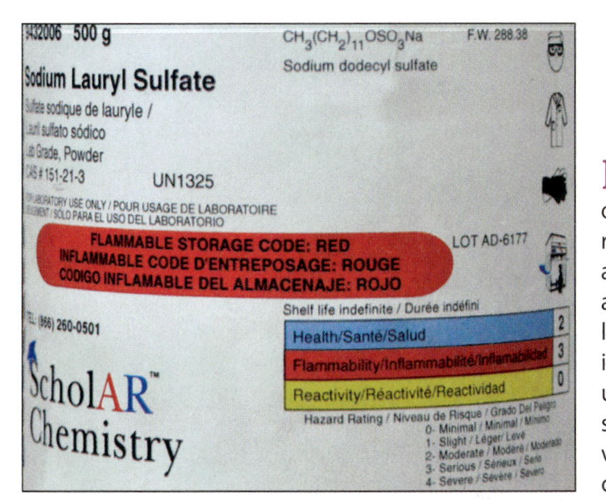


**I.2** **Chemical Hazard Label** ■ The blue, red, and yellow boxes on this label indicate the health, flammability, and reactivity levels, respectively, of the compound sodium dodecyl sulfate (SDS). The colors are standardized (regardless of the chemical) and each is assigned a number between "0" (minimal) and "4" (severe) as shown in the legend at the lower right. SDS presents a moderate health hazard, is very flammable, but minimally reactive. Also notice the icons in the upper right, which indicate (from top to bottom) that this compound should be handled with eye protection, a lab coat, gloves, and adequate ventilation. Be aware of these warnings and comply with the associated cautions for all chemicals you handle in lab.

## Student Conduct

- To reduce the risk of infection, do not smoke, eat, drink, or bring food or drinks into the laboratory room—even if lab work is not being done at the time.
- Do not apply cosmetics or handle contact lenses in the laboratory.
- Wash your hands *thoroughly* with soap and water after handling living microbes and before leaving the laboratory each day. Also, wash your hands after removing gloves.
- Do not remove any organisms or chemicals from the laboratory.
- Lab time is precious, so come to lab prepared for that day's work. Figuring out what to do as you go is likely to produce confusion and accidents.
- Work carefully and methodically. Do not hurry through any laboratory procedure.

## Basic Laboratory Safety

- Wear protective clothing (i.e., a lab coat) in the laboratory when handling microbes. Remove the coat prior to leaving the lab and autoclave it regularly (Fig. I.3). Do not take it home for washing until it has been autoclaved.
- Wear only closed-toe shoes in the laboratory.
- Wear eye protection whenever you are handling microbes or chemicals (especially during heating) even if you wear glasses or contacts (Fig. I.3).
- Turn off your Bunsen burner when it is not in use. In addition to being a fire and safety hazard, it is an unnecessary source of heat in the room.
- Tie back long hair, because it is a potential source of contamination as well as a likely target for fire.
- If you are feeling ill, go home. A microbiology laboratory is not a safe place if you are ill.
- If you are pregnant, immune compromised, or are taking immunosuppressant drugs, please see the instructor. It may be in your best long-term interests to postpone taking this class. Discuss your options with your instructor.
- If it is your lab's practice to wear disposable gloves while handling microorganisms (and this is becoming the norm), be sure to remove them each time you leave the laboratory. The proper method for removal is with the thumb under the cuff of the other hand's glove and turning it inside out without snapping it (Fig. I.4). Gloves should then be disposed of in the container for contaminated materials. Finally, wash your hands.
- Wear disposable gloves while staining microbes and handling blood products—plasma, serum, antiserum, or whole blood. Handling blood can be hazardous, even if you are wearing gloves. Consult your instructor before attempting to work with any blood products.
- Use an antiseptic (e.g., Betadine®) on your skin if it is exposed to a spill containing microorganisms. Your instructor will tell you which antiseptic you will be using. Have a lab partner turn on the water for you if your hands have been contaminated.
- Never pipette by mouth. Always use a mechanical pipettor (Fig. C.1, Appendix C).
- Dispose of broken glass or any other item that could puncture an autoclave bag (contaminated or not) in an appropriate sharps or broken-glass container (Fig. I.5).
- Use a fume hood to perform any work involving highly volatile chemicals or stains that need to be heated.
- Find the first-aid kit, and make a mental note of its location.
- Find the fire blanket, shower, and fire extinguisher, note their locations, and develop a plan for how to access them in an emergency.
- Find the eyewash basin, learn how to operate it, and remember its location.

**I.3 Safety First** ■ This student is prepared to work safely with microorganisms. The lab area is uncluttered, tubes are upright in a test tube rack, and the flame is accessible but not in the way. The student is wearing a protective lab coat, gloves, and goggles, all of which are to be removed prior to leaving the laboratory. Not all procedures require gloves and eye protection. Your instructor will advise you as to the standards in your laboratory.

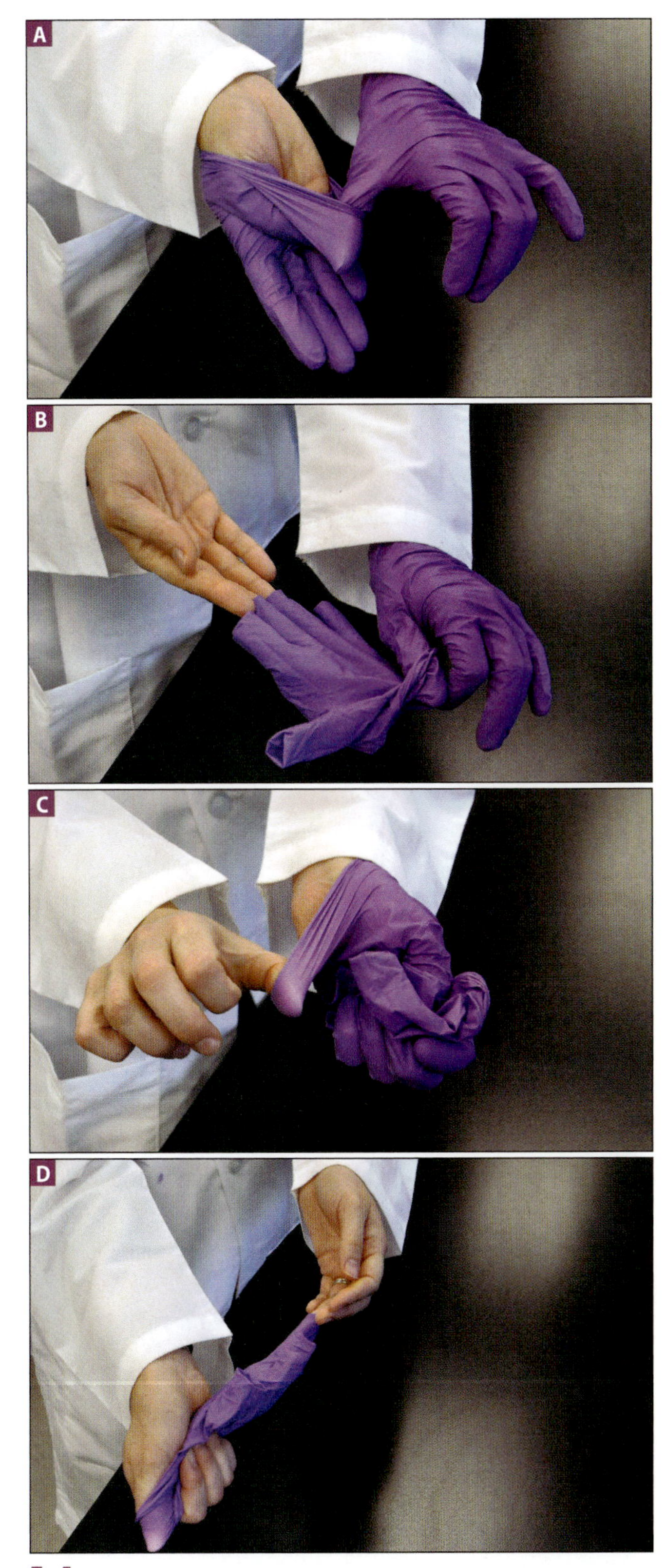


**I.4 Proper Glove Removal** ■ Protective gloves fit tightly around the hands and must be removed in such a way to limit their snapping, which produces aerosols. (**A**) Begin by hooking the thumb of one hand under the cuff of the glove on the other hand. (**B**) Next, roll the glove off the hand so that it turns inside out. (**C**) Take the removed glove in the hand that still has a glove on it and repeat the process. (**D**) As you roll the second glove off your hand, the first glove will end up inside the second glove, which also ends up inside out. Upon completion, all contaminated surfaces are inside the second glove.

## Reducing Contamination of Self, Others, Cultures, and the Environment

- Wipe the desktop with a disinfectant (e.g., Coverage® or 10% chlorine bleach) before *and* after each lab period. Never assume that the class before you disinfected the work area. An appropriate disinfectant will be supplied. Allow the disinfectant to evaporate; do not wipe it dry.
- Never lay culture tubes on the table; they always should remain upright in a tube holder (Fig. I.3). Even solid media tubes contain moisture or condensation that may leak out and contaminate everything it contacts.
- Cover any culture spills with paper towels. Soak the towels immediately with disinfectant and allow them to stand for 20 minutes. Report the spill to your instructor. When you are finished, place the towels in the container designated for autoclaving.
  - If the culture spills on you, remain where you are, do not touch anything, and have your lab partner notify the instructor. Your instructor will advise your group on how to handle the spill.
  - If you get a microbial culture in your eyes IMMEDIATELY have a lab partner lead you to the eyewash basin and rinse your eyes for at least 15 minutes. Time is of the essence for eye contamination. Your lab partners must notify the instructor of the spill and your situation, and follow his/her instructions for handling your care and the spill.

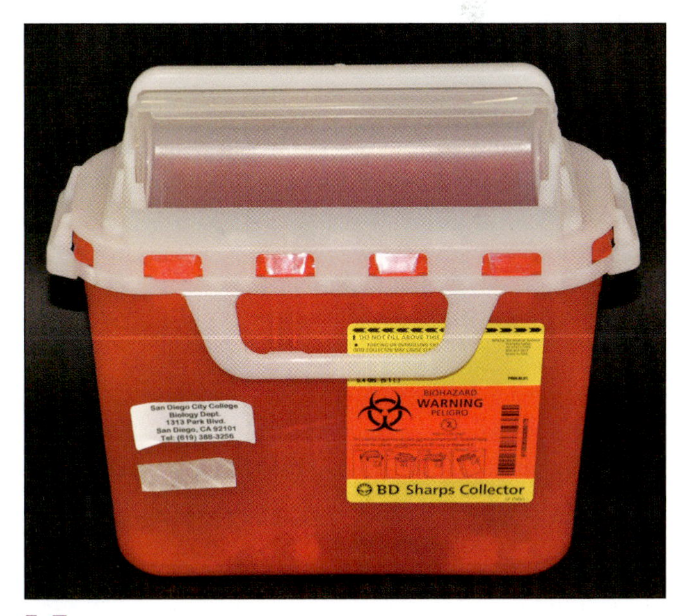


**I.5 Sharps Container** ■ Needles, glass, and other contaminated items that can penetrate the skin or an autoclave bag should be disposed of in a sharps container. Do not fill above the dashed black line. Notice the autoclave tape in the lower left. The white stripes will turn black after proper autoclaving. Above the autoclave tape is the address of the institution that produced the biohazardous waste.

- Keep all nonessential books and papers off the desk. A cluttered lab table is an invitation for an accident that may contaminate your expensive school supplies. Your instructor will advise you where to store these items.
  - Cell phones, tablets, computers, and other electronic devices must never be on the lab table when working with microbes. Contamination of these items can be a health hazard as well as expensive.
- When pipetting microbial cultures, place a disinfectant-soaked towel on the work area. This reduces contamination and possible aerosols if a drop escapes from the pipette and hits the tabletop.

## Guidelines Governing Handling of BSL-2 Organisms

Following is a list of precautions for handling BSL-2 organisms. A number of these were covered previously, but are repeated here along with the new ones and now are considered "highly recommended." It is also highly recommended that before students are allowed to handle BSL-2 organisms, they should demonstrate skill in handling BSL-1 organisms safely.

Your college may have other guidelines for handling BSL-2 organisms and other standards of practice may be adopted during the lifetime of this book, so pay attention to any announcements your lab instructor may give. And, as always, it is the instructor's responsibility to choose organisms and lab exercises appropriate to the skill level of his/her students.

- The lab instructor should announce to the class when BSL-2 precautions are in effect. Your college may also institute a mechanism for identifying BSL-2 cultures, such as using red caps on culture tubes. Be aware of any conventions used by your lab.
- Access to the lab must be limited once work with the BSL-2 organisms has begun. Doors and windows must be closed.
- All unnecessary materials (books, backpacks, etc.—in other words, anything you plan on taking home) must be kept off the work surface and safe from aerial contamination and spills. Your lab will have a designated place for their storage.
- Disinfect your lab table top before and after the activity. Allow the disinfectant to air-dry.
- A lab coat, gloves, and eye protection must be worn throughout the handling of BSL-2 organisms. Do not touch any items not directly involved in the lab activity. Properly remove and dispose of the gloves when finished. (See Fig. I.4 for proper glove removal.)
- Writing utensils must be "lab dedicated," that is, they do not leave the lab until they have been disinfected. It is best to wipe them down with disinfectant after use but keep them in your lab drawer for the entire semester. There is no sense in risking contamination of multiple pens and pencils by using different ones each day.
- Electronic devices (e.g., laptops, tablets, cell phones) must be kept off the work surface and stored in an area protected from spills and aerial contamination.
- Use page covers to protect your lab exercise pages while performing the activity. Once your work is completed, disinfect the page covers, wash your hands, and remove the pages. When dry, the page covers can be stored in your locker and reused. The lab book pages can be taken home.
- It is best to use a photocopy of your data sheet to record results. Once you have decontaminated the area, it can be photographed and the results can be transcribed to the original data sheet.[2] Dispose of the photocopy in an appropriate biohazard container. It must not leave the lab.
- Minimizing aerosol production during the transfer of microorganisms is important and the instructions in Section 1 address methods for doing so. However, here is a summary of recommended practices specific to BSL-2 organisms.
  - Wire loops and needles should be incinerated in an enclosed electric incinerator (refer to Section 1, Fig. 1.15, p. 34). Sterilization with an open flame should be avoided because of its potential for aerosol production.
  - An alternative to the electric incinerator is to perform transfers with sterile disposable loops or sterile wooden sticks. Both will work for transferring from solid or liquid media. After each use, they should be put in a disposal container (such as a can) contaminated end down, submerged in about 2 cm of disinfectant. When finished with the exercise, the can should be put in a bin for autoclaving or other disposal receptacle your lab uses. Alternatively, sticks and disposable loops can be put in a sharps container if that is your lab's practice.
  - Pipetting must be done in such a way that the last drop is *not* "blown out" unless a Petri dish lid is covering a plate. Appendices C and D give specific instructions for pipetting BSL-2 organisms.

[2] Alternatively, if your lab has a dedicated document scanner, your completed data sheet can be scanned and emailed to your instructor for grading, with the original being retained in the lab. Thanks to Dr. Brian Gray, York College of Pennsylvania, for sharing this idea with me at the 2015 ASMCUE conference in Austin, TX.

- Clean up after the activity.
  - For incubation, place tubed cultures in a rack. For plate cultures, it is best to tape the lids on each plate individually by taping around the edge of the plate rather than taping around the base and lid (Fig. I.6). This allows viewing of the agar surface without having to remove the tape.
  - Wipe down the work surface, writing utensils, and page covers (if any) with disinfectant and let them air-dry. Only after the disinfectant has dried is it okay to place items to be removed from lab on your table.
  - Remove your gloves, eye protection, and lab coat and store them according to your lab's conventions.
  - Wash your hands thoroughly.
- Photograph or scan your data sheet and store or dispose of it properly.

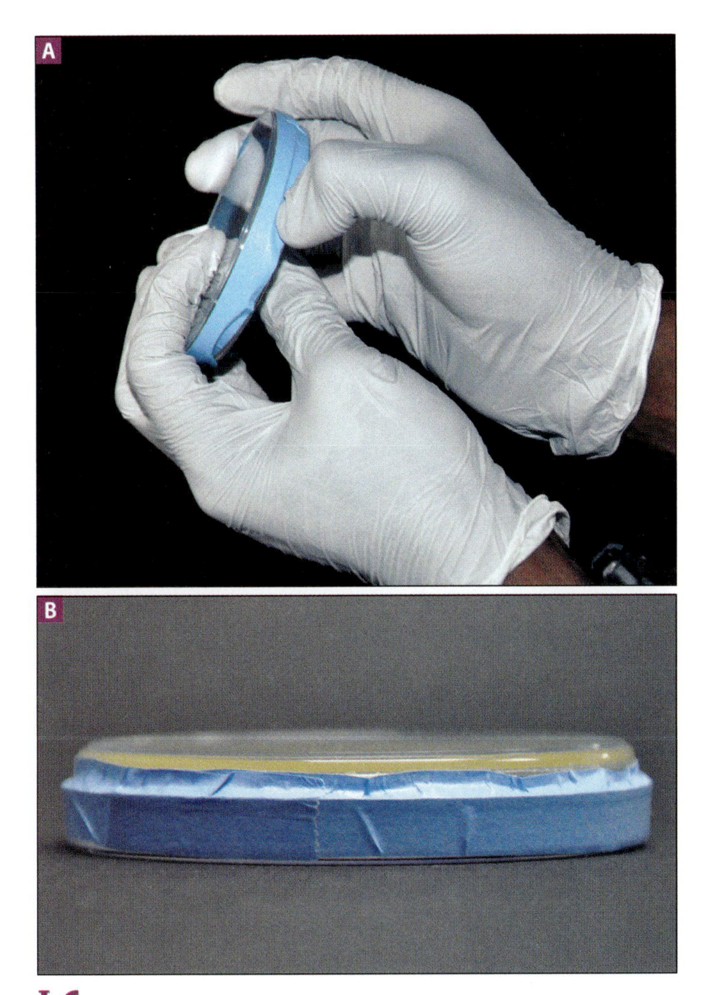


**I.6 Taping Plates** ■ It is a good idea to tape the lid of a Petri dish to its base when incubating or storing it, especially if it contains a BSL-2 organism or a culture of unknown microbes. (**A**) One method is to wrap a length of tape around the lid and then use your thumb to press it firmly to the base. (**B**) When finished, the entire agar surface is visible and not obscured by any tape, making lid removal unnecessary to view colonies.

## Disposing of Contaminated Materials

In most instances, the preferred method of decontaminating microbiological waste and reusable equipment is the autoclave (Fig. I.7).

- Remove all labels from tube cultures and other contaminated *reusable* items and place them in the designated autoclave container. This will likely be an open autoclave pan to enable cleaning the tubes and other items following sterilization.
- Dispose of plate cultures (if plastic Petri dishes are used) and other contaminated non-sharp *disposable* items in the designated autoclave container, such as an autoclave bag (Fig. I.8). Petri dishes should be taped closed. (***Note:*** Autoclave containers are designed to be autoclaved, permanently closed, and discarded. Therefore, do not place reusable and nonreusable items in the same container.)
- Dispose of all blood product samples and disposable gloves in the container designated for autoclaving.
- Place used microscope slides of bacteria in a sharps container designated for autoclaving, or soak them in disinfectant solution for at least 30 minutes before cleaning or discarding them. Follow your laboratory guidelines for disposing of glass.
- Place contaminated broken glass and other sharp objects (anything likely to puncture an autoclave bag) in a sharps container designated for autoclaving (Fig. I.5). Uncontaminated broken glass does not need to be autoclaved, but should be disposed of in a specialized broken glass container.

**I.7 An Autoclave** ■ Media, cultures, and equipment to be sterilized are placed in the basket of the autoclave. Steam heat at a temperature of 121°C (produced at atmospheric pressure plus 15 psi) for 15 minutes is effective at killing even bacterial spores. Some items that cannot withstand the heat, or have irregular surfaces that prevent uniform contact with the steam, are sterilized by other means.

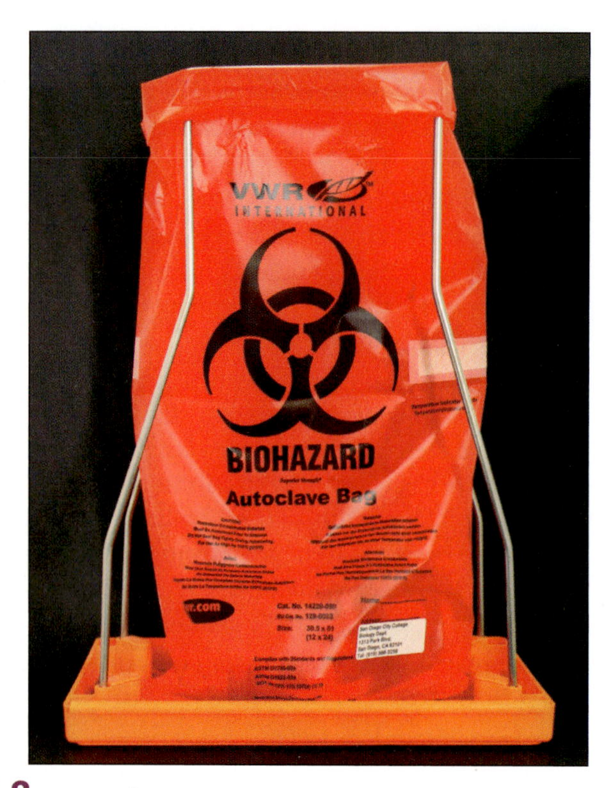


**I.8 An Autoclave Bag** ■ Nonreusable items (such as plastic Petri dishes) are placed in an autoclave bag for decontamination. Petri dishes should be taped closed. Do not overfill or place sharp objects in the bag. Notice the autoclave tape at the middle right. The white stripes will turn black after proper autoclaving. At the lower right is the address of the institution that produced the biohazardous waste.

## Student Compliance with Laboratory Safety Regulations

Your institution may augment the safety regulations outlined above or compile a list tailored to your laboratory. In either case, these constitute a Laboratory Safety Statement. It is highly recommended that students verify that they have read and understand these safety regulations, agree to comply with them, and are aware of noncompliance consequences by signing a statement to that effect. The American Society for Microbiology has produced a publication (Emmert et al., 2012) that includes a sample Laboratory Safety Statement and a student signature page. The signature portion in a slightly modified form is reproduced on page 13. Your instructor may have you sign it or sign his/her own version of it.

### References


Barkley, W. Emmett and John H. Richardson. Chap. 29 in *Methods for General and Molecular Bacteriology*, Philipp Gerhardt, R. G. E. Murray, Willis A. Wood, and Noel R. Krieg, eds. Washington, DC: American Society for Microbiology, 1994.

Collins, C. H., Patricia M. Lyne, and J. M. Grange. Chaps. 1 and 4 in *Collins and Lyne's Microbiological Methods*, 7th ed. Oxford, UK: Butterworth-Heinemann, 1995.

Darlow, H. M. Chap. VI in *Methods in Microbiology,* Volume 1, J. R. Norris and D. W. Ribbins, eds. London, UK: Academic Press, Ltd., 1969.

Emmert, Elizabeth A. B., Jeffrey Byrd, Ruth A. Gyure, Diane Hartman, and Amy White et al. "Biosafety Guidelines for Handling Microorganisms in the Teaching Laboratory: Development and Rationale." *Journal of Microbiology & Biology Education* (May 2013): 78–83. DOI: http://dx.doi.org/10.1128/jmbe.v14i1.531.

Emmert, Elizabeth A. B., Jeffrey Byrd, Ruth A. Gyure, Diane Hartman, and Amy White et al. *Appendix to the Guidelines for Biosafety in Teaching Laboratories*. Washington, DC: American Society for Microbiology, 2012.

Fleming, Diane O. and Debra L. Hunt, eds. *Laboratory Safety— Principles and Practices*, 3rd ed. Washington, DC: American Society for Microbiology, 2000.

Koneman, Elmer W., Stephen D. Allen, William M. Janda, Paul C. Schreckenberger, and Washington C. Winn, Jr. *Color Atlas and Textbook of Diagnostic Microbiology*, 5th ed. Philadelphia and New York: Lippincott-Raven Publishers, 1997.

Power, David A. and Peggy J. McCuen. Pages 2–3 in *Manual of BBL™ Products and Laboratory Procedures*, 6th ed. Cockeysville, MD: Becton Dickinson Microbiology Systems, 1988.

Wilson, Deborah E. and L. Casey Chosewood. U.S. Department of Health and Human Services, *Biosafety in Microbiological and Biomedical Laboratories,* 5th ed. Washington, DC: U.S. Government Printing Office, 2007.


## A Word About Experimental Design

Like most sciences, microbiology has descriptive and experimental components. Here we are concerned with the latter. Science is a philosophical approach to finding answers to questions. Despite what you may have been taught in grade school about *THE* "Scientific Method," science can approach problems in many ways, rather than in any *single* way. The nature of the problem, personality of the scientist, intellectual environment at the time, and good, old-fashioned luck all play a role in determining which approach is taken. Nevertheless, in experimental science, one component that is always present is a **control** (or controls).

A controlled experiment is one in which all **variables** except one—the **experimental variable**—are maintained without change. Frequently, we are looking for the cause of some phenomenon. By limiting variables to only one we can draw a provisional conclusion about whether or not that variable causes the phenomenon.

If changing the variable causes a change in the phenomenon, then we can provisionally conclude that the variable and the phenomenon are causally related; that is, we have demonstrated a **cause and effect relationship** between the phenomenon and the experimental variable. Alternatively, if there is no observed change, we can eliminate the experimental variable from involvement with the phenomenon.

For example, if we want to determine the effect of increasing temperature on microbial growth rate, we could grow the same microorganism in two test tubes containing the same nutrient source at two different temperatures. One would be grown at "normal" temperature (whatever it is for that organism) and the other would be grown at a higher temperature. Everything but temperature would be the same, so that if we see a difference in growth rate we can attribute it to the experimental variable temperature.

But how would we know there is a difference in growth rate? Simple—by comparing the growth rate at the higher temperature to the growth rate at the "normal" temperature. The "normal" temperature tube is the control; it provides a baseline growth rate against which the experimental growth rate is compared. Growth rate in the experimental tube will be faster, slower, or the same as the control.

Without the control we would only be able to measure growth rate in the experimental tube, but we couldn't answer the question about the effect of higher temperature on growth rate. Had we grown different microorganisms with different nutrient sources at different temperatures and we saw a difference, we wouldn't know if the difference was due to the organism, the nutrients available, and/or the temperature. This is why we test one variable at a time.

Controls are an essential and integral part of all experiments, but there are many types. Two commonly used controls are **positive** and **negative controls.** A positive control is one that is set up to produce a positive result. Alternatively, a negative control is set up to demonstrate a negative result or no change.

As you work your way through the exercises in this book, pay attention to the various ways positive, negative, and other controls are used to improve the **validity** (effectiveness) of the experiment and the **reliability** (accuracy) of the results. In fact, there are many questions on the data sheets that ask some form of the question: "What is the role of the control in this experiment?" We are pretty sure that your professor won't be satisfied with the answer, "It's a control."

Microbiological experimentation often involves tests that determine the ability of an organism to use or produce some chemical, or to determine the presence or absence of a specific organism in a sample. Ideally, a positive result in the test indicates that the microbe has the ability or is present in the sample, and a negative result indicates a lack of that ability or absence in the sample (Fig. I.9).

The tests we run, however, have limitations and occasionally may give **false-positive** or **false-negative** results. An inability to detect small amounts of the chemical or organism in question would yield a false negative result and would be the result of inadequate **sensitivity** of the test (Fig. I.9). An inability to discriminate between the

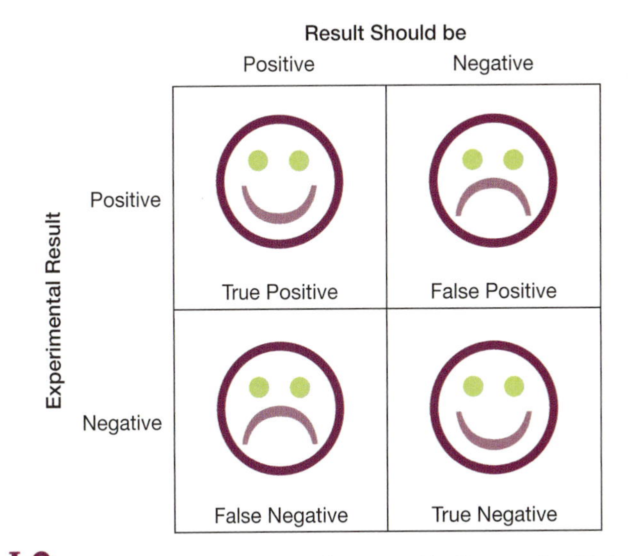


**I.9 Limitations of Experimental Tests** ■ Ideally, tests should give a positive result for specimens that are positive, and a negative result for specimens that are negative. False positive and false negative results do occur, however, and these are attributed to inadequate specificity and inadequate sensitivity, respectively, of the test system.

chemical or organism in question and similar chemicals or organisms would yield false-positive results when the similar chemical or organism was being tested and would be the result of inadequate **specificity** of the test (Fig. I.9). Sensitivity and specificity can be quantified using the following equations:

$$\text{Sensitivity} = \frac{\text{True Positives}}{\text{True Positives} + \text{False Negatives}}$$

$$\text{Specificity} = \frac{\text{True Negatives}}{\text{True Negatives} + \text{False Positives}}$$

The closer sensitivity and specificity are to a value of one, the more useful the test. As you perform the tests in this book, be mindful of each test's limitations, and be open to the possibility of false positive and false negative results.

## References


Forbes, Betty A., Daniel F. Sahm, and Alice. S. Weissfeld. Chap. 5 in *Bailey & Scott's Diagnostic Microbiology*, 10th ed. St. Louis, MO: Mosby-Year Book, 1998.

Lilienfeld, David E. and Paul D. Stolley. Page 118 in *Foundations of Epidemiology*, 3rd ed. New York: Oxford University Press, 1994.

Mausner, Judith S. and Shira Kramer. Pages 217–220 in *Epidemiology: An Introductory Text*, 2nd ed. Philadelphia: W.B. Saunders Company, 1985.

## Data Presentation: Tables and Graphs

In microbiology, we perform experiments and collect data, but it is often difficult to know what the data mean without some method of organization. Tables and graphs allow us to summarize data in a way that makes interpretation easier.

### Tables

A table is often used as a preliminary means of organizing data. As an example, Table I-2 shows the winning times for each male and female age division in a half-marathon race. Again, the aim of a table is to provide information to the reader. Notice the meaningful title, the column labels, and the appropriate measurement units. Without these, the reader cannot completely understand the table and your work will go unappreciated! Data tables are provided for you on the data sheets for each exercise in this book, but you may be required to fill-in certain components (units, labels, etc.) in addition to the data.

TABLE **I-2** **Winning Half-Marathon Times by Sex and Age Division**

| Male Runners | | Female Runners | |
|---|---|---|---|
| Winner's Age (years) | Winning Time (minutes) | Winner's Age (years) | Winning Time (minutes) |
| 15 | 73 | 15 | 88 |
| 25 | 67 | 24 | 82 |
| 31 | 67 | 30 | 82 |
| 35 | 71 | 39 | 84 |
| 40 | 71 | 42 | 85 |
| 52 | 78 | 50 | 109 |
| 62 | 95 | 62 | 108 |
| 70 | 123 | 70 | 126 |

### Graphs

Table I-2 does give the information, but what it is telling us may not be entirely clear. It appears that the half-marathon times increase as the runners get older, but we have difficulty determining if this is truly a pattern. That is why data also are presented in graphic form at times; a graph usually shows the relationship between variables better than a table of numbers.

**X–Y Scatter Plot** The type of graph you will be using in this manual is an "*X–Y* Scatter Plot," in which two variables are graphed against each other. Figure I.10 shows the same data as Table I-2, but in an *X–Y* Scatter Plot form.

Notice the following important features of the graph in Figure I.10:

- *Title:* The graph has a meaningful title—which should tell the reader what the graph is about. A title of "Age vs. Winning Time" is vague and inadequate.
- *Dependent and independent variables:* The graph is read from left to right. In our example, we might say for the male runners, "As runners get older, winning times get longer." *Winning time* depends on *age*, so winning time is the *dependent* variable and age is the *independent* variable. (Age does *not* depend on the winning time!) By convention, the independent variable is plotted on the *x*-axis and the dependent variable is plotted on the *y*-axis. By way of comparison, notice the consequence of plotting age on the *y*-axis and winning time on the *x*-axis: "As runners get slower, they get older"—which doesn't reflect the actual relationship between the variables and worse yet, is nonsense.
- *Axis labels:* Each axis is labeled, including the appropriate units of measure. "Age" without units is meaningless. Does the scale represent months? Years? Centuries?

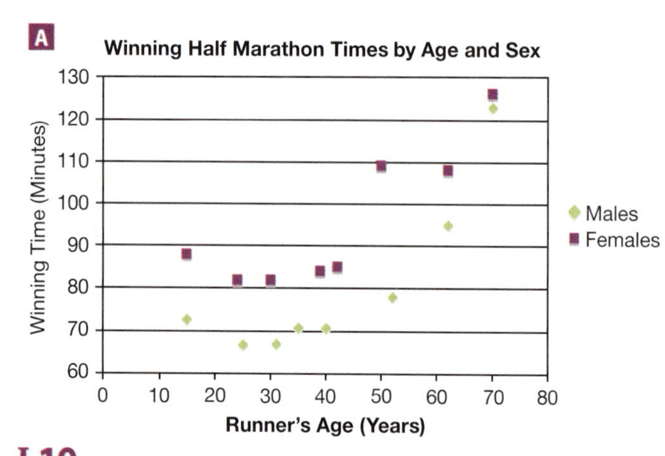


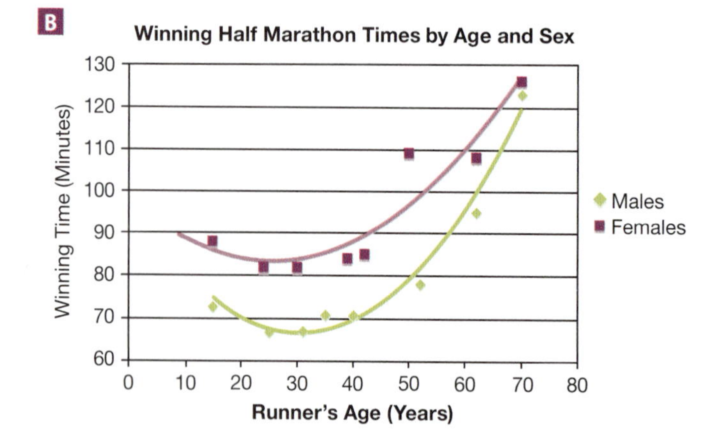


**I.10** **Sample X-Y Scatter Plot** ■ A graph often shows the relationship between variables better than a table of numbers. Examine this sample and identify the essential components of a quality graph (see text). (**A**) Presentation of data without a best-fit line is acceptable if there are not enough data points to justify illustrating a trend. (**B**) Shown here are the same data but with a trend line. Notice that the points do not fall directly on the line but, rather, that the line gives the general trend of the data. "Connecting the dots" is not appropriate.

- *Axis scale:* The scale on each axis is uniform. The distance between marks on the axis is always the same and represents the same amount of that variable. (But increments on the $x$-axis don't have to equal those on the $y$-axis, as shown.) The size of each increment is up to the person making the graph and is dictated by the magnitude and range of the data. Most of the time, we choose a length for the axis that fills the available graphing space.
- *Axis range:* The scale for an axis does not have to begin with "0." Use a scale that best presents the data. In this case, the smallest $y$-value was 67 minutes, so the scale begins at 60 minutes.
- *Multiple data sets and the legend:* The two data series (male and female times) are plotted on the same set of axes, but with different symbols that are defined in the legend at the right. The symbols shown differ in color *and* shape, but one of these is adequate.
- *Best-fit line:* If a line is to be drawn at all, it should be an average line for the data points, not one that "connects the dots" (Fig. I.10B). Notice that the points are not necessarily *on* the line. The purpose of a best-fit line is to illustrate the general trend of the data, not the specifics of the individual data points. (Be assured that most graphs in your textbooks where a smooth line is shown were experimentally determined and the lines are derived from points scattered around the line.) There is a mathematical formula that allows one to compute the slope and $y$-intercept of the **trend line** if the relationship is linear, or a **best-fit line** if the relationship is nonlinear (as in the half-marathon times example), but this is beyond our needs. For our purposes, a hand-drawn trend line that looks good is good enough. (If you use a computer graphing program, then it will produce the trend line without you doing any of the math—the best of all situations!)

### Bar Graphs

Bar graphs are used to illustrate one variable. Often, the variable is considered to be categorical. That is, the data are in distinct groups (rather than continuous, where the breaks are arbitrary). Using a bar graph to show the relationship between winning times and ages is inappropriate. Examine Figure I.11A. Notice that the space each bar fills is meaningless; that is, the only important part of the bar is the top—which is the value used in the $X$–$Y$ scatter plot. Also notice that the $x$-axis scale is not uniform—the gaps between bars range from 4 years (31 to 35) to 12 years (40 to 52). An appropriate use of a bar graph would be the distribution of student performance on an exam, even though exam scores are not categorical (Fig. I.11B). Notice that the space each bar fills has meaning. Each student in a particular group adds height to the bar.

## Data Presentation: Be Creative, But Complete

There is no single correct way to produce a graph for a particular data set. Actually, most people working independently would graph the same data set in different ways (e.g., different scales, colors, wording of the title, and axis labels), but the essential components listed would have to be there. You will be asked to graph some of the data you collect. Be sure your graphs tell a complete and clear story of what you have done.

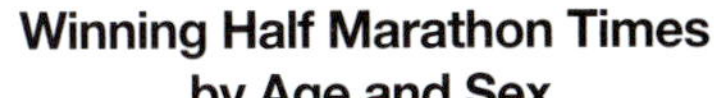


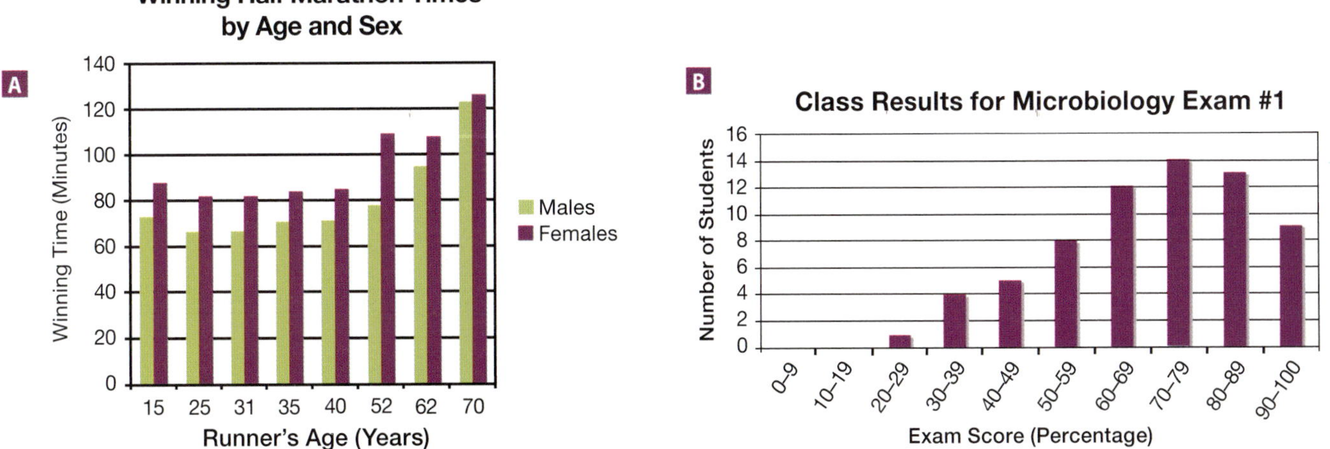


**I.11 Bar Graphs** ■ A bar graph is appropriate to present data involving a single variable, especially if that variable is easily divided into distinct categories or groups. (**A**) Plotting the winning half-marathon times from Table I-2 using a bar graph is inappropriate because the only meaningful point is at the top. (**B**) A bar graph is useful in presenting data of a single variable, such as the number of students earning a specific score on their microbiology exam.

# Student Safety Contract

### Student Agreement on Laboratory Safety

I have read the Laboratory Safety Statement as written in *Microbiology: Laboratory Theory & Application, Brief,* 3rd edition, and/or the equivalent document supplied by the Department of Biological Sciences,

_______________________________________________,
(Institution Name)

and I understand its content. I agree to abide by all laboratory rules set forth by the instructor. I understand that my safety is entirely my own responsibility and that I may be putting myself and others in danger if I do not abide by all the rules set forth by the instructor.

COURSE: _______________________________________________

STUDENT NAME (PRINT): _______________________________________________

STUDENT SIGNATURE: _______________________________________________

DATE: _______________________________________________

SECTION 1

# Fundamental Skills for the Microbiology Laboratory

A necessary skill for safely working in a laboratory, handling foods, and just living in a world full of microbes, is effective hand washing. In Exercises 1-1 and 1-2, you will have the opportunity to evaluate your hand-washing technique to correct any deficiencies you observe and to compare the effectiveness of several hand-cleansing products.

A second skill microbiologists should at the very least be familiar with, and at most be able to do as a matter of routine, is prepare the **media** used to grow bacterial and fungal **cultures**. Media must be sterile prior to use and must also supply the organisms to be grown in or on them with the chemical nutrients necessary for the organisms' growth. Preparation of these media involves weighing ingredients, measuring liquid volumes, calculating proportions, handling basic laboratory glassware, and operating a pH meter and an autoclave. In Exercise 1-3 you will learn and practice these fundamental skills by preparing a couple of simple growth media. When you have completed the exercise, you will have the skills necessary to prepare almost any medium if given the recipe.

A third fundamental skill necessary for any microbiologist is the ability to transfer microbes from one place to another without contaminating the original culture, the new medium, or the environment (including the microbiologist and others in the laboratory). This **aseptic** ("without contamination") transfer technique is required for virtually all procedures in which living microbes are handled, including isolations, staining, and differential testing. Exercises 1-4 through 1-6 present descriptions of common transfer and inoculation methods. Less frequently used methods are covered in Appendices B through D.

## A Word About Hand Hygiene

The concept of good hand hygiene has gone from a controversial beginning (in the early 1800s) to an accepted practice that is still problematic. Current studies designed to test the efficacy of various hand-cleansing agents often have subjects wash their hands for unrealistic lengths of time (that is, longer than workers routinely wash on the job), test artificially contaminated hands (or not), and use different standards of evaluation, making comparison difficult. We are still left with the question, "What works best?"

While hand washing has been identified as an important, easily performed behavior that minimizes transfer of pathogens to others, uniform compliance with hand-washing standards has been difficult to achieve. Heavy workloads, skin reactions to the agent (e.g., plain or antimicrobial soap, iodine compounds, alcohol), skin dryness due to frequent washing, and many other factors contribute to noncompliance (Boyce and Pittet, 2002, and The Joint Commission, 2009).

Alcohol-based hand rubs have, in many instances, replaced conventional hand-washing agents because they are more effective than soap and water (disputed), require less time, produce fewer skin reactions, and have been shown to result in a higher level of compliance by health-care workers.

In Exercises 1-1 and 1-2 you will have the opportunity to evaluate your ability to remove artificial "germs" from your own hands and then compare the relative effectiveness of commercial agents used to cleanse hands (or skin in general). ■

# Glo Germ™ Hand Wash Education System

## Theory

The Glo Germ™ Hand Wash Education System was developed as a training aid for people to learn to wash their hands more effectively. The lotion (a powder is also available) contains minute plastic particles (artificial germs) that fluoresce when illuminated with ultraviolet (UV) radiation but are invisible with normal lighting.

Initially the hands are covered with the lotion, but the location and density of the "germs" is unknown because of the normal room lighting. After washing, a UV lamp is shined on the hands. Wherever the "glowing germs" remain, hand washing was not effective. This provides immediate feedback to the washer as to the effectiveness of their hand washing and provides information about where they need to concentrate their efforts in the future.

## Application

Effective hand washing to minimize direct person-to-person and indirect contact transmission of pathogens by health-care professionals and food handlers is essential. It also is critical to laboratorians handling pathogens to minimize transmission to others, inoculation of oneself, and contamination of cultures.

## In This Exercise

You will cover your hands with nontoxic synthetic fluorescent "germs" and compare the degree of contamination before and after hand washing to evaluate your hand-washing technique and demonstrate the difficulty in removing hand contaminants.

## Materials[1]

### Per Student Group

- □ One bottle of Glo Germ™ lotion-based simulated germs
- □ One ultraviolet penlight
- □ (Optional) fingernail brush

[1] Available from Glo Germ™, 1101 S. Murphy Ln., Moab, UT 84532. 1-800-842-6622 (USA); www.glogerm.com/

## PROCEDURE

1. Shake the lotion bottle well.
2. Have your lab partner apply 2–3 drops of gel on the palms of both your hands. Be careful not to get the gel on your clothing, or in your eyes or mouth.
3. Rub your hands together, thoroughly covering your hand surfaces, including the backs and between the fingers (Fig. 1.1). Spread the lotion up to your wrists on both sides. Also, scratch your palms with all fingernails.
4. Have your lab partner shine the UV light on your hands to see the extent of coverage with the lotion. *Do not look directly at the lamp*. This works best in an area with limited ambient light. Do *not* handle the light yourself because you will contaminate it with the artificial germs.
5. Have your lab partner turn on warm water at a sink for you. Then wash your hands with soap and warm water as thoroughly as you can for at least 20 seconds. (If you don't have a watch handy, sing the "Happy Birthday Song" to yourself twice. Or, sing it out loud twice!) Use a fingernail brush if you have one. When you are finished, have your lab partner turn off the water and hand you a fresh paper towel. Dry your hands.
6. Have your lab partner shine the UV light on your hands once more. *Do not look directly at the lamp*. Examine the hand surfaces contaminated by the artificial germs. Then, turn off the lamp.
7. Now that you know where the artificial germs remain, wash your hands once more to remove as many as possible. As before, have your lab partner turn the water on and off for you.
8. Repeat the experiment with your lab partner, but with roles reversed.
9. Record your results on the data sheet on page 19 and answer the questions.
10. After recording your results, shine the UV lamp once more on your data sheet, desk top, and pen/pencil to see how much of the lotion was transferred to these. *Do not look directly at the lamp*.

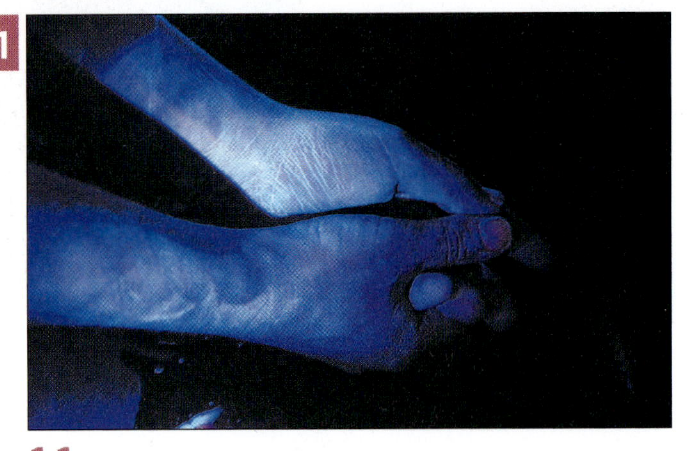

**1.1 Hands Covered With Glo Germ™ Prior to Washing** ■ Shown are properly prepared hands covered with the fluorescent Glo Germ™ lotion prior to washing. Note the thorough coverage, including the back of the hands and under the fingernails.

## References

Boyce, John M., and Didier Pittet. Centers for Disease Control and Prevention. *Guideline for Hand Hygiene in Health-Care Settings: Recommendations of the Healthcare Infection Control Practices Advisory Committee and the HICPAC/SHEA/APIC/IDSA Hand Hygiene Task Force*. MMWR 51 (No. RR-16) (2002): pages 1–45.

Glo Germ™. Package insert for the Glo Germ™ Hand Wash Education System. www.glogerm.com/

Name ______________________________

Date ______________________________

Lab Section ______________________________

I was present and performed this exercise (initials) ______________________________

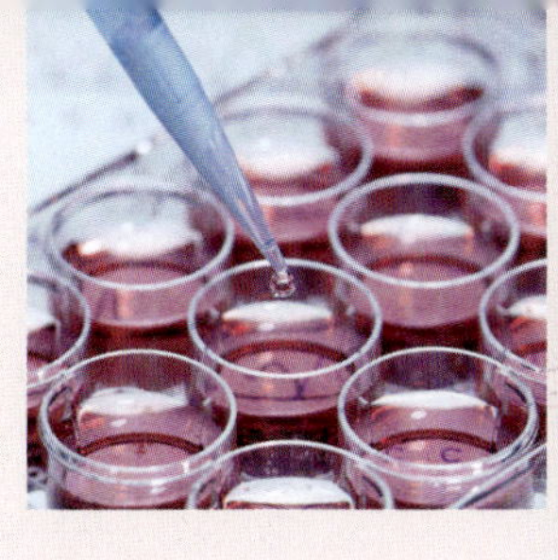

DATA SHEET 1-1

# Glo Germ™ Hand Wash Education System

## OBSERVATIONS AND INTERPRETATIONS

1 Record the degree of hand contamination before and after washing in the table below. Use this qualitative scale for evaluation

+++ means "a lot of contamination"

++ means "moderate contamination"

+ means "little contamination"

0 means "no contamination"

2 There is no absolute cutoff between any of these categories, and what you call "moderate contamination" might be called "little contamination" by another student. Just try to be consistent within your evaluation.

| Body Region | Left Hand | | Right Hand | |
|---|---|---|---|---|
| | Before Washing | After Washing | Before Washing | After Washing |
| Palm | | | | |
| Fingers | | | | |
| Between Fingers | | | | |
| Tops of Fingernails | | | | |
| Under Fingernails | | | | |
| Back of Hand | | | | |
| Front of Wrist | | | | |
| Back of Wrist | | | | |

## QUESTIONS

1 *What areas were most thoroughly cleaned by your washing technique?*

______________________________

______________________________

2 *What areas were most difficult for you to clean with your washing technique?*

______________________________

______________________________

**3** *In general, were your two hands cleaned an equal amount, or was one cleaned more than the other? What could account for any differences?*

**4** *How do your answers to questions 1, 2, and 3 compare to your lab partner's answers? Why might they differ?*

**5** *Why were you instructed to have your lab partner turn the water on and off and operate the UV lamp rather than you doing these actions yourself?*

**6** *Why might it be advisable to modify the procedure and use the UV light to check your hands prior to application of lotion and the paper towels prior to drying?*

**7** *Using the same qualitative scale as before, record the amount of Glo Germ™ that was transferred to this data sheet, your table top, and to your writing instrument. What does this tell you about the ease of transferring the "unseen" by contact.*

# A Comparison of Hand-Cleansing Agents[1]

EXERCISE 1-2

## Theory

Now that you have honed your hand-washing skills in Exercise 1-1, you can compare the relative effectiveness of several hand-cleansing agents. Many hand-cleansing agents are commercially available and each has its appropriate application. Some are intended for routine home use, such as "washing up before dinner." Others are available to the general public, but can be used in clinical settings where prevention of infection is a matter of life and death. And others fall somewhere in-between. Examples of products that can be tested in this exercise are shown in Figure 1.2.

Each product has an **active ingredient** that either kills *(–cidal*, e.g., "bactericidal") or stops the growth of certain microbes *(–static*, e.g., "bacteriostatic"). Their effectiveness comes from an ability to interact with one or more essential cellular components and make them nonfunctional. The cell component could be as small as an enzyme or as large as the entire cytoplasmic membrane, and anything in-between. Thus the active ingredient dictates the utility of the product, because the more damage it does the higher level of decontamination it produces. Question 5 in the data sheet assigns you the task of looking up common active ingredients and reporting on their mechanism(s) of action and their target microbe(s).

[1] Thanks to reviewers Johana Melédez-Santiago and Janice Yoder Smith for their suggestions on developing this laboratory exercise.

## Application

Effective hand washing to minimize direct person-to-person transmission, and indirect contact of pathogens by health-care professionals and food handlers, is essential. It also is critical to laboratorians handling pathogens to minimize transmission to others, inoculation of oneself, and contamination of cultures.

## In This Exercise

You will evaluate the relative effectiveness of a variety of hand-washing agents.

## ▼ Materials

### Per Student

- □ Lab coat
- □ Disposable gloves
- □ Chemical eye protection
- □ One nutrient agar or tryptic soy agar plate
- □ Access to a sink and paper towels
- □ Permanent marking pen

### Per Table

- □ Paper towels
- □ Tap water

**1.2 Hand-Washing Agents** ■ From left to right are: Betadine® scrub (10% povidone iodine), alcohol-based hand sanitizer (65% ethyl alcohol, 0.13% benzalkonium), chloride-based hand sanitizer, antibacterial soap (0.15% triclosan), and Hibiclens® (4% chlorhexidine gluconate). As written, this exercise uses these agents, but others can be substituted, including tap water.

- ☐ Bottles of as many as five of the following or suitable replacements:
  - Antibacterial liquid soap (0.15% triclosan)
  - Hand sanitizer (65% ethyl alcohol = 65% ethanol)
  - Hy5 brand of alcohol-free hand sanitizer (0.13% benzalkonium chloride)
  - Betadine® scrub (10% povidone iodine)
  - Hibiclens® (4% w/v chlorhexidine gluconate)

**Note**

Students should not ingest or get any of these products in their eyes. Follow package inserts for treatments. Students with allergies or sensitivities to the active ingredients in any of these products should be assigned an alternate product.

## PROCEDURE

### Lab One

1. Wear a lab coat, gloves, and chemical eye protection when performing this procedure.
2. Label the base of a nutrient agar or tryptic soy agar plate with your name. Divide the plate in half and label one half "Before" and the other half "After." Then, number 1 through 5 on both sides of the dividing line so the numbers line up. That is, number 1 on the "Before" side should be opposite number 1 on the "After" side, and so on. Use the entire diameter of the plate for the five numbers and spread them out as much as you can. You will be pressing your fingers at each number in steps 4 and 7.
3. Rub your left-hand fingertips (all five of them) on the area to be sampled. (Good choices are the floor, shoe soles, tabletop near a sink, backpack, and your forehead or feet. Be creative!)
4. Then, one at a time, gently press your thumb tip next to the "Before" number 1, your index finger next to the "Before" number 2, and so on. Your middle, ring, and pinky fingers are numbered 3, 4, and 5, respectively.
5. Wash your hands with soap and water.
6. Record the hand-cleansing agents you will be using on the data sheet, page 23.
7. Now sample the same site used in step 2 with the fingertips of your right hand, but this time one fingertip at a time. Then, you will rub that fingertip in a few milliliters of the hand-cleansing agent in your left palm (see steps a and b for further directions). If the agent requires rinsing and drying, perform this at a sink. Agents that only require air-drying can be done at your table.
   a. Place a quarter-size drop of the first hand-cleansing agent in the palm of your left hand. Rub your right thumb on the area to be sampled. Then, rub your thumb into the agent for 10 seconds.[2] (If you don't have a watch, sing the "Happy Birthday Song" once to yourself.) Once your thumb is dry, gently press it next to the number 1 on the "After" side. Wash both hands with soap and water, and dry them.
   b. Repeat step a with each finger of the right hand and the remaining four hand-cleansing agents. Be sure to press each finger onto the agar next to its correct number and to wash your hands in between exposures.
8. Tape the lid on the plate and incubate at 25°C until the next lab period.

### Lab Two

1. Examine your plates for growth. (***Note:*** do not remove the plate's lid unless given permission to do so.) Each "blob" is probably the product of a single microbial species that has reproduced from one or a few cells to such an extent that it is now visible to the naked eye. This is called a "colony." Different microbial species often produce distinctly different colonies.
2. Fill in the table provided on the data sheet, page 24, as you evaluate the relative amount of growth and microbial diversity (based on colony differences) on the "Before" and "After" sides of the plate for each agent. Recognize that you are comparing "apples and oranges" with respect to their intended uses. That is, the products tested range from surgical scrubs to over-the-counter soaps intended for household use. A surgical scrub may have better antimicrobial activity than hand soap, but that doesn't make it "better" for household use!
3. Dispose of all plates in the appropriate autoclave container when finished.

---

[2] The recommended time for hand washing is 15–20 seconds, but because you are only washing the fingertips and concentrating on those, that time has been reduced in order to see representative results.

### References


Boyce, John M. and Didier Pittet. Centers for Disease Control and Prevention. *Guideline for Hand Hygiene in Health-Care Settings: Recommendations of the Healthcare Infection Control Practices Advisory Committee and the HICPAC/SHEA/APIC/IDSA Hand Hygiene Task Force.* MMWR 51 (No. RR-16) (2002): 1-45.

The Joint Commission, 2009. *Measuring Hand Hygiene Adherence: Overcoming the Challenges.* Available online. URL: http://www.jointcommission.org/assets/1/18/hh_monograph.pdf.

Name ______________________________

Date ______________________________

Lab Section ______________________________

I was present and performed this exercise (initials) ______________________________

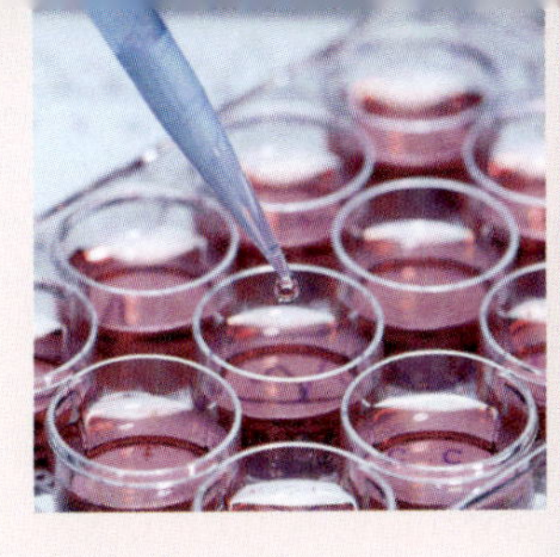

# DATA SHEET 1-2

## A Comparison of Hand-Cleansing Agents

### OBSERVATIONS AND INTERPRETATIONS

**1** Record your results in the table below. For **relative amount of growth**, use this qualitative scale for evaluation:

+++ means "a lot of contamination"
++ means "moderate contamination"
+ means "little contamination"
0 means "no contamination"

**2** There is no absolute cutoff between any of these categories. What you call "moderate contamination" might be called "little contamination" by another student. Just try to be consistent within your evaluation. For **diversity**, look for different colors, shapes, textures, sizes, and anything else that is indicative of a different organism growing, and record the number of different organisms present. **Interpretation** would be relative effectiveness of the agent (excellent, good, fair, poor).

| Number on Plate | Hand-washing Agent | "Before" | | "After" | | Interpretation |
|---|---|---|---|---|---|---|
| | | Relative Growth | Diversity | Relative Growth | Diversity | |
| 1 | | | | | | |
| 2 | | | | | | |
| 3 | | | | | | |
| 4 | | | | | | |
| 5 | | | | | | |

## QUESTIONS

**1** *Which agent seemed to be most effective at removing microbes from the fingertips?*

______________________________

**2** *Which agent seemed to be least effective at removing microbes from the fingertips?*

______________________________

**3** *What other variables besides the cleansing agent might be responsible for any differences noted between fingers or students?*

**4** *Did you or any other students notice some microbes persisting more than others after washing? What factor(s) might account for this?*

**5** *Do some research and fill in the following table. Your answer to "Antibacterial Effect" should address the structural and/or functional part of the cell affected, as well as if the agent is bactericidal (kills bacteria) or bacteriostatic (stops bacterial growth, but doesn't kill them). Under "Other Antimicrobial Effects," note any antiviral, antiprotozoal, and/or antifungal properties of the agent.*

| Active Ingredient | Antibacterial Effect | Other Antimicrobial Effects |
|---|---|---|
| Triclosan | | |
| Isopropanol/ethanol (isopropyl/ethyl alcohol) | | |
| Benzalkonium chloride | | |
| Povidone iodine scrub | | |
| Chlorhexidine gluconate | | |

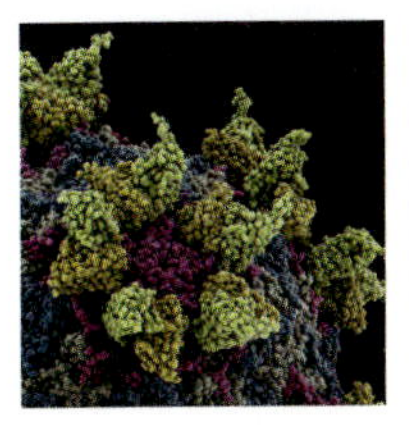

## A Word About Basic Growth Media

To cultivate microbes, microbiologists use a variety of growth media. Although these media may be formulated from scratch, they more typically are produced by rehydrating commercially available powdered media. Media that are routinely encountered in the microbiology laboratory range from the widely used, general-purpose growth media, to the more specific selective and differential media used in identification of microbes (Sections 4 and 5). In Exercise 1-3 you will learn how to prepare simple general growth media. ■

# Nutrient Broth and Nutrient Agar Preparation

EXERCISE **1-3**

## Theory

Nutrient broth and nutrient agar are common media used for maintaining bacterial cultures[1]. To be of practical use, they have to meet the diverse nutrient requirements of routinely cultivated bacteria. As such, they are formulated from sources that supply carbon and nitrogen in a variety of forms—amino acids, purines, pyrimidines, monosaccharides to polysaccharides, and various lipids.

Generally, these are provided in digests of plant material (phytone) or animal material (peptone and others). Because the exact composition and amounts of carbon and nitrogen in these ingredients are unknown, these media are considered to be **undefined**. They are also known as **complex media**[2].

In most classes (because of limited time), media are prepared by a laboratory technician or by the instructor. Still, it is instructive for novice microbiologists to at least gain exposure to what is involved in media preparation. Your instructor will provide specific instructions on how to execute this exercise using the equipment in your laboratory.

## Application

Microbiological growth media are prepared to cultivate microbes. These general growth media are used to maintain bacterial stock cultures.

[1] This is a true statement but a bit misleading. The implication is that these will support growth of most bacteria. In fact, recent estimates indicate we have yet to find appropriate media and laboratory conditions to cultivate over 90% of bacteria in the environment!

[2] In contrast, **chemically defined** media are composed of ingredients with a known chemical structure and in known quantities. For an example, see the recipe for Citrate Agar (Simmons) on page 612. In it, the only carbon source is the six-carbon sugar citrate ($C_6H_8O_7$) and the only nitrogen source is ammonium ($NH_4^+$).

## In This Exercise

You will prepare 1-liter batches of two general growth media: nutrient broth and nutrient agar. During the course of the semester, a laboratory technician will probably do this for you, but it is good to gain firsthand appreciation for the work done behind the scenes!

## ▼ Materials

### Per Student

- □ Lab coat
- □ Disposable gloves
- □ Chemical eye protection

### Per Student Group

- □ One 2-liter Erlenmeyer flask for each medium made
- □ Three or four 500 mL Erlenmeyer flasks and covers (can be aluminum foil)
- □ Three or four stirring hot plates
- □ Three or four magnetic stir bars
- □ All ingredients listed in the following recipes (or commercially prepared dehydrated media)
- □ Sterile Petri dishes
- □ Test tubes (16 mm × 150 mm) and caps
- □ Balance
- □ Weighing paper or boats
- □ Spatulas

## Medium Recipes

### Nutrient Broth

| | |
|---|---|
| □ Beef extract | 3.0 g |
| □ Peptone | 5.0 g |
| □ Distilled or deionized water | 1.0 L |

*pH 6.6–7.0 at 25°C*

1

### Nutrient Agar

- □ Beef extract 3.0 g
- □ Peptone 5.0 g
- □ Agar 15.0 g
- □ Distilled or deionized water 1.0 L

*pH 6.6–7.0 at 25°C*

Note that the only difference between nutrient broth and nutrient agar is the agar. Agar is indigestible by most bacteria so it is not considered a nutrient source in the medium. It acts solely as a solidifying agent in which to suspend the nutrients, which makes the name "nutrient agar" a bit misleading. Also note the agar concentration: 15 g/L is 15 g/1,000 mL = 1.5 g%. You will notice different agar concentrations in other media you encounter during the course.

## PREPARATION OF THE MEDIA

### Lab One

To minimize contamination while preparing media clean the work surface, turn off all fans, and close any doors that might allow excessive air movements. Wear a lab coat, gloves, and chemical eye protection when performing these procedures.

**Note**

Your instructor will determine how many tubes and plates of each medium individual students will prepare. Recipes are given in 1 L volumes and will be enough to prepare approximately 140 tubes at 7 mL and 50 plates at 20 mL. Thus, in a class of 24 students 1 L of broth and 1 L of agar would permit each student to prepare 5 or 6 tubes and 2 plates, respectively. Smaller batches can be made if greater emphasis is placed on more students experiencing the weighing and mixing aspect of medium preparation. For example, pairs of students in a class of 24 could prepare 100 mL batches, which would allow each student to make approximately 7 tubes or 2 to 3 plates.

### Nutrient Agar Tubes

1. Weigh the ingredients on a balance (Fig. 1.3).
2. Suspend the ingredients in 1 L of distilled or deionized water in the 2-liter flask, mix well, and boil until fully dissolved (Fig. 1.4).
3. Dispense 7 mL portions into test tubes and cap loosely (Fig. 1.5). If your tubes are smaller than those listed in Materials, adjust the volume to fill 20% to 25% of the tube. Fill to approximately 50% for agar deeps.
4. Sterilize the medium by autoclaving for 15 minutes at 121°C (Fig. 1.6).
5. After autoclaving, cool to room temperature with the tubes in an upright position for agar deep tubes. Cool with the tubes on an angle for agar slants (Fig. 1.7).
6. Incubate the slants and/or deep tubes at 35 ± 2°C for 24 to 48 hours.

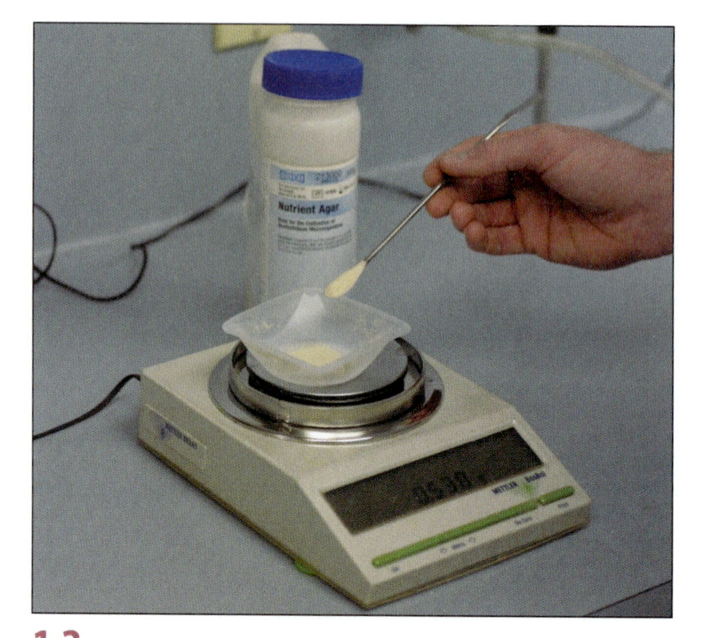


**1.3 Weighing Medium Ingredients** ■ Solid ingredients are weighed with an analytical balance. A spatula is used to transfer the powder to a tared weighing boat. Shown here is dehydrated nutrient agar, but the weighing process is the same for any powdered ingredient.

**1.4 Mixing the Medium** ■ The powder is added to a flask of distilled or deionized water on a hot plate. A magnetic stir bar mixes the medium as it is heated to dissolve the powder.

## Nutrient Agar Plates

1 Weigh the ingredients on a balance (Fig. 1.3).

2 Suspend the ingredients in 1 L of distilled or deionized water in the 2-liter flask, mix well, and boil until fully dissolved (Fig. 1.4).

3 Divide into three or four 500 mL flasks for pouring. Smaller flasks are easier to handle when pouring plates. Don't forget to add a magnetic stir bar and to cover each flask before autoclaving.

4 Autoclave for 15 minutes at 121°C to sterilize the medium.

5 Remove the sterile agar flasks from the autoclave and allow them to cool to 50°C while stirring on a hot plate.

6 Dispense approximately 20 mL into sterile Petri plates (Fig. 1.8). **Be careful! The flask will still be hot, so wear an oven mitt.** While you pour the agar, shield the Petri dish with its lid to reduce the chance of introducing airborne contaminants. If necessary, *gently* swirl each plate so the agar completely covers the bottom; do not swirl the agar up into the lid. Allow the agar to cool and solidify before moving the plates (Fig. 1.9).

7 Store these plates on a countertop for 24 hours to allow them to dry prior to use.

## Nutrient Broth

1 Weigh the ingredients on a balance (Fig. 1.3).

2 Suspend the ingredients in 1 L of distilled or deionized water in the 2-liter flask. Agitate and heat slightly (if necessary) to dissolve them completely (Fig. 1.4).

3 Dispense 7 mL portions into test tubes (or less, depending on your lab customs) and cap loosely (Fig. 1.5). As with agar slants, if your tubes are smaller than those recommended in Materials, add enough broth to fill them approximately 20% to 25%.

4 Sterilize the medium by autoclaving for 15 minutes at 121°C (Fig. 1.6).

**1.6 Autoclaving the Tubed Media** ■ Media are sterilized in an autoclave for 15 minutes at 121°C. Shown are four racks of tubed media being removed from an autoclave. The tubes on the left are brain-heart infusion broth (BHI or BHIB) and they will be allowed to stand and cool. Once cool, they will remain liquid. The tubes on the right contain brain-heart infusion agar (BHIA), a solid medium, which is liquid when it is removed from the autoclave but will solidify as it cools due to the agar in it. The position of the tube as it cools will determine whether an agar deep tube or an agar slant is produced (Fig. 1.7). Note the angled sides of the racks that allow them to be tipped so the agar will solidify in a slant. Plated media are autoclaved in a flask and then dispensed into sterile Petri dishes (Fig. 1.8).

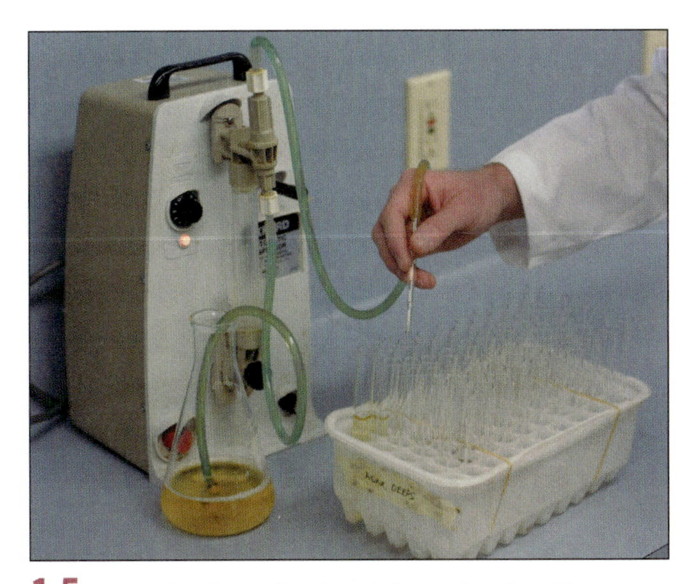

**1.5 Dispensing the Medium into Tubes** ■ An adjustable pump can be used to dispense the appropriate volume (usually 7 mL–10 mL) into tubes. Then, loosely cap the tubes.

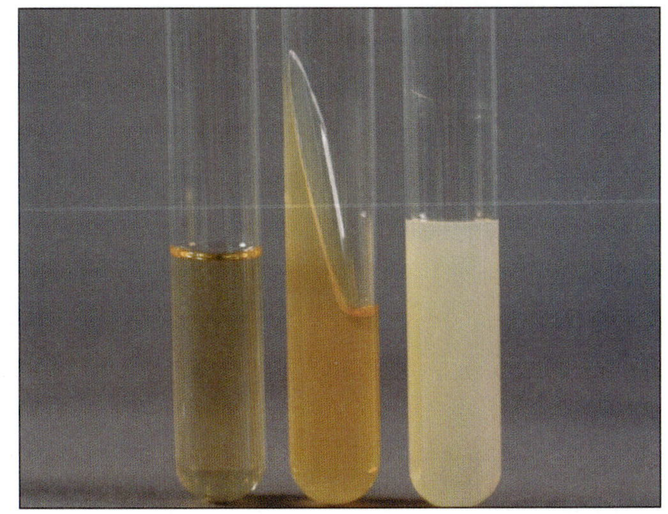

**1.7 Tubed Media** ■ From left to right: a broth, an agar slant, and an agar deep tube. The solid media are liquid when they are removed from the autoclave. Agar deeps are allowed to cool and solidify in an upright position, whereas agar slants are cooled and solidified on an angle.

## Lab Two

1. Examine the tubes and plates for evidence of growth.
2. Record your observations on the data sheet, page 29.
3. Save the plates and tubes as directed by your instructor.

**1.8 Pouring Agar Plates** ■ Agar plates are made by pouring sterilized medium into sterile Petri dishes. The lid is used as a shield to prevent airborne contamination. Once poured, the dish is gently swirled so the medium covers the base. Plates are then cooled and dried to eliminate condensation.

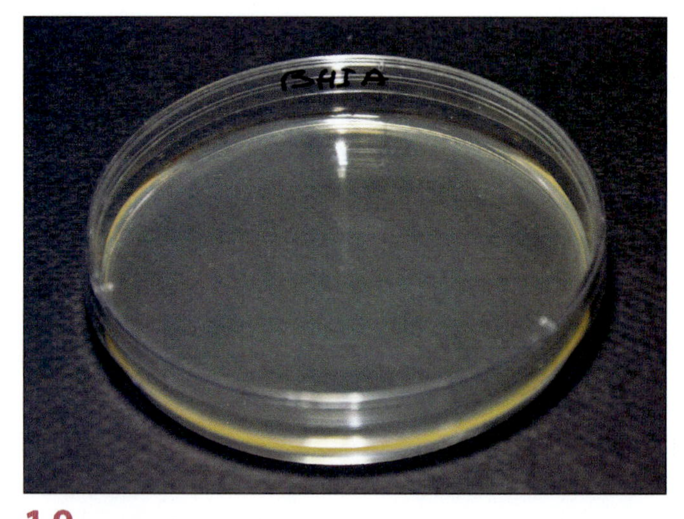


**1.9 An Agar Plate** ■ Plated media are often used for isolating individual species from a mixed culture (Exercises 1-5 and 1-6) or for counting the number of cells in a diluted sample (Exercise 6-2). Some differential tests also use plated media (e.g., milk agar, DNase agar, starch agar). Shown is a BHIA plate, which stands for "brain-heart infusion agar." The 1.5%–2% agar in the medium acts as a solidifying agent to suspend the nutrients—extracts of brain and heart tissues. Brain and heart provide carbon, nitrogen, and other essential nutrients for growth, as well as energy.

### Reference

Zimbro, Mary Jo and David A. Power. Pages 404–405 and 408 in *DIFCO™ & BBL™ Manual—Manual of Microbiological Culture Media*. Sparks, MD: Becton Dickinson and Company, 2003.

Name ______________________________

Date ______________________________

Lab Section ______________________________

I was present and performed this exercise (initials) ______________________________

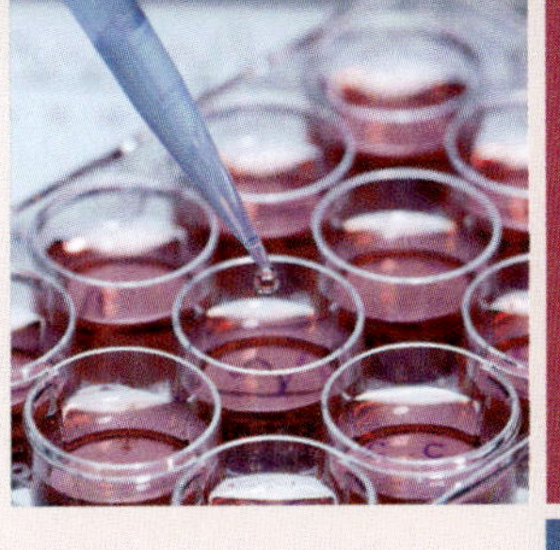

DATA SHEET

1-3

# Nutrient Broth and Nutrient Agar Preparations

## OBSERVATIONS AND INTERPRETATIONS

**1** In the following table, record the number of each medium you prepared, and then record the number of apparently sterile ones. Calculate your percentage of successful preparations for each. In the last column, speculate as to probable/possible sources of contamination (even if all your media are sterile).

| Medium | Total Number Prepared | Number of Sterile Preparations | Percentage of Successful Preparations | Probable Sources of Contamination |
|---|---|---|---|---|
| Nutrient agar tubes (slant or deep) | | | | |
| Nutrient agar plates | | | | |
| Nutrient broths | | | | |

## QUESTIONS

**1** *Which medium was most difficult to prepare without contamination? Why do think this might be so?*

1

## 2 *For each of the following types of contamination suggest the most likely point in preparation (or later) at which the contaminant was introduced.*

**a.** *Growth in all broth tubes.*

**b.** *Growth in one broth tube.*

**c.** *Growth only on the surface of a plate.*

**d.** *Growth throughout the agar's thickness on a plate.*

**e.** *Growth only in the upper 1 cm of agar in an agar deep tube.*

**f.** *All plates in a batch have the same type and density of contaminants.*

**g.** *Only a few plates in a batch are contaminated, and each looks different.*

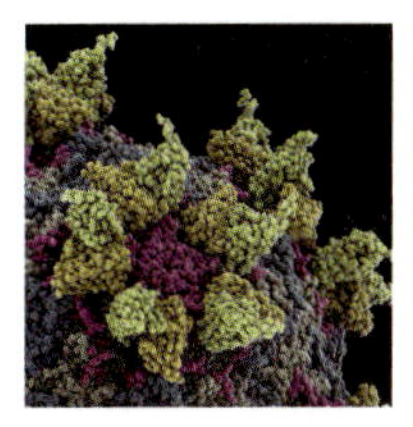

## A Word About Aseptic Transfers and Inoculation Methods

Microbiologists must be able to transfer living microbes from one place to another aseptically (i.e., without contamination of the culture, the sterile medium, or the surroundings). Microbiology students must also acquire this skill and do so very early on. While you won't be expected to master all transfer methods right now, you will be expected to perform most of them over the course of the semester. Refer to Exercises 1-4 through 1-6 and Appendices B, C, and D as needed.

To prevent contamination of the sample, inoculating instruments (Fig. 1.10) must be sterilized prior to use. Wire inoculating loops and needles are sterilized immediately before use in an incinerator or Bunsen burner flame. The mouths of tubes or flasks containing cultures or media are also incinerated at the time of transfer by passing their openings through a flame. Instruments that are not conveniently or safely incinerated, such as Pasteur pipettes, cotton applicators, glass pipettes, and digital pipettor tips, are sterilized inside wrappers or containers by autoclaving prior to use.

Aseptic transfers are not difficult; however, a little preparation will help assure a safe and successful procedure. Before you begin, you will need to know where the sample is coming from, its destination, and the type of transfer instrument to be used. These exercises provide step-by-step descriptions of routine transfer methods. Certain less-routine transfer methods are discussed in Appendices B through D. ■

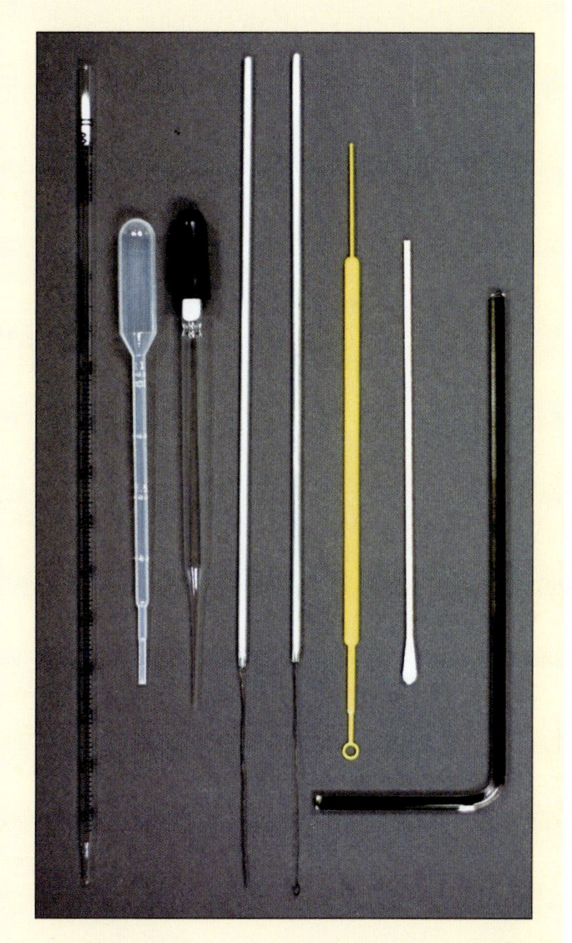

**1.10 Inoculating Instruments** ■ Any of several different instruments may be used to transfer a microbial sample, the choice of which depends on the sample source, its destination, and any special requirements imposed by the specific protocol. Shown here are several examples of transfer instruments. From left to right: serological pipette (see Appendix C), disposable transfer pipette, Pasteur pipette, inoculating needle, inoculating loop, disposable inoculating needle/loop, cotton swab (see Appendix B and Exercise 1-5), and glass spreading rod (see Exercise 1-6). (***Note:*** the glass rod is not an inoculation instrument, but it is used to spread an inoculum introduced to an agar plate by another instrument. As such, it is an instrument used in an inoculation process.) When transferring BSL-2 organisms, we advise using a sterile disposable loop or wooden stick (not shown). Neither of these requires incineration after use and each minimizes the threat of aerosol production.

# Common Aseptic Transfers and Inoculation Methods

EXERCISE **1-4**

## ■ Theory

A medium that contains living microbes is called a **culture**. If a culture contains a single species it is said to be a **pure culture**. It is essential to transfer microbes from their pure culture to a sterile medium **aseptically**, that is, without contamination of yourself, others, the environment, the source culture, or the medium being inoculated. In other words, you want your pure culture to stay pure, your new culture to be pure, and the surroundings to remain uninoculated.

### General Techniques and Practices for Aseptic Transfers

Following are general techniques and practices that improve your chances of successfully making an aseptic transfer. Equally important are the techniques related to safety. It has been demonstrated that while "...the causative incident for most LAIs (laboratory acquired infections) is unknown...A procedure's potential to release microorganisms into the air as aerosols and droplets is the most important operational risk factor..." (Chosewood and Wilson, 2009). In other words, limiting aerosol production is a *safety issue and not an issue of keeping pure cultures pure*. Please adhere to these practices for your safety and the safety of those around you.

- **Minimize the potential of contamination.** Do not perform any transfers over your books and papers because you may inadvertently and unknowingly contaminate them with droplets or aerosols that settle. Put them safely away. Some labs advise performing transfers over a disinfectant-soaked paper towel.
- **Be organized.** Arrange all media in advance and clearly label them with your name, the date, the medium, and the inoculum (Fig. 1.11). Tubes may be labeled with tape, paper held on with a rubber band, or by writing directly on the glass (this option presumes you will clean the writing off before you dispose of the culture for decontamination). You *should* write directly on the base (not the lid) of plastic Petri plates because they are disposable. Be sure not to place any labels in such a way as to obscure or obstruct your view of the tube's or plate's interior.
- **Place all media tubes in a test tube rack when not in use whether they are sterile or not.** Tubes should never be laid on the table surface because they may leak (Fig. 1.12).
- **Take your time.** Work efficiently, but do not hurry. You are handling potentially dangerous microbes. Working at a frenzied pace leads to carelessness and accidents.
- **Never hold a tube culture by its cap.** Caps are generally loose to allow aeration and are not secure enough to be used as a handle. Even screw caps can be loose enough so as not to be secure.
- **Hold the inoculating loop or needle like a pencil in your dominant hand and relax** (Fig. 1.13)!
- **Adjust your Bunsen burner so its flame has an inner and outer cone** (Fig. 1.14).

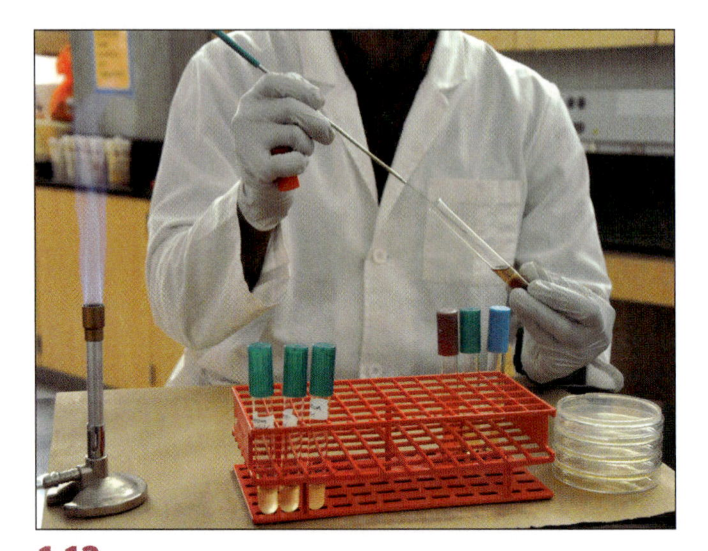

**1.12 Microbiologist at Work** ■ Materials are neatly positioned and not in the way, and the Bunsen burner is accessible, but not so close as to be a major fire hazard. To prevent spills, culture tubes are stored upright in a test tube rack. They are never laid on the table. The microbiologist is relaxed and ready for work. Notice he is holding the loop like a pencil, not gripping it like a dagger.

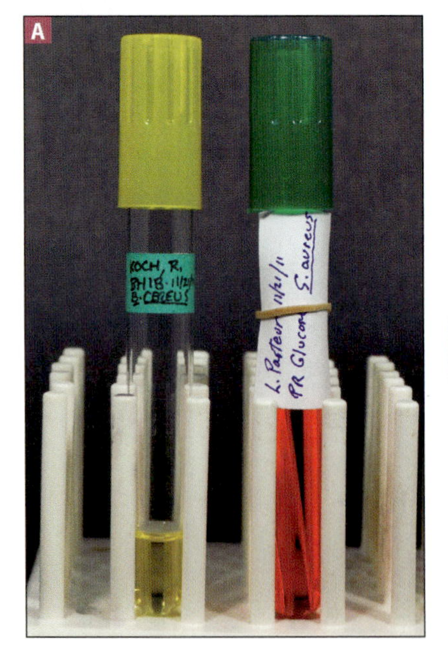

**1.11 Label the Media** ■ To avoid confusion after-the-fact, it is best to label sterile media prior to inoculating it. (**A**) Tubed media can be labeled with tape, paper labels, or directly on the glass with a marking pen. Labels must be removed when tubes are put in the autoclave bin for sterilization. (**B**) Plastic Petri dishes should be labeled on their base, not on their lid, because the lid may get separated from its base during reading or rotated from its correct orientation. Write the information at the edge to avoid obscuring growth on the plate. Because most labs use disposable Petri dishes, the labels do not have to be removed prior to autoclaving.

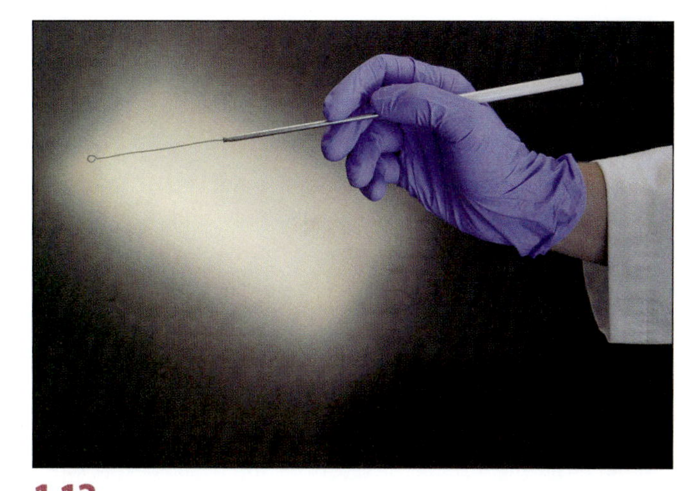

**1.13 Hold the Loop like a Pencil** ■ Holding the loop as shown puts the hand in a convenient position to hold tube caps with the "pinky" finger.

**1.14 Bunsen Burner Flame** ■ When properly adjusted, a Bunsen burner produces a flame with two cones. Sterilization of inoculating instruments is done in the hottest part of the flame—the tip of the inner cone (red arrow). Heat-fixing bacterial smears on slides and incinerating the mouths of open glassware items are done in the outer cone (white arrow).

### Types of Media

Media come in many forms, each with specific applications. **Broths** are used to grow microbes when fresh cultures or large numbers of cells are required. Broths of differential media are also used in microbial identification (Section 5). **Agar slants** are generally used to grow stock cultures that can be refrigerated after incubation and maintained for several weeks. In addition, many differential media are agar slants. **Plated media** are typically used for obtaining isolation of species (Exercises 1-5 and 1-6), differential testing, and quantifying bacterial densities (Exercise 6-1). In all cases, using these media requires aseptic inoculation in which a portion of an existing pure culture is transferred to a sterile medium to start a new pure culture. (A streak plate and sometimes a spread plate are exceptions to this. See Exercises 1-5 and 1-6.)

Transfers can be made between all forms of media—slants, broths, and plates—depending on the intended use of the new culture. The following is organized into transfers from broth culture to sterile broth, agar slant culture to sterile agar slant, and plate culture to sterile broth. If you can do these, then you have the skills to transfer between most any combination of media. (Inoculation of sterile agar plates is covered in Exercises 1-5 and 1-6 and will complete your skill set.)

### Transfer Instruments

The instruments usually used for transfers are either **inoculating loops** or **inoculating needles**. (Pipettes are also sometimes used and these are covered in Appendices C and D.) For simplicity, the following instructions only refer to inoculating loops, but the same apply to inoculating needles.

### A Special Note about Transferring BSL-2 Organisms

Most college biology teaching laboratories have eliminated or reduced the use of BSL-2 organisms, as we have in this edition of the lab manual. However, use of some BSL-2 organisms is unavoidable for some tests. In other exercises, they are included as optional test organisms.

The primary concern with BSL-2 organisms is aerosol production, which can lead to contamination of the environment or infection due to inhalation. Aerosols are problematic because we generally are unaware of their production and they remain suspended in the air long after the procedure has been completed.

Throughout this section we emphasize techniques that minimize aerosol production, but they are even more essential when handling BSL-2 organisms. Your lab will have specific guidelines on how to handle BSL-2 organisms and you should take these seriously. They may include any or all of the following precautions, depending on the exercise and the equipment you have available.

- Performing tests in a Class I or II biosafety cabinet (Fig. I.1) is recommended, but not all teaching laboratories have these installed.
- Using an electric incinerator (Fig. 1.15) for decontaminating wire loops and needles, which simultaneously contains aerosols within the ceramic interior and decontaminates them.
- Using sterile disposable loops/needles or wooden sticks that don't require flaming after use and can be disposed of in sharps containers or other appropriate receptacles for autoclaving.
- Wearing gloves and eye protection.
- Identifying BSL-2 organisms with red caps on tube cultures or BSL-2 labels on plate cultures.

Whatever precautions your college's guidelines dictate, take them seriously. Your health and the health of others in your lab are at stake.

### Transfer from a Broth Culture to a Sterile Broth

As you read these instructions, also follow the procedural diagram in Figure 1.16 to get a summary view of the process. **Make appropriate adjustments if handling a BSL-2 organism.**

1. Label the sterile broth tube with your name, the date, the medium, and the organism you are inoculating it with.
2. Make sure your loop is a closed circle. If it isn't, pinch it closed. Hold it like a pencil, and then flame it from base to tip as shown in Figure 1.17. Be sure the entire wire becomes orange-hot at some point.
3. Suspend the bacteria in the broth culture with a vortex mixer prior to transfer (Fig. 1.18). Be sure not to mix so vigorously that broth gets into the cap or that you lose control of the tube. Start slowly, and then gently increase the speed until the tip of the vortex reaches the bottom of the tube. Alternatively, the culture may be agitated by drumming your fingers along the length of the tube several times (Fig. 1.19). Again, be careful not to splash broth into the cap or lose control of the tube.
4. Loosen the cap of the culture tube (this is especially important if you are using a screw-cap tube). Move the culture tube to your loop hand. Remove and hold the tube's cap with the little finger of your loop hand (Fig. 1.20). Moving the tube and not the loop prevents excessive movement of the loop that might result in aerosols or droplets.
5. Incinerate the lip of the tube by passing it quickly through the flame two or three times (Fig. 1.21). Do not wave it through the flame so fast that broth sloshes out of the open end.
6. Hold the tube on an angle to minimize the opportunity for airborne contamination (Fig. 1.22).
7. This step is important to minimize the production of aerosols. Hold the loop hand still and move the tube up the wire until the loop's tip is in the broth. Continue holding the loop hand still while you remove the tube from over the loop (Fig. 1.23). *Be careful not to catch the loop on the lip of the tube or you will produce contaminated droplets and aerosols.* At this point, there should be a visible film of broth in your loop (Fig. 1.24). If there is not a visible film of broth, replace the cap, flame the loop, let it cool, and then pinch the loop so it is a closed circle. Then, start over.
8. While holding your loop hand still, flame the tube and replace the cap. Set the tube in the rack and pick up the sterile broth tube in your free hand.
9. Repeat the process of removing the cap with the little finger of the loop hand and flaming the lip of the sterile broth tube while holding the loop hand still. Hold the tube on an angle. *Don't forget you have living microbes exposed to the environment on the loop at this point and careless movements can spread them.*
10. Move the tube over the loop until it is submerged in the sterile broth and mix by gently swirling the loop.
11. Before you remove the loop from the tube, tap the loop's face on the inside of the tube to remove the

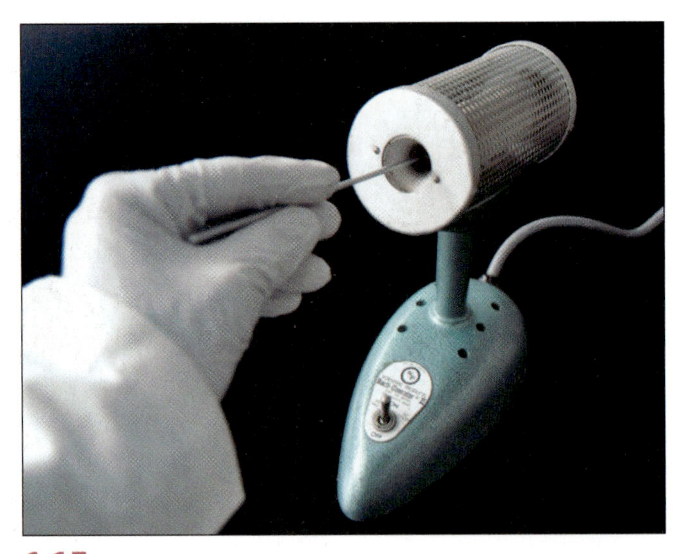

**1.15 Bacteriological Incinerator** ■ Bacterial incinerators use infrared heat and reach temperatures over 800°C. A wire loop/needle is inserted into the incinerator and heated for 5–7 seconds. (The handle may also get hot, so be careful. You may wish to wrap the handle in several layers of tape for insulation.) The loop/needle is then removed and allowed to cool without touching anything. It may then be used to transfer microbes, or if this is done at the completion of a transfer, it may be set-aside in a holder (often in the base of the incinerator, as shown) until needed again.

film within the loop (Fig. 1.25). Be persistent in this—don't give up until you are successful at its removal. If you were to flame the loop with the film present, you would produce aerosols.

**12** Carefully remove the tube from over the loop and avoid catching the loop on the lip of the tube.

**13** Keeping your loop hand still, flame the tube, replace the cap, and set it in the rack.

**14** Flame the loop from base to tip until it is uniformly orange-hot.

**15** Incubate the inoculated culture at the assigned temperature for the assigned time.

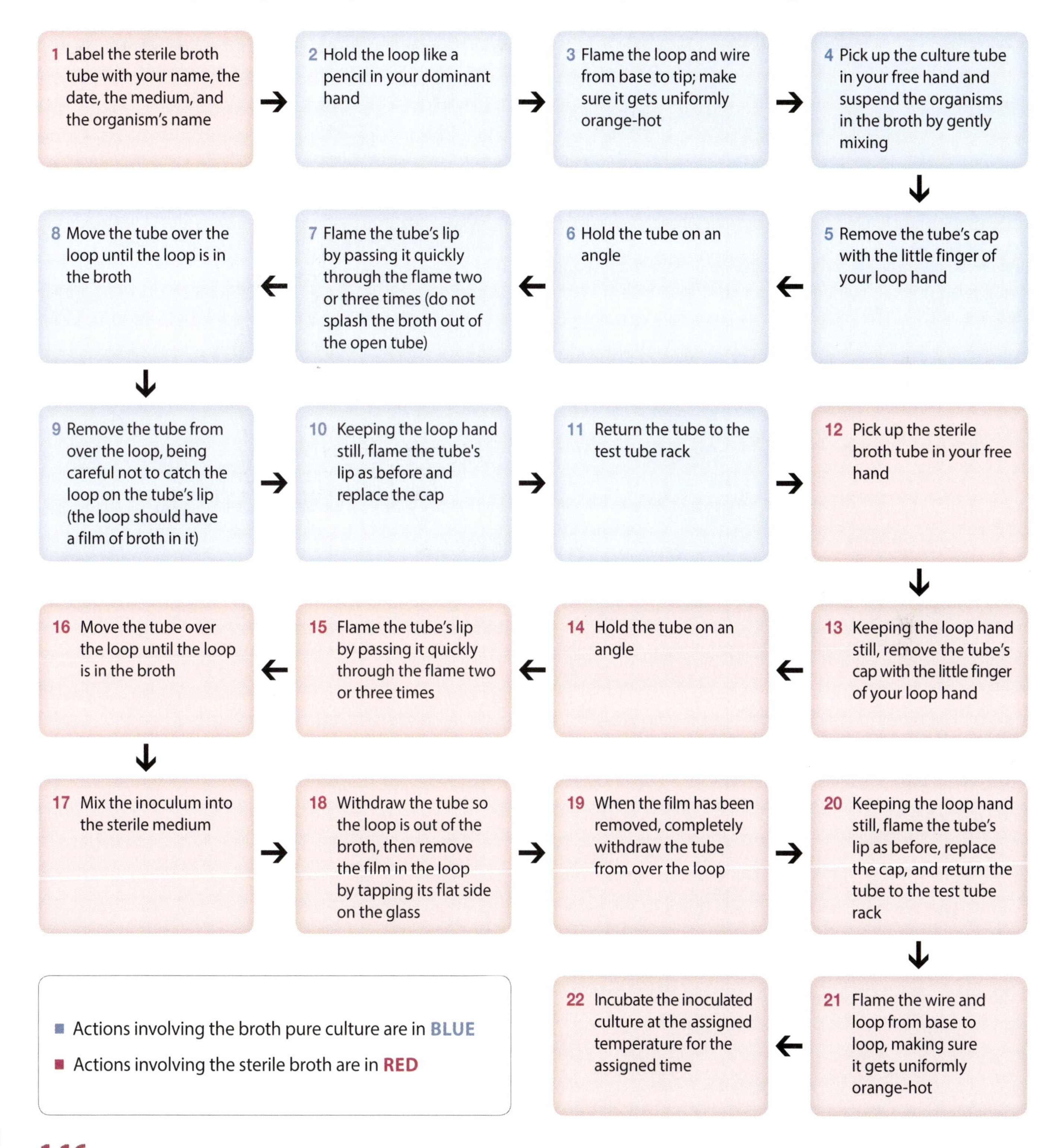


**1.16 Procedural Diagram: Aseptic Transfer from a Broth Pure Culture to a Sterile Broth Tube** ■ This is a summary of the procedure. Make every effort to keep your loop hand as still as possible throughout the transfer. Details can be found in the text. Make appropriate adjustments if transferring a BSL-2 organism.

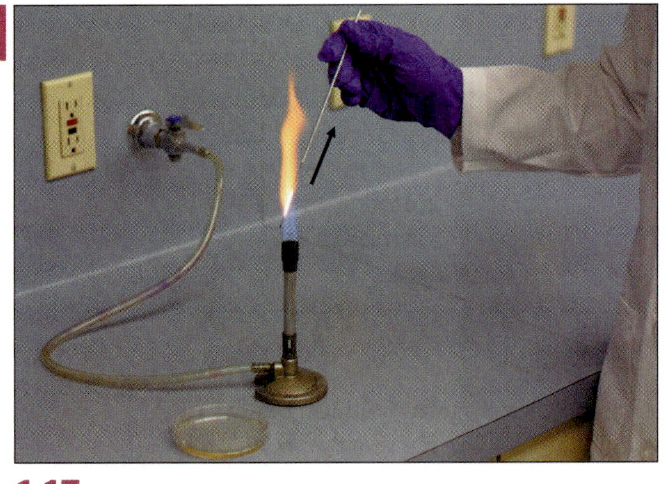

**1.17 Flaming the Loop** ■ Incineration of an inoculating loop's wire is done by passing it through the tip of the flame's inner cone. Begin at the wire's base and continue to the end, making sure that all parts are heated to a uniform orange color. Allow the wire to cool before touching it or placing it on/in a culture. The former will burn you; the latter will cause aerosols of microorganisms.

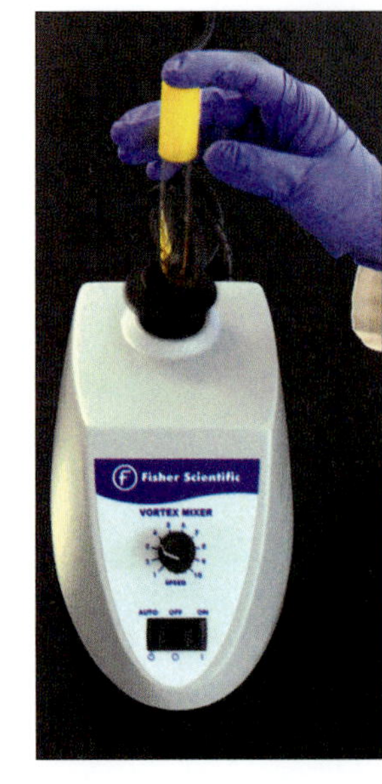


**1.18 Vortex Mixer** ■ Bacteria may be suspended in a broth using a vortex mixer. The switch on the bottom has three positions: "auto" (left), "off" (center), and "on" (right). The rubber boot is activated when touched only if the "auto" position is used; "on" means the boot is constantly vibrating. Above the on/off/auto switch is a variable speed knob. The slowest speed that allows the vortex to reach the bottom of the tube is used. Caution must be used to prevent broth from getting into the cap or losing control of the tube and causing a spill (note the hand position around the tube, ready to grab it). Short bursts of vortexing can be used if the glassware is too full to allow vortexing to the bottom.

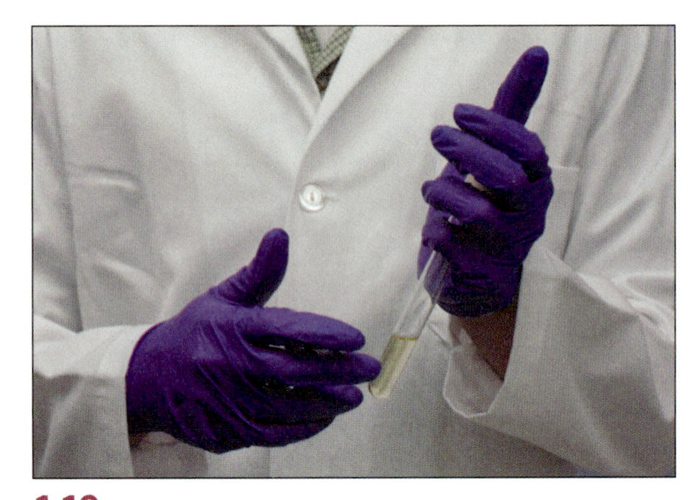

**1.19 Mixing Broth by Hand** ■ A broth culture always should be mixed prior to transfer. Tapping the tube with your fingers gets the job done safely and without special equipment.

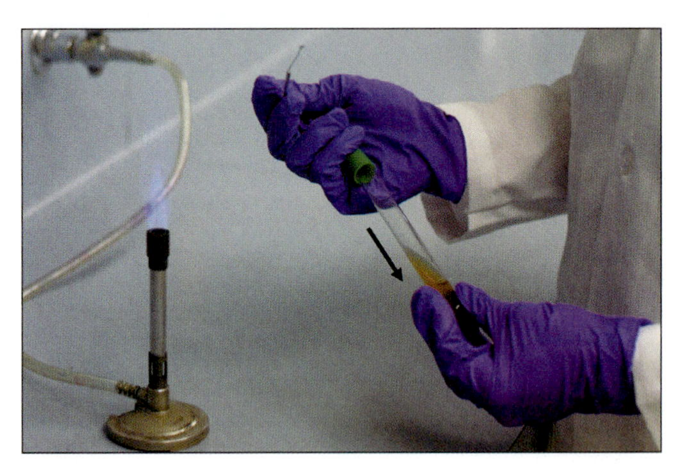

**1.20 Removing the Tube Cap** ■ The loop is held in the dominant hand and the tube in the other hand. Remove the tube's cap with the little finger of your loop hand by pulling the tube away with the other hand; keep your loop hand still. Hold the cap in your little finger during the transfer. When replacing the cap, move the tube back to the cap to keep your loop hand still. The replaced cap does not have to be on firmly at this time—just enough to cover the tube.

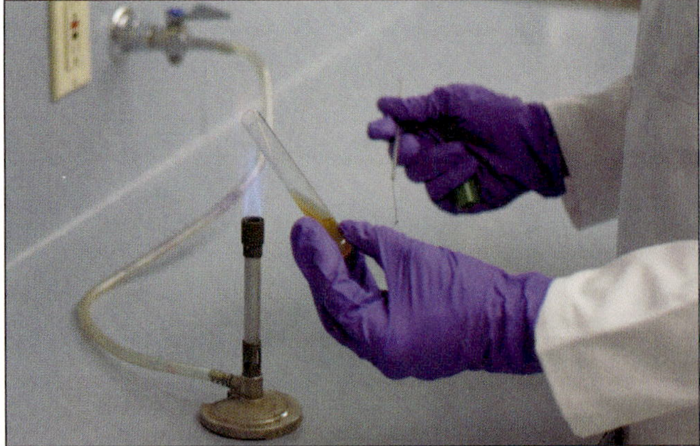

**1.21 Flaming the Tube** ■ The tube's mouth is passed quickly through the flame a couple of times to sterilize the tube's lip and the surrounding air. Do not move the tube so quickly that broth sloshes out the opening. Notice that the tube's cap is held in the loop hand.

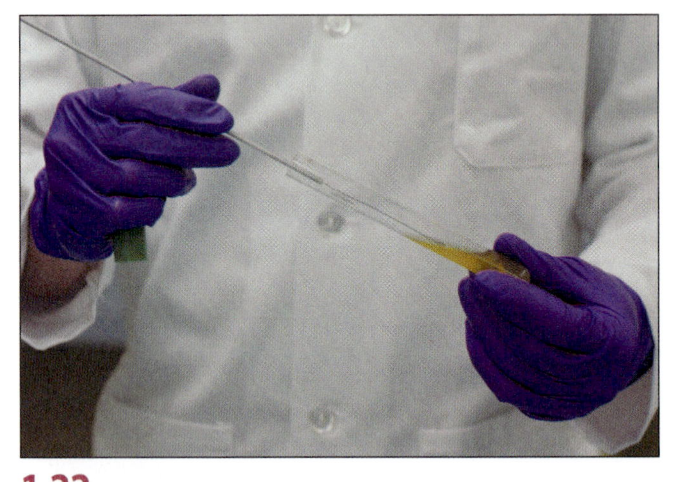

**1.22 Holding the Tube at an Angle** ■ The tube is held at an angle to minimize the chance that airborne microbes will drop into it. Notice that the tube's cap is held in the loop hand.

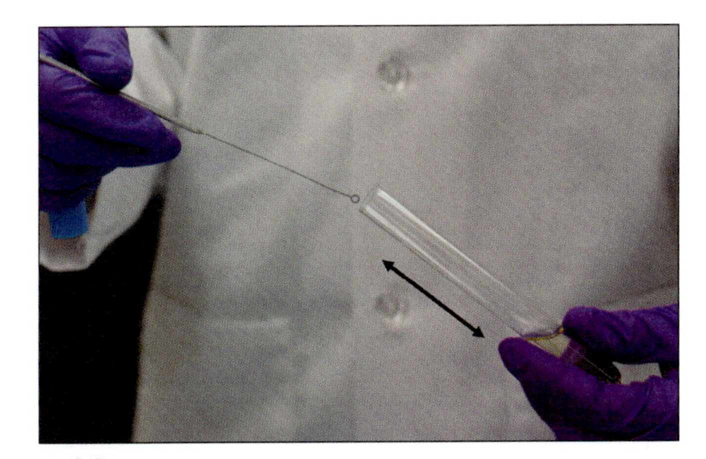

**1.23 Move the Tube, Not the Loop** ■ The open tube is held at an angle to minimize airborne contamination of it. When placing a loop into a broth tube or removing it, keep the loop hand still and move the tube. ***Be careful not to catch the loop on the tube's lip when removing it.*** This produces aerosols that can be dangerous or produce contamination.

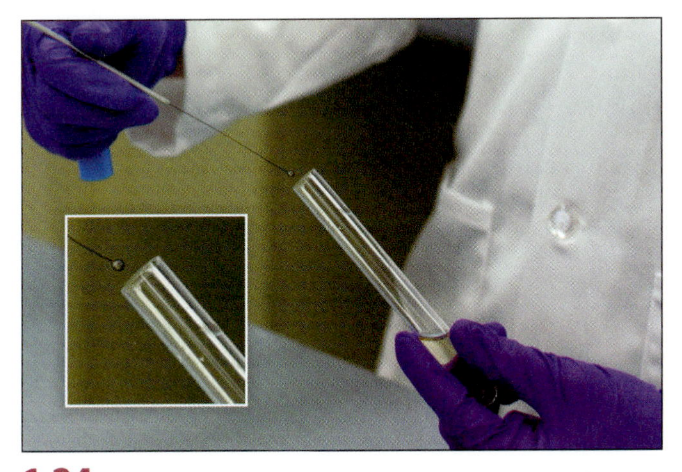

**1.24 Removing the Loop from Broth** ■ Notice the film of broth in the loop (see inset). Be careful not to catch the loop on the lip of the tube when removing it. This would produce aerosols and droplets that can be dangerous or produce contamination.

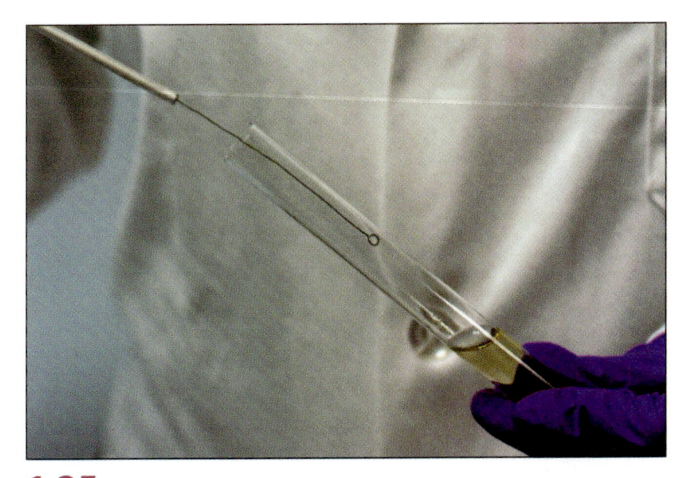

**1.25 Removing Excess Broth from Loop** ■ Before removing it from the new culture tube, tap the *face* of the loop on the glass to remove the broth film (Fig. 1.24). Failing to do so will result in splattering and aerosols when sterilizing the loop in a flame.

### Transfer from an Agar Slant Culture to a Sterile Agar Slant

As you read these instructions, also follow the procedural diagram in Figure 1.26 to get a summary view of the process. This transfer has a lot in common with the broth-to-broth transfer and you are referred back to it at relevant points. **Make appropriate adjustments if handling a BSL-2 organism.**

1. Label the sterile agar slant with your name, the date, the medium, and the inoculum.
2. Hold the loop like a pencil and then flame it from base to tip as in a broth transfer (Fig. 1.17). Be sure the entire wire becomes orange-hot at some point.
3. Loosen and remove the cap on the culture tube as in a broth transfer (Fig. 1.20).
4. Hold the culture tube on an angle with the agar surface facing upward.
5. Flame the tube's lip as in a broth transfer (Fig. 1.21).
6. Move the culture tube up the wire of the loop, and then gently touch the loop's tip to the growth on the agar's surface (Fig. 1.27). You don't need to dig into the agar, nor do you need to scoop up a glob of growth. Just touch the loop to the growth and pick up the smallest amount you can see with your naked eye.
7. Holding your loop hand still, carefully remove the tube from over the wire, flame the culture tube's lip, and replace its cap. Place the culture in the test tube rack.
8. Pick up the sterile agar slant in your free hand, remove the cap, and flame the tube as before.
9. Holding the tube on an angle with the agar surface upward, move the tube over the wire so the loop is near the bottom of the slant.
10. Touch the tip of the loop (where there is organism) to the agar. Then, as you withdraw the tube move the loop back and forth (Fig. 1.28). Be careful not to cut the agar with the loop. This is called a **fishtail inoculation** or **fishtail streak** because you are seeding the agar surface in a wavy pattern resembling the movement of a fish tail.
11. Be careful not to catch the loop on the tube's lip as you remove the tube. Then, keeping the loop hand still, flame the tube's lip, replace its cap, and put it in the test tube rack.
12. Flame the loop from base to tip as before.
13. Incubate the inoculated culture at the assigned temperature for the assigned time.

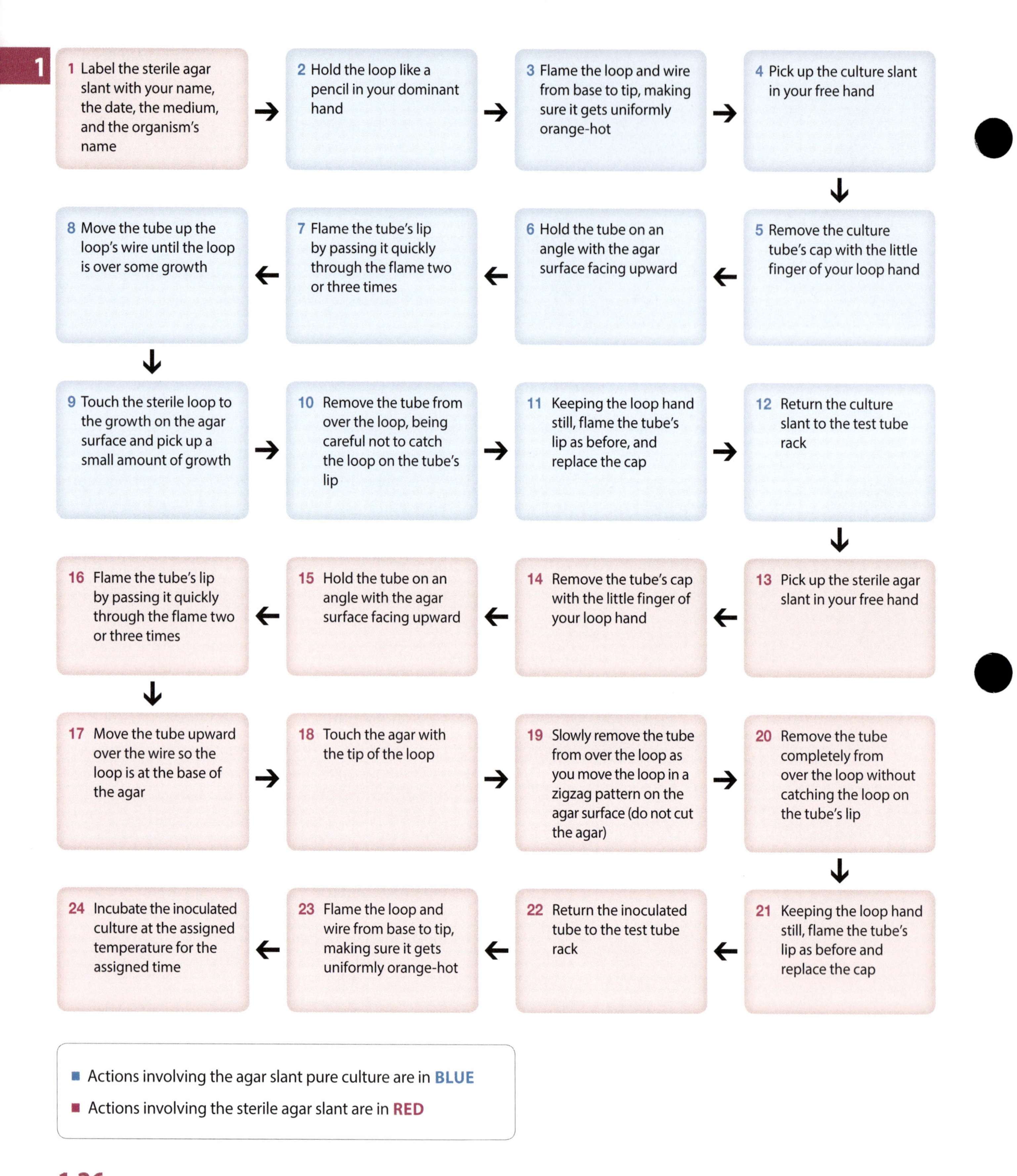


**1.26 Procedural Diagram: Aseptic Transfer from a Nutrient Agar Slant to a Sterile Nutrient Agar Slant** ■ This is a summary of the procedure. Make every effort to keep your loop hand as still as possible throughout the transfer. Details can be found in the text. Make appropriate adjustments if transferring a BSL-2 organism.

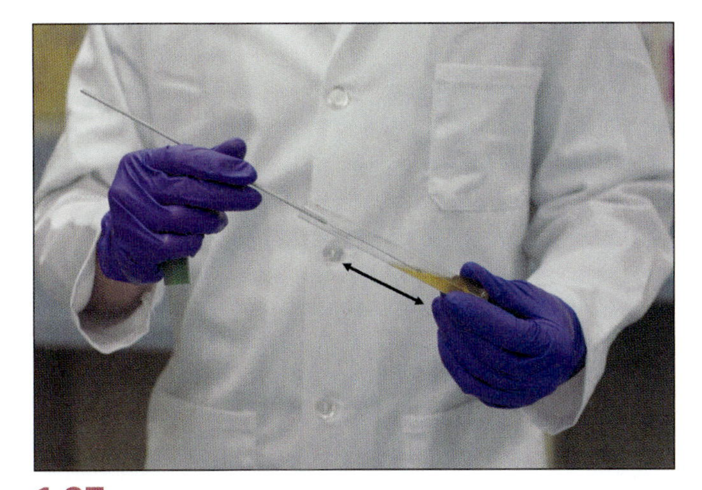

**1.27 A Loop and an Agar Slant** ■ When placing a loop into a slant tube or removing it, the loop hand is kept still while the tube is moved. Hold the tube so the agar is facing upward. To pick up the inoculum, you only need to gently touch the growth on the agar surface.

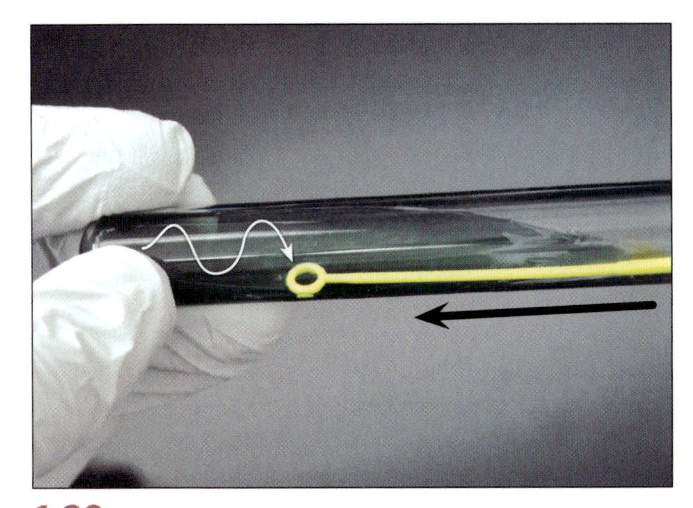

**1.28 Fishtail Inoculation of a Slant** ■ Begin at the base of the slant and gently move the loop back and forth as you withdraw the tube. Use the end of the loop and be careful not to cut the agar. After completing the transfer, sterilize the loop or dispose of it properly. (A disposable loop was used in this photo.)

### Transfers from a Plate Culture to a Sterile Broth or Agar Slant

As you read these instructions, also follow the procedural diagram in Figure 1.29 to get a summary view of the process. **Make appropriate adjustments if transferring a BSL-2 organism.**

1. Label the sterile broth tube with your name, the date, the medium, and the organism you are inoculating it with.
2. Flame the loop from base to tip (Fig. 1.17).
3. Lift the lid of the Petri dish and use it as a shield from airborne contamination (Fig. 1.30).
4. Touch the loop to an uninoculated portion of the plate to cool it. (Loop wires can get very hot if a series of transfers are made in a short period of time. Placing a hot wire on growth may cause the growth to spatter and create aerosols. If you are not doing a lot of successive transfers, this is probably unnecessary.) Gently touch the loop to the center of an isolated colony on the agar surface and collect the smallest amount you can see. As with the slant, you don't need to dig into the agar or scoop up a glob of growth.
5. Remove the loop and replace the lid.
6. Pick up the sterile broth and continue with step 9 in the broth-to-broth transfer (p. 34) or step 12 in Figure 1.16. If transferring to a slant, continue with step 8 in the slant-to-slant transfer (p. 37) and step 13 in Figure 1.26.

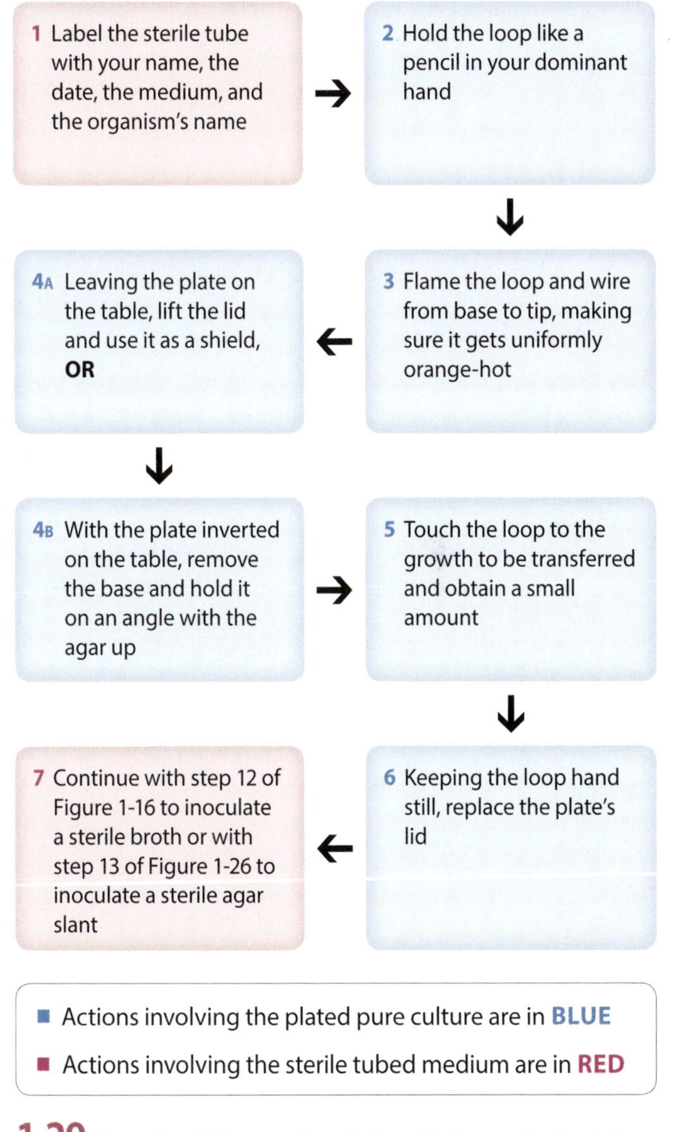


**1.29 Procedural Diagram: Aseptic Transfer from a Nutrient Agar Plate Pure Culture to a Sterile Tubed Medium** ■ This is a summary of the procedure. Make every effort to keep your loop hand as still as possible throughout the transfer. Details can be found in the text. Inoculation of a sterile broth is the same as in Figure 1.16, whereas inoculation of a sterile slant is the same as in Figure 1.26. Make appropriate adjustments if transferring a BSL-2 organism.

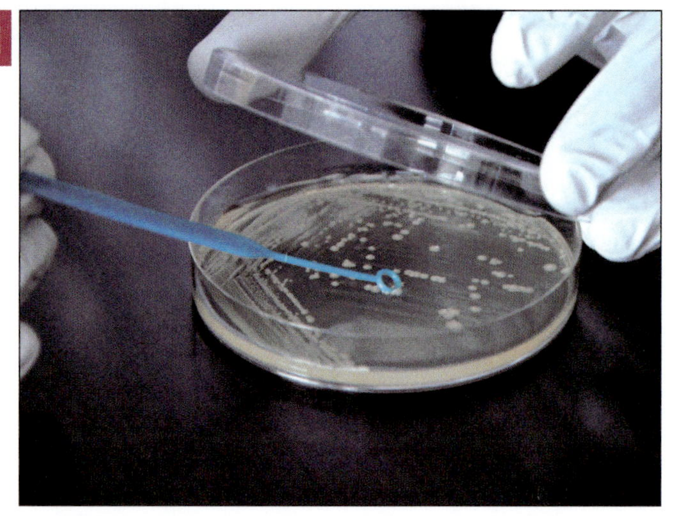

**1.30 "Picking" a Colony for Transfer** ■ Touch the tip of the loop to the center of an isolated colony and get a small amount of growth. Use the lid as a shield from airborne contamination.

## Application

To be a successful microbiologist, you must be able to transfer microorganisms from one place to another aseptically.

## In This Exercise

You now have the procedures for obtaining growth from a broth culture, an agar slant culture, and a plated culture. You also have the procedures for inoculating a sterile broth and a sterile agar slant. These can be performed in any combination necessary. Today, you will begin by testing your baseline dexterity by using sterile agar and sterile broth as the "culture" tubes *and* the tubes to be inoculated. After a little practice with these "blanks," you will do transfers with real cultures and sterile media in the following combinations: agar slant culture to sterile agar slant and sterile broth, broth culture to sterile agar slant and sterile broth, and plate culture to sterile broth (Fig. 1.31).

## Materials

### Per Student

- □ Lab coat
- □ Disposable gloves
- □ Chemical eye protection
- □ Inoculating loop
- □ Bunsen burner
- □ Four sterile nutrient broth tubes
- □ Three sterile nutrient agar slants
- □ Marking pens and labeling tape (or materials for the preferred labeling method in your laboratory)
- □ (Optional) vortex mixer

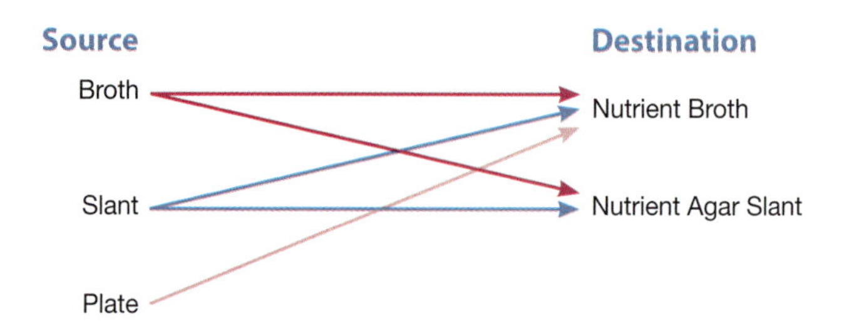


**1.31 Procedural Diagram for this Lab** ■ In today's lab, you will make transfers from broth and slant cultures of *S. epidermidis* to sterile nutrient broth and nutrient agar slant media. You will also transfer from a plate culture of *S. epidermidis* to a nutrient broth.

### Per Student Group

- □ Nutrient agar (NA) slant culture of *Staphylococcus epidermidis*
- □ Nutrient broth (NB) culture of *Staphylococcus epidermidis*
- □ Nutrient agar streak plate of *Staphylococcus epidermidis*

## PROCEDURE

### Lab One

1. Wear a lab coat, gloves, and chemical eye protection when performing this procedure.
2. Label one nutrient broth (NB) and one nutrient agar (NA) tube with your name and the word "sterile."
3. Using the sterile NB tube and the sterile NA tube, practice making transfers between them in all possible combinations. Your lab partner has the same assignment using a second set of tubes. Work with her/him and alternate transferring and evaluating each other's technique. (Not only will you learn by doing and evaluating another person, alternating will allow the tubes to cool.)
4. Incubate the "sterile" practice tubes at 35 ± 2°C until the next lab period. (You may wait to do this until you have completed step 5 and then transfer all the tubes to the incubator at one time.)
5. Once you have been "cleared" by your instructor to make transfers using real cultures, each student should perform the following, using Figure 1.31 as a guide:
   a. Label a sterile nutrient agar slant and a sterile nutrient broth with your name, the medium in the tube (either NA or NB), the source of inoculum (NA slant), and the organism. Then, aseptically transfer from the *S. epidermidis* NA slant culture to the sterile NA slant and the sterile NB.

b Label a sterile NA slant and a sterile NB with your name, the medium in the tube, the source of inoculum (NB), and the organism. Then, transfer from the *S. epidermidis* NB to the sterile NA slant and the sterile NB.

c Label a sterile NB with your name, the medium in the tube, the source of inoculum (NA plate), and the organism. Then, transfer from the *S. epidermidis* NA plate culture to the sterile NB. Choose a well-isolated colony and touch the center with the loop as in Figure 1.30.

d Incubate these five tubes at 35 ± 2°C until the next lab period.

6. Save or dispose of the original cultures as directed by your instructor.

### Lab Two

1. Remove your cultures and "sterile" practice tubes from the incubators and examine them for growth. Record your observations and answer the questions on the data sheet, page 43.
2. Your instructor may ask you to save your cultures for later use. If so, put them in the refrigerator. Otherwise, remove the labels and dispose of them in the appropriate autoclave container.

## References


Barkley, W. Emmett and John H. Richardson. Chap. 29 in *Methods for General and Molecular Bacteriology*. Washington, DC: American Society for Microbiology, 1994.

Chosewood, L. Casey and Deborah E. Wilson, eds. Page 4 in *Biosafety in Microbiological and Biomedical Laboratories*, 5th ed. U.S. Department of Health and Human Services Publication No. (CDC) 21-1112, December 2009.

Claus, G. William. Chap. 2 in *Understanding Microbes—A Laboratory Textbook for Microbiology*. New York: W. H. Freeman and Company, 1989.

Darlow, H. M. Chap. VI in *Methods in Microbiology*, Vol. 1. J. R. Norris and D. W. Ribbins, eds. London, UK: Academic Press, Ltd., 1969.

Emmert, Elizabeth A. B., Jeffrey Byrd, Ruth A. Gyure, Diane Hartman, and Amy White et al. "Biosafety Guidelines for Handling Microorganisms in the Teaching Laboratory: Development and Rationale." *Journal of Microbiology & Biology Education* (May 2013): 78–83. DOI: http://dx.doi.org/10.1128/jmbe.v14i1.531.

Estridge, Barbara and Anna Reynolds. Page 712 in *Basic Clinical Laboratory Techniques,* 6th ed. Independence, KY: Cengage, 2012.

Fleming, Diane O. Chap. 13 in *Laboratory Safety—Principles and Practices*, 2nd ed. Diane O. Fleming, John H. Richardson, Jerry J. Tulis, and Donald Vesley, eds. Washington, DC: American Society for Microbiology, 1995.

Koneman, Elmer W., Stephen D. Allen, William M. Janda, Paul C. Schreckenberger, and Washington C. Winn, Jr. Chap. 2 in *Color Atlas and Textbook of Diagnostic Microbiology*, 5th ed. Philadelphia: Lippincott-Raven Publishers, 1997.

Murray, Patrick R., Ellen Jo Baron, Michael A. Pfaller, Fred C. Tenover, and Robert H. Yolken. *Manual of Clinical Microbiology*, 6th ed. Washington, DC: American Society for Microbiology, 1995.

Power, David A. and Peggy J. McCuen. *Manual of BBL™ Products and Laboratory Procedures*, 6th ed. Cockeysville, MD: Becton Dickinson Microbiology Systems, 1988.

Name ______________________________

Date ______________________________

Lab Section ______________________________

I was present and performed this exercise (initials) ______________________________

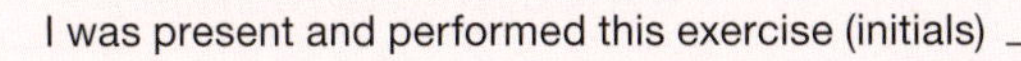

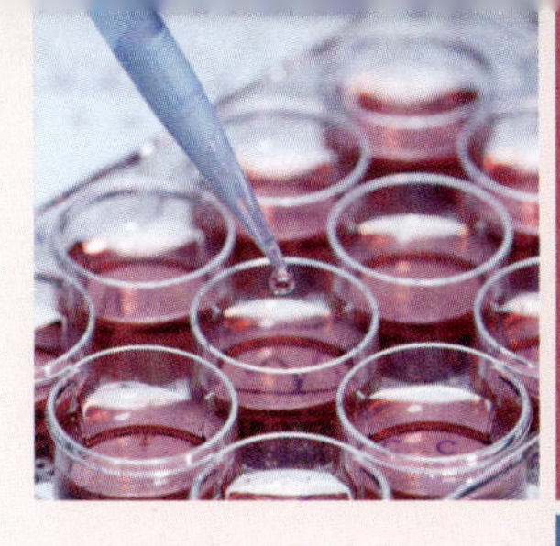

DATA SHEET 1-4

## Common Aseptic Transfers and Inoculation Methods

### OBSERVATIONS AND INTERPRETATIONS

1 Describe the appearance of growth on/in each medium. Growth on a solid medium could be described by color and amount (abundant, sparse, absent). Draw representative samples of each growth type. Growth in broth can be described by its degree of cloudiness (turbidity), using a qualitative scoring system:

+++ means "very turbid"
++ means "somewhat turbid"
+ means "barely turbid"
0 means "not turbid"

| Source | Medium Inoculated | |
|---|---|---|
| | Nutrient Broth | Nutrient Agar Slant |
| Sterile practice media | | |
| *Staphylococcus epidermidis* on NA slant | | |
| *Staphylococcus epidermidis* in NB | | |
| *Staphylococcus epidermidis* colony on sterile NA plate | | NA |

## QUESTIONS

**1** *Did you get growth on/in the sterile NB and NA slant tubes you practiced with? If not, congratulations! If so, where did you see it and what might have been its source(s)?*

**2** *Considering the cultures used to inoculate each medium in this exercise, how many different microbial types should you expect to see on/in each medium? Explain your answer.*

**3** *Which medium was most difficult for you to transfer from? Which medium was most difficult for you to inoculate? Explain your difficulties. (**Note:** There are no correct answers to these questions. They are based on evaluation of your personal experience.)*

**4** *Did you notice a difference in density (turbidity) of growth in NB tubes inoculated from NB and NA slants? Suggest possible reasons why a difference might occur.*

**5** *Did you notice a difference in density of growth on NA slants inoculated from NA slants and NB? Suggest possible reasons why a difference might occur.*

# Streak Plate Methods of Isolation

EXERCISE 1-5

## Theory

A microbial culture consisting of two or more species is said to be a **mixed culture**, whereas a **pure culture** contains only a single species. Obtaining isolation of individual species from a mixed sample is generally the first step in identifying an organism. A commonly used **isolation technique** is the **streak plate** (Fig. 1.32).

In the streak plate method of isolation, a bacterial sample (always assumed to be a mixed culture) is streaked over the surface of a plated agar medium. During streaking, the cell density decreases, eventually leading to individual cells being deposited separately on the agar surface. Cells that have been sufficiently isolated will grow into **colonies** consisting only of the original cell type (assuming the medium supports their growth). Because some colonies form from individual cells and others from pairs, chains, or clusters of cells, the term **colony-forming unit** (**CFU**) is a more correct description of the colony origin.

Several patterns are used in streaking an agar plate, the choice of which depends on the source of inoculum and microbiologist's preference. Although streak patterns range from simple to more complex, all are designed to separate deposited cells (CFUs) on the agar surface so individual cells (CFUs) grow into isolated colonies. A quadrant streak or a T-streak is generally used with samples suspected of high cell density, whereas a simple zigzag (continuous streak) pattern may be used for samples containing lower cell densities.

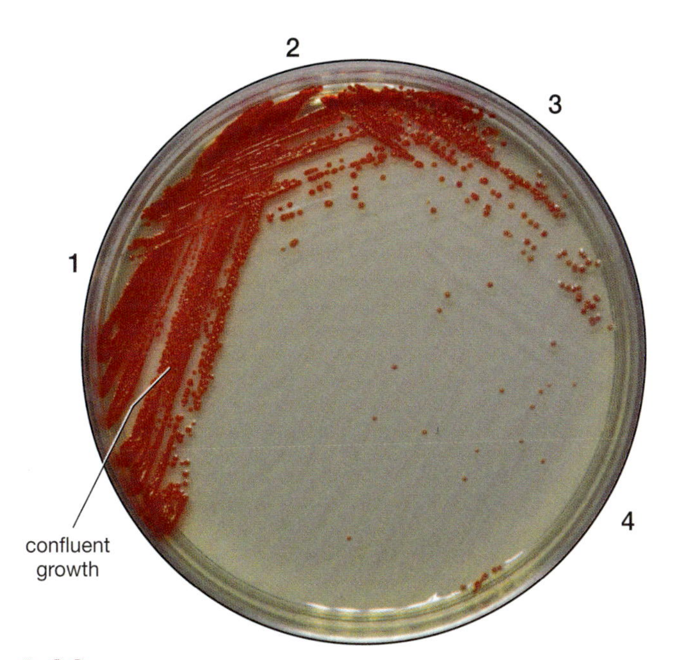


**1.32 Quadrant Streak Plate of *Serratia marcescens*** ■ Note the decreasing density of growth in the four streak patterns (indicated by numerals). On this plate, isolation is first achieved in the second streak, but the microbiologist would not know that at the time of streaking, so all four streaks are performed in the hope that isolation will occur in at least one of them. Cells from an isolated colony (one that is not touching another colony) can be transferred to a sterile medium to start a pure culture.

## Application

The identification process of an unknown microbe relies on obtaining a pure culture of that organism. The streak plate method produces individual colonies on an agar plate. A portion of an isolated colony then may be transferred to a sterile medium to start a pure culture.

Following are descriptions of streak techniques.

### Inoculation of Agar Plates Using the Quadrant Streak Method

This inoculation pattern is usually performed as the initial streak for isolation of two or more bacterial species in a mixed culture with suspected high cell density.

1. Label the plate's base with your name, date, and sample inoculated.
2. Obtain the sample of mixed culture with a sterile loop.
3. You have two options at this point. Use whichever is more comfortable for you or is required by your instructor.
   - a Leave the sterile agar plate on the table and lift the lid slightly, using it as a shield from airborne contamination (Fig. 1.33).

     or
   - b Place the plate lid down on the table (Fig. 1.34A). Then remove the base and hold it in the air on an angle (Fig. 1.34B).
4. Starting at the edge of the plate lightly drag the loop back and forth across the agar surface as shown in Figure 1.35A. Be careful not to cut the agar surface. The loop should contact the agar as shown in Figure 1.36.
5. Remove the loop and replace the lid.
6. Sterilize your loop as before. It is especially important to flame it from base to tip now because the loop has bacteria on it.
7. Rotate the plate a little less than 90°.
8. Let the loop cool for a few moments (or you can touch an open part of the agar), then perform

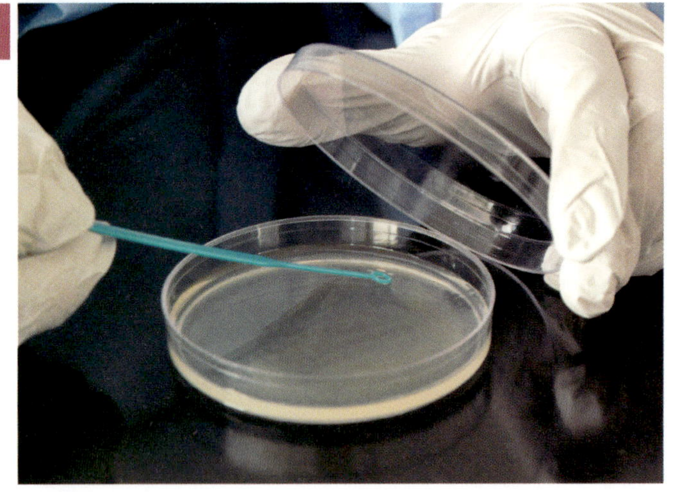

**1.33 Streak Plate Inoculation; Plate on the Table** ■ The streak plate may be performed with the plate's base resting on the table while holding the lid over it to prevent airborne contamination. Perform the streak plate as described in the text and as shown in Figure 1.35 or Figure 1.37.

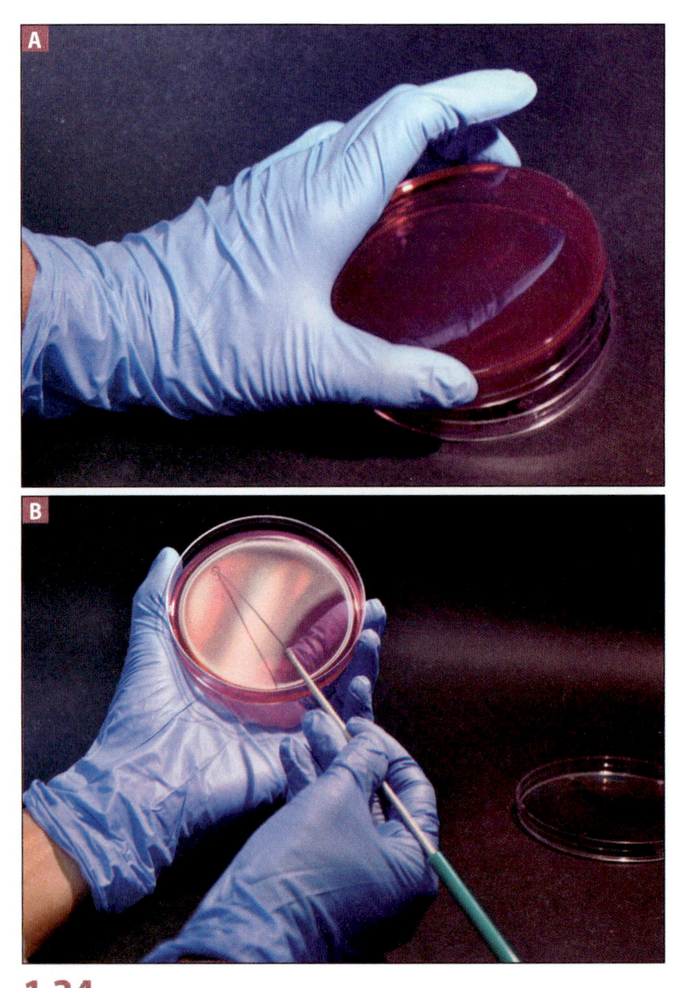

**1.34 Streak Plate Inoculation; Plate Held in the Hand** ■ (A) Some microbiologists prefer to hold the Petri dish in the air when performing a streak plate. To do this, place the plate lid down on the table and lift the base from it, holding it on an angle. (B) Perform the streak as described in the text and as shown in Figures 1.35 and 1.37.

another streak with the sterile loop beginning at one end of the first streak pattern (Fig. 1.35B). Intersect the first streak only two or three times.

9 Sterilize the loop, and then repeat with a third streak beginning in the second streak (Fig. 1.35C).

10 Sterilize the loop, and then perform a fourth streak beginning in the third streak and extending into the middle of the plate. Be careful not to enter any streaks but the third (Fig. 1.35D).

11 Sterilize the loop.

12 Incubate the plate in an inverted position for the assigned time at the appropriate temperature.

### Inoculation of Agar Plates Using the T-Streak Method

The T-streak method is a variation on the quadrant streak, but only three streakings are done (Fig. 1.37). There is no particular advantage of one method over the other. It basically comes down to personal preference.

1 Label the plate's base with your name, date, and sample inoculated.

2 With a marking pen, draw one line across the plate's base about one-third of the way down the plate. Then, draw a vertical line in the larger of the two regions roughly dividing it in half. The two lines make a "T."

3 Obtain the sample of mixed culture with a sterile loop.

4 You have two options at this point. Use whichever is more comfortable for you or is required by your instructor.

   a Leave the sterile agar plate on the table and lift the lid slightly, using it as a shield from airborne contamination (Fig. 1.33).

   or

   b Place the plate lid down on the table (Fig. 1.34A). Then remove the base and hold it in the air on an angle (Fig. 1.34B).

5 Streak the sample across the large region several times. Be careful not to cut the agar. The loop should contact the agar as shown in Figure 1.36.

6 Flame the loop from base to tip and let it cool in the air or touch it to an uninoculated region of the agar.

7 Make two or three streaks out of the first region into the second region, and then continue with an additional four or five streaks exclusively in the second region.

8 Flame the loop from base to tip and let it cool in the air or touch it to an uninoculated region of the agar.

9 Make two or three streaks out of the second region into the third region, and then continue with an additional four or five streaks exclusively in the third region.

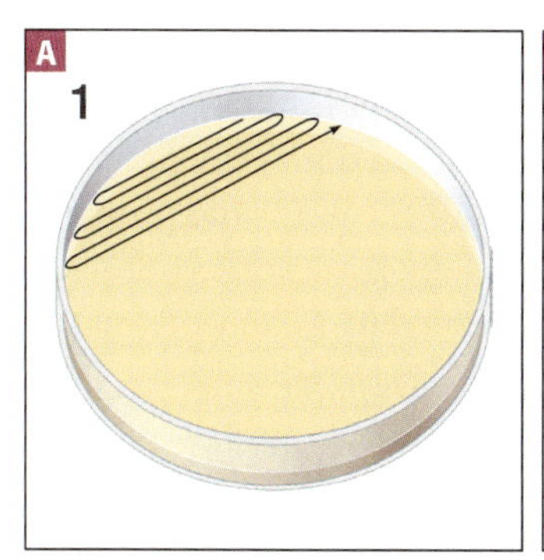


**1.35A Beginning the Quadrant Streak Pattern** ■ Streak the mixed culture back and forth in one quadrant of the agar plate. Stay close to the plate's edge and make the streaks long. Do not cut the agar with the loop. Flame the loop, and then proceed.

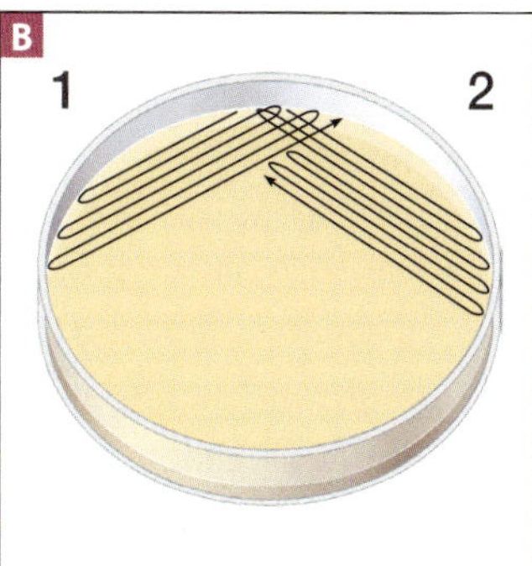


**1.35B Second Streak** ■ Rotate the plate nearly 90° and touch the agar in an uninoculated region to cool the loop. Streak again, using the same wrist motion. Flame the loop afterward. (***Note:*** In these illustrations, the plate is not rotated.)

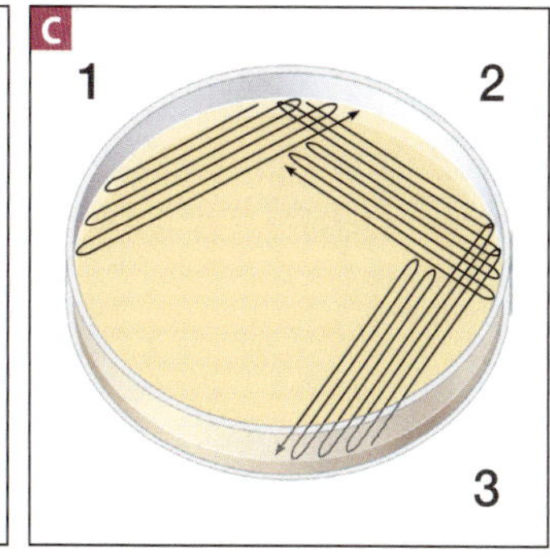


**1.35C Third Streak** ■ Rotate the plate nearly 90° and streak again, using the same wrist motion. Be sure to cool the loop prior to streaking and flame it afterward.

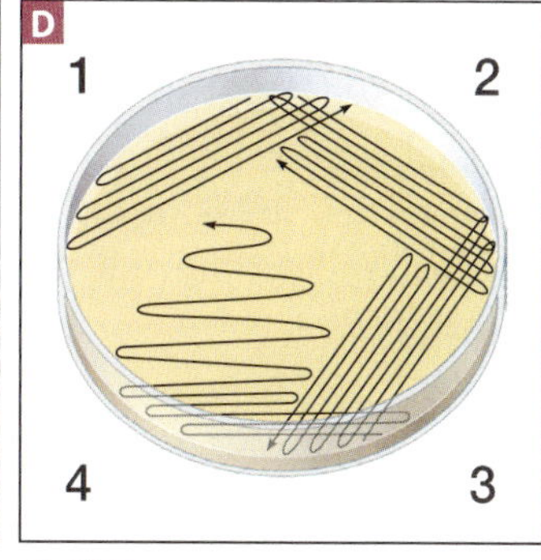


**1.35D Fourth Streak into the Center** ■ After cooling the loop, streak one last time into the center of the plate. Flame the loop, and incubate the plate in an inverted position for the assigned time at the appropriate temperature.



10 Flame the loop from base to tip.

11 Incubate the plate in an inverted position for the assigned time at the appropriate temperature.

### Zigzag (Continuous) Inoculation of Agar Plates Using a Cotton Swab

This inoculation pattern is usually performed when the sample does not have a high cell density and with pure cultures when isolation is not necessary.

1. Label the base of the plate with your name, date, and sample.
2. Hold the swab comfortably in your dominant hand and lift the lid of the Petri dish with the other. Use the lid as a shield to protect the agar from airborne contamination (Fig. 1.33). Alternatively, the plate can be held in the air as shown in Figures 1.34A and 1.34B.
3. Lightly drag the cotton swab across the agar surface in a zigzag pattern, rolling it as you do so. Be careful not to cut the agar surface (Fig. 1.38).
4. Replace the lid.
5. Dispose of the swab according to your lab's practices (generally in a sharps or biohazard container).
6. Incubate the plate in an inverted position for the assigned time at the appropriate temperature.

### Inoculation of Agar Plates with a Cotton Swab in Preparation for a Quadrant Streak Plate

This inoculation pattern is usually performed as the initial streak for isolation of two or more bacterial species in a mixed culture with suspected high cell density.

1. Label the plate's base with your name, date, and sample.

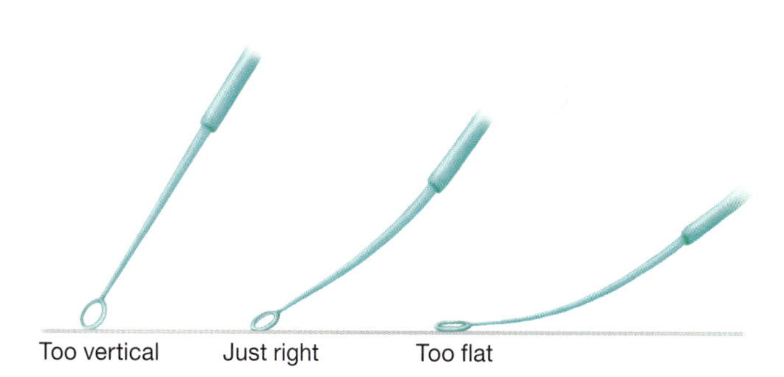


**1.36 Proper Pressure on the Loop** ■ The loop should not be held too vertically because it will cut the agar (left drawing). It also should not be held too flatly against the agar because the streaks will be too wide (right drawing). Moderate pressure should be applied to the loop so only the outer third or so of the loop's face contacts the agar (middle drawing).

2. Hold the swab comfortably in your dominant hand and lift the lid of the Petri dish with the other. Use the lid as a shield to protect the agar from airborne contamination (Fig. 1.33). Alternatively, the plate can be held in the air as shown in Figures 1.34A and 1.34B.
3. Lightly drag the cotton swab back and forth across the agar surface in one quadrant of the plate (Fig. 1.39). This replaces the first streak as shown in Figure 1.35A.
4. Dispose of the swab according to your lab's practices (generally in a sharps container).
5. Further streaking is performed with a loop as shown in Figures 1.35B through 1.35D.
6. Incubate the plate in an inverted position for the assigned time at the appropriate temperature.

1

## In This Exercise

You will learn how to isolate individual organisms from a mixed culture, the first step in producing a pure culture. Three related streaking techniques will be used, the choice of which is determined by the anticipated cell density of the sample.

## Materials

### Per Student

- □ Lab coat
- □ Disposable gloves
- □ Chemical eye protection
- □ Inoculating loop
- □ Three tryptic soy agar (TSA) plates
- □ One sterile cotton swab in sterile distilled water

### Per Student Group

- □ Fresh broth cultures of these recommended organisms:
  - *Micrococcus luteus*
  - *Staphylococcus epidermidis*

### Lab One

1. Wear a lab coat, gloves, and chemical eye protection when performing this procedure.
2. Using a pencil, practice quadrant-streaking the "plate" on the data sheet, page 51, before trying it with living bacteria. Rotate the paper so that each streaking motion is the same even though you are streaking a different part of the plate. ***Hint***: Keep your wrist relaxed.
3. Each student should transfer a loopful of *M. luteus* to one sterile tryptic soy agar plate and follow the diagrams in Figure 1.35 to perform a quadrant streak for isolation. Avoid digging into or cutting the agar, which ruins the plate and may create dangerous aerosols. See Figure 1.36 for the proper pressure on the loop. Label the plate with your name, the date, and the organism.
4. Each student should transfer *S. epidermidis* to one sterile TSA plate and perform a T-streak for isolation (Fig. 1.37). As with the quadrant streak, avoid digging into or cutting the agar, which ruins the plate and may create dangerous aerosols. Label the plate with your name, the date, and the organism.
5. Each student should use the cotton swab to sample an environmental source (see Appendix B), and then do a simple zigzag streak on a tryptic soy agar plate (Fig. 1.38). Dispose of the swab in a sharps or biohazard container to be autoclaved. Label the plate with your name, the date, and the sample source.
6. Tape the three plates together, invert them (the plates' bases with the agar should be "up"), and incubate them at 25°C for 24 to 48 hours.
7. Save or dispose of the original cultures as directed by your instructor.

### Lab Two

1. After incubation, examine the plates for isolation.
2. Compare your streak plates with your lab partner's plates and critique each other's technique. Remember, a successful streak plate is one that has isolated colonies; the pattern does not have to be textbook quality—it's just that textbook quality provides you with a greater chance of getting isolation.
3. Dispose of all plates in the appropriate autoclave container when finished.
4. Complete the data sheet on page 51.

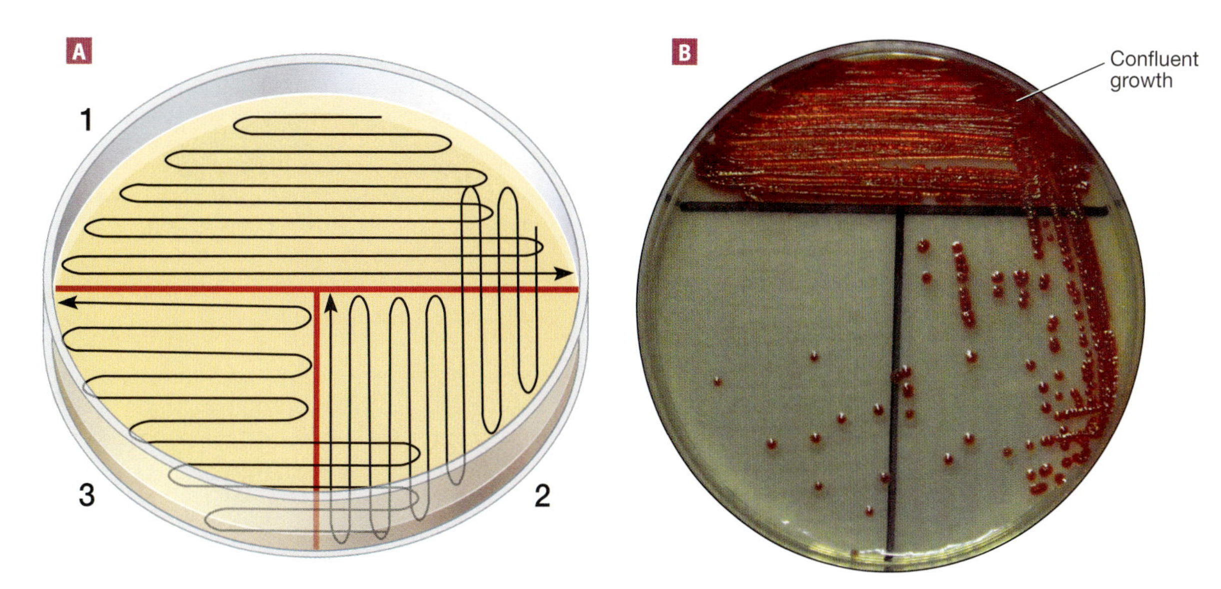


**1.37 T-Streak Pattern** ■ (**A**) This method uses only three streaks, each occupying a region designated by drawing lines (shown in red) in the shape of a "T" (hence, the procedure's name) on the plate's bottom. The first region (I) is the largest (about one-third of the agar's surface) and is streaked with the original sample. After flaming the loop and allowing it to cool, a second streak is made by entering the first streak two or three times, followed by four or five streaks just in the second region (II). After flaming and cooling the loop, a third streak is made from the second region into the third region (III) as before. (**B**) *Serratia marcescens* was streaked on the plate using the T-streak method. As in Figure 1.32, isolation was achieved on the second streak, which is not surprising because both plates were made from the same culture and had the same cell density.

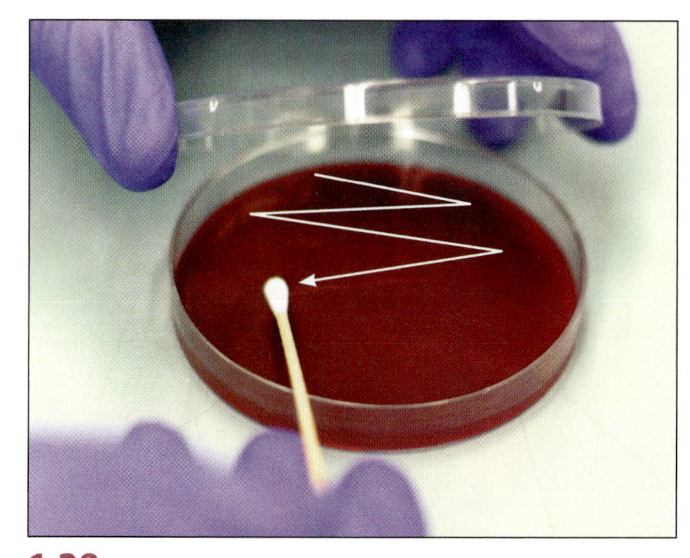

**1.38 Zigzag (Continuous) Inoculation with a Swab** ■ Use the swab to streak the agar surface to get isolated colonies after incubation. Be careful not to cut the agar as you rotate the swab. Properly dispose of the swab in a sharps or biohazard container.

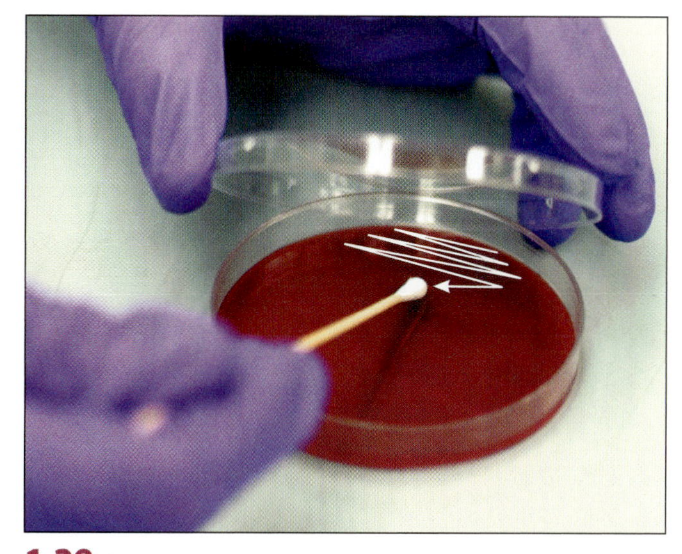

**1.39 Inoculation in Preparation for a Quadrant Streak Using a Swab** ■ If the sample is expected to have a high density of organisms, use the swab to perform the first streak toward one edge of the plate. Then continue the quadrant streak using a loop. Be careful not to cut the agar. Properly dispose of the swab in a sharps or biohazard container.

## References

Collins, C. H. and Patricia M. Lyne. Chap. 6 in *Collins and Lyne's Microbiological Methods*, 7th ed. Oxford, UK: Butterworth-Heinemann, 1995.

DeMers, Marlene. Exercise 12 in *Fundamentals of Microbiology Laboratory Manual*, 8th ed. Dubuque, IA: Kendall Hunt Publishing, 2010.

Forbes, Betty A., Daniel F. Sahm, and Alice S. Weissfeld. Chap. 1 in *Bailey & Scott's Diagnostic Microbiology*, 11th ed. St. Louis, MO: Mosby-Yearbook, 2002.

Koneman, Elmer W., Stephen D. Allen, William M. Janda, Paul C. Schreckenberger, and Washington C. Winn, Jr. Chap. 2 in *Color Atlas and Textbook of Diagnostic Microbiology*, 5th ed. Philadelphia: J. B. Lippincott Company, 1997.

Power, David A. and Peggy J. McCuen. Pages 2–3 in *Manual of BBL™ Products and Laboratory Procedures*, 6th ed. Cockeysville, MD: Becton Dickinson Microbiology Systems, 1988.

Tille, Patricia M. Pages 90–91 in *Bailey & Scott's Diagnostic Microbiology*, 13th ed. St. Louis, MO: Mosby, 2014.

Name ____________________

Date ____________________

Lab Section ____________________

I was present and performed this exercise (initials) ____________________

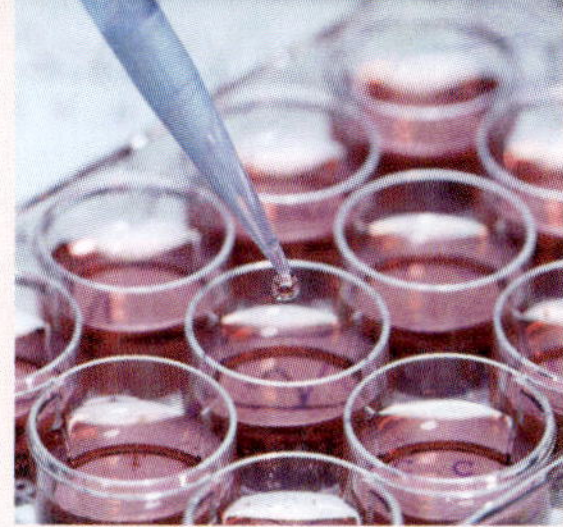

DATA SHEET

1-5

## Streak Plate Methods of Isolation

### RESULTS AND INTERPRETATIONS

**1** *Using your pencil, perform a quadrant streak on the "practice plate" below.*

**2** *Examine the quadrant streak and T-streak plates. Have your lab partner write a critique of your isolation technique in the space below. The following should be addressed: Was isolation produced on one or both plates? On the quadrant streak, were the first three streaks near the edges of the plate? On both plates, did any streaks intersect streaks they should not have? Was the whole surface of the agar used? Was the agar cut by the loop?*

**3** *Did you achieve isolation using the quadrant streak? If so, in which streak (1, 2, 3, or 4) did it occur? If you did not achieve isolation, what might you do differently next time to improve your results?*

**4** *Did you achieve isolation using the T–streak? If so, in which streak (1, 2, or 3) did it occur? If you did not achieve isolation, what might you do differently next time to improve your results?*

**5** *Examine the environmental sample. Did you achieve isolation?*

**6** *Were the three different streak methods appropriate to the cell densities recovered?*

**7** *Most colonies on streak plates grow from isolated colony-forming units (CFUs). On rare occasions, however, a colony can be a mixture of two different organisms. If a culture is started from this colony (thinking it is pure), correct identification will be next to impossible because the extra organism could confound the identifying test results. How could you verify the purity of a colony? (The answers may vary depending on what experience you have had prior to performing this exercise.) If you found the colony to be a mixture of organisms, what could you do to purify it?*

# Spread Plate Method of Isolation

## Theory

The spread plate technique is a method of isolation in which a diluted microbial sample is deposited on an agar plate and spread uniformly across the surface with a glass rod. With a properly diluted sample, cells (CFUs) will be deposited far enough apart on the agar surface to grow into individual colonies.

## Application

After incubation, a portion of an isolated colony can be transferred to a sterile medium to begin a pure culture. The spread plate technique also has applications in quantitative microbiology (see Section 6).

Following is a description of the spread plate technique.

### Spread Plate Technique

Generally, multiple plates are made from successively more dilute samples. What follows is a description of how each plate is inoculated. The process is repeated for each plate.

1. Label the plate's base with your name, date, organism, and any other relevant information (such as dilution).
2. Arrange the alcohol jar, Bunsen burner, and agar plate as shown in Figure 1.40. This arrangement minimizes the chances of catching the alcohol on fire.
3. Lift the plate's lid and use it as a shield to protect from airborne contamination.
4. Using an appropriate pipette, deposit the designated inoculum volume on the agar surface (Fig. 1.41). (Please see Appendices C and D for use of pipettes.) From this point, the remaining steps should be completed within about 15 seconds to prevent the inoculum from soaking into the agar.
5. Properly dispose of the pipetting instrument used to inoculate the medium, because it is contaminated. Each lab has its own specific procedures and your instructor will advise you what to do.
6. Remove the glass spreading rod from the alcohol and pass it through the flame to ignite the alcohol (Fig. 1.42). Remove the rod from the flame and allow the alcohol to burn off completely. Do not leave the rod in the flame; the combination of the alcohol and brief flaming are sufficient to sterilize it. **Be careful not to drop any flaming alcohol on the work surface. Be especially careful not to drop flaming alcohol back into the alcohol jar. If the jar catches on fire, place the lid over it to smother the flame.**
7. After the flame has gone out on the glass rod, lift the lid of the plate and use it as a shield from airborne contamination. Then touch the rod to the agar surface away from the inoculum to cool it.

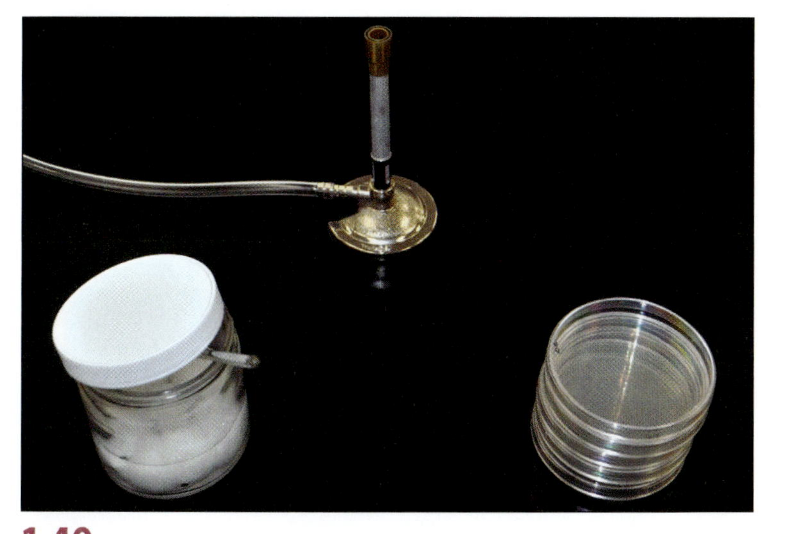

**1.40 Spread Plate Set-up** ■ The spread plate technique requires a Bunsen burner, a screw-cap jar with alcohol, a glass spreading rod, and the plate. Position these components in your work area as shown: isopropyl alcohol, flame, and plate. This arrangement reduces the chance of accidentally catching the alcohol on fire. Notice the cotton in the jar's bottom to reduce the chance of breaking the glass rod.

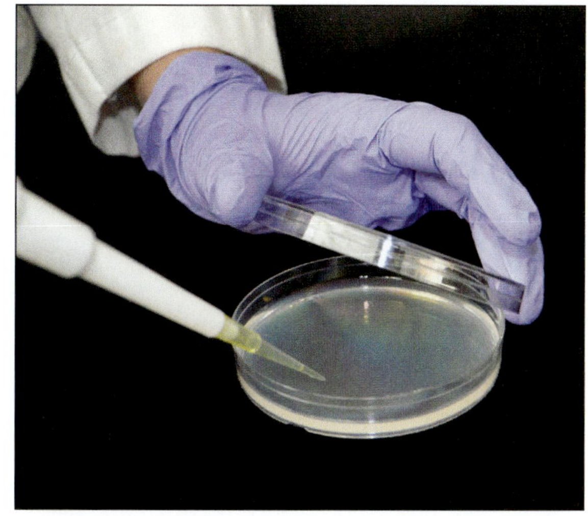

**1.41 Delivering the Inoculum** ■ Deposit the inoculum near one side of the agar surface. Use the lid as a shield and properly dispose of the pipette or pipette tip in a sharps or other biohazard container.

8 To spread the inoculum, hold the plate lid with the base of your thumb and index finger and use the tip of your thumb and middle finger to rotate the base (Fig. 1.43). At the same time, move the rod in a back-and-forth motion across the agar surface. After a couple of turns, do one last turn with the rod next to the plate's edge. Alternatively, place the plate on a rotating platform and spread the inoculum (Fig. 1.44).

9 Remove the rod from the plate and replace the lid.

10 Return the rod to the alcohol in preparation for the next inoculation.

11 Repeat until all dilutions have been inoculated.

12 Tape the plates into stacks of no more than four. Make sure they all face the same direction.

13 Incubate the plates in an inverted position (with the agar "up") at the appropriate temperature for the assigned time. (If you plated a volume of inoculum greater than 0.5 mL, wait a few minutes and allow it to soak in before inverting the plates.)

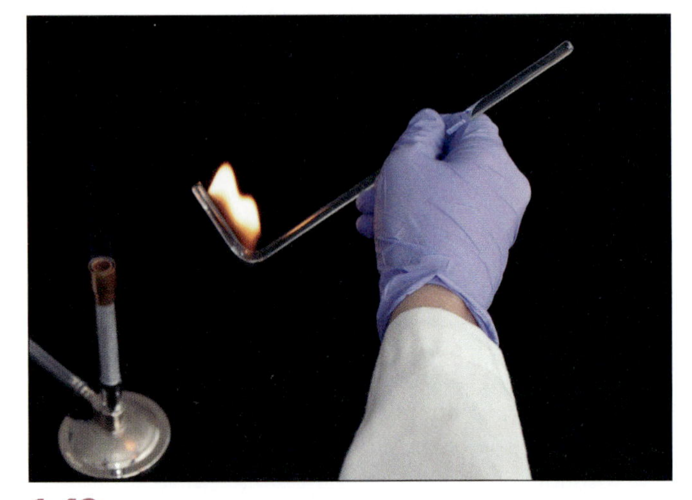

**1.42 Flaming the Glass Rod** ■ Remove the glass spreading rod from the alcohol jar, tap off any excess alcohol, and then pass it through the flame *away from the alcohol jar* to ignite the alcohol on it. Allow the alcohol to burn off completely. Do not leave the rod in the flame; the combination of the alcohol and brief flaming are sufficient to sterilize it. *Be careful not to drop any flaming alcohol on the work surface or back into the alcohol jar. If the alcohol catches on fire, smother the flame by replacing the cap on the jar.*

## ■ In This Exercise

You will perform a spread plate inoculation. In the context of this exercise, it is used as an isolation procedure, but it also can be used in quantifying cell densities in broth samples.

## ▼ Materials

### Per Student

- □ Lab coat
- □ Disposable gloves
- □ Chemical eye protection

### Per Student Pair

- □ Inoculating loop (each student)
- □ Six sterile plastic transfer pipettes
- □ Glass spreading rod (alternatively, sterile disposable plastic spreaders are available)

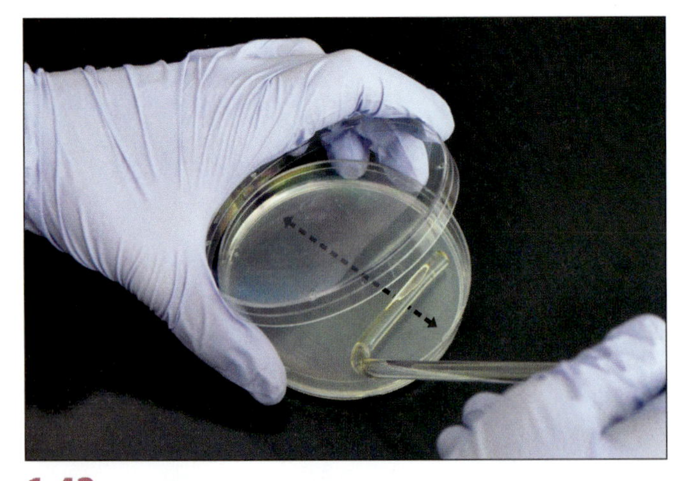

**1.43 Spreading the Inoculum** ■ After the flame has gone out on the rod, lift the lid of the plate and use it as a shield from airborne contamination. Then, touch the rod to the agar surface away from the inoculum in order to cool it. To spread the inoculum, hold the plate lid with the base of your thumb and index finger, and use the tip of your thumb and middle finger to rotate the base. At the same time, move the rod in a back-and-forth motion across the agar surface. After a couple of turns, do one last turn with the rod next to the plate's edge.

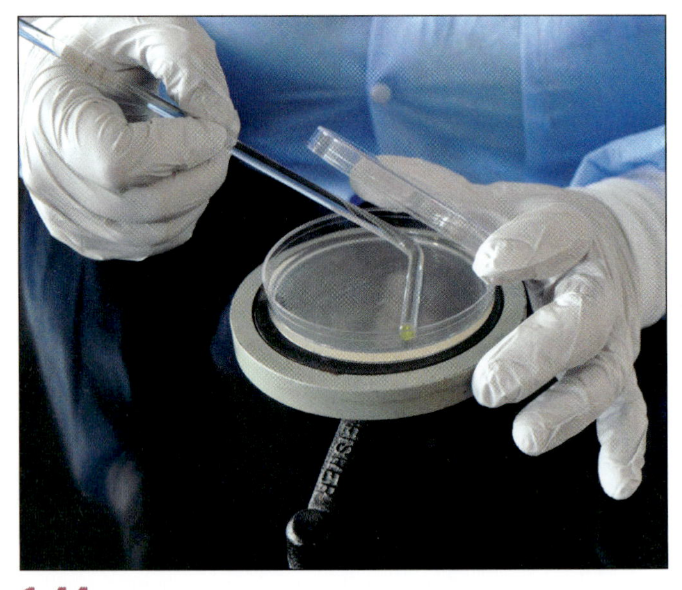

**1.44 Spreading with a Turntable** ■ An inoculating turntable makes it easier to rotate the plate during the spread plate technique.

- ☐ Glass, screw-cap jar with cotton in the bottom, and enough isopropyl alcohol to only cover the spreading part of the glass rod (not necessary if disposable spreaders are used)
- ☐ Bunsen burner and striker
- ☐ Four nutrient agar plates
- ☐ One sterile microtube
- ☐ Four capped microtubes with about 1 mL sterile distilled or deionized water ($dH_2O$)
- ☐ (Optional) vortex mixer
- ☐ Fresh broth cultures of these recommended organisms:
  - *Escherichia coli*
  - *Serratia marcescens*

## PROCEDURE

### Lab One

1. Wear a lab coat, gloves, and chemical eye protection when performing this procedure.
2. Using a different pipette for each, transfer a few drops of *E. coli* and *S. marcescens* to the same microtube. Cap the tube and mix well with a vortex mixer. Or use the second pipette to mix well by gently drawing and dispensing the mixture in and out of the tube a couple of times. Do not spray the mixture!
3. Label the four microtubes containing sterile $dH_2O$ "A," "B," "C," and "D."
4. Label the four nutrient agar plates "A," "B," "C," and "D."
5. Transfer a loopful of the mixture from the microtube to Tube A and mix well with the loop or a vortex mixer. If using a vortex mixer, be sure to cap the tube.[1]
6. Transfer a loopful of the mixture in Tube A to Tube B and mix well with the loop or a vortex mixer.
7. Transfer a loopful of the mixture in Tube B to Tube C and mix well with the loop or a vortex mixer.
8. Transfer a loopful of the mixture in Tube C to Tube D and mix well with the loop or a vortex mixer.
9. Using a sterile transfer pipette, place a couple of drops of sample from Tube A on Plate A. Spread the inoculum with a glass rod as described in "Spread Plate Technique" and as shown in Figures 1.41 through 1.43. Let the plate sit for a few minutes.
10. Repeat step 9 for Tubes B, C, and D and Plates B, C, and D, respectively. Do your best to transfer the same volume to each plate. Dispose of the microtubes in an appropriate autoclave container.
11. Tape the four plates together (be sure they are facing the same direction), invert them (agar "up") and incubate them at 25°C for 24 to 48 hours.
12. Save or dispose of the original broth cultures as directed by your instructor.

### Lab Two

1. After incubation, examine the plates for isolation. *S. marcescens* produces reddish-orange colonies and *E. coli* produces buff-colored colonies.
2. Complete the data sheet on page 57.
3. Dispose of all plates in the appropriate autoclave container when finished.

[1] Because this is not a quantitative procedure, it is not necessary to flame the loop between transfers.

## References

Clesceri, Lenore S. WEF, Chair; Arnold E. Greenberg, APHA; Andrew D. Eaton, AWWA; and Mary Ann H. Franson. Pages 9–38 in *Standard Methods for the Examination of Water and Wastewater*, 20th ed. Washington, DC: Joint publication of American Public Health Association, American Water Works Association, and Water Environment Federation. APHA Publication Office, 1998.

Downes, Frances Pouch, and Keith Ito. Page 57 in *Compendium of Methods for the Microbiological Examination of Foods,* 4th ed. Washington, DC: American Public Health Association, 2001.

Gerhard, Philipp, R. G. E. Murray, Willis A. Wood, and Noel R. Kreig. Pages 255–257 in *Methods for General and Molecular Bacteriology.* Washington, DC: American Society for Microbiology, 1994.

Name ______________________

Date ______________________

Lab Section ______________________

I was present and performed this exercise (initials) ______________________

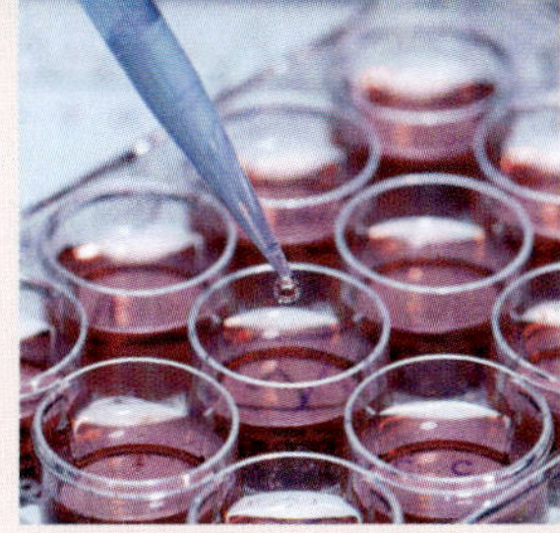

DATA SHEET

1-6

## Spread Plate Method of Isolation

### OBSERVATIONS AND INTERPRETATIONS

**1** Record your observations in the table below.

| Organism | Plate(s) with Isolation | Comments |
|---|---|---|
| *E. coli* | | |
| *S. marcescens* | | |

## QUESTIONS

**1** *Think about your results in attempting to isolate* E. coli *and* S. marcescens.

**a.** *On which plate did you first obtain isolation with* E. coli?

______________________

**b.** *On which plate did you first obtain isolation with* S. marcescens?

______________________

**c.** *Do you have reason to suspect that they should become isolated on the same dilution plate? Why or why not?*

______________________

______________________

______________________

______________________

______________________

______________________

**2** *What is the primary negative consequence of not spreading the inoculum evenly over the agar surface?*

**3** *To get isolated colonies on a plate, only about 300 cells can be in the inoculum. What will happen if the cell density of the inoculum significantly exceeds this number?*

**4** *Suppose you have two organisms in a mixture and Organism A is 1,000 times more abundant than Organism B. Will you (without counting on good luck!) be able to isolate Organism B using the spread plate technique? Explain your answer.*

SECTION 2

# Microbial Growth

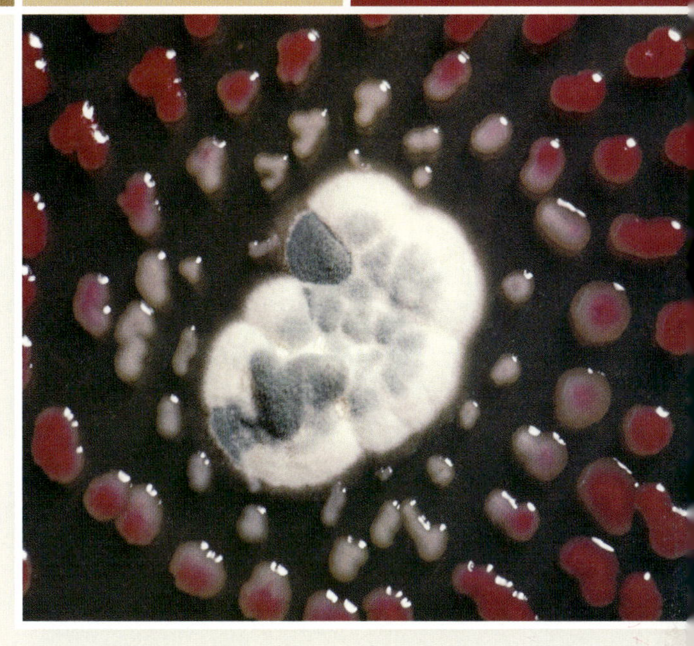

Microorganisms are extraordinarily diverse, and every species demonstrates a unique combination of characteristics, some of which can be easily observed. In this section we illustrate some of those characteristics and factors that affect them.

You will begin this section with an exercise intended to sensitize you to the diversity of microbial populations living all around us. Allowing for variables, such as the growth medium and incubation conditions, much can be determined about an organism by simply looking at the colonies it produces, or its appearance on slants or in broths. Distinguishing growth patterns on or in different media is an important skill—one that you can use as you progress through the semester.

Note the growth characteristics of all the organisms provided for your laboratory exercises, and jot them down or even sketch them. When the time comes to identify your unknown species, you may find your records very useful.

Next, you will examine microbial nutritional diversity by growing bacteria on media with varying amounts of carbon and nitrogen resources. Following that, you will look at some environmental factors affecting microbial growth, such as oxygen, temperature, pH, and osmotic pressure. Finally, you will examine some physical and chemical microbial control agents and systems, that is, ways in which humans can control bacterial growth.

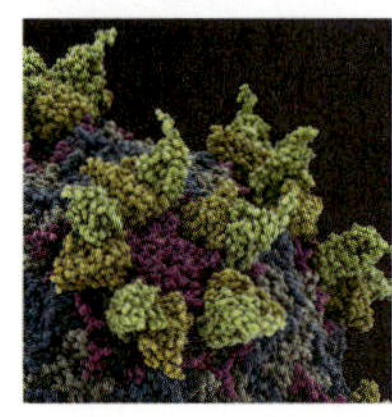

## Ubiquity and Diversity of Microorganisms

2

Microorganisms are found everywhere that other forms of life exist. They can be isolated from soil, bodies of water, and even from the air. As unwanted parasites or colonizers, some microorganisms cause diseases or infections. Most, however, are harmless saprophytes that live in, on, or around plants and animals and decompose dead organic matter. In so doing, they perform the essential function of nutrient recycling in ecosystems.

In Exercise 2-1, you will transfer microorganisms from, in some cases, seemingly uninhabited sources and grow them on agar plates. Each source is likely to have multiple species growing on or in it and represents a **mixed culture**. When you inoculate the plates you will be transferring untold numbers of unknown cells to them. Any cells that are able to grow on the plate's medium will divide and produce visible **colonies** of identical cells.

In Exercise 2-2, you will learn to identify some of the various growth characteristics produced by the "invisible" cohabitants from Exercise 2-1, as well as examine plates of known species. An ability to recognize differences in **colony morphology** is often the first clue to a microbiologist that two organisms are different species. If a colony is not contacting other colonies, it is said to be **isolated** and a portion of it can be transferred to a sterile medium to start a **pure culture** of the species, which is then frequently used in tests designed to identify it. Pure cultures are usually grown in a broth or on a slanted medium. In Exercises 2-3 and 2-4 you will examine growth characteristics of bacteria on slants and in broth, respectively. ■

# Ubiquity of Microorganisms

EXERCISE 2-1

## Theory

Microorganisms have a long, rich history on Earth and have successfully adapted to a wide range of habitats. The literature on microorganisms often describes them as being "ubiquitous in nature." More specifically, this means that microorganisms of all sorts can be isolated from soil (of all sorts), water (over a large range of salinities), plants, and animals (including humans). Microbes are even found in apparently uninhabitable sites such as hot acid pools (Fig. 2.1).

Many microorganisms are **free-living**—they do not reside on or in a specific plant or animal **host** and are not known to cause disease; they are **nonpathogenic.** Frequently they are **saprophytes** and perform the important ecosystem role of decomposing organic matter. Others perform important conversions of inorganic molecules and participate in **biogeochemical cycles**, such as the sulfur or nitrogen cycles. Other microorganisms reside on or in another species and benefit from the symbiotic association with their host(s)—they have a place to live and reproduce successfully. If they cause damage to their host, that is, they cause disease, they are **pathogens** of it. In other instances the microbe may actually benefit their host, an example of **mutualism.** Lastly, some microbes are **commensals,** where they benefit but have no significant effect on their host. However, even many of the commensal or mutualistic strains inhabiting our bodies are **opportunistic pathogens.** That is, they are capable of producing a disease state if introduced into a suitable part of the body. Any area, including sites outside of the host organism, where a microbe resides and serves as a potential source of infection is called a **reservoir.**

**2.1** ***Sulfolobus* in the Sulfur Caldron, Yellowstone National Park** This geologic feature of Yellowstone National Park produces copious amounts of $H_2S$ gas. *Sulfolobus* (domain Archaea) uses the $H_2S$ as an energy source and oxidizes it to sulfuric acid ($H_2SO_4$), lowering the pH of its environment to 2 or less (about the equivalent of battery acid!). The high acidity breaks down rock and soil to produce the muddy conditions seen in the photo. The steam you see rising from the caldron is due to its high temperature. *Sulfolobus* species grow between 65°C and 85°C—that's roughly 150°F to 185°F! *Sulfolobus* usually gets its carbon from $CO_2$, just like a plant.

## Application

This exercise is designed to demonstrate the ubiquitous nature of microorganisms and the ease with which many can be cultivated. (It should be noted that although we can find living microorganisms virtually everywhere and confirm their presence by cultivation, molecular techniques developed over the last two or three decades demonstrate that the organisms successfully grown in the lab represent a minute fraction of those still uncultivable.)

## In This Exercise

Today, you will work in small groups to sample and culture several locations in your laboratory. Your instructor may have other locations outside of the lab to sample as well. Remember that even relatively "harmless" bacteria, when cultivated on a growth medium, are in sufficient enough numbers to constitute a health hazard. Treat them with care after incubation.

## Materials

### Per Student

- ☐ Lab coat
- ☐ Disposable gloves
- ☐ Chemical eye protection

### Per Student Group

- ☐ Eight nutrient agar plates
- ☐ One sterile cotton swab

## PROCEDURE

### Lab One

**1** Wear a lab coat, gloves, and chemical eye protection when performing this procedure.

**2** Students in each group should share responsibility for inoculating the plates in this exercise.

**3** Number the plates 1 through 8.

**4** Open plate number 1 and expose it to the air for 30 minutes or longer. Set it aside and out of the way of the other plates.

**5** Use the cotton swab to sample your desk area and then streak plates 2 and 3 in the pattern shown in Figure 2.2. Roll the swab on the table top as you sample it. Then, press it lightly on the agar surface (don't cut into the agar) and roll it as you streak. Be sure the contaminated part of the swab contacts the agar so you actually transfer any organisms collected. Dispose of the swab as directed by your instructor.

**6** With the lid off, hold plate 4 directly in front of your mouth and cough several times on the agar surface. Be sure you aren't facing anyone when you cough. In fact, it is best to cough downward onto the plate to minimize spreading germs throughout the classroom. Send them to the floor instead!

**7** Rub your hands together, and then touch the agar surface of plate 5 lightly with your fingertips. A light touch is sufficient; touching too firmly will crack the agar. Your fingers will feel gooey afterward, but the agar is sterile so there should actually be *fewer* microbes on your fingers than before you touched it. Wash your hands, if you must!

**8** Remove the lid of plate 6 and vigorously scratch your head above it. It is best to bend your head forward so your hair is actually above the plate. When inoculating, do so away from plate 1 to avoid cross-contamination.

**9** Leave plates 7 and 8 covered; do not open them.

**10** Label the base of each plate with the date, type of exposure it has received, incubation temperature (see step 11), and the name or number of your group.

**11** Invert all plates and incubate them for 24 to 48 hours at the following temperatures:

| | |
|---|---|
| Plates 1, 2, and 8: | 25°C |
| Plates 3, 4, 5, 6, and 7: | 35 ± 2°C |

**2.2 Simple Streak Pattern on an Agar Plate** ■ Roll the swab as you collect the sample from the table and then again as you inoculate the plate. Do not press so hard that you cut the agar. Dispose of the swab as directed by your instructor.

### Lab Two

**1** Using the plate diagrams on the data sheet, page 63, draw two representative *colonies* that are growing on each of your agar plates. Be sure to label them according to incubation time, temperature, and source of inoculum. It is best to select one colony from each plate to draw before doing the second one. That way you will have observed each plate in case time runs short. ***Note***: Do not remove the plate's lid if there is fuzzy growth on it or your instructor has told you not to. If you are in doubt about "fuzzy growth," ask your instructor.

**2** Save these plates in a refrigerator for use in Exercise 2-2.

### References


Holt, John G., ed. *Bergey's Manual of Determinative Bacteriology,* 9th ed. Baltimore: Lippincott Williams & Wilkins, 1994.

"Sulphur Caldron." Available online. URL: http://www.yellowstoneparknet.com/geothermal_features/sulphur_caldron.php.

Tille, Patricia M. Chap. 7 in *Bailey & Scott's Diagnostic Microbiology,* 13th ed. St. Louis, MO: Mosby, 2014.

Varnam, Alan H. and Malcolm G. Evans. *Environmental Microbiology.* Washington, DC: ASM Press, 2000.

Winn, Washington C. et al. *Koneman's Color Atlas and Textbook of Diagnostic Microbiology,* 6th ed. Baltimore: Lippincott Williams & Wilkins, 2006.

Name ______________________________

Date ______________________________

Lab Section ______________________________

I was present and performed this exercise (initials) ______________________________

# DATA SHEET 2-1

## Ubiquity of Microorganisms

### OBSERVATIONS AND INTERPRETATIONS

**1** Use the circles below as Petri dishes. Then, for each plate, choose two different colonies and draw each as seen from above and from the side. Simple line drawings are acceptable, but do them with care. Chicken scratches are not very useful! Label the plates according to incubation time, temperature, and source of inoculum. Also include other useful colony information, such as color and relative abundance.

**2** Save the plates for Exercise 2-2.

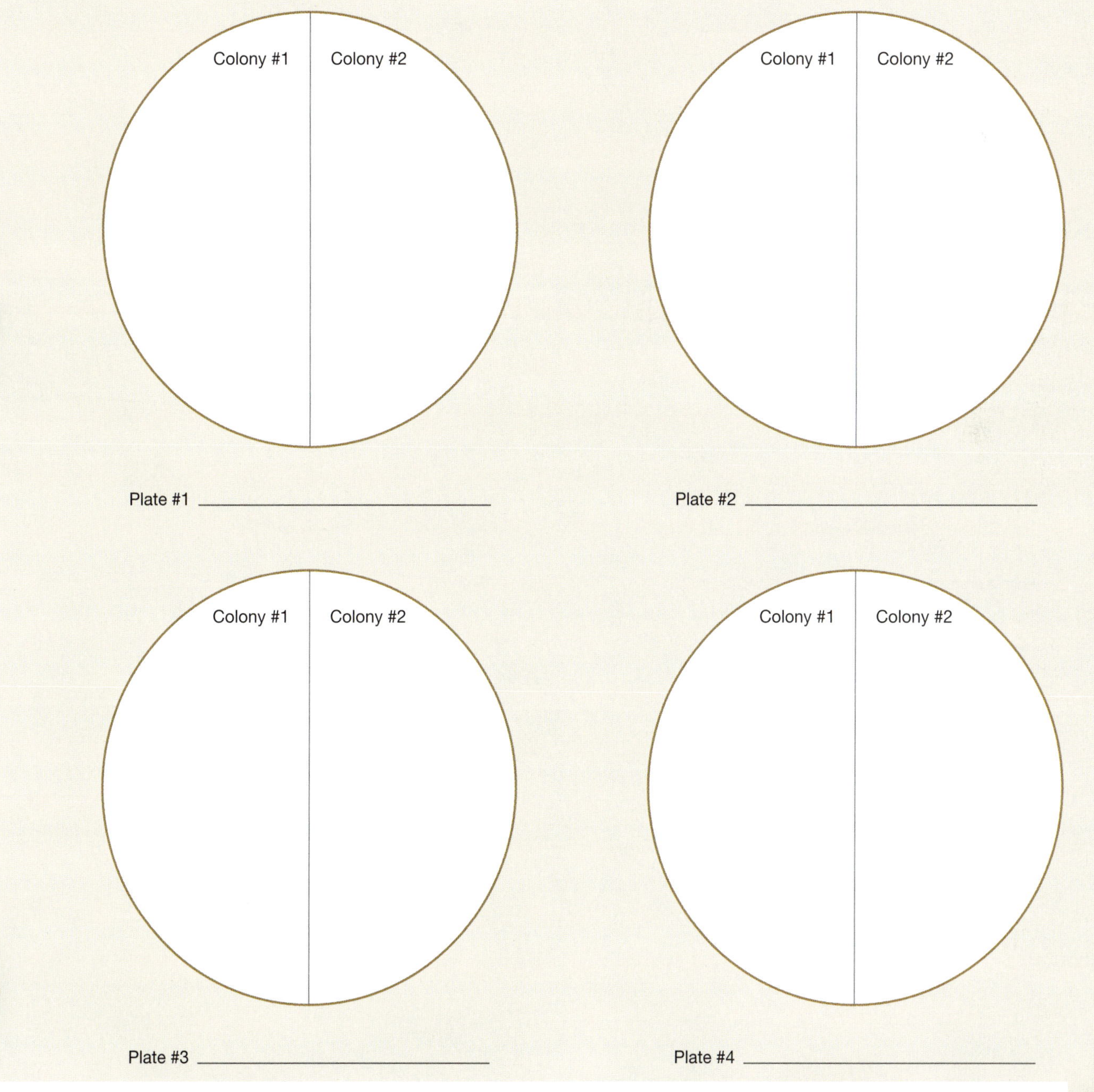


Plate #1 ______________________________

Plate #2 ______________________________

Plate #3 ______________________________

Plate #4 ______________________________

Colony #1
Colony #2
Plate #5
Colony #1
Colony #2
Plate #6
Colony #1
Colony #2
Plate #7
Colony #1
Colony #2
Plate #8

Name ______________________________

Date ______________________________

Lab Section ______________________________

I was present and performed this exercise (initials) ______________

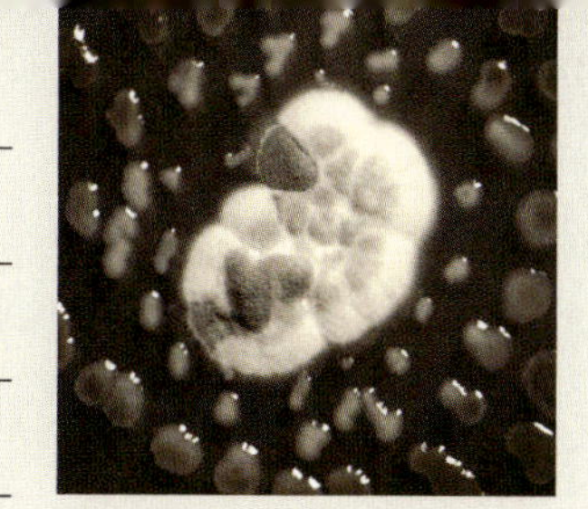

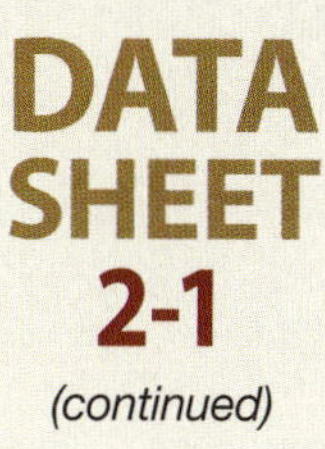

# DATA SHEET 2-1

*(continued)*

## QUESTIONS

**1** *Consider plates 7 and 8.*

**a.** *What was the purpose of incubating the uninoculated plates? Be specific.*

**b.** *What is an appropriate name for these plates?*

**c.** *If growth appears on both uninoculated plates, what are some likely explanations?*

**d.** *What if growth appears on only one plate?*

**e.** *How does growth on the uninoculated plates affect your interpretation of the other plates?*

**2** *Why do you think the specific types of exposure (air, hair, tabletop, etc.) were chosen for this exercise?*

**3** *Consider plates 2 and 3 (table top).*

**a.** *Did you get different-appearing colonies on plates 2 and 3? If so, explain why.*

**2**

**b.** *What is the likely source (reservoir) of organisms that grew best at 35°C, and how do they survive at room temperature without nutrients?*

**4** *Suppose plate 4 (cough) has no growth after incubation. It is highly unlikely the "cougher" has sterile coughs! Suggest reasons why no growth was recovered on the plate.*

**5** *The plates you are using for this lab will be autoclaved to completely sterilize them. The measures taken to disinfect the tabletops (the source of the organisms on plates 2 and 3) are not as extreme. Why?*

EXERCISE
2-2

# Colony Morphology

## Theory

When a single bacterial cell is deposited on an appropriate solid nutrient medium, it begins to divide. One cell makes two, two make four, four make eight . . . one million make two million, and so on. Eventually a visible mass of cells—a **colony**—appears where the original cell was deposited. Color, size, shape, and texture of microbial growth are determined by the genetic makeup of the organism (in many cases by yet unknown mechanisms), but are also greatly influenced by environmental factors, including nutrient availability, temperature, and incubation time.

Colony morphological characteristics may be viewed with the naked eye, a hand lens, a stereo (dissecting) microscope, or a colony counter (Fig. 2.3). The seven basic categories include colony size, shape, margin (edge), surface, elevation, texture, and optical properties (Fig. 2.4).

1. *Size* is simply a measurement of the colony's dimensions—the diameter if circular or length and width if shaped otherwise.
2. *Shape* may be described as **round (circular)**, **irregular**, or **punctiform** (tiny, pinpoint).
3. The *margin* may be **entire** (**smooth**, with no irregularities), **undulate** (wavy), **lobate** (lobed), **filamentous** (unbranched strands), or **rhizoid** (branched like roots).
4. The *surface* may be **smooth**, **rough**, **wrinkled** (**rugose**), **shiny**, or **dull**.
5. The *texture* may be **moist**, **mucoid** (sticky), **butyrous** (buttery), or **dry**.
6. *Elevations* include **flat**, **raised**, **convex**, **pulvinate** (very convex), and **umbonate** (raised in the center).
7. Other useful features include **color** and optical properties such as **opaque** (you can't see through it) and **translucent** (light passes through).

Features such as colony shape, margin, surface, texture (shiny or dull), and color are best viewed by observing from above while holding the plate level with the lid off (if it is safe to do so), but rocking it back and forth slightly so reflected light hits it at different angles. If allowed to do so, you may also check texture by touching the growth with an inoculating loop or wooden stick. Be sure to flame the loop afterward or dispose of the wooden stick properly.

Elevations are best viewed with the plate tilted slightly at eye level. Opacity and translucence are best viewed by placing the plate on a colony counter or holding it (lid on) so it is illuminated from behind (transmitted light). Colony dimensions are best measured from the plate's base rather than through the lid.

When reporting colony morphology, it is important to include the medium and the incubation time and temperature, all of which can affect a colony's appearance.

## Application

Recognizing different bacterial growth morphologies on agar plates is a useful step in the identification process. It is often the first indication that one organism is different from another. Once purity of a colony has been confirmed by an appropriate staining procedure (this is not always done), cells can be transferred to a sterile medium, grown, and maintained as a pure culture, which then acts as a source of that microbe for identification or other purposes.



**2.3 Colony Counter** ■ Subtle differences in colony shape and size can best be viewed with magnification, such as is provided by a colony counter. The **transmitted light** and magnifying glass allow observation of greater detail; however, colony color and many other features are best determined with reflected light. The grid in the background is a counting aid; each big square is 1 square centimeter.

## In This Exercise

Today you will be viewing colony characteristics on the plates saved from Exercise 2-1 and (if available) prepared streak plates provided by your instructor. Figures 2.4 through 2.30 show a variety of bacterial colony forms and characteristics. Where applicable, contrasting environmental factors are indicated.

2

## Materials

### Per Student

- ◻ Lab coat
- ◻ Disposable gloves
- ◻ Chemical eye protection

### Per Student Group

- ◻ (Optional) colony counter, stereo (dissecting) microscope, or hand lens
- ◻ Metric ruler
- ◻ Plates from Exercise 2-1
- ◻ (Optional) tryptic soy agar or brain-heart infusion agar streak plate cultures of any of the following:
  - *Bacillus subtilis*
  - *Corynebacterium xerosis*
  - *Kocuria rosea*
  - *Lactobacillus plantarum* or *Lactobacillus acidophilus*
  - *Micrococcus luteus*
  - (Optional) tryptic soy agar or brain-heart infusion agar streak plate cultures of these BSL-2 organisms:
    - *Mycobacterium smegmatis* (BSL-2)
    - *Proteus mirabilis* (BSL-2)

## PROCEDURE

1. Working with your group, use the terms in Figure 2.4 and in the text to describe some representative colonies on your plates from Exercise 2-1 (if not already described) and the pure cultures supplied. Figures 2.5 through 2.30 may also be useful. Measure colony diameters (in mm) with a ruler and include them with your descriptions in the table on the data sheet, page 77. If you see a distinctive feature that has not been given a name, make up one! Just make it descriptive and easily understood by others. That's what the early microbiologists did to compile the list you have been given. (***Note:*** Remember that many microorganisms are opportunistic pathogens, so be sure to handle the plates carefully. **Do not open plates with BSL-2 organisms on them or those containing fuzzy growth,** because a fuzzy appearance suggests fungal growth containing spores that can spread easily and contaminate the laboratory and other cultures. If you are in doubt, check with your instructor.)
2. Unless you have been instructed to save today's cultures for future exercises (such as Exercise 2-3), discard all plates in an appropriate autoclave container.

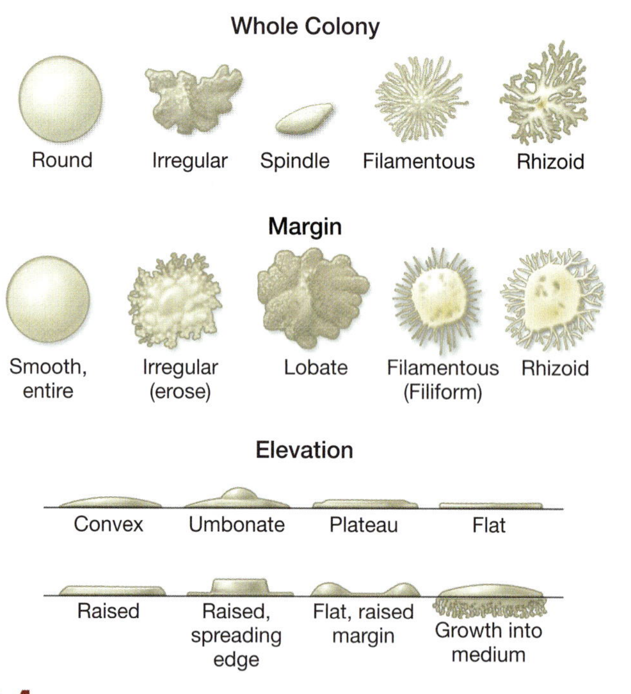


**2.4 A Sampling of Bacterial Colony Features** ■ These terms are used to describe colony morphology. Descriptions also should include color, size, surface characteristics, texture, and optical properties (opaque or translucent). See the text for details.

## Diversity of Colony Morphologies in Mixed Cultures (Figures 2.5–2.6)

**2.5 Two Mixed Soil Cultures on Nutrient Agar** ■ These plates show the morphological diversity present in two diluted soil samples incubated for 48 hours. If two colonies look different when grown under the same conditions, they most likely are different species. The opposite is not always true, however. Two different species can produce colonies that are virtually identical.

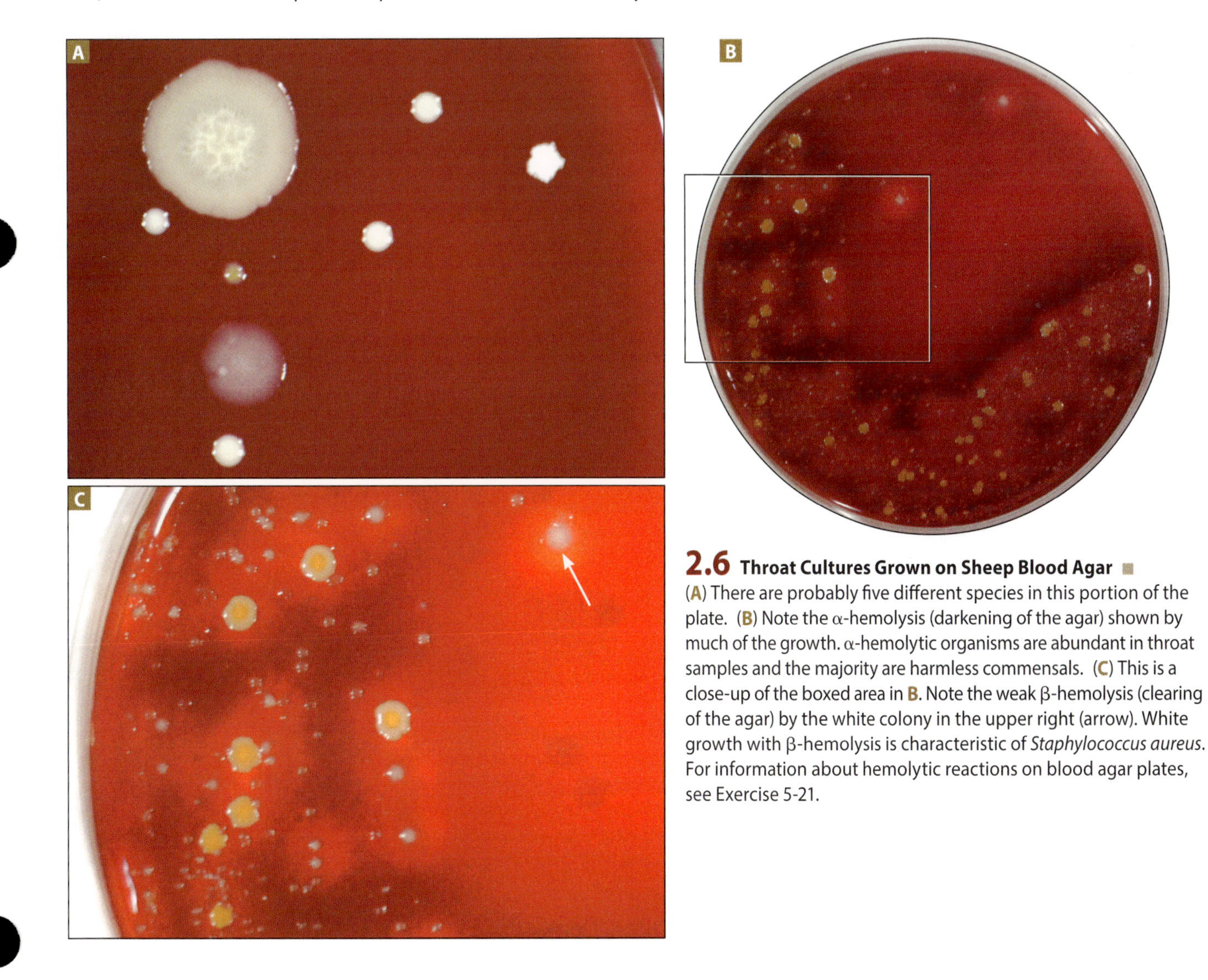


**2.6 Throat Cultures Grown on Sheep Blood Agar** ■ (**A**) There are probably five different species in this portion of the plate. (**B**) Note the $\alpha$-hemolysis (darkening of the agar) shown by much of the growth. $\alpha$-hemolytic organisms are abundant in throat samples and the majority are harmless commensals. (**C**) This is a close-up of the boxed area in **B**. Note the weak $\beta$-hemolysis (clearing of the agar) by the white colony in the upper right (arrow). White growth with $\beta$-hemolysis is characteristic of *Staphylococcus aureus*. For information about hemolytic reactions on blood agar plates, see Exercise 5-21.

## Common Colony Morphologies and Colors (Figures 2.7–2.13)

**2.7 Round, Shiny, Convex Colonies** ■ (**A**) These buff-colored colonies of *Providencia stuartii* grew on nutrient agar in 48 hours. *P. stuartii* is a frequent isolate in urine samples obtained from hospitalized and catheterized patients. It is highly resistant to antibiotics. (**B**) The colonies of the soil and water bacterium *Chromobacterium violaceum* grown on sheep blood agar are purple (hence "violaceum"). The effect of nutrient availability on pigment production by *C. violaceum* is shown in Figure 2.30.

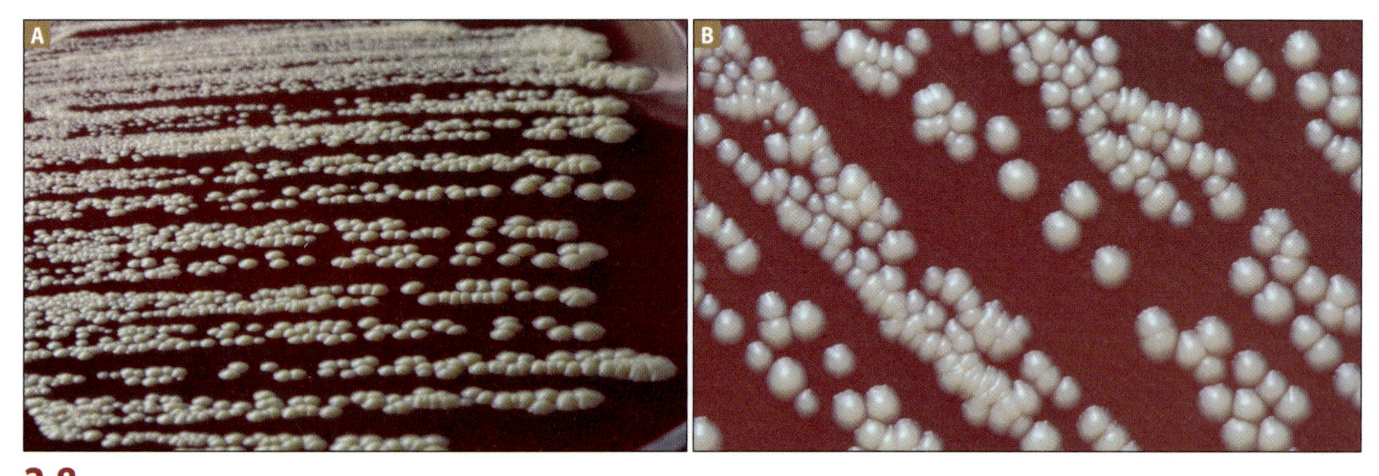

**2.8 Round, Dull, Convex Colonies** ■ (**A**) Dry, buff-colored *Corynebacterium xerosis* colonies on sheep blood agar photographed from the side. *C. xerosis* is rarely an opportunistic pathogen. (**B**) Close-up of the same *C. xerosis* colonies, but from above.

**2.9 Irregular Colony Shape** ■ This unidentified contaminant grew on tryptic soy agar and was isolated from a laboratory tabletop. In addition to its irregular shape, it has a lobed margin and wrinkled (rugose) surface. Its widest dimension was approximately 5 mm. This photo was taken through a stereo microscope.

**2.10 Punctiform Colonies** ■ (**A**) Shown are *Mycobacterium smegmatis* colonies grown on sheep blood agar. The colonies of this slow-growing relative of *M. tuberculosis* are less than 1 mm in diameter. (**B**) These punctiform colonies of the rose-colored *Kocuria rosea* are <0.5 mm in diameter and were grown for 96 hours. (**C**) These are the same colonies as in (**B**) but viewed with a stereoscopic microscope. Notice their irregular shape and wrinkled surface. *K. rosea* is an inhabitant of water, dust, and salty foods.

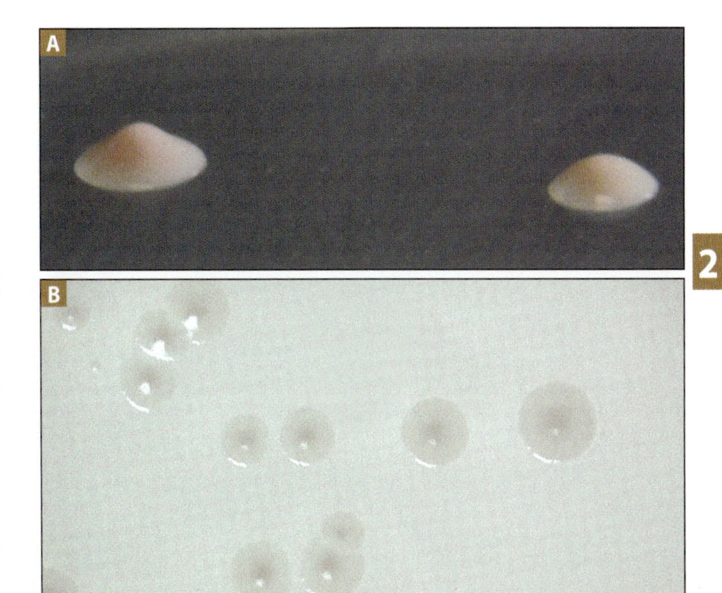

**2.11 Umbonate Colonies** ■ (**A**) The colony on the left of this anaerobic lab contaminant is truly umbonate. The one on the right is getting there. Their diameters are about 3 mm. (**B**) These *Enterococcus faecium* colonies were grown on tryptic soy agar for 48 hours and were photographed using a stereo microscope. The colonies are 1–2 mm in diameter and are white, circular, and umbonate (note the thicker center) with an entire margin. Notice, though, how flat the colonies are except for the central bump that makes them umbonate. *E. faecium* (formerly known as *Streptococcus faecium*) is found in human and animal feces.

**2.12 Flat Colony** ■ This unknown soil isolate is flat only around the edge. The center is concave (but mostly flat, not concave like a bowl) and at the very center is another small, raised ring surrounding another concavity. It's almost like an automobile tire around the wheel and hub. Very unusual! Also note the dull surface.

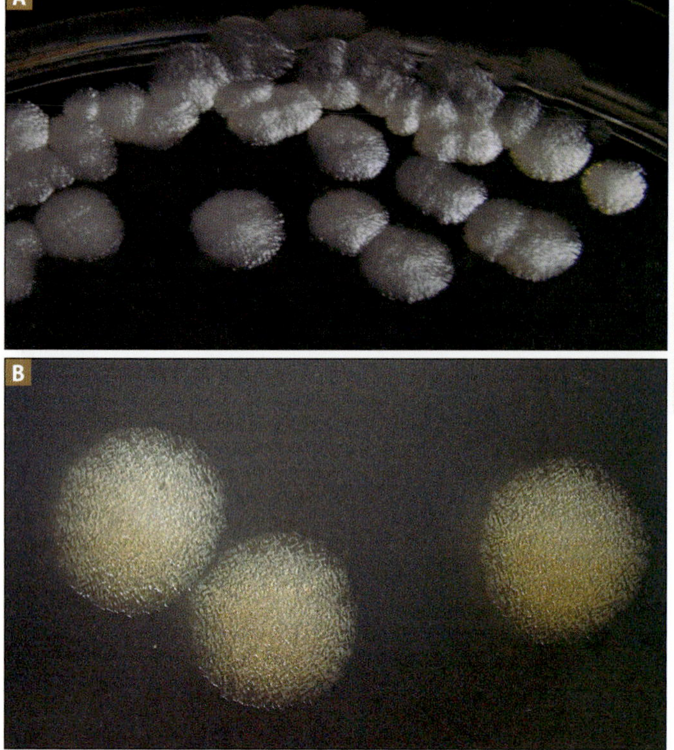

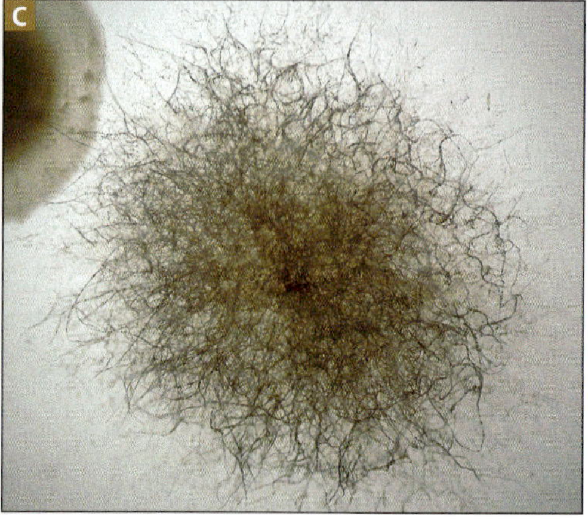

**2.13 Filamentous Colonies** ■ (**A**) These colonies of *Rhizobium leguminosarum* were grown on brain-heart infusion agar and are about 5 mm in diameter. In addition to being filamentous, they are convex, circular, and mucoid with translucent edges. *R. leguminosarum* is capable of causing root nodule formation (*rhiz* means "root") in many legumes and subsequently fixing atmospheric nitrogen. (**B**) This micrograph of three *R. leguminosarum* colonies on brain-heart infusion agar was taken through a stereo microscope. (**C**) Fungi other than yeasts grow as filaments called hyphae, which collectively are referred to as a mycelium. Shown is an unidentified mold mycelium obtained from the microbiology lab viewed with a stereo microscope.

## Common Colony Margins (Figures 2.14–2.18)

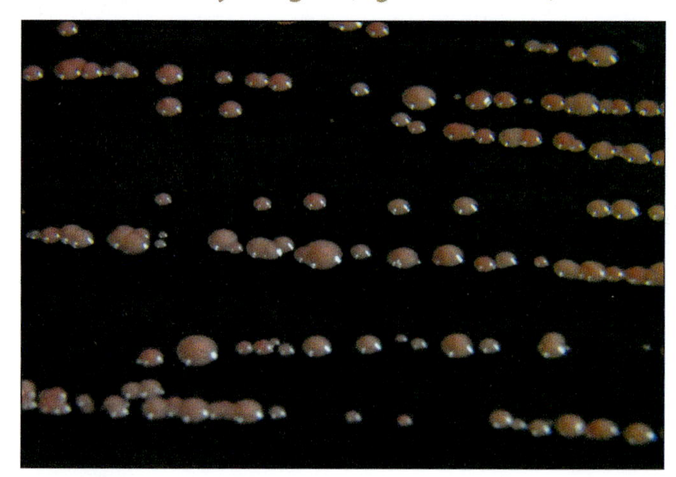

**2.14 Entire or Smooth Margin** ■ These circular and convex *Rhodococcus rhodochrous* colonies were grown on brain-heart infusion agar for 48 hours. They are about 1 mm in diameter with a smooth margin and a shiny, pink surface (hence "*rhodo*"). *Rhodococcus* species are soil organisms.

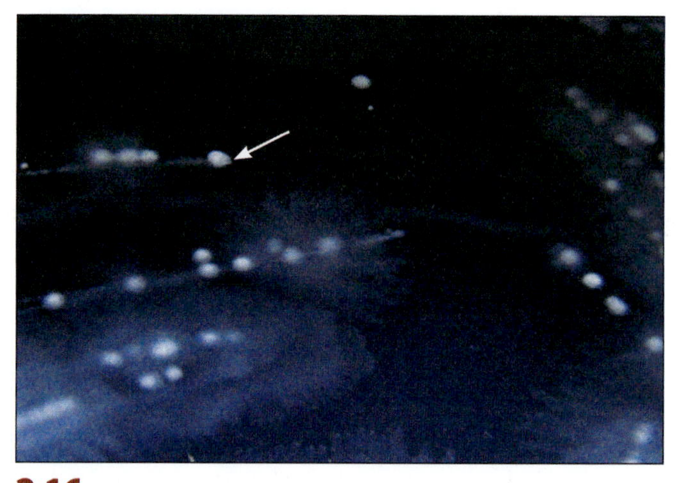

**2.16 Spreading Margin** ■ *Erwinia amylovora* colonies show an irregular spreading margin. Notice that not all colonies exhibit this feature (arrow). These colonies were grown for 24 hours on tryptic soy agar at 25°C. *E. amylovora* is a plant pathogen. Its scientific name literally means "Erwin (Smith's) starch devourer."

**2.15 Lobed Margin** ■ This lab contaminant colony has a lobed margin and a rugose surface. Notice that even though the margin is lobed, the overall colony shape is circular. It was approximately 1 cm in diameter at 48 hours of incubation at 35°C.

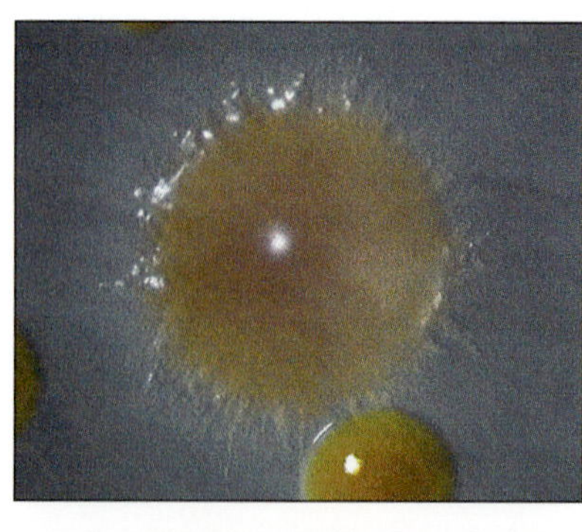

**2.17 Filamentous Margin** ■ Viewed with the stereo microscope, the filamentous margin of this unidentified lab contaminant is visible. The colony is also circular, convex, shiny, and about 2 mm in diameter.

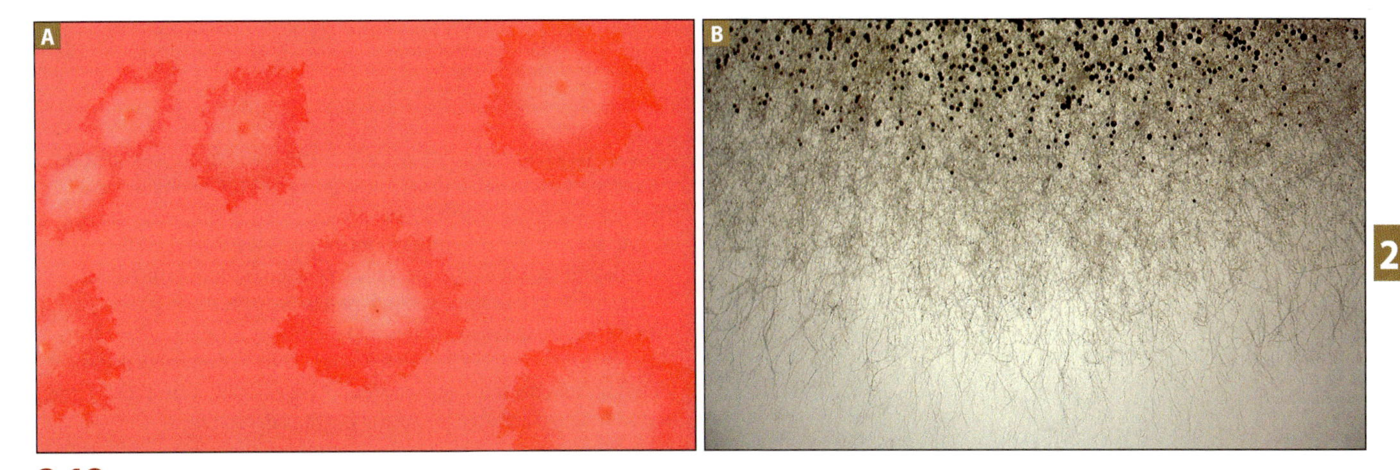


**2.18 Rhizoid Margins** ■ (**A**) These irregular colonies of *Clostridium sporogenes* were grown anaerobically on sheep blood agar and are viewed through a stereo microscope. They have a raised center and a flat, spreading edge of branched, tangled filaments (reminiscent of the mythological creature Medusa, who had snakes for hair!). They vary in size from 2 mm to 6 mm. *C. sporogenes* is found in soils worldwide. (**B**) Fungi (other than yeast) are naturally filamentous and sometimes the filaments are branched, as in this lab contaminant (probably *Aspergillus niger*). It takes magnification to see that the margin is rhizoid and not simply filamentous. The black spheres are asexual spores called conidia.

## Colony Textures (Figures 2.19–2.22)

**2.19 Mucoid Colonies** ■ (**A**) These *Klebsiella pneumoniae* colonies grown on nutrient agar are mucoid, raised, and shiny. While it is a normal inhabitant of the human intestinal tract, it is associated with community-acquired pneumonia and nosocomial urinary tract infections. (**B**) *Pseudomonas aeruginosa* grown on Endo agar illustrates a mucoid texture. *P. aeruginosa* is found in soil and water and can cause infections in burn patients.

**2.20 Butyrous Colony** ■ This unidentified 12 mm colony was found on a glycerol yeast extract plate inoculated with a diluted soil sample. Butyrous (*butyrum* means "buttery") colonies have the consistency of melted butter. This colony was almost liquid in texture, something that is demonstrated by its contact with the yellow colony to its right.

**2.21 Granular Colony** ■ These colonies of *Streptomyces griseus* grown on brain-heart infusion agar are circular, entire, and granular with a ridged surface. At a later stage of development, they produce yellow reproductive spores. Growth of streptomycetes is associated with an "earthy" smell. This one plate fragranced the entire incubator!

**2.22 Dry Colony** ■ Shown are colonies of *Streptomyces violaceus* viewed with the stereo microscope after 3 weeks of incubation at 25°C. They are umbonate, granular, and dry. Members of this genus share a growth pattern that resembles fungi (but they are Gram-positive bacteria and the similarities are only coincidental) and some fungal terminology has traditionally been used to describe them. For instance, their filaments are called hyphae and the colony is called a mycelium. They also grow aerial hyphae that produce chains of reproductive spores (different from bacterial endospores). These colonies were about 4 mm in diameter and adhered tenaciously to the agar surface.

## Optical Properties of Colonies (Figure 2.23)

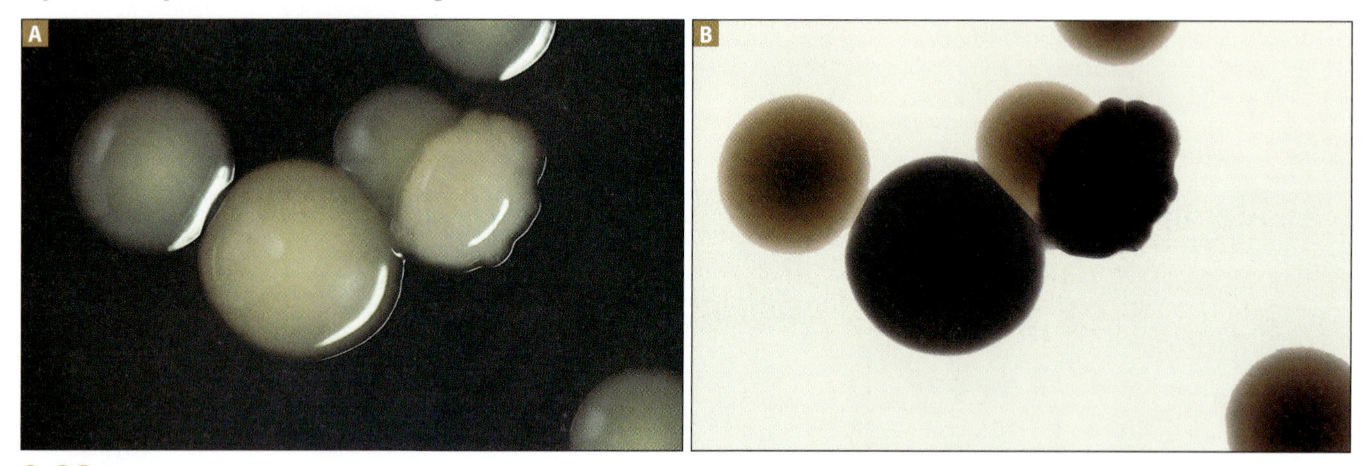


**2.23 Opaque and Translucent Colonies** ■ (A) These colonies were photographed with a stereo microscope using reflected light. There are slight differences in size and color, and one is irregular in shape, whereas the others are circular. But there is another difference not visible with reflected light. (B) Here, the same colonies were photographed with transmitted light and it becomes obvious that the medium-sized round colonies are translucent, whereas the others are opaque.

## Other Less Common, But Distinctive, Colony Features (Figures 2.24–2.25)

**2.24 Swarming Growth** ■ Members of the genus *Proteus* will swarm at certain intervals and produce a pattern of concentric rings because of their motility. This photograph demonstrates the swarming behavior of *P. vulgaris* on DNase agar.

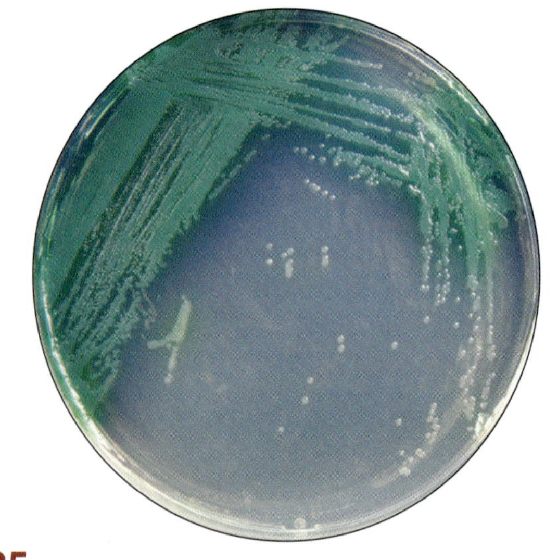

**2.25 Diffusible Pigment** ■ The blue-green pigment pyocyanin diffusing from the growth is distinctive of *Pseudomonas aeruginosa*. Here *P. aeruginosa* is growing on tryptic soy agar.

## Species Diversity (Figure 2.26)

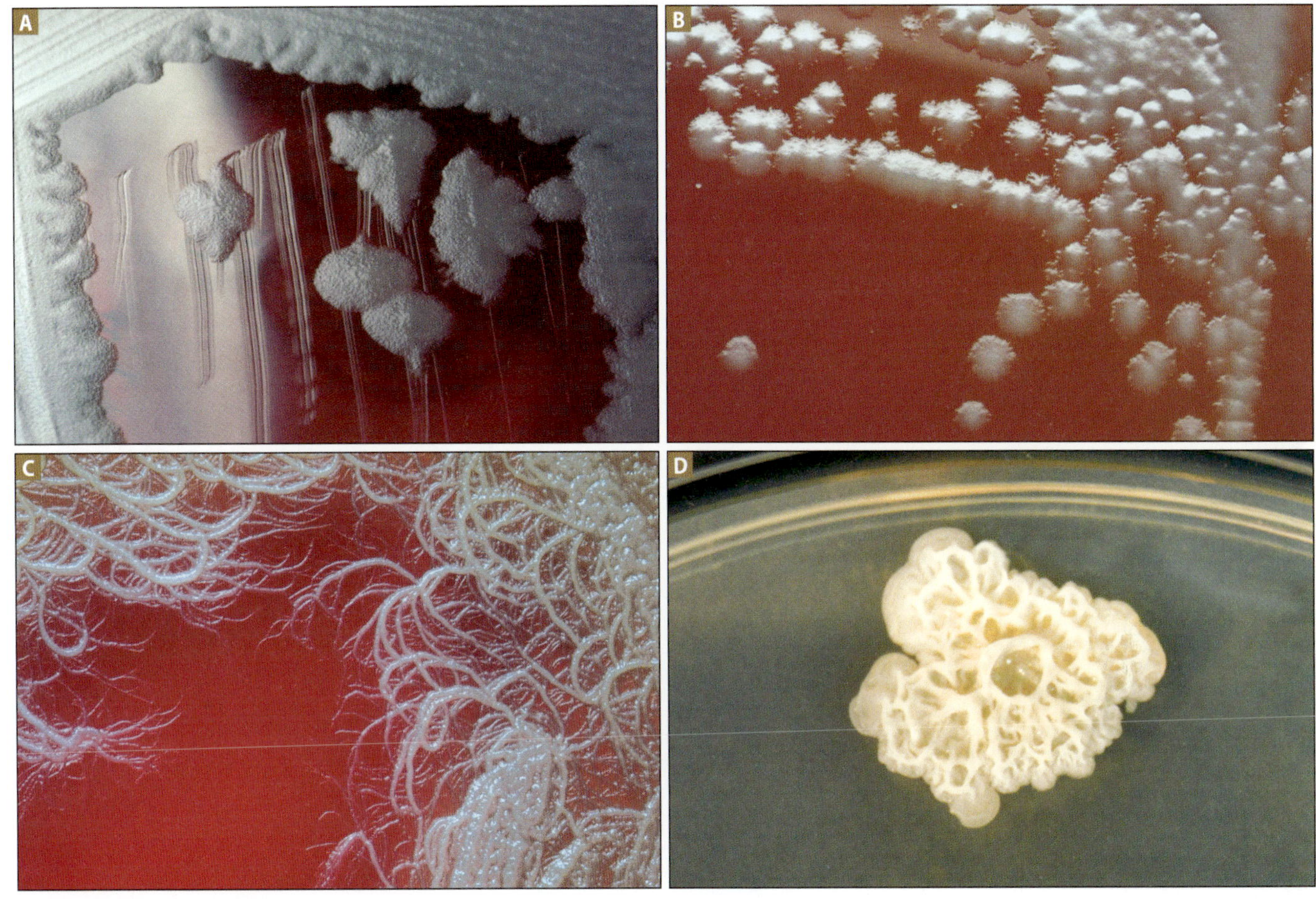


**2.26 Comparison of Four *Bacillus* Species Colonies** ■ (**A**) *B. cereus* grown on sheep blood agar (SBA) produces distinctively large (up to 7 mm), gray, granular, irregular colonies. They often produce a "mousy" smell. Also note the distinctive extensions of growth along the streak line. (**B**) *B. anthracis* colonies on SBA resemble *B. cereus*, but are usually smaller and adhere to the medium more tenaciously. It must be handled at *least* in a BSL-2 laboratory, and sometimes BSL-3 if the cell density is high enough, so it is unlikely you will be testing its tenacity! (**C**) *B. mycoides* produces distinctive, rapidly spreading, rhizoid (note the branching) colonies. It is a common isolate from soil and is shown here on SBA. (**D**) This unknown *Bacillus* isolated as a laboratory contaminant produced a wrinkled, irregular colony with an irregular (wavy) margin on tryptic soy agar.

## Extrinsic Factors Affecting Colony Morphology (Figures 2.27–2.30)

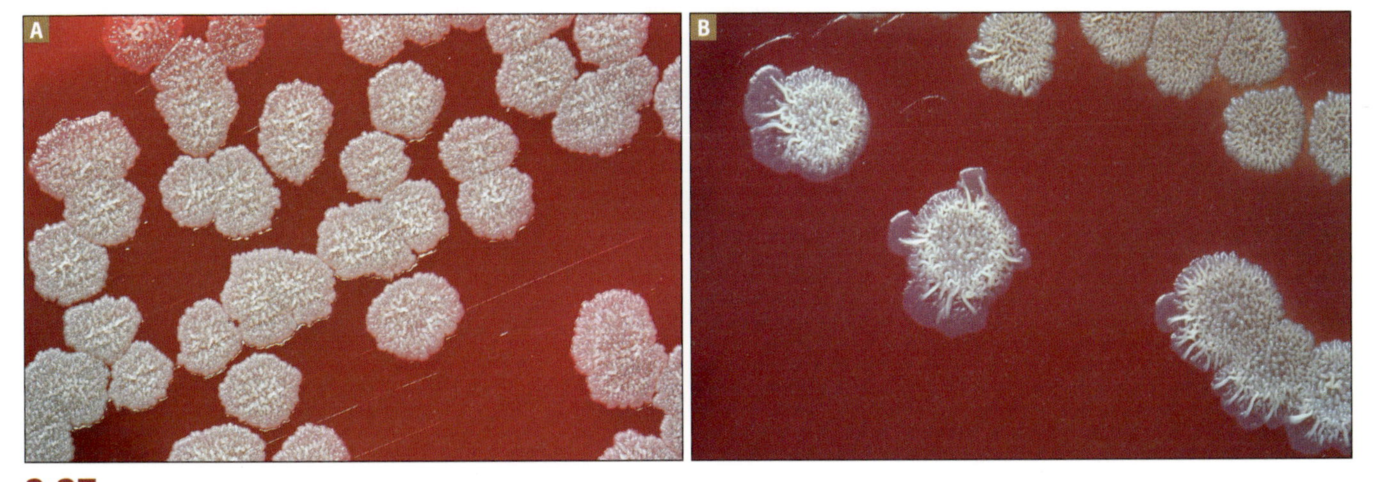


**2.27 Effect of Incubation Time on Colony Morphology** ■ (**A**) Close-up of *Bacillus subtilis* on sheep blood agar after 24 hours of incubation. (**B**) Close-up of the same *Bacillus subtilis* culture after 48 hours of growth. Note the wormlike extensions.

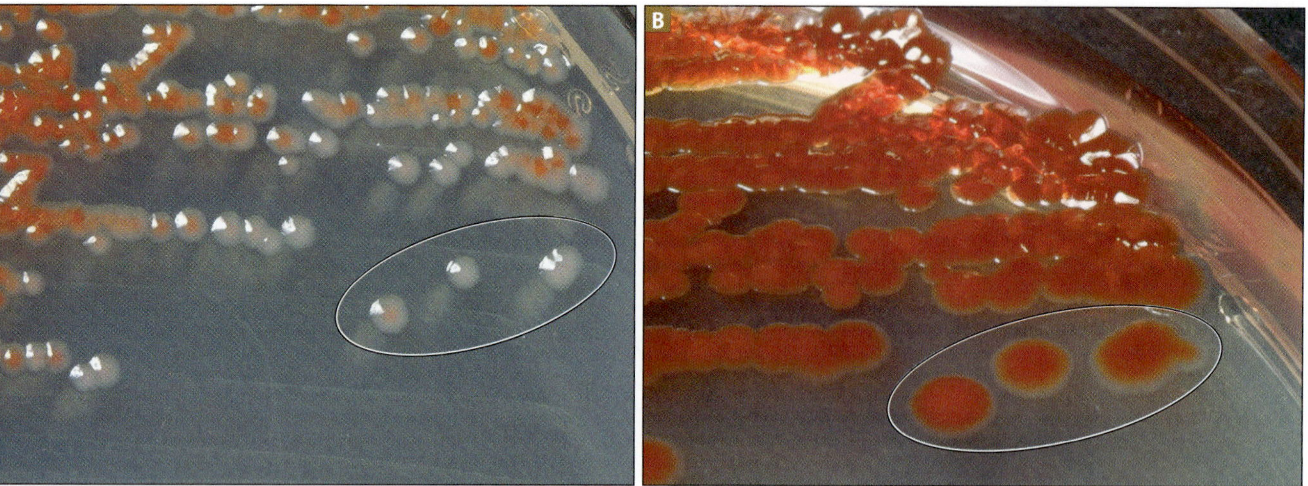


**2.28 Effect of Incubation Time on Pigment Production** ■ (**A**) *Serratia marcescens* grown on nutrient agar after 24 hours. (**B**) The same plate of *S. marcescens* after 48 hours. Note in particular the change in the three colonies in the lower right (circled).

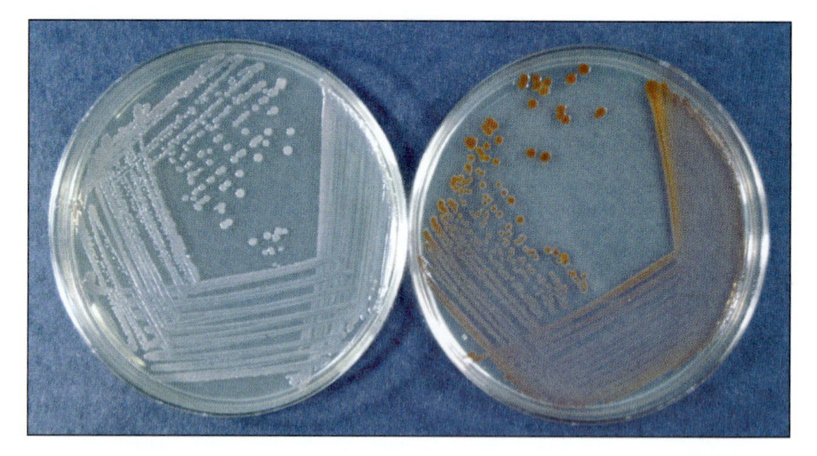

**2.29 Effect of Incubation Temperature on Pigment Production** ■ Pigment production may be influenced by temperature. *Serratia marcescens* produces less orange pigment when grown at 37°C (left) than when grown at 25°C (right).

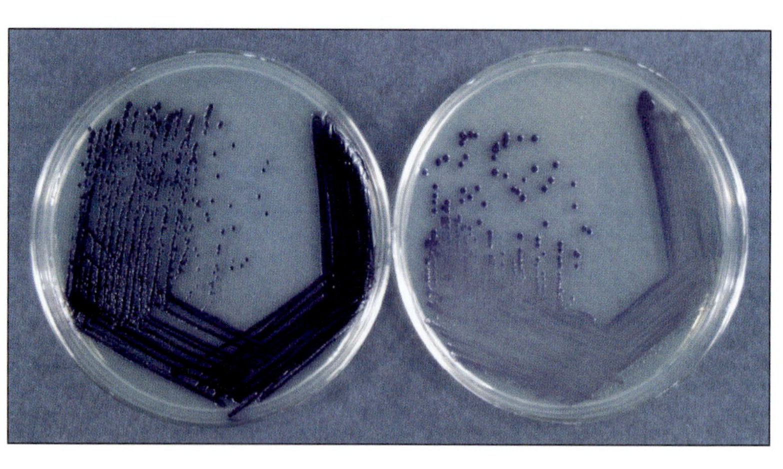

**2.30 Effect of Nutrient Availability on Pigment Production** ■ Pigment production may be influenced by environmental factors such as nutrient availability. *Chromobacterium violaceum* produces a much more intense purple pigment when grown on tryptic soy agar (left) than when grown on nutrient agar (right), a less nutritious medium.

## References


Claus, G. William. Chap. 14 in *Understanding Microbes—A Laboratory Textbook for Microbiology*. New York: W. H. Freeman and Co., 1989.

Collins, C. H., Patricia M. Lyne, and J. M. Grange. Chap. 6 in *Collins and Lyne's Microbiological Methods*, 7th ed. Oxford, England: Butterworth-Heinemann, 1995.

Tille, Patricia M. Pages 92–93 in *Bailey & Scott's Diagnostic Microbiology*, 13th ed. St. Louis, MO: Mosby, 2014.

Winn, Washington C. et al. *Koneman's Color Atlas and Textbook of Diagnostic Microbiology*, 6th ed. Baltimore: Lippincott Williams & Wilkins, 2006.

Name ______________________________

Date ______________________________

Lab Section ______________________________

I was present and performed this exercise (initials) ______________

DATA SHEET

2-2

# Colony Morphology

## OBSERVATIONS AND INTERPRETATIONS

**1** In the table below, use the terms provided in Figure 2.4 and the text to describe and carefully sketch representative colonies on the plates you examined. Draw separate sketches for the colonies as seen from above and in elevation (from the side). Plates from Exercise 2-1 can also be used, if not already examined for colony morphology. Use a colony counter for magnification if necessary. Measure colony diameters, and include them with your descriptions along with incubation times and temperatures, and culture sources.

| Organism/Plate | Colony Description and Sketch |
|---|---|
| | |
| | |
| | |
| | |
| | |

## QUESTIONS

**1** *A description of colony morphology (shape) provides important information about an organism. What other colony features should you include?*

**2** *Three critical aspects of a description of bacterial growth are colony size, color, and shape. At least three other important factors—not descriptions of the organism itself—typically are included when describing bacterial growth. What are they and why are they important?*

# Growth Patterns on Slants

EXERCISE **2-3**

## Theory

Agar slants are useful primarily as media for cultivation and maintenance of stock cultures. Organisms cultivated on slants, however, do display a variety of growth characteristics. We offer these more as an item of interest than one of diagnostic value.

Many of the organisms you will see in this class produce **filiform** growth (dense and opaque with a smooth edge), but you will see differences in other features. Growth on slants can vary in texture (e.g., moist or **friable** [dry, crusty]), optical properties (e.g., **opaque, translucent**), and margin (e.g., **smooth, lobed, spreading/effuse, filamentous, rhizoid** [branched], or **echinulate** [spiny]). Be advised that margins can be difficult to evaluate or even observe if the slant is not streaked in a straight line or if the organism covers the entire slant with its edge butting up against the glass. Figure 2.31 illustrates some of these features. Color is also a variable feature of growth on slants. All the organisms in Figure 2.32 are naturally **pigmented** (an obvious difference) and all but one are filiform (the second from the left is effuse).

Growth patterns of virtually all microorganisms can be influenced by environmental factors. Two important factors influencing bacterial growth are incubation time and temperature. Figure 2.33 illustrates the effect of temperature on pigment production by *Serratia marcescens*. See "Environmental Factors Affecting Microbial Growth" on page 87 for further discussion.

## Application

Slants are usually inoculated with pure cultures, so a streak plate has already been done and colony morphology has been described. Still, growth characteristics on slants can provide useful information when attempting to identify an organism.

**2.32 Pigment Production on Slants** ■ All of these organisms naturally produce a pigment. From left to right: *Staphylococcus epidermidis* (white), *Pseudomonas aeruginosa* (green), *Chromobacterium violaceum* (violet), *Serratia marcescens* (red/orange), *Kocuria rosea* (rose), *Micrococcus luteus* (yellow). All but *Pseudomonas*, which is effuse, have an entire (smooth) edge. The waviness you see is due to the inoculation pattern. Bacteria grow where you put them!

**2.31 Growth Patterns on Slants** ■ From left to right: filiform (smooth, even growth) and translucent; friable and opaque; effuse and translucent; echinulate and opaque; and rhizoid and opaque.

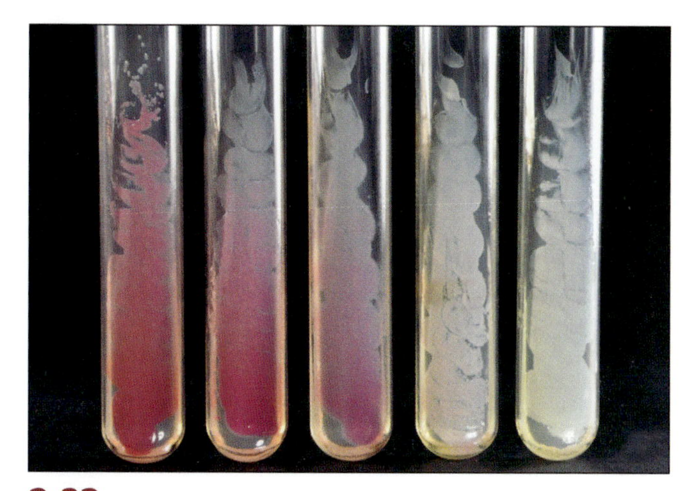

**2.33 Influence of Temperature on Pigment Production in *Serratia marcescens*** ■ *Serratia marcescens* was grown for 48 hours on tryptic soy agar slants at five different temperatures. From left to right: 25°C, 30°C, 33°C, 35°C, and 37°C. A difference of 2°C makes the difference between being pigmented or not!

## In This Exercise

In Exercise 2-2, you examined colony growth on agar plates. Today you will be looking at the growth patterns of up to seven organisms on prepared agar slants. If your instructor had you save your plates from Exercise 2-2, you will compare the growth patterns on the two different types of media.

2

## Materials

### Per Student Group

- ☐ Fresh agar slant cultures of these recommended organisms:
  - *Bacillus subtilis*
  - *Corynebacterium xerosis*
  - *Kocuria rosea*
  - *Lactobacillus plantarum* or *Lactobacillus acidophilus*
  - *Micrococcus luteus*
- ☐ (Optional)
  - *Mycobacterium smegmatis* (BSL-2)
  - *Proteus mirabilis* (BSL-2)
- ☐ Uninoculated tube of the medium used for cultures.

## PROCEDURE

1. Wear a lab coat, gloves, and chemical eye protection when performing this procedure.
2. Working with your group, examine the slants and describe the different growth patterns on the data sheet, page 81. Include a sketch of a representative portion of each. Use the uninoculated tube for comparison if growth is difficult to see on any of the slants.
3. (Optional) Compare the slant growth to the colonies on the corresponding agar plates from Exercise 2-2.
4. Dispose of all tubes (and plates) in the appropriate autoclave container when finished.

### Reference

Chan, E. C. S., Michael J. Pelczar, Jr., and Noel R. Krieg. Chap. 22 in *Laboratory Exercises in Microbiology*, 6th ed. New York: McGraw-Hill, Inc., 1993.

Claus, G. William. Chap. 17 in *Understanding Microbes—A Laboratory Textbook for Microbiology*. New York: W. H. Freeman and Co., 1989.

Name ____________________

Date ____________________

Lab Section ____________________

I was present and performed this exercise (initials) ____________________

DATA SHEET

2-3

# Growth Patterns on Slants

## OBSERVATIONS AND INTERPRETATIONS

**1** In the table below, describe the growth on your slants, including margin, texture, and color. Draw a picture of a short segment and include other information about the conditions of incubation, such as time, temperature, and the medium used.

| Organism | Description and Sketch |
|---|---|
| Control | |
| | |
| | |
| | |
| | |
| | |
| | |
| | |

## QUESTIONS

**1** *Match the following:*

| | | |
|---|---|---|
| ______ Filiform | **1.** | Produces colored growth |
| ______ Spreading edge | **2.** | Smooth texture with solid edge |
| ______ Transparent | **3.** | Solid growth seeming to radiate outward |
| ______ Friable | **4.** | Almost invisible, or easy to see light through |
| ______ Pigmented | **5.** | Rough texture with crusty appearance |

**2** *List some reasons why growth characteristics are more useful on agar plates than on agar slants.*

**3** *Why are agar slants better suited than agar plates to maintain stock cultures? (Think in practical terms, not necessarily in biological terms.)*

**4** *Suggest possible reasons for how temperature affects pigment production (as in Fig. 2.33).*

# Growth Patterns in Broth

EXERCISE **2-4**

2

## Theory

Microorganisms cultivated in broth display a variety of growth characteristics. Some organisms float on top of the medium and produce a type of surface membrane called a **pellicle**. Others sink to the bottom as **sediment**. Some bacteria produce **uniform fine turbidity,** and others appear to clump in what is called **flocculent** growth. Others show pigmentation or form a ring at the surface. Refer to Figures 2.34 and 2.35.

## Application

Bacterial genera, and frequently individual species within a genus, demonstrate characteristic growth patterns in broth that provide useful information when attempting to identify an organism.

## In This Exercise

Today, you will be examining the growth characteristics of up to seven different bacteria in broth.

## Materials

### Per Student Group

- ☐ One sterile uninoculated brain–heart infusion (BHI) broth control
- ☐ Fresh broth cultures of these recommended organisms:
  - *Bacillus subtilis*
  - *Corynebacterium xerosis*
  - *Kocuria rosea*
  - *Lactobacillus plantarum* or *Lactobacillus acidophilus*
  - *Micrococcus luteus*
- ☐ (Optional)
  - *Mycobacterium smegmatis* (BSL-2)
  - *Proteus mirabilis* (BSL-2)
- ☐ Your slants from Exercise 2-3

**2.34 Growth Patterns in Trypticase Soy Broth (TSB)** ■ From left to right: Uniform fine turbidity—UFT; Flocculent growth—notice the flakes within the turbidity; Pellicle—a film floating on surface. This pellicle's consistency is illustrated by the portions hanging down into the broth, an artifact of handling during the photo shoot...as is the sediment at the bottom, which fell from the pellicle at the surface; Uninoculated TSB for comparison; Ring at surface, with fine turbidity and a small sediment. Also note the purple coloration of the organism; Sediment.

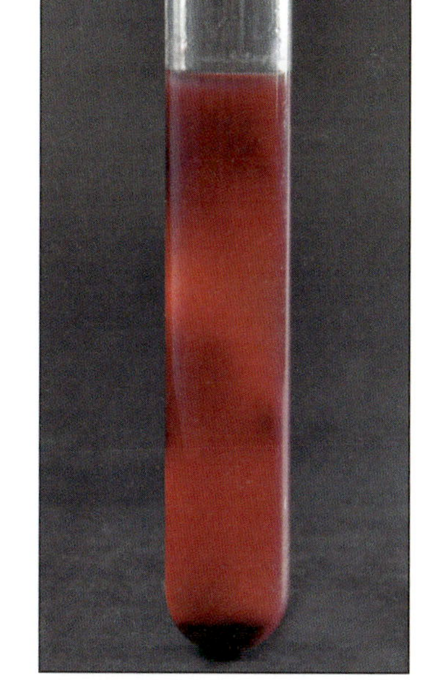

**2.35 Pigmentation in Broth** ■ *Rhodospirillum rubrum* has a red color due to carotenoid pigments. It grows as a photoheterotroph in the presence of light and the absence of oxygen.

1. Working with your group, examine the tubes and compare them with the control. Describe the different growth patterns in the table on the data sheet, page 85.
2. Dispose of all tubes in the appropriate autoclave container when finished.

2

### References


Claus, G. William. Chap. 17 in *Understanding Microbes—A Laboratory Textbook for Microbiology*. New York: W. H. Freeman and Co., 1989.

Grant, William D. Page 301 in Genus I. *Halobacterium* in *Bergey's Manual of Systematic Bacteriology*, 2nd ed. Vol. 1, The Archaea and The Deeply Branching and Phototrophic Bacteria. David R. Boone and Richard W. Castenholz, eds. New York: Springer, 2001.

Name ______________________________

Date ______________________________

Lab Section ______________________________

I was present and performed this exercise (initials) ______________________________

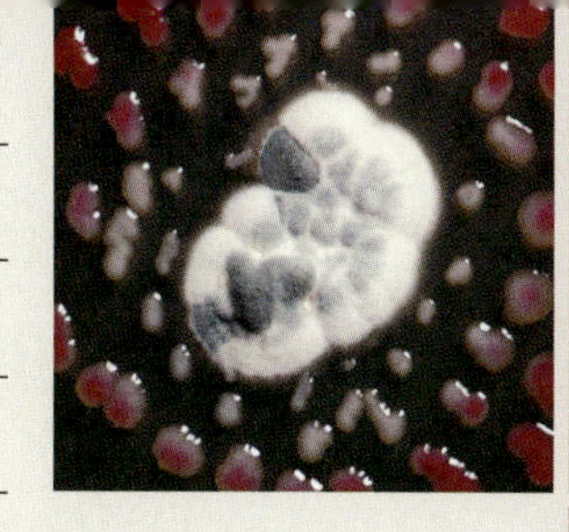

DATA SHEET 2-4

# Growth Patterns in Broth

## OBSERVATIONS AND INTERPRETATIONS

**1** Examine the broths and enter the description and a sketch of each in the table below. Include other information about incubation conditions, such as time, temperature, and the medium used.

| Organism | Sketch and Description of Growth in Broth |
|---|---|
| Control | |
| | |
| | |
| | |
| | |
| | |
| | |
| | |

## QUESTIONS

**1** *Match the following:*

| | | |
|---|---|---|
| ______ | Flocculent | **1.** Evenly cloudy throughout |
| ______ | Sediment | **2.** Growth at top around the edge |
| ______ | Ring | **3.** Growth on the bottom |
| ______ | Pellicle | **4.** Membrane at the top |
| ______ | Uniform fine turbidity | **5.** Suspended chunks or pieces |

**2** *What factors besides physical growth characteristics of the organism are important when recording data about an organism? Why?*

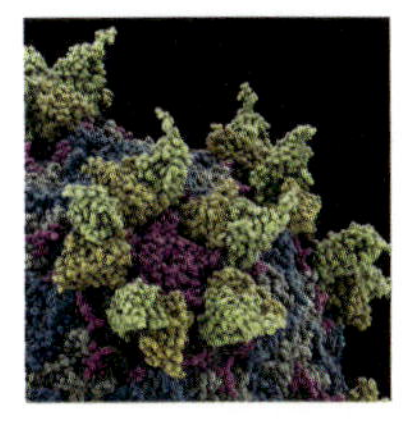

## Environmental Factors Affecting Microbial Growth

All organisms rely on their environment to provide them with the resources and conditions necessary for their growth. That is actually another way of saying, "Organisms are adapted to their environment." However, not all organisms are equally dependent on their environment, and not all are equally finicky about their environmental conditions.

With the disclaimer that absence of *any* nutritional requirement is enough to prevent microbial growth, three elements are of major importance. These are carbon, nitrogen, and oxygen. As you go through this lab course you will be using media and incubation conditions that attempt to mimic the organisms' natural habitats, if not perfectly, then well enough to support their growth. Be sure to examine the media recipes in each exercise and identify the source(s) of these elements. It will help you understand the exercise as well as get a better understanding of the organisms themselves.

Probably the most obvious environmental need organisms have is colloquially referred to as "food." However, the concept of "food" is more complex than you might think, because organisms need energy *and* they need a carbon source, along with numerous other elements (recall the acronym CHOPKNSCaFe—read as "C. HOPK(i)NS café—for starters). But, we will focus on energy and carbon right now. Some organisms get energy from chemicals and are classified as **chemotrophs**, whereas others get their energy from light and are called **phototrophs**—and most people wouldn't consider light as "food."

Organisms able to survive with carbon dioxide ($CO_2$) as the only carbon source are classified as **autotrophs**. Again, most people wouldn't consider $CO_2$ as "food," yet it is a necessary resource for autotrophs. Organisms requiring organic carbon are classified as **heterotrophs**. For many organisms (called **chemoheterotrophs**) the same organic molecule(s) provide both carbon and energy. (If you are feeling self-conscious about now, you should: *you* are a chemoheterotroph!)

In Exercise 2-5 you will be examining the role of medium composition in determining what organism(s) it can support and how effectively it can do so. You also will be examining how finicky different organisms can be with respect to dependence on their environment. Optimal growth conditions, as might be expected, result in faster growth and greater cell density (as evidenced by greater turbidity) than do less than optimal conditions.

Understand that growth rate, as it is used in this manual, is synonymous with reproductive rate, and while inherent growth rates may differ for each organism, we can legitimately compare growth rates of the same organism in different media. Growth rate comparisons between different organisms in the *same* medium require the assumption that the inherent growth rates are the same or similar, which we will make with the understanding that it may not be true. ■

# Evaluation of Media

EXERCISE 2-5

## Theory

Living things are composed of compounds from four biochemical families:

1. proteins
2. carbohydrates
3. lipids
4. nucleic acids.

Even though all organisms share this fundamental biochemical composition, they differ greatly in their ability to make these molecules. Some are capable of making them out of the simple carbon compound—carbon dioxide—and a nitrogen source. These organisms, called autotrophs, require the least "assistance" from the environment to grow. The remaining organisms, called heterotrophs, require preformed organic compounds from the environment.

Some heterotrophs are metabolically flexible and require only a few simple organic compounds from which to make all their biochemicals. Others require the environment to supply a greater portion of their organic compounds. An organism that relies heavily on the environment to supply ready-made organic compounds is referred to as **fastidious**. Fastidious heterotrophs vary greatly in their dependence on the environment and, as such, range from highly fastidious to **nonfastidious**. Autotrophs are less fastidious than the most nonfastidious heterotrophs.

Successful cultivation of a microbe in the laboratory requires an ability to satisfy its nutritional needs. The absence of a single required chemical resource prevents its growth. In general, the more fastidious the organism, the more ingredients a medium must have. **Undefined media** (also known as **complex media**) are composed of extracts from, or digests of, plant or animal matter and are usually quite rich in nutrients. Even though the exact composition of the medium and the amount of each ingredient are unknown, undefined media are useful in growing the greatest variety of culturable microbes.

A **defined medium** (or **chemically defined medium**) is one in which the amount and identity of every ingredient are known, which means it must be formulated from purified chemicals. As a result, defined media typically support a narrower range of organisms and are usually used with the growth of a particular organism or group of organisms in mind.

## Application

The ability of a microbiologist to cultivate a microorganism requires some knowledge of its metabolic needs. One quick way to make this determination is to transfer it to a variety of media containing different nutritional components and observe how well it grows.

## In This Exercise

Today you will evaluate the ability of three media—brain–heart infusion broth, nutrient broth, and glucose salts medium—to support bacterial growth. You will do this by visually comparing the density of growth (turbidity) of organisms in the various broths. It is important to make your inoculations as uniform as possible, so you will be adding the inocula with sterile transfer pipettes.

## Materials

### Per Student

- ☐ Lab coat
- ☐ Disposable gloves
- ☐ Chemical eye protection

### Per Student Group

- ☐ Five tubes each of:
  - Brain–heart infusion broth
  - Glucose salts broth
  - Nutrient broth
- ☐ Fresh broth cultures of these recommended organisms:
  - *Escherichia coli*
  - *Lactococcus lactis*
  - *Moraxella catarrhalis* (some strains are BSL-2)
  - *Staphylococcus epidermidis*
- ☐ Four sterile transfer pipettes

## Medium Recipes

### Nutrient Broth

|  |  |
|---|---|
| ☐ Beef extract | 3.0 g |
| ☐ Peptone | 5.0 g |
| ☐ Distilled or deionized water | 1.0 L |

### Brain–Heart Infusion Broth

|  |  |
|---|---|
| ☐ Calf brains, infusion from 200 g | 7.7 g |
| ☐ Beef heart, infusion from 250 g | 9.8 g |
| ☐ Proteose peptone | 10.0 g |
| ☐ Dextrose (glucose) | 2.0 g |
| ☐ Sodium chloride | 5.0 g |
| ☐ Disodium phosphate | 2.5 g |
| ☐ Distilled or deionized water | 1.0 L |

### Glucose Salts Broth

|  |  |
|---|---|
| ☐ Glucose | 5.0 g |
| ☐ Sodium chloride | 5.0 g |
| ☐ Magnesium sulfate | 0.2 g |
| ☐ Ammonium dihydrogen phosphate | 1.0 g |
| ☐ Dipotassium phosphate | 1.0 g |
| ☐ Distilled or deionized water | 1.0 L |

# PROCEDURE

## Lab One

1. Wear a lab coat, gloves, and chemical eye protection when performing this procedure.
2. Working with your group, obtain five tubes of each medium (a total of 15 tubes) and label each tube with your group name, the date, and the medium it contains.
3. Label one set of three tubes (one of each medium) with the name of one organism. Repeat the labeling process with the remaining three sets of medium tubes and organisms, using a different organism per set of tubes. Label the fifth set of tubes "uninoculated."
4. Each student should choose one or two cultures to inoculate from (whatever divides the work load evenly). Mix them thoroughly, but be careful not to splash into the cap. **Use BSL-2 precautions when transferring *Moraxella catarrhalis*.**
5. Inoculate one tube of each medium with one drop of the organism (be consistent) using a sterile transfer pipette. You may fill the pipette with about 1 mL of broth and deliver the drops with the same pipette to the tubes without flaming them. In your group, leave one of each tube uninoculated. When finished, each organism will have been inoculated into each medium.

**Caution!**

- Insert the pipette well into the tube (but not touching the glass or the medium) before delivering the inoculum.
- Be careful not to spatter as you deliver the drop(s) to each tube, or accidentally drip organism between tubes. Clean up the spill if you do.
- If you don't have enough organism left in the pipette to inoculate a tube, get more organism. You must not empty ("blow-out") the pipette because you will create aerosols.
- Properly dispose of the pipette when finished.

6. Incubate all tubes (including the uninoculated ones) at 35 ± 2°C for 24–48 hours.
7. Save or dispose of the original cultures as directed by your instructor.

## Lab Two

1. Wipe the outside of all tubes with a tissue or towel, and place them in a test tube rack organized into groups by organism in one direction and by medium in the other.
2. Mix all tubes well and examine them for turbidity. Be careful not to splash broth into the caps. Score relative amounts of growth using "0" for no growth, and "1," "2," and "3" for successively greater degrees of growth. Choose the tube (regardless of medium) with the most turbidity to act as a "3." What will you use to act as "0"?

____________________________________

____________________________________

____________________________________

3. Record your results on the data sheet, page 91. (***Note:*** Results may vary. Record what you see, not what you expect or what other groups are getting for results.)
4. Dispose of all tubes in the appropriate autoclave container when finished.

## References


Delost, Maria Dannessa. Page 144 in *Introduction to Diagnostic Microbiology: A Text and Workbook*. St. Louis, MO: Mosby, 1997.

Tille, Patricia M. Chap. 7 in *Bailey & Scott's Diagnostic Microbiology*, 13th ed. St. Louis, MO: Mosby, 2014.

Zimbro, Mary Jo and David A. Power, eds. *Difco™ and BBL™ Manual—Manual of Microbiological Culture Media*. Sparks, MD: Becton Dickinson and Co., 2003.


2

Name ______________________________

Date ______________________________

Lab Section ______________________________

I was present and performed this exercise (initials) ______________

DATA SHEET 2-5

# Evaluation of Media

## OBSERVATIONS AND INTERPRETATIONS

**1** In the table below, score the relative amount of growth in each tube. Compare the five tubes of each medium with each other, as well as the amount of growth for each organism in the various media. Use "0" for no growth, "3" for abundant growth, and "1" and "2" for degrees of growth in between. Use the most turbid of the 15 broths as your "3."

| Specimen | Nutrient Broth | Brain-Heart Infusion Broth | Glucose Salts Broth | Interpretation (relative fastidiousness) |
|---|---|---|---|---|
| Uninoculated control | | | | NA |
| | | | | |
| | | | | |
| | | | | |
| | | | | |

**2** In the table below, identify the carbon and nitrogen source(s) in each medium and then characterize each as "defined" or "undefined" ("complex"). If undefined, list the ingredient(s) that make it undefined. Finally, use your results to characterize each medium based on its ability to support growth of a wide range of organisms (poor, fair, good).

| Medium Description | Nutrient Broth | Brain-Heart Infusion Broth | Glucose Salts Broth |
|---|---|---|---|
| Carbon source(s) | | | |
| Nitrogen source(s) | | | |
| Defined/undefined | | | |
| Ingredient(s) making it "undefined" | | | |
| Ability to support a wide range of organisms (poor, fair, good) | | | |

Name ______________________________

Date ______________________________

Lab Section ______________________________

I was present and performed this exercise (initials) ______________________________

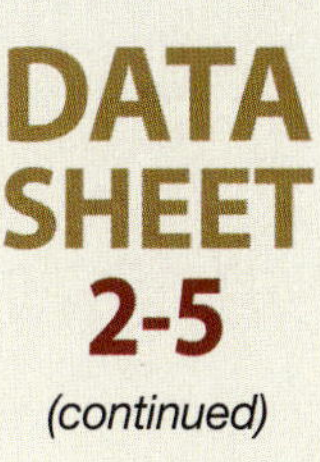


*(continued)*

## QUESTIONS

**1** *Why is it recommended that you use the broth (regardless of medium) with the most growth to serve as the "standard" for scoring a "3"?*

**2** *Why is it important to use the uninoculated controls* for each medium *to serve as the "standard" for scoring a "0" in their respective tubes?*

**3** *What does comparing growth of a given organism in the three media tell you? (That is, are you gaining information primarily about the organism or about the medium?)*

**4** *What does comparing growth of the four organisms in a given medium tell you? (Again, are you gaining information primarily about the organism or about the medium?)*

**5** *Suppose after performing this experiment, you examine the growth of two different organisms in Medium X. Both show growth and Organism 1 has produced less turbidity than Organism 2. Explain how it could be possible that Organism 1's needs are actually better supplied by the medium than Organism 2's, yet there is less turbidity.*

**6** *Evaluate the media.*

**a.** *Which medium supports growth of the widest range of organisms?*

**b.** *Which medium supports the fewest organisms?*

2

**c.** *Is there a correlation between your answers to questions 6a and 6b and the terms defined medium and undefined medium? If so, what is it? If not, attempt to explain the relationship you observe.*

**7** *Evaluate the organism.*

**a.** *Which organism appears to be most fastidious? How can you tell?*

**b.** *Which organism appears to be least fastidious? How can you tell?*

**8** *What is the biochemical basis for the spectrum of fastidiousness seen in the microbial world? (That is, why are some organisms fastidious and others are nonfastidious?)*

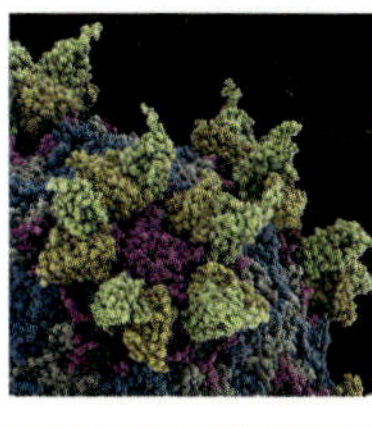

## Aerotolerance

Most microorganisms can survive within a range of environmental conditions, but not surprisingly, tend to produce growth with the greatest density in the areas where conditions are most favorable. One important resource influencing microbial growth is oxygen ($O_2$)[1]. Some organisms require oxygen for their metabolic needs (e.g., in aerobic respiration). Remarkably (from our perspective) other organisms don't use it and cannot even survive in its presence! Still other organisms don't use it and are not affected by it at all. This ability or inability to live in the presence of oxygen is called **aerotolerance**.

Most growth media are sterilized in an autoclave during preparation. This process not only kills unwanted microbes, but removes most of the free oxygen from the medium as well. After the medium is removed from the autoclave and allowed to cool, oxygen begins to diffuse back in from the air (which is approximately 21% $O_2$).

In tubed media (both liquid and solid) this process creates a gradient of oxygen concentrations, ranging from **aerobic** at the top, nearest the source of oxygen, to **anaerobic** at the bottom. Because of microorganisms' natural tendency to proliferate where the oxygen concentration best suits their metabolic needs, differing degrees of population density will develop in the medium over time that can be used to visually examine their aerotolerance.

**Obligate** (**strict**) **aerobes**, organisms that require oxygen for aerobic respiration, grow at the top where oxygen is most plentiful. **Facultative anaerobes** grow in the presence *or* absence of oxygen. When oxygen is available, they respire aerobically. When oxygen is not available, they either respire anaerobically (reducing sulfur or nitrate instead of oxygen) or ferment an available substrate. Refer to Appendix A and Section 5 for more information on anaerobic respiration and fermentation.

Where an oxygen gradient exists, facultative anaerobes grow throughout the medium but appear denser near the top because of the higher ATP yield from aerobic respiration compared to fermentation or anaerobic respiration. **Aerotolerant anaerobes**, organisms that don't require oxygen and are not adversely affected by it, live uniformly throughout the medium. Aerotolerant anaerobes ferment even in the presence of free oxygen.

**Microaerophiles**, as the name suggests, survive only in environments containing lower than atmospheric levels of oxygen. Some microaerophiles called **capnophiles** can survive only if carbon dioxide levels are elevated. Microaerophiles will be seen somewhere near the middle or upper middle region of the medium. Finally, **obligate** (**strict**) **anaerobes** are organisms for which even small amounts of oxygen are lethal and, therefore, will be seen only in the lower regions of the medium, depending on how far into the medium oxygen has diffused. ■

[1] Most biochemicals and many inorganic compounds contain oxygen, so all organisms require oxygen as part of other molecules. The use of the term *oxygen* in most microbiological contexts refers to molecular oxygen, $O_2$.

# EXERCISE 2-6 Fluid Thioglycollate Broth

## Theory

Fluid thioglycollate broth (FTB) is prepared as a basic medium (as used in this exercise) or with a variety of supplements, depending on the specific needs of the organisms being cultivated. As such, this medium can support growth of a broad variety of aerobic and anaerobic, fastidious and nonfastidious organisms. It is particularly well adapted for cultivation of strict anaerobes and microaerophiles.

Key components of the medium are yeast extract, pancreatic digest of casein, dextrose (glucose), sodium thioglycollate, L-cystine, and resazurin. Yeast extract and pancreatic digest of casein provide nutrients; sodium thioglycollate and L-cystine reduce oxygen to water; and resazurin (pink when oxidized, colorless when reduced) acts as an $O_2$ indicator. A small amount of agar is included to slow oxygen diffusion.

As mentioned in the introduction to this exercise, "Aerotolerance," oxygen removed during autoclaving will diffuse back into the medium as the tubes cool to room temperature. This produces a gradient of concentrations from aerobic (comparable to air) at the top to anaerobic at the bottom. Thus, fresh media will appear clear to straw-colored with a pink region at the top where the dye has become oxidized (Fig. 2.36). Figure 2.37 shows some basic bacterial growth patterns in the medium as influenced by the oxygen gradient.

## Application

Fluid thioglycollate broth is a liquid medium that can be modified to promote growth of a wide variety of fastidious to nonfastidious microorganisms. It can be used to illustrate microbial growth representing all levels of oxygen tolerance; however, its primary use is cultivation of anaerobic and microaerophilic bacteria.

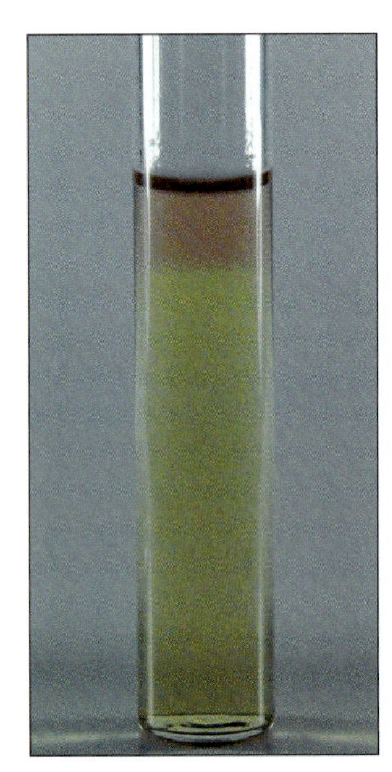

**2.36 Aerobic and Anaerobic Zones in Thioglycollate Broth** ■ Note the pink region in the upper, aerobic portion of the broth resulting from oxidation of the indicator resazurin. In the lower, anaerobic portion, the dye has been reduced and is colorless, leaving the medium its typical straw color.

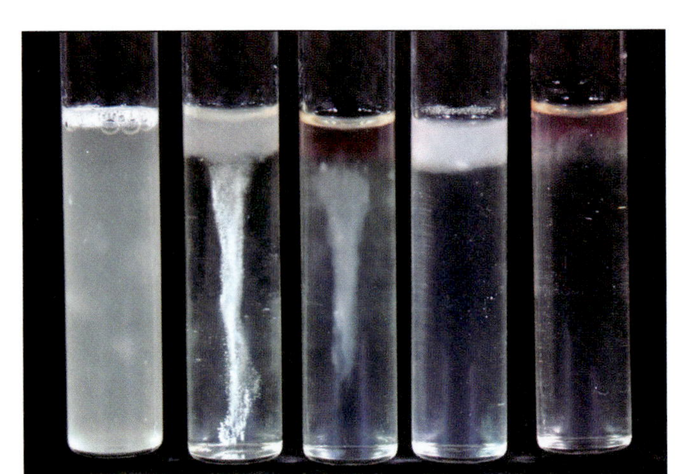

**2.37 Growth Patterns in Thioglycollate Broth** ■ Growth patterns of a variety of organisms are shown in these fluid thioglycollate broths. Pictured from left to right are: aerotolerant anaerobe, facultative anaerobe, obligate (strict) anaerobe, obligate (strict) aerobe, and microaerophile. Compare these tubes with the uninoculated broth in Figure 2.36.

## In This Exercise

Today you will be inoculating fluid thioglycollate broth to determine the aerotolerance categories of three bacteria. Your instructor may have additional organisms or may have you collect environmental samples to test.

## Materials

### Per Student

- ☐ Lab coat
- ☐ Disposable gloves
- ☐ Chemical eye protection

### Per Student Group

- ☐ Four fluid thioglycollate broth (FTB) tubes
- ☐ Fresh broth cultures of these recommended organisms:
  - *Alcaligenes faecalis* subsp. *faecalis*
  - *Clostridium butyricum*
  - *Staphylococcus epidermidis*

## Medium Recipe

### Fluid Thioglycollate Broth

|  |  |
|---|---|
| ☐ Yeast extract | 5.0 g |
| ☐ Pancreatic digest of casein | 15.0 g |
| ☐ Dextrose (glucose) | 5.5 g |
| ☐ Sodium chloride | 2.5 g |
| ☐ Sodium thioglycollate | 0.5 g |
| ☐ L-cystine | 0.5 g |
| ☐ Agar | 0.75 g |
| ☐ Resazurin | 0.001 g |
| ☐ Distilled or deionized water | 1.0 L |

## PROCEDURE

### Lab One

1. Wear a lab coat, gloves, and chemical eye protection when performing this procedure.
2. Working with your group, obtain four fluid thioglycollate tubes and label them with your name, the date, and the medium. Label three of the tubes with the name of an organism, using a different organism on each tube; label the fourth tube "uninoculated."

3. Using your loop, inoculate three broths with the organisms provided. (***Note:*** when inoculating thioglycollate broth, it helps to dip the loop all the way to the bottom of the tube and gently mix the broth with the loop as you remove it. Do not shake the tube. Finish mixing by gently rolling the tube between your hands.) Do not inoculate the fourth tube.
4. Incubate the tubes at 35 ± 2°C for 24 to 48 hours.
5. Save or dispose of the original cultures as directed by your instructor.

### Lab Two

1. Check the uninoculated tube for growth to assure sterility of the medium. Note any changes that may have occurred as a result of incubation, especially in the colored region at the surface.
2. Using the control as a comparison, examine and note the location of the growth in all tubes.
3. Enter your observations and interpretations in the table provided on the data sheet, page 99.
4. Dispose of all tubes in the appropriate autoclave container when finished.

## References

Allen, Stephen D., Christopher L. Emery, and David M. Lyerly. Chap. 54 in *Manual of Clinical Microbiology*, 8th ed. Patrick R. Murray, Ellen Jo Baron, James H. Jorgensen, Michael A. Pfaller, and Robert H. Yolken, eds. Washington, DC: ASM Press, American Society for Microbiology, 2003.

Atlas, Ronald M. and James W. Snyder. Page 299 in *Manual of Clinical Microbiology,* 10th ed., James Versalovic, Karen C. Carroll, Guido Funke, James H. Jorgensen, Marie Louise Landry, and David W. Warnock, eds. Washington, DC: ASM Press, 2011.

Ryan, Kenneth J. and C. George Ray. Pages 64–65 in *Sherris Medical Microbiology*, 6th ed. New York: McGraw-Hill, 2014.

Tille, Patricia M. Pages 85 and 87–88 in *Bailey & Scott's Diagnostic Microbiology,* 13th ed. St. Louis, MO: Mosby, 2014.

Winn, Washington C. et al. *Koneman's Color Atlas and Textbook of Diagnostic Microbiology,* 6th ed. Baltimore: Lippincott Williams & Wilkins, 2006.

Zimbro, Mary Jo and David A. Power, eds. *Difco™ and BBL™ Manual—Manual of Microbiological Culture Media.* Sparks, MD: Becton Dickinson and Company, 2003.

Name ______________________________

Date ______________________________

Lab Section ______________________________

I was present and performed this exercise (initials) ______________

DATA SHEET

2-6

# Fluid Thioglycollate Broth

## OBSERVATIONS AND INTERPRETATIONS

**1** Draw a sketch of each culture showing the location of growth. Indicate the amount of growth in the aerobic and anaerobic regions of the broth using the following system:

"0" = no growth  "3" = abundant growth  "1" and "2" = degrees of growth in between

Then, place each organism into an aerotolerance category.

| Organism | Location of Growth in Medium | Aerotolerance Category |
|---|---|---|
| Control | | Not Applicable |
| | | |
| | | |
| | | |

## QUESTIONS

**1** *Consider aerotolerance categories and their growth in FTB.*

**a.** *Where would you expect to see growth of a strict aerobe?*

______________________________

**b.** *Anaerobe?*

______________________________

**c.** *Microaerophile?*

______________________________

**d.** *Facultative anaerobe?*

______________________________

**e.** *Aerotolerant anaerobe?*

______________________________

**2** *Consider resazurin dye.*

**a.** *What purpose does the dye serve in FTB?*



**b.** *Where in the broth column is resazurin dye located?*

**c.** *Why is the broth pink at its surface?*

**d.** *Considering the primary application of FTB, which is more desirable, a thick or a thin colored band at the surface? Explain your answer.*

**3** *Consider medium freshness.*

**a.** *Why is it important that this medium be fresh?*

**b.** *Which type of organism (obligate aerobe, obligate anaerobe, microaerophile, facultative anaerobe, aerotolerant anaerobe) would most likely be affected negatively by the use of old media? Explain your answer.*

**c.** *Which would most likely be affected positively by the use of old media? Explain your answer.*

**d.** *Which would be least affected by old media? Explain your answer.*

# Anaerobic Jar

EXERCISE 2-7

## Theory

Various airtight jars and pouches have been developed that allow cultivation of obligately anaerobic, microaerophilic, or capnophilic microorganisms. They vary in size, but can carry one or more agar plates or test tubes and still fit in a standard incubator, thus making cultivation of these "environmentally demanding" organisms a relatively simple task. The GasPak® 100 Anaerobic Jar by BD™ is an example (Fig. 2.38). It holds up to 12 Petri dishes.

Each system can be used with a variety of gas generating sachets (packets) that produce the desired atmosphere. For instance, BD GasPak® EZ Anaerobe Sachets contain water, ascorbic acid, activated carbon, and inorganic carbonate. When removed from their wrapper and sealed in the jar, the contents are exposed to air and ascorbic acid reacts with $O_2$ to form dehydroascorbic acid and water (Fig. 2.39), thus making the interior of the jar anaerobic (<1% $O_2$). The inorganic carbonate reacts to produce an atmosphere with roughly 13% carbon dioxide (air is 0.04% $CO_2$).[1]

With a slightly different composition, the BD GasPak® EZ Campy Sachets ("Campy" stands for *Campylobacter jejuni*, a microaerophilic pathogen) produce microaerophilic conditions (6–16% $O_2$). The inorganic carbonate produces an atmosphere of 2–10% $CO_2$.

Oxoid AnaeroGen Atmospheric Generation Systems are an alternative to the BD systems and work in a similar way. The advantages of both systems over previous ones are that they don't require addition of water or a palladium catalyst, and most importantly don't produce flammable $H_2$ gas as a byproduct.

**2.38 An Anaerobic Jar** The large white envelope is the gas generator packet. It is inserted between the jar and the plates. The small white strip at the left is the indicator. It is white because the gas generator has successfully removed all $O_2$, making conditions inside the jar anaerobic and leaving the methylene blue dye in its colorless oxidized state. This jar is large enough to hold 12 Petri dishes.

An indicator strip is used with the anaerobic systems to ensure that anaerobic conditions have been created. The indicator strip is soaked with methylene blue or resazurin dye that is blue or pink, respectively, when oxidized and colorless when reduced.

## Application

This procedure provides a means of conveniently cultivating anaerobic and microaerophilic bacteria in a standard incubator. The jar can accommodate either plated or broth media.

## In This Exercise

Today you will use the jar to experience a common way to grow anaerobes (or microaerophiles), but you will also take advantage of the opportunity to demonstrate aerotolerance categories of the organisms grown. To do the latter, you will inoculate two nutrient agar plates with three organisms (your instructor may have additional organisms for you to test) and incubate one of the plates inside the jar and the other outside the jar. Both plates will be incubated at the same temperature for the same amount of time. Following incubation, growth on the two plates will be examined and compared to determine aerotolerance.

## Materials

### Per Student

- ☐ Lab coat
- ☐ Disposable gloves
- ☐ Chemical eye protection

### Per Class

- ☐ One anaerobic jar with anaerobic gas generator packet[2,3]

---

[1] The Chemical Company. https://www.thechemco.com/chemical/ascorbic-acid/

[2] Available from Becton Dickinson Microbiology Systems, Sparks, MD http://www.bd.com.

[3] Available from Thermo Fisher Scientific, Oxoid Microbiology Products, Waltham, MA, http://www.oxoid.com/UK/blue/prod_detail/prod_detail.asp?pr=AN0035&c=UK&lang=EN.

### Per Student Group

- ☐ Two nutrient agar plates
- ☐ Fresh broth cultures of these recommended organisms:
  - *Alcaligenes faecalis* subsp. *faecalis*
  - *Clostridium butyricum*
  - *Staphylococcus epidermidis*

2

## PROCEDURE

### Lab One

1. Wear a lab coat, gloves, and chemical eye protection when performing this procedure.
2. Working with your group, obtain two nutrient agar plates. Using a marking pen, divide the bottom of each plate into three sectors (or more, if your instructor has additional organisms for you to use).
3. Label each plate with your name, the date, and organism by sector.
4. Using your loop, inoculate corresponding sectors of both plates with the organisms provided. (***Note:*** inoculate with single streaks about one centimeter long.) Tape the lids in place.
5. Place one plate in the anaerobic jar in an inverted position. Depending on your class size, more than one jar may be necessary.
6. When all groups have placed their plates in the jar, discharge the packet as follows.

   If using a BD system, follow the packet instructions, but they should be something like this:

   **a** Stick the methylene blue strip on the wall of the jar.

   **b** Open the packet and place it in the jar with the label facing inward.

   **c** Immediately close the jar.

   **d** Examine for condensation within about 30 minutes. If none appears, start over with a new packet.

   If using an Oxoid system, follow the packet instructions, but they should be something like this:

   **a** Stick the resazurin indicator strip on the wall of the jar.

   **b** Open the foil packet and remove the sachet. Place the sachet in the jar.

   **c** Close the jar within 1 minute.

   **d** Examine for condensation within about 30 minutes. If none appears, start over with a new packet.

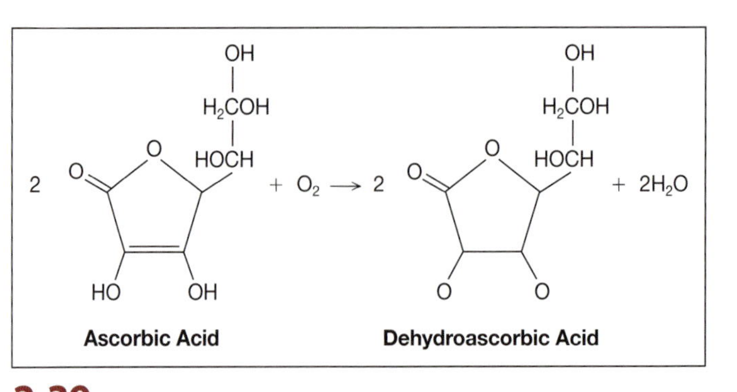


**2.39 Reaction of Ascorbic Acid and Oxygen** ■ When exposed to air in the sealed anaerobic jar, ascorbic acid reacts with oxygen to produce dehydroascorbic acid and water, thus making the atmosphere within the jar anaerobic.

7. Place the second (aerobic) plate and the anaerobic jar in the 35 ± 2°C incubator for 24 to 48 hours. Be sure the plate is inverted.
8. Save or dispose of the original cultures as directed by your instructor.

### Lab Two

1. Examine and compare the growth on the plates. (***Note:*** density of growth will be the most useful basis for comparison.)
2. Properly dispose of the packet in laboratory waste. The Oxoid packet may still be warm to the touch. Let it cool before disposing of it.
3. Record your results and interpretations in the table provided on the data sheet, page 103.
4. Dispose of all plates in the appropriate autoclave container when finished.

### *References*


Ryan, Kenneth J. and C. George Ray. Pages 64–65 in *Sherris Medical Microbiology*, 6th ed. New York: McGraw-Hill, 2014.

Thermo Fisher Scientific, Oxoid Microbiology Products, Waltham, MA, http://www.oxoid.com/UK/blue/prod_detail/prod_detail.asp?pr=AN0035&c=UK&lang=EN.

Tille, Patricia M. Pages 460–464 in *Bailey & Scott's Diagnostic Microbiology*, 13th ed. St. Louis, MO: Mosby, 2014.

Winn, Washington C. et al. *Koneman's Color Atlas and Textbook of Diagnostic Microbiology*, 6th ed. Baltimore: Lippincott Williams & Wilkins, 2006.

Zimbro, Mary Jo and David A. Power, eds. *Difco™ and BBL™ Manual—Manual of Microbiological Culture Media*. Sparks, MD: Becton Dickinson and Company, 2003.

Name ______________________________

Date ______________________________

Lab Section ______________________________

I was present and performed this exercise (initials) ______________________________

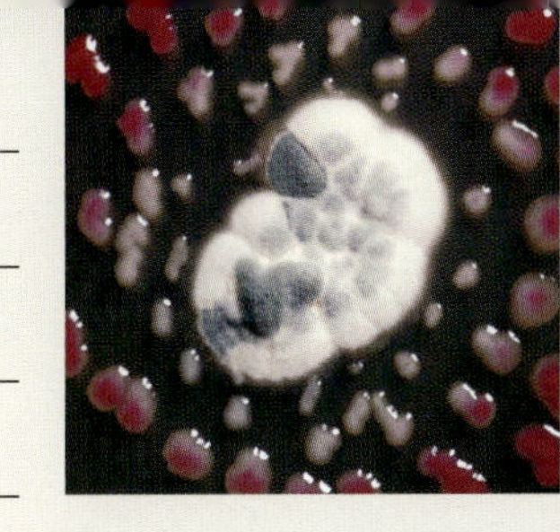

# DATA SHEET 2-7

## Anaerobic Jar

### OBSERVATIONS AND INTERPRETATIONS

**1** Indicate the amount of growth of each organism on each plate.

"0" = no growth "3" = dense growth "1" and "2" = degrees of growth in between

| Organism | Growth on Aerobic Plate | Growth on Anaerobic Plate | Aerotolerance Category |
|---|---|---|---|
| | | | |
| | | | |
| | | | |

### QUESTIONS

**1** *Suppose that after incubation you examine the jar before opening it and see that the methylene blue strip is blue (or the resazurin strip is pink).*

**a.** *What would you guess the internal environment to be—aerobic or anaerobic?*

______________________________

**b.** *How would you expect the growth on the plates inside the jar to differ from the plates incubated outside the jar?*

______________________________

______________________________

______________________________

______________________________

**2** *Based on your results, which of the three organisms would be most affected by the conditions described in question 1?*



**3** *An alternative to the anaerobic jar is a candle jar, in which a candle is placed in the jar, lit, and the lid closed to enable the flame to use the available oxygen. Typically, in this system, not all of the oxygen is used. Which types of organisms would most likely benefit from this environment?*

**4** *Considering the gaseous changes resulting from the burned candle, which group would benefit the most from a functional candle jar? (**Hint:** Think about a gas not involved in aerotolerance.)*

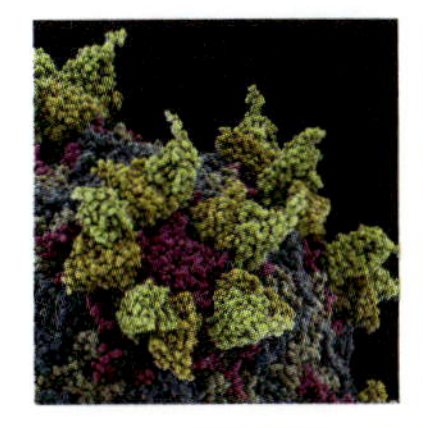

## Effect of Physical and Chemical Environmental Factors on Microbial Growth

Bacteria and other microbes have limited control over their internal environments. Whereas many eukaryotes have evolved sophisticated internal control mechanisms, microbes are almost completely dependent on external factors to provide conditions suitable for their existence. Minor environmental changes can dramatically change a microorganism's ability to transport materials across the membrane, perform complex enzymatic reactions, and maintain critical cytoplasmic pressure.

One way to observe microbial responses to environmental changes is to artificially manipulate an external factor and measure its effect on growth rate; that is, cell density after a given incubation time. In this series of laboratory exercises, you will examine the effects of temperature, pH, and osmotic pressure on growth rate. When appropriate, you will attempt to classify organisms based on your results. ■



# The Effect of Temperature on Microbial Growth

EXERCISE 2-8

## Theory

Bacteria and Archaea have been discovered living in habitats ranging from –10°C to more than 110°C. The temperature range of any single species, however, is a small portion of this overall range. As such, each species is characterized by a minimum, maximum, and optimum temperature—collectively known as its **cardinal temperatures** (Fig. 2.40). Minimum and maximum temperatures are, simply, the temperatures below and above which the organism will not survive. Optimum temperature is the temperature at which an organism grows the fastest—its highest growth rate.

Organisms that only grow below 20°C are called **psychrophiles.** These are common in ocean, Arctic, and Antarctic habitats where the temperature remains permanently cold with little or no fluctuation. Organisms adapted to cold habitats that fluctuate from about 0°C to above 30°C are called **psychrotrophs.** Bacteria adapted to temperatures between 15°C and 45°C are known as **mesophiles.** Most bacterial residents in the human body, as well as numerous human pathogens, are mesophiles.

**Thermophiles** are organisms adapted to temperatures above 40°C. Thermophiles that will not grow at temperatures below 40°C are called **obligate thermophiles;** those that will grow below 40°C are known as **facultative thermophiles.** Environments in which thermophilic Bacteria and Archaea are found include composting organic material, soil surfaces subjected to direct sunlight, and silage. Bacteria and Archaea isolated from ocean floor hydrothermal vents and other geothermal sites (Fig. 2.1) are called **extreme thermophiles** because they can survive temperatures in the 65°C and 110°C range (up to 122°C for *Methanopyrus*). Extreme thermophiles grow best above 80°C. Figure 2.41 illustrates typical temperature ranges and classifications of Bacteria and Archaea.

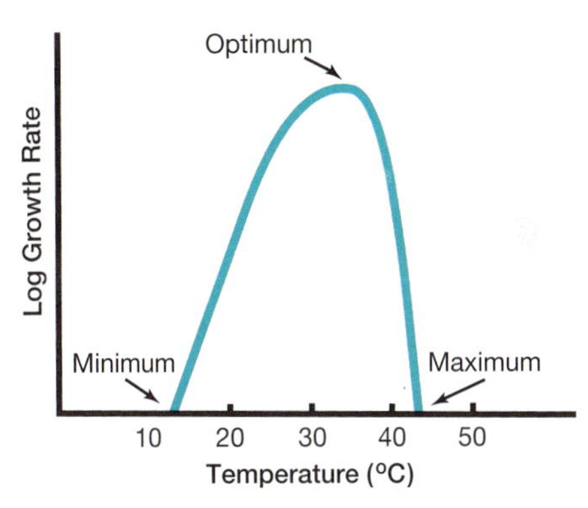


**2.40 Typical Growth Range of a Mesophile** ■ The "minimum" and "maximum" are temperatures beyond which no growth takes place. The "optimum" is the temperature at which growth rate is highest. These are the "cardinal temperatures."

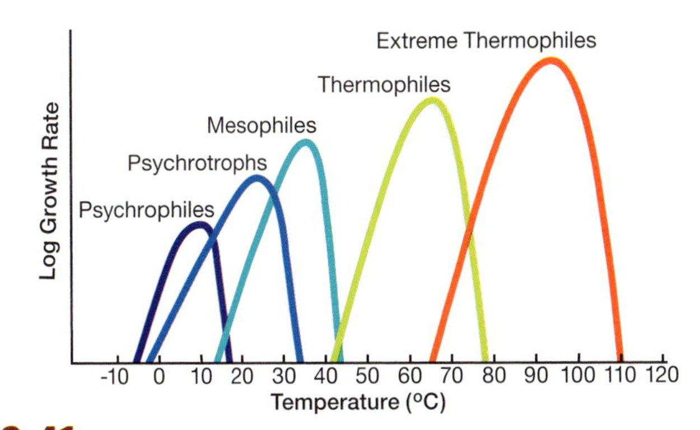


**2.41 Thermal Classifications of Bacteria** ■ These are generic cardinal temperature graphs for each category and represent a composite of all the individual graphs for each species. Refer to the text for a description of each.

## Application

This is a qualitative procedure designed for observing the effect of temperature on microbial growth. It allows an estimation of the cardinal temperatures for individual species (rather than for an entire group, as shown in Fig. 2.41).



## In This Exercise

Today you will examine the growth characteristics of four organisms at five different temperatures. In addition, you will observe the influence of temperature on pigment production.

### Materials

#### Per Student

- □ Lab coat
- □ Disposable gloves
- □ Chemical eye protection

#### Per Student Group

- □ Twenty-one sterile tryptic soy broths (or nutrient broths)
- □ Two tryptic soy agar (TSA) plates
- □ Four sterile transfer pipettes
- □ Fresh broth cultures of these recommended organisms:
  - *Escherichia coli*
  - *Geobacillus* (*Bacillus*) *stearothermophilus*
  - *Serratia marcescens*
  - *Pseudomonas fluorescens*

#### Per Class

- □ Five incubating devices set at 10°C, 25°C, 35°C, 45°C, and 60°C. These devices may be any combination of the following:
  - refrigerator
  - incubator
  - hot-water bath
  - cold-water bath

## PROCEDURE

### Lab One

1. Wear a lab coat, gloves, and chemical eye protection when performing this procedure.
2. Working with your group, obtain 21 tryptic soy broths (TSB)—one broth for each organism at each temperature plus one for an uninoculated control. Label them accordingly. Also obtain two TSA plates and label them 20°C and 35°C, respectively.
3. Each student should choose one or two cultures to inoculate from (whatever divides the work load evenly). Mix them thoroughly, but be careful not to splash into the cap.
4. Inoculate five TSB tubes with one drop of organism (be consistent) using a sterile transfer pipette. You may fill the pipette with about 1 mL of broth and deliver the drops with the same pipette to the tubes without flaming them. In your group, leave one tube uninoculated.

**Caution!**

- Insert the pipette well into the tube (but not touching the broth) before delivering the inoculum.
- Be careful not to spatter as you deliver the drop to each tube, or accidentally drip organism between tubes. Clean up the spill if you do.
- If you don't have enough organism left in the pipette to inoculate a tube, get more organism. You must not empty ("blow-out") the pipette because you will create aerosols.
- Properly dispose of the pipette when finished.

5. Using a simple zigzag pattern with a loop, as in Exercise 1-5, inoculate each TSA plate with *Serratia marcescens*.
6. Incubate all tubes at their appropriate temperatures for 24 to 48 hours.
7. Incubate the *Serratia marcescens* plates in the 20°C and 35°C incubators in an inverted position.
8. Save or dispose of the original cultures as directed by your instructor.

## Lab Two

1. Clean the outside of all tubes with a tissue, and place them in a test tube rack organized into groups by organism.
2. Mix each broth gently until uniform turbidity is achieved. Be careful not to splash into the caps.
3. Compare all tubes of a species to each other. Rate each as 0, 1, 2, or 3, according to its turbidity (0 is clear—the uninoculated tube—and 3 is highly turbid and is established by the most turbid tube out of the 20). Record these in the Broth Data table on the data sheet, page 109.
4. Examine the *Serratia marcescens* plates incubated at two different temperatures, compare the growth characteristics, and enter your results in the Plate Data table on the data sheet.
5. Using the data from the Broth Data table, circle the estimated cardinal temperatures of each of the four organisms. Then, based on your results, classify each organism.
6. On the graph paper provided with the data sheet, plot the relative turbidity values versus temperature for the four organisms. If you need help with graphing, see pages 10–11.
7. Dispose of both plates and all tubes in their appropriate autoclave containers when finished.

## References


Holt, John G., ed. *Bergey's Manual of Determinative Bacteriology,* 9th ed. Baltimore: Lippincott Williams & Wilkins, 1994.

Moat, Albert G., John W. Foster, and Michael P. Spector. Pages 597–601 in *Microbial Physiology,* 4th ed. New York: Wiley-Liss, 2002.

Prescott, Lansing M., John P. Harley, and Donald A. Klein. Chap. 6 in *Microbiology,* 6th ed. Boston: WCB McGraw-Hill, 2005.

Tille, Patricia M. Page 89 in *Bailey & Scott's Diagnostic Microbiology,* 13th ed. St. Louis, MO: Mosby, 2014.

Varnam, Alan H. and Malcolm G. Evans. *Environmental Microbiology.* Washington, DC: ASM Press, 2000.

White, David. Pages 384–387 in *The Physiology and Biochemistry of Prokaryotes,* 2nd ed. New York: Oxford University Press, 2000.

Winn, Washington C. et al. *Koneman's Color Atlas and Textbook of Diagnostic Microbiology,* 6th ed. Baltimore: Lippincott Williams & Wilkins, 2006

Name ___________________________________________

Date ___________________________________________

Lab Section ___________________________________________

I was present and performed this exercise (initials) ___________________

# DATA SHEET 2-8

## The Effect of Temperature on Microbial Growth

### OBSERVATIONS AND INTERPRETATIONS

**1** For each species, record a "3" for the temperature(s) with maximum growth and "0" for no growth. Use "1" and "2" for intermediate amounts of growth.

| Broth Data | | | | | | |
|---|---|---|---|---|---|---|
| **Organism** | **10°C** | **25°C** | **35°C** | **45°C** | **60°C** | **Classification** |
| | | | | | | |
| | | | | | | |
| | | | | | | |
| | | | | | | |

**2** Record the growth characteristics of the *Serratia marcescens* incubated at two temperatures.

| Plate Data | |
|---|---|
| **Incubation Temperature** | **Description of Growth** |
| 20°C | |
| 35°C | |

### QUESTIONS

**1** *Using the data from the Broth Data table in step 1, determine the cardinal temperatures for each of the four organisms. Circle the optimum temperature for each organism. Use brackets to designate the range for each. Is there any overlap between species? Describe any examples.*

___________________________________________

___________________________________________

___________________________________________

**2** *Plot relative turbidity (numeric values) versus temperature on the graph paper provided. See pages 10–11 for proper graphing technique.*

**3** *Why do different temperatures produce different growth rates?*

**4** *Why is it not advisable to connect the data points for each organism in your graph?*

**5** *Consider the shape of each curve. Where is the optimum temperature relative to the range for each organism?*

**6** *In what way(s) could you adjust the incubation temperature to grow an organism at less than its optimal growth rate?*

# The Effect of pH on Microbial Growth

2

## Theory

The conventional means of expressing the concentration (or activity) of hydrogen ions in a solution is "pH." The term *pH*, which stands for *pondus hydrogenii* (variably defined as hydrogen power or hydrogen potential), was invented in 1909 by the Danish biochemist, Søren Peter Lauritz Sørensen. The 0–14 pH range he developed is a logarithmic scale designed to simplify acid and base calculations that otherwise would be expressed as molar values. Sørensen's formula for the calculation of pH is expressed as follows:

$$pH = -\log [H^+]$$

For example, an aqueous solution containing $10^{-6}$ moles of disassociated hydrogen ions per liter would be converted using Sørensen's formula as follows:

$$pH = -\log [H^+]$$
$$pH = -\log 10^{-6} M\ H^+$$
$$pH = -(-6)$$
$$pH = 6$$

Pure water contains $10^{-7}$ moles of hydrogen ions per liter and has a pH of 7. As hydrogen ions increase, the solution becomes more acidic and the pH decreases (Table 2-1). Bacteria live in habitats throughout the pH spectrum; however, the range of most individual species is small. Like temperature and salinity, pH tolerance is used as a means of classification. The three major classifications are

1. **acidophiles:** organisms adapted to grow well in environments below about pH 5.5 (Fig. 2.1),
2. **neutrophiles:** organisms that prefer pH levels between 5.5 and 8.5, and
3. **alkaliphiles:** organisms that live above pH 8.5.

Under natural circumstances, the majority of bacteria maintain a near-neutral internal environment regardless of their habitat because they have cytoplasmic buffers. Changes to pH outside an organism's range may destroy necessary membrane potential (used in ATP production) and damage vital enzymes beyond repair. This **denaturing** of cellular enzymes may be as minor as conformational changes in the proteins' tertiary structure, but usually is lethal to the cell.

Acids from carbohydrate fermentation and alkaline products from protein metabolism are sufficient to disrupt microbial enzyme integrity when grown in vitro. This is why buffers made from weak acids such as hydrogen phosphate are added to bacteriological growth media.

TABLE **2-1** **pH Scale**

| Solution Classification Acidity/Alkalinity | pH | H+ Concentration in Moles/Liter | Common Examples | Organismal Classification |
|---|---|---|---|---|
| ↑ | 0 | $10^{0}$ | Nitric acid | ↑ |
| | 1 | $10^{-1}$ | Stomach acid | |
| | 2 | $10^{-2}$ | Lemon juice | |
| | 3 | $10^{-3}$ | Vinegar, cola | |
| | 4 | $10^{-4}$ | Tomatoes, orange juice | |
| | 5 | $10^{-5}$ | Black coffee | Acidophiles |
| Acidic | 6 | $10^{-6}$ | Urine | |
| Neutral | 7 | $10^{-7}$ | Pure water | Neutrophiles |
| Alkaline | 8 | $10^{-8}$ | Seawater | |
| | 9 | $10^{-9}$ | Baking soda | Alkaliphiles |
| | 10 | $10^{-10}$ | Soap, milk of magnesia | |
| | 11 | $10^{-11}$ | Ammonia | |
| | 12 | $10^{-12}$ | Lime water [$Ca(OH)_2$] | |
| | 13 | $10^{-13}$ | Household bleach | |
| ↓ | 14 | $10^{-14}$ | Drain cleaner | ↓ |

In solution, buffers are able to alternate between weak acid ($H_2PO_4^-$) and conjugate base ($HPO_4^{2-}$) to maintain $H^+/OH^-$ equilibrium.

$$H^+ + HPO_4^{2-} \rightarrow H_2PO_4^-$$

$$OH^- + H_2PO_4^- \rightarrow HPO_4^{2-} + H_2O$$

2

## Application

This is a qualitative procedure used to estimate the pH minimum, maximum, and optimum for growth of microbial species.

## In This Exercise

Today you will cultivate four organisms and observe the effect of pH on them. Then you will classify them based on your results.

## Materials

### Per Student

- ☐ Lab coat
- ☐ Disposable gloves
- ☐ Chemical eye protection

### Per Student Group

- ☐ Five of each pH adjusted tryptic soy broth (or nutrient broth) as follows: pH 2, pH 4, pH 6, pH 8, and pH 10
- ☐ Four sterile transfer pipettes
- ☐ Fresh broth cultures of these recommended organisms:
  - *Alcaligenes faecalis*
  - *Lactobacillus plantarum* or *Lactobacillus acidophilus*
  - *Lactococcus lactis*
  - *Staphylococcus saprophyticus*

## Medium Recipe

### pH-Adjusted Nutrient Broth

| | |
|---|---|
| ☐ Beef extract | 3.0 g |
| ☐ Peptone | 5.0 g |
| ☐ Distilled or deionized water | 1.0 L |
| ☐ NaOH or HCl as needed to adjust pH | |

## PROCEDURE

### Lab One

1. Wear a lab coat, gloves, and chemical eye protection when performing this procedure.
2. Working with your group, obtain five tubes of each pH broth—one of each pH per organism plus a control (25 tubes total). Label them accordingly.
3. Each student should choose one or two cultures to inoculate from (whatever divides the work load evenly). Mix them thoroughly, but be careful not to splash into the cap.
4. Inoculate one tube of each pH with one drop of the organism (be consistent) using a sterile transfer pipette. You may fill the pipette with about 1 mL of broth and deliver the drops with the same pipette to the tubes without flaming them. **Use BSL-2 precautions when transferring *A. faecalis*, if used.** Leave one tube of each pH uninoculated.

**Caution!**

- Insert the pipette well into the tube (but not touching the glass or the medium) before delivering the inoculum.
- Be careful not to spatter as you deliver the drop to each tube, or accidentally drip organism between tubes. Clean up the spill if you do.
- If you don't have enough organism left in the pipette to inoculate a tube, get more organism. You must not empty ("blow-out") the pipette because you will create aerosols.
- Properly dispose of the pipette when finished.

5. It will make reading the results easier if you arrange the tubes in a rack, with the organisms in vertical rows and the same pHs in horizontal rows. Then, incubate all tubes at 35 ± 2°C for 48 hours.
6. Save or dispose of the original broth cultures as directed by your instructor.

### Lab Two

1. Clean the outside of all tubes with a tissue and place them in a test tube rack organized into groups by organism in one direction and pH in the other (if not previously done)
2. Mix each broth gently until uniform turbidity is achieved.

3. Compare all tubes of a species to each other and the appropriate control. Rate each one as 0, 1, 2, or 3 according to its turbidity (0 is clear—the uninoculated tube for each pH—and 3 is highly turbid, established by the most turbid of the 20 inoculated tubes). Enter your observations in the table on the data sheet, page 115. (***Note:*** some color variability may exist between the different pH broths; therefore, base your conclusions solely on turbidity, not color.)
4. Dispose of all tubes in an appropriate autoclave container when finished.
5. Determine the optimum pH and pH range for each organism. Record the pH category for each on the data sheet.
6. On the graph paper provided with the data sheet, plot the relative turbidity versus pH of the four organisms. If you need help with graphing, see pages 10–11.

## References

Forbes, Betty A., Daniel F. Sahm, and Alice S. Weissfeld. Chap. 10 in *Bailey & Scott's Diagnostic Microbiology*, 11th ed. St. Louis, MO: Mosby, 2002.

Holt, John G., ed. *Bergey's Manual of Determinative Bacteriology*, 9th ed. Baltimore: Lippincott Williams & Wilkins, 1994.

Varnam, Alan H. and Malcolm G. Evans. *Environmental Microbiology*. Washington, DC: ASM Press, 2000.

Winn, Washington C. et al. *Koneman's, Color Atlas and Textbook of Diagnostic Microbiology*, 6th ed. Baltimore: Lippincott Williams & Wilkins, 2006.

Name ______________________________

Date ______________________________

Lab Section ______________________________

I was present and performed this exercise (initials) ______________________________

DATA SHEET

2-9

# The Effect of pH on Microbial Growth

## OBSERVATIONS AND INTERPRETATIONS

1 For each species, record a "3" for the pH(s) with maximum growth and "0" for no growth. Use "1" and "2" for intermediate amounts of growth.

| Organism | pH 2 | pH 4 | pH 6 | pH 8 | pH 10 | Classification |
|---|---|---|---|---|---|---|
| | | | | | | |
| | | | | | | |
| | | | | | | |
| | | | | | | |

## QUESTIONS

1 *Circle the pH optimum for each organism. Place brackets around the range. Is there any overlap between species? Describe any examples.*

______________________________

______________________________

______________________________

______________________________

2 *Account for the inability of organisms to grow outside their pH ranges. Why, for instance, are alkaliphiles able to survive at high pHs when neutrophiles cannot?*

______________________________

______________________________

______________________________

______________________________

**3** *Plot relative turbidity (numeric values) versus pH for each organism on the graph paper provided. See pages 10–11 for proper graphing technique.*

2

**4** *Why is it not advisable to connect the data points for each organism in your graph?*

**5** *Consider the shape of the four curves.*

**a.** *Where is the pH optimum relative to the pH range for each organism?*

**b.** *Do you see any parallels between these data and the data produced in Exercise 2-8? Explain.*

# The Effect of Osmotic Pressure on Microbial Growth

EXERCISE 2-10

## Theory

Water is essential to all forms of life. It is not only the principal component of cellular cytoplasm, but also an essential source of electrons and hydrogen ions. Species of Bacteria and Archaea[1], like plants, require water to maintain cellular **turgor pressure**, that is, internal pressure against the cell wall. Whereas animal cells burst with a constant influx of water, cells with a wall require water to prevent shrinking of the cytoplasm and the associated separation of the cytoplasmic membrane from the cell wall—an occurrence known as **plasmolysis**.

Many bacteria regulate turgor pressure by transporting in and maintaining a relatively high cytoplasmic potassium or sodium ion concentration, thereby creating a concentration gradient that promotes inward **diffusion** of water by **osmosis** (see below). For bacteria living in saline habitats, the job of maintaining turgor pressure is continuous because of the constant efflux of water.

Irrespective of a cell's efforts to control its internal environment, natural forces will cause water to move through its semipermeable (selectively permeable) membrane from an area of low **solute** concentration to an area of high solute concentration. Where solute concentration is low in a solution, water concentration is high, and vice versa. Therefore, water moves through a cytoplasmic membrane from where *its* concentration is high to where *its* concentration is low. This process is called osmosis, and the force that controls it is called **osmotic pressure.**

[1] For convenience, further references will only be to Bacteria, but the same applies to Archaea.

Osmotic pressure is a quantifiable term and refers, specifically, to the ability of a solution to *pull water toward itself* through a semipermeable membrane. If a bacterial cell is placed into a solution that is **hyposmotic** (a solution having low osmotic pressure), there will be a *net* movement of water into the cell—at least until internal hydrostatic pressure equals osmotic pressure. At this point, net water movement stops because a dynamic equilibrium is reached.

If an organism is placed into a **hyperosmotic** solution (a solution having high osmotic pressure), there will be a net movement of water out of the cell. For a bacterial cell in an **isosmotic** solution (a solution having osmotic pressure equal to that of the cell), water will tend to move in both directions equally; that is, there is no net movement (Fig. 2.42).

Bacteria and Archaea constitute a diverse group of organisms and, as such, have evolved many adaptations for survival. Microorganisms tend to have a distinct range of salinities that are optimal for growth, with little or no survival outside that range. For example, some bacteria called **halophiles** grow optimally in NaCl concentrations of 3% or higher. **Extreme halophiles** are organisms with specialized cell membranes and enzymes that require salt concentrations from 15% up to about 25% and will not survive where salinity is lower (Fig. 2.43). Except for a few **osmotolerant** Bacteria and Archaea, which will grow over a wide range of salinities, most live where NaCl concentrations are less than 3%.

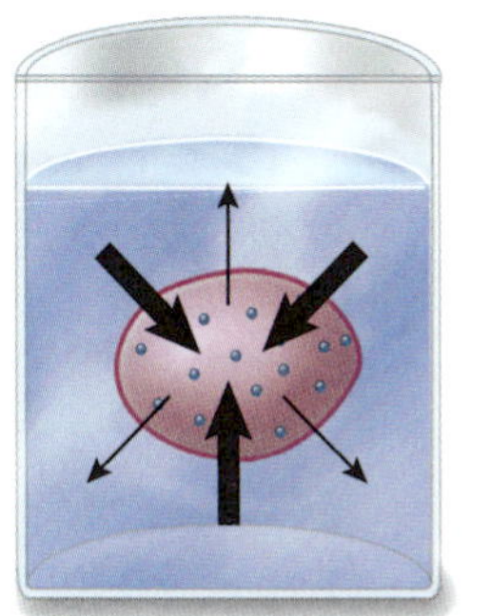


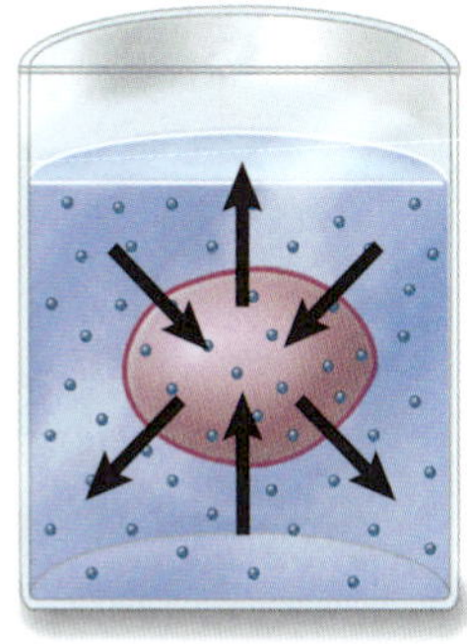


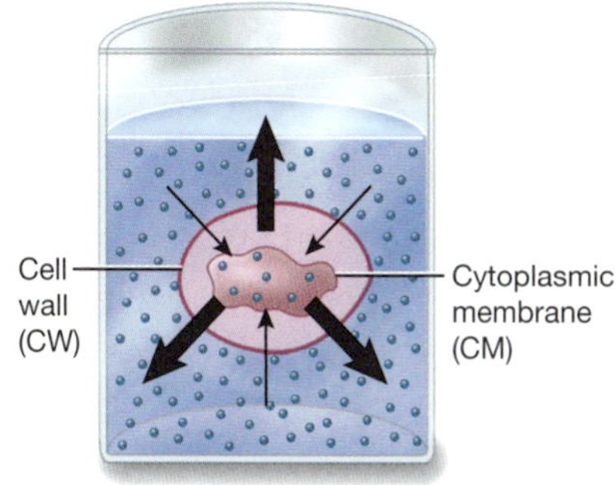


**2.42 The Effect of Osmotic Pressure on Bacterial Cells** ■ This osmosis diagram illustrates the movement of water into and out of cells. The labels refer to the solution outside the cell. In a **hyposmotic** environment, the cell has greater osmotic pressure, so the net movement of water (heavy arrows) will be into the cell. In an **isosmotic** environment, there is no net movement because the osmotic pressure of the cell and that of the environment are equal. (Actually, water molecules are moving equally in both directions.) In a **hyperosmotic** environment, the osmotic pressure of the environment is greater, so the net water movement is outward and results in plasmolysis. Note the smaller cytoplasmic volume as evidenced by the cytoplasmic membrane (CM) that has pulled away from the rigid cell wall (CW) in the hyperosmotic solution.

**2.43 Salterns in San Diego Bay** ■ Salterns are low pools of saltwater used in the harvesting of salt. As water evaporates, the saltwater becomes saltier and saltier, until only salt remains. This can then be purified and sold. The colors in the pools result from different communities of halophilic microorganisms that are associated with different salinities as the pools dry out.

## Application

This is a qualitative procedure used to demonstrate microbial tolerances to NaCl.

## In This Exercise

You will be growing three microorganisms at a variety of NaCl concentrations to determine the salinity tolerance range and optimum salinity for each organism.

## Materials

### Per Student

- □ Lab coat
- □ Disposable gloves
- □ Chemical eye protection

### Per Student Group

- □ Four tubes each of saline medium prepared with 0%, 5%, 10%, 15%, 20%, and 25% NaCl
- □ Three sterile transfer pipettes
- □ Fresh broth cultures of these recommended organisms:
  - *Escherichia coli*
  - *Halobacterium salinarum*
  - *Staphylococcus epidermidis*

## Medium Recipe

### Modified *Halobacterium* Broth

- □ Sodium chloride — 0 g, 50 g, 100 g, 150 g, 200 g, or 250 g (for 0%, 5%, 10%, 15%, 20%, and 25% saline broths)
- □ Magnesium sulfate, heptahydrate — 20.0 g
- □ Trisodium citrate, dihydrate — 3.0 g
- □ Potassium chloride — 2.0 g
- □ Casamino acids — 5.0 g
- □ Yeast extract — 5.0 g
- □ Deionized water — 1.0 L

*Adjust pH to 7.2 using 5* M *or concentrated HCl*

## PROCEDURE

### Lab One

1. Wear a lab coat, gloves, and chemical eye protection when performing this procedure.
2. Working with your group, obtain four tubes of each *Halobacterium* broth—one of each concentration per organism plus a control (24 tubes total). Label them accordingly.
3. Each student should choose one or two cultures to inoculate from (whatever divides the work load evenly). Mix them thoroughly, but be careful not to splash into the cap.
4. Inoculate one tube of each salinity with one drop of the organism (be consistent) using a sterile transfer pipette. You may fill the pipette with about 1 mL of broth and deliver the drops with the same pipette to the tubes without flaming them. In your group, leave one tube of each salinity uninoculated.

**Caution!**

- Insert the pipette well into the tube (but not touching the glass or the medium) before delivering the inoculum.
- Be careful not to spatter as you deliver the drop to each tube, or accidentally drip organism between tubes. Clean up the spill if you do.
- If you don't have enough organism left in the pipette to inoculate a tube, get more organism. You must not empty ("blow-out") the pipette because you will create aerosols.
- Properly dispose of the pipette when finished.

5 It will make reading the results easier if you arrange the tubes in a rack, with the organisms in vertical rows and the same salinities in horizontal rows. Then, incubate all tubes at $35 \pm 2°C$ for 48 hours (or longer, if necessary).

6 Save or dispose of the original cultures as directed by your instructor.

### Lab Two

1 Clean the outside of all tubes with a tissue, and place them in a test tube rack organized into groups by organism in one direction and by salinity in the other (if not previously done)

2 Mix each broth gently until uniform turbidity is achieved.

3 Compare all tubes of a species to each other and to the control. Rate each one as 0, 1, 2, or 3 according to its turbidity (0 is clear—the uninoculated tube for each salinity—and 3 is the most turbid of the 18 inoculated tubes). Enter your observations in the table on the data sheet, page 121. (***Note:*** the different broths may show some color variability. Check the uninoculated controls and, thus, base your conclusions solely on turbidity, not color.)

4 Dispose of all tubes in an appropriate autoclave container when finished.

5 From your data, determine the optimum salinity and salinity range for each organism.

6 On the graph paper provided with the data sheet, plot the relative turbidity versus the salt concentration for all three organisms. If you need help with graphing, see pages 10–11.

## References


Forbes, Betty A., Daniel F. Sahm, and Alice S. Weissfeld. Chap. 10 in *Bailey & Scott's Diagnostic Microbiology,* 11th ed. St. Louis, MO: Mosby, 2002.

Hauser, Juliana T. *Techniques for Studying Bacteria and Fungi*. Burlington, NC: Carolina Biological Supply Company, 2006.

Holt, John G., Ed. *Bergey's Manual of Determinative Bacteriology,* 9th ed. Baltimore: Lippincott Williams & Wilkins, 1994.

Koneman, Elmer W., Stephen D. Allen, William M. Janda, Paul C. Schreckenberger, and Washington C. Winn, Jr. *Color Atlas and Textbook of Diagnostic Microbiology,* 5th ed. Philadelphia: J. B. Lippincott Co., 1997.

Moat, Albert G., John W. Foster, and Michael P. Spector. Pages 582–587 in *Microbial Physiology,* 4th ed. New York: Wiley-Liss, 2002.

Varnam, Alan H. and Malcolm G. Evans. *Environmental Microbiology.* Washington, DC: ASM Press, 2000.

White, David. Pages 388–394 in *The Physiology and Biochemistry of Prokaryotes,* 2nd ed. New York: Oxford University Press, 2000.

Name ______________________________

Date ______________________________

Lab Section ______________________________

I was present and performed this exercise (initials) ______________________________

# DATA SHEET 2-10

## The Effect of Osmotic Pressure on Microbial Growth

### OBSERVATIONS AND INTERPRETATIONS

**1** For each species, record a "3" for the salt concentration(s) with maximum growth and "0" for no growth. Use "1" and "2" for intermediate amounts of growth.

| Organism | NaCl Concentration | | | | | |
|---|---|---|---|---|---|---|
| | 0% | 5% | 10% | 15% | 20% | 25% |
| Control | | | | | | |
| | | | | | | |
| | | | | | | |
| | | | | | | |

## QUESTIONS

**1** *Circle the optimum salinity for each organism. Place brackets around the range. Is there any overlap between species? Describe any examples.*

**2** *Plot relative turbidity (numeric values) versus salt concentration on the graph paper provided. See pages 10–11 for proper graphing techniques.*

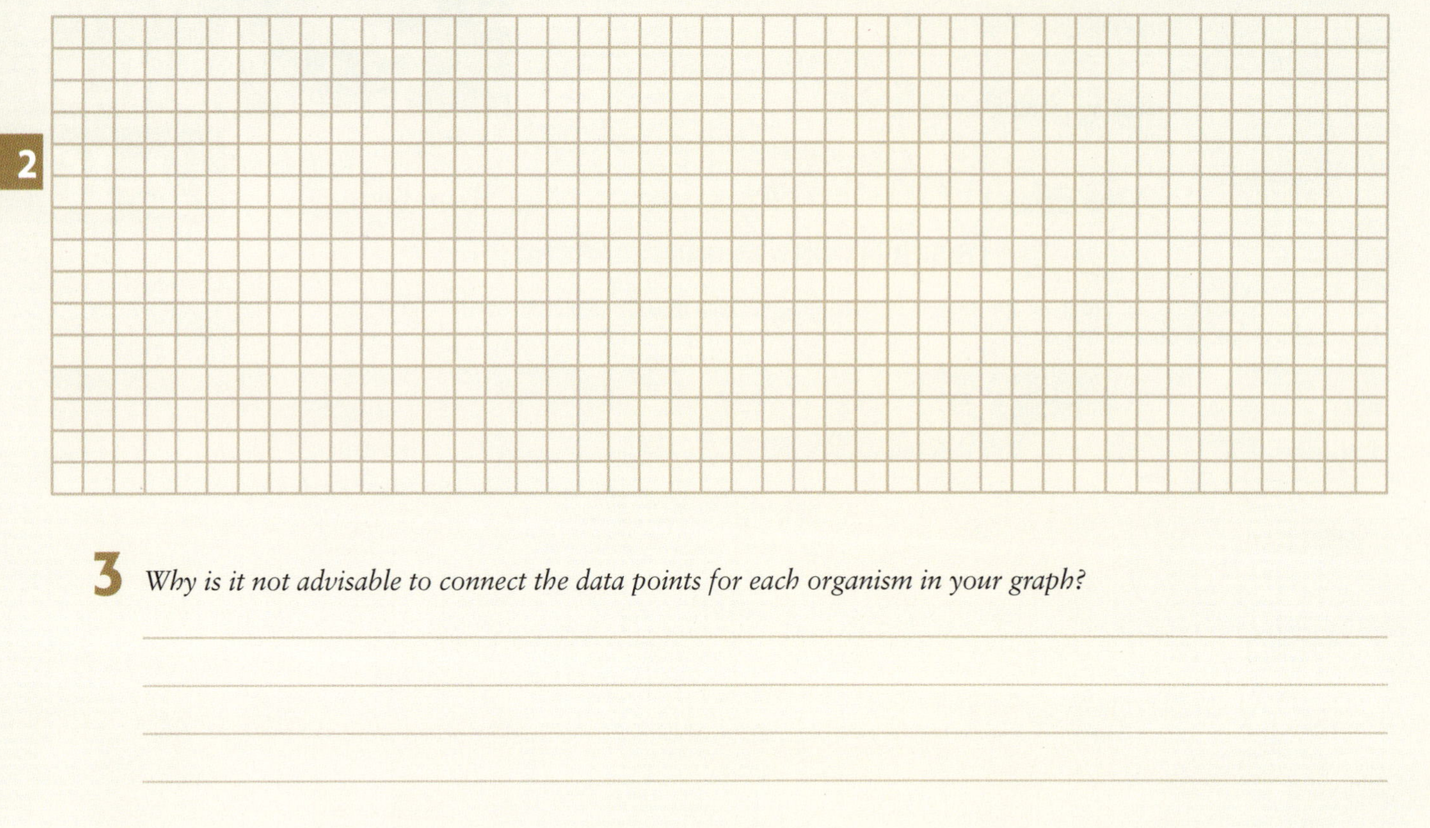

2

**3** *Why is it not advisable to connect the data points for each organism in your graph?*

**4** *Which organism(s) exhibit(s) the greatest tolerance range?*

**5** *Use your textbook or other available references to look up the habitat of each organism. Do the habitats make sense in light of the tolerance ranges you obtained? Explain your answers.*

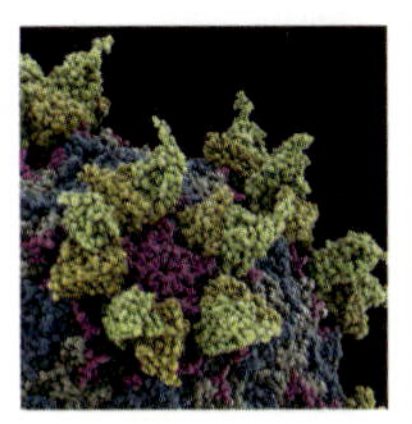

## Physical and Chemical Methods of Pathogen Control

Every patient in a hospital or other clinical setting has the right to expect that he or she will not contract a disease or infection while in that institution's care. Every person donating blood at a blood bank or mobile center has the right to expect that all materials and surfaces they come in contact with will be free of pathogens. Workers in health clinics, hospitals, medical laboratories, and public health laboratories have the right to assume that reasonable precautions have been and are being taken to protect their safety while in the workplace.



These are only a few of the many reasons why the importance of understanding and use of pathogen control systems cannot be overemphasized. Fortunately, with relatively few exceptions, the above-described conditions exist in this and other developed countries largely because of the dedication of thousands of employees and the oversight of dozens of international, governmental, and private organizations such as the World Health Organization (WHO), Centers For Disease Control and Prevention (CDC), Food and Drug Administration (FDA), Environmental Protection Agency (EPA), American Public Health Association (APHA), and Association of Official Analytical Chemists (AOAC). These and many other federal and private organizations are responsible for the proper testing, registration, and classification of the substances or systems used to prevent the spread of pathogens.

These substances or systems, both chemical and physical, are referred to broadly as germicides. Some germicides are specific in nature and typically include the name of the target pathogen, such as "tuberculocide," "virucide," or "sporocide." Most germicides are broad-spectrum and, thus, target a wide variety of pathogens. Although some overlap occurs, germicidal systems fall into three categories: decontamination, disinfection, or sterilization.

1. **Decontamination** is the lowest level of control and is defined as "reduction of pathogenic microorganisms to a level at which items are safe to handle without protective attire." Decontamination usually includes physical cleaning with soaps or detergents, and removal of all (ideally) or most organic and inorganic material. Proper cleaning of all instruments and surfaces is considered the critical first step toward disinfection or sterilization because, to be fully effective, a disinfectant or sterilant must come in direct contact with all pathogens present. Materials left to dry on a surface or apparatus can actually shield pathogens from a disinfecting or sterilizing agent or otherwise neutralize it.

2. **Disinfection** is the next level of control and is divided into three sublevels—low, medium, and high—based on effectiveness against specific control pathogens or their surrogates. All sublevels kill large numbers, if not all, of the targeted pathogens but typically do not kill large numbers of bacterial endospores. Some high-level disinfectants are called chemical sterilants because they have the ability to kill all vegetative cells and endospores.

   Disinfectants typically are liquid chemical agents but can also be solid or gaseous. Other disinfection methods include dry heat, moist heat, and ultraviolet light. Disinfectants that are designed to reduce or eliminate pathogens on or in living tissue are called **antiseptics**. For obvious safety reasons, antiseptics are subject to additional testing to minimize the risks of side effects. Some antiseptics are considered drugs and, therefore, are regulated by the FDA.

3. **Sterilization** is the complete elimination of viable organisms including bacterial endospores and, as such, is the highest level of pathogen control. Sterilization can be achieved by some chemicals, some gases, incineration, dry heat, moist heat, ethylene oxide gas, ionizing radiation (gamma, X-ray, and electron beam), low-temperature plasma (utilizing a combination of chemical sterilants and ultraviolet radiation in a vacuum chamber), or low-temperature ozone (utilizing bottled oxygen, water, and electricity in a chamber to produce a lethal level of ozone).

In this last portion of Section 2 you will examine one final aspect of microbial growth: its control, by both physical and chemical means. The following exercises illustrate the germicidal effects of steam (moist heat) sterilization, UV radiation, and disinfection and antisepsis. (For more information on microbial control, refer to Exercise 7-2 Antimicrobial Susceptibility Test.) ■

# Steam Sterilization

## Theory

Of the many methods or agents that have been developed for sterilizing surgical and dental instruments, microbiological media, infectious waste, and other materials not harmed by moisture or heat, steam is still the most effective and most common. The device used most commonly for this purpose is called a steam sterilizer, or **autoclave**. Autoclaves are relatively safe, easy to operate, and, if used properly, effective at killing all microbial vegetative cells and bacterial endospores.[1]

Under atmospheric pressure, water boils at 100°C (212°F). At pressures above atmospheric pressure, water must be heated above 100°C before it will boil. Similar to home pressure cookers, which create pressure and high temperatures to shorten cooking times, autoclaves use super-heated steam under pressure to kill heat-resistant organisms. Examples of heat-resistant organisms include members of the spore-producing genera—*Bacillus*, *Geobacillus*, and *Clostridium*.

In the microbiology laboratory, sterilizing temperature usually is set at between 121°C and 127°C (250°F and 260°F); however, sterilizing time can vary according to the size and consistency of the material being sterilized.[2] At a minimum, to be sure that all vegetative cells and endospores have been killed, items being processed must reach optimum temperature for at least 15 minutes. This includes items deep inside the autoclave container that may be partially insulated from the steam by surrounding items.

Understandably, larger loads take longer to process than smaller loads. (Certain sensitive applications, such as microbiological media preparation, in which formula integrity must be maintained and when specific growth inhibiting ingredients are included, lower times and temperatures are acceptable.)

To maintain laboratory safety and comply with laws regarding infectious waste disposal, sterilizers must be checked regularly for operating effectiveness. Special thermometers placed in an autoclave can record the maximum temperature reached inside the chamber but do not measure how low the temperature dips during the normal cycling of the heating elements. Specialized color-coded autoclave tape (Fig. 2.44) can be a fairly good indicator that sterilization is complete, but the only way, with certainty, to determine that sterilization has been achieved is by using a device called a **biological indicator** (Fig. 2.45).

Biological indicators, as the name suggests, are test systems that contain something living. A typical biological indicator that is particularly useful for testing autoclaves is one that contains **bacterial endospores**. Bacterial endospores, the dormant form of an organism, are highly resistant to both chemical and physical means of control.

Therefore, if an autoclave kills the endospores in the test system, it is safe to assume that it has destroyed other microbes as well. This is important not only for safety reasons but is of legal importance as well. Public health and safety agencies maintain compliance with hazardous waste disposal regulations by requiring regular testing of autoclaves used to process biohazardous material.

A typical system, and the one selected for today's lab, includes a small, heat-resistant plastic vial containing a glass ampule of sterile fermentation broth. Also inside the vial, but outside of the ampule, is a strip of filter paper containing bacterial endospores. The vial is placed inside the autoclave and heated at 121°C for 15 minutes. After autoclaving, the vial is cooled and crushed with a special device that breaks the inner ampule without damaging the plastic vial. Breaking the ampule allows the fermentation broth to come in contact with the bacterial endospores in the filter paper.

The vials are then incubated at 55°C for 48 hours. If the endospores have been killed in the autoclave,

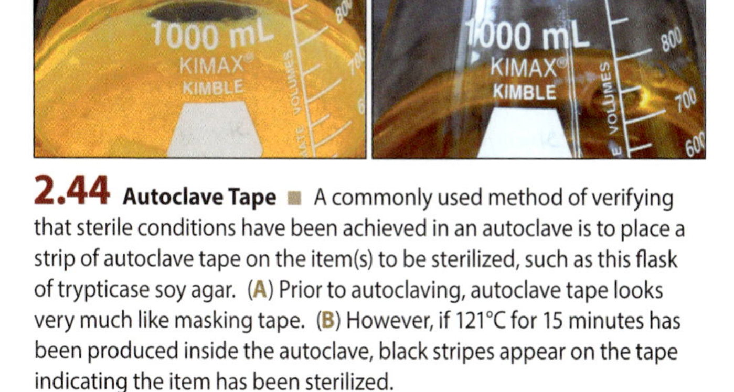


**2.44 Autoclave Tape** ■ A commonly used method of verifying that sterile conditions have been achieved in an autoclave is to place a strip of autoclave tape on the item(s) to be sterilized, such as this flask of trypticase soy agar. (**A**) Prior to autoclaving, autoclave tape looks very much like masking tape. (**B**) However, if 121°C for 15 minutes has been produced inside the autoclave, black stripes appear on the tape indicating the item has been sterilized.

[1] Steam sterilization is not yet considered reliable at inactivating prion proteins, such as those that cause bovine spongiform encephalopathy, the so-called mad cow disease, and its variant, Creutzfeldt-Jakob disease.

[2] In clinics, hospitals, or other locations where surgical instruments are being processed, the World Health Organization (WHO) recommends a minimum processing time and temperature of 134°C for 18 minutes.

incubation will produce no growth. If they have not been killed, they will germinate and ferment the substrate in the broth. A pH-indicating dye, included in the broth, will reveal any acid produced (during fermentation) with a distinctive color change. No color change during incubation is, thus, an indication that sterilization is complete, the endospores have been killed, and the autoclave is operating properly. (For more information on fermentation, refer to Section 5 and Appendix A.)

## Application

Biological indicators are available in many forms and commonly used to test the efficiency of steam sterilizers.

## In This Exercise

You are going to use bacterial endospores and their resistance to steam sterilization to test the effectiveness of your lab's autoclave. This procedure is written for a product called BTSure Biological Indicator[3], but can be applied to other brands as well. BTSure Biological Indicators include a pH indicator that turns the broth from purple to yellow if acid has been produced by fermentation.

[3] BTSure Biological Indicators are available from Fisher Science, http://www.fishersci.com/shop/products/thermo-scientific-biological-indicators-incubator-pouches-5/1449027.

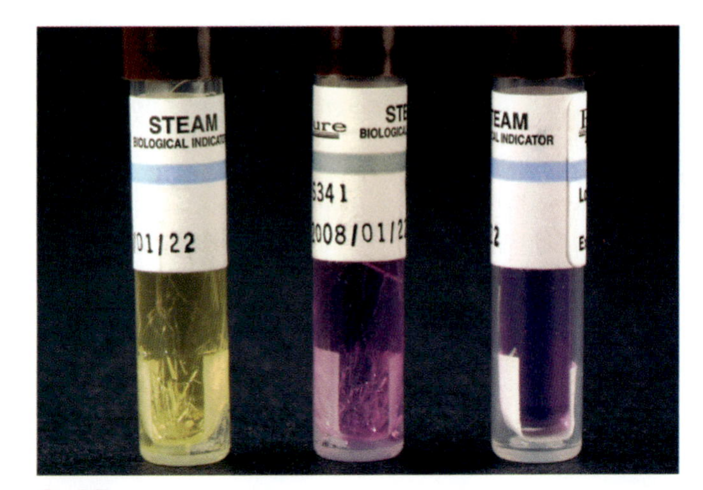


**2.45 Autoclave Biological Indicator** ■ These indicator vials contain an ampule of fermentation broth and a filter paper strip containing endospores of *Geobacillus stearothermophilus*—an endospore-forming organism capable of withstanding high temperatures. The vial in the center was autoclaved for 15 minutes at 121°C, cooled, pinched to crush the inner glass ampule, and incubated. The purple color (compared to the uncrushed negative control on the right) indicates that an acidic pH from fermentation does not exist. This suggests that the endospores have been killed by the autoclaving. Note the gray-colored band on the label. This chemical indictor changes from blue to gray upon autoclaving. The ampule in the negative control was not crushed, so the endospores in the filter paper never made contact with the broth and, thus, provides a color example of unfermented broth. The ampule on the left is a positive control to verify the viability of the organism used in the system. It was not autoclaved (as evidenced by the blue band on the label) but was crushed and incubated. The development of yellow color indicates that the organism in the system is viable and that the lack of yellow color in the center vial is a result of autoclaving.

## ▼ Materials

### Per Student

- ☐ Lab coat
- ☐ Disposable gloves
- ☐ Chemical eye protection

### Per Student Group

- ☐ Four BTSure Biological Indicators
- ☐ One autoclave pan

### Per Class

- ☐ One (or more) steam autoclave(s)
- ☐ Incubator set at 55°C

## PROCEDURE

### Lab One

1. Wear a lab coat, gloves, and chemical eye protection when performing this procedure.
2. Working as a group, obtain four BTSure Biological Indicator vials and label them #1, #2, #3, and #4 with a Sharpie® or other permanent marker. Do not label the vial with tape or paper, because this will insulate it from the steam.
3. Place vial #1 on its side uncovered in an autoclave pan. If the size of your class and the size of the pan allow it, it is most efficient if all vials are put in the same autoclave pan.
4. Place vial #2 inside a container (or multiple containers), as well insulated as can be achieved with materials provided by your instructor. We recommend placing the vial inside a screw-capped test tube inside other tubes—two, three, or four layers deep. Place this vial in the autoclave pan with vial #1.
5. Do nothing with vials #3 and #4 as yet.
6. Place the autoclave pan in the autoclave.
7. Follow your instructor's guidelines to add water to the chamber, set the temperature at 121°C (250°F), and set the timer at 15 minutes. When instructed to do so, close and start the autoclave.

8. When autoclaving is complete, all of the steam has been vented, and the machine has been allowed to cool slightly, remove your pan (while protecting your hands with appropriate gloves) and allow the contents to cool to room temperature.
9. Keeping the vials in an upright position, use the crushing device to squeeze vials #1, #2, and #3 (one at a time) until you hear the glass ampule inside the vial break.
10. Place these vials along with vial #4, again in an upright position, into the 55°C incubator for 48 hours.

### Lab Two

1. After incubation, examine the vials for color changes.
2. Using Table 2-2 as a guide, record and interpret your results in the table provided on the data sheet, page 129.

**TABLE 2-2 Autoclave Biological Indicator Test Results and Interpretations**

| Broth Color | Interpretation (assume the ampules inside the vials have been crushed) |
|---|---|
| Purple | No fermentation or acid production in the medium. The organism is dead. |
| Yellow | Fermentation with acid production in the medium. The organism is alive. |

## References

McDonnell, Gerald E. *Antisepsis, Disinfection, and Sterilization: Types, Action, and Resistance*. Washington, DC: ASM Press, American Society for Microbiology, 2007.

Widmer, Andreas F. and Reno Frei. Chap. 7 in *Manual of Clinical Microbiology*, 9th ed. Patrick R. Murray, Ellen Jo Baron, James H. Jorgensen, Marie Louise Landry, and Michael A. Pfaller, eds. Washington, DC: ASM Press, American Society for Microbiology, 2007.

Name ____________________

Date ____________________

Lab Section ____________________

I was present and performed this exercise (initials) ____________________

# DATA SHEET 2-11

## Steam Sterilization

### OBSERVATIONS AND INTERPRETATIONS

**1** Examine all vials and record your results in the table below.

| Indicator Vial | Autoclaved? (Yes or No) | Ampule Crushed? (Yes or No) | Color Result | Interpretation |
|---|---|---|---|---|
| Vial #1 | | | | |
| Vial #2 | | | | |
| Vial #3 | | | | |
| Vial #4 | | | | |

### QUESTIONS

**1** *Why are the bacterial endospores placed on the paper strip and not directly in the fermentation broth?*

**2** *What is the purpose of the unautoclaved/unbroken vial (#4)?*

2

**3** *How would you interpret the following combinations of results? Vial numbers indicate the treatment as in this experiment.*

**a.** *Vial #1—purple, vial #2—purple, vial #3—purple, vial #4—purple?*

**b.** *Vial #1—purple, vial #2—purple, vial #3—yellow, vial #4—purple?*

**c.** *Vial #1—purple, vial #2—yellow, vial #3—yellow, vial #4—purple?*

**d.** *Vial #1—yellow, vial #2—yellow, vial #3—yellow, vial #4—purple?*

**e.** *Vial #1—yellow, vial #2—yellow, vial #3—yellow, vial #4—yellow?*

**4** *What changes would you make to avoid repeating the faulty scenarios illustrated in question 3?*

# The Effect of Ultraviolet Radiation on Microbial Growth

EXERCISE 2-12

2

## Theory

Ultraviolet radiation (UV light) is a type of **electromagnetic energy**. Like all electromagnetic energy, UV travels in waves and is distinguishable from all others by its **wavelength**. Wavelength is the distance between adjacent wave crests and is typically measured in nanometers (nm) (Fig. 2.46).

Ultraviolet light is divided into three groups categorized by wavelength:

1. UV-A, the longest wavelengths, ranging from 315 nm to 400 nm
2. UV-B, wavelengths between 280 nm and 315 nm
3. UV-C, wavelengths ranging from 100 nm to 280 nm. (These wavelengths—more specifically, 240 nm–280 nm—are most detrimental to bacteria. Bacterial exposure to UV-C for more than a few minutes usually results in irreparable DNA damage and death of the organism. For a discussion on the mutagenic effects of UV and DNA repair, refer to Exercise 8-2.)

The germicidal effect of UV-C is related to time of exposure, lamp intensity, and distance to the target (the inverse square law applies here—intensity diminishes by the reciprocal of square of the distance: $\frac{1}{\text{distance}^2}$ ). It also must have "line of sight" to the surface being decontaminated. That is, it doesn't penetrate well, bend around corners, or trickle into crevices. Lastly, its effectiveness is diminished by dust, organic material, etc. Taken together, this means that UV-C as a germicidal agent has its limited applications.

## Application

Ultraviolet light is commonly used to disinfect laboratory and health-care environment work surfaces and surrounding air.

## In This Exercise

Today you will compare the effect of UV exposure on four cultures—*Bacillus subtilis* (24-hour and 7-day cultures), *Deinococcus radiophilus*, and *Escherichia coli*. Because of the large number of plates to be treated, the work will be divided among six groups of students. Refer to Table 2-3 for assignments. ***Note:*** if the class doesn't divide conveniently into six groups, then the groups assigned to the shorter exposure times can do more than one set of plates.

## Materials

### Per Student

- ☐ Lab coat
- ☐ Disposable gloves
- ☐ Chemical eye protection

### Per Student Group

- ☐ Short wavelength ultraviolet lamp (UV-C) with appropriate shielding and support
- ☐ Cardboard to cover plates (see Fig. 2.47)
- ☐ Four tryptic soy agar (TSA) plates (eight for group one)
- ☐ Stopwatch or electronic timer
- ☐ Sterile cotton swabs (four per group; eight for group 1)
- ☐ Fresh broth cultures of these recommended organisms:
  - *Bacillus subtilis*
  - *Bacillus subtilis* 7-day culture
  - *Deinococcus radiophilus*
  - *Escherichia coli*

TABLE **2-3** **Group Assignments by Number**

| Organism | No UV | 1 minute | 2 minutes | 4 minutes | 8 minutes | 12 minutes | 16 minutes |
|---|---|---|---|---|---|---|---|
| *B. subtilis* (24-hour culture) | 1 | 1 | 2 | 3 | 4 | 5 | 6 |
| *B. subtilis* (7-day culture) | 1 | 1 | 2 | 3 | 4 | 5 | 6 |
| *Deinococcus radiophilus* | 1 | 1 | 2 | 3 | 4 | 5 | 6 |
| *E. coli* | 1 | 1 | 2 | 3 | 4 | 5 | 6 |

$10^{-4}$ nm | 10 nm | 380 nm | 750 nm | 1 mm | 100 mm

Gamma rays | X-rays | UV-C | UV-B | UV-A | Visible light | Infrared | Microwaves | Radio waves

**2.46 Electromagnetic Spectrum** ■ The shortest and highest energy wavelengths are those of gamma rays with wavelengths less than $10^{-4}$ nm. Radio waves, at the other end of the spectrum, can be 1 kilometer or longer. Between about 100 nm and 380 nm (just shorter than visible light) is the portion known as ultraviolet light.

## PROCEDURE

### Lab One

1. Wear a lab coat, gloves, and chemical eye protection when performing this procedure.
2. Enter your group number and exposure time (from Table 2-3) on the data sheet, page 133.
3. Obtain four TSA plates and label the bottom of each with the names of the organisms to be inoculated and your group number. Draw a line to divide the plates in half, and label the sides "A" and "B."
4. Dip a sterile cotton swab into the broth of one culture and wipe the excess on the inside of the tube. Inoculate the appropriate plate by spreading the organism over the entire surface of the agar. Do this by streaking the plate surface completely three times, rotating it one-third turn between streaks. There should be no gaps between streaks. When incubated, this will form a bacterial lawn.
5. Repeat step 4 with the other organisms and plates.
6. Place a paper towel on the table next to the UV lamp and soak it with disinfectant.
7. Place your plates under the UV lamp with the agar surface exposed. Set the covers open side down on the disinfectant-soaked towel. Cover each plate's B side with the cardboard as shown in Figure 2.47.
8. Be sure the lamp is the same distance from the plates for all groups (we recommend between 8 and 12 inches). Then, turn on the lamp for the prescribed time. ***Caution:* Be sure the protective shield is in place and do not look at the light while it is on!**
9. Immediately replace the covers, invert the plates and incubate them as follows: *B. subtilis* and *E. coli*, $35 \pm 2$°C for 24 to 48 hours; *D. radiophilus*, 30°C for 72 to 96 hours.
10. Save or dispose of the original cultures as directed by your instructor.

### Lab Two

1. Remove your plates from the incubator and observe for growth. Side B should be covered with a bacterial lawn (confluent growth). If this is not the case, see your instructor.
2. Record the growth on side A of each plate in the table on the data sheet, page 133. Record as "heavy confluent growth," "sparse confluent growth," "individual colonies," and "no growth."
3. Dispose of all plates in an appropriate autoclave container when finished.
4. Using the results from other groups, complete the class data table on the data sheet.
5. On the graph paper provided with the data sheet, construct a graph representing growth versus UV exposure time for all three organisms. If you need help with graphing, see pages 10–11.

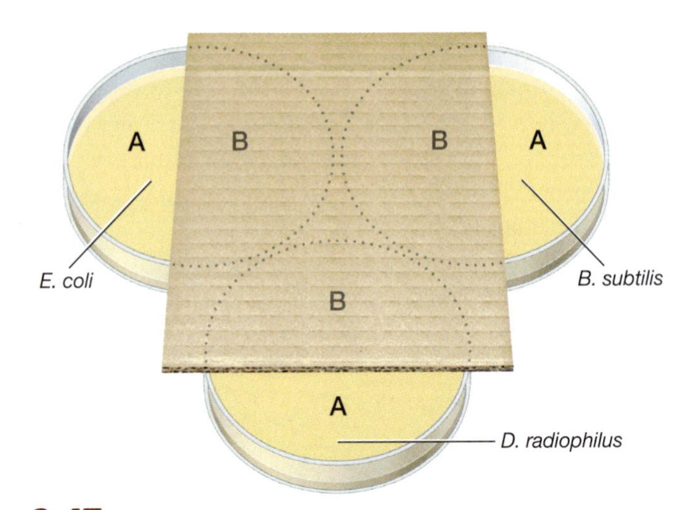


**2.47 Plates Shielded for UV Exposure** ■ Place the plates under the UV lamp with the lids removed and the cardboard shield covering half of each plate as shown. Make sure the Petri dish lids are placed open side down on a disinfectant-soaked towel.

### References

"Guideline for Disinfection and Sterilization in Healthcare Facilities, 2008," Centers for Disease Control and Prevention, Healthcare Infection Control Practices Advisory Committee (HICPAC), http://www.cdc.gov/hicpac/Disinfection_Sterilization/10_0MiscAgents.html#2.

Krebs, Jocelyn E., Elliott S. Goldstein, and Stephen T. Kilpatrick. Chap. 16 in *Lewin's Genes XI*. Burlington, MA: Jones & Bartlett Learning, 2012.

Varnam, Alan H., and Malcolm G. Evans. *Environmental Microbiology*. Washington, DC: ASM Press, 2000.

Name ______________________________

Date ______________________________

Lab Section ______________________________

I was present and performed this exercise (initials) ______________________________

DATA SHEET

2-12

# The Effect of Ultraviolet Radiation on Microbial Growth

## OBSERVATIONS AND INTERPRETATIONS

**1** Enter your class data in the table below using the following descriptions:

0 = no growth  1 = individual colonies  2 = sparse confluent growth  3 = heavy confluent growth

| **UV Exposure** (minutes) | ***B. subtilis*** (24-hour culture) | | ***B. subtilis*** (7-day culture) | | ***D. radiophilus*** | | ***E. coli*** | |
|---|---|---|---|---|---|---|---|---|
| | **Side A** | **Side B** | **Side A** | **Side B** | **Side A** | **Side B** | **Side A** | **Side B** |
| | | | | | | | | |
| | | | | | | | | |
| | | | | | | | | |
| | | | | | | | | |
| | | | | | | | | |
| | | | | | | | | |

## QUESTIONS

**1** *The purpose of this exercise is to demonstrate the comparative effect of UV on four bacterial populations. This could have been accomplished without the cardboard cover. Why was the cover used?*

______________________________

______________________________

______________________________

______________________________

______________________________

______________________________

**2** *This is not a quantitative exercise. Keeping this in mind, what is the general trend between bacterial death and UV exposure time?*

______________________________

______________________________

______________________________

______________________________

______________________________

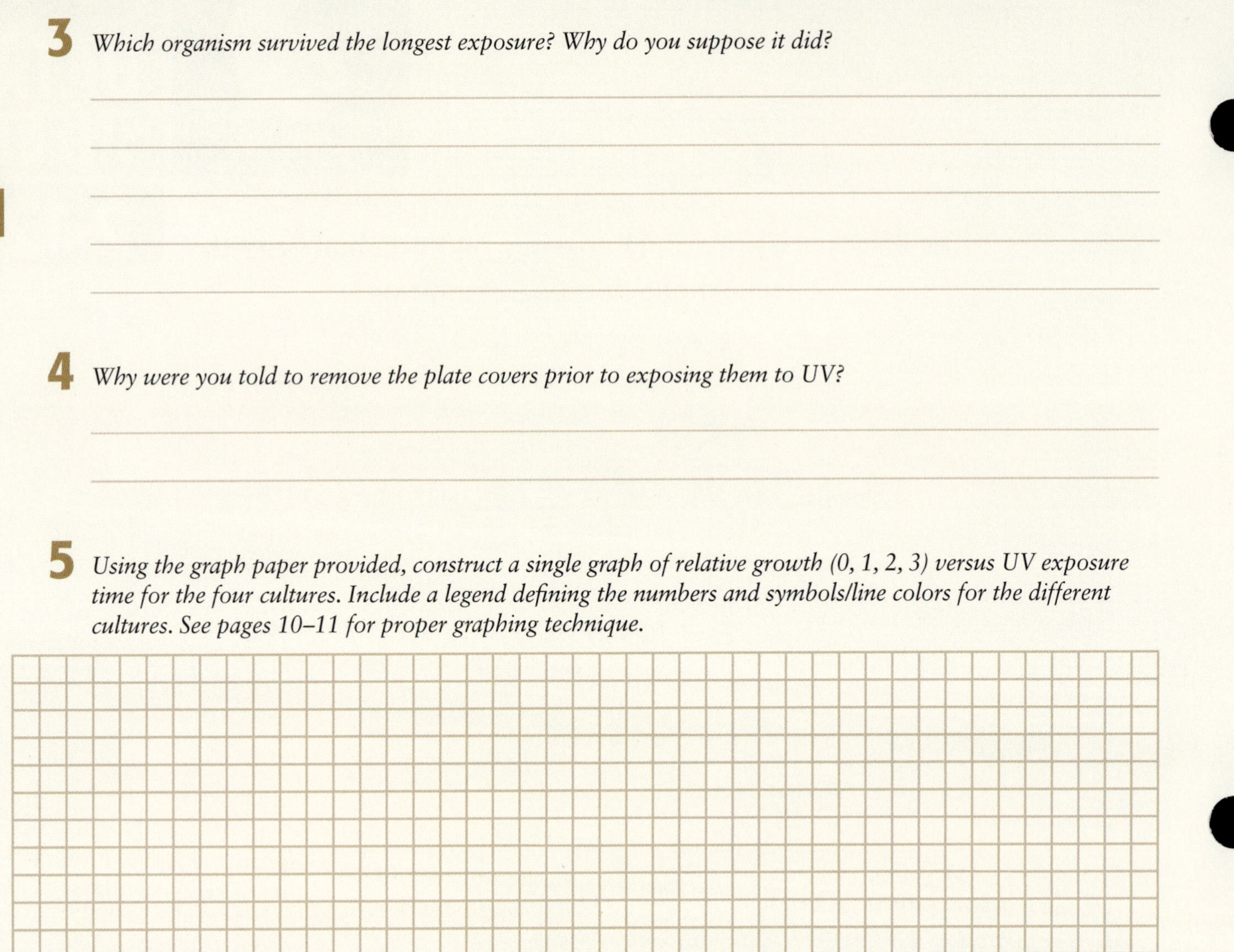

**3** *Which organism survived the longest exposure? Why do you suppose it did?*

**4** *Why were you told to remove the plate covers prior to exposing them to UV?*

**5** *Using the graph paper provided, construct a single graph of relative growth (0, 1, 2, 3) versus UV exposure time for the four cultures. Include a legend defining the numbers and symbols/line colors for the different cultures. See pages 10–11 for proper graphing technique.*

# Effectiveness of Chemical Germicides: The Use-Dilution Test for Disinfectants and Antiseptics

EXERCISE **2-13**



## Theory

Chemical germicides are substances designed to reduce the number of pathogens on a surface, in a liquid, or on or in living tissue. Germicides designed for use on surfaces (floors, tables, sinks, countertops, surgical instruments, etc.) or in liquids are called **disinfectants**. Germicides designed for use on or in living tissue are called **antiseptics**.

Before a new substance can be registered by either the FDA or EPA and allowed on the market, it must be tested and classified according to its effectiveness against pathogens. The Use-Dilution Test, published by the Association of Official Analytical Chemists (AOAC), is one of many commonly used tests for this purpose.

The Use-Dilution Test is a standard procedure used to measure the effectiveness of disinfectants specifically against *Staphylococcus aureus*, *Salmonella enterica* serovar Choleraesuis, and *Pseudomonas aeruginosa*. In the standard procedure, glass beads or stainless steel cylinders coated with living bacteria are exposed to varying concentrations (dilutions) of test germicides and then transferred to a growth medium.

After a period of incubation, the medium is examined for growth. If a solution is sufficient to prevent microbial growth at least 95% of the time, it meets the required standards and is considered a usable dilution of that germicide for a specific application. Today's exercise is an adaptation of this method.

## Application

This procedure is used to test the effectiveness of germicides against *Staphylococcus aureus*, *Salmonella enterica* serovar Choleraesuis, and *Pseudomonas aeruginosa*.

## In This Exercise

Today you will examine the effectiveness of four germicides—two common household disinfectants and two over-the-counter antiseptics. The disinfectants selected for the exercise are household bleach and Lysol® Brand II Disinfectant. The antiseptics are hydrogen peroxide and isopropyl alcohol. The organisms used for the test are *Staphylococcus epidermidis* and *Escherichia coli*. (The usual test organisms are BSL-2 and we can demonstrate the protocol without using them.)

You will first coat the beads with bacteria, expose them to three concentrations of your assigned germicide, and then use them to inoculate sterile nutrient broth. If all of the bacteria on the bead are killed during exposure to the germicide, the broth inoculated with that bead will remain clear. If any of the bacteria survive the germicide exposure, they will reproduce during incubation and make the broth turbid. You will use the results to determine the effective concentration (dilution) of your assigned germicide.

The tasks for the exercise are divided among eight groups of students (or a convenient number for your lab size). Each group will be responsible for one organism and three dilutions of one germicide. Refer to Table 2-4 for your assignments.

Finally, this is an interesting exercise with moderate amount of work involved. If you do not hurry and are careful to use aseptic technique, you will be rewarded with reliable data at the end.

TABLE **2-4** **Group Assignments**

| Germicide | *Staphylococcus epidermidis* | *Escherichia coli* |
|---|---|---|
| Bleach | Group 1 | Group 2 |
| Lysol® | Group 3 | Group 4 |
| Hydrogen peroxide | Group 5 | Group 6 |
| Isopropyl alcohol | Group 7 | Group 8 |

## Materials

### Per Student

- ☐ Lab coat
- ☐ Disposable gloves
- ☐ Chemical eye protection

### Per Student Group

- ☐ 100 mL flask of sterile deionized water
- ☐ Three concentrations of one germicide (listed above)
- ☐ Five sterile 60 mm Petri dishes
- ☐ One sterile glass 100 mm Petri dish containing filter or bibulous paper
- ☐ One container of sterile ceramic or glass beads[1]
- ☐ Sterile transfer pipette
- ☐ Seven sterile nutrient broth tubes

[1] Sterilized #8 seed beads from a craft store will work for this purpose.

- ☐ Needle-nose forceps (or appropriate device for aseptically picking up beads)
- ☐ Small screw-cap jar with alcohol (for flaming forceps)
- ☐ Small beaker with 10 mL of disinfectant (for disposal of the broth culture)
- ☐ Fresh broth cultures of these recommended organisms (only one per group):
  - *Escherichia coli*
  - *Staphylococcus epidermidis*

2

### Per Class (see Table 2-4)

Disinfectants

- ☐ 0.01%, 0.1%, and 1% household bleach
- ☐ 25%, 50%, and 100% Lysol® Brand II Disinfectant

Antiseptics

- ☐ 0.03%, 0.3%, and 3% hydrogen peroxide (3% is full strength as purchased at the pharmacy)
- ☐ 10%, 30%, and 50% isopropyl alcohol (70% or 90% is full strength as purchased at the pharmacy)

## PROCEDURE[2]

Timing is important in this procedure. Read through it and make a plan before you begin so your transfers and soaking times are done uniformly and are consistent with those of other groups.

### Lab One

1. Wear a lab coat, gloves, and chemical eye protection when performing this procedure.
2. Enter the name of your organism here:

   ______________________________

3. Enter the name of your germicide here:

   ______________________________

4. Obtain all of the necessary items for your group as listed in Materials.
5. Place the materials properly on your workspace as shown in Figure 2.48. Label all seven broths with your group name or number. Label three of the broths "Broth #1, #2, and #3" and label the last four broths "Control #1, #2, #3, and #4" as shown in the diagram in Figure 2.48.
6. Label three of the 60 mm plates with the name and concentration of your germicide. Label the other two plates "#4 sterile water" and "#5 sterile water."
7. Add enough of each germicide concentration to its respective plate (Plates #1 through #3) to cover the beads (approximately 15 mL). Add an equal volume of sterile water into Plates #4 and #5.
8. Carefully mix your bacterial culture until uniform turbidity is achieved. Take care not to splash into the cap.
9. Aseptically transfer one loopful of culture broth to Control #2.
10. Alcohol-flame your forceps and aseptically drop six beads into the broth culture. (You will only use four beads; the other two are extras in case you drop one during the procedure.)

**Caution!**

Store your forceps in the alcohol jar between transfers. When it is time to make a transfer, pinch the forceps and remove them from the alcohol. Then pass them through the flame to burn off the alcohol, holding them away from the alcohol jar while doing so. If the alcohol jar catches on fire, smother the flame with the lid. When finished, return the forceps directly to the jar without flaming.

11. After 1 minute, decant the broth into a beaker of disinfectant. Remove as much of the broth as possible without losing the beads in the disinfectant.
12. Dispense the beads onto the sterile filter paper in the glass Petri dish. This can be done by tapping the mouth of the tube on the paper. If this doesn't work, remove the beads with a sterile inoculating loop and flame it afterward
13. Using alcohol-flamed forceps spread the beads apart on the paper and allow them to dry for 10 minutes. Do not roll them around because this may remove bacteria.
14. After 10 minutes, place one bead in each of the three germicide plates using alcohol-flamed forceps. Mark the time here:

    ______________________________

15. With alcohol-flamed forceps, immediately place the fourth bead in plate #4 (sterile water).
16. With alcohol-flamed forceps, immediately place a sterile bead in plate #5 (sterile water).
17. After 10 minutes from the time marked in step 14 (step 7 in the procedural diagram), remove the five beads from the solutions in the same order as they were added, and place them in their respective nutrient broths. Carefully mix the broths immediately to disperse any residual disinfectant on the beads. Do not splash the broth into the cap.

[2] This procedure has been modified from its original form and is to be used for instructional purposes only.

18 Incubate all seven broths at 35 ± 2°C for 48 hours.

19 Save or dispose of the original culture as directed by your instructor.

### Lab Two

1 Remove all broth tubes from the incubator. Gently mix the controls and examine them for evidence of growth. Enter your results on the data sheet, page 139. Control #1 should have no growth and Control #2 should show turbidity. If both of these conditions have been met, you may proceed. If not, see your instructor.

2 Using Controls #1 and #2 as comparisons, examine broths containing beads exposed to germicide (Broths #1, #2, and #3). Using "G" to indicate growth and "NG" to indicate no growth, enter your results in both the individual data table and the class data table on the data sheet.

3 Again using Controls #1 and #2 as comparisons, examine Controls #3 and #4. Using "G" to indicate growth and "NG" to indicate no growth, enter your results on the data sheet.

4 Dispose of all plates in an appropriate autoclave container when finished.

5 Your instructor will provide you with a means to share data obtained by all eight groups. Record these data on your data sheet.

6 Answer the questions on the data sheet.

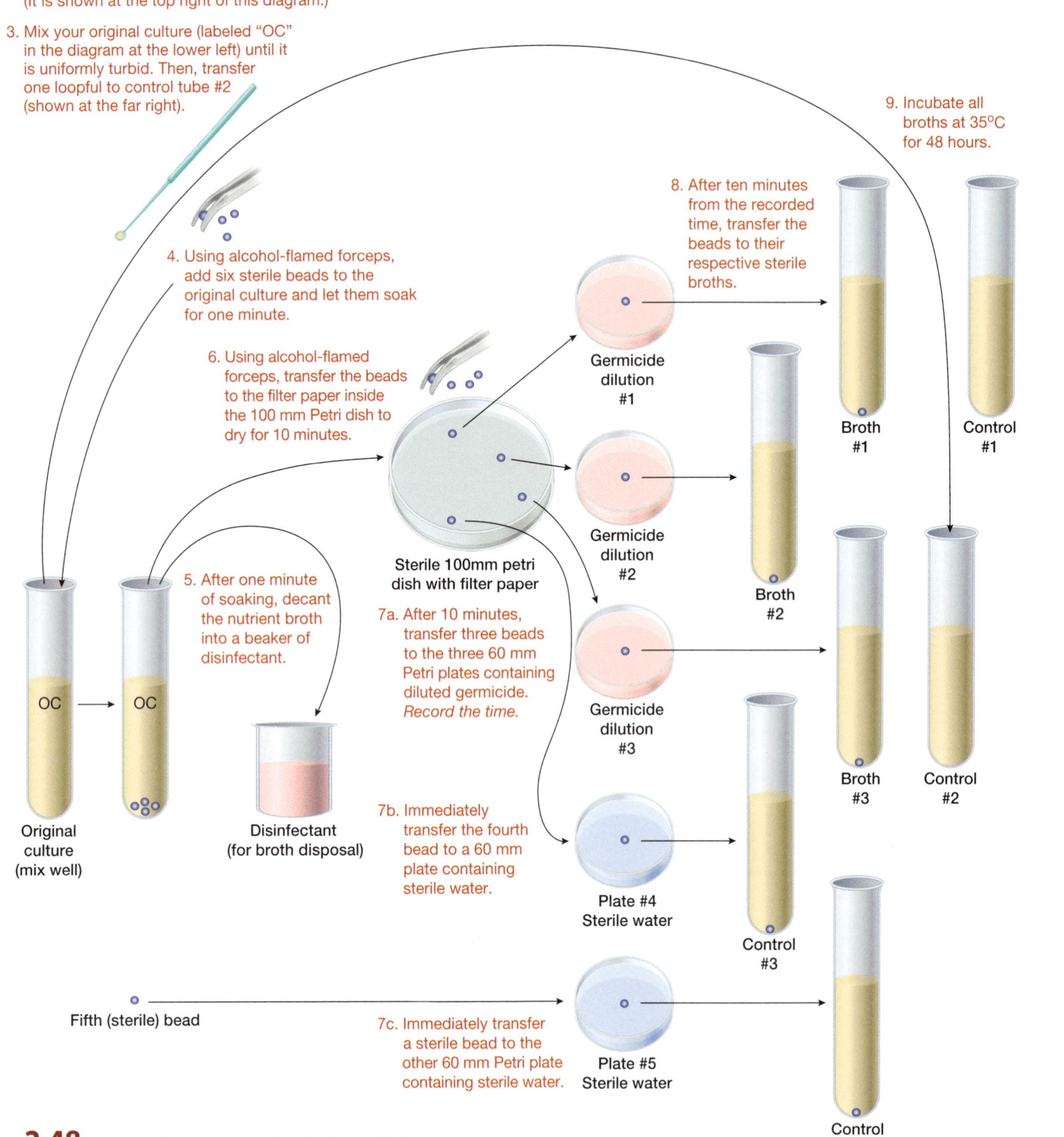


**2.48 Procedural Diagram for Chemical Germicides** At first glance, this procedural diagram looks pretty imposing. Read it from left to right and follow the steps in sequence. With a little preparation and familiarity, it really isn't very complex.

## References

McDonnell, Gerald E. *Antisepsis, Disinfection, and Sterilization: Types, Action, and Resistance.* Washington, DC: ASM Press, American Society for Microbiology, 2007.

Widmer, Andreas F. and Reno Frei. Chap. 7 in *Manual of Clinical Microbiology,* 9th ed. Patrick R. Murray, Ellen Jo Baron, James H. Jorgensen, Marie Louise Landry, and Michael A. Pfaller, eds. Washington, DC: ASM Press, American Society for Microbiology, 2007.

Name ______________________________

Date ______________________________

Lab Section ______________________________

I was present and performed this exercise (initials) ______________________________

DATA SHEET

2-13

# Effectiveness of Chemical Germicides: The Use-Dilution Test for Disinfectants and Antiseptics

## OBSERVATIONS AND INTERPRETATIONS

**1** Enter your individual data in the tables below.

| Controls | Growth (G) or No Growth (NG) |
|---|---|
| #1 | |
| #2 | |
| #3 | |
| #4 | |

**2** Enter the class data in the table below. Use "G" for "Growth" and "NG" for "No Growth."

| Organism | Household Bleach | | | Hydrogen Peroxide | | | Lysol® Brand II Disinfectant | | | Isopropyl Alcohol | | |
|---|---|---|---|---|---|---|---|---|---|---|---|---|
| | 0.01% | 0.1% | 1% | 0.03% | 0.3% | 3% | 25% | 50% | 100% | 10% | 30% | 50% |
| *S. epidermidis* | | | | | | | | | | | | |
| *E. coli* | | | | | | | | | | | | |

## QUESTIONS

**1** *Compare your results with the class data.*

**a.** *Which germicide was most effective and at what concentration? Defend your choice.*

______________________________

______________________________

______________________________

______________________________

______________________________

**b.** *Which was least effective? Defend your choice.*

______________________________

______________________________

______________________________

______________________________

______________________________

**2** *Which organism seemed to be more resistant to the germicides?*

2

**3** Consider the controls.

**a.** *Control #1: Was this a positive or a negative control? What purpose did it serve in the experiment?*

**b.** *Control #2: Was this a positive or a negative control? What purpose did it serve in the experiment?*

**c.** *Control #3: Was this a positive or a negative control? What purpose did it serve in the experiment?*

**d.** *Control #4: Was this a positive or a negative control? What purpose did it serve in the experiment?*

SECTION 3

# Microscopy and Staining

Microbiology as a biological discipline would not be what it is today without microscopes and cytological stains. Our ability to visualize, sometimes in great detail, the form and structure of microbes too small or transparent to be seen otherwise is attributable to developments in microscopy and staining techniques. In this section you will learn (or refine) your microscope skills. Then you will learn simple and more sophisticated bacterial staining techniques.

The earliest microscopes used visible light to create images and were little more than magnifying glasses. Today, more sophisticated compound light microscopes (Fig. 3.1) are used routinely in microbiology laboratories. The various types of light microscopy include bright-field, dark-field, fluorescence, and phase contrast microscopy (Fig. 3.2). Although each method has specific applications and advantages, the one used most commonly in introductory classes and clinical laboratories is bright-field microscopy. Many research applications use electron microscopy because of its ability to produce higher-quality images of greater magnification.

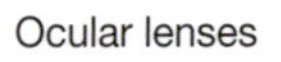
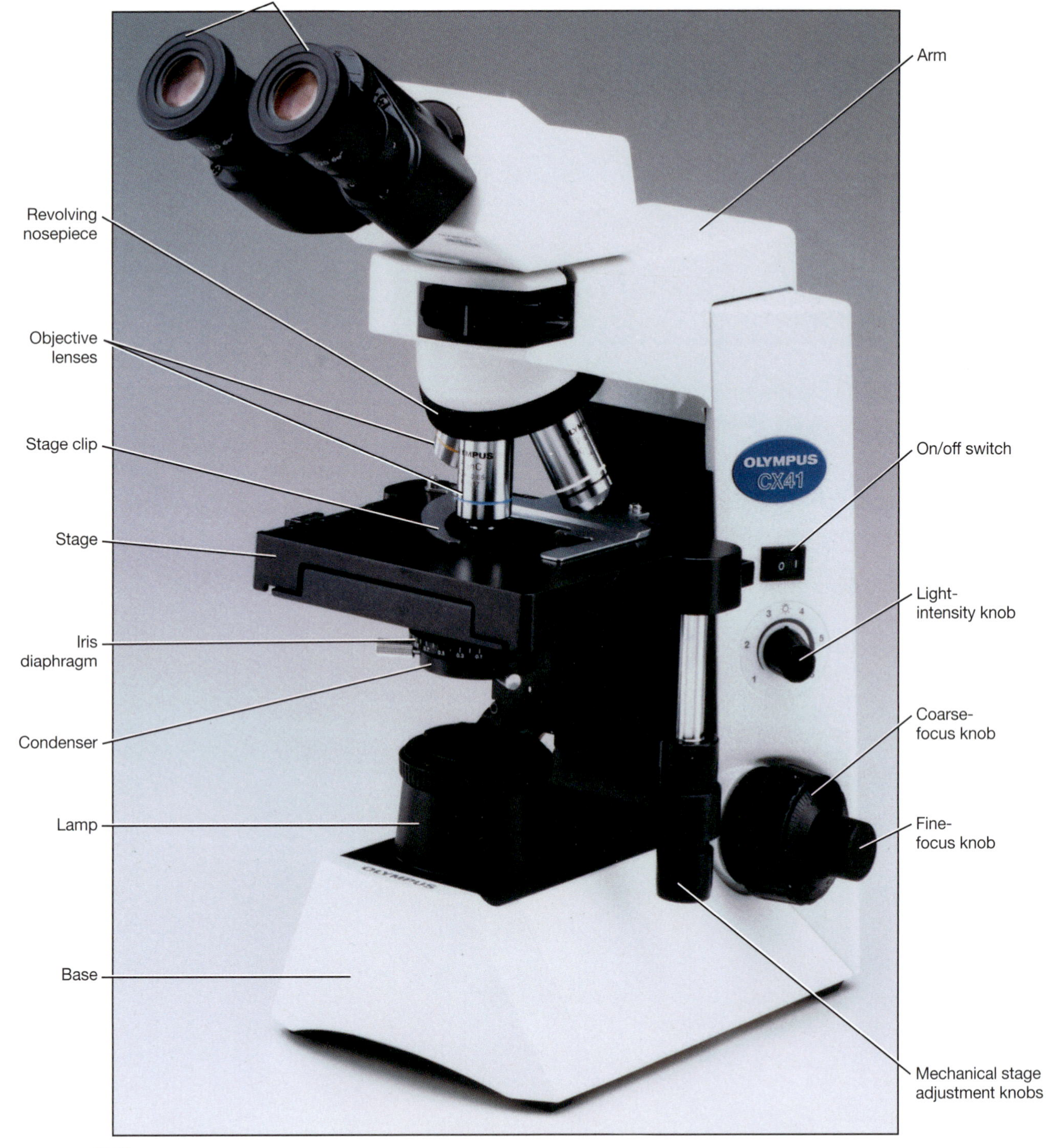


**3.1 Binocular Compound Microscope** ■ A quality microscope is an essential tool for microbiologists. Most are assembled with exchangeable component parts and can be customized to suit the user's specific needs.

Photograph courtesy of Olympus America Inc.

EXERCISE 3-1

# Introduction to the Light Microscope

## Theory

Bright-field microscopy produces an image made from light that is transmitted through a specimen (Fig. 3.2A). The specimen restricts light transmission and appears "shadowy" against a bright background (where light enters the microscope unimpeded). Because most biological specimens are transparent, the contrast between the specimen and the background can be improved with the application of stains to the specimen (Exercises 3-4 through 3-9 and 3-11). The "price" of the improved contrast is that the staining process usually kills cells. This is especially true of bacterial-staining protocols.

Image formation begins with light coming from an internal or an external light source (Fig. 3.3). It passes through the **condenser** lens, which concentrates the light and makes illumination of the specimen more uniform. **Refraction** (bending) of light as it passes through the **objective lens** from the specimen produces a magnified **real image.** This image is magnified again as it passes through the **ocular lens** to produce a **virtual image** that appears below or within the microscope. The amount of magnification that each lens produces is marked on the lens (Fig. 3.4). Total magnification of the specimen can be calculated by using the following formula:

$$\text{Total Magnification} = \text{Magnification by the Objective Lens} \times \text{Magnification by the Ocular Lens}$$

The practical limit to magnification with a light microscope is around 1300×. Although higher magnifications are possible, image clarity is more difficult to maintain as the magnification increases. Clarity of an image is called **resolution** (Fig. 3.5). The **limit of resolution** (or **resolving power**) is an actual measurement of how far apart two points must be for the microscope to view them as being separate. Notice that resolution *improves* as the limit of resolution is made *smaller*.

The best limit of resolution achieved by a light microscope is about 0.2 µm. (That is, at its absolute best, a light microscope cannot distinguish between two points closer together than 0.2 µm.) For a specific microscope, the actual limit of resolution can be calculated using the following formula:

$$D = \frac{\lambda}{NA_{condenser} + NA_{objective}}$$

where D is the minimum distance at which two points can be resolved, λ is the wavelength of light used, and $NA_{condenser}$ and $NA_{objective}$ are the numerical apertures of

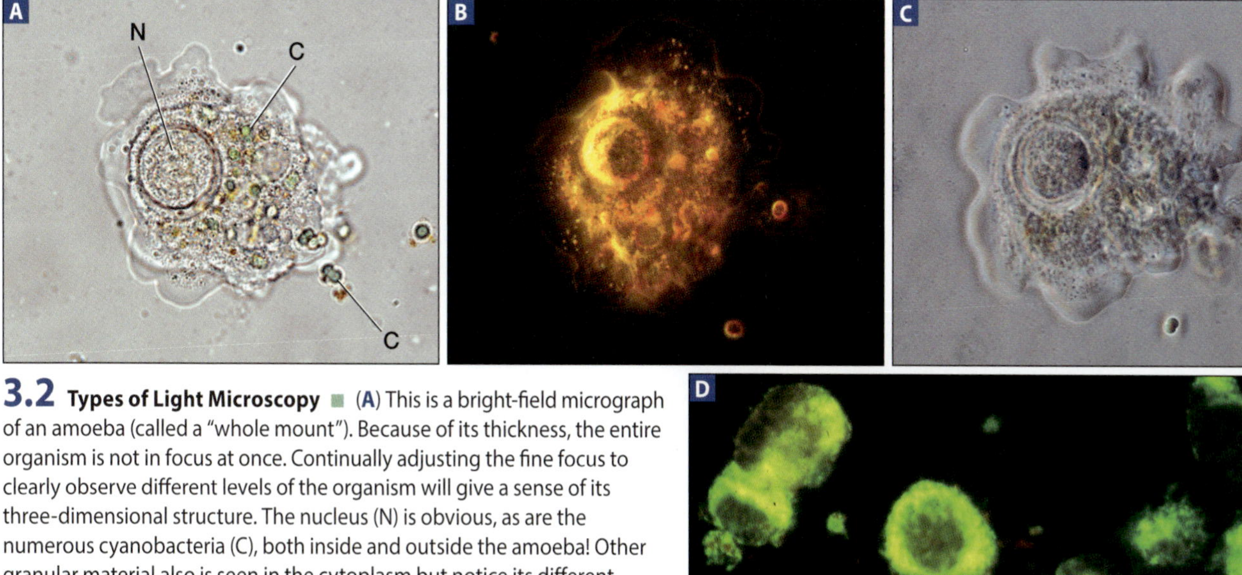


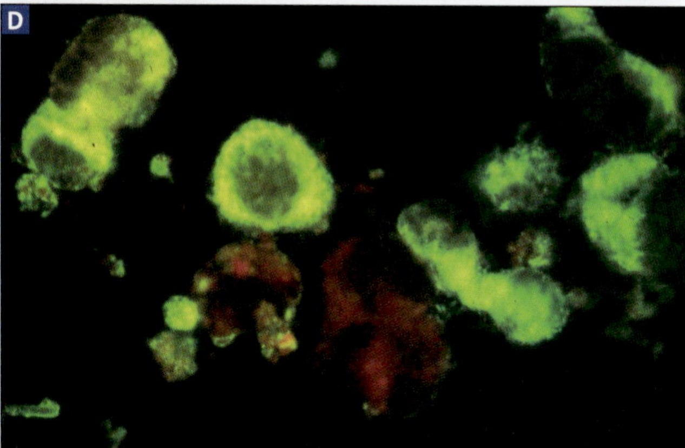


**3.2 Types of Light Microscopy** ■ **(A)** This is a bright-field micrograph of an amoeba (called a "whole mount"). Because of its thickness, the entire organism is not in focus at once. Continually adjusting the fine focus to clearly observe different levels of the organism will give a sense of its three-dimensional structure. The nucleus (N) is obvious, as are the numerous cyanobacteria (C), both inside and outside the amoeba! Other granular material also is seen in the cytoplasm but notice its different texture toward the periphery. **(B)** This is a dark-field micrograph of the same amoeba. Notice the more three-dimensional image and that the peripheral cytoplasm is barely visible. **(C)** This is a phase contrast image of the same amoeba. Different parts of the interior and its detail are visible than what is seen in the other two micrographs. **(D)** This is a fluorescence micrograph of *Mycobacterium kansasii.* The apple green is one of the characteristic colors of fluorescence microscopy.

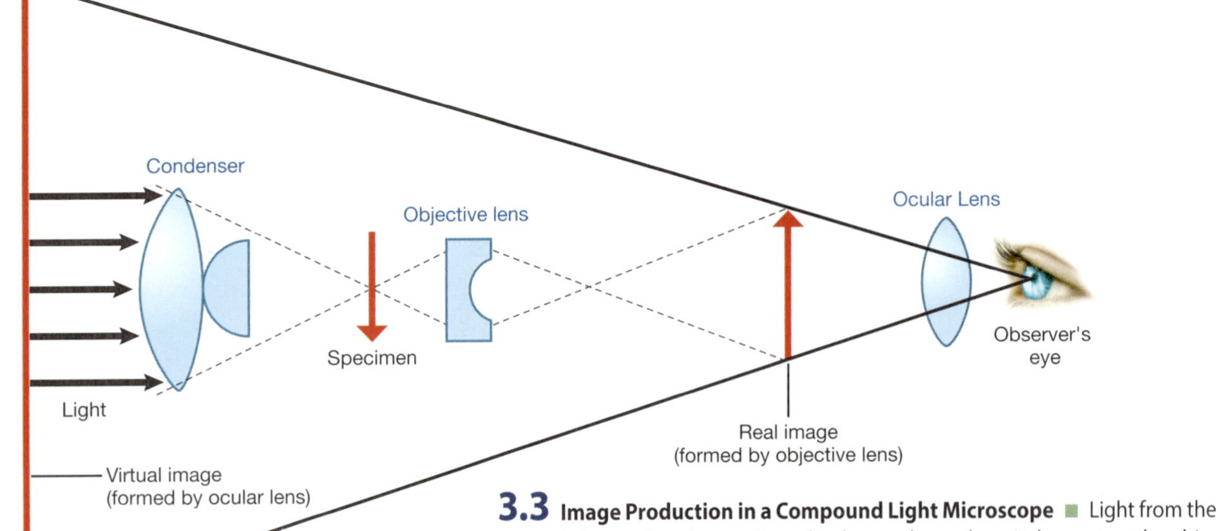


**3.3 Image Production in a Compound Light Microscope** ■ Light from the source is focused on the specimen by the condenser lens. It then enters the objective lens, where it is used to produce a magnified real image. The real image is magnified again by the ocular lens to produce a virtual image that is seen by the eye as being below or within the microscope. (After Chan, et al., 1986)

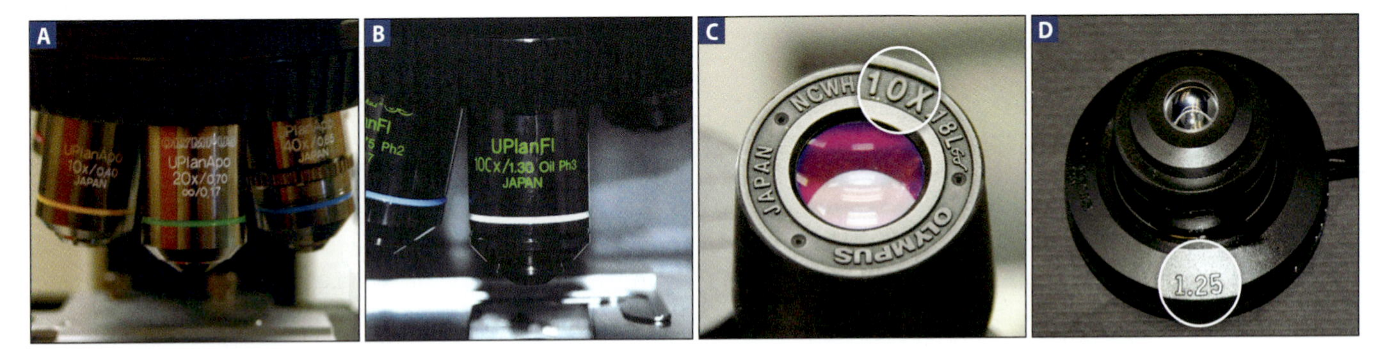


**3.4 Markings of Magnification and Numerical Aperture on Microscope Components** ■ **(A)** Three plan apochromatic objective lenses on the nosepiece of a light microscope. *Plan* means the lens produces a flat field of view. Apochromatic lenses are made in such a way that chromatic aberration is reduced to a minimum. From left to right, the lenses magnify 10×, 20×, and 40×, and have numerical apertures of 0.40, 0.70, and 0.85. The 20× lens has other markings on it. The mechanical tube length is the distance from the nosepiece to the ocular and has become standardized at 160 mm. However, this 20× lens has been corrected so the light rays are made parallel, effectively creating an infinitely long mechanical tube length (∞). This allows insertion of accessories into the light path without decreasing image quality. The thickness of cover glass to be used is also given (0.17±0.01 mm). Also notice the standard colored rings for each objective: yellow for 10×, green for 20× (or 16×), and light blue for 40× (or 50×). **(B)** This is an oil-immersion lens, indicated by the black ring (below the white ring, which is used to indicate a 100× or greater lens). It is the only lens constructed in such a way as not to be damaged by oil and, as such, is the only one with which oil is to be used. This particular lens is also constructed for phase contrast microscopy. It indicates it is to be used with the #3 setting on the phase condenser. **(C)** A 10× ocular lens. **(D)** A condenser (removed from the microscope) with a numerical aperture of 1.25. The lever at the right is used to open and close the iris diaphragm and adjust the amount of light entering the specimen.

the condenser lens and objective lens, respectively. Because numerical aperture has no units, the units for D are the same as the units for wavelength, which typically are in nanometers (nm).

**Numerical aperture** is the measure of a lens's ability to "capture" light coming from the specimen and use it to make the image. As with magnification, it is marked on the lens (Figs. 3.4A, 3.4B, and 3.4D). Using immersion oil between the specimen and the oil-immersion lens increases its numerical aperture and, in turn, makes its limit of resolution smaller. (If necessary, oil also may be placed between the condenser lens and the slide.) The result is better resolution.

The light microscope may be modified to improve its ability to produce images with contrast without staining, which often distorts or kills the specimen. In **dark-field microscopy** (Fig. 3.2B), a special condenser is used so only the light reflected off the specimen enters the objective. The appearance is of a brightly lit specimen against a dark background, often with better resolution than that of the bright-field microscope.

**Phase contrast microscopy** (Fig. 3.2C) uses special optical components to exploit subtle differences in the refractive indices of water and cytoplasmic components to produce contrast. Light waves that are in phase (that is, their peaks and valleys exactly coincide) reinforce one

**3.5 Resolution and Limit of Resolution** ■ The headlights of most automobiles are around 1.5 m apart. As you look at the cars in the foreground of the photo, it is easy to see both headlights as separate objects. The automobiles in the distance appear smaller (but really aren't) as does the apparent distance between the headlights. When the apparent distance between automobile headlights reaches about 0.1 mm, they blur into one because that is the limit of resolution of the human eye.

another, and their total intensity (because of the summed amplitudes) increases. Light waves that are out of phase by exactly one-half wavelength cancel each other and result in no intensity—that is, darkness.

Wavelengths that are out of phase by any amount will produce some degree of cancellation and result in brightness that is less than maximum but more than darkness. Thus, contrast is provided by differences in light intensity that result from differences in refractive indices in parts of the specimen that put light waves more or less out of phase. As a result, the specimen appears as various levels of "darks" against a bright background.

**Fluorescence microscopy** (Fig. 3.2D) uses a fluorescent dye that emits fluorescence when illuminated with ultraviolet radiation. Use of multiple fluorescent dyes that emit different colors and have an affinity for certain cellular structures allows differential staining of specimens. Advances in production of different fluorescent proteins (first derived from jellyfish green fluorescent protein; see Fig. 8.5, p. 507) have extended the utility of fluorescence microscopy in observing living organisms. In some cases, specimens possess naturally fluorescing chemicals and no dye is needed.

## ■ Application

Light microscopy (used in conjunction with cytological stains) is used to identify microbes from patient specimens or the environment. It also may be used to visually examine a specimen for the presence of more than one type of bacteria, or for the presence of other cell types that indicate tissue inflammation or contamination by a patient's cells.

## ■ In This Exercise

Today you will become familiar with the operation and limitations of your light microscope. You also will examine two practice slides to learn about microscope functioning.

### ▼ Materials

**Per Student**

- □ Lab coat
- □ Disposable gloves
- □ Chemical eye protection
- □ Compound light microscope
- □ Lens paper
- □ Non-sterile cotton swabs
- □ Lens-cleaning solution or 95% ethanol
- □ Letter "e" slide
- □ Colored-threads slide

## ■ Instructions for Using the Microscope

Proper use of the microscope is essential for your success in microbiology. Fortunately, with practice and by following a few simple guidelines, you can achieve satisfactory results quickly. Because student labs may be supplied with a variety of microscopes, your instructor may supplement the following procedures and guidelines with instructions specific to your equipment. Refer to Figure 3.1 as you read the following (if working independently), or follow along on your microscope as your instructor guides you. (*Note:* This is a thorough treatment of microscope use and not all parts may be immediately relevant to your laboratory. Refer back to this exercise as necessary.)

### Transport

1. Carry your microscope to your workstation using both hands—one hand grasping the microscope's arm and the other supporting the microscope beneath its base.
2. Gently place the microscope on the table.

### Cleaning

1. Lens paper is used for gently cleaning the condenser and objective lenses. Light wiping is usually enough. If that still doesn't clean the lens, call your instructor.

2 To clean an ocular, moisten a cotton swab with cleaning solution and gently wipe in a spiral motion starting at the center of the lens and working outward. Follow with a dry swab in the same pattern.[1]

### Basic Operation

1 Raise the substage condenser to a couple of millimeters below its maximum position nearly even with the stage (be sure not to raise it too high if you have already placed a slide on the stage) and open the iris diaphragm.

2 Plug in the microscope and turn on the lamp. Adjust the light intensity slowly to its maximum.

3 Using the nosepiece ring move the scanning objective (usually 4×) or low-power objective (10×) into position. Do not rotate the nosepiece by the objectives because this can damage the objective lenses and cause them to unscrew from the nosepiece.

4 Place a slide on the stage in the mechanical slide holder and center the specimen over the opening in the stage.

5 If using a binocular microscope, adjust the distance between the two oculars to match your own interpupillary distance as you examine the specimen. Position your eyes above the oculars so the images from the two oculars fuse into one.

6 Adjust the iris diaphragm to produce optimum illumination, contrast, and image. In the simplest sense, this means opening the iris diaphragm as you increase magnification because a smaller portion of the light beam is entering the lens. (More specifically, use the maximum light intensity combined with the smallest aperture in the iris diaphragm that produces optimum illumination. Remember: This is bright-field microscopy, so don't close down the iris diaphragm too much unless necessary to see detail, as in unstained specimens.)

7 Use the coarse-focus adjustment knob to bring the image into focus. (*Note:* For most microscopes, the distance from the nosepiece opening to the focal plane of each lens has been standardized at 45 mm. This makes the lenses **parfocal** and gives the user an idea of where to begin focusing.) Bring the image into sharpest focus using the fine-focus adjustment knob. Then observe the specimen with your eyes relaxed and slightly above the oculars to allow the images to fuse into one. If you are using a monocular microscope, keep both eyes open anyway to reduce eye fatigue.

8 If you are using a binocular microscope, adjust the oculars' focus to compensate for differences in visual acuity of your two eyes. Close the eye with the adjustable ocular and bring the image into focus with the coarse- and fine-focus knobs. Then, while using only the eye with the adjustable ocular, focus the image using the ocular's focus ring.

9 Scan the specimen to locate a promising region to examine in more detail.

10 If you are observing a nonbacterial specimen, progress through the objectives until you see the degree of structural detail necessary for your purposes. You will have to adjust the fine focus (but not the coarse focus because the lenses are parfocal—all that should be necessary after changing lenses is a slight adjustment) and illumination for each objective. Before advancing to the next objective, be sure to position a desirable portion of the specimen in the center of the field or you will risk "losing" it at the higher magnification.

11 If you are working with a bacterial smear, you will have to use the oil-immersion lens.

12 Follow these instructions to use the oil-immersion lens.

- Work through the low (10×), then high-dry (40×) objectives, adjusting the fine focus and illumination for each. Before advancing to the next objective, be sure to position a desirable portion of the specimen in the center of the field or you risk "losing" it at the higher magnification.
- When the specimen is in focus under high dry, rotate the nosepiece to a position midway between the high-dry and oil-immersion lenses. Then place a drop of immersion oil on the specimen. *Be careful not to get any oil on the microscope or its lenses, and be sure to clean it up if you do.* Rotate the oil lens so its tip is submerged in the oil drop, pass through it, and then return the oil lens into the oil. This minimizes the occurrence of air bubbles.
- *Note*: Do not move the stage down to add oil to the slide or the specimen will no longer be in focus. On a properly adjusted microscope, the oil and the high-dry lenses have the same focal plane. Therefore, when a specimen is in focus on high dry, the oil lens, although longer, will also be in focus and won't touch the slide when rotated into position.
- Focus and adjust the illumination to maximize the image quality.

13 When you are finished, lower the stage (or raise the objective) and remove the slide. Dispose of the freshly prepared slides in a jar of disinfectant or a sharps container; return permanent slides to storage.

[1] Lens paper can be gently used on ocular lenses, but even lens paper can scratch a lens if excessive pressure is applied.

### Storage

When you are finished using your microscope for the day:

1. Move the scanning objective into position.
2. Center and lower the mechanical stage.
3. Lower the light intensity to its minimum, and then turn off the light.
4. Wrap the electrical cord according to your particular lab rules.
5. Clean any oil off the lenses, stage, etc. Be sure to use only cotton swabs or lens paper for cleaning any of the optical surfaces of the microscope (see "Cleaning," pages 145–146).
6. Return the microscope to its appropriate storage place.

## PROCEDURE

1. Get out your microscope and record the magnifications and numerical aperture values in the chart on the data sheet, page 149.
2. Clean your microscope lenses as outlined in "Instructions for Using the Microscope."
3. Plug in the microscope and position the scanning objective over the stage. Make condenser and lamp adjustments appropriate for scanning power.
4. Pick up the letter "e" slide and examine it without the microscope. Record the orientation of the letter when the slide label is on the left. Sketch the orientation of the letter "e" on your data sheet.
5. Place the slide on the stage in the same position as you examined it with your naked eyes. Now, center the "e" in the field and examine it with the scanning objective. After focusing, sketch the orientation of the letter "e" image as viewed with the microscope on your data sheet.
6. Now move the *stage* to the right and record the direction the *image* moves.[2]

**3.6 Challenge of the Threads** ■ Even with the microscope, determining the order of threads from top to bottom is a challenge. This will require patience and use of the fine focus! Making it worse, not all the slides will be the same. Good luck!

7. Position the "e" in the center of the field again. Move the *stage* toward you and record the direction the *image* moves on your data sheet. Then remove the slide from the microscope and return it to its box.
8. Examine the colored-threads slide without the microscope (Fig. 3.6). See if you can tell where in the stack of three threads each color resides. That is, is the red thread on the top, bottom, or middle? Do the same for the yellow and blue threads.
9. Now, place the slide on the microscope and determine the order of the threads using the low- and high-power objectives. Record your observations on the data sheet. Then, return the slide to its box.
10. Return your microscope to storage.

[2] If your microscope doesn't have a mechanical stage, move the slide with your hands in the appropriate direction.

## References


Abramowitz, Mortimer. *Microscope Basics and Beyond.* Olympus America Inc. Melville, NY: Scientific Equipment Group, 2003.

Ash, Lawrence R. and Thomas C. Orihel. Pages 187–190 in *Parasites: A Guide to Laboratory Procedures and Identification.* Chicago: American Society for Clinical Pathology (ASCP) Press, 1991.

Bradbury, Savile and Brian Bracegirdle. Chap. 1 in *Introduction to Light Microscopy.* Oxford, UK: BIOS Scientific Publishers Limited, 1998.

Chan, E.C.S., Michael J. Pelczar, Jr., and Noel R. Krieg. Page 26 in *Laboratory Exercises in Microbiology*, 5th ed. New York, NY: McGraw-Hill Book Company, 1986.

Forbes, Betty A., Daniel F. Sahm, and Alice S. Weissfeld. Pages 119–121 in *Bailey & Scott's Diagnostic Microbiology,* 11th ed. St. Louis, MO: Mosby, 2002.

Piston, David W., George H. Patterson, Jennifer Lippincott-Schwartz, Nathan S. Claxton, and Michael W. Davidson, contributing authors. *Introduction to Fluorescent Proteins.* Nikon Microscopy U. Available online. URL: https://www.microscopyu.com/articles/livecellimaging/fpintro.html.

Wiedbrauk, Danny L. "Microscopy." Chap. 2 in *Manual of Clinical Microbiology,* 11th ed. James H. Jorgensen, Michael A. Pfaller, Karen C. Carroll, Guido Funke, Marie Louise Landry, Sandra S. Richter, and David W. Warnock, eds. Washington, DC: ASM Press, American Society for Microbiology, 2015.

Name ______________________________

Date ______________________________

Lab Section ______________________________

I was present and performed this exercise (initials) ______________________________

# DATA SHEET 3-1

## Introduction to the Light Microscope

### DATA AND CALCULATIONS

**1** Record the relevant values off your microscope and perform the calculations of total magnification for each lens.

| Lens System | Magnification of Objective Lens | Magnification of Ocular Lens | Total Magnification | Numerical Aperture | Calibration of Ocular Micrometer (from Ex. 3-2) |
|---|---|---|---|---|---|
| Scanning | | | | | |
| Low power | | | | | |
| High dry | | | | | |
| Oil immersion | | | | | |
| Condenser lens | NA | NA | NA | | NA |

**2** Sketch your observations of the letter "e" slide in the table below. Be sure the slide is right side up with the label at the left.

| Appearance of the "e" with the naked eye | Appearance of the "e" under the microscope | When the stage moves to the right, the image moves to the... | When the stage moves toward you, the image moves... |
|---|---|---|---|
| | | | |

**3** Record your observations of the colored-threads slide below. Check it under low and high power and see if your answer changes.

| | |
|---|---|
| Low power (usually the 10× objective) | Top: |
| | Middle: |
| | Bottom: |
| High power (usually the 40× objective) | Top: |
| | Middle: |
| | Bottom: |

## QUESTIONS

**1** *Why aren't the magnifications of both ocular lenses of a binocular microscope used to calculate total magnification?*

**2** *What is the total magnification for each lens setting on a microscope with 15× oculars and 4×, 10×, 45×, and 97× objectives lenses?*

3

**3** *Assuming that all other variables remain constant, explain why light of shorter wavelengths will produce a clearer image than light of longer wavelengths.*

**4** *Why is wavelength the main limiting factor on limit of resolution in light microscopy?*

**5** *On a given microscope, the numerical apertures of the condenser and low-power objective lenses are 1.25 and 0.25, respectively. You are supplied with a filter that selects a wavelength of 520 nm.*

**a.** *What is the limit of resolution on this microscope?*

**b.** *Will you be able to distinguish two points that are 300 nm apart as being separate, or will they blur into one?*

Name ______________________________

Date ______________________________

Lab Section ______________________________

I was present and performed this exercise (initials) ______________________________

# DATA SHEET 3-1

*(continued)*

**6** *On the same microscope as in question 5, the high-dry objective lens has a numerical aperture of 0.85.*

**a.** *What is the limit of resolution on this microscope?*

______________________________

______________________________

______________________________

______________________________

**b.** *Will you be able to distinguish two points that are 300 nm apart as being separate, or will they blur into one?*

______________________________

**7** *Calculate the limit of resolution for the oil lens of your microscope. Assume an average wavelength of 500 nm.*

______________________________

______________________________

______________________________

______________________________

**8** *Examine Figure 3.3 and explain the results you observed with the letter "e" slide.*

______________________________

______________________________

______________________________

**9** *With which objective was it easier to determine the sequence of colored threads?*

______________________________

**10** *Why should closing the iris diaphragm improve your ability to determine thread order?*

______________________________

**11** *What does the colored-threads slide demonstrate about specimens you will be observing later in the class?*

3

# Calibration of the Ocular Micrometer

EXERCISE 3-2

## Theory

An **ocular micrometer** is a type of ruler installed in the microscope eyepiece, composed of uniform but unspecified graduations (Fig. 3.7). Therefore, it must be calibrated before any viewed specimens can be measured. The device used to calibrate ocular micrometers is called a **stage micrometer**, which is illustrated in Figure 3.8. A stage micrometer is a type of microscope slide with a ruler divided into 10 µm and 100 µm increments. Other measuring instruments may be used in place of a stage micrometer, as shown in Figure 3.9.

When the stage micrometer is placed on the stage, it is magnified by the objective being used; therefore, the apparent distance between lines increases as magnification increases, but the ocular micrometer is unchanged. Consequently, the *value* of ocular micrometer divisions decreases as magnification increases. For this reason, calibration must be done for each objective.

Figure 3.10 illustrates how the stage micrometer is placed on the stage and brought into focus with the ocular micrometer superimposed over it. Then the first (left) line of the ocular micrometer is aligned with one of the marks

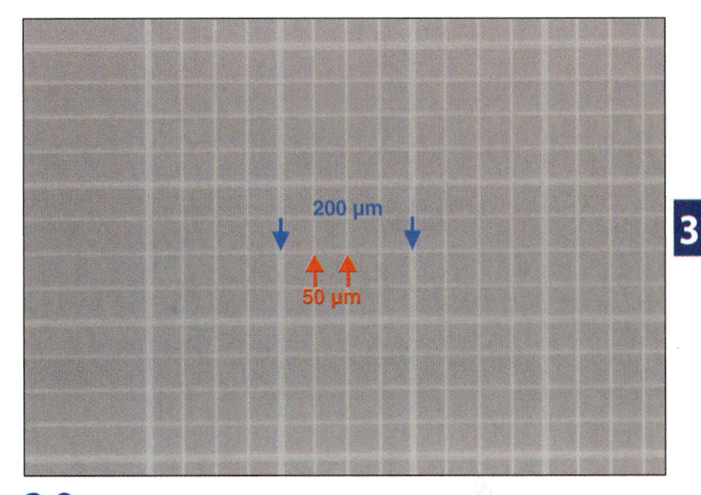


**3.9 Hemacytometer** ■ Any instrument with markings of known distance apart may be used as a stage micrometer. The hemacytometer is a grid with lines 50 µm apart (red arrows). A larger grid is formed by lines 200 µm apart (blue arrows). Use any horizontal line as the micrometer, with the smallest divisions 50 µm apart.

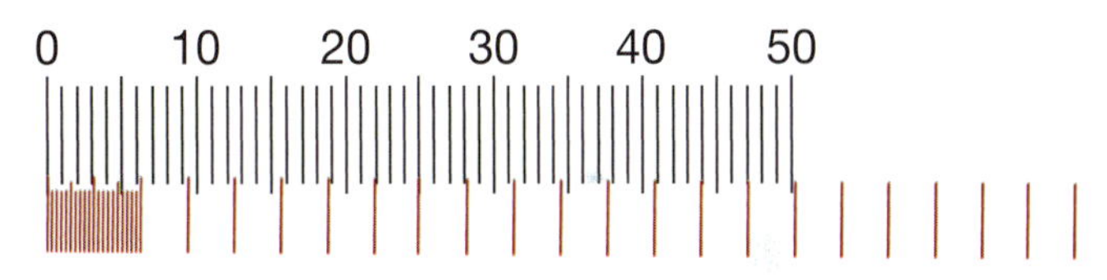


**3.10 What They Look Like in Use** ■ When properly aligned, the ocular micrometer scale is superimposed over the stage micrometer scale. Notice that they line up at their left ends. These two micrometers have intentionally not been drawn to the correct proportions relative to each other because they are intended only to show how to use them. When you work with your microscope and stage micrometer, expect different numbers to align.

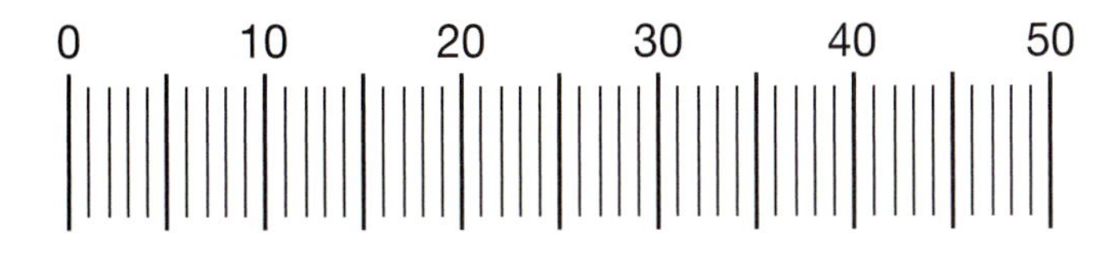


**3.7 Ocular Micrometer** ■ This illustration representing an ocular micrometer shows a scale with uniform increments of unknown size. It has to be calibrated for each objective lens on the microscope.

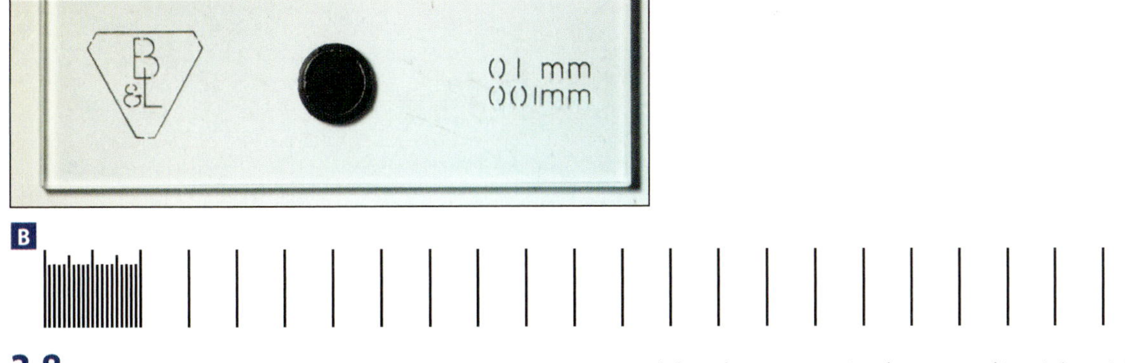


**3.8 Stage Micrometer** ■ (**A**) The stage micrometer is a microscope slide with a microscopic ruler engraved into it (not visible in the dark center of the slide). The number markings at the right on this micrometer indicate that the major increments are 0.1 mm (100 µm) apart. There is also a section of the scale that is marked off in 0.01 mm (10 µm) increments. (**B**) This drawing represents what the stage micrometer on the slide in (**A**) looks like. The micrometer is 2,200 µm long. The 22 major divisions are 100 µm apart. The 200 µm at the left are divided into 10 µm increments.

on the stage micrometer. (The line chosen on the stage micrometer depends on the power of the lens being calibrated. Larger increments can be used for lower magnifications, but higher magnifications will require using the smaller ones. Figure 3.10 depicts proper alignment with the scanning objective.)

Notice in Figure 3.10 that line 25 of the ocular micrometer and the eighth major line of the stage micrometer are perfectly aligned. (***Note***: This is a hypothetical situation. Your stage and ocular micrometers won't align in the same way.) This indicates that 25 ocular micrometer divisions (also called ocular units, or OU) span a distance of 800 μm because the stage micrometer lines are 100 μm apart. Notice also that line 47 of the ocular micrometer is aligned with the fifteenth major stage micrometer line. This means that 47 ocular units span 1,500 μm. These values have been entered for you in Table 3-1.



To determine the value of an ocular unit on a given magnification, divide the distance (from the stage micrometer) by the corresponding number of ocular units.

$$\frac{800\ \mu m}{25\ \text{ocular units}} = 32\ \frac{\mu m}{OU}$$

$$\frac{1{,}500\ \mu m}{47\ \text{ocular units}} = 31.9\ \frac{\mu m}{OU} \approx 32\ \frac{\mu m}{OU}$$

It is customary to record more than one measurement, with each being calculated separately. If the calculated ocular unit values differ, use their arithmetic mean as the calibration for that objective lens.

As mentioned previously, each objective must be calibrated. Because of its short working distance, calibrating the oil-immersion lens may be difficult to accomplish using the stage micrometer. It also may be difficult because the magnified distance between stage micrometer lines is too large. If this is the case, its value can be calculated using the calibration value of one of the other lenses. Refer to Table 3-2 for the total magnifications of each objective lens on a typical microscope.

TABLE **3-1** **Sample Data from Figure 3.10**

| Stage Micrometer | Ocular Micrometer | Calibration |
|---|---|---|
| 800 μm | 25 OU | $32\ \frac{\mu m}{OU}$ |
| 1,500 μm | 47 OU | $31.9\ \frac{\mu m}{OU} \approx 32\ \frac{\mu m}{OU}$ |

Notice that the magnification of the oil-immersion lens is 10 times greater than the low-power lens. This means that objects viewed on the stage (stage micrometer *or* specimens) appear 10 times larger when changing from low power to oil immersion. But, because the magnification of the ocular micrometer does not change, an ocular division now covers only one-tenth the distance. Thus, the size of an ocular unit using the oil-immersion lens can be calculated by dividing the calibration for low power by 10.

Ocular micrometer values can be calculated for any lens using values from any other lens, and they provide a good check of measured values. The calibration of the scanning objective in the hypothetical microscope in Figure 3.10 was 32 μm/OU. For practice, use the scanning calibration to calculate the low, high-dry and oil-immersion calibrations. Write the values in Table 3-2.

Once you have determined the ocular unit values for each objective lens on your microscope, use the ocular micrometer as a ruler to measure specimens. For instance, if you determine that under the scanning objective a cell is 5 ocular units long, the cell's actual length would be determined as follows (using the sample values from Table 3-2):

$$\text{Cell Dimension} = \text{Ocular Units} \times \text{Calibration}$$
$$\text{Cell Dimension} = 5\ \text{Ocular Units} \times 32\ \mu m/OU$$
$$\text{Cell Dimension} = 160\ \mu m$$

Be sure to include the proper units in your answer!

## Application

The ability to measure microbes is useful in their identification and characterization.

## In This Exercise

This lab exercise involves calibrating the ocular micrometer on your microscope. Actual measurement of specimens will be done in subsequent lab exercises as assigned by your instructor.

## ▼ Materials

### Per Student

- ☐ Compound microscope equipped with an ocular micrometer
- ☐ Stage micrometer

TABLE **3-2** **Using one Calibration to Calculate the Others—A Sample Problem**

| Power | Total Magnification | Magnification Factor | Calibration (µm/OU) |
|---|---|---|---|
| Scanning | 40× | | 32 |
| | | $\frac{100}{40} = 2.5$ | |
| Low Power | 100× | | |
| | | | |
| High-Dry Power | 400× | | |
| | | | |
| Oil Immersion | 1000× | | |

## PROCEDURE

Following is the general procedure for calibrating the ocular micrometer on your microscope. Your instructor will notify you of any specific details unique to your laboratory.

1. Check your microscope and determine which ocular has the micrometer in it (it will be in only one).
2. Move the scanning objective into position.
3. Place the stage micrometer on the stage and position it so its image is superimposed by the ocular micrometer and the left-hand marks line up.
4. Examine the two micrometers and, as previously described, record two or three points where they line up exactly. Record these values on the data sheet, page 157, and calculate the value of an ocular unit.
5. Change to low power and repeat the process.
6. Change to high-dry power and repeat the process.
7. Change to the oil-immersion lens and repeat the process. Be sure to look at the oil lens from the side as you rotate it into position. If it looks like it is going to contact the stage micrometer, **stop**! Return the high-dry lens into position and calculate the calibration using the value of another lens. You will also need to calculate the oil lens' calibration if the stage micrometer lines are too far apart for direct measurement.
8. Compute average calibrations for each objective lens and record these on the data sheet.
9. As long as you keep this microscope throughout the term, you may use the calibrations you recorded without recalibrating the microscope. You may wish to write the calibrations in Table 1 of the Exercise 3-1 data sheet (p. 149) so all the relevant features of your microscope are recorded in one place.

## References

Abramoff, Peter and Robert G. Thompson. Pages 5–6 in *Laboratory Outlines in Biology—III*. San Francisco: W. H. Freeman, 1982.

Ash, Lawrence R. and Thomas C. Orihel. Pages 187–190 in *Parasites: A Guide to Laboratory Procedures and Identification*. Chicago: American Society for Clinical Pathology (ASCP) Press, 1991.

Wiedbrauk, Danny L. Page 11 in *Manual of Clinical Microbiology*, 11th ed. James H. Jorgensen, Michael A. Pfaller, Karen C. Carroll, Guido Funke, Marie Louise Landry, Sandra S. Richter, and David W. Warnock, eds. Washington, DC: ASM Press, American Society for Microbiology, 2015.

Name ______________________________

Date ______________________________

Lab Section ______________________________

I was present and performed this exercise (initials) ______________________________

# DATA SHEET 3-2

## Calibration of the Ocular Micrometer

### DATA AND CALCULATIONS

**1** Record two or three values where the ocular micrometer and the stage micrometer line up for the scanning, low, high-dry, and oil-immersion objective lenses. Then calculate the calibration for each. Be sure to include proper units in your calibrations. ***Note:*** Use caution when moving the oil-immersion lens into position. If it looks like it will hit the stage micrometer, stop. You will have to calculate its calibration using the value of one of the other lenses.

#### Scanning Objective Lens

| Stage Micrometer (μm) | Ocular Micrometer (OU) | Calibration |
|---|---|---|
| | | |
| | | |
| | | |

#### Low-Power Objective Lens

| Stage Micrometer (μm) | Ocular Micrometer (OU) | Calibration |
|---|---|---|
| | | |
| | | |
| | | |

### High-Dry Objective Lens

| Stage Micrometer (µm) | Ocular Micrometer (OU) | Calibration |
|---|---|---|
| | | |
| | | |
| | | |

### Oil-Immersion Objective Lens

| Stage Micrometer (µm) | Ocular Micrometer (OU) | Calibration |
|---|---|---|
| | | |
| | | |
| | | |

**2** Calculate the average value for each calibration and record them in the table below. If necessary, use one of the values to calculate the calibration of the oil lens. Be sure to include proper units in your answer. You may also want to record these values in Table 1 on the Exercise 3-1 data sheet (p. 149).

### Average Calibrations for My Microscope

| Objective Lens | Average Calibration |
|---|---|
| Scanning | |
| Lower power | |
| High-dry power | |
| Oil immersion | |

# Microscopic Examination of Eukaryotic Microbes

EXERCISE 3-3

## Theory

Over the last several decades, cells have been divided into two major groups—the **prokaryotes** and the **eukaryotes**—based primarily on size and complexity. However, the validity of the label *prokaryote* has been questioned because it only describes what prokaryotes are *not* rather than describing what they *are*. The initial distinction was made based on the absence of a nucleus, but using the same reasoning the argument could be made that sequoia redwoods and humans belong in the same taxonomic group because they don't have feathers!

It might be a few more years before the scientific community settles on whether the term *prokaryote* should be retained or not. However, we are convinced by the arguments in favor of abandoning it and won't use it in this book. We will refer to the three domains: *Bacteria*, *Archaea*, and *Eukarya* (Fig. 3.11).

The eukaryotes are further subdivided into four "supergroups" until kingdom associations can reasonably be determined. Figure 3.12 shows a provisional phylogenetic tree of the three domains based on RNA comparisons and other data.

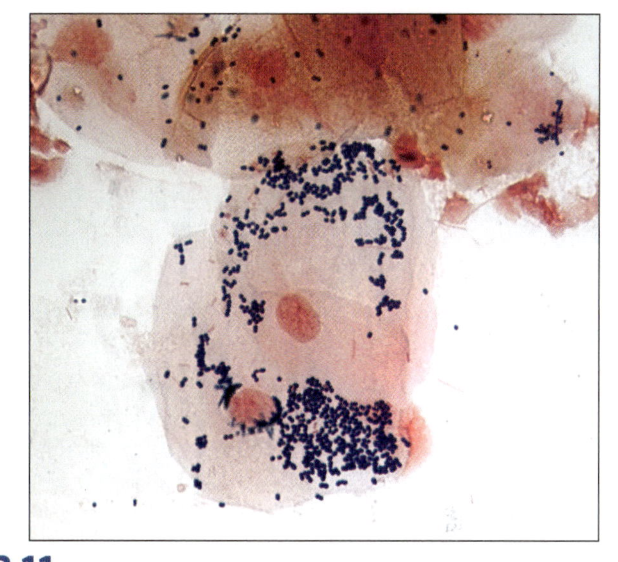

**3.11 Bacterial and Eukaryotic Cells (Gram Stain, Oil Immersion)** This is a direct smear specimen taken from around the base of the teeth below the gum line. The large, pink cells are human epithelial cells and are eukaryotic (notice the prominent nuclei). The small, purple and pink cells are bacteria, which are typically smaller than eukaryotic cells (as are *Archaea*)–roughly 1 μm to 5 μm in length versus 10 μm to 100 μm for eukaryotic cells.

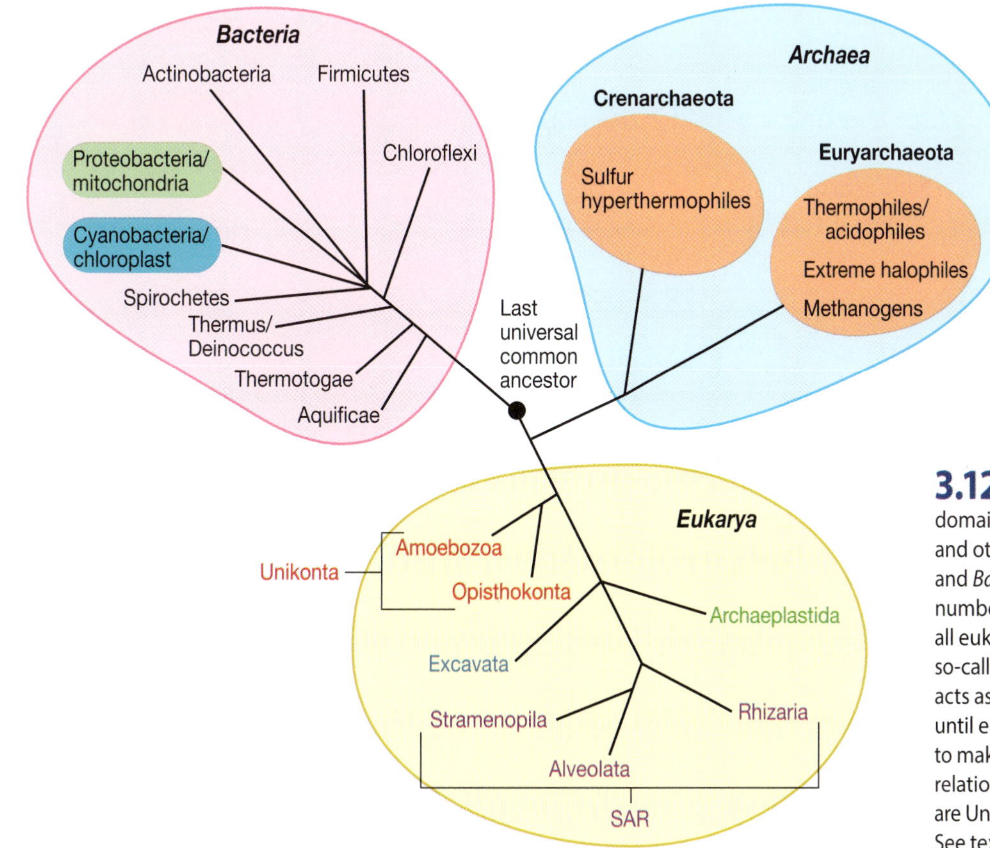


**3.12 The Three Domains of Life** The domains are based on rRNA sequencing results and other molecular comparisons. The *Archaea* and *Bacteria* comprise an as yet undetermined number of kingdoms. Domain *Eukarya* includes all eukaryotic organisms and is divided into so-called supergroups, an informal category that acts as a place to temporarily group organisms until enough evidence has been accumulated to make more informed decisions about their relationships and placement. The supergroups are Unikonta, Excavata, SAR, and Archaeplastida. See text for details.

## Eukaryotic Supergroups

Only a brief introduction to the eukaryotic supergroups—Excavata, SAR (formerly Chromalveolata and Rhizaria), Archaeplastida, and Unikonta—will be given here, with an emphasis on microscopic species.

- **Supergroup Unikonta.** This group is composed of heterotrophs and includes amoebas (of all sorts), animals, fungi, and others. There is evidence that unikonts diverged from the other eukaryotic lineages fairly early in eukaryote evolution. This is represented by their branch point being closer to the "root" of the phylogenetic tree in Figure 3.12.
  - Amoebas belong to two main groups. "Classical" amoebas belong to a subgroup called **gymnamoebas** (Fig. 3.13A). These are free-living in marine, freshwater, and soil environments. The other group comprises the **entamoebas** (Fig. 3.13B), some of which are parasitic. All amoebas move by forming cytoplasmic extensions called **pseudopods,** which are also used to engulf their food. Division is by binary fission and they do not have a cell wall.
  - Fungi are nonmotile eukaryotes. Their cell wall is usually made of the polysaccharide **chitin**, not cellulose as in plants. Unlike animals and many other unikonts (that ingest then digest their food), fungi are **absorptive heterotrophs**. That is, they secrete **exoenzymes** into the environment, and then absorb the digested nutrients. Most are saprophytes that decompose dead organic matter, but some are parasites of plants, animals, or humans. Fungi are informally divided into **filamentous molds** and **unicellular yeasts** based on their overall appearance.

    Molds grow as filaments of cells. These filaments are called **hyphae** and collectively form a **mycelium**. Molds produce a variety of sexual and asexual reproductive spores, which are used as a basis for classification. Three molds will be covered here: *Rhizopus*, *Aspergillus*, and *Penicillium*.

    The dozen or so *Rhizopus* are fast-growing molds that produce white or grayish, cottony growth. The mycelium becomes darker with age as sporangia are produced, giving it a "salt and pepper" appearance (Fig. 3.14A). *Rhizopus* species produce broad (10 µm) surface and aerial hyphae with irregular diameters. Anchoring **rhizoids** (Fig. 3.14B) are produced where the surface hyphae (**stolons**) join the bases of the long, unbranched **sporangiophores**.

    The *Rhizopus* life cycle has both sexual and asexual phases. Asexual **sporangiospores** are produced by large, circular sporangia (Fig. 3.14C) borne at the ends of the elevated sporangiophores. A hemispherical **columella** supports the sporangium. After release, the sporangiospores develop into hyphae identical to those that produced them.

    On occasion, sexual reproduction occurs when hyphae of different mating types (designated + and − strains) make contact. Initially, **progametangia** (Fig. 3.14D) extend from each hypha and nuclei and cytoplasm flow into them. As a consequence, they enlarge and contact one another. Upon contact, a septum separates the end of each progametangium into a **gametangium** and a **suspensor** (Fig. 3.14E).

    The walls between the two gametangia dissolve, and a thick-walled **zygosporangium** develops (Figs. 3.14F and 3.14G). Fusion of nuclei (karyogamy) occurs within the zygosporangium, resulting in multiple diploid zygote nuclei. After a dormant period, zygote meiosis occurs and produces haploid **zygospores** which, when released, develop into new hyphae and the life cycle is completed.

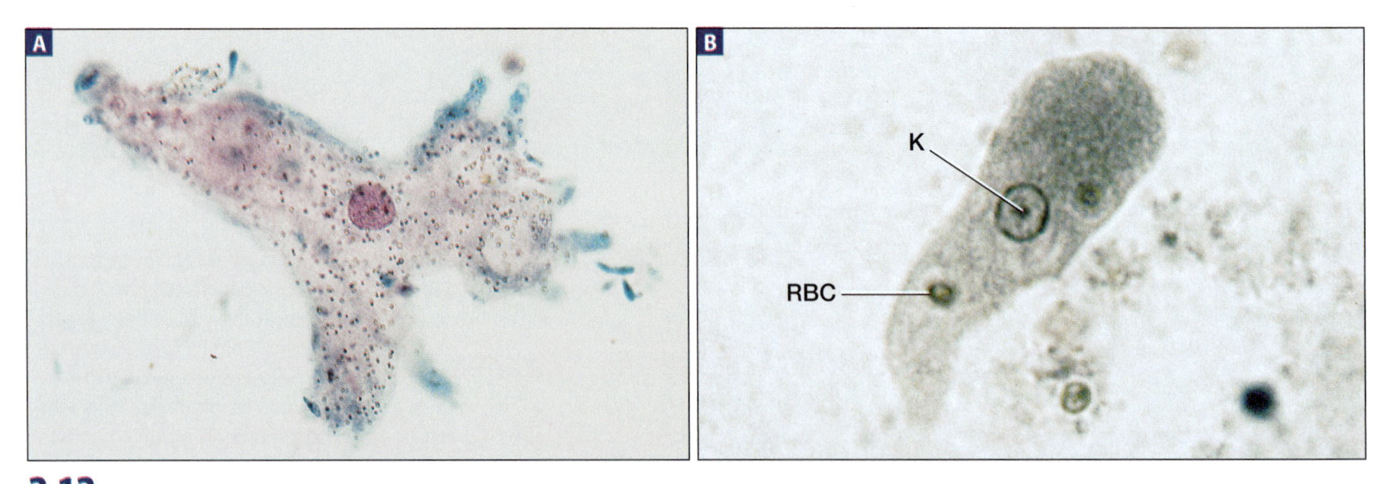


**3.13 Supergroup Unikonta—Amoebas** ■ **(A)** This *Amoeba* belongs to the gymnamoeba subgroup. Visible are the numerous pseudopods, which are used for movement as well as engulfing prey. (Commercially prepared slide.) **(B)** *Entamoeba histolytica* is the causative agent of amebic dysentery. Trophozoites range in size from 15 to 50 µm. Notice the small, central karyosome (K), the dark-staining, beaded chromatin at the periphery of the nucleus, and the ingested red blood cell (RBC), all of which are characteristic of *E. histolytica*. (Iron hematoxylin stain of fecal smear.)

Two *Rhizopus* species (*R. oryzae* and *R. microsporus*) are primarily responsible for producing **zygomycosis** (mucormycosis), a condition found most frequently in diabetics, organ transplant recipients, and immunocompromised patients. Portal of entry and overall health of the patient are critical factors in determining the location and extent of infection, but the mortality rate is high—between 47–84% worldwide. On a less severe note, most species are harmless saprophytes. *R. stolonifer* is the common bread mold.

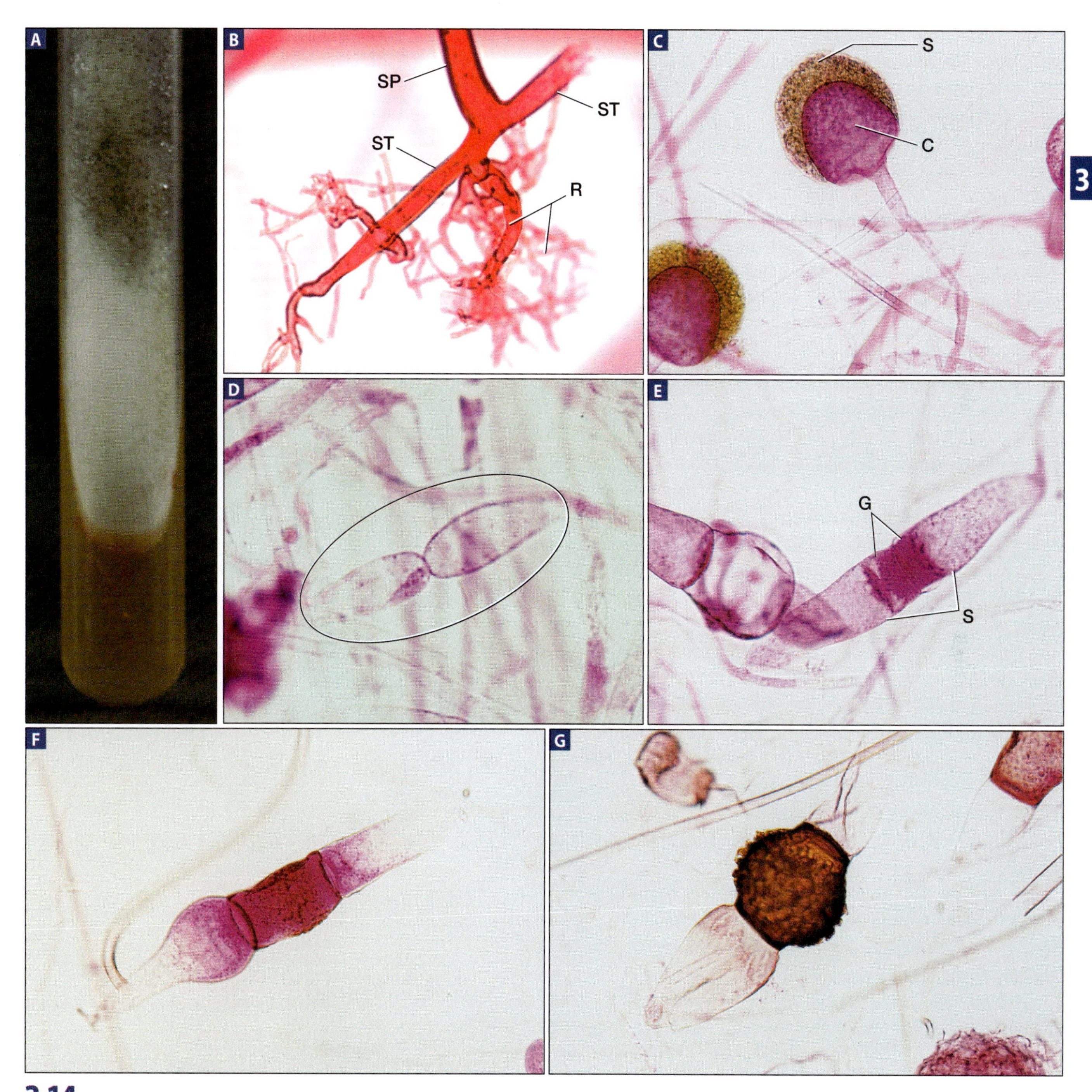


**3.14 Supergroup Unikonta—Mold *Rhizopus*** ■ **(A)** Black asexual sporangia *Rhizopus stolonifer*, the bread mold, have begun to form at the top, giving the growth a "salt and pepper" appearance. **(B)** Anchoring rhizoids (R) form at the junction of each sporangiophore (SP) and the stolon (ST). Note the absence of the septa. **(C)** The sporangium is found at the end of a long, unbranched, and nonseptate sporangiophore. The haploid asexual sporangiospores (S) in the sporangium cover the surface of the columella (C), which has a flattened base. **(D)** The black circle encloses progametangia from different hyphae. Contact between them results in each forming a gamete. **(E)** Gametangia (G) and suspensors (S) are shown in the center of the field. Gametangia contain one or more haploid nuclei from each mating type. **(F)** The zygosporangium forms when the cytoplasm from the two mating strains fuse (plasmogamy). **(G)** Haploid nuclei from each strain fuse within the zygosporangium (karyogamy) to produce many diploid zygote nuclei. Meiosis follows to produce numerous haploid zygospores. (Commercially prepared slides.)

*Aspergillus* species are commonly found in soil and on plants. Roughly 180 species have been identified. They are characterized by green to yellow or brown granular colonies with a white edge. One species, *A. niger*, produces distinctive black colonies. The *Aspergillus* "fruiting body" is distinctive, with chains of spore (called conidia) arising from one or two rows of cells attached to a swollen **vesicle** at the end of an unbranched stalk (Fig. 3.15A). Fruiting body structure and size, and conidia color are useful in species identification. Since their discovery it was thought that *Aspergillus* species were incapable of sexual reproduction, but evidence is mounting that under the right environmental conditions viable ascospores can be produced.

*A. fumigatus* and other species are opportunistic pathogens that cause aspergillosis, an umbrella term covering many diseases. Immunocompromised patients are at highest risk of infection. Allergic aspergillosis may occur in individuals who are in frequent contact with the spores and become sensitized to them. Subsequent contact produces symptoms similar to asthma. Aspergilloma (fungus ball) involves colonization of the paranasal sinuses or lungs, resulting in abscess formation. It is frequently asymptomatic and resolves without treatment, but in serious cases it can be fatal.

Some *Aspergillus* species are of commercial importance. Fermentation by *A. oryzae* is used in the production of sake (made from rice); and soy paste (made from soybeans), which may be further fermented with *A. oryzae* and *A. soyae* into soy sauce. Aspergilli are also used in commercial production of citric acid, various enzymes, and many other products. Further, *A. niger* and *A. oryzae* have been designated as GRAS (generally regarded as safe) by the Food and Drug Administration and World Health Organization, which has opened the door for using them as hosts in biotechnological applications.

Members of the genus *Penicillium* are ubiquitous, being found in the air, soil, and decaying organic matter worldwide. Over 200 species have been identified. They produce distinctive green, powdery, radially furrowed colonies with a white edge. *Penicillium* fruiting bodies are distinctive, looking like a brush (*penicillus* means "paintbrush") at the end of an elevated stalk (Fig. 3.15B).

*Penicillium* is best known for its production of the antibiotic penicillin (discovered by Alexander Fleming in 1928), but it is also a common contaminant. One pathogen, *P. marneffei*, is endemic to Asia and is responsible for disseminating opportunistic infections of the lungs, liver, and skin in immunosuppressed and immunocompromised patients. *P. chrysogenum* is common in damp, indoor habitats and their spores can cause allergic reactions. Other species are important because they cause spoilage of fruits and vegetables. Some *Penicillium* species are of commercial importance for fermentations used in cheese production. Examples include *P. roquefortii* (Roquefort cheese) and *P. camembertii* (Camembert and Brie cheeses).

Yeasts are generally unicellular, but may form short chains of cells. Two will be covered here: *Saccharomyces cerevisiae* and *Candida albicans*.

*Saccharomyces cerevisiae* is used in the production of bread, wine, and beer but is not an important human pathogen. It does not form a mycelium but, rather, produces a colony similar to bacteria (Fig. 3.16A). Vegetative cells are generally oval to round in shape, and asexual reproduction occurs by budding (Fig. 3.16B). Short **pseudohyphae** are sometimes produced when the budding cells fail to separate. In an *S. cerevisiae* population, haploid and diploid cells

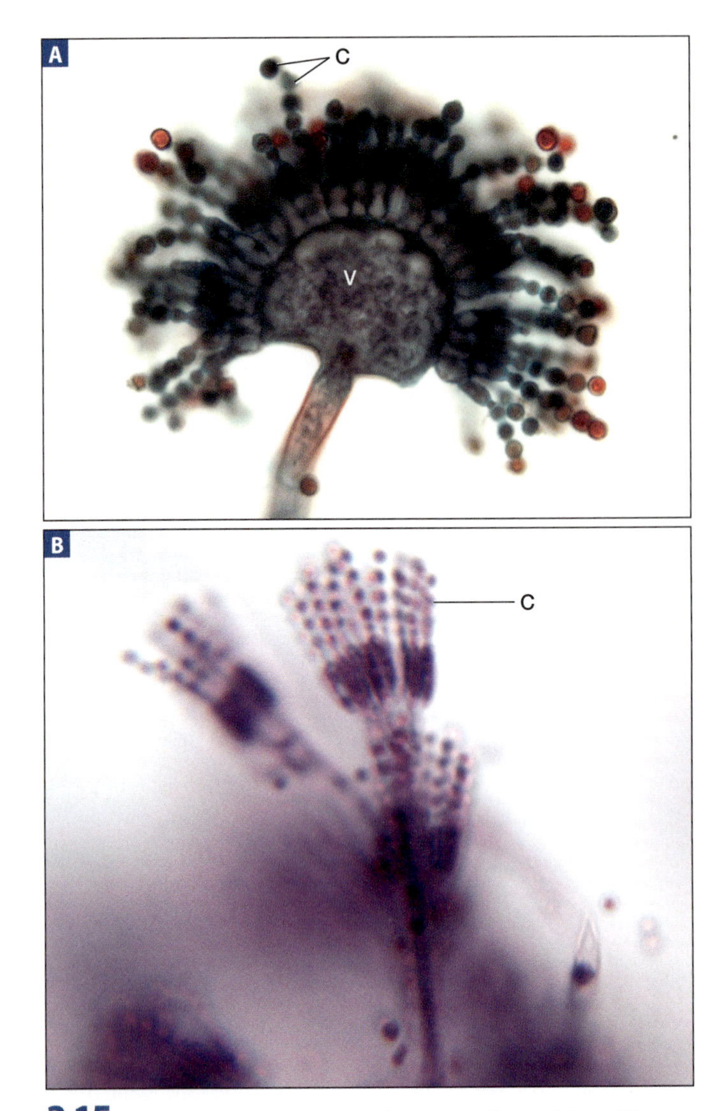


**3.15 Supergroup Unikonta—Molds *Aspergillus* and *Penicillium***
(**A**) Shown is a section of an *Aspergillus niger* conidiophore. Note the chains of spores (conidia–C) extending from the surface of the vesicle (V). (**B**) *Penicillium* species produce a characteristic brush-shaped fruiting body with chains of conidia (C). (Commercially prepared slides.)

are indistinguishable from one another and both reproduce asexually by budding. Diploid cells occasionally undergo meiosis and produce one to four haploid **ascospores**. When released, ascospores bud to produce more haploid cells. Occasionally, haploid cells of opposite mating types combine to create a new diploid cell, thus acting like gametes.

*Candida albicans* (Fig. 3.16C) is part of the normal respiratory, gastrointestinal, and female urogenital tract floras, but it is also the most common fungal opportunistic pathogen. It is the causative agent of thrush in the oral cavity, vulvovaginitis of the female genitals, and cutaneous candidiasis of the skin. Individuals most susceptible to *Candida* infections are diabetics, those with immunodeficiency (e.g., AIDS), catheterized patients, and individuals taking antimicrobial medications.

*Candida albicans* are round to oval cells that reproduce by budding and are 3–6 μm in diameter. Sometimes the cells separate after budding, but other times they remain attached, forming branched chains of two types: **pseudohyphae**, which are narrower at their junctions, and true hyphae which are not. Both types are asexual spores. The ability to switch growth pattern between pseudohyphae and true hyphae may increase virulence. The sexual life cycle is similar to *S. cerevisiae*.

- **Supergroup Excavata.** These are unicellular species that usually have a feeding groove *excavated* from one side of the cell and possess one or more flagella. Many species have a life cycle in which there is an active feeding stage called the **trophozoite** and an inert resting stage called a **cyst**. Three subgroups are recognized: **parabasalids**, **diplomonads**, and **euglenozoans**. Only the euglenozoans and parabasalids are considered here.
  - Euglenozoans are characterized by a crystalline structure and a rod within their flagella in addition to the usual 9 + 2 arrangement of microtubules. Most euglenozoans are unicellular, green, photosynthetic autotrophs when light is available, but can live heterotrophically when it is not (making them **mixotrophic**). They usually have two flagella, or more rarely only one. The red "eyespot" is a distinctive characteristic that is easily seen with the light microscope. A euglenid is shown in Figure 3.17A.

    Trypanosomes are heterotrophic members of the Euglenozoa. Two *Trypanosoma* species (*T. brucei gambiense* and *T. brucei rhodesiense*) are pathogens, causing West African trypanosomiasis and East African trypanosomiasis, respectively (Fig. 3.17B). The former is generally a mild, chronic disease that may last for years before the nervous system is affected, whereas the latter is more acute and results in death within a year.

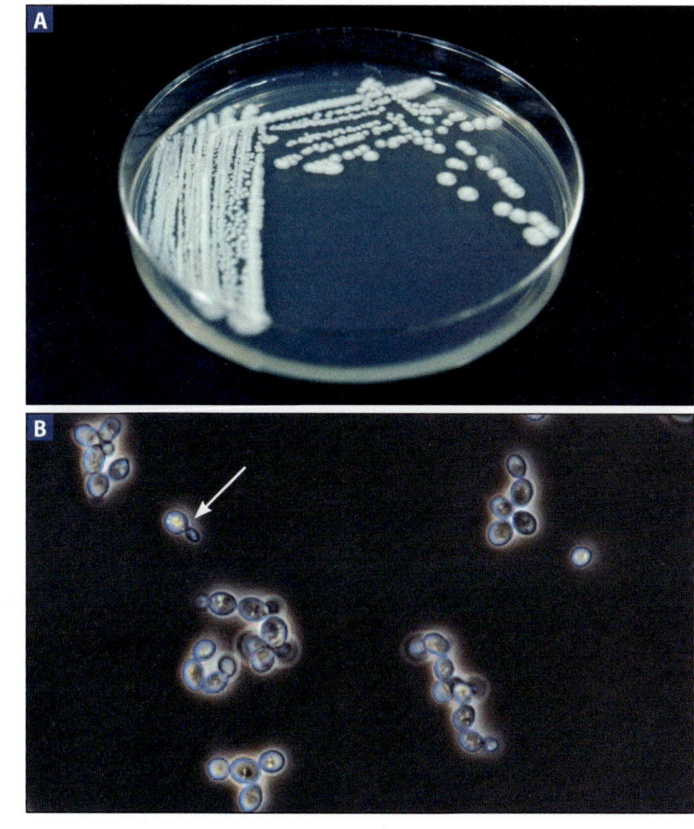


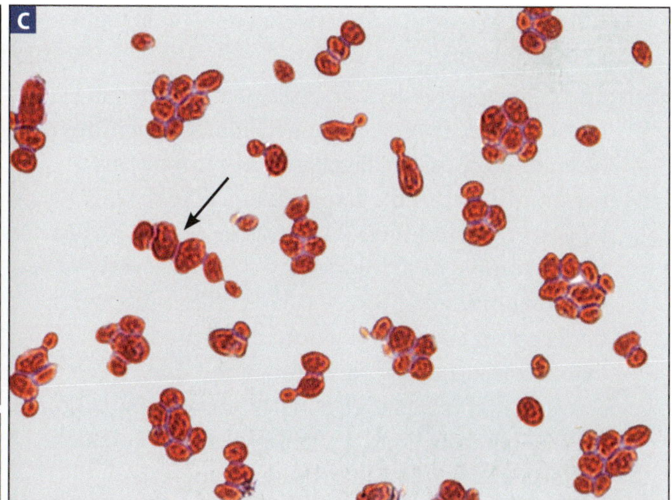


**3.16 Supergroup Unikonta—Yeasts *Saccharomyces* and *Candida*** (**A**) The brewer's yeast *Saccharomyces cerevisiae* colonies resemble bacterial colonies. Mold colonies present a more "fuzzy" appearance. (**B**) *S. cerevisiae* is shown in a wet mount preparation using phase contrast microscopy. The cells are oval with dimensions of 3–8 μm by 5–10 μm. One budding cell is indicated by the arrow. (**C**) *Candida albicans* cells grown on a solid medium don't show natural arrangements seen in tissue samples. However, in this micrograph the oval cell shape and a pseudohypha (arrow) with a budding cell at its end are visible.

Trypanosomes have a complex life cycle involving the tsetse fly (genus *Glossina*) as an intermediate host, which transmits the pathogen to the human host during a blood meal. The symptoms of sleeping sickness—sleepiness, emaciation, and unconsciousness—begin when the central nervous system becomes infected. Depending on the infecting strain, the disease may last for months or years, but the mortality rate is high.

- Parabasalids possess a group of anterior flagella and an undulating membrane, an extension of the plasma membrane associated with the posterior flagellum. They also have degenerated mitochondria (hydrogenosomes) that produce ATP with $H_2$ gas as the end product. *Trichomonas vaginalis* (Fig. 3.18) is the causative agent of trichomoniasis in humans. It causes inflammation of genitourinary mucosal surfaces—typically the vagina, vulva, and cervix in females and the urethra, prostate, and seminal vesicles in males—and infection in females is more common. Most infections are asymptomatic or mild.

- **Supergroup Archaeplastida.** There is strong evidence to support the conclusion that members of this group are all descended from an ancestor that engulfed a cyanobacterium (by **primary endosymbiosis**), which ultimately evolved into the chloroplasts of green algae, red algae, and true plants. Thus, all members are autotrophic. Most also possess walls made of cellulose. Two groups are covered here: **chlorophytes** and **charophytes.**
  - The chlorophytes are commonly called green algae. Most are freshwater and unicellular, filamentous, or colonial species. Flagella are commonly used for motility, as in the haploid, unicellular *Chlamydomonas* (Fig. 3.19A). *Volvox* (Fig. 3.19B) is an example of a colonial species made up of *Chlamydomonas*-like cells. *Volvox* colonies can undergo sexual reproduction; there are male colonies that produce sperm bundles and female colonies that produce eggs. They also can undergo asexual reproduction by forming daughter colonies within, which are ultimately released as new individuals.
  - Closely related to the chlorophytes, but even more closely related to true plants (because of certain aspects of cell ultrastructure and physiology), are the charophytes, which include *Spirogyra* (Fig. 3.20).

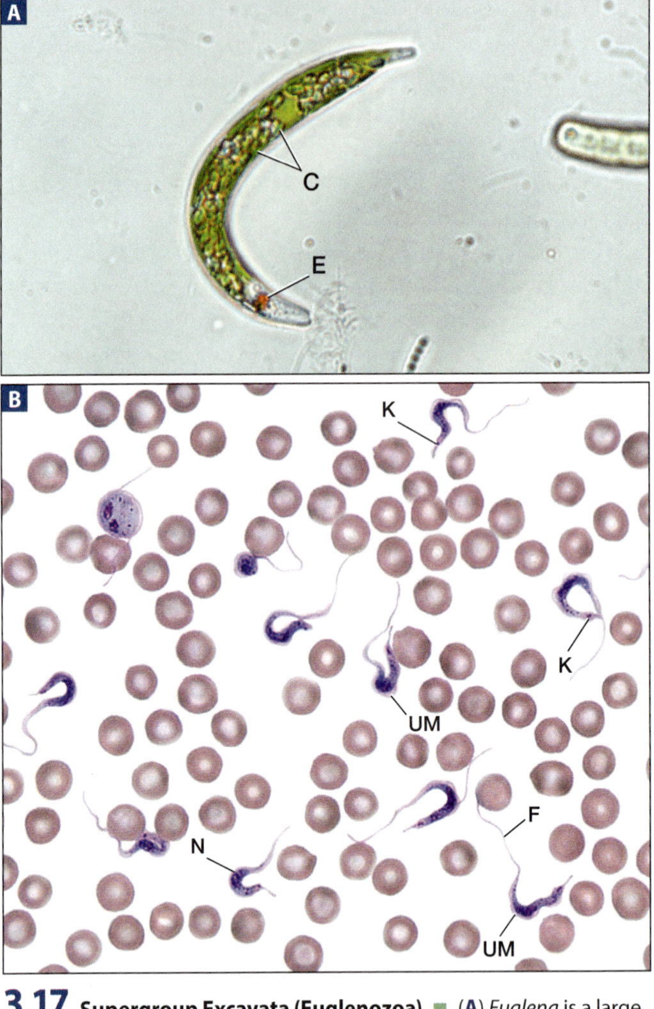


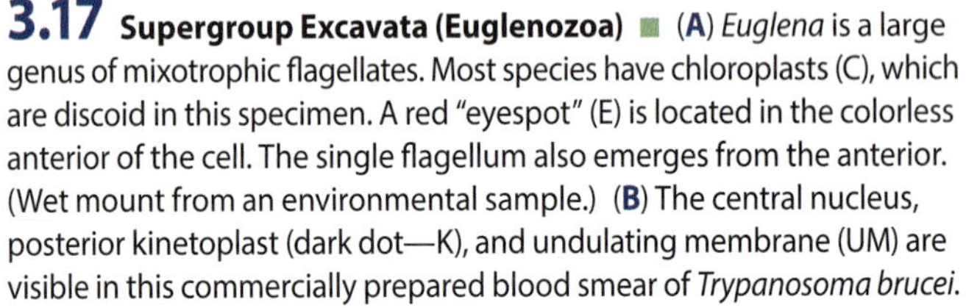

**3.17 Supergroup Excavata (Euglenozoa)** ■ (**A**) *Euglena* is a large genus of mixotrophic flagellates. Most species have chloroplasts (C), which are discoid in this specimen. A red "eyespot" (E) is located in the colorless anterior of the cell. The single flagellum also emerges from the anterior. (Wet mount from an environmental sample.) (**B**) The central nucleus, posterior kinetoplast (dark dot—K), and undulating membrane (UM) are visible in this commercially prepared blood smear of *Trypanosoma brucei*.

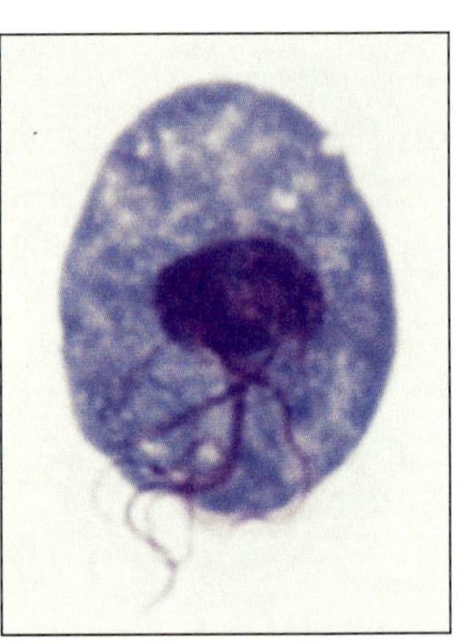

**3.18 Supergroup Excavata (Parabasalid—*Trichomonas*)** ■ The *Trichomonas vaginalis* trophozoite is the only stage of the life cycle and is 7–23 µm long by 5–15 µm wide. The four anterior flagella are visible. (Commercially prepared slide.)

- **SAR Supergroup.** This is a very diverse group recently joined primarily because of strong similarities in DNA sequences of the entire genome. Subsumed into the SAR supergroup are stramenopiles, alveolates, and rhizarians (from which the acronym SAR is derived). Only alveolates and stramenopiles are considered here.

  The main unifying feature of stramenopiles and alveolates (though not demonstrated in all groups) is evidence of being descendants of an ancestor that engulfed a red algal cell that evolved into a plastid with red algal characteristics. (Because the red alga already had a chloroplast from primary endosymbiosis, the plastids derived from it are said to have been formed by **secondary endosymbiosis**.) In some modern species, the plastid is still present; in others, only red algal plastid DNA remains in the genome. More work needs to be done on this group before we have a clear picture.

  - Alveolates may be either heterotrophic or autotrophic, but they all have small, membranous sacs beneath their cytoplasmic membrane. Alveolates are divided into three subgroups, of which only the ciliates and apicomplexans will be covered.

    Ciliates are characterized by cilia covering their outer surface that provide motility when they sweep back-and-forth. Cilia also line the **oral groove** and sweep food particles inward toward the **cytostome** where it is engulfed. They are represented here by three species—*Paramecium* (Fig. 3.21A), *Stentor* (Fig. 3.21B), and *Balantidium coli* (Fig. 3.21C).

    *Paramecium* and *Stentor* are nonpathogenic ciliates that feed on organic matter in their environments. *Paramecium* species are found in marine and freshwater environments and are active swimmers as they search for food. *Stentor* species are found in freshwater and, while they can swim, spend most of their time anchored to a surface as they feed.



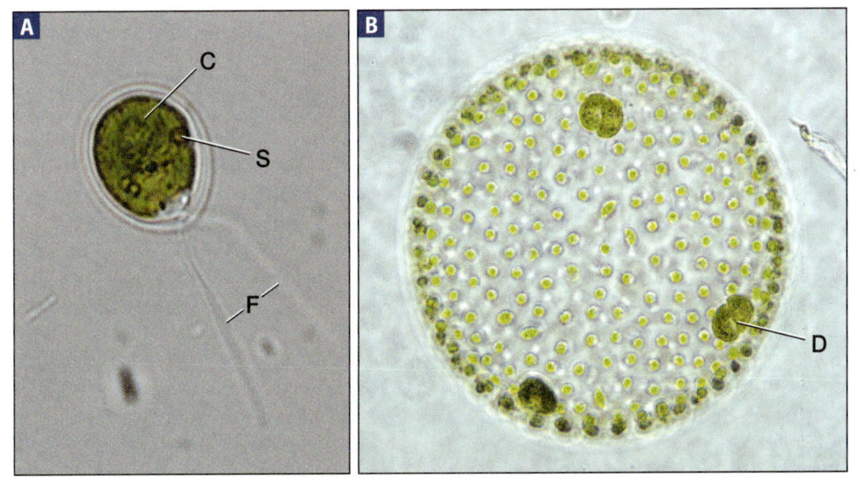


**3.19 Supergroup Archaeplastida (Chlorophyte Subgroup)** ■ (**A**) *Chlamydomonas* is a unicellular green alga with a single, cup-shaped chloroplast (C). A red-orange stigma (S) is found within the chloroplast and is involved in phototaxis (attraction to light). It also has two flagella (F) for locomotion. Cells are approximately 30 µm long and 20 µm wide. (Wet mount of an environmental sample.) (**B**) *Volvox* is a spherical colonial green alga made of cells similar in shape to *Chlamydomonas*. Daughter colonies (D) form asexually from special cells in the parent colony. Release of a daughter colony involves its eversion as it exits through an enzymatically produced pore. Mature colonies can reach a size visible to the naked eye. (Wet mount from culture.)

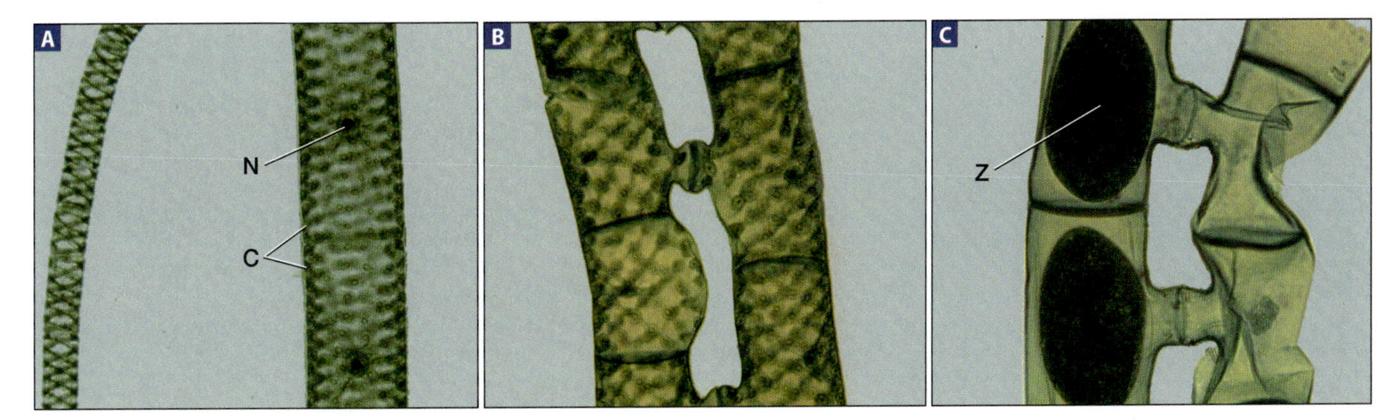


**3.20 Supergroup Archaeplastida (Charophyte Subgroup)** ■ *Spirogyra* is a filamentous alga with distinctive spiral chloroplasts (C). The cells are haploid, that is, they have only one complete set of chromosomes. (**A**) Two vegetative filaments are shown here. Also note the nuclei (N) in the cells on the right. (**B**) These two filaments have begun **scalariform conjugation**, a sexual process that occurs between different filaments. The protrusions are called **conjugation tubes** and serve as a connection through which the cytoplasm of one cell (acting as a gamete) can move and join with the cytoplasm of a cell in the other filament. (**C**) The cells and their two haploid nuclei fuse to form a diploid **zygospore** (Z) and conjugation is complete. When conditions are favorable, the zygospore germinates by undergoing meiosis. Only one haploid nucleus survives and the zygospore cell develops into a new vegetative filament. Note the absence of spiral chloroplasts in the empty cells and in the zygospores. (Commercially prepared slides.)

*Balantidium coli* is an alveolate pathogen and causes balantidiasis, a disease found worldwide. *B. coli* exists in two forms: a vegetative trophozoite (Fig. 3.21C) and a cyst. Laboratory diagnosis is made by identifying either the cyst or the trophozoite in feces, with the latter being more commonly found. A major risk factor is contact with pigs.

The ciliate trophozoite is highly motile and has a macronucleus and a micronucleus. Cysts in sewage-contaminated water are the infective form and after ingestion become trophozoites in the intestines. Trophozoites reproduce and live in the colon, generally feeding on bacteria but they may cause mucosal ulcerations. Symptoms of acute infection include bloody and mucoid feces, nausea, vomiting, and abdominal tenderness, among others. Diarrhea alternating with constipation may occur in chronic infections. Most infections probably are asymptomatic.

Apicomplexans have a complex life cycle, often involving more than one host and multiple tissues. For instance, species of *Plasmodium* (Fig. 3.21D) cause malaria and infect both liver and red blood cells of humans, and the gut and salivary glands of the insect vector, the female *Anopheles* mosquito.

- The stramenopiles are characterized by a unique flagellum with three-parted lateral hairs emerging from it as well as some other physiological features. Even though diatoms (Fig. 3.22) lack flagella, they are so successful and are so frequently seen in aquatic samples that we have chosen them to represent the group.

Diatoms are single-celled or colonial autotrophs. They have a distinctive golden-brown color due to the abundance of the pigment **fucoxanthin**. Their highly ornamented cell wall is made of silica embedded in an organic matrix and consists of two halves with one half overlapping the other in the same way the lid of a Petri dish overlaps its base. **Centric diatoms** have a round shape, whereas **pennate diatoms** are bilaterally symmetrical. Oil droplets are frequently visible.

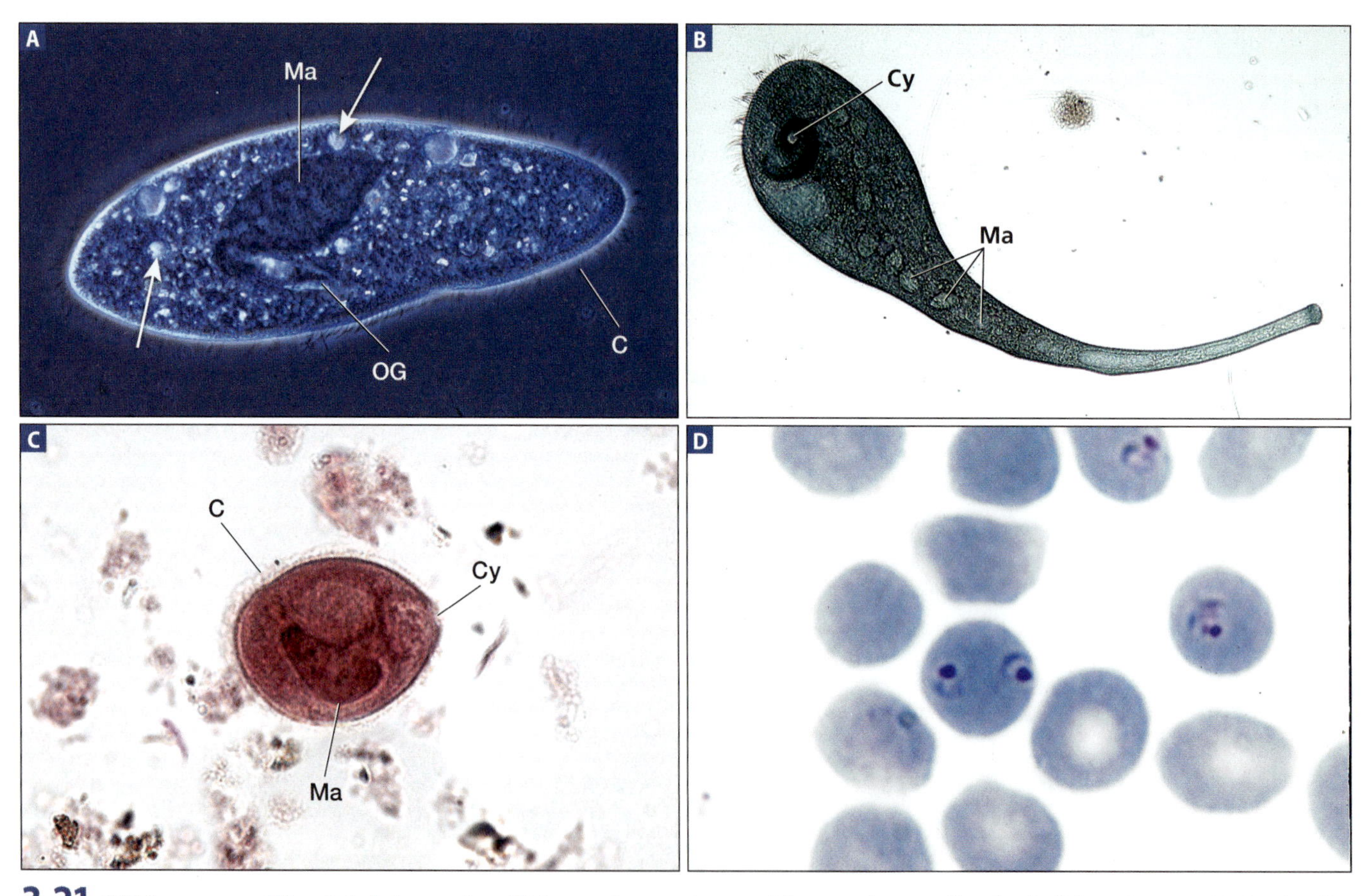


**3.21 SAR Supergroup (Alveolate Subgroup)** ■ (**A**) *Paramecium* may be one of the most famous ciliated unicellular eukaryotes. Notice the cilia (C), the oral groove (OG), and macronucleus (Ma). The white circles (arrows) are food vacuoles. (Phase contrast wet mount.) (**B**) Its trumpet shape, size (up to 2 mm), and beaded macronucleus (Ma) make *Stentor* an easy ciliate to identify. The white spot inside the dark coiled region is the cytostome (Cy). This species is naturally green; this is not a stained specimen. (**C**) *Balantidium coli* trophozoites are oval in shape with dimensions of 50 to 100 μm long by 40 to 70 μm wide. Cilia (C) cover the cell surface. Internally, the macronucleus (Ma) is prominent; the adjacent micronucleus is not. An anterior cytostome (Cy) is usually visible. (**D**) Double infections of red blood cells are commonly seen in *Plasmodium falciparum* infections (center). A single infection is seen at the right. Young trophozoites are said to be in the "ring stage." (B through D commercially prepared slides.)

In this lab you will observe representative simple eukaryotes from the four supergroups. Some specimens are on commercially prepared slides, whereas others need to be prepared as wet mounts (Fig. 3.23) of living cultures. In a wet mount, a drop of water is placed on the slide and the organisms are introduced into it. Or, if the organism is already in a liquid medium, then a drop of medium is placed on the slide. A cover glass is placed over the preparation to flatten the drop and keep the objective lens from getting wet. A stain may or may not be applied to add contrast.

## Application

Familiarity with eukaryotic cells not only rounds out your microbiological experience, it is important to be able to differentiate eukaryotic from bacterial cells when examining environmental or clinical specimens.

## In This Exercise

Most of this manual is devoted to bacteria, but in this exercise you will be given the opportunity to examine various eukaryotic microorganisms. This will not only

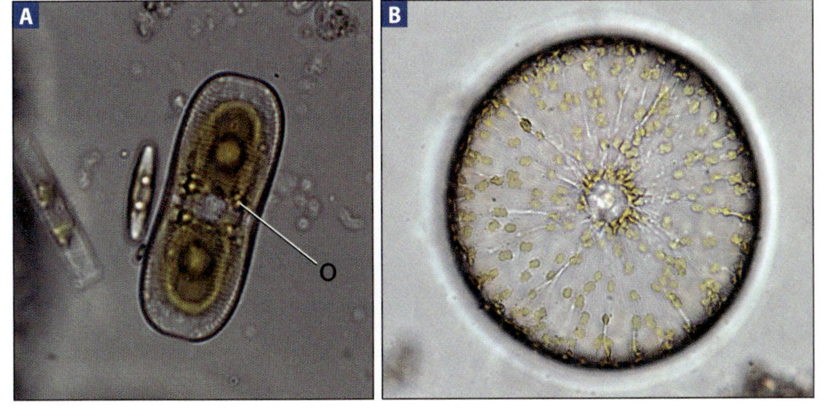


**3.22 SAR Supergroup (Stramenopile Subgroup)** While diatoms lack the flagella characteristic of most stramenopiles, they are so ubiquitous we have chosen them to represent the group. Their cell wall is divided into two halves, one of which fits into the other like the lid and base of a Petri dish, and are composed of silica and organic material. Diatoms are divided into two groups based on their symmetry. Those with bilateral symmetry (**A**) are said to be pennate and those with radial symmetry (**B**) are said to be centric. Note the distinctive golden chloroplasts, the cell wall ornamentation, and the oil droplets (O). (Wet mount preparations of environmental samples.)

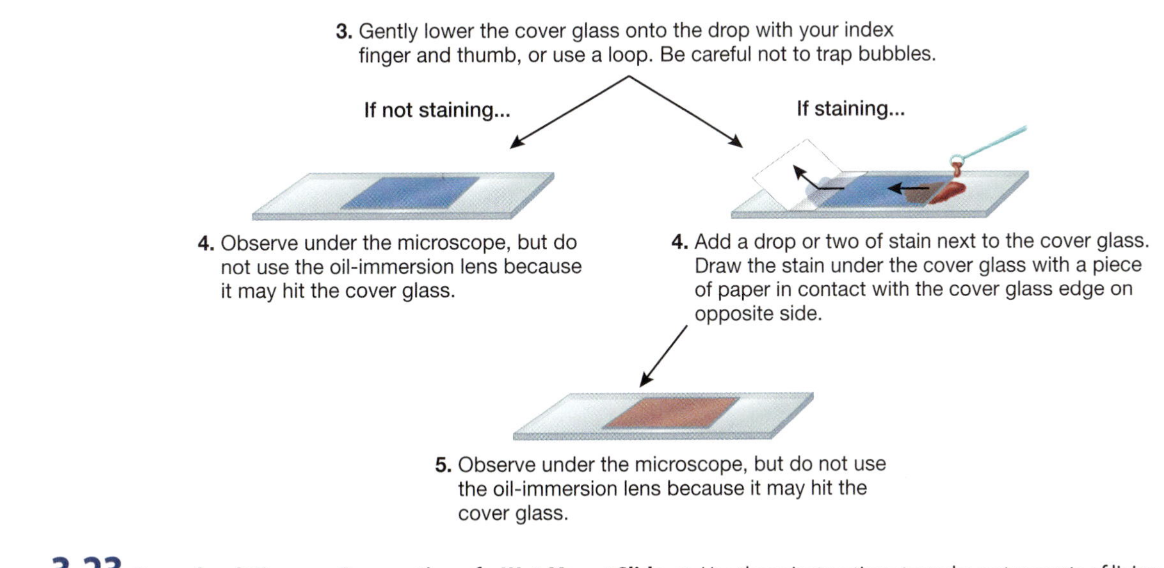


**3.23 Procedural Diagram: Preparation of a Wet-Mount Slide** Use these instructions to make wet mounts of living specimens. The preparation may be stained or not, but staining will eventually kill the organisms. Of course, so will drying. You can expect a wet mount to last only 15–20 minutes.

serve to familiarize you with simple eukaryotes, but also give you practice using the microscope, measuring specimens, and making wet-mount preparations.

## ▼ Materials

### Per Class

- ☐ Living cultures (as available) of a variety of simple eukaryotes and fungi:
  - *Amoeba*
  - *Saccharomyces cerevisiae*
  - *Euglena*
  - *Chlamydomonas*
  - *Volvox*
  - *Spirogyra*
  - *Paramecium*
  - *Stentor*
- ☐ Prepared slides (as available) of:
  - *Entamoeba histolytica* trophozoites
  - *Rhizopus* sporangia and gametangia
  - *Aspergillus* conidiophore
  - *Penicillium* conidiophore
  - *Candida albicans*
  - *Trypanosoma* sp. trypomastigotes in blood
  - *Trichomonas vaginalis*
  - *Balantidium coli* trophozoites
  - *Plasmodium* sp. (thick blood smear)
  - Diatoms (centric and pennate)

### Per Student

- ☐ Clean glass slides and cover glasses
- ☐ Compound microscope with ocular micrometer
- ☐ Cytological stains (e.g., methylene blue, $I_2KI$)
- ☐ Methyl cellulose
- ☐ Immersion oil
- ☐ Cotton swabs
- ☐ Lens paper
- ☐ Lens cleaning solution

### General Instructions for Prepared Slides

1. Obtain a microscope and place it on the table or workspace. Check to be sure the stage is all the way down and the scanning objective is in place.
2. Begin with a prepared slide. Clean it with a tissue if it is dirty, and then place it on the microscope stage. Center the specimen under the scanning objective.
3. Follow the instructions given in Exercise 3-1 to bring the specimen into focus at the highest magnification that allows you to see the entire structure you want to view.
4. Practice scanning with the mechanical stage until you're satisfied that you have seen everything interesting to see. Sketch what you see in the table provided on the data sheet, page 171.
5. Measure cellular dimensions and record these in the table provided on the data sheet.
6. Repeat with as many slides as you have time for. Use the appropriate figures in this exercise to help you identify the organisms and their structures. Following is a list of what to look for. ***Note:*** You may not be able to find all items in every specimen. You may have to look at more than one slide. Your instructor will advise you as to what his/her expectations of you are.

### Prepared Slides

#### Unikonta

- ☐ *Entamoeba histolytica* trophozoites: identify pseudopods; nucleus with uniform, beaded chromatin at periphery and small, central karysome; ingested red blood cells
- ☐ *Rhizopus* sporangia and gametangia
  - Sporangia: identify sproangiophores, sporangia, and spores
  - Gametangia: identify progametangia, gametangia, young zygosporangia, and mature zygosporangia
- ☐ *Aspergillus* conidiophore: identify hyphae, conidiophores, and chains of conidia
- ☐ *Penicillium* conidiophore: identify hyphae, conidiophores, and chains of conidia
- ☐ *Candida albicans*: identify vegetative cells and budding cells

#### Excavata

- ☐ *Trypanosoma* in blood: identify trypomastigotes with flagella, undulating membrane, kinteoplast, and nucleus; red blood cells
- ☐ *Trichomonas vaginalis*: identify nucleus and flagella

#### SAR—Alveolata

- ☐ *Balantidium coli* trophozoites: identify cilia, macronucleus, micronucleus (maybe), and cytostome
- ☐ *Plasmodium falciparum* (or other species) thick smear: identify red blood cells, ring stage of trophozoites

#### SAR—Stramenopiles

- ☐ Diatoms: identify centric and pennate shapes, chloroplasts, oil droplets, and cell walls with sculpturing

### General Instructions for Wet-Mount Slides

1. Prepare wet mounts of available specimens by following the procedural diagram in Figure 3.23. A drop of methyl cellulose may be added to the wet mount if you have fast swimmers.
2. Sketch what you see and record cellular dimensions in the table provided on the data sheet (p. 171).
3. Repeat with as many specimens as are available and you have time for. ***Note***: Your instructor may provide you with prepared slides rather than living specimens. These features should still be recognizable.

#### Unikonta

- □ *Amoeba*: identify nucleus, pseudopods, and ingested material; do not stain
- □ *Saccharomyces cerevisiae*: identify vegetative and budding cells, and nuclei; stain with methylene blue or $I_2KI$

#### Excavata

- □ *Euglena*: identify chloroplast, flagellum, eyespot; do not stain

#### Archaeplastida

- □ *Chlamydomonas*: identify chloroplast, flagella, and stigma; do not stain
- □ *Volvox*: identify colony and daughter colonies; do not stain
- □ *Spirogyra*: identify nuclei, spiral chloroplasts, and cell wall; do not stain

#### SAR—Alveolata

- □ *Paramecium*: identify macronucleus, cilia, oral groove, and contractile vacuole; do not stain
- □ *Stentor*: identify cilia, beaded macronucleus, and cytostome

### General Instructions for Clean Up

1. When you are finished observing the specimens, blot any oil from the oil-immersion lens (if used) with a lens tissue. Also, check the high-dry lens for oil and clean it if necessary.
2. Return all lenses and adjustments to their storage positions before putting the microscope away.
3. Return prepared slides to their proper storage box. Dispose of wet-mount slides as directed by your instructor.

## References

Brown, James W. Pages 22–23 in *Principles of Microbial Diversity*. Washington, DC: ASM Press, 2015.

Reece, Jane B., Lisa A. Urry, Michael L. Cain, Steven A. Wasserman, Peter V. Minorsky, and Robert B. Jackson. Chap. 28 in *Campbell Biology* 10th ed. Boston: Pearson, 2014.

Madigan, Michael T., John M. Martinko, Kelly S. Bender, Daniel H. Buckley, and David A. Stahl. Chap. 12 in *Brock's Biology of Microorganisms,* 14th ed. San Francisco: Pearson/Benjamin Cummings, 2015.

Name ______________________________

Date ______________________________

Lab Section ______________________________

I was present and performed this exercise (initials) ______________________________

# DATA SHEET 3-3

## Microscopic Examination of Eukaryotic Microbes

### OBSERVATIONS AND INTERPRETATIONS

**1** Fill in the table for each eukaryotic microbe you observe.

| **Organism** (include wet mount or prepared slide) | **Sketch** (include magnification and stain, if any) | **Dimensions** | **Identifying Characteristics** (list only those you observed) |
|---|---|---|---|
| | | | |
| | | | |
| | | | |
| | | | |
| | | | |
| | | | |
| | | | |
| | | | |
| | | | |

| Organism (include wet mount or prepared slide) | Sketch (include magnification and stain, if any) | Dimensions | Identifying Characteristics (list only those you observed) |
|---|---|---|---|
| | | | |
| | | | |
| | | | |
| | | | |
| | | | |

3

## QUESTIONS

**1** *What features did the cells you observed have in common? How were they different?*

______________________________

______________________________

______________________________

**2** *If you observed the same organism on a prepared slide and a wet mount, how did the images compare?*

______________________________

______________________________

______________________________

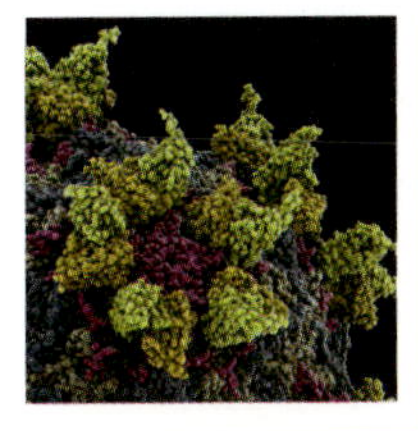

## Bacterial Structure and Simple Stains

### Importance of Contrast

In Exercise 3-1 you were introduced to two of the three important features of a microscope and microscopy: magnification and resolution. A third feature is **contrast**. To be visible, the specimen must contrast with the background of the microscope field. Because cytoplasm is essentially transparent, viewing cells with the standard light microscope is difficult without stains to provide that contrast. In the upcoming exercises, you will learn how to correctly prepare a bacterial smear for staining and how to perform simple and negative stains. Details of cell structure (cell morphology, size, and arrangement) then may be seen more easily. In a medical laboratory these usually are determined with a Gram stain (Exercise 3-6), but you will be using simple stains as an introduction to the staining process, as well as an introduction to these cellular characteristics.



### Bacterial Cell Morphologies

Bacterial cells are much smaller than eukaryotic cells (Fig. 3.11) and come in a variety of **morphologies** (shapes) and **arrangements**. Determining cell morphology is an important first step in identifying a bacterial species. Cells may be spheres (**cocci**, singular **coccus**), rods (**bacilli**, singular **bacillus**), or spirals (**spirilla**, singular **spirillum**). Variations of these shapes include slightly curved rods (**vibrios**), short rods (**coccobacilli**), and flexible spirals (**spirochetes**). Examples of cell morphologies are shown in Figures 3.24 through 3.38. In Figure 3.31, *Corynebacterium xerosis* illustrates pleomorphism, where a variety of cell shapes—slender, ellipsoidal, or ovoid rods—may be seen in a given sample.

### Bacterial Cell Arrangements

Cell arrangement, determined by the number of planes in which division occurs and whether the cells separate after division, is also useful in identifying bacteria (Table 3-3). Spirilla rarely are seen as anything other than single cells, but cocci and bacilli do form multicellular associations. Because cocci exhibit the most variety in arrangements, they are used for illustration in Figure 3.32. If the two daughter cells remain attached after a coccus divides, a **diplococcus** is formed. The same process happens in bacilli that produce **diplobacilli**. If the cells continue to divide in the same plane and remain attached, they form a chain and exhibit a **streptococcus** or **streptobacillus** arrangement.

TABLE **3-3** **Cell Arrangements Found Associated with Different Cell Morphologies**

| Cell Arrangement | Cell Morphologies with the Cell Arrangement |
|---|---|
| Single cells | Cocci, bacilli, spirilla, spirochaetes, and vibrios |
| Pairs of cells ("diplo–") | Cocci and bacilli |
| Chains of cells ("strepto–") | Cocci and bacilli |
| Tetrads | Cocci |
| Cube (sarcina) | Cocci |
| Irregular cluster ("staphylo–") | Cocci |
| Palisade and angular | Bacilli |

If a second division occurs in a plane perpendicular to the first, a **tetrad** is formed. A third division plane perpendicular to the other two produces a cube-shaped arrangement of eight cells called a **sarcina**. Tetrads and sarcinae are seen only in cocci. If the division planes of a coccus are irregular, a cluster of cells is produced to form a **staphylococcus**. Figures 3.33 through 3.38 illustrate common cell arrangements.

Arrangement and morphology often are easier to see when the organisms are grown in a broth rather than a solid medium, or are observed from a direct smear. If you have difficulty identifying cell morphology or arrangement, consider transferring the organism to a broth culture, incubating it (even for an hour or so), and trying again.

One last bit of advice: Do not expect nature to conform perfectly to our categories of morphology and cell arrangement. These are convenient descriptive categories that will not be applied easily in all cases. When examining a slide, look for the most common morphology and most complex arrangement. Do not be afraid to report what you see. For instance, it is okay to say, "Cocci in singles, pairs, and chains."

## Surface Area and Volume of Cells

Cells are three-dimensional objects with a surface that contacts the environment and a volume made up of cytoplasm. The ability to transport nutrients into the cell and wastes out of the cell is proportional to the amount of surface area doing the transport. The demand for nutrients and production of wastes is proportional to a cell's volume. It is a mathematical fact of life that as an object gets bigger, its volume increases more rapidly than its surface area. Therefore, a cell can achieve a size at which its surface area is not adequate to supply the nutrient needs of its cytoplasm. That is, its **surface-to-volume** ratio is too small. At this point, a cell usually divides its volume to increase its surface area. This phenomenon is a major factor in limiting cell size and determining a cell's habitat.

Bacilli, cocci, and spirilla with the same volume have different amounts of surface area. A sphere has a lower surface to volume ratio than a bacillus or spirillum of the same volume. A streptococcus, however, would have approximately the same surface to volume ratio as a bacillus of the same volume. ■



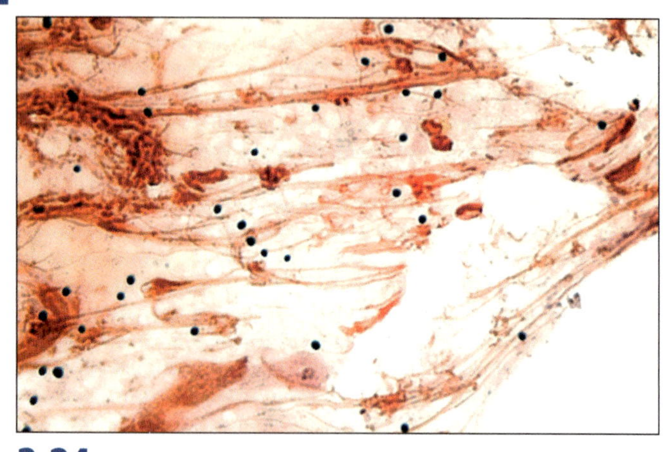

**3.24 Single Cocci from a Nasal Swab (Gram Stain)** ■ This direct smear of a nasal swab illustrates unidentified cocci (dark circles) stained with crystal violet. Each cell is about 1 µm in diameter. The red-orange background material is mostly mucus.

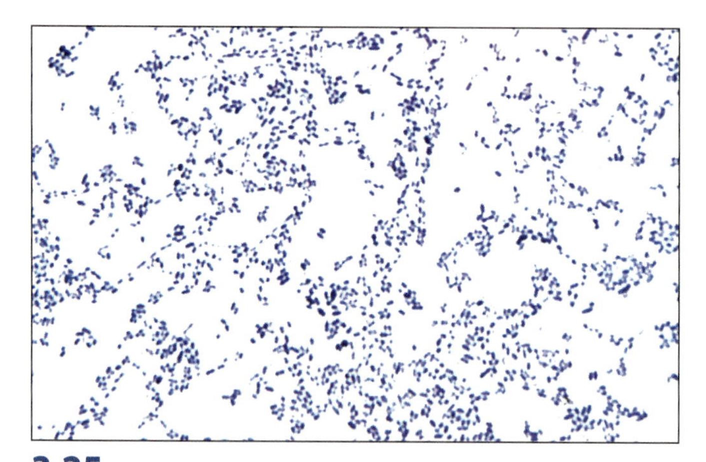

**3.25 Ovoid Cocci (Gram Stain)** ■ *Lactococcus lactis* is an elongated coccus that a beginning microbiologist might confuse with a rod. Notice the slight elongation of the cells, and also that most cells are not more than twice as long as they are wide. *L. lactis* is found naturally in raw milk and milk products, but these cells were grown in culture. Cell dimensions are about 1–2 µm long by 1 µm wide.

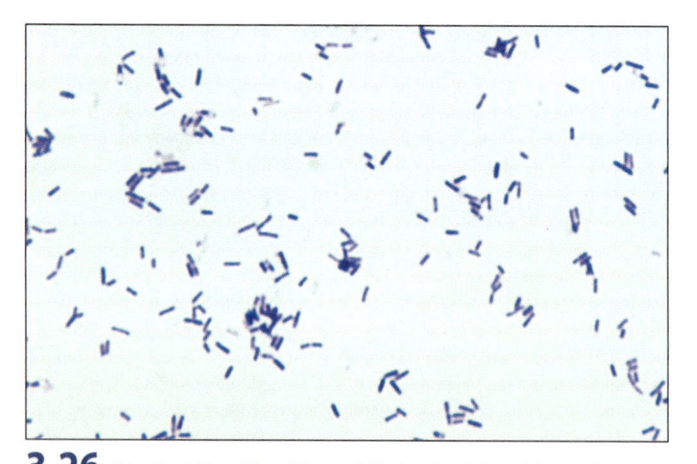

**3.26 "Typical" Bacillus (Crystal Violet Stain)** ■ Notice the variability in rod length (because of different ages of the cells) in this stain of the soil organism *Bacillus subtilis* grown in culture. Cell dimensions are 2–3 µm long by 0.7 µm wide.

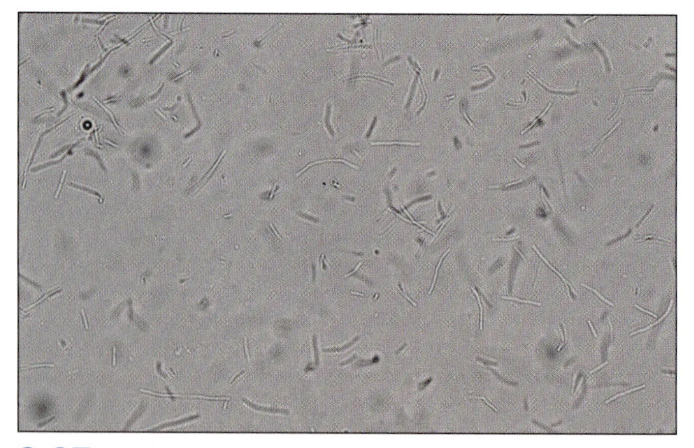

**3.27 Long, Thin Bacillus (Phase Contrast Wet Mount)** ■ This *Bacillus subtilis* culture was grown in trypticase soy broth for 24 hours and is shown in a wet mount using high-dry magnification. Notice that the cells are longer and thinner than in Figure 3.26 (which was viewed with the oil-immersion lens). Cell morphologies can vary depending on the medium used (solid vs. liquid) and age of the culture. The longer "individuals" are actually short chains of two or three cells, as evidenced by transverse constrictions or bends where the cells are attached. The water of the wet mount has considerable thickness relative to the cell sizes, so many of the cells are above or below the focal plane and are out of focus.

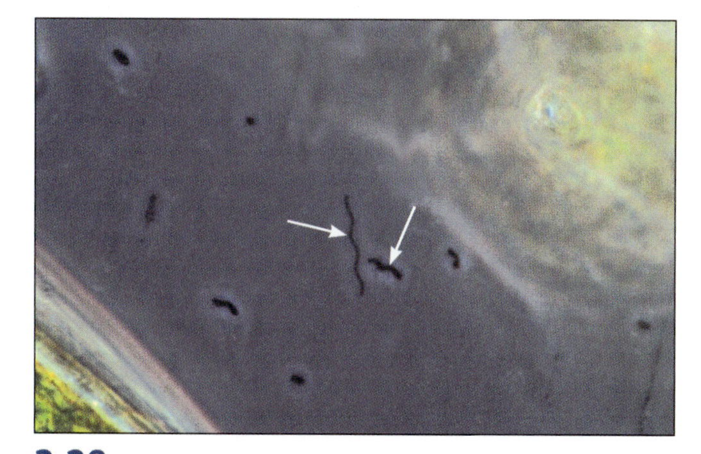

**3.28 Two Different Spirilla (Phase Contrast Wet Mount)** ■ The two different spirilla (arrows) are undoubtedly different species based on their different morphologies: one is long and slender with loose spirals; the other is shorter and fatter with tighter coils. Cell dimensions are less than 1 µm wide and up to 35 µm long.

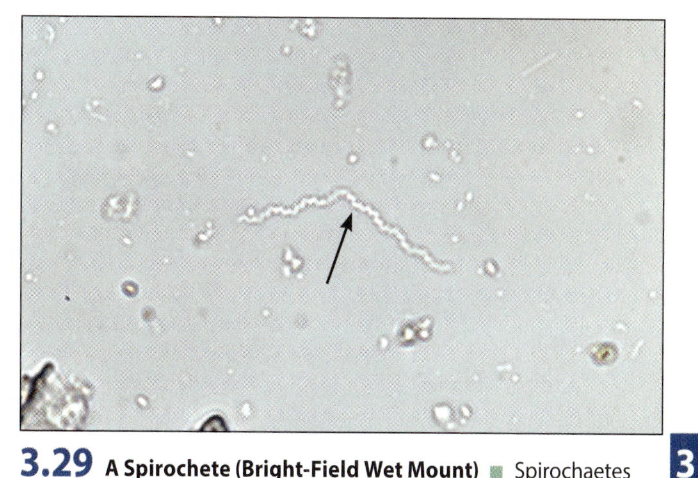

**3.29 A Spirochete (Bright-Field Wet Mount)** ■ Spirochaetes are tightly coiled, flexible rods. This bright-field micrograph shows a marine species of *Spirochaeta* (arrow). Notice the bend in the center of the cell. Cell dimensions of the genus are 5–500 µm long (!) by less than 1 µm wide.



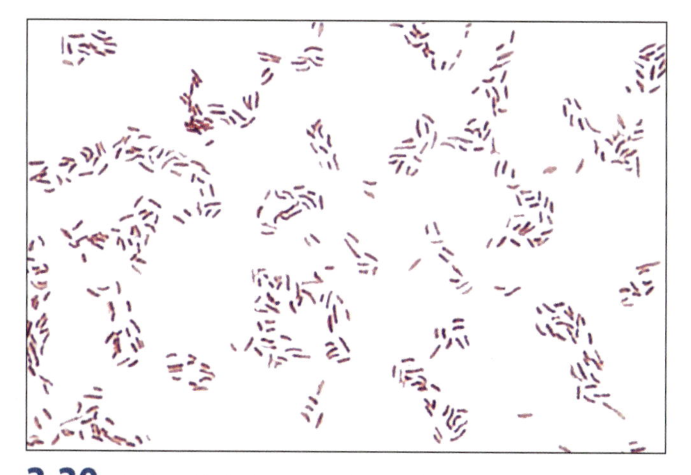

**3.30 Marine Vibrio (Gram Stain)** ■ Slightly curved rods are demonstrated by this unidentified species isolated from a Southern California estuary. Notice that not all rods are curved. Cells are approximately 2–3 µm long by 0.5 µm wide.

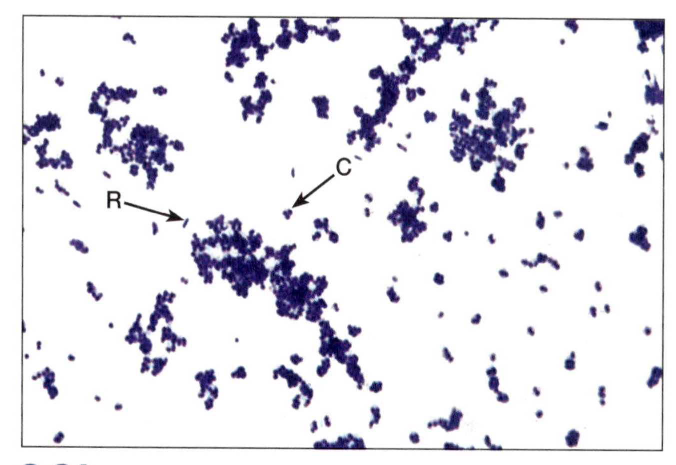


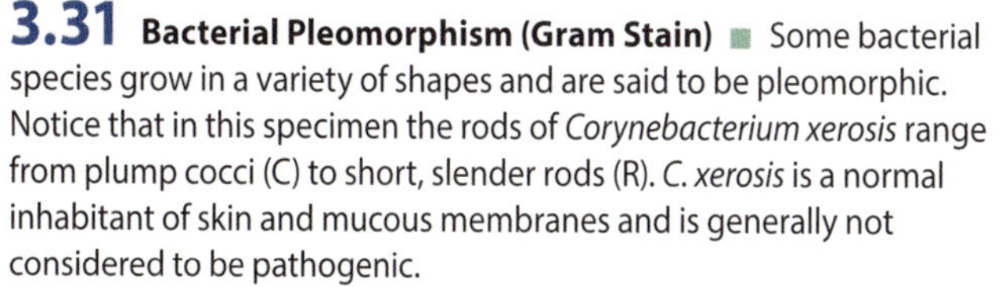

**3.31 Bacterial Pleomorphism (Gram Stain)** ■ Some bacterial species grow in a variety of shapes and are said to be pleomorphic. Notice that in this specimen the rods of *Corynebacterium xerosis* range from plump cocci (C) to short, slender rods (R). *C. xerosis* is a normal inhabitant of skin and mucous membranes and is generally not considered to be pathogenic.

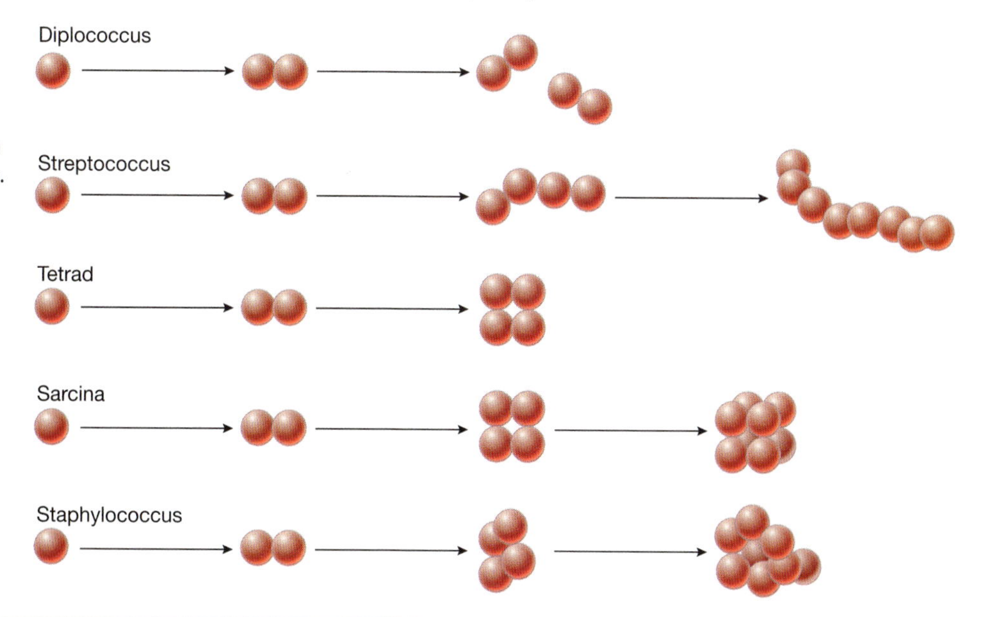


**3.32 Division Patterns Among Cocci** ■ Diplococci have a single division plane and the cells generally occur in pairs. Streptococci also have a single division plane, but the cells remain attached to form chains of variable length. If there are two perpendicular division planes, the cells form tetrads. Sarcinae have divided in three perpendicular planes to produce a regular cuboidal arrangement of cells. Staphylococci have divided in more than three planes to produce a characteristic grapelike cluster of cells. (***Note:*** Rarely will a sample be composed of just one arrangement. Report what you see, and emphasize the most complex arrangement.)

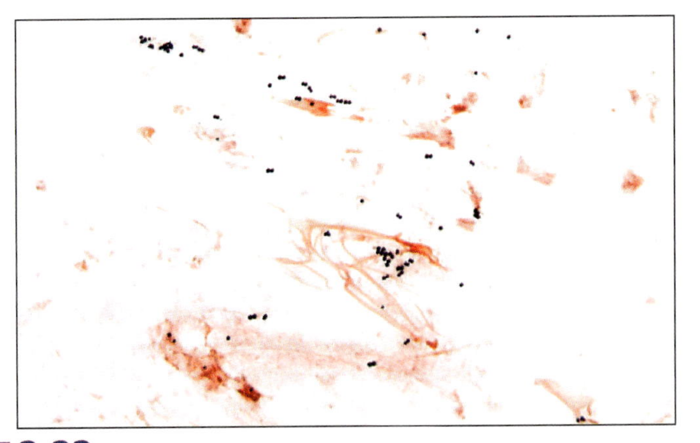

**3.33 Diplococcus Arrangement (Gram Stain)** ■ Most cells are in pairs, but a few are singles because they have yet to divide. Expect to see a mixture of cell arrangements representing stages before the "final" one.

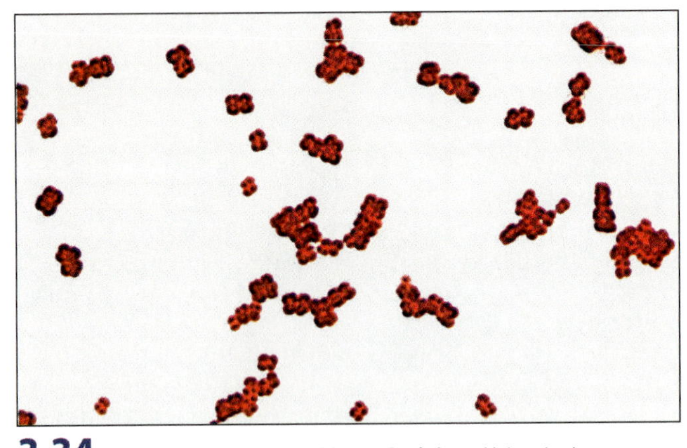

**3.34 Tetrad Arrangement (Gram Stain)** ■ *Neisseria sicca* frequently grows in tetrads, as shown in this Gram stain of a trypticase soy agar culture. Cell arrangement, especially with cocci, is frequently more easily determined if the specimen comes from a broth culture because groups of cells separate from one another in the liquid, but tetrads are pretty easy to spot even when taken from a solid medium. Look at the bigger clumps and you will see tetrads within them. *N. sicca* is a normal inhabitant of the human nasopharynx, saliva, and sputum, and is an opportunistic pathogen. Cells are between 1 and 2 µm in diameter.

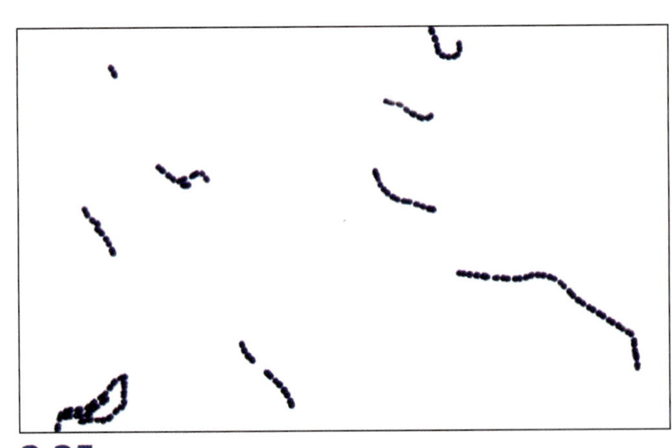

**3.35 Streptococcus Arrangement (Gram Stain)** ■ As its name suggests, *Streptococcus salivarius* forms chains of spherical cells and it inhabits the oral cavity, but primarily the saliva and tongue. This specimen is from a broth culture (which enables the cells to form long chains). Individual cells are approximately 1 µm in diameter.

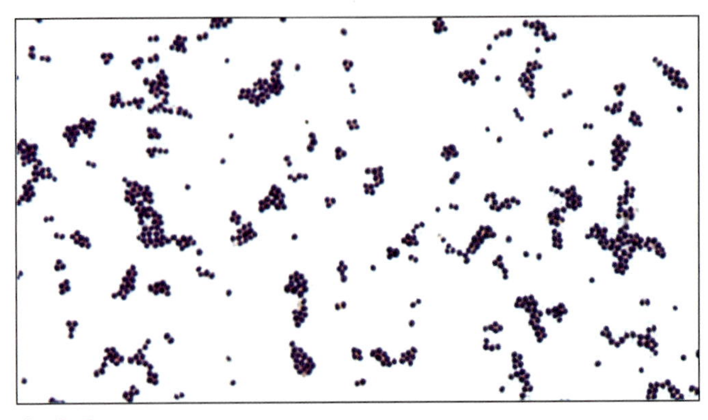

**3.36 Staphylococcus Arrangement (Gram Stain)** ■ *Staphylococcus aureus* cells grown in broth culture are shown. Note that the staphylococcal groups are rarely bigger than a dozen cells or so. Larger clusters are probably two or more groups that happen to be next to each other on the slide. This "bunching" can be especially problematic when making a preparation of cells grown on a solid medium. Be sure to emulsify them thoroughly to spread clusters out as much as possible. *S. aureus* is a common opportunistic pathogen of humans. Cells are approximately 1 µm in diameter.

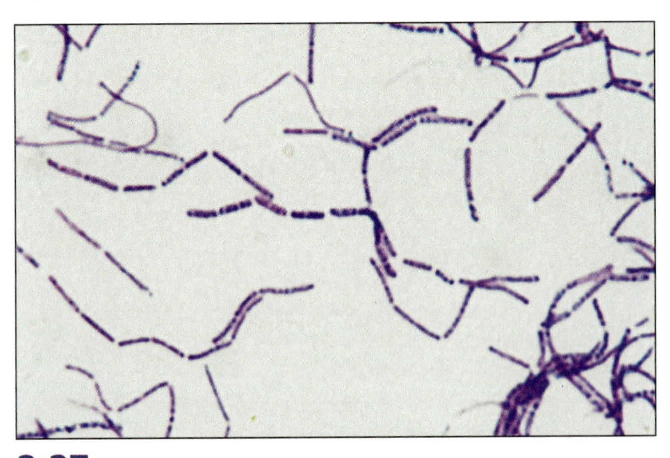

**3.37 Streptobacillus Arrangement (Crystal Violet Stain)** ■ *Bacillus megaterium* is a streptobacillus. These cells were obtained from culture and are 2–5 µm long by 1.2–1.5 µm wide.

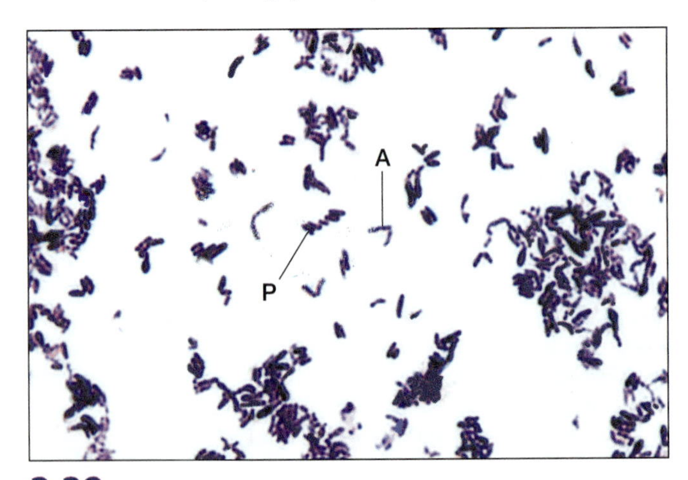


**3.38 Palisade and Angular Arrangements of *Arthrobacter* (Gram Stain)** ■ Stacking of rod-shaped cells side-by-side is the palisade arrangement (P). An angular arrangement (A) is when a pair of rods is bent where the cells join. Both arrangements are due to "snapping division" in which the cells divide lengthwise and remain more or less attached. *Corynebacterium* also shows these arrangements.

# Simple Stains

EXERCISE **3-4**

## Theory

Stains are solutions consisting of a solvent (usually water or ethanol) and a colored molecule (often a benzene derivative), the **chromogen.** The portion of the chromogen that gives it its color is the **chromophore.** A chromogen may have multiple chromophores, with each chromophore adding intensity to the color. The **auxochrome** is the charged portion of a chromogen and allows it to act as a dye through ionic or covalent bonds between the chromogen and the cell. **Basic stains**[1] (where the auxochrome becomes positively charged as a result of picking up a hydrogen ion or losing a hydroxide ion) are attracted to the negative charges on the surface of most bacterial cells. Thus, the cell becomes colored (Fig. 3.39). Common basic stains include methylene blue, crystal violet, and safranin. Examples of basic stains may be seen in Figures 3.26, 3.37, and 3.40.

Basic stains are applied to bacterial smears that have been **heat-fixed.** Heat-fixing kills the bacteria, makes them adhere to the slide, and coagulates cytoplasmic proteins to make them more visible. It also distorts the cells to some extent.

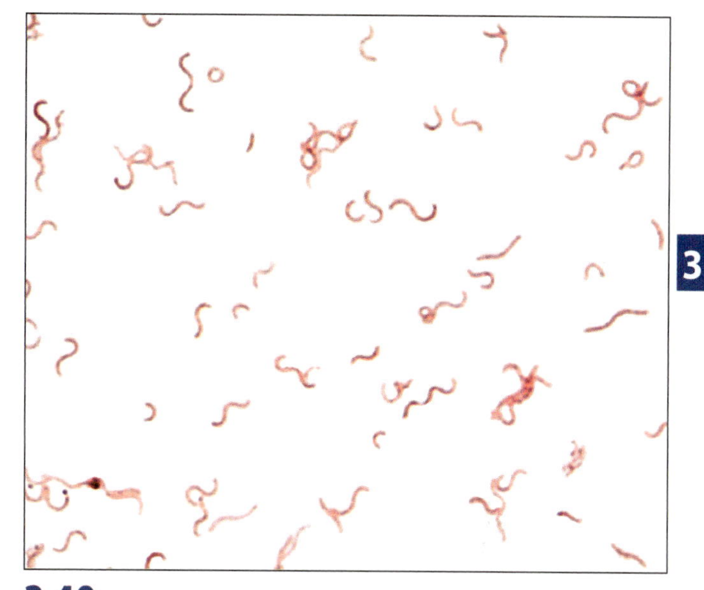

**3.40 Safranin Dye in a Simple Stain** ■ This is a simple stain using safranin, a basic stain. Notice that the stain is associated with the cells and not the background. The organism is *Rhodospirillum rubrum* grown in broth culture and photographed with the oil-immersion lens. Compare cell size and morphology in this preparation with the negative stain of *R. rubrum* in Figure 3.45.

## Application

Because cytoplasm is transparent, cells usually are stained with a colored dye to make them more visible under the microscope. Then cell morphology, size, and arrangement can be determined. In a medical laboratory, these are usually determined with a Gram stain (Exercise 3-6), but you will be using simple stains as an introduction to these.

[1] Notice that the term *basic* means "alkaline," not "elementary"; however, coincidentally *basic* stains can be used for *simple* staining procedures.

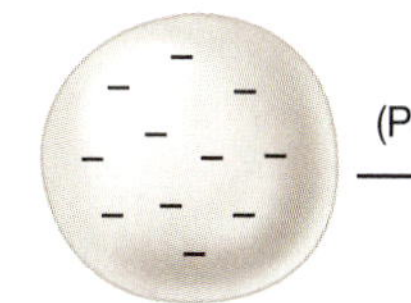

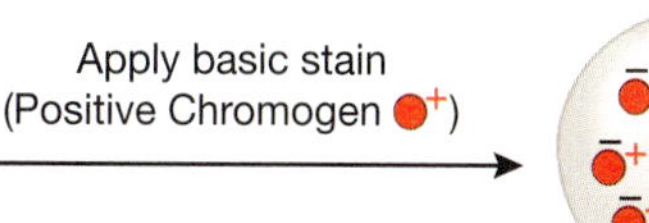

**3.39 Chemistry of Basic Stains** ■ Basic stains have a positively charged chromogen (●+), which forms an ionic bond with the negatively charged bacterial cell, thus colorizing the cell.

## In This Exercise

Today you will learn how to prepare a bacterial emulsion (smear) and perform simple stains. Several different organisms will be supplied so you can begin to see the variety of cell morphologies and arrangements in the bacterial world. You may wish to divide the workload with your lab partners so all specimens are stained by someone within your group. Then, if you don't have time to stain all the specimens yourself, you can examine slides prepared by your lab partner.

### ▼ Materials

**Per Student**

- ☐ Lab coat
- ☐ Disposable gloves
- ☐ Chemical eye protection

**Per Student Group**

- ☐ Clean glass microscope slides
- ☐ Methylene blue stain
- ☐ Safranin stain
- ☐ Crystal violet stain

- □ Squirt bottle with water
- □ Staining tray
- □ Staining screen
- □ Bibulous paper (or paper towels)
- □ Slide holder
- □ Compound microscope with oil-immersion lens and ocular micrometer
- □ Immersion oil
- □ Lens paper
- □ Fresh broth or slant cultures of these recommended organisms:
  - *Bacillus cereus* or any *Bacillus* species
  - *Micrococcus luteus*
  - *Neisseria sicca* or *Moraxella catarrhalis*
  - *Rhodospirillum rubrum* or *Aquaspirillum* species
  - *Staphylococcus epidermidis*
  - *Vibrio* species

## PROCEDURE

1. Wear a lab coat, gloves, and chemical eye protection when staining. Remove your gloves and eye protection when using the microscope.
2. A bacterial emulsion (smear) is made prior to most staining procedures. Follow the procedural diagram in Figure 3.41 to prepare smears of two organisms on a slide. Be sure not to mix them. Make a second and third slide with the remaining four organisms at the same time so they can be air-drying simultaneously. Emulsions should be about the size of a dime. If you are short on time, divide the workload with your lab partners so all specimens get stained by someone, then examine each other's slides so you see all six organisms.
3. Heat-fix each slide as described in Figure 3.41. If you are using a bacterial incinerator, hold the slide (with a slide holder) near the opening for a few seconds (Fig. 3.42).
4. Following the basic staining procedure illustrated in the procedural diagram in Figure 3.43, stain one of your slides with each stain using the following times:

| | |
|---|---|
| crystal violet: | stain for 30 to 60 seconds |
| safranin: | stain for up to 1 minute |
| methylene blue: | stain for 30 to 60 seconds |

   Record your actual staining times in the table provided in the data sheet, page 181, so you can adjust for over- or under-staining.
5. Using the oil-immersion lens, observe each slide. Record your observations of cell morphology, arrangement, and size in the chart provided on the data sheet.
6. Dispose of the slides and used stain according to your laboratory's policy.

**1.** Wearing gloves and eye protection, place a small drop of water (not too much) on a clean slide using an inoculating loop. If you are staining from a broth culture, begin with Step 2.

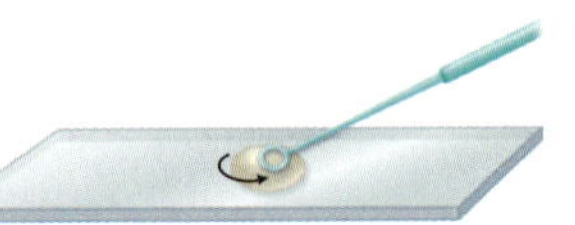

**2.** Aseptically add bacteria to the water. ***Note:*** If you find that you are transferring too much organism with a loop, use an inoculating needle. If you are transferring a BSL-2 organism, use a sterile wooden stick or a disposable loop and dispose of it properly when finished. Up to three emulsions can comfortably be made on a single slide. Mix in the bacteria and spread the drop out. Avoid spattering the emulsion as you mix. Flame your loop when done.

**3.** Allow the smear to air dry. If prepared correctly, the smear should be slightly cloudy.

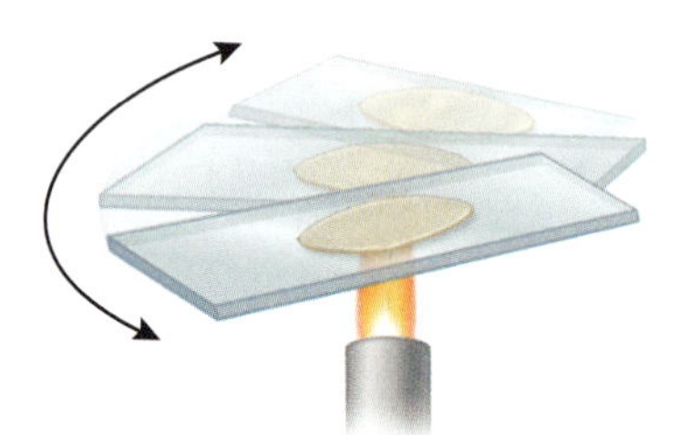

**4.** Using a slide holder, pass the smear through the upper part of a flame two or three times.This heat-fixes the preparation. Avoid overheating the slide because aerosols may be produced.

**5.** Allow the slide to cool, then continue with the staining protocol.

**3.41 Procedural Diagram: Making a Bacterial Smear (Emulsion)** You will apply this technique as the first step in performing most stain procedures (the negative and capsule stains are the exceptions). It is an important skill to learn, because preparation of uniform bacterial smears will make it easier to obtain consistent staining results. ***Caution:*** Wear gloves and eye protection and avoid producing aerosols. Do not spatter the smear as you mix it, do not blow on or wave the slide to speed up air-drying, and do not overheat when heat-fixing. If working with a BSL-2 organism, use a wooden stick or disposable loop to transfer the organism to the slide and heat-fix by holding the slide near the opening of a bacterial incinerator for 5-10 seconds with the smear facing *away* from the opening (Fig. 3.42).

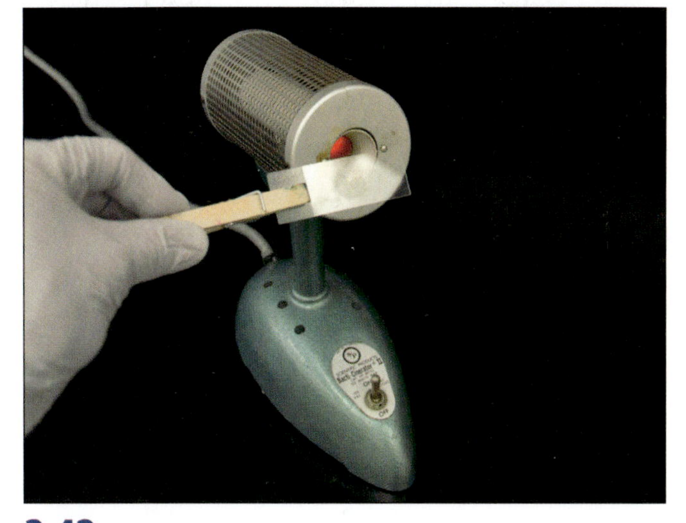

**3.42 Heat-fixing Using a Bacterial Incinerator** ■ In order to reduce aerosol production, heat-fixing a slide can be done by holding it near the opening of a microincinerator (with the smear facing away from the opening) for 5–10 seconds. This is an absolute requirement if staining a BSL-2 organism.

1. Begin with a heat-fixed emulsion (see Fig. 3.41). More than one organism can be put on a slide. In this exercise, we recommend putting three organisms on each slide. Be sure to wear gloves and eye protection.

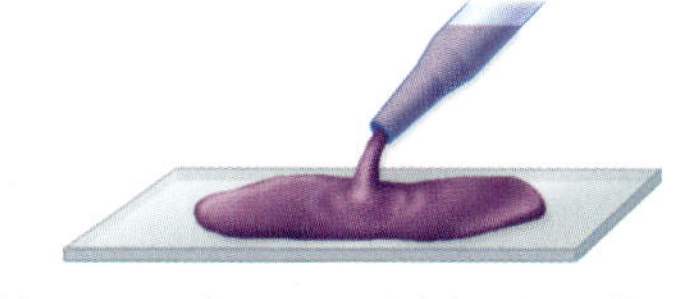

2. Place the slide on a rack over a staining tray. Cover the smear(s) with the stain. Make sure any excess stain falls into the staining tray.

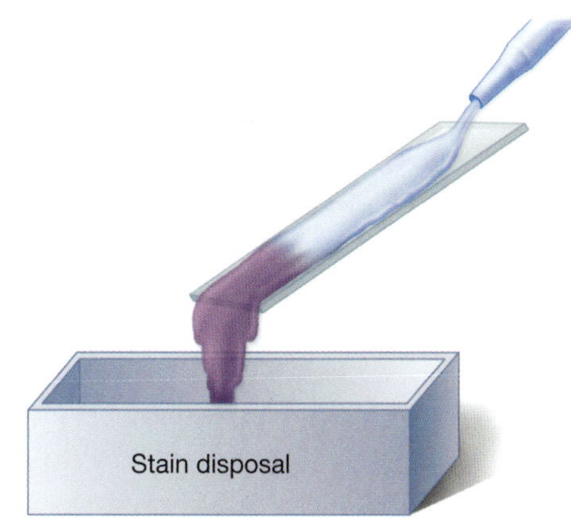


3. Grasp the slide with a slide holder and hold it on an angle. Gently rinse the slide with distilled water into the staining tray. Dispose of stain in the tray at the end of lab according to your lab's practices.

4. Gently blot dry in a tablet of bibulous paper or paper towels. (Alternatively, a page from the tablet can be removed and used for blotting.) Do not rub. When dry, observe under oil immersion.

**3.43 Procedural Diagram: Simple Stain** ■ Wear gloves and eye protection. Staining times differ for each stain, but cell density of your smear also affects staining time. Strive for consistency in making your smears. You also only need to cover the emulsion(s) on the slide with stain, not the slide's whole surface. ***Caution***: Be sure to flame your loop after cell transfer and properly dispose of the slide when you are finished observing it.

## References


Atlas, Ronald and James Snyder. Page 323 in *Manual of Clinical Microbiology,* 11th ed. James H. Jorgensen, Michael A. Pfaller, Karen C. Carroll, Guido Funke, Marie Louise Landry, Sandra S. Richter, and David W. Warnock, eds. Washington, DC: ASM Press, American Society for Microbiology, 2015.

Chapin, Kimberle. Chap. 4 in *Manual of Clinical Microbiology,* 6th ed. Patrick R. Murray, Ellen Jo Baron, Michael A. Pfaller, Fred C. Tenover, and Robert H. Yolken, eds. Washington, DC: American Society for Microbiology, 1995.

Chapin, Kimberle C. and Patrick R. Murray. Pages 257–259 in *Manual of Clinical Microbiology,* 8th ed. Patrick R. Murray, Ellen Jo Baron, James H. Jorgensen, Michael A. Pfaller, and Robert H. Yolken, eds. Washington, DC: American Society for Microbiology, 2003.

Murray, R. G. E., Raymond N. Doetsch, and C. F. Robinow. Page 27 in *Methods for General and Molecular Bacteriology*. Philipp Gerhardt, R. G. E. Murray, Willis A. Wood, and Noel R. Krieg, eds. Washington, DC: American Society for Microbiology, 1994.

Norris, J. R. and Helen Swain. Chap. II in *Methods in Microbiology,* Vol. 5. A. J. R. Norris and D. W. Ribbons, eds. London, UK: Academic Press, Ltd., 1971.

Power, David A. and Peggy J. McCuen. Page 4 in *Manual of BBL™ Products and Laboratory Procedures*, 6th ed. Cockeysville, MD: Becton Dickinson Microbiology Systems, 1988.

Tille, Patricia M. Pages 70–71 in *Bailey & Scott's Diagnostic Microbiology*, 13th ed. St. Louis, MO: Mosby, 2014.

Name ______________________________

Date ______________________________

Lab Section ______________________________

I was present and performed this exercise (initials) ______________

**DATA SHEET**

**3-4**

# Simple Stains

## OBSERVATIONS AND INTERPRETATIONS

**1** Record your observations in the table below.

| Organism | Stain and Duration | Cellular Morphology and Arrangement (include a detailed sketch of a few representative cells) | Cell Dimensions |
|---|---|---|---|
| | | | |
| | | | |
| | | | |
| | | | |
| | | | |
| | | | |

## QUESTIONS

**1** *What are some consequences of leaving a stain on a bacterial smear too long (over-staining)?*

**2** *What are some consequences of not leaving a stain on a smear long enough (under-staining)?*

3

**3** *Choose a coccus and a bacillus from the organisms you observed and calculate their surface-to-volume ratios. Consider the coccus to be a perfect sphere and the bacillus to be a cylinder. Use the equations supplied.*

| Cell Morphology | Surface Area | Volume |
|---|---|---|
| Coccus | $SA = 4\pi r^2$ | $V = \frac{4}{3}\pi r^3$ |
| Bacillus | $SA = 2\pi rh + 2\pi r^2 = \pi d(r+h)$ | $V = \pi r^2 h$ |

r = radius, d = diameter, h = height, $\pi = 3.14$

### Surface-to-Volume Ratio of Sample Cells

| Organism | Cell Morphology | Surface Area ($\mu m^2$) | Volume ($\mu m^3$) | Surface-to-Volume Ratio |
|---|---|---|---|---|
| | | | | |
| | | | | |

**4** *Consider a coccus and a rod of equal volume.*

**a.** *Which is more likely to survive in a dry environment? Explain your answer.*

**b.** *Which would be better adapted to a moist environment? Explain your answer.*

# Negative Stains

EXERCISE 3-5

## Theory

The negative staining technique uses a dye solution in which the chromogen is acidic and carries a negative charge. (An acidic chromogen gives up a hydrogen ion, which leaves it with a negative charge.) The negative charge on the bacterial surface repels the negatively charged chromogen, so the cell remains unstained against a colored background (Fig. 3.44). A specimen stained with the acidic stain nigrosin is shown in Figure 3.45.

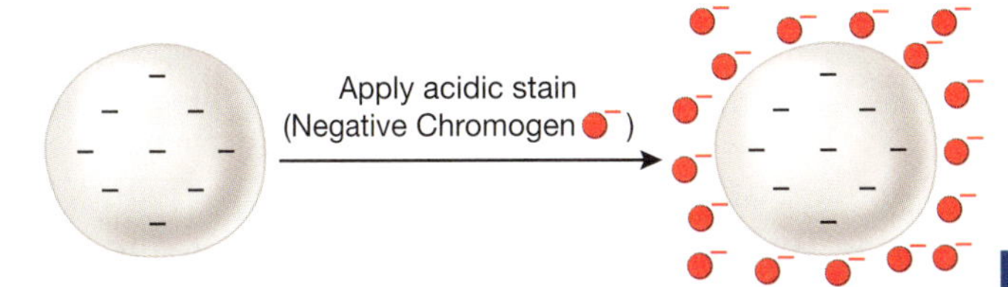


**3.44 Chemistry of Acidic Stains** ■ Acidic stains have a negatively charged chromogen (•⁻) that is repelled by negatively charged cells. Thus, the background is colored and the cell remains transparent.

## Application

The negative staining technique is used to determine morphology and cellular arrangement in bacteria that are too delicate to withstand heat-fixing. A primary example is the spirochete *Treponema*, which is distorted by the heat-fixing of other staining techniques. Also, where determining the accurate size is crucial, a negative stain can be used because it produces minimal cell shrinkage.

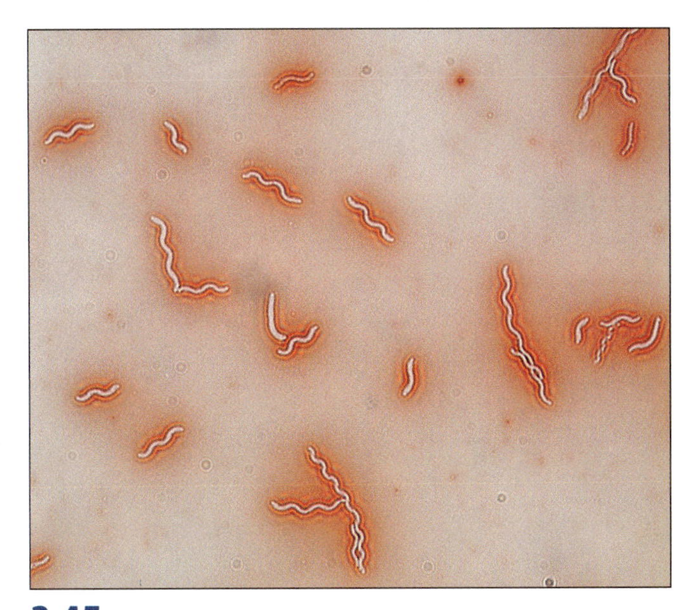

**3.45 Congo Red Negative Stain** ■ A negative stain procedure is so-called because it is the background that is colorized, not the cells. Heat-fixing is used in most staining procedures and it causes some degree of cell distortion. Because cells in a negative stain are not heat-fixed, their appearance is closer to what they really look like. These *Rhodospirillum rubrum* cells were grown in broth culture and were photographed with the oil-immersion lens. Compare their size and morphology with the simple stain shown in Figure 3.40.

## In This Exercise

Today you will perform negative stains on three different organisms. You also will have the opportunity to compare the sizes of *Bacillus cereus*, *Micrococcus luteus*, and *Rhodospirillum rubrum* (or their substitutes) measured with the negative stain to their sizes as determined using a simple stain. You may wish to divide the workload with your lab partners so all specimens are stained by someone within your group. Then, if you don't have time to stain all the specimens yourself, you can examine slides prepared by your lab partner.

### ▼ Materials

**Per Student**

- ☐ Lab coat
- ☐ Disposable gloves
- ☐ Chemical eye protection

**Per Student Group**

- ☐ Nigrosin stain or eosin stain
- ☐ Clean glass microscope slides
- ☐ Compound microscope with oil-immersion lens and ocular micrometer
- ☐ Immersion oil
- ☐ Lens paper
- ☐ Bibulous paper or paper towel
- ☐ Fresh broth or slant cultures of these recommended organisms:
  - *Bacillus cereus* or any *Bacillus* species (which should match the one used in Exercise 3-4)
  - *Micrococcus luteus*
  - *Rhodospirillum rubrum* or *Aquaspirillum* species (which should match the one used in Exercise 3-4)

## PROCEDURE

1. Wear a lab coat, gloves, and chemical eye protection when staining. Remove your gloves and eye protection when using the microscope.
2. Follow the procedural diagram in Figure 3.46 to prepare a negative stain of each organism. Notice that only one organism can be put on a slide. Do NOT heat-fix your slides after air-drying them. If you are short on time, divide the workload with your lab partners so all specimens get stained by someone, then examine each other's slides so you see all three organisms.
3. Dispose of the spreader slide in a disinfectant jar or sharps container immediately after use.
4. Observe using the oil-immersion lens. Record your observations in the chart on the data sheet, page 185.
5. Dispose of the specimen slide in a disinfectant jar or sharps container after use.

### References


Claus, G. William. Chap. 5 in *Understanding Microbes—A Laboratory Textbook for Microbiology*. New York: W. H. Freeman and Co., 1989.

Murray, R. G. E., Raymond N. Doetsch, and C. F. Robinow. Page 27 in *Methods for General and Molecular Bacteriology*. Philipp Gerhardt, R. G. E. Murray, Willis A. Wood, and Noel R. Krieg, eds. Washington, DC: American Society for Microbiology, 1994.


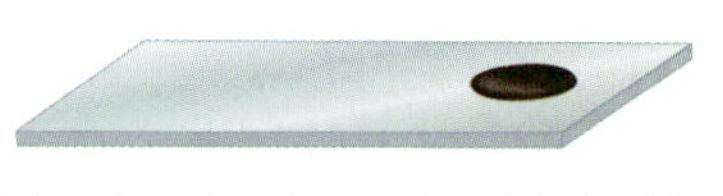

**1.** Begin with a drop or loopful of acidic stain at one end of a clean slide. Be sure to wear gloves and eye protection.

**2.** Working on the table top covered with a paper towel, aseptically add organisms and emulsify with the loop. Don't over-inoculate and avoid spattering the mixture. Sterilize the loop after emulsifying. ***Note:*** If you find you are transferring too much organism with a loop, use an inoculating needle. If you are transferring a BSL-2 organism, use a sterile wooden stick or disposable loop/needle and dispose of it properly when finished.

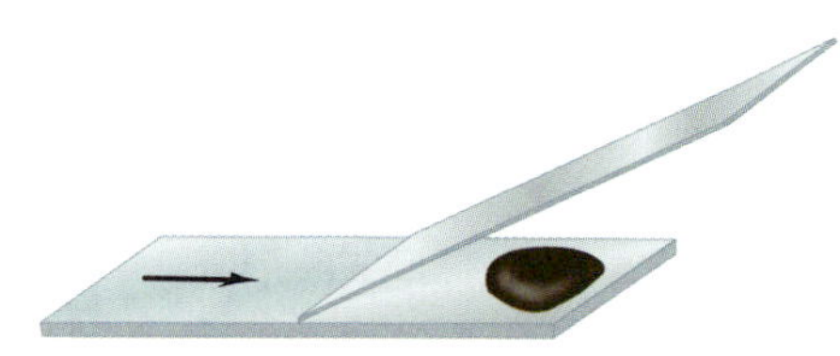

**3.** Take a second clean slide, place it on the surface of the first slide and hold it at a 30° angle, then draw it back into the drop.

**4.** When the drop flows across the width of the spreader slide...

**5.** ...push the spreader slide to the other end. Dispose of the spreader slide in a jar of disinfectant or sharps container.

**6.** Air-dry and observe under the microscope. Do NOT heat-fix.

**3.46 Procedural Diagram: Negative Stain** ■ Wear gloves and eye protection. Be sure to sterilize your loop after transfer and to appropriately dispose of the spreader slide immediately after use; do not set it on the table. Once air-dried, the slide is ready for viewing. No heat-fixing is required in this procedure.

Name ______________________________

Date ______________________________

Lab Section ______________________________

I was present and performed this exercise (initials) ______________________________

# DATA SHEET 3-5

## Negative Stains

### OBSERVATIONS AND INTERPRETATIONS

**1** Record your observations in the table below.

| Organism | Stain | Cellular Morphology and Arrangement (include a detailed sketch of a few representative cells) | Cell Dimensions from Negative Stain | Cell Dimensions from Simple Stain (Exercise 3-4) |
|---|---|---|---|---|
| | | | | |
| | | | | |
| | | | | |

### QUESTIONS

**1** *Why doesn't a negative stain colorize the cells in the smear?*

______________________________

______________________________

______________________________

______________________________

______________________________

______________________________

______________________________

______________________________

*Eosin is a red stain and methylene blue is blue. What should be the result of staining a bacterial smear with a mixture of eosin and methylene blue?*

3

**3** *Compare the sizes of* B. cereus, M. luteus, *and* R. rubrum *cells (or their substitutes) as measured using a basic stain (Exercise 3-4) and an acidic stain. What might account for any difference?*

## Differential and Structural Stains

**Differential stains** allow a microbiologist to detect differences between organisms or differences between parts of the same organism. In practice, these are used much more frequently than simple and negative stains because they not only allow determination of cell size, morphology, and arrangement (as with simple or negative stains) but they provide information about other features as well.

The Gram stain is the most commonly used differential stain in bacteriology. Other differential stains are used for organisms not distinguishable by the Gram stain and for those that have other important cellular attributes, such as acid-fastness, a capsule, spores, or flagella. With the exception of the acid-fast stain, these other stains sometimes are referred to as **structural stains.** ■



# Gram Stain

EXERCISE 3-6

## Theory

The Gram stain is a differential stain in which a **decolorization** step occurs between the application of two basic stains. There are several minor variations of the Gram stain procedure, but they all work in basically the same way (Fig. 3.47). The **primary stain** is crystal violet. Iodine is added as a **mordant** to enhance crystal violet staining by forming a **crystal violet–iodine complex**.

Decolorization follows and is the most critical step in the procedure. Gram-negative cells are decolorized by the solution (generally an alcohol or alcohol/acetone mixture of varying proportions) whereas Gram-positive cells are not. Gram-negative cells can thus be colorized by the red **counterstain** safranin, but Gram-positive cells are already violet and cannot. Upon successful completion of a Gram stain, Gram-positive cells appear purple and Gram-negative cells appear reddish-pink (Fig. 3.48).

Electron microscopy and other evidence indicate that the ability to resist decolorization or not is based on the different wall constructions of Gram-positive and Gram-negative cells. Gram-negative cell walls have a higher lipid content (because of the outer membrane) and a thinner peptidoglycan layer than Gram-positive cell walls (Fig. 3.49).

The alcohol/acetone in the decolorizer extracts the lipid, making the Gram-negative wall more porous and incapable of retaining the crystal violet–iodine complex,

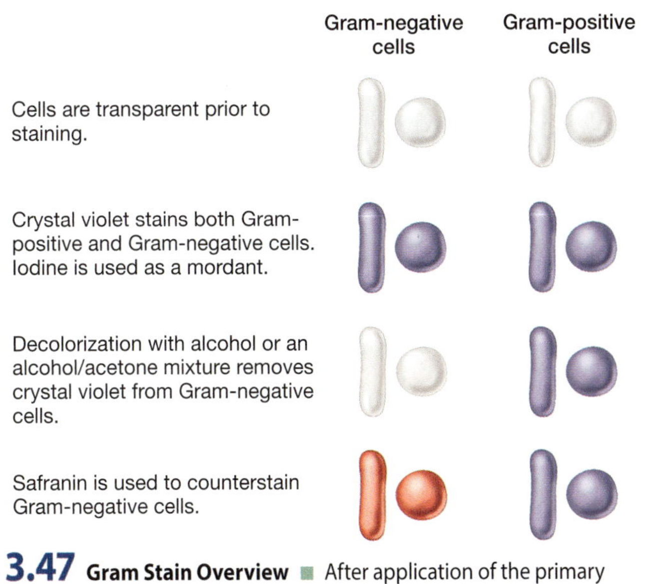


**3.47 Gram Stain Overview** ■ After application of the primary stain (crystal violet), decolorization, and counterstaining with safranin, Gram-positive cells stain violet and Gram-negative cells stain pink/red. Notice that crystal violet and safranin are both basic stains, and that the decolorization step is what makes the Gram-stain differential.

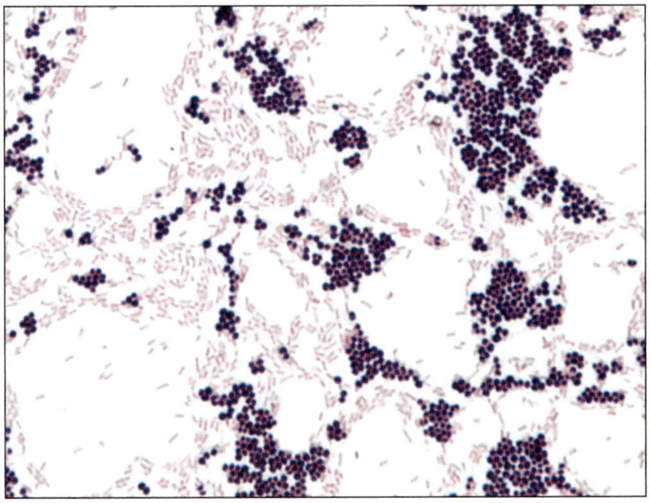

**3.48 Gram Stain of Staphylococcus (+) and Proteus (−)** ■ The violet staphylococcal cells are Gram positive; the pink rods are Gram negative. Depending on your Gram-stain kit, the safranin may be anywhere from a light pink (as in this micrograph) to a more intense reddish color. In either case, it should be easily distinguishable from the crystal violet color.

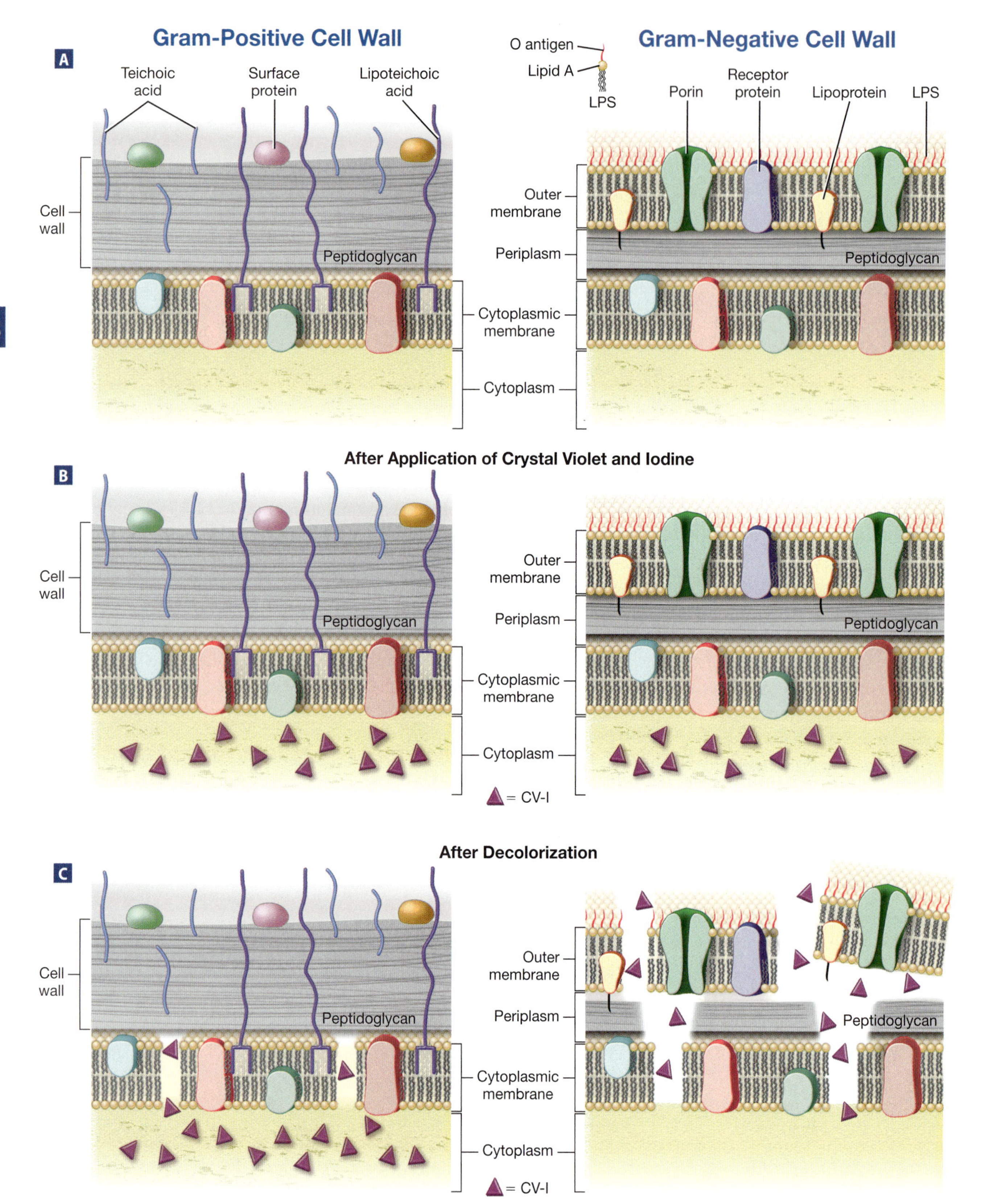


**3.49 Bacterial Cell Walls** ■ (**A**) The Gram-negative wall (**right**) is composed of less peptidoglycan (as little as a single layer) and more lipid (due to the outer membrane) than the Gram-positive wall (**left**). Though it is shown as a solid layer, the peptidoglycan is actually quite porous, kind of like a three-dimensional chain-link fence. (**B**) Relatively large crystal violet-iodine complexes (purple triangles) form in the cytoplasm of both Gram-positive and Gram-negative cells. (**C**) The thicker Gram-positive peptidoglycan layer dehydrates when decolorizer is added, trapping the crystal violet-iodine complexes. The lipid in the Gram-negative cell wall is dissolved by the decolorizer and the thinner peptidoglycan is incapable of preventing extraction of the crystal violet–iodine complex.

thereby decolorizing it. Dehydration of the Gram-positive wall by the decolorizer coupled with its thicker peptidoglycan and greater degree of cross-linking (because of teichoic acids) trap the crystal violet–iodine complex more effectively, making the Gram-positive wall less susceptible to decolorization.

Although some organisms give Gram-variable results, most variable results are a consequence of poor technique. The decolorization step is the most crucial and most likely source of Gram-stain inconsistency. It is possible to **over-decolorize** by leaving the decolorizer on too long and get reddish Gram-*positive* cells. It also is possible to **under-decolorize** and produce purple Gram-*negative* cells. Neither of these situations changes the actual Gram reaction of the organism being stained. Rather, these are false results because of poor technique.

A second source of poor Gram stains is inconsistency in preparation of the emulsion. Remember, a good emulsion is about the size of a dime and dries to a faint haze on the slide.

A third source of inconsistent Gram stains may be the organisms themselves. Some Gram positives, especially *Bacillus* and *Staphylococcus* species, lose their ability to retain the crystal violet–iodine complex in as little as 24 hours of incubation. Figure 3.50 shows a *Bacillus* culture Gram stained at 24 hours of growth. Always plan on doing your Gram stains on cultures no older than 24 hours for best results.

Until correct results are obtained consistently, it is recommended that control smears of Gram-positive and Gram-negative organisms be stained along with the organism in question (Fig. 3.51). As an alternative control, a direct smear made from the gumline may be Gram stained (Fig. 3.52) with the expectation that both Gram-positive and Gram-negative organisms will be seen. Over-decolorized and under-decolorized gumline direct smears are shown for comparison (Figs. 3.53 and 3.54). Positive controls also should be run when using new reagent batches.

Interpretation of Gram stains can be complicated by nonbacterial elements. For instance, stain crystals from an old or improperly made stain solution can disrupt the field (Fig. 3.55) or stain precipitate may be mistakenly identified as bacteria (Fig. 3.56).

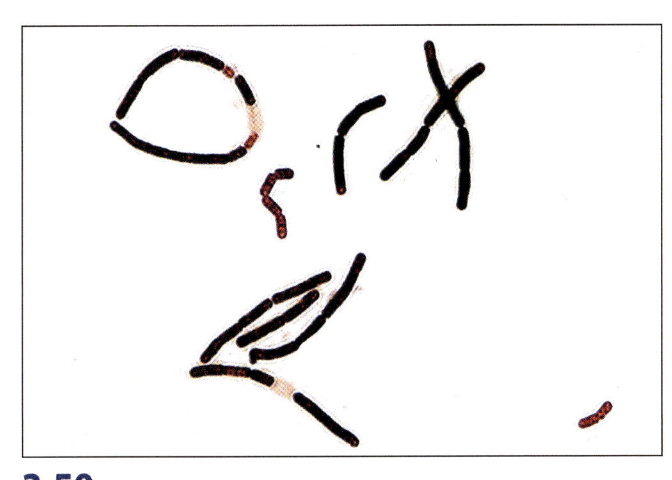

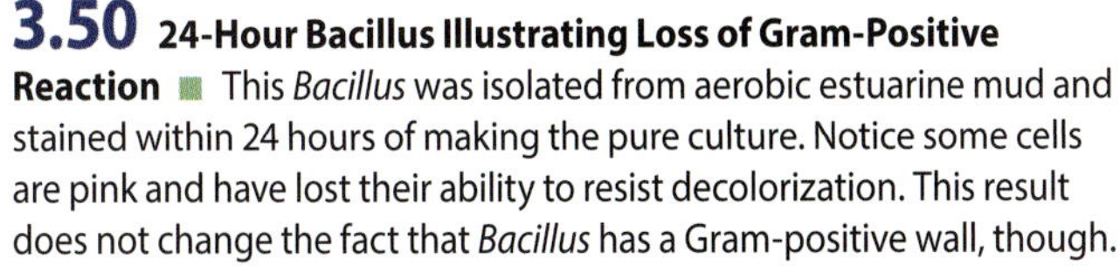

**3.50 24-Hour Bacillus Illustrating Loss of Gram-Positive Reaction** ■ This *Bacillus* was isolated from aerobic estuarine mud and stained within 24 hours of making the pure culture. Notice some cells are pink and have lost their ability to resist decolorization. This result does not change the fact that *Bacillus* has a Gram-positive wall, though.

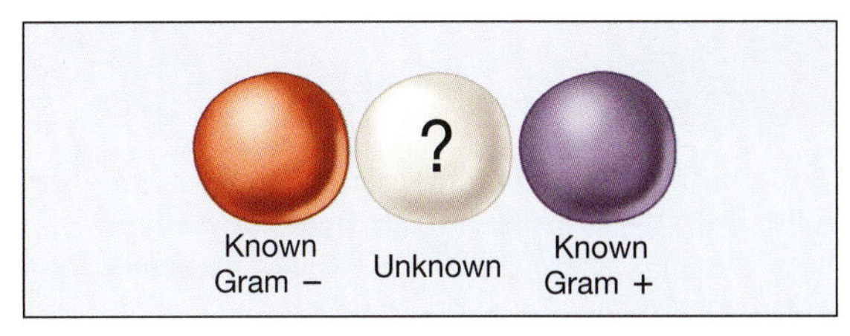


**3.51 Positive Controls to Check Your Technique** ■ Staining known Gram-positive and Gram-negative organisms on either side of your unknown organism acts as positive controls for your technique. Try to make the emulsions as close to one another as possible. Spreading them out across the slide makes it difficult to stain and decolorize them equally. It is also beneficial to put the Gram-negative control at the end where you will be applying the decolorizer. This will make it easier to see when it is running clear. Putting it at the other end means it will have runoff from the Gram-positive control and the unknown obscuring its runoff.

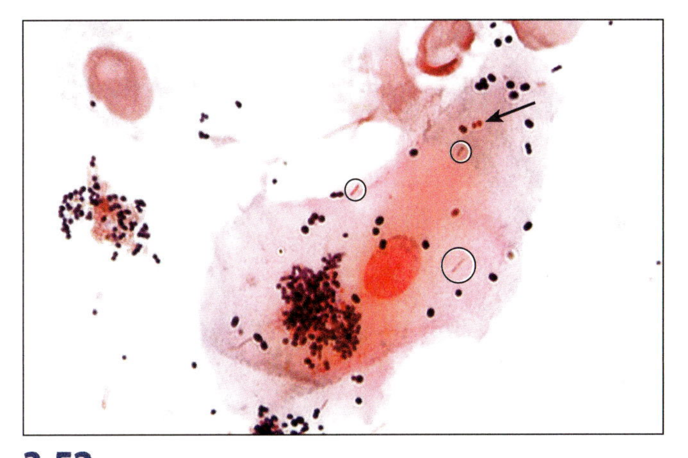

**3.52 Direct Smear Positive Control (Gram Stain)** ■ A direct smear made from the gumline may also be used as a Gram-stain control. Expect numerous Gram-positive bacteria (especially cocci) and some Gram-negative cells, including your own epithelial cells. In this slide, Gram-positive cocci predominate, but a few Gram-negative cells are visible, including Gram-negative rods (circled) and a Gram-negative diplococcus (arrow). Notice that most bacterial cells are on the epithelial cell's surface, which is where they reside when in the mouth.

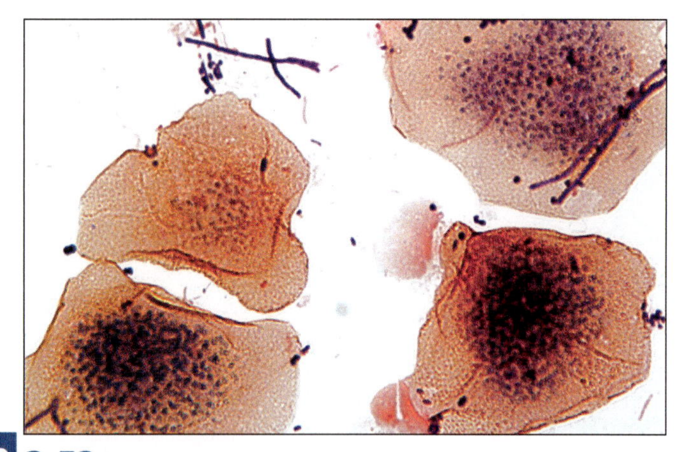

**3.53 Under-Decolorized Gram Stain** ■ This is a direct smear from the gumline. Notice the purple patches of stain on the epithelial cells. Also notice the variable quality of this stain—the epithelial cell to the left of center is stained better than the others.

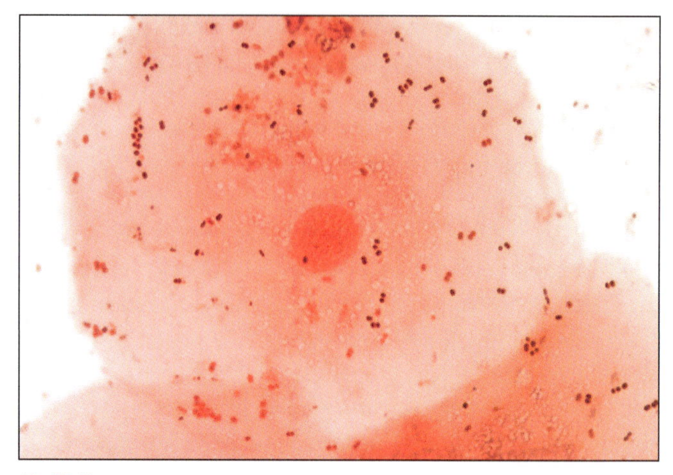

**3.54 Over-Decolorized Gram Stain** ■ This also is a direct smear from the gumline. Notice the virtual absence of any purple cells, a certain indication of over-decolorization.

**3.55 Crystal-Violet Crystals (Gram Stain)** ■ If the staining solution is not adequately filtered or is old, crystal-violet crystals may appear. Although they are pleasing to the eye, they obstruct your view of the specimen. Crystals from two different Gram stains are shown here. (**A**) This is a gumline direct smear. The bacterial cells are the pink material in the background (this emulsion was way too thick) and the crystals are the large, purple "snowflakes." (**B**) In this Gram stain of *Micrococcus roseus* grown in culture, the cells are the lighter purple in the background (can you detect any tetrads?) and the crystals are the darker spikes.

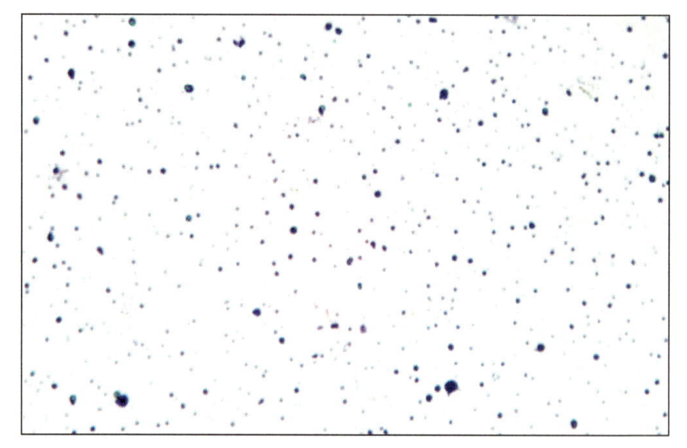

**3.56 Stain Precipitate (Gram Stain)** ■ If the slide is not rinsed thoroughly or the stain is allowed to dry on the slide, spots of stain precipitate may form and may be confused with bacterial cells. Their variability in size is a clue that they are not bacteria.

## Application

The Gram stain, used to distinguish between Gram-positive and Gram-negative cells, is the most important and widely used microbiological differential stain. In addition to Gram reaction, this stain allows determination of cell morphology, size, and arrangement. It typically is the first differential test run on a specimen brought into the laboratory for identification. In some cases, a rapid, presumptive identification of the organism or elimination of a particular organism is possible.

## In This Exercise

As stated above (but worth repeating), the Gram stain is the single most important differential stain in bacteriology. Therefore, you will have to practice it and practice it some more to become proficient in its execution. The organisms should be used in the combinations given so you will have

one Gram positive and one Gram negative on each slide. We recommend that you stain one slide, observe it, evaluate it, and then stain another slide incorporating corrections indicated by the first slide's results (Fig. 3.57).

### ▼ Materials

#### Per Student

- ☐ Lab coat
- ☐ Disposable gloves
- ☐ Chemical eye protection

#### Per Student Group

- ☐ Compound microscope with oil-immersion lens and ocular micrometer
- ☐ Clean glass microscope slides
- ☐ Sterile toothpick
- ☐ Gram-stain solutions (commercial kits are available):
  - Gram crystal violet
  - Gram iodine
  - 95% ethanol (or ethanol/acetone solution)
  - Gram safranin
- ☐ Squirt bottle with water
- ☐ Bibulous paper or paper towels
- ☐ Staining tray
- ☐ Staining screen
- ☐ Slide holder
- ☐ Recommended organisms (overnight cultures grown on agar slants or turbid broth cultures):
  - *Staphylococcus epidermidis* or *Micrococcus luteus*
  - *Escherichia coli* or *Rhodospirillum rubrum*
  - *Neisseria sicca* or *Moraxella catarrhalis*
  - *Corynebacterium xerosis*

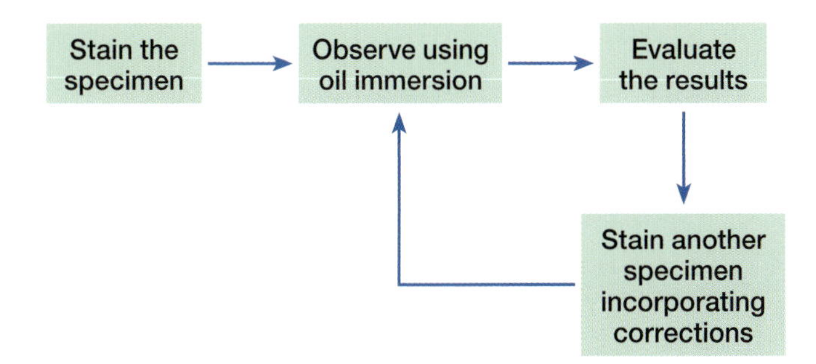


**3.57 Evaluating Staining Technique** ■ It is best when learning a staining technique to prepare several slides and stain them one at a time. Observe the first slide, evaluate its quality (good or poor) and if poor, determine what steps to take to improve the quality. Then, stain another slide incorporating the changes suggested by your previous attempt. This technique is applicable to any stain, not just the Gram stain.

## PROCEDURE

1. Wear a lab coat, gloves, and chemical eye protection when staining. Remove your gloves and eye protection when using the microscope.
2. Of the listed organisms, *Staphylococcus epidermidis*, *Micrococcus luteus*, and *Corynebacterium xerosis* have the same Gram reaction. Call these "Group A." *Escherichia coli, Rhodospirillum rubrum*, *Neisseria sicca*, and *Moraxella catarrhalis* have the other Gram reaction. Call these "Group B."
3. Follow the procedure illustrated in Figure 3.41 to prepare and heat-fix smears of one Group A and one Group B organism *immediately next to one another* on the same clean glass slide. (If you make the emulsions at opposite ends of the slide, you may find it difficult to stain and decolorize each equally.) Strive to prepare smears of uniform thickness, because thick smears risk being under-decolorized.
4. Repeat step 3 for a different pair of organisms on a second slide.
5. Because Gram stains require much practice, we recommend that you prepare several slides of each combination and let them air-dry simultaneously. As long as you heat-fix them, they can be stained during another lab period if you don't have time today. Then they will be ready when you need them.
6. Wash your hands, and then use the sterile toothpick to obtain a sample from your teeth at the gumline. (Do not draw blood! What you want is easily removed from your gingival pockets—that space between your teeth and gums.) Transfer the sample to a drop of water on a clean glass slide, air-dry, and heat-fix.
7. Follow the basic staining procedure illustrated in Figure 3.58. We recommend staining the pure cultures first, one at a time, and evaluating your technique before doing another (Fig. 3.57). After your technique is consistent, stain the gumline sample. Be sure to wear gloves and eye protection. Also, pay special attention to the decolorization "tips" in the caption and steps 6 and 7 of Figure 3.58.
8. Observe using the oil-immersion lens. Record your observations of cell morphology and arrangement, dimensions, and Gram reactions in the chart provided on the data sheet, page 193.
9. Dispose of the specimen slides in a jar of disinfectant or a sharps container after use.

## References


Atlas, Ronald and James Snyder. Page 322 in *Manual of Clinical Microbiology*, 11th ed. James H. Jorgensen, Michael A. Pfaller, Karen C. Carroll, Guido Funke, Marie Louise Landry, Sandra S. Richter, and David W. Warnock, eds. Washington, DC: ASM Press, American Society for Microbiology, 2015.

Chapin, Kimberle C. and Patrick R. Murray. Pages 258–260 in *Manual of Clinical Microbiology*, 8th ed. Patrick R. Murray, Ellen Jo Baron, James H. Jorgensen, Michael A. Pfaller, and Robert H. Yolken, eds. Washington, DC: American Society for Microbiology, 2003.

Koneman, Elmer W., Stephen D. Allen, William M. Janda, Paul C. Schreckenberger, and Washington C. Winn, Jr. Chap. 14 in *Color Atlas and Textbook of Diagnostic Microbiology*, 5th ed. Philadelphia: J. B. Lippincott Co., 1997.

Murray, R. G. E., Raymond N. Doetsch, and C. F. Robinow. Pages 31 and 32 in *Methods for General and Molecular Bacteriology*. Philipp Gerhardt, R. G. E. Murray, Willis A. Wood, and Noel R. Krieg, eds. Washington, DC: American Society for Microbiology, 1994.

Norris, J. R., and Helen Swain. Chap. II in *Methods in Microbiology*, Vol. 5A. J. R. Norris and D. W. Ribbons, eds. London, UK: Academic Press, Ltd., 1971.

Power, David A. and Peggy J. McCuen. Page 261 in *Manual of BBL™ Products and Laboratory Procedures*, 6th ed. Cockeysville, MD: Becton Dickinson Microbiology Systems, 1988.

Tille, Patricia M. Pages 70–73 in *Bailey & Scott's Diagnostic Microbiology*, 13th ed. St. Louis, MO: Mosby, 2014.

**1.** Begin with up to three heat-fixed emulsions on one slide. (See Fig. 3.41.) Wearing gloves, place the slide on a rack over a staining tray.

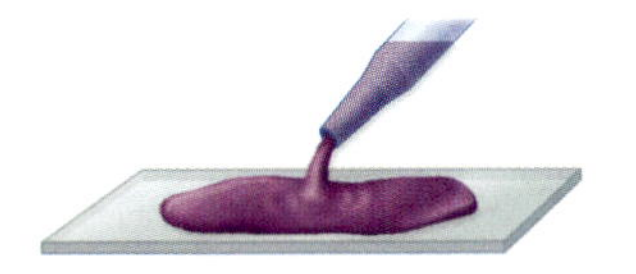

**2.** Cover the smear(s) (not the whole slide) with crystal violet stain for 1 minute. Be sure the staining tray catches all excess stain.

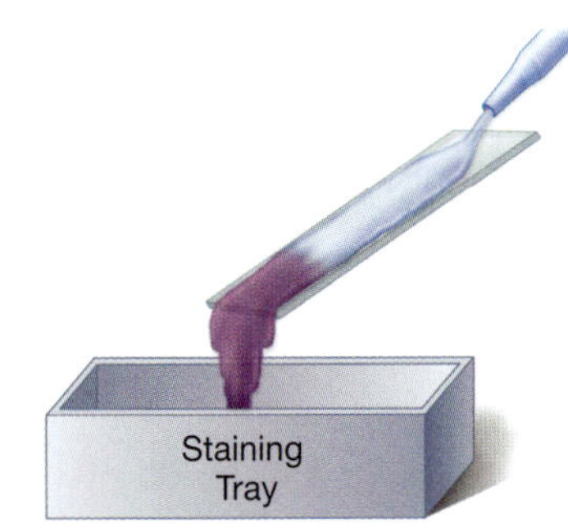


**3.** Grasp the slide with a slide holder and hold it on an angle. Gently rinse the slide with distilled water into the staining tray. Alternatively, see steps 4 and 5.

**4.** Cover the smear(s) with Gram iodine stain for 1 minute. Be sure the staining tray catches all excess stain.

Staining Tray

**5.** Grasp the slide with a slide holder and hold it on an angle. Gently rinse the slide with distilled water into the staining tray.

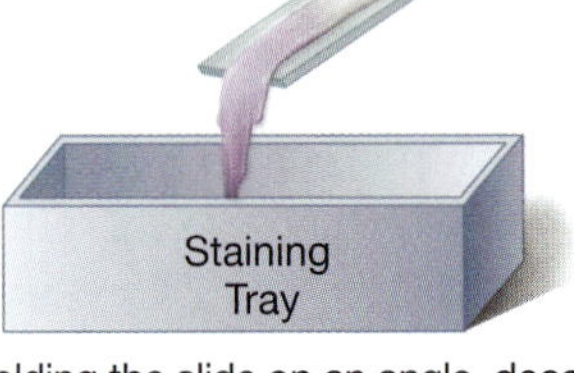


**6.** Holding the slide on an angle, decolorize with Gram decolorizer by allowing it to trickle down the slide until the runoff is clear. Catch the runoff in the staining tray. ***Note:*** Decolorizers differ depending on the kit's manufacturer. The more acetone in the decolorizer, the faster it will decolorize.

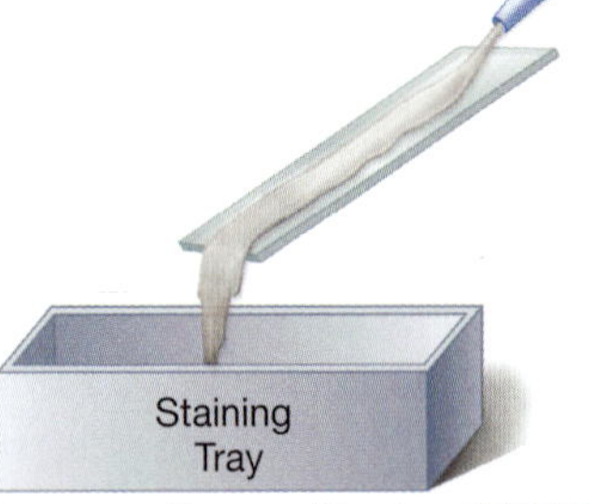


**7.** As soon as the runoff is clear, IMMEDIATELY rinse the slide with distilled water into the staining tray (***Note:*** It will still be decolorizing while you're grabbing the water bottle! Plan ahead for that transition.).

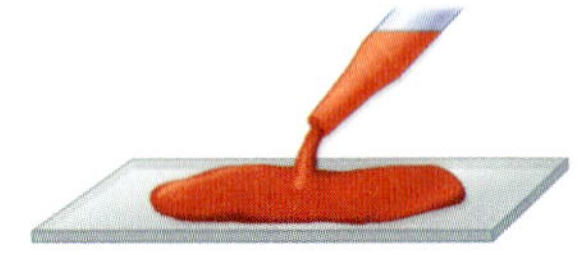

**8.** Counterstain the smear(s) with safranin for 1 minute. Be sure the staining tray catches all excess stain.

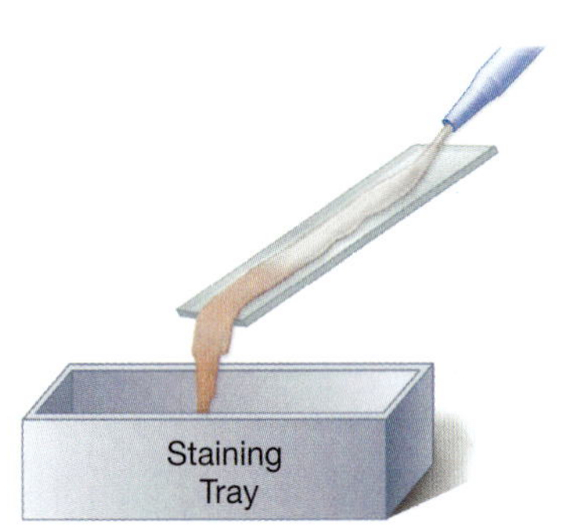


**9.** Grasp the slide with a slide holder and hold it on an angle. Gently rinse the slide with distilled water into the staining tray.

**10.** Gently blot dry in a tablet of bibulous paper or paper towels. (Alternatively, a page from the tablet can be removed and used for blotting.) Do not rub. When dry, observe under oil immersion.

**3.58 Procedural Diagram: Gram Stain** ■ Pay careful attention to the staining times. If your preparations do not give "correct" results, the most likely source of error is in the decolorization step. When the runoff is clear (generally within 10 seconds, shorter if the acetone concentration is high), IMMEDIATELY rinse with $dH_2O$. *The decolorizer continues decolorizing as you put its bottle down and pick up the water bottle, so anticipate the end point of decolorization to compensate for this.*

Name ______________________________

Date ______________________________

Lab Section ______________________________

I was present and performed this exercise (initials) ______________________________

DATA SHEET 3-6

# Gram Stain

## OBSERVATIONS AND INTERPRETATIONS

1 Record your observations in the table below. Use separate lines for different organisms found in the Gram stain of your gumline. Include a drawing of your own epithelial cells.

| Organism or Source | Cellular Morphology and Arrangement (include a detailed sketch of a few representative cells) | Cell Dimensions | Color | Gram Reaction (+/−) |
|---|---|---|---|---|
| | | | | |
| | | | | |
| | | | | |
| | | | | |
| | | | | |
| | | | | |

## QUESTIONS

**1** *Predict the effect on Gram-positive and Gram-negative cells of the following "mistakes" made when performing a Gram stain. Consider each mistake independently.*

**a.** *Failure to add the iodine.*

**b.** *Failure to apply the decolorizer.*

3

**c.** *Failure to apply the safranin.*

**d.** *Reversal of crystal violet and safranin stains.*

**2** *Both crystal violet and safranin are basic stains and may be used to do simple stains on Gram-positive and Gram-negative cells. This being the case, explain how they end up staining Gram-positive and Gram-negative cells differently in the Gram stain.*

**3** *If you saw large, eukaryotic cells in the preparation made from your gumline, they were most likely your own epithelial cells. Are you Gram-positive or Gram-negative? (You can make a good guess about this even if you didn't see your cells.)*

**4** *One of your lab partners has followed the recommended procedure of running Gram-positive and Gram-negative control organisms on her Gram stain of an unknown species. Her choices of controls were* Escherichia coli *and* Bacillus subtilis. *She tries several times and each time concludes she is decolorizing too long because both controls have pink cells (one more than the other). What might you suggest she try and why?*

# Acid-Fast Stains

EXERCISE 3-7

## Theory

The presence of mycolic acids in the cell walls of acid-fast organisms is the cytological basis for the acid-fast differential stain. Mycolic acid is a waxy substance that gives acid-fast cells a higher affinity for the primary stain and resistance to decolorization by an acid alcohol solution (3% HCl in 95% ethanol).

A variety of acid-fast staining procedures are employed, two of which are the Ziehl-Neelsen (ZN) method and the Kinyoun (K) method. These differ primarily in that the ZN method uses heat as part of the staining process (this is in addition to heat-fixing the preparation), whereas the K method is a "cold" stain (which is still heat-fixed). In both protocols the bacterial smear may be prepared in a drop of serum to help the "slippery" acid-fast cells (they are waxy, after all) adhere to the slide. The two methods provide comparable results and the choice of method comes down to lab conventions and personal preference.

The waxy wall of acid-fast cells repels typical aqueous stains. As a result, most acid-fast positive organisms are only weakly Gram positive. The general sequence of events in an acid-fast stain are shown in Figure 3.59. In both the ZN and K methods, the phenolic compound carbolfuchsin is used as the primary stain because it is lipid soluble and penetrates the waxy cell wall. Staining by carbolfuchsin is further enhanced in the ZN method by steam heating the preparation to melt the wax and allow the stain to move into the wall.

The K method compensates for not heating the smear by using a more concentrated and lipid-soluble carbolfuchsin solution as the primary stain. But as a consequence of not heating, the K method is slightly less sensitive than the ZN method. In both, acid alcohol is used to decolorize nonacid-fast cells; acid-fast cells resist this decolorization. A counterstain, such as methylene blue, then is applied. Acid-fast cells are reddish-purple, often with a beaded appearance; nonacid-fast cells are blue (Fig. 3.60).

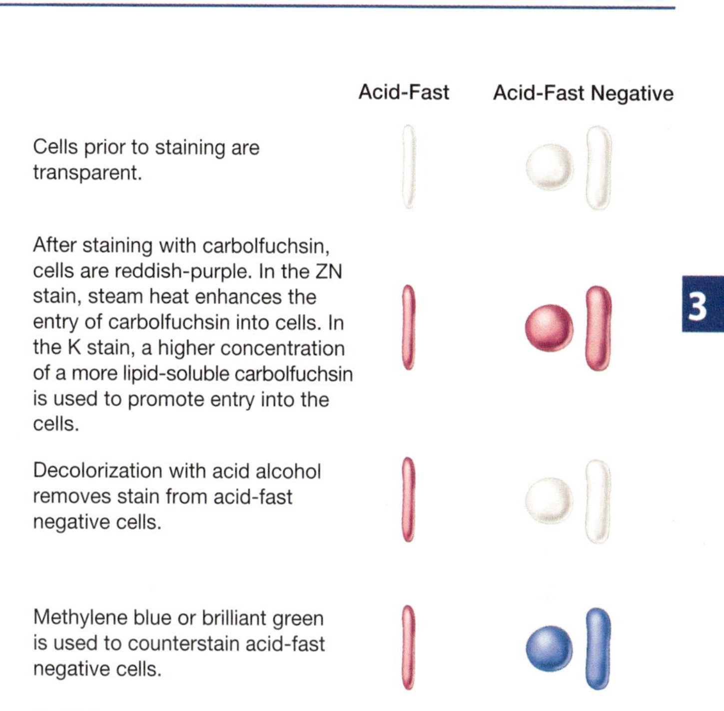


**3.59 Acid-Fast Stain Overview of Both ZN and K Methods** Acid-fast cells stain reddish purple; acid-fast negative cells stain blue or the color of the counterstain if a different one is used.

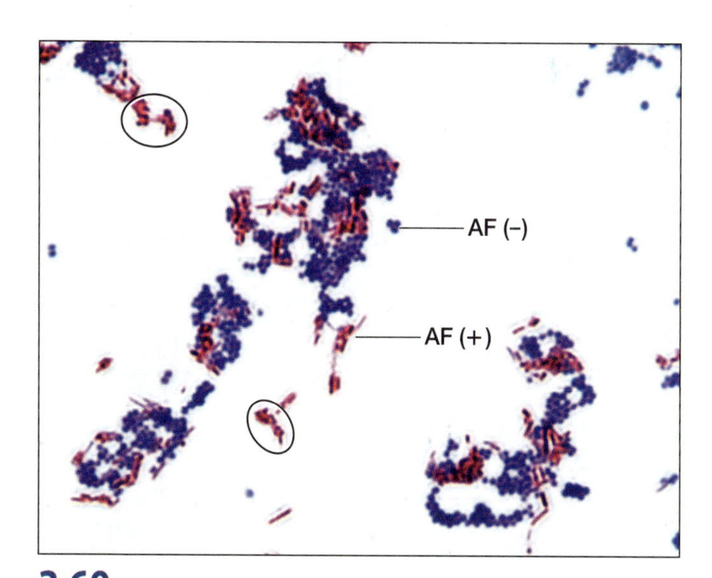


**3.60 Acid-Fast Stain Results** Notice how most of the *Mycobacterium phlei* (AF+) cells are in clumps, an unusual state for most rods. They do this because their waxy cell walls make them sticky. Try carefully and gently mixing them a little longer when preparing the slide than you would for other stains. A few individual cells are visible, however, and they clearly are rods that measure 0.5 by 2–3 μm. The *Staphylococcus epidermidis* cells (AF−) are also in clumps, but that is because they grow as grapelike clusters. Each cell's diameter is approximately 1 μm. Also notice the characteristic beaded appearance of some AF+ cells (circled).

## Application

The acid-fast stain is a differential stain used to detect cells capable of retaining a primary stain when treated with an acid alcohol. It is an important differential stain used to identify bacteria in the genus *Mycobacterium*, some of which are pathogens (e.g., *M. leprae* and *M. tuberculosis*, causative agents of leprosy and tuberculosis, respectively).

Members of the actinomycete genus *Nocardia* (*N. brasiliensis* and *N. asteroides* are opportunistic pathogens) are partially acid-fast. Oocysts of coccidian parasites, such as *Cryptosporidium* and *Isospora,* are also acid-fast. Because so few organisms are acid-fast, the acid-fast stain is run only when infection by an acid-fast organism is suspected.

Acid-fast stains are useful in identifying **acid-fast bacilli** (**AFB**) and in rapid, preliminary, and provisional diagnosis of tuberculosis (with greater than 90% predictive value from sputum samples). It also can be performed on patient samples to track the progress of antibiotic therapy and determine the degree of contagiousness. A prescribed number of microscopic fields are examined and the number of AFB is determined and reported using a standard scoring system.

3

## In This Exercise

Today you will perform an acid-fast stain designed primarily to identify members of the genus *Mycobacterium*. Because you know ahead of time which organisms should give a positive result and which should give a negative result, it is okay to mix them into a single emulsion.

## Materials

### Per Student

- □ Lab coat
- □ Disposable gloves
- □ Acid-Fast Stain Using the ZN Method

### Per Student Group

- □ Compound microscope with oil-immersion lens and ocular micrometer
- □ Clean glass microscope slides
- □ Staining tray
- □ Staining screen
- □ Bibulous paper or paper towel
- □ Slide holder
- □ Sterile wooden sticks or disposable plastic loops
- □ Ziehl-Neelsen Stains (complete kits are commercially available)
  - Methylene blue stain
  - Ziehl's carbolfuchsin stain
  - Acid alcohol (95% ethanol + 3% HCl)
- □ Kinyoun Stains (complete kits are commercially available)
  - Kinyoun carbolfuchsin
  - Acid alcohol (95% ethanol + 3% HCl)
  - Methylene blue stain or brilliant green stain (some kits come with this counterstain)
- □ Sheep serum
- □ Squirt bottle with water
- □ Heating apparatus (steam or hot plate)—for ZN only
- □ Nonsterile Petri dish for transporting slides
- □ Fresh agar slant cultures of these recommended organisms:
  - *Mycobacterium phlei* (BSL-2) or *M. smegmatis* (BSL-2)
  - *Staphylococcus epidermidis*

## PROCEDURE

### Ziehl-Neelsen (ZN) Method

1. Wear a lab coat, gloves, and chemical eye protection when staining. Remove your gloves and eye protection when using the microscope.
2. **Using BSL-2 precautions,** prepare a smear of each organism on a clean glass slide as illustrated in Figure 3.41, substituting a drop of sheep serum for the drop of water. Gently mix the *Mycobacterium* smear thoroughly because the cells tend to stick together in clumps (Fig. 3.60), but use care not to produce aerosols. Air-dry and then heat-fix the smears using a microincinerator. ***Note***: Because you know which organism is acid-fast positive and which is acid-fast negative, you may make two separate smears right next to one another on the slide or mix the two organisms in one smear. This is not recommended if you are staining an unknown.
3. Follow the staining protocol shown in the procedural diagram (Fig. 3.61). Use a steaming apparatus (such as the one in Fig. 3.62) to heat the slide. If the slide must be carried to and from the steaming apparatus, put it in a covered Petri dish. ***Caution!*** Be sure to have adequate ventilation (such as a fume hood or biosafety cabinet) while staining.
4. Observe using the oil-immersion lens. Record your observations of cell morphology and arrangement, dimensions, and acid-fast reaction in the table on the data sheet, page 199.
5. When finished, dispose of slides in a disinfectant jar or sharps container.

## PROCEDURE

### Kinyoun Method

1 Wear a lab coat, gloves, and chemical eye protection when staining. Remove your gloves and eye protection when using the microscope.

2 **Using BSL-2 precautions,** prepare a smear of each organism on a clean glass slide as illustrated in Figure 3.41, substituting a drop of sheep serum for the drop of water. Gently mix the *Mycobacterium* smear thoroughly because the cells tend to stick together in clumps (Fig. 3.60), but use care not to produce aerosols. Air-dry and then heat-fix the smears using a microincinerator. ***Note:*** Because you know which organism is acid-fast positive and which is acid-fast negative, you may make two separate smears right next to one another on the slide or mix the two organisms in one smear. This is not recommended if you are staining an unknown.

3 Follow the staining protocol shown in the procedural diagram (Fig. 3.63). ***Caution!*** Be sure to wear gloves, eye protection, and perform the stain with adequate ventilation. A fume hood or a biosafety cabinet is recommended.

4 Observe using the oil-immersion lens. Record your observations of cell morphology and arrangement, dimensions, and acid-fast reaction on the data sheet.

5 When finished, dispose of slides in a disinfectant jar or a sharps container.

**1.** Prepare a slide with up to three emulsions, each in a drop of serum (optional). Use a small inoculum and mix each smear thoroughly to separate the sticky mycobacterial cells, but do not spatter (Fig. 3.41). Air dry and then heat-fix. **Use BSL-2 precautions**. Wearing gloves and eye protection, place the slide on the steaming apparatus (Fig. 3.62). If you need to transport the slide to another part of the lab for staining, put it in a covered Petri dish.

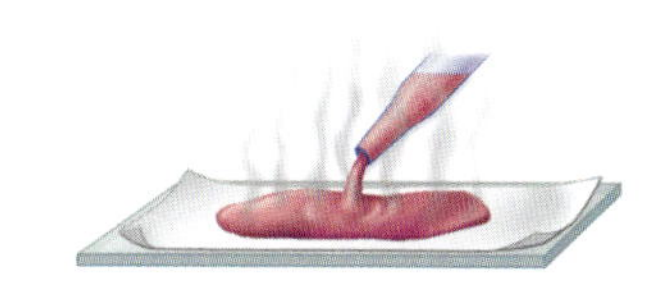

**2.** Perform this step with adequate ventilation (e.g., in a fume hood). Wearing gloves and chemical eye protection, and the slide on the steaming apparatus, cover the smear(s) with a strip of bibulous paper cut slightly smaller than the slide and apply ZN carbolfuchsin stain. Heat the slide to *steaming* for 5 minutes as shown in Figure 3.62. Keep the paper moist with stain and do not boil it.

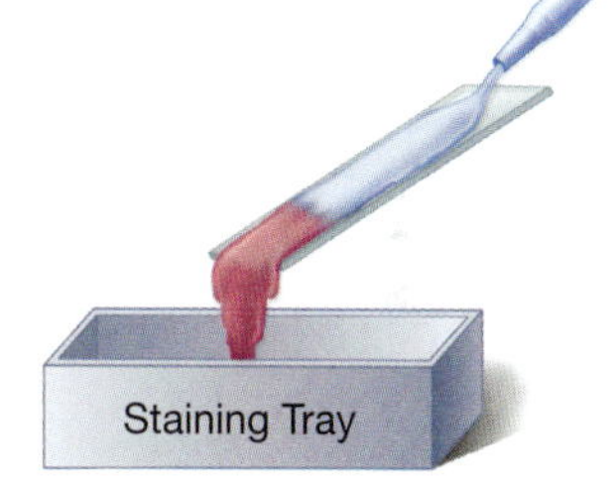


**3.** Remove the bibulous paper (with forceps, if available, to keep the stain off your gloves) and dispose of it properly. Grasp the slide with a slide holder and hold it on an angle. Gently rinse both sides of the slide with distilled water into the staining tray.

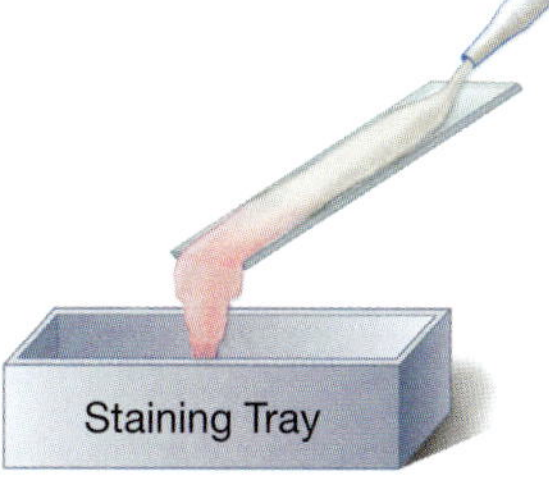


**4.** Continue holding the slide with a slide holder. Decolorize with acid-alcohol until the runoff is clear. **Use caution when handling the acid-alcohol.**

**5.** Still holding the slide on an angle, gently rinse with distilled water into a staining tray.

**6.** If not already there, return to your lab station carrying the slide in a Petri dish. Then, place the slide on the staining tray and counterstain the smear(s) with methylene blue for 1 minute. Be sure the staining tray catches any excess stain.

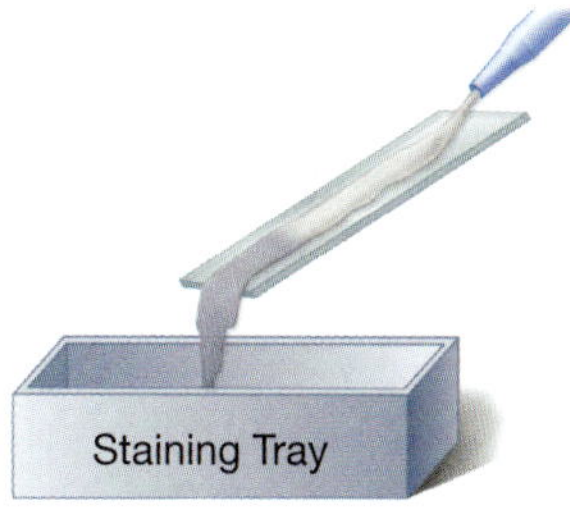


**7.** Grasp the slide with a slide holder and hold it on an angle. Gently rinse the slide with distilled water into the staining tray.

**8.** Gently blot dry in a tablet of bibulous paper or paper towels. (Alternatively, a page from the tablet can be removed and used for blotting.) Do not rub. When dry, observe under oil immersion.

**3.61 Procedural Diagram: ZN Acid-Fast Stain** ■ **Use BSL-2 precautions** throughout this procedure and perform the staining in a fume hood or well-ventilated area. Carry the slide to and from the steaming apparatus in a covered Petri dish.

**3.62 Steaming the Slide During the Ziehl-Neelsen Stain** ■ Carefully steam the slide to melt the waxy wall so the carbolfuchsin can get into acid-fast cells. Do not boil the slide or let it dry out. Keep the paper moist with stain for the entire 5 minutes of steaming. ***Caution***: This must be performed in a fume hood or a biosafety cabinet with hand, clothing, and eye protection.

## References


Atlas, Ronald and James Snyder. Pages 321–322 in *Manual of Clinical Microbiology,* 11th ed. James H. Jorgensen, Michael A. Pfaller, Karen C. Carroll, Guido Funke, Marie Louise Landry, Sandra S. Richter, and David W. Warnock, eds. Washington, DC: ASM Press, American Society for Microbiology, 2015.

Chapin, Kimberle C. and Patrick R. Murray. Pages 259–261 in *Manual of Clinical Microbiology,* 8th ed. Patrick R. Murray, Ellen Jo Baron, James H. Jorgensen, Michael A. Pfaller, and Robert H. Yolken, eds. Washington, DC: American Society for Microbiology, 2003.

Doetsch, Raymond N. and C. F. Robinow. Page 32 in *Methods for General and Molecular Bacteriology*. Philipp Gerhardt, R. G. E. Murray, Willis A. Wood, and Noel R. Krieg, eds. Washington, DC: American Society for Microbiology, 1994.

Norris, J. R. and Helen Swain. Chap. II in *Methods in Microbiology,* Vol. 5A. J. R. Norris and D. W. Ribbons. eds. London, UK: Academic Press, Ltd., 1971.

Power, David A. and Peggy J. McCuen. Page 5 in *Manual of BBL™ Products and Laboratory Procedures,* 6th ed. Cockeysville, MD: Becton Dickinson Microbiology Systems, 1988.

Tille, Patricia M. Pages 73–76 and 497–499 in *Bailey & Scott's Diagnostic Microbiology*, 13th ed. St. Louis, MO: Mosby, 2014.


**1.** Prepare a slide with up to three emulsions, each in a drop of serum (optional). Use a small inoculum and mix each smear thoroughly to separate the sticky mycobacterial cells, but do not spatter (Fig. 3.41). Air dry and then heat-fix. **Use BSL-2 precautions.**

**2.** Perform this step with adequate ventilation (e.g., in a fume hood). Wearing gloves and eye protection, place the slide on a rack over a staining tray and apply Kinyoun carbolfuchsin for 5–10 minutes. DO NOT heat the slide. Be sure the staining tray catches any excess stain.

Staining Tray

**3.** Grasp the slide with a slide holder and hold it on an angle. Gently rinse the slide with distilled water into the staining tray.

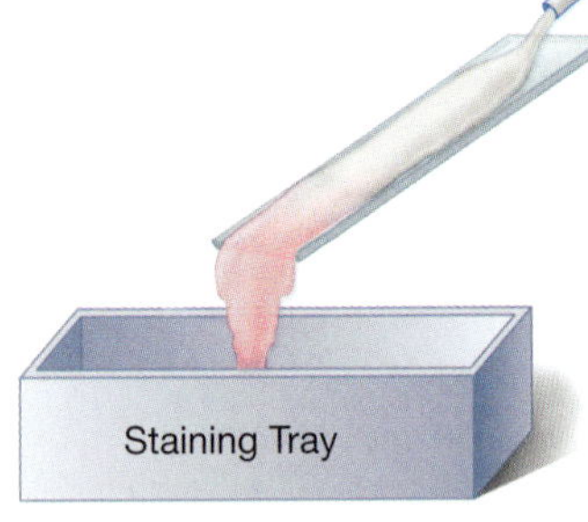


**4.** Continue holding the slide with a slide holder. Decolorize with acid-alcohol until the run-off is clear. **Use caution when handling the acid-alcohol.**

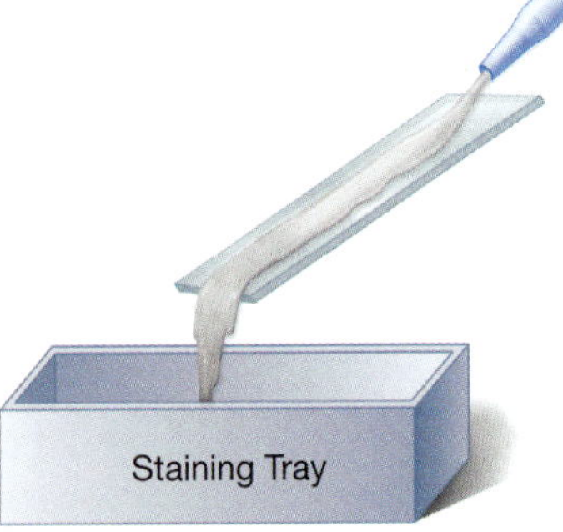


**5.** Still holding the slide on an angle, gently rinse with distilled water into the staining tray.

**6.** Return the slide to the staining tray and counterstain the smear(s) with methylene blue or brilliant green stain for 1 minute. Be sure the staining tray catches any excess stain.

**3.63 Procedural Diagram: Kinyoun Acid-Fast Stain** ■ Use BSL-2 precautions throughout this procedure and perform the stain in a fume hood or well-ventilated area. Carry the slide to and from the fume hood in a covered Petri dish.

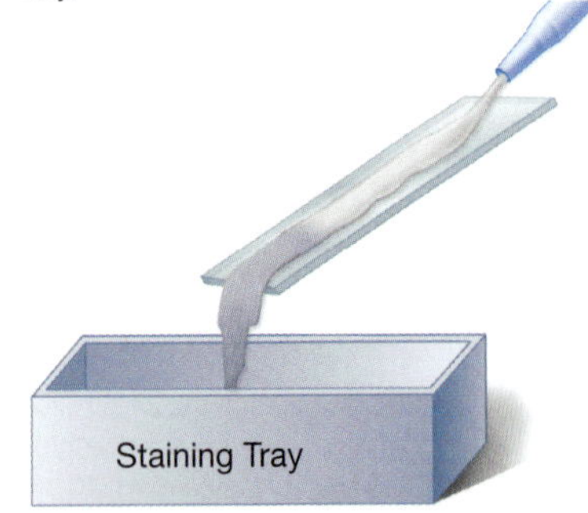


**7.** Grasp the slide with a slide holder and hold it on an angle. Gently rinse the slide with distilled water into the staining tray.

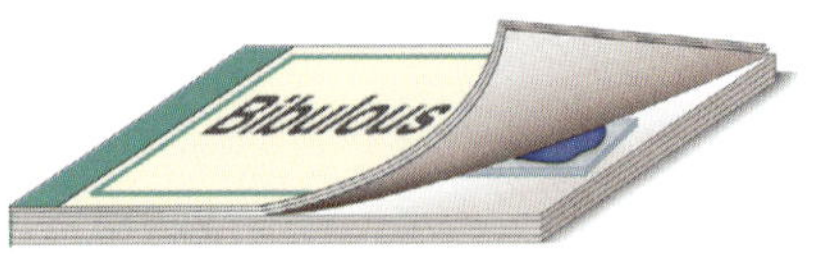


**8.** Gently blot dry in a tablet of bibulous paper or paper towels. (Alternatively, a page from the tablet can be removed and used for blotting.) Do not rub. When dry, observe under oil immersion.

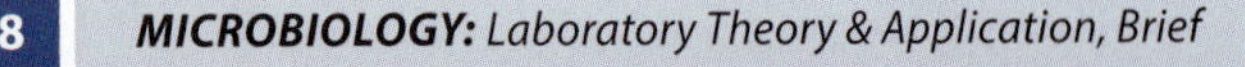

Name ______________________________

Date ______________________________

Lab Section ______________________________

I was present and performed this exercise (initials) ______________________________

# DATA SHEET 3-7

## Acid-Fast Stains

### OBSERVATIONS AND INTERPRETATIONS

**1** Record your observations in the table below.

| Organism | Staining Method (ZN or K) | Cellular Morphology and Arrangement (include a detailed sketch of a few representative cells) | Cell Dimensions | Color | Acid-Fast Reaction (+/−) |
|---|---|---|---|---|---|
| | | | | | |
| | | | | | |

## QUESTIONS

**1** *How does heating the bacterial smear during a ZN stain promote entry of carbolfuchsin into the acid-fast cell wall?*

**2** *Are acid-fast negative cells stained by carbolfuchsin? If so, how can this be a differential stain?*

**3** *Why do you suppose the acid-fast stain is not as widely used as the Gram stain? When is it more useful than the Gram stain?*

# Capsule Stain

EXERCISE **3-8**

## Theory

**Capsules** are composed of mucoid polysaccharides or polypeptides that repel most stains because of their neutral charge. The capsule stain technique takes advantage of this characteristic by staining *around* the cells. Typically, an acidic stain such as Congo red or nigrosin, which stains the background, and a basic stain that colorizes the cell proper are used in combination. The capsule remains unstained and appears as a white halo between the cells and the colored background (Fig. 3.64).

As in a negative stain, cells are spread in a film of an acidic stain and are not heat-fixed, but for a different reason. Heat-fixing causes the cells to shrink and leaves an artifactual white halo around them that might be interpreted as a capsule. In place of heat-fixing, cells may be emulsified in a drop of serum to promote their adhering to the glass slide.

## Application

The capsule stain is a differential stain used to detect cells capable of producing an extracellular capsule. Capsule production increases virulence in some microbes (such as the anthrax bacillus *Bacillus anthracis* and the pneumococcus *Streptococcus pneumoniae*) by making them less vulnerable to phagocytosis.

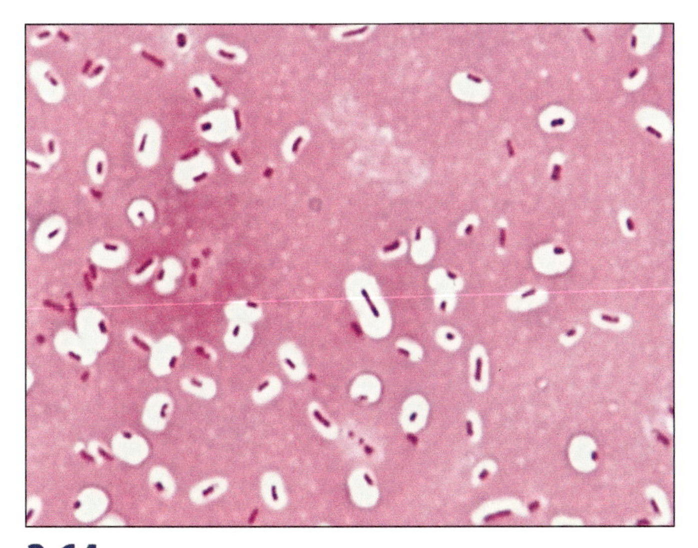

**3.64 Capsule Stain** ■ The acidic stain colorizes the background while the basic stain colorizes the cell, leaving the capsules as unstained, white clearings around the cells. Notice the lack of uniform capsule size, and even the absence of a capsule in some cells. There are several variations of the capsule stain. The capsule stain protocol provided in this exercise will produce different colors than in this micrograph, but the principle is the same—capsules will appear as white clearings.

## In This Exercise

The capsule stain allows you to visualize an extracellular capsule, if present. Be careful to distinguish between a tiny white halo (as a result of cell shrinkage) and a true capsule.

## Materials

### Per Student

- ☐ Lab coat
- ☐ Disposable gloves
- ☐ Chemical eye protection
- ☐ Clean glass slides
- ☐ Sheep serum
- ☐ Maneval's stain
- ☐ Congo red stain
- ☐ Squirt bottle with water
- ☐ Staining tray
- ☐ Staining screen
- ☐ Coplin staining jar (or a beaker) with distilled water
- ☐ Bibulous paper or paper towel
- ☐ Slide holder
- ☐ Sterile toothpicks
- ☐ Compound microscope with oil-immersion lens and ocular micrometer
- ☐ Immersion oil
- ☐ Lens paper
- ☐ Recommended organisms (18- to 24-hour skim milk or tryptic soy agar slant pure cultures):
  - *Aeromonas sobria*
  - *Rhizobium leguminosarum*
  - (Optional) *Klebsiella pneumoniae* (BSL-2)

## PROCEDURE

1. Wear a lab coat, gloves, and chemical eye protection when staining. Remove your gloves and eye protection when using the microscope.
2. Follow the protocol in the procedural diagram (Fig. 3.65) to make stains of the organisms supplied. **Use BSL-2 precautions if *K. pneumoniae* is stained.** Use a separate slide for each specimen and do not heat-fix them.

3 Remove your gloves and wash your hands. Then, using a sterile toothpick, obtain a sample from below the gumline in your mouth. (Do not draw blood!) Mix the sample into a drop of water or serum on a slide, and then perform a capsule stain on it.

4 Observe, using the oil-immersion lens. Record your observations of cell morphology and arrangement, cell dimensions, and presence or absence of a capsule in the table on the data sheet, page 203.

5 Dispose of the specimen slides in a jar of disinfectant or sharps container after use.

## References


Hughes, Roxana B. and Ann C. Smith. "Capsule Stain Protocols," ASM MicrobeLibrary (September 29, 2007). Updated July 22, 2013. Available online. URL: http://www.microbelibrary.org/library/laboratory-test/3041-capsule-stain-protocols.

Murray, R. G. E., Raymond N. Doetsch, and C. F. Robinow. Page 35 in *Methods for General and Molecular Bacteriology*. Philipp Gerhardt, R. G. E. Murray, Willis A. Wood, and Noel R. Krieg, eds. Washington, DC: American Society for Microbiology, 1994.

Norris, J. R. and Helen Swain. Chap. II in *Methods in Microbiology*, Vol. 5A. J. R. Norris and D. W. Ribbons, eds. London, UK: Academic Press, Ltd. 1971.

**1.** Working on the table top covered with a paper towel, place a loopful of sheep serum at one end of a clean glass slide. Add a drop or loopful of Congo red stain. Be sure to wear gloves.

**2.** Aseptically add organisms and emulsify with a loop. **Be careful not to spatter.** Sterilize the loop after emulsifying. ***Note:*** If you are transferring a BSL-2 organism, use a sterile wooden stick or disposable loop/needle and dispose of it properly when finished.

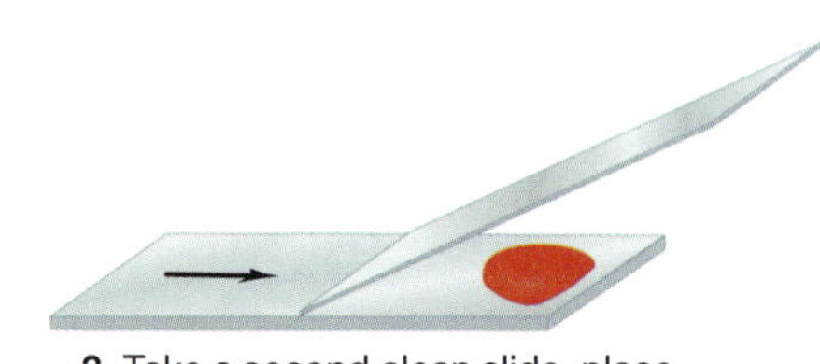

**3.** Take a second clean slide, place it on the surface of the first slide, and draw it back into the drop while holding it at a 30° angle.

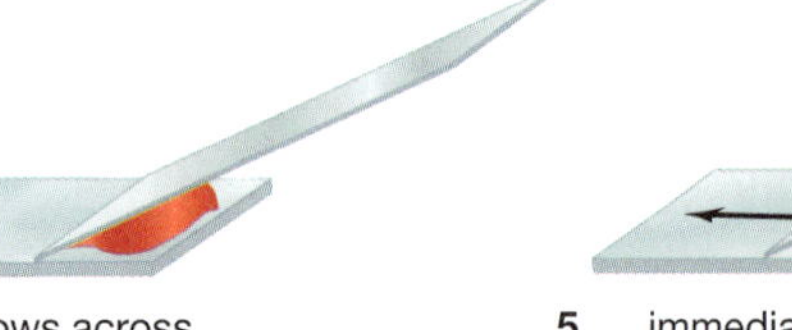

**4.** When the drop flows across the width of the spreader slide (or slightly before) ...

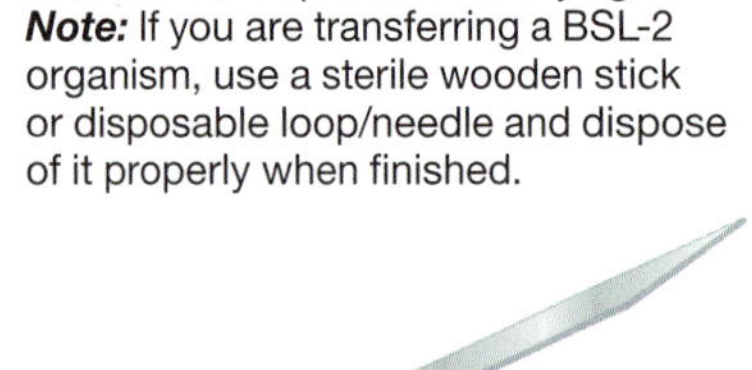

**5.** ... immediately push the spreader slide to the other end. Dispose of the spreader slide in a jar of disinfectant or sharps container.

**6.** Air-dry and do NOT heat-fix.

**7.** Place the slide on a rack over a staining tray. Cover the spread with Maneval's stain for 1 minute. Make sure the staining tray catches all excess stain.

**8.** Grasp the slide with a slide holder and dip it into a jar of distilled water two or three times to rinse off the Maneval's stain. Change the water in the jar frequently.

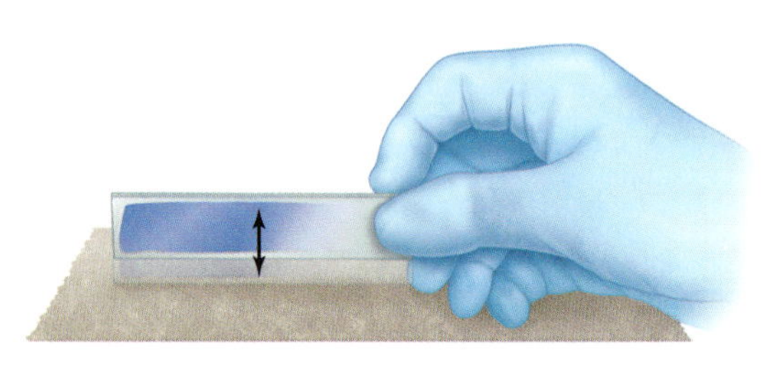

**9.** Gently tap the edge of the slide on the paper towel to remove excess water, then set it down to air-dry. When dry, observe under oil immersion.

**3.65 Procedural Diagram: Capsule Stain** ■ Be sure to sterilize your loop after transfer and to appropriately dispose of the spreader slide immediately after use. Do not set it on the table. Once air-dried, the slide is ready for counterstaining. No heat-fixing is required. Gently rinsing the Maneval's stain and not blotting the slide dry will help with the quality of your preparations. Properly dispose of the slide after viewing.

Name ____________________

Date ____________________

Lab Section ____________________

I was present and performed this exercise (initials) ____________________

# DATA SHEET 3-8

## Capsule Stain

### OBSERVATIONS AND INTERPRETATIONS

**1** Record your observations in the table below. Use a separate line for each organism from your gumline sample that illustrates a capsule or an unusual cell morphology or arrangement. Measure cell dimensions and the width of the capsule from the cell's surface to the capsule's edge.

| Organism | Cellular Morphology and Arrangement (include a detailed sketch of a few representative cells) | Capsule (+/−) | Cell Dimensions and Width of Capsule (if any) |
|---|---|---|---|
| | | | |
| | | | |
| | | | |
| | | | |
| | | | |
| | | | |

## QUESTIONS

**1** *Capsules are neutrally charged. This being the case, what is the purpose of emulsifying the sample in serum in this staining procedure?*

**2** *Some oral bacteria produce an extracellular "capsule." Of what benefit is a capsule to these cells?*

**3** *Sketch any cells from your mouth sample that display an unusual morphology or arrangement.*

# Endospore Stain

## Theory

Some bacterial species are able to differentiate into dormant cells called **endospores** when environmental conditions, such as nutrient depletion or high temperatures, are unsuitable for growth. Endospores are highly resistant to heat and chemicals, which allows them to survive in this state for long periods of time. The total absence of ATP within endospores is an indication of how dormant they are!

In addition to nutrient depletion, sporulation has also been shown to be dependent on population density. With increased density, a secreted peptide (called "competence and sporulation factor," or CSF) reaches a critical concentration and results in derepression of sporulation genes, which is a fancy way of saying, "Taking the brakes off" of them. Endospores' resistance is due to a combination of factors, including a tough outer covering made of the protein **keratin**, its dehydrated state, DNA protective proteins, and other adaptations. When conditions are suitable again, endospores germinate into metabolically active **vegetative cells.**

The keratin in the spore coat also resists staining, so extreme measures must be taken to stain an endospore. In the Schaeffer-Fulton method (Fig. 3.66), a primary stain of malachite green is forced into the spore by steaming the bacterial emulsion. Alternatively, malachite-green can be left on the slide for 15 minutes or more to stain the spores. Malachite green is water soluble and has a low affinity for cellular material, so vegetative cells and **spore mother cells** (which are responsible for producing the endospore) can be decolorized with water and counterstained with safranin (Fig. 3.67).

Endospores may be located in the middle of the cell (**central**), at the end of the cell (**terminal**), or between the end and middle of the cell (**subterminal**). Endospore location in some species is variable and cells may exhibit a combination of terminal and subterminal, for instance. Endospores also may be differentiated based on shape—either **spherical** or **elliptical** (**oval**)—and size relative to the cell, that is, whether they cause the cell to look swollen or not. These structural features are shown in Figures 3.68 through 3.70.

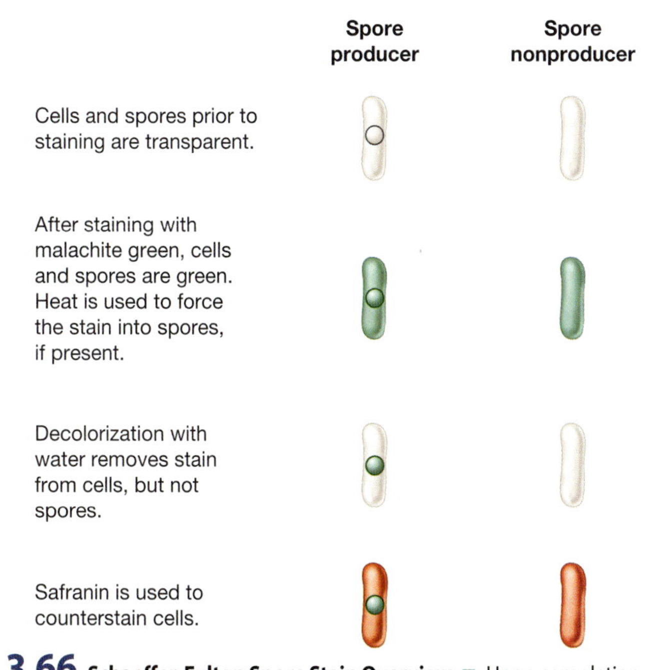


**3.66 Schaeffer-Fulton Spore Stain Overview** ■ Upon completion, endospores are green, vegetative and spore mother cells are red.

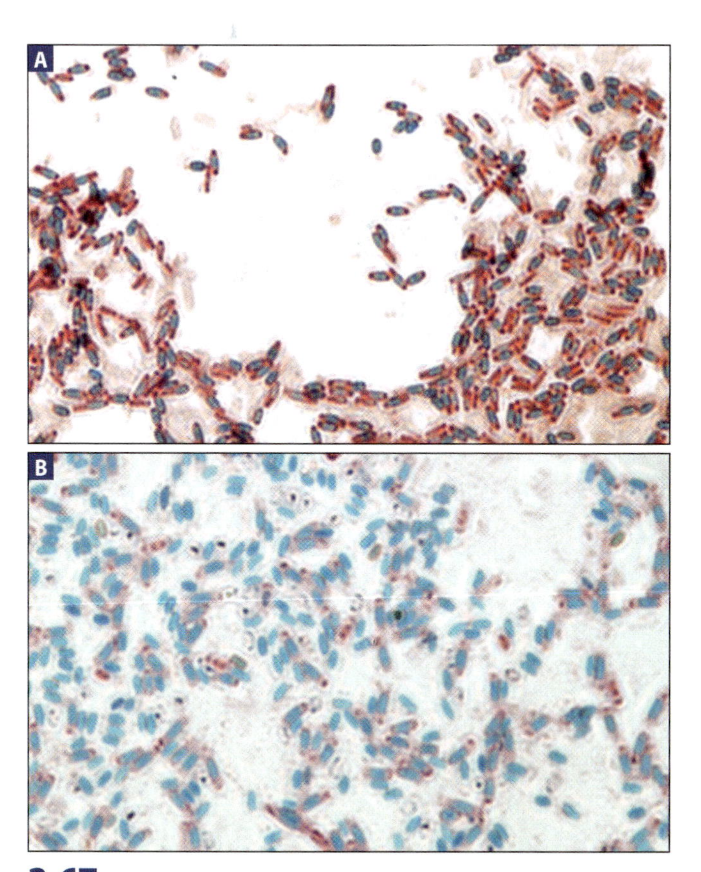


**3.67 Culture Age Can Affect Sporulation** ■ Bacteria capable of producing endospores do not do so uniformly during their culture's growth. Sporulation is done in response to nutrient depletion, and so is characteristic of older cultures. These two cultures illustrate different degrees of sporulation. (**A**) Most cells in this specimen contain endospores; very few have been released. The light pink rods are probably spore mother cells that have released their spore and have died. (**B**) This specimen consists mostly of released endospores.

Given enough time, *Bacillus* species will begin to produce endospores in most media that support their growth. However, sporulation can be accelerated by including manganese in the medium in excess of the amount required for vegetative growth. Manganese is required as a cofactor by several enzymes during sporulation (and germination). Your instructor may supply you with cultures grown in sporulating agar (see recipe, p. 207).

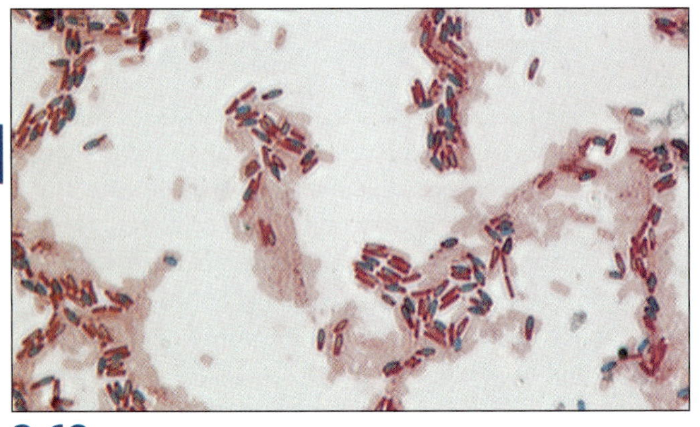

**3.68 Central Elliptical Endospores** ■ Most of these *Bacillus megaterium* endospores are centrally positioned, but some are subterminal. All are elliptical, but they do not distend the mother cell.

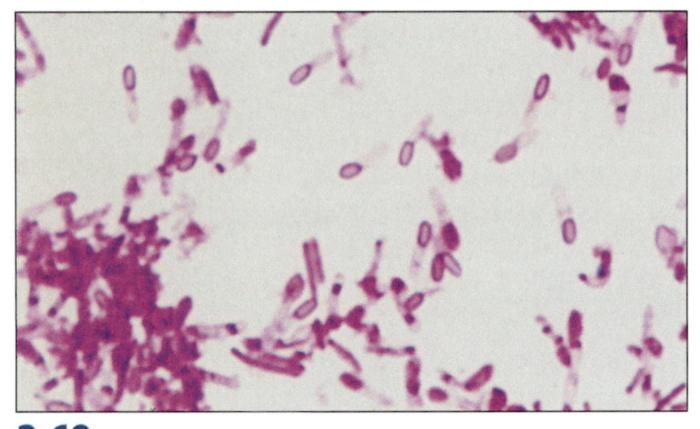

**3.69 Terminal Elliptical Endospores** ■ Shown is *Clostridium tetani* stained by a different endospore stain protocol using carbolfuchsin. Notice how the endospores have caused the ends of the cells to distend (swell).

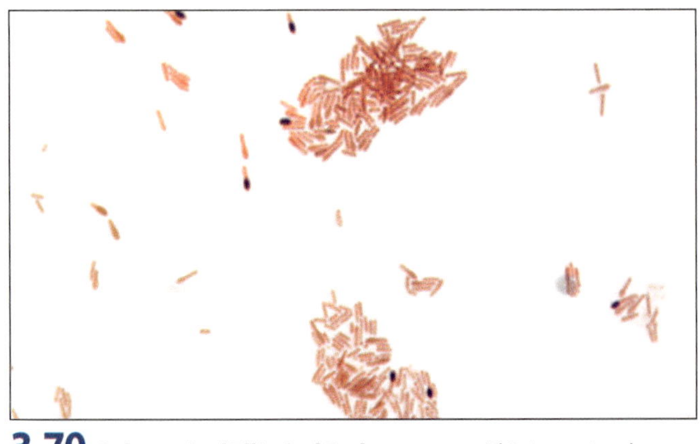

**3.70 Subterminal Elliptical Endospores** ■ This is a stained preparation of *Clostridium botulinum* using an alternative procedure. The black, elliptical endospores slightly distend the cell.

## Application

The endospore stain is a differential stain used to detect the presence, shape, and location of endospores in bacterial cells. Only a few genera produce spores. Most common are the genera *Bacillus* (over 100 species) and *Clostridium* (over 160 species). In addition, several "new" genera formerly classified as *Bacillus* species are also endospore producers. These include *Brevibacillus*, *Geobacillus*, *Paenibacillus*, and *Virgibacillus*, among others.

Most *Bacillus* species are soil, freshwater, or marine **saprophytes**, but two are pathogens. *B. anthracis* is the causal agent of anthrax and *B. cereus* causes two kinds of food poisoning: emetic and diarrheal. Most members of *Clostridium* are soil or aquatic saprophytes, or inhabitants of human intestines, but four pathogens are fairly well known: *C. tetani* (tetanus), *C. botulinum* (botulism), *C. perfringens* (gas gangrene), and *C. difficile* (pseudomembranous colitis).

## In This Exercise

Today you will perform the endospore stain on as many as four species of *Bacillus* to see variations in endospore shapes, locations, and sizes. In addition, you will be supplied with two cultures of different ages of each species to see if there is a relationship between culture age and endospore production. You won't have to stain each species. Divide the work within your lab group, but be sure to look at each other's slides.

## ▼ Materials

### Per Student

- ☐ Lab coat
- ☐ Disposable gloves
- ☐ Chemical eye protection

### Per Student Group

- ☐ Clean glass microscope slides
- ☐ Malachite green stain
- ☐ Safranin stain
- ☐ Squirt bottle with water
- ☐ Heating apparatus (steam apparatus or hot plate)
- ☐ Fume hood
- ☐ Bibulous paper or paper towel
- ☐ Staining tray
- ☐ Staining screen
- ☐ Slide holder
- ☐ Compound microscope with oil-immersion lens and ocular micrometer
- ☐ Immersion oil
- ☐ Lens paper

- ☐ Nonsterile Petri dish for transporting slides
- ☐ Recommended organisms (per student group):
  - 24-hour and 48-hour sporulating agar slant pure cultures of *Bacillus cereus*
  - 24-hour and 48-hour sporulating agar slant pure cultures of *Bacillus coagulans*
  - 24-hour and 48-hour sporulating agar slant pure cultures of *Bacillus megaterium*
  - 24-hour and 48-hour sporulating agar slant pure cultures of *Bacillus subtilis*

**Note**

Tryptic soy agar can be substituted for sporulating agar, but incubation times should be adjusted to 48 hours and 5 days.

## Medium Recipe

### Sporulating Agar

| | |
|---|---|
| ☐ Pancreatic digest of gelatin | 6.0 g |
| ☐ Pancreatic digest of casein | 4.0 g |
| ☐ Yeast extract | 3.0 g |
| ☐ Beef extract | 1.5 g |
| ☐ Dextrose | 1.0 g |
| ☐ Agar | 15.0 g |
| ☐ Manganous sulfate | 0.3 g |
| ☐ Distilled or deionized water | 1.0 L |

*Final pH 6.6 ± 0.2 at 25°C*

## PROCEDURE

1. Wear a lab coat, gloves, and chemical eye protection when staining. Remove your gloves and eye protection when using the microscope.
2. For each organism prepare and heat-fix two smears on the same slide as illustrated in Figure 3.41. One smear should be the 24-hour culture, and the other the 48-hour culture (or, 48-hour and 5-day cultures if TSA is the medium). Divide the work of slide preparation within your lab group. Minimally, each student should prepare one slide with smears of the 24-hour and 48-hour cultures of one species. ***Note*: Use BSL-2 precautions if you have subcultured soil isolates for endospore staining.**
3. Follow the instructions in the procedural diagram in Figure 3.71. Use a steaming apparatus like the one shown in Figure 3.72. Be sure to have adequate ventilation (a fume hood is highly recommended), eye protection, and gloves. If you must carry the slide to and from the steaming apparatus, put it in a covered Petri dish.
4. Observe using the oil-immersion lens. Record your observations of cell morphology and arrangement, cell dimensions, and endospore presence, position and shape in the table provided on the data sheet, page 209.

**1.** Begin with up to three heat-fixed emulsions on one slide. (Fig. 3.41.) **Use BSL-2 precautions, if appropriate**. Wearing gloves and chemical eye protection, cover the smear with a strip of bibulous paper cut slightly smaller than the slide. If you need to transport the slide to another part of the lab for staining, put it in a covered Petri dish.

**2.** Set the slide on a steaming apparatus and saturate the bibulous paper with malachite green stain. Heat it to steaming for 5 to 7 minutes (as shown in Fig. 3.72). Be sure to keep the paper moist with stain, but don't have it so wet that stain runs off the slide. Perform this step with adequate ventilation, preferably in a fume hood.

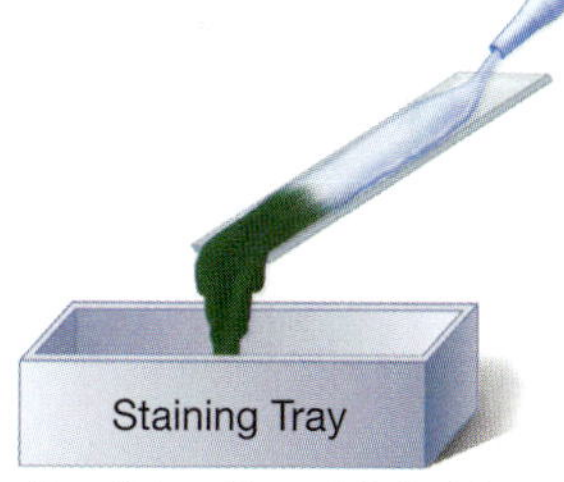


**3.** Grasp the slide with a slide holder and properly dispose of the bibulous paper with forceps (to prevent stain from getting on your gloves). Then, hold the slide on an angle over a stain tray and gently rinse both sides with distilled water until the runoff is clear. Then, if not already there, return to your lab station carrying your slide in a Petri dish.

**4.** Place the slide on the staining rack and counterstain with safranin for 1 minute. Be sure excess stain falls into the staining tray.

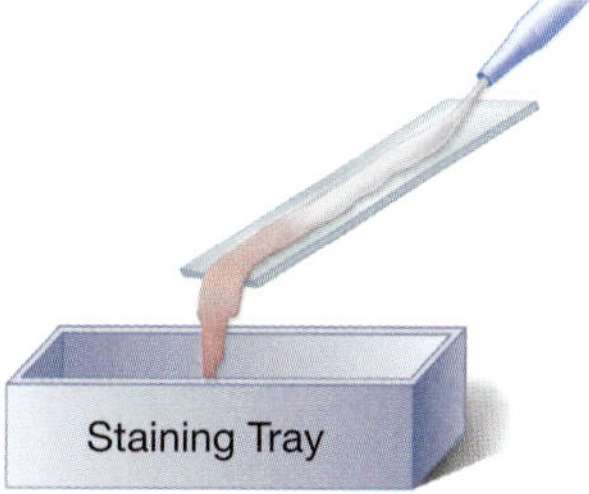


**5.** Grasp the slide with a slide holder and hold it on an angle. Gently rinse the slide with distilled water into the staining tray.

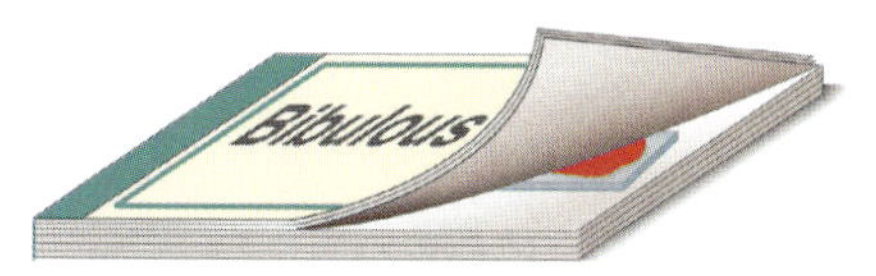


**6.** Gently blot dry in a tablet of bibulous paper or paper towels. (Alternatively, a page from the tablet can be removed and used for blotting.) Do not rub. When dry, observe under oil immersion.

**3.71 Procedural Diagram: Schaeffer-Fulton Endospore Stain** ■ Steam the staining preparation; do not boil it. Be sure to perform this procedure with adequate ventilation (preferably a fume hood), eye protection, a lab coat, and gloves. **Employ BSL-2 precautions** if you have subcultured a soil isolate for endospore staining.

**5** Dispose of specimen slides in a disinfectant jar or a sharps container.

**3.72 Steaming the Slide During the Endospore Stain** ■ Carefully steam the slide to force the malachite green into the endospores. Do not boil the slide or let it dry out. Keep it moist with stain for up to 7 minutes of steaming (not 7 minutes on the apparatus—7 minutes of steaming!). ***Caution:*** This should be performed in a well-ventilated area (preferably a fume hood) with hand, clothing, and eye protection.

## References


Atlas, Ronald and James Snyder. Page 323 in *Manual of Clinical Microbiology,* 11th ed. James H. Jorgensen, Michael A. Pfaller, Karen C. Carroll, Guido Funke, Marie Louise Landry, Sandra S. Richter, and David W. Warnock, eds. Washington, DC: ASM Press, American Society for Microbiology, 2015.

Claus, G. William. Chap. 9 in *Understanding Microbes—A Laboratory Textbook for Microbiology*. New York: W. H. Freeman and Co., 1989.

Jakubovics, Nicholas S. and Howard F. Jenkinson. "Out of the iron age: new insights into the critical role of manganese homeostasis in bacteria." *Microbiology* 147 (2001): 1709–1718.

Logan, Niall A. and Paul De Vos. Genus I. Bacillus in *Bergey's Manual of Systematic Bacteriology*, 2nd ed., Vol. III The *Firmicutes.* Paul De Vos, George M. Garrity, Dorothy Jones, Noel R. Krieg, Wolfgang Ludwig, Fred A. Rainey, Karl-Heinz Schleifer, and William B. Whitman, eds. New York: Springer, 2009.

Murray, R. G. E., Raymond N. Doetsch, and C. F. Robinow. Page 34 in *Methods for General and Molecular Bacteriology*. Philipp Gerhardt, R. G. E. Murray, Willis A. Wood, and Noel R. Krieg, eds. Washington, DC: American Society for Microbiology, 1994.

Turenne, Christine Y., James W. Snyder, and David C. Alexander. Pages 448-449 in *Manual of Clinical Microbiology,* 11th ed. James H. Jorgensen, Michael A. Pfaller, Karen C. Carroll, Guido Funke, Marie Louise Landry, Sandra S. Richter, and David W. Warnock, eds. Washington, DC: ASM Press, American Society for Microbiology, 2015.

Weinberg, Eugene D. "Manganese Requirement for Sporulation and Other Secondary Biosynthetic Processes of *Bacillus*." *Applied Microbiology* 12, no. 5 (1964): 436–441.

Name ______________________________

Date ______________________________

Lab Section ______________________________

I was present and performed this exercise (initials) ______________

# DATA SHEET 3-9

## Endospore Stain

### OBSERVATIONS AND INTERPRETATIONS

**1** Record your observations in the table below.

| **Organism** (include culture age) | **Cellular Morphology and Arrangement** (include a detailed sketch of a few representative cells and spores) | **Cell Dimensions** | **Spores** (present or absent) | **Spore Shape, Position, and Dimensions** (if present) |
|---|---|---|---|---|
| | | | | |
| | | | | |
| | | | | |
| | | | | |
| | | | | |
| | | | | |
| | | | | |
| | | | | |

## QUESTIONS

**1** *Why does this exercise call for an older (48-hour or 5-day) culture of* Bacillus?

3

**2** *Consider the possible results of an endospore stain.*

**a.** *What does a positive result for the endospore stain indicate about the organism?*

**b.** *What does a negative result for the endospore stain indicate about the organism?*

**3** *Why is it not necessary to include a negative control for this stain procedure?*

**4** *Endospores do not stain easily. Perhaps you have seen them as unstained white objects inside* Bacillus *species in other staining procedures. If they are visible as unstained objects in other stains, of what use is the endospore stain?*

# Bacterial Motility: Wet Mount and Hanging Drop Preparations

EXERCISE 3-10

## Theory

A wet mount preparation is made by placing the specimen in a drop of water on a microscope slide and covering it with a cover glass. Because no stain is used and most cells are transparent, viewing is best done with as little illumination as possible by adjusting the iris diaphragm (Fig. 3.73).

Motility often can be observed at low or high-dry magnification, but viewing must be done quickly because the preparation will dry out in 15–20 minutes. Swimming bacteria will move independently of one another, either in a straight line or in a zigzag, random path. Cell size, morphology, arrangement, and binary fission may also be observed at higher magnifications.

A hanging drop preparation allows longer observation of the specimen because it does not dry out as quickly. A thin ring of petroleum jelly is applied around the well of a depression slide. A drop of water then is placed in the center of the cover glass and living microbes are transferred into it. The depression microscope slide is carefully placed over the cover glass in such a way that the drop is received into the depression and is undisturbed. The petroleum jelly causes the cover glass to stick to the slide.

The preparation then may be picked up, inverted so the cover glass is on top, and placed under the microscope for examination. As with the wet mount, viewing is best done with as little illumination as possible, so adjust the iris diaphragm. The petroleum jelly forms an airtight seal that slows drying of the drop, allowing a longer period for observation of cell size, morphology, arrangement, binary fission, and motility.

If these techniques are done to determine motility, the observer must be careful to distinguish between true motility and the **Brownian motion** created by collisions between cells and water molecules. In the latter, cells will appear to vibrate in place. Nonmotile cells should exhibit Brownian motion. Cells that actually swim will exhibit independent movement over greater distances and the effects of collisions with water molecules will be insignificant.

In addition to Brownian motion, other complications can present themselves when observing wet mounts or hanging drop preparations for motility. Following is a list you can use to troubleshoot if you do not see motility.

- Observation: Cells are in one focal plane and not moving, even by Brownian motion. Interpretation and solution: The cells are stuck to the glass and may or may not be motile if freed (which you won't be able to do). Focus on cells in the water between the coverslip and the slide.
- Observation: Cells are streaming across the field in the same direction. Interpretation and solution: There is a current carrying the cells and they may or may not be motile. The water is either receding due to evaporation or the coverslip was put on too vigorously. Try finding a calmer place to observe the cells or simply wait until the current dissipates. If the water is evaporating, make another slide.
- Observation: Cells are packed together and not moving except, perhaps, by Brownian motion. Interpretation and solution: You have got a traffic jam and the cells may or may not be motile. Look around for a part of the slide where the cells are not so densely packed. If you can't find such a region, make a new slide with fewer cells.

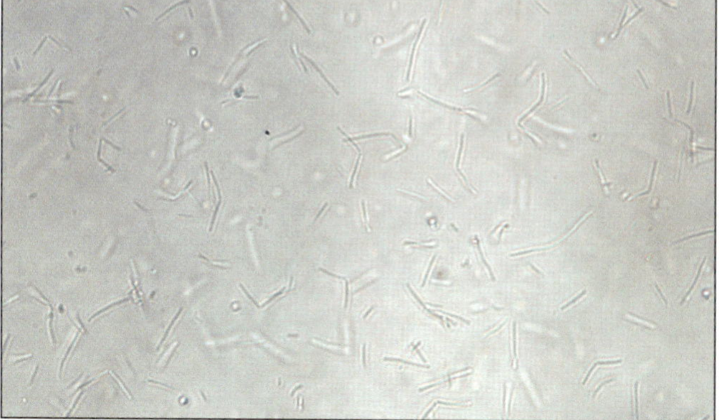

**3.73 Wet Mount** ■ Shown is an unstained wet-mount preparation of a motile Gram-negative rod using the high-dry lens. Because of the thickness of the water in the wet mount, cells show up in many different focal planes and are mostly out of focus. To get the best possible image, adjust the condenser height and reduce the light intensity with the iris diaphragm (as described in Exercise 3-1). Then, use the fine focus adjustment to scan for motile cells. Just be careful not to hit the slide with the objective lens.

One last complication presents itself, and you won't know that it has occurred because most bacterial flagella are too thin to see with the light microscope. Flagella are very delicate and it is possible to damage them when transferring cells to the slide. It is best to check for motility using a broth culture rather than trying to transfer cells from a solid medium to the water drop. Also, do not emulsify the loopful of broth on the slide. Just gently allow the broth to merge with the water drop already on the slide.

## Application

Most bacterial microscopic preparations result in death of the microorganisms as a result of heat-fixing and staining. Simple **wet mounts** and the **hanging drop technique** allow observation of living cells to determine motility. They also are used to see natural cell size, arrangement, and shape, as well as to study binary fission. These characteristics may be useful in identification of a microbe.

## In This Exercise

Today you will have the opportunity to view living bacterial cells swimming. The hanging drop preparation allows longer viewing, whereas the simple wet mount can be used to view motility, and it is the starting point for some flagella stain protocols.

3

## Materials

### Per Student

- ☐ Lab coat
- ☐ Disposable gloves
- ☐ Chemical eye protection
- ☐ Compound microscope with oil-immersion lens and ocular micrometer
- ☐ Depression slide and cover glass
- ☐ Clean microscope slides and cover glasses
- ☐ Petroleum jelly
- ☐ Toothpick
- ☐ Immersion oil
- ☐ Lens paper
- ☐ Recommended organisms (overnight cultures grown in broth media):
  - *Aeromonas sobria*
  - *Rhodospirillum rubrum*
  - *Staphylococcus epidermidis* or *Lactobacillus acidophilus*
  - (Optional) pond water

## PROCEDURE

### Wet Mount Preparation

1. Wear a lab coat, gloves, and chemical eye protection when staining. Remove your gloves (no staining was done, so follow your instructor's directions) and eye protection when using the microscope.
2. If you are short on time, divide the workload of slide preparation with your lab partners. Just make sure all specimens are viewed.
3. Place a loopful of water on a clean glass slide. If inoculating from a broth culture, the water drop may be omitted if desired.
4. Gently add bacteria to the drop by placing the loop over the drop for several seconds. Do not over-inoculate and do not vigorously emulsify. Flame the loop after transfer.
5. Using a loop, gently lower a cover glass over the drop of water (Fig. 3.74). Avoid trapping air bubbles.
6. Observe under high-dry power and record your results in the table on the data sheet, page 215. Avoid hitting the cover slip with the high-dry lens. Use of the oil-immersion lens is not recommended because in most instances it will hit the cover glass.
7. Dispose of the slide and cover glass in a disinfectant jar or a sharps container when finished.

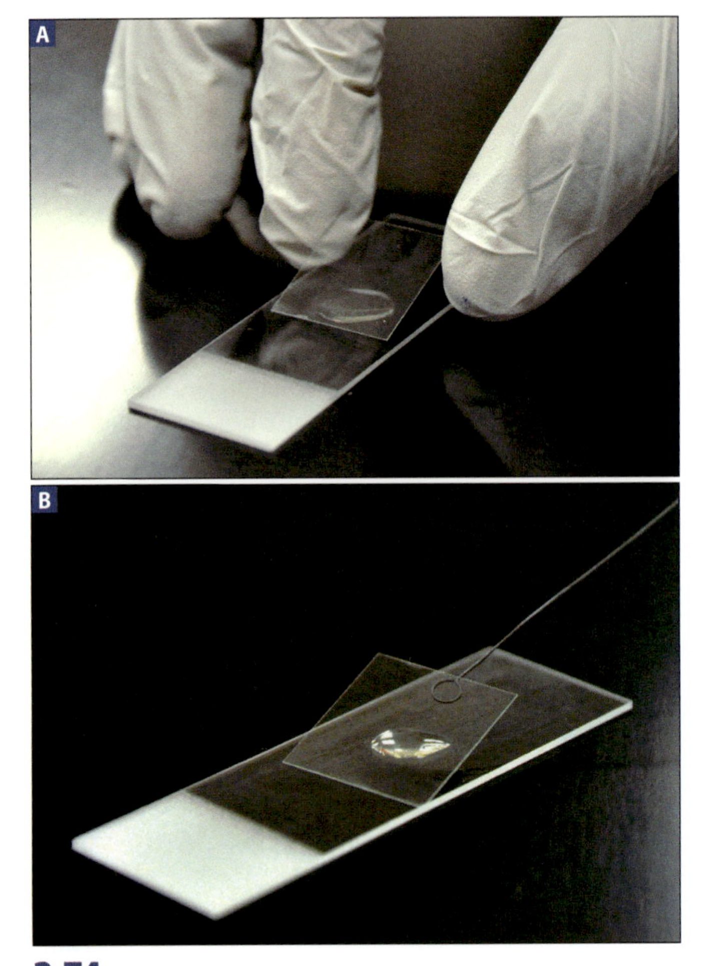


**3.74 Applying the Cover Glass** ■ There are many ways to put the cover glass onto a slide. (**A**) In this photo, the microbiologist is using his fingers to place the cover glass over the culture. Notice the cover glass is at one edge of the culture and is being (gently) lowered over it. Do not just drop it onto the culture from above. This can cause splashing of the culture. (**B**) The culture will spread out to the edges under the cover glass (and sometimes beyond) and gloves can get contaminated when placing the cover glass by hand. For this reason, we prefer an alternative method. Place one edge of the cover glass next to the specimen and gently lower it with an inoculating loop. Again, don't drop it onto the specimen. Be sure to flame the loop afterward. Whichever method you use, if you do contact the culture dispose of the gloves and wash your hands. Also, clean up any splatters if the cover glass lands too hard.

# PROCEDURE

## Hanging Drop Preparation

1 Wear a lab coat, gloves, and chemical eye protection when staining. Remove your gloves (no staining was done, so follow your instructor's directions) and eye protection when using the microscope.

2 Follow the procedure illustrated in Figure 3.75 for each specimen. Divide the workload within your lab group and observe each other's preparations.

3 Observe under high dry and record your results in the table on the data sheet. Avoid hitting the cover slip with the high-dry objective. Use of the oil-immersion lens is not recommended because in most instances it will hit the cover glass.

4 When finished, put on at least one glove and push the cover glass from the depression slide into a disinfectant jar. Then, put the depression slide in the disinfectant. Soak both for at least 15 minutes. Flame the loop. After soaking, rinse the slide with 95% ethanol to remove the petroleum jelly, then with water to remove the alcohol. Dry the depression slide for reuse. (Don't dispose of it! They are expensive!)

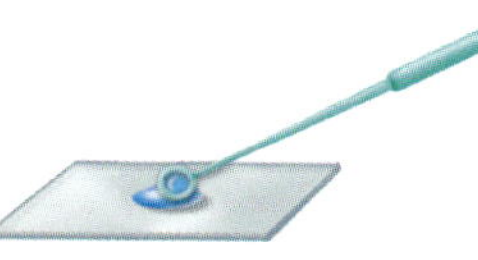

**1.** Apply a light ring of petroleum jelly around the well of a depression slide with a toothpick.

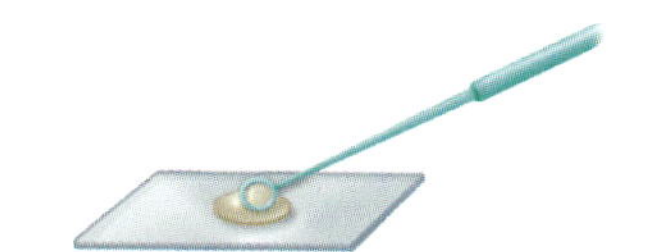

**2.** Apply a drop or a loopful of water to a cover glass. Do not use too much water. If using a broth culture, water may be omitted.

**3.** Aseptically, gently transfer a loopful of culture into the drop of water. (If you are transferring from a slant, get a loopful of sterile water and touch it to the growth for 15 seconds.) Then, without mixing, hold the loop in the water. Flame your loop (or dispose of it properly) after the transfer.

**4.** Invert the depression slide so the drop is centered in the well. Gently press until the petroleum jelly has created a seal between the slide and cover glass. Then, turn the slide right side up and place it on the stage of your microscope.

**5.** From the side, the preparation should look like this. Notice that the drop is "hanging," and is not in contact with the depression slide. Observe under high dry with the iris diaphragm closed down to reduce glare. Be careful not to hit the cover glass with the high–dry objective when using the fine-focus adjustment.

**3.75 Procedural Diagram: Hanging Drop Preparation** ■ The hanging drop method is used for long-term observation of a living specimen. Gently transfer the cells into the drop, and then observe with reduced light intensity using the high-dry lens.

## References

Iino, Tetsuo, and Masatoshi Enomoto. Chap. IV in *Methods in Microbiology*, Vol. 1. J. R. Norris and D. W. Ribbins, eds. London, UK: Academic Press, Ltd., 1969.

Murray, R. G. E., Raymond N. Doetsch, and C. F. Robinow. Page 26 in *Methods for General and Molecular Bacteriology*. Philipp Gerhardt, R. G. E. Murray, Willis A. Wood, and Noel R. Krieg, eds. Washington, DC: American Society for Microbiology, 1994.

Quesnel, Louis B. Chap. X in *Methods in Microbiology*, Vol. 1. J. R. Norris and D. W. Ribbins, eds. London, UK: Academic Press, Ltd., 1969.

Name ____________________

Date ____________________

Lab Section ____________________

I was present and performed this exercise (initials) ____________________

DATA SHEET 3-10

# Bacterial Motility: Wet Mount and Hanging Drop Preparations

## OBSERVATIONS AND INTERPRETATIONS

**1** Record your observations in the table below.

| Organism | Procedure (wet mount or hanging drop) | Cellular Morphology and Arrangement (include a detailed sketch of a few representative cells) | Cell Dimensions | Motility (+/−) |
|---|---|---|---|---|
| | | | | |
| | | | | |
| | | | | |
| | | | | |
| | | | | |
| | | | | |

## QUESTIONS

**1** *In a wet mount, each of the following complications could lead to a false interpretation of motility. For each, write "false positive," or "false negative," or "no effect," depending on how it could interfere with your reading of a motile organism and a nonmotile organism.*

| Complication | Motile Organism | Nonmotile Organism |
|---|---|---|
| Over-inoculation of the slide with organisms | | |
| Cells attaching to the glass slide or cover glass | | |
| Receding water line | | |
| Using an old culture | | |



**2** *You are told that viewing is best done with as little illumination as possible. Why will transparent cells be easier to view with less light?*

**3** *Why should Brownian motion increase the longer you observe a hanging drop or wet mount preparation?*

**4** *If you examined pond water, did you notice any bacteria swimming in "runs and tumbles"? Did you notice any swimming in more-or-less straight lines? If yes, describe the cell morphology(-ies).*

**5** *Motility because of flagella is the most common form of bacterial locomotion, but some bacteria glide. If you examined pond water, did you notice any gliding cyanobacteria? If yes, describe the morphology(-ies)*

EXERCISE **3-11**

# Bacterial Motility: Flagella

## Theory

Bacterial flagella typically are too thin to be observed with the light microscope and ordinary stains, but, there are exceptions. Figure 3.76 shows a wet mount preparation of a spirillum viewed with phase contrast and its flagella are visible without staining. However, in most instances staining will be necessary. Staining flagella requires more than just adding color to them.

Various special flagella stains have been developed that use a **mordant** to assist in coating flagella with stain to a visible thickness. Most require experience and advanced techniques, and generally are not performed in beginning microbiology classes. Instead, you will be examining a variety of commercially prepared slides.

The number and arrangement of flagella may be observed with a flagella stain. A single flagellum is said to be **polar** and the cell has a **monotrichous** arrangement (Fig. 3.77). Other arrangements (shown in Figs. 3.78 through 3.80) include **amphitrichous**, with flagella at both ends of the cell; **lophotrichous**, with tufts of flagella at the end of the cell; and **peritrichous**, with flagella emerging from the entire cell surface.

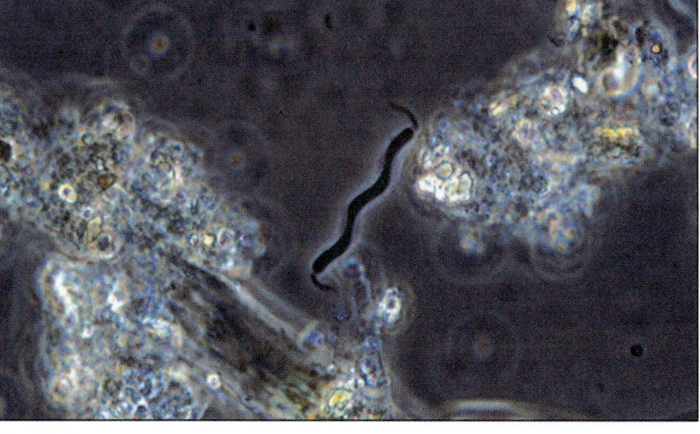

**3.76 Bacterial Flagella Without Staining—Phase Contrast** ■ The flagella of some larger bacteria are visible without staining, as in this wet mount of an unidentified spirillum viewed with phase contrast (they are not nearly as visible with bright field). A single flagellum occupies each end of the cell. They don't look particularly long because much of their length is out of the focal plane of the objective lens.

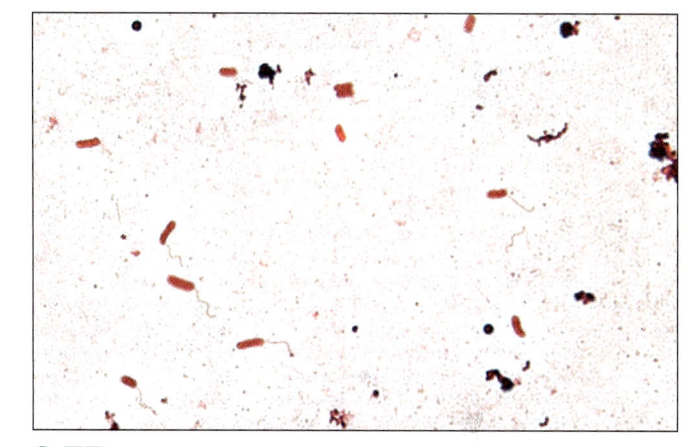

**3.77 Polar Flagella** ■ *Pseudomonas aeruginosa* is often suggested as a positive control for flagella stains, but it is a BSL-2 organism. Notice the single flagellum emerging from the ends of many (but not all) cells. This is a result of the fragile nature of flagella, which can be broken from the cells during slide preparation.

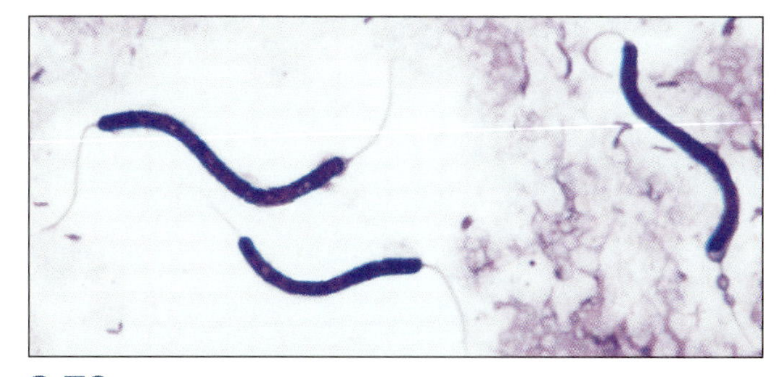

**3.78 Amphitrichous Flagella** ■ *Spirillum volutans* has a flagellum at each end. Some spirilla are large enough that the flagella are visible with phase contrast microscopy, but they are the exceptions. As a rule, bacterial flagella must be stained to be seen with the light microscope. The spirillum in Figure 3.76 also has amphitrichous flagella.

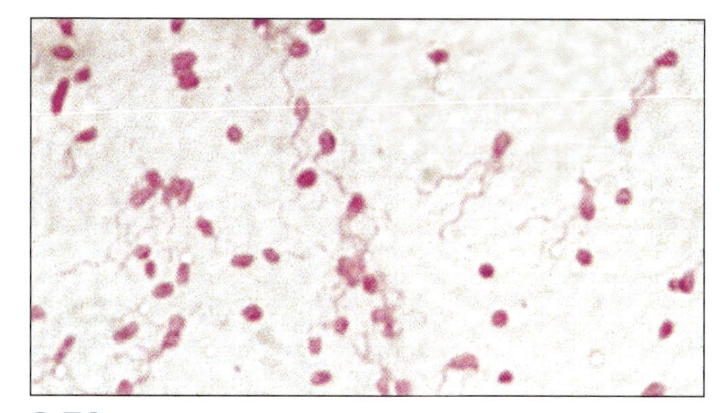

**3.79 Lophotrichous Flagella** ■ Several flagella emerge from one end of this *Pseudomonas* species. Not all cells have flagella because they were too delicate to stay intact during the staining procedure.

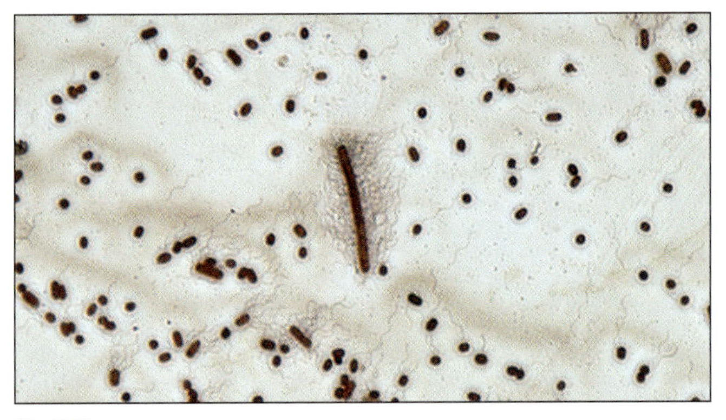

**3.80 Peritrichous Flagella** ■ Notice the flagella emerging from the entire cell surface. *P. vulgaris* is capable of swarming motility (Fig. 2.24), in which the cells spread across the agar surface at specific intervals, and then remain in place until the next swarm. The smaller cells are called "swimmers" and are the form seen when grown in a liquid medium. When transferred to a solid medium or under certain environmental conditions, swimmers differentiate into "swarmers." A swarmer is seen in the center of this micrograph. Swarmers are larger, contain multiple nucleoids (the site of DNA), and produce an extracellular slime or capsule that assists in swarming. After swarming, they break up into swimmer cells. *P. vulgaris* is an intestinal inhabitant of humans and animals, but can cause urinary tract infections. Swarmers are the more virulent form.

## Application

The flagella stain allows direct observation of flagella. The presence and arrangement of flagella may be useful in identifying bacterial species.

## In This Exercise

You will examine commercially prepared slides illustrating various flagellar arrangements.

## Materials

### Per Student

- ☐ Lab coat
- ☐ Compound microscope with ocular micrometer
- ☐ Immersion oil
- ☐ Lens paper
- ☐ Commercially prepared slides of motile organisms showing flagella

## PROCEDURE

1. Examine the available prepared microscope slides of bacteria with flagella. You will need to search carefully, because flagella are extraordinarily thin even in the best preparations (see Figs. 3.76 through 3.79) and not all cells will have them. Using the oil-immersion lens is essential.
2. Record your observations of flagellar length arrangement and number on the data sheet, page 219. Also record cell morphology, arrangement, and dimensions.
3. When finished, use a tissue to clean the oil off the slide and return it to its appropriate box.

### References


Heimbrook, Margaret E., Wen Lan L. Wang, and Gail Campbell. "Staining Bacterial Flagella Easily." *J. Clin. Microbiol* 27 (1989): 2612–2615.

Iino, Tetsuo and Masatoshi Enomoto. Chap. IV in *Methods in Microbiology*, Vol. 5A. J. R. Norris and D. W. Ribbins, eds. London, UK: Academic Press, Ltd, 1971.

Murray, R. G. E., Raymond N. Doetsch, and C. F. Robinow. Page 35 in *Methods for General and Molecular Bacteriology*. Philipp Gerhardt, R. G. E. Murray, Willis A. Wood, and Noel R. Krieg, eds. Washington, DC: American Society for Microbiology,1994.

Penner, John L. Genus *XXIX*. Proteus in *Bergey's Manual of Determinative Bacteriology*, 2nd ed. Vol. Two, The *Proteobacteria*, Part B, The *Gammaproteobacteria*. Don J. Brenner, Noel R. Krieg, and James T. Stanley, eds., 746. New York: Springer, 2005.

Name ______________________________

Date ______________________________

Lab Section ______________________________

I was present and performed this exercise (initials) ______________

DATA SHEET 3-11

# Bacterial Motility: Flagella

## OBSERVATIONS AND INTERPRETATIONS

**1** Record your observations in the table below.

| Organism | Cellular Morphology and Arrangement (include a detailed sketch of a few representative organisms with their flagella) | Cell Dimensions | Flagellar Length |
|---|---|---|---|
| | | | |
| | | | |
| | | | |
| | | | |

## QUESTIONS

**1** *Why can't flagella be observed in action?*

**3**

**2** *Flagella have a diameter of about 1 nm. A hypothetical question: To resolve flagella, what is the maximum wavelength of the electromagnetic spectrum that would have to be used to create the image (given numerical apertures of 1.25 for both the condenser and the oil lens)? Refer to Exercise 3-1 for help with this question.*

**3** *Based on the samples of motile bacteria you have observed in Exercise 3-10 and this exercise, what general conclusion can you draw about cell morphology and the possession of flagella?*

# Morphological Unknown

EXERCISE 3-12



## Theory

In this exercise you will be given one pure bacterial culture selected from the organisms listed in Table 3-4. Your job will be to identify it using only the staining techniques covered in Section 3.

Your first task is to convert the information organized in Table 3-4 into flowchart form. A flowchart is simply a visual tool to illustrate the process of elimination that is the foundation of unknown identification. We have started it for you on the data sheet, page 225, by giving you a few of the branches and listing appropriate organisms. You must complete it by adding necessary branches until you have shown a path to identify each of the organisms. The following Procedure contains a detailed explanation of the process.

Once you have designed your flowchart, you will run one stain at a time on your organism. As you match your staining results with those in the flowchart, you will follow a path to its identification, with each staining result informing you of the next stain to run.

A final stain will be performed as a **confirmatory test**. It will serve as further evidence that you have identified your unknown correctly. The only hard and fast rule about confirmatory tests is that they can't have been used during the identification process. Beyond that, it is *nice* if the confirmatory test also serves to differentiate the identified organism from the other organisms on the last branch of the flowchart, but that is not a requirement.

It is also *nice* if the confirmatory test is anticipated to give a positive result because we generally have more confidence in positive results than negative results. However, in this exercise you don't have many stains to choose from as confirmatory tests, so do your best. Another consideration is that you might want to choose a particular stain for the confirmatory test simply for practice because it was difficult for you.

## Application

Schemes employing differential tests are the main strategy for microbial identification.

## In This Exercise

You will practice the stains you have learned to this point, as well as familiarize yourself with the standard process of elimination used in bacterial identification.

## Materials

### Per Class

- ☐ Unknown organisms in numbered tubes (one per student). Fresh broth or slant cultures of[1]:
  - *Bacillus subtilis*
  - *Corynebacterium xerosis*
  - *Lactococcus lactis*
  - *Micrococcus luteus*
  - *Mycobacterium phlei* (BSL-2) or *Mycobacterium smegmatis* (BSL-2)
  - *Neisseria sicca*
  - *Pseudomonas putida*
  - *Rhizobium leguminosarum*
  - *Rhodospirillum rubrum*
  - *Shigella flexneri* (BSL-2)
  - *Staphylococcus epidermidis*

**Note**

Your instructor will choose organisms appropriate to your specific microbiology course and facilities.

- ☐ Sterile tryptic soy broth tubes (for subculturing)
- ☐ Sterile tryptic soy agar slants (for subculturing)
- ☐ (Optional) sporulating agar

### Per Student Group

- ☐ Overnight TSA slant cultures of one Gram-positive and one Gram-negative control organism, labeled only with Gram reaction
- ☐ 24-hour tryptic soy agar slant culture of an acid-fast positive organism to be used as a control (BSL-2)
- ☐ Gram stain kit
- ☐ Acid-fast stain kit
- ☐ Capsule stain kit
- ☐ Spore stain kit

### Per Student

- ☐ Lab coat
- ☐ Disposable gloves
- ☐ Chemical eye protection
- ☐ Compound microscope with oil-immersion lens
- ☐ Immersion oil

[1] Cultures should have abundant growth.

- □ Clean microscope slides
- □ Clean cover glasses
- □ Lens paper
- □ Bunsen burner
- □ Striker
- □ Disposable inoculating loops or sterile wooden sticks

## PROCEDURE

3

1. Using the information contained in Table 3-4, complete the flowchart on the data sheet, page 225. Each of the 11 organisms should occupy a solitary position at the end of a branch. Do not include stains in a branch that do not differentiate any organisms. Also, be sure the names and number of organisms on any level are equal to the names and number of organisms leading into that branch. Have your instructor check your flowchart before you begin staining.
2. Obtain one unknown slant or broth culture. Record its number in the space labeled "Unknown number" in step 2 on the data sheet.
3. ***NOTE*: Because some of the unknowns are BSL-2 organisms, all procedures must be done with BSL-2 precautions.**
4. Gloves, chemical eye protection, and a lab coat must be worn while performing all stains. Remove gloves when using the microscope and recording results.
5. Perform a Gram stain of the unknown organism. The stain should be run with known Gram-positive and Gram-negative controls to verify your technique (Fig. 3.51). In addition to providing information on Gram reaction, the Gram stain will allow observation of cell size, morphology, and arrangement. Enter all these and the date in the appropriate boxes of the table provided on the data sheet. (*Note*: Cell size for the unknown candidates is not given and should not be used to differentiate the species. To give you an idea of typical cell sizes, the cocci should be about 1 μm in diameter, and the rods 1 to 5 μm in length and about 1 μm in width.)
6. If you have cocci, or are unclear about cell arrangement from the Gram stain, aseptically transfer your unknown to a sterile tryptic soy broth and examine it again in 24–48 hours. Cell arrangement often is easier to interpret using cells grown in broth.
7. Based on the Gram stain and cell morphology results, follow the appropriate branches in the flowchart until you reach the list of remaining organisms that matches (and hopefully, includes!) your unknown.
8. Perform the next stain and determine to which new branch of the flowchart your unknown belongs. Record the result and date of this stain in the table provided on the data sheet. (Because a major purpose of this exercise is to practice staining, your instructor may prohibit you from using cell arrangement until it is the only remaining differential characteristic.)
9. Repeat step 8 until you eliminate all but one organism in the flowchart—this is your unknown! (If you are not doing all the stains on the day the unknowns are handed out, be sure to inoculate appropriate media so you will have fresh cultures to work with. If you need to do an endospore stain, incubate your original culture for another 24–48 hours or inoculate sporulation agar and incubate it for 24–48 hours, and then perform the stain.) If checking for motility, inoculate a broth and incubate it for 24–48 hours.
10. After you identify your unknown, perform one more stain to confirm your result. This confirmatory test must not be one you used in the identification process and should be added to your flowchart. The result of this test should agree with the predicted result (as given in Table 3-4). If your confirmatory test result does not match the expected result, repeat any suspect stains and find the source of error. (*Note:* The source of error may be your technique or bacterial strain variability. The results in Table 3-4 are typical for each organism, but some strains may vary. By rerunning suspect stain(s) *with controls*, if appropriate, you probably will be able to identify the source of error.)
11. Use a colored marker to highlight the path on the flowchart that leads to your unknown. Have your instructor check your work.

TABLE **3-4** **Table of Results for Organisms Used in the Exercise** These results are typical for the species listed. Your specific strains may vary because of their genetics, their age, or the environment in which they are grown. It is important that you record your results in case strain variability leads to misidentification. Also, pay attention to the footnotes!

| Organism | Gram Stain | Cell Morphology | Cell Arrangement[1] | Acid-Fast Stain | Motility (wet mount)[2] | Capsule[3] | Spore Stain[4] |
|---|---|---|---|---|---|---|---|
| *Bacillus subtilis* | + | rod | usually single cells | – | + | ± | + |
| *Corynebacterium xerosis* | + | irregular rods | single cells or multiples in angular or palisade arrangement | – | – | – | – |
| *Lactococcus lactis* | + | ovoid cocci (appearing stretched in the chain or pair) | chains and some pairs | – | – | + | – |
| *Micrococcus luteus* | + | cocci | tetrads (some singles and pairs) | – | – | – | – |
| *Mycobacterium phlei* (BSL-2) or *Mycobacterium smegmatis* (BSL-2) | weak + | thin rods | single cells or branched; sometimes in dense clusters | + | – | – | – |
| *Neisseria sicca* | – | cocci | pairs and/or tetrads with adjacent sides flattened | – | – | + | – |
| *Pseudomonas putida* | – | straight to slightly curved rods | single cells | – | + | – | – |
| *Rhizobium leguminosarum* | – | short rods | usually single rods | – | – | + | – |
| *Rhodospirillum rubrum* | – | thick spirilla or bent rods | single | – | + | – | – |
| *Shigella flexneri* (BSL-2) | – | rods | single | – | – | – | – |
| *Staphylococcus epidermidis* | + | cocci | irregular clusters (some singles, pairs, and tetrads) | – | – | – | – |

[1] It may be necessary to grow your unknown in broth culture to see the arrangement clearly. Because the purpose of this lab is to practice the stains, use cell arrangement in the identification flowchart only as a last resort.

[2] You might have better luck looking for motility if you transfer some culture to a small amount of sterile broth (in a microtube, for instance) and incubate it for an hour or so at 35°C. Then, make your wet mount from that culture.

[3] For purposes of this exercise, to be considered a capsule, the white halo needs to be at least the thickness of the cell. If you suspect cells have shrunk, measure them and compare their size and the size of the "capsule" to the cell dimensions as measured in your Gram stain.

[4] If spores are not readily visible, continue incubating your culture for another 24–48 hours and check again. You may also inoculate sporulation agar and incubate for 24–48 hours.

Name ______________________

Date ______________________

Lab Section ______________________

I was present and performed this exercise (initials) ______________________

DATA SHEET 3-12

# Morphological Unknown

## OBSERVATIONS AND INTERPRETATIONS

**1** Complete the flowchart using the information in Table 3-4, page 223. Each path should end with a single organism.

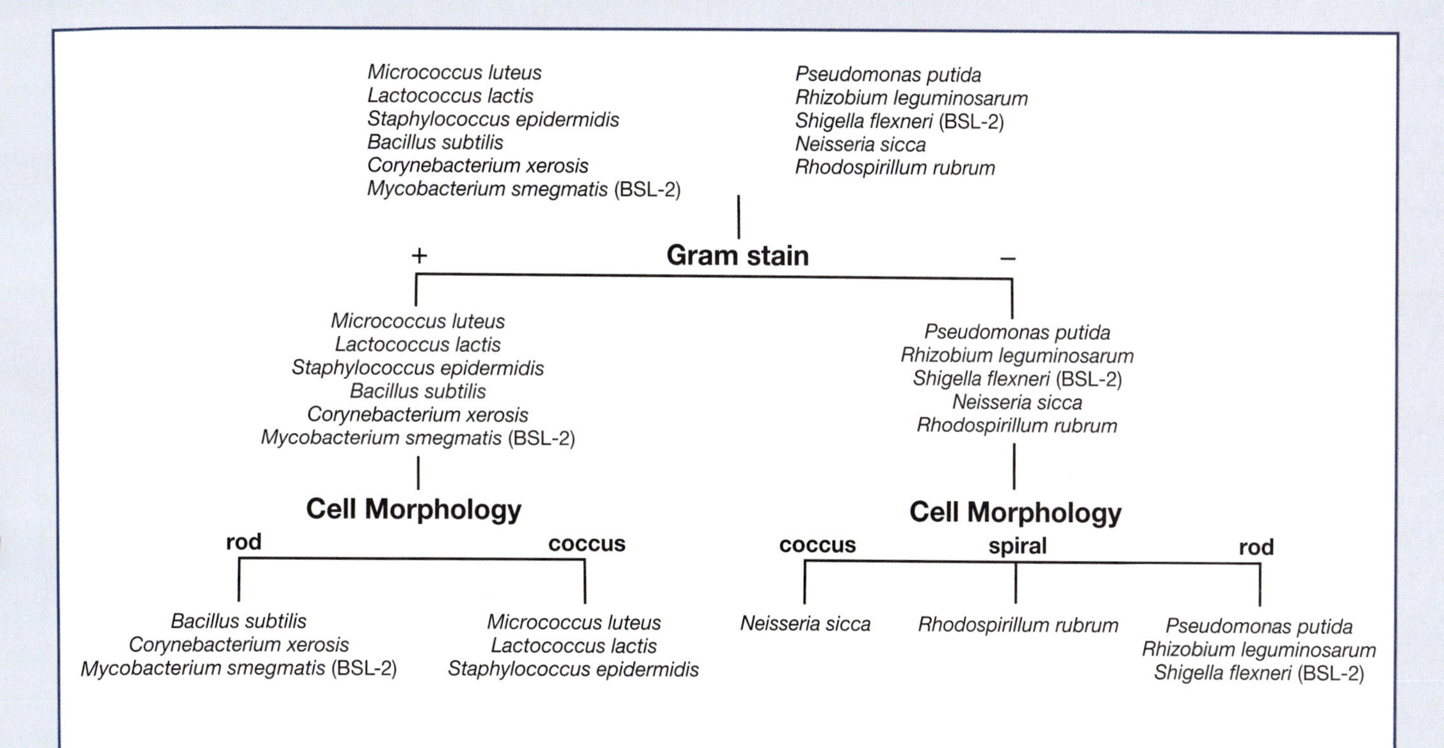

**2** Record your unknown number. Then, record your stain results and the date each was run in the table below. Include sketches (as appropriate) along with your written descriptions. Sketch the cells carefully—do not scribble a bunch of dots or squiggles.

Unknown number: ________________

| | Gram Reaction Cell Morphology and Arrangement Cell Dimensions | Acid-Fast Stain | Motility (wet mount) | Capsule Stain | Spore Stain (shape and location) |
|---|---|---|---|---|---|
| Date(s) Run | | | | | |
| Results | Gram Reaction:<br>Cell Morphology and Arrangement<br>Cell Dimensions:<br>Sketch | Result: | Result: | Result:<br>Sketch: | Result:<br>Sketch: |

**3** Match your results with those in the flowchart. Highlight the path on the flowchart that leads to identification of your unknown.

**4** Highlight your confirmatory test in the results table.

**5** Write the identity of your unknown organism in the space below.

My unknown is: ________________

SECTION 4

# Selective Media

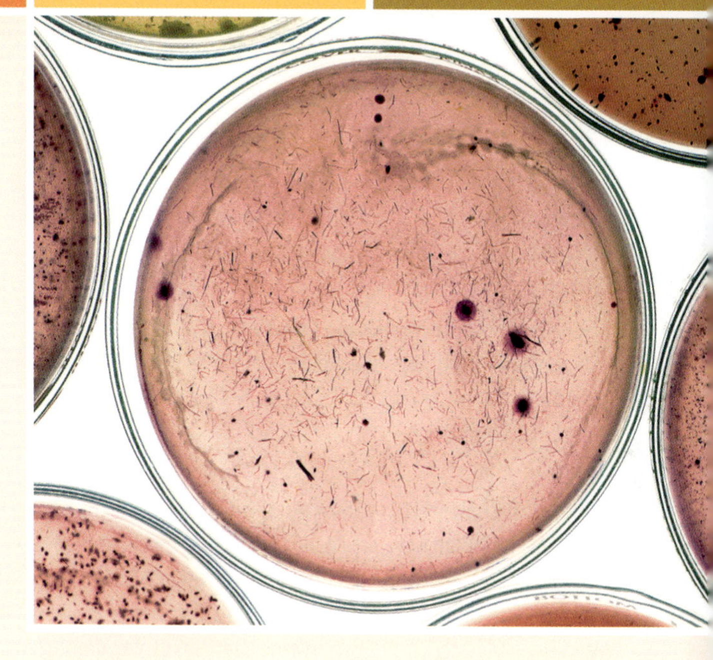

Individual microbial species in a **mixed culture** must be isolated and cultivated as **pure cultures** before they can be tested and properly identified. The most common means of isolating an organism from a mixed culture is to streak for isolation. Refer to Exercise 1-5 for a description of the streak-plate isolation method.

The **media** introduced in this section are designed to enhance the isolation procedure by inhibiting growth of some organisms while encouraging the growth of others. Thus, they are referred to as **selective media**. Many types of selective media contain indicators to expose differences between organisms, which also make them **differential media**. The media illustrated in this section are used specifically to isolate pathogenic Gram-negative bacilli and Gram-positive cocci from human or environmental samples containing a mixture of microorganisms.

Clinical microbiologists, who are familiar with human pathogens and the types of infections they cause, choose selective media that will screen out normal flora that also are likely to be in a sample. Environmental microbiologists often choose selective and differential media to detect **coliform** bacteria. Coliforms are a subgroup of *Enterobacteriaceae* (*Escherichia coli* being the most prominent member) that produce acid and gas from lactose fermentation. Fermentation will be discussed more fully in Section 5. Most coliforms are normal inhabitants of the human intestinal tract; therefore, their presence in the environment may be evidence of fecal contamination.

All of the exercises contain a Table of Results identifying the various reactions produced on a specific medium. These include color results, interpretations of results, symbols used to quickly identify the various reactions, and presumptive identification of typical organisms encountered with these media. Presumptive identification is not final identification. It is an "educated guess" based on evidence provided by the selective/ differential medium coupled with information about the origin of a sample.

## A Word about Selective Media

As you read the Theory of each of the following media and learn a little about how the tests actually perform on a biochemical level, you will see that each medium has six fundamental features. For a fuller understanding of the media and tests included in this section, watch for these six features and make sure you can identify them. (***Note:*** In the following, a distinction is made between nutritional components and substrates. Since, in the strictest sense all reactants in a biochemical reaction are substrates, this is an artificial distinction. We separate the terms to illustrate the importance of the specific substance added that allows observations of biochemical distinctions to be made.)

### 1. Selective, Differential, Defined, or Undefined—Which Is It?

The medium may be most any combination of the above. If it is selective, it encourages growth of some organisms and discourages (inhibits) growth of others. If it is differential, it allows us to distinguish between different microbes. If it is defined, or "chemically defined," each of its ingredients is known and in exactly what amounts. If it is undefined, or "complex," it contains one or more ingredient(s) of unknown composition and/or amount, such as yeast extract, beef extract, digest of gelatin, etc.

### 2. Nutritional Components

Nutritional components are selected to obtain the optimum growth of the organisms being tested for or suspected of being in the sample. Although many ingredients are suitable for many different types of microorganisms, typically, slight variations are made to tailor the medium for a specified group. For instance, yeast extract can be added for fastidious organisms requiring specific vitamins.



### 3. Inhibitors

Inhibitors make the medium selective. They are designed to exploit weaknesses in specific groups of organisms and thus prevent or inhibit their growth, while allowing other organisms to grow. Some inhibitors function by interrupting DNA synthesis or expression of a gene. Other inhibitors function at the enzymatic level, interfering with a critical reaction. Still other inhibitors interfere with membrane permeability, thus upsetting homeostasis and starting a cascade of catastrophic changes inside the cell. Sodium chloride can be added to inhibit organisms that can't tolerate high osmotic pressures. The precise "menu" of nutritional components can enhance selectivity by excluding certain ingredients required by undesired organisms in the sample. (In the absence of an inhibitor, this strategy would make it an enrichment medium.)

### 4. Substrates

Substrates are almost always what make the medium differential. Differentiation and identification of organisms frequently relies on their differing abilities to perform a specific chemical reaction or set of reactions and to do them in a way that can be observed. This can all be staged in an artificial environment (differential medium) by providing the organisms all the required components for growth and by including at least one substrate that only one organism or group of organisms can utilize. Once the reaction has taken place, presumably, organisms that before the test were indistinguishable are now (with the help of indicators—see #5) easily differentiated. For example, if a fecal sample is being tested for the presence of a pathogen known not to ferment lactose, then lactose is an important substrate needed to identify and isolate *E. coli* and other coliforms, all of which ferment lactose and all of which are likely to be found in the sample. Colonies of lactose non-fermenters will be visibly different and would be grown in pure culture for further testing.

### 5. Indicators

Indicators make a desired or expected reaction visible. An indicator is virtually always included in differential media because bacterial growth, by itself, is usually not enough to reliably differentiate between microbial groups. The medium must change fairly dramatically for us to be able to see it. Subtle changes can lead to false readings. The indicator frequently is a dye that changes color as pH changes. All pH dyes have a specific functional range and are chosen based on the starting pH of the medium, which in turn is based on the optimum pH of the organisms being tested.

Indicators can also be chemicals that react with products of a reaction and produce a color change. For example, if the organism being tested reduces the medium substrate, an indicator that becomes oxidized can produce a pretty spectacular change. Oxidized iron is a common reactant used to form a visible precipitate upon its reduction.

### 6. Positive or Negative?

The last item covered is a description of the positive and negative reactions. We tell you what to look for. ■

## Selective Media for Isolation of Gram-Positive Cocci

Gram-negative organisms frequently inhibit growth of Gram-positive organisms when they are cultivated together. Therefore, when looking for staphylococci or streptococci in a clinical sample, it may be necessary to begin by streaking the unknown mixture onto a selective medium that inhibits Gram-negative growth. The three selective media introduced in the first part of Section 4—Phenylethyl Alcohol Agar, Columbia CNA Agar, and Mannitol Salt Agar—are all used to isolate streptococci, enterococci, or staphylococci in human samples. ■

# Phenylethyl Alcohol Agar

EXERCISE 4-1



## ■ Theory

Phenylethyl alcohol agar (PEA) is an undefined, selective medium that inhibits the growth of many Gram-negative and some Gram-positive organisms. It is not a differential medium because no distinction is made between organisms that successfully grow on it.

Digests of casein and soybean meal provide nutrition while sodium chloride provides a stable osmotic environment suitable for the addition of sheep blood if desired. Phenylethyl alcohol is the selective agent and may be bacteriostatic or bactericidal, depending on concentration.

The major effect on susceptible bacteria is physical disruption of the cytoplasmic membrane, which affects its ability to act as a permeability barrier. One noteworthy consequence of this is a rapid loss of potassium ions by the cell, which in turn affects the cell's ability to maintain pH and osmotic homeostasis. A second effect is inhibition of DNA synthesis, which is likely due to loss of molecules essential for this process.

## ■ Application

PEA is used to isolate staphylococci and streptococci (including enterococci and lactococci) from specimens containing mixtures of bacterial flora. Typically, it is used to screen out the common contaminants *Escherichia coli* and *Proteus* species. When prepared with 5% sheep blood, it is used for cultivation of Gram-positive anaerobes.

## ■ In This Exercise

You will spot inoculate one PEA plate and one nutrient agar (NA) plate with three test organisms. (In a clinical application, PEA plates would be streaked for isolation with a patient or environmental sample.) The NA plate will serve as a comparison for growth quality on the PEA plate.

## ▼ Materials

### Per Student

- □ Lab coat
- □ Disposable gloves
- □ Chemical eye protection

### Per Student Group

- □ One PEA plate
- □ One NA plate
- □ Fresh broth cultures of these recommended organisms:
  - *Escherichia coli*
  - *Staphylococcus epidermidis*
  - (Optional) *Streptococcus gallolyticus*

## ■ Medium Recipes

### Phenylethyl Alcohol Agar

|  |  |
|---|---|
| □ Pancreatic digest of casein | 15.0 g |
| □ Papaic digest of soybean meal | 5.0 g |
| □ Sodium chloride | 5.0 g |
| □ β-Phenylethyl alcohol | 2.5 g |
| □ Agar | 15.0 g |
| □ Distilled or deionized water | 1.0 L |

*pH 7.1–7.5 at 25°C*

### Nutrient Agar

|  |  |
|---|---|
| □ Beef extract | 3.0 g |
| □ Peptone | 5.0 g |
| □ Agar | 15.0 g |
| □ Distilled or deionized water | 1.0 L |

*pH 6.6–7.0 at 25°C*

## PROCEDURE

### Lab One

1. Wear a lab coat, gloves, and chemical eye protection when performing this procedure.
2. Gently mix each culture according to your lab's standards.
3. Using a permanent marker, divide the bottom of each plate into three sectors.
4. Label the plates with the organisms' names, your name, and the date. Use the same positions for each specimen on both plates.
5. Spot inoculate (refer to Appendix B, "Spot Inoculation of an Agar Plate," p. 601) the sectors on the PEA plate with the test organisms.
6. Repeat step 5 with the NA plate. Be sure to inoculate the specimens in the sectors that correspond to their position on the PEA plate.
7. Invert and incubate the plate at 35 ± 2°C for 24 to 48 hours.
8. Save or dispose of the original cultures as directed by your instructor.

### Lab Two

1. Examine and compare the plates for quality of growth (no growth, poor growth, or good growth). See Figure 4.1 for an example.
2. Record your results on the data sheet, page 231. Refer to Table 4-1 as needed.
3. Dispose of all plates in the appropriate autoclave container when finished.

### References


Atlas, Ronald M. and James W. Snyder. Page 297 in *Manual of Clinical Microbiology,* 10th ed. James Versalovic, Karen C. Carroll, Guido Funke, James H. Jorgensen, Marie Louise Landry, and David W. Warnock, eds. Washington, DC: ASM Press, 2011.

Corre, J, J. J. Lucchini, G. M. Mercier, and A. Cremieux. "Antibacterial activity of phenethyl alcohol and resulting membrane alterations." *Res Microbiol* 141, no. 4 (1990): 483–497.

Forbes, Betty A., Daniel F. Sahm, and Alice S. Weissfeld. Chap. 10 in *Bailey & Scott's Diagnostic Microbiology,* 11th ed. St. Louis, MO: Mosby, 2002.

Silva, M. T., J. C. Sousa, M. A. Macedo, J. Polónia, and A. M Parente. "Effects of phenylethyl alcohol on Bacillus and Streptococcus." *J. Bacteriol* 127, no. 3 (1976): 1359–1369.

Silver, S. and L. Wendt. "Mechanism of Action of Phenethyl Alcohol: Breakdown of the Cellular Permeability Barrier." *J. Bacteriol* 93, no. 2 (1967): 560–566.

Tille, Patricia M. Page 236 in *Bailey & Scott's Diagnostic Microbiology,* 13th ed. St. Louis, MO: Mosby, 2014.

Zimbro, Mary Jo and David A. Power, eds. Page 443 in *Difco™ & BBL™ Manual–Manual of Microbiological Culture Media.* Sparks, MD: Becton Dickinson and Co., 2003.


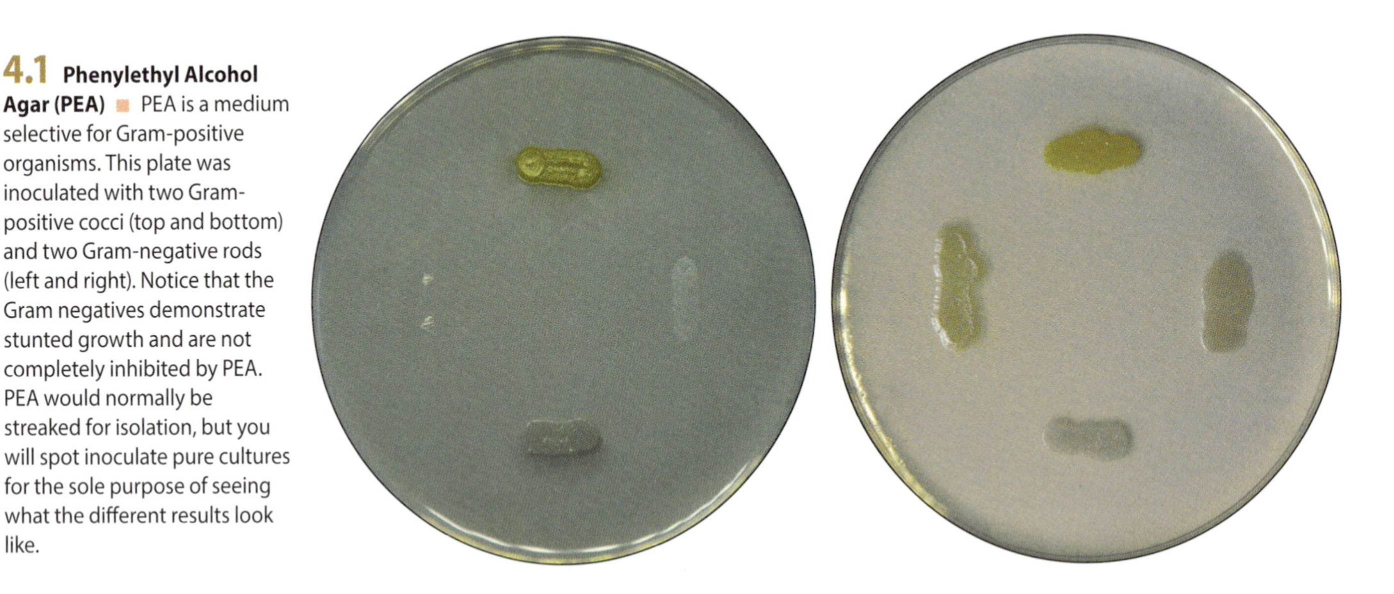

**4.1 Phenylethyl Alcohol Agar (PEA)** ■ PEA is a medium selective for Gram-positive organisms. This plate was inoculated with two Gram-positive cocci (top and bottom) and two Gram-negative rods (left and right). Notice that the Gram negatives demonstrate stunted growth and are not completely inhibited by PEA. PEA would normally be streaked for isolation, but you will spot inoculate pure cultures for the sole purpose of seeing what the different results look like.

TABLE **4-1** **PEA Results and Interpretations**

| Result | Interpretation | Presumptive ID |
|---|---|---|
| Poor growth (P) or no growth (N) | Organism is inhibited by phenylethyl alcohol | Probable Gram-negative organism |
| Good growth (G) | Organism is not inhibited by phenylethyl alcohol | Probable *Staphylococcus, Streptococcus, Enterococcus,* or *Lactococcus* |

Name ______

Date ______

Lab Section ______

I was present and performed this exercise (initials) ______

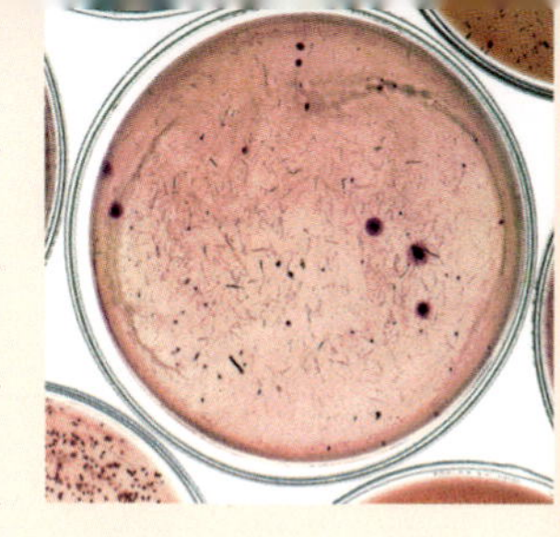

DATA SHEET

4-1

# Phenylethyl Alcohol Agar

## OBSERVATIONS AND INTERPRETATIONS

**1** Refer to Table 4-1, page 230, when recording your results and interpretations in the table below.

| Organism | Growth (N/P/G) | | Interpretation and Presumptive ID |
|---|---|---|---|
| | PEA | NA | |
| | | | |
| | | | |
| | | | |

## QUESTIONS

**1** *In your own words, what is the application (purpose) of PEA?*

**2** *Which ingredient(s) in PEA supply(ies)*

**a.** *Carbon?*

**b.** *Nitrogen?*

**3** *Is PEA a defined or an undefined medium? Provide the reasoning behind your choice and explain why this formulation is desirable.*

**4** *In your own words, what is the role of* β*-phenylethyl alcohol in PEA and how does it work?*

4

**5** *Think about the results you recorded.*

**a.** *Growth on the PEA and NA plates was recorded as "good growth," "poor growth," or "no growth." These are qualitative and, at least for the first two, subjective terms. What did you use to establish what constituted "good growth?"*

**b.** *Why wouldn't it be advisable to compare growth of the organisms on each plate to each other? There are at least two answers to this question!*

Name ______

Date ______

Lab Section ______

I was present and performed this exercise (initials) ______

# DATA SHEET 4-1

*(continued)*

**6** *Would removing β-phenylethyl alcohol from PEA alter the medium's sensitivity or specificity? (See pages 8 and 9, "A Word About Experimental Design," for assistance with these terms.)*

**7** *If you observed growth of Gram-negative organisms on the PEA plate you inoculated, does this negate the usefulness of PEA as a selective medium? Why or why not? (**Hint:** Compare the growth on the NA and PEA plates.)*

**8** *PEA contains only 0.25% phenylethyl alcohol because high concentrations inhibit both Gram-positive and Gram-negative organisms. Describe some possible reasons why this occurs.*

EXERCISE

4-2

# Columbia CNA with 5% Sheep Blood Agar

## Theory

Columbia CNA with 5% sheep blood agar is an undefined, differential, and selective medium that allows growth of Gram-positive organisms (especially staphylococci, streptococci, and enterococci) and stops or inhibits growth of most Gram-negative organisms (Fig. 4.2). Casein, digest of animal tissue, beef extract, yeast extract, corn starch, and sheep blood provide a range of carbon and energy sources to support a wide variety of organisms.

In addition, sheep blood supplies the X factor (heme) and yeast extract provides B vitamins. The antibiotics colistin and nalidixic acid (CNA) act as selective agents against Gram-negative organisms by affecting membrane integrity and interfering with DNA replication, respectively. They are particularly effective against *Klebsiella, Proteus,* and *Pseudomonas* species. Further, sheep blood makes possible differentiation of Gram-positive organisms based on hemolytic reaction (see Exercise 5-21).

**4.2 Columbia CNA with 5% Sheep Blood Agar** ■ This medium is selective for Gram positives and also allows determination of hemolytic reaction (Exercise 5-21). Shown is a plate inoculated with four organisms: three Gram-positive cocci and one Gram-negative rod. Only the Gram-positive organisms grew well on the Columbia CNA agar. The Gram negative at the right didn't grow. Further, the top Gram-positive coccus is β-hemolytic (arrow indicates clearing), whereas the lower one is α-hemolytic (slightly green) and the one on the left is nonhemolytic. This medium would normally be streaked for isolation, but you will spot inoculate pure cultures for the sole purpose of seeing what the different results look like.

## Application

Columbia CNA agar is used to isolate staphylococci, streptococci, and enterococci, primarily from clinical specimens. The inclusion of 5% sheep blood allows differentiation based on hemolysis reaction.

## In This Exercise

You will spot inoculate one Columbia CNA with 5% sheep blood agar plate and one nutrient agar (NA) plate with four test organisms. (In a clinical application, CNA plates would be streaked for isolation with a patient's sample.) The NA plate will serve as a comparison for growth quality on the Columbia CNA with 5% sheep blood agar plate.

## Materials

### Per Student

- □ Lab coat
- □ Disposable gloves
- □ Chemical eye protection

### Per Student Group

- □ One Columbia CNA with 5% sheep blood agar plate[1]
- □ One nutrient agar plate
- □ Fresh broth cultures of these recommended organisms:
  - *Enterococcus faecalis* (BSL-2) or *Enterococcus faecium* (BSL-2)
  - *Escherichia coli* (or other Gram negative, as available)
  - *Staphylococcus epidermidis*
  - *Streptococcus gallolyticus*
  - A β-hemolytic streptococcus may be added if desired (probable BSL-2)
- □ (Optional) candle jar or anaerobic jar with 3%–5% $CO_2$

[1] Columbia CNA agar without sheep blood may be substituted, but the ability to differentiate organisms based on hemolysis will be sacrificed.

4

## Medium Recipes

### Columbia CNA with 5% Sheep Blood Agar

- □ Pancreatic digest of casein 12.0 g
- □ Peptic digest of animal tissue 5.0 g
- □ Yeast extract 3.0 g
- □ Beef extract 3.0 g
- □ Corn starch 1.0 g
- □ Sodium chloride 5.0 g
- □ Colistin 10.0 mg
- □ Nalidixic acid 10.0 mg
- □ Agar 13.5 g
- □ Distilled or deionized water 1 L

  *pH 7.1–7.5 at 25°C*
- □ Sheep blood (defibrinated) 50 mL added to 950 mL of above

### Nutrient Agar

- □ Beef extract 3.0 g
- □ Peptone 5.0 g
- □ Agar 15.0 g
- □ Distilled or deionized water 1.0 L

  *pH 6.6–7.0 at 25°C*

## PROCEDURE

### Lab One

1. Wear a lab coat, gloves, and chemical eye protection when performing this procedure.
2. Gently mix each culture according to your lab's standards.
3. Using a permanent marker, divide the bottom of each plate into four sectors (five if a β-hemolytic streptococcus is added).
4. Label the plates with the organisms' names, your name, and the date. Use the same positions for each specimen on both plates.
5. Spot inoculate (refer to Appendix B, "Spot Inoculation of an Agar Plate," p. 601) the sectors on the CNA plate with the test organisms. **Use BSL-2 precautions when transferring *E. faecalis* (or *E. faecium*) and the β-hemolytic streptococcus, if used.**
6. Repeat step 5 with the nutrient agar plate. Be sure to inoculate the specimens in the sectors that correspond to their position on the CNA plate.
7. Invert and incubate the plate at 35 ± 2°C for 24 to 48 hours. If possible, incubate the plates in a candle jar or anaerobic jar with 3%–5% $CO_2$.
8. Save or dispose of the original cultures as directed by your instructor.

### Lab Two

1. Referring to Table 4-2, examine and compare the plates for quality of growth (none, poor, or good) and hemolytic reaction. For more information on hemolysis, see Exercise 5-21.
2. Record your results in the space provided on the data sheet, page 237.
3. Dispose of all plates in the appropriate autoclave container when finished.

### References


Chapin, Kimberle C. and Tsai-Ling Lauderdale. Chap. 21 in *Manual of Clinical Microbiology*, 9th ed. Patrick R. Murray, Ellen Jo Baron, James H. Jorgensen, Marie Louise Landry, and Michael A. Pfaller, eds. Washington, DC: ASM Press, 2007.

Tille, Patricia M. Pages 86 and 236 in *Bailey & Scott's Diagnostic Microbiology*, 13th ed. St. Louis, MO: Mosby, 2014.

Zimbro, Mary Jo and David A. Power, eds. Page 156 in *Difco™ & BBL™ Manual–Manual of Microbiological Culture Media*. Sparks, MD: Becton, Dickinson and Co., 2003.


TABLE **4-2** **Columbia CNA with 5% Sheep Blood Agar Results and Interpretations**

| Result | Interpretation | Presumptive ID |
|---|---|---|
| Poor growth (P) or no growth (N) | Organism is inhibited by colistin and/or nalidixic acid | Probable Gram-negative organism |
| Good growth (G) with clearing of medium (β-hemolysis) | Organism is not inhibited by colistin and nalidixic acid and completely hemolyzes RBCs | Probable β-hemolytic *Staphylococcus*, *Streptococcus*, or rarely *Enterococcus* (***Note:*** Artifactual greening may occur with some β-hemolytic organisms) |
| Good growth (G) with greening of medium (α-hemolysis) | Organism is not inhibited by colistin and nalidixic acid and partially hemolyzes RBCs | Probable α-hemolytic *Staphylococcus*, *Streptococcus*, or *Enterococcus* |
| Good growth (G) with no change of medium's color (γ-hemolysis) | Organism is not inhibited by colistin and nalidixic acid and does not hemolyze RBCs | Probable γ-hemolytic *Staphylococcus*, *Streptococcus*, or *Enterococcus* |

Name ______________________________

Date ______________________________

Lab Section ______________________________

I was present and performed this exercise (initials) ______________

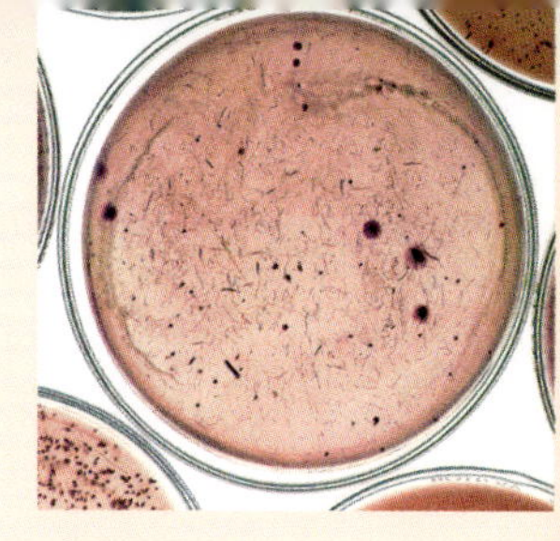

# DATA SHEET 4-2

## Columbia CNA with 5% Sheep Blood Agar

### OBSERVATIONS AND INTERPRETATIONS

1 Refer to Table 4-2, page 236, when recording your results and interpretations in the table below.

| Organism | Growth (N/P/G) | | Hemolysis (α,β,γ) | Interpretation and Presumptive ID |
|---|---|---|---|---|
| | Columbia CNA Agar | NA | | |
| | | | | |
| | | | | |
| | | | | |
| | | | | |
| | | | | |

## QUESTIONS

1 *In your own words, what is the application (purpose) of Columbia CNA with 5% sheep blood agar?*

**2** *Which ingredient(s) in Columbia CNA plus 5% sheep blood agar supply(ies)*

**a.** *Carbon?*

**b.** *Nitrogen?*

**3** *Is Columbia CNA agar a defined or an undefined medium? Provide the reasoning behind your choice and explain why this formulation is desirable.*

4

**4** *In your own words, what are the roles of colistin and nalidixic acid in CNA and how does each work?*

**5** *In your own words, what is the role of sheep blood in CNA agar?*

Name ______________________

Date ______________________

Lab Section ______________________

I was present and performed this exercise (initials) ______________________

# DATA SHEET 4-2

*(continued)*

**6** *Think about the results you recorded.*

**a.** *Growth on the Columbia CNA and NA plates was recorded as "good growth," "poor growth," or "no growth." These are qualitative and, at least for the first two, subjective terms. What did you use to establish what constituted "good growth?"*

**b.** *Why wouldn't it be advisable to compare growth of the organisms on each plate to each other? There are at least two answers to this question!*

**7** *Would removing colistin and nalidixic acid from CNA alter the medium's sensitivity or specificity? (See pages 8 and 9, "A Word About Experimental Design," for assistance with these terms.)*

**8** *Why might colistin affect Gram-negative bacteria more severely than Gram-positive bacteria? (**Hint:** Consider their structural differences.)*

**9** *Compare the recipes of nutrient agar and Columbia CNA agar. If an organism can grow on both media, on which would you expect it to grow better? Explain your answer.*

4

# Mannitol Salt Agar

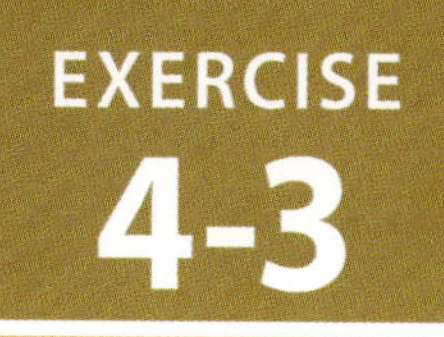


## Theory

Mannitol salt agar (MSA) contains the carbohydrate mannitol, 7.5% sodium chloride (NaCl), and the pH indicator phenol red. Phenol red is yellow below pH 6.8, red at pH 7.4 to 8.4, and pink at pH 8.4 and above. Mannitol provides the substrate for fermentation and makes the medium differential. Sodium chloride makes the medium selective because its concentration is high enough to dehydrate and kill most bacteria.

Staphylococci thrive on the medium, largely because of their adaptation to salty habitats such as human skin. Phenol red indicates whether fermentation with an acid end-product has taken place by changing color as the pH changes. (See Section 5, p. 268, "A Word About Biochemical Tests and Acid-Base Reactions.")

Most staphylococci are able to grow on MSA, but do not ferment mannitol, so their growth appears pink or red and the medium remains unchanged. *Staphylococcus aureus* ferments mannitol, which produces acids and lowers the pH of the medium (Fig. 4.3). The result is formation of bright yellow colonies usually surrounded by a yellow halo (Fig. 4.4). For more information on fermentation, refer to Exercise 5-2 and Appendix A.

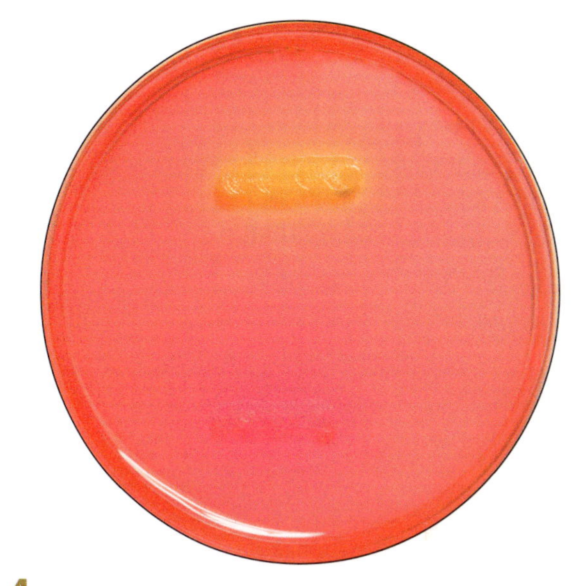

**4.4 Mannitol Salt Agar** ■ This medium is selective for members of the genus *Staphylococcus* due to their salt tolerance. It further allows differentiation between *Staphylococcus* species based on the ability to ferment mannitol to acid end products (top) and those that do not (bottom). MSA would normally be streaked for isolation, but you will spot inoculate pure cultures for the sole purpose of seeing what the different results look like.

## Application

Mannitol salt agar is used for isolation and differentiation of *Staphylococcus aureus* from other *Staphylococcus* species. It is not used to determine the ability of an isolate to ferment mannitol. A fermentation broth, such as phenol red mannitol, would be used instead (see Exercise 5-2).

## In This Exercise

Today you will spot inoculate one MSA plate and one nutrient agar (NA) plate with three test organisms. (In a clinical application, MSA plates would be streaked for isolation with a patient's sample.) The NA will serve as a comparison for growth quality on the MSA plate.

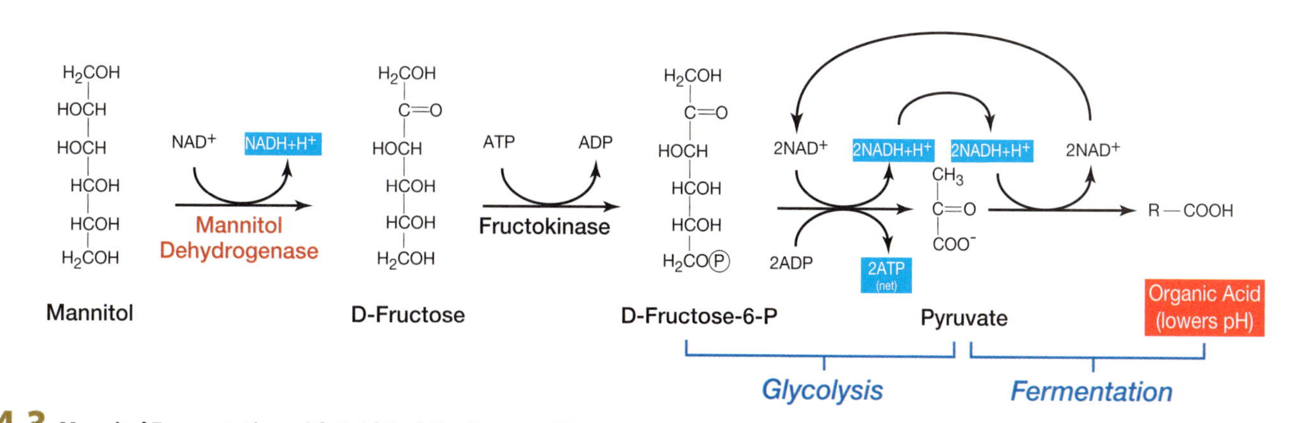


**4.3 Mannitol Fermentation with Acid End-Products** ■ There are two important points to note here: one, that the fermentation provides a mechanism for oxidizing NADH back to $NAD^+$ so it can be used again in glycolysis and two, that the end product of the fermentation is an acid—the acid detected by the phenol red pH indicator in MSA. For simplicity, only the relevant parts of glycolysis are shown. See Appendix A, Figure A.1 for details.

## ▼ Materials

### Per Student

- ☐ Lab coat
- ☐ Disposable gloves
- ☐ Chemical eye protection

### Per Student Group

- ☐ One MSA plate
- ☐ One NA plate
- ☐ Fresh broth cultures of these recommended organisms:
  - *Staphylococcus aureus* (BSL-2)
  - *Staphylococcus epidermidis*
  - *Escherichia coli*

## ■ Medium Recipes

### Mannitol Salt Agar

| | |
|---|---|
| ☐ Beef extract | 1.0 g |
| ☐ Peptone | 10.0 g |
| ☐ Sodium chloride | 75.0 g |
| ☐ D-Mannitol | 10.0 g |
| ☐ Phenol red | 0.025 g |
| ☐ Agar | 15.0 g |
| ☐ Distilled or deionized water | 1.0 L |

*pH 7.2–7.6 at 25°C*

### Nutrient Agar

| | |
|---|---|
| ☐ Beef extract | 3.0 g |
| ☐ Peptone | 5.0 g |
| ☐ Agar | 15.0 g |
| ☐ Distilled or deionized water | 1.0 L |

*pH 6.6–7.0 at 25°C*

## PROCEDURE

### Lab One

1. Wear a lab coat, gloves, and chemical eye protection when performing this procedure.
2. Gently mix each culture according to your lab's standards.
3. Using a permanent marker, divide the bottom of each plate into three sectors.
4. Label the plates with the organisms' names, your name, and the date. Use the same positions for each specimen on both plates.
5. Spot inoculate (refer to Appendix B, "Spot Inoculation of an Agar Plate," p. 601) the sectors on the MSA plate with the test organisms. **Use BSL-2 precautions when transferring *S. aureus*.**
6. Repeat step 5 with the nutrient agar plate. Be sure to inoculate the specimens in the sectors that correspond to their position on the MSA plate.
7. Invert and incubate the plate at 35 ± 2°C for 24 to 48 hours.
8. Save or dispose of the original cultures as directed by your instructor.

### Lab Two

1. Examine and compare the plates for color and quality of growth.
2. Record your results on the data sheet, page 243. Refer to Table 4-3 as needed.
3. Dispose of all plates in the appropriate autoclave container when finished.

### References


Delost, Maria Dannessa. Page 112 in *Introduction to Diagnostic Microbiology, a Text and Workbook*. St. Louis, MO: Mosby, 1997.

Tille, Patricia M. Page 199 in *Bailey & Scott's Diagnostic Microbiology*, 13th ed. St. Louis, MO: Mosby, 2014.

Zimbro, Mary Jo and David A. Power, eds. Page 349 in *Difco™ & BBL™ Manual–Manual of Microbiological Culture Media*. Sparks, MD: Becton Dickinson and Co., 2003.


TABLE **4-3** **Mannitol Salt Agar Results and Interpretations**

| Result | Interpretation | Presumptive ID |
|---|---|---|
| Poor growth (P) or no growth (N) | Organism is inhibited by NaCl | Not *Staphylococcus* |
| Good growth (G) | Organism is not inhibited by NaCl | *Staphylococcus* |
| Yellow growth or halo (Y) | Organism produces acid from mannitol fermentation | Possible pathogenic *Staphylococcus aureus* |
| Red growth (no halo) (R) | Organism does not ferment mannitol | *Staphylococcus* other than *S. aureus* |

Name ______________________________

Date ______________________________

Lab Section ______________________________

I was present and performed this exercise (initials) ______________________________

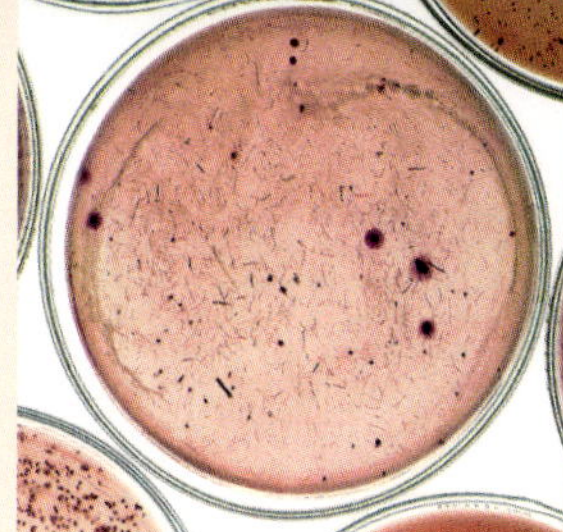

# DATA SHEET 4-3

## Mannitol Salt Agar

### OBSERVATIONS AND INTERPRETATIONS

**1** Refer to Table 4-3, page 242, when recording your results and interpretations in the table below. Use abbreviations or symbols as needed.

| Organism | Growth (N/P/G) | | MSA Growth Color (Y/R) | Interpretation and Presumptive ID |
|---|---|---|---|---|
| | MSA | NA | | |
| | | | | |
| | | | | |
| | | | | |

### QUESTIONS

**1** *In your own words, what is the application (purpose) of MSA?*

**2** *Which ingredient(s) in MSA supply(ies)*

**a.** *Carbon?*

**b.** *Nitrogen?*

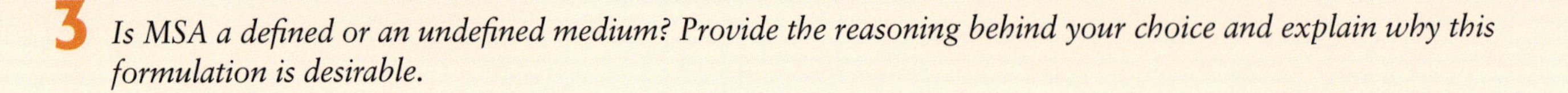

**3** *Is MSA a defined or an undefined medium? Provide the reasoning behind your choice and explain why this formulation is desirable.*

**4** *In your own words, what is the role of sodium chloride in MSA and how does it work?*

4

**5** *In your own words, what are the roles of mannitol and phenol red in MSA?*

**6** *Think about the results you recorded.*

**a.** *Growth on the MSA and NA plates was recorded as "good growth," "poor growth," or "no growth." These are qualitative and, at least for the first two, subjective terms. What did you use to establish what constituted "good growth?"*

**b.** *Why wouldn't it be advisable to compare growth of the organisms on each plate to each other? There are at least two answers to this question!*

Name

Date

Lab Section

I was present and performed this exercise (initials)

# DATA SHEET 4-3

*(continued)*

**7** *Would removal of sodium chloride from MSA alter the medium's sensitivity or specificity? Explain your answer. (See pages 8 and 9, "A Word About Experimental Design," for assistance with these terms.)*

**8** *Suppose a mistake is made in preparing a batch of MSA and the starting pH is 7.8 instead of 7.2–7.6. Would that affect the medium's sensitivity or specificity? Explain your answer. (See pages 8 and 9, "A Word About Experimental Design," for assistance with these terms.)*

**9** *With the diversity of microorganisms in the world, how can a single test such as MSA be used to "confidently" identify* Staphylococcus aureus*? (**Hint:** Consider the sample source and assumptions being made.)*

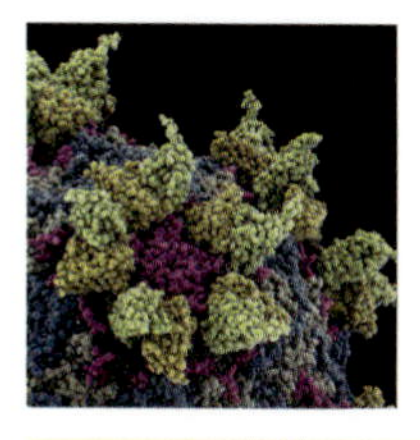

## Selective Media for Isolation of Gram-Negative Rods

Members of the family *Enterobacteriaceae*—the enteric "gut" bacteria—are commonly found in clinical samples. Depending on the circumstances, some organisms in a mixed sample are contaminants and relatively benign, whereas others are potentially harmful and must be isolated and identified. The media in this part of Section 4 are designed to isolate and differentiate these organisms from each other and to discourage growth of other organisms that are not currently of interest to the microbiologist.

The three examples selected for this unit are MacConkey Agar, Eosin Methylene Blue (EMB) Agar, and Hektoen Enteric (HE) Agar. MacConkey agar and EMB agar are selective for Gram-negative organisms and both contain indicators to differentiate lactose fermenters from lactose non-fermenters. EMB agar is commonly used to test for the presence of coliforms in environmental samples. As mentioned in the introduction to this section, the presence of coliforms in the environment suggests fecal contamination and the possible presence of enteric pathogens.

HE agar differentiates *Salmonella* and *Shigella* from each other and from other enterics based on their ability to overcome the inhibitory effects of bile; reduce sulfur to $H_2S$; and ferment lactose, sucrose, and/or salicin. (For more information about fermentation or decarboxylation reactions, refer to Section 5, "A Word About Biochemical Tests and Acid-Base Reactions," "Introduction to Energy Metabolism Tests," and Exercises 5-2, 5-3, and 5-8. Also see Appendix A.) ■

4

# MacConkey Agar

EXERCISE 4-4

## ■ Theory

MacConkey agar is a commonly used selective and differential medium containing lactose, bile salts, neutral red, and crystal violet. Bile salts and crystal violet inhibit growth of Gram-positive bacteria. Neutral red dye is a pH indicator that is colorless above a pH of 6.8 and red at a pH less than 6.8. Organisms that ferment lactose to acid end-products lower the pH (Fig. 4.5) and their colonies turn a pink to red color.

On occasion, a pink to red bile precipitate will also form in the agar due to the lowered pH. Lactose non-fermenters retain their normal color or the color of the medium (Fig. 4.6). Formulations without crystal violet allow growth of *Enterococcus* and some species of *Staphylococcus*, which ferment the lactose and appear pink on the medium.

## ■ Application

MacConkey agar is used to isolate and differentiate members of the *Enterobacteriaceae* based on the ability to ferment lactose. Variations on the standard medium include MacConkey agar w/o CV (without crystal violet) to allow growth of Gram-positive cocci, or MacConkey agar CS to control swarming bacteria (*Proteus*–see Fig. 2.24) that interfere with interpretation of results and isolation of other species.

## ■ In This Exercise

You will spot inoculate one MacConkey agar plate and one nutrient agar (NA) plate with three test organisms. (In a clinical application, MacConkey plates would be streaked for isolation with a patient's sample.) The NA plate will serve as a comparison for growth quality on the MacConkey agar plate.

## ▼ Materials

### Per Student

- □ Lab coat
- □ Disposable gloves
- □ Chemical eye protection

### Per Student Group

- □ One MacConkey agar plate
- □ One nutrient agar plate
- □ Fresh broth cultures of these recommended organisms:
  - *Escherichia coli*
  - *Providencia stuartii*
  - *Staphylococcus epidermidis*

## Medium Recipes

### MacConkey Agar

|  |  |
|---|---|
| □ Pancreatic digest of gelatin | 17.0 g |
| □ Pancreatic digest of casein | 1.5 g |
| □ Peptic digest of animal tissue | 1.5 g |
| □ Lactose | 10.0 g |
| □ Bile salts | 1.5 g |
| □ Sodium chloride | 5.0 g |
| □ Neutral red | 0.03 g |
| □ Crystal violet | 0.001 g |
| □ Agar | 13.5 g |
| □ Distilled or deionized water | 1.0 L |

*pH 6.9–7.3 at 25°C*

### Nutrient Agar

|  |  |
|---|---|
| □ Beef extract | 3.0 g |
| □ Peptone | 5.0 g |
| □ Agar | 15.0 g |
| □ Distilled or deionized water | 1.0 L |

*pH 6.6–7.0 at 25°C*

## PROCEDURE

### Lab One

1. Wear a lab coat, gloves, and chemical eye protection when performing this procedure.
2. Gently mix each culture according to your lab's standards.
3. Using a permanent marker, divide the bottom of each plate into three sectors.
4. Label the plates with the organisms' names, your name, and the date. Use the same position for each specimen on both plates.
5. Spot inoculate (refer to Appendix B, "Spot Inoculation of an Agar Plate," p. 601) the sectors on the MacConkey agar plate with the test organisms.
6. Repeat step 5 with the nutrient agar plate. Be sure to inoculate the specimens in the sectors that correspond to their position on the MacConkey plate.
7. Invert and incubate the plate at 35 ± 2°C for 24 to 48 hours.
8. Save or dispose of the original cultures as directed by your instructor.

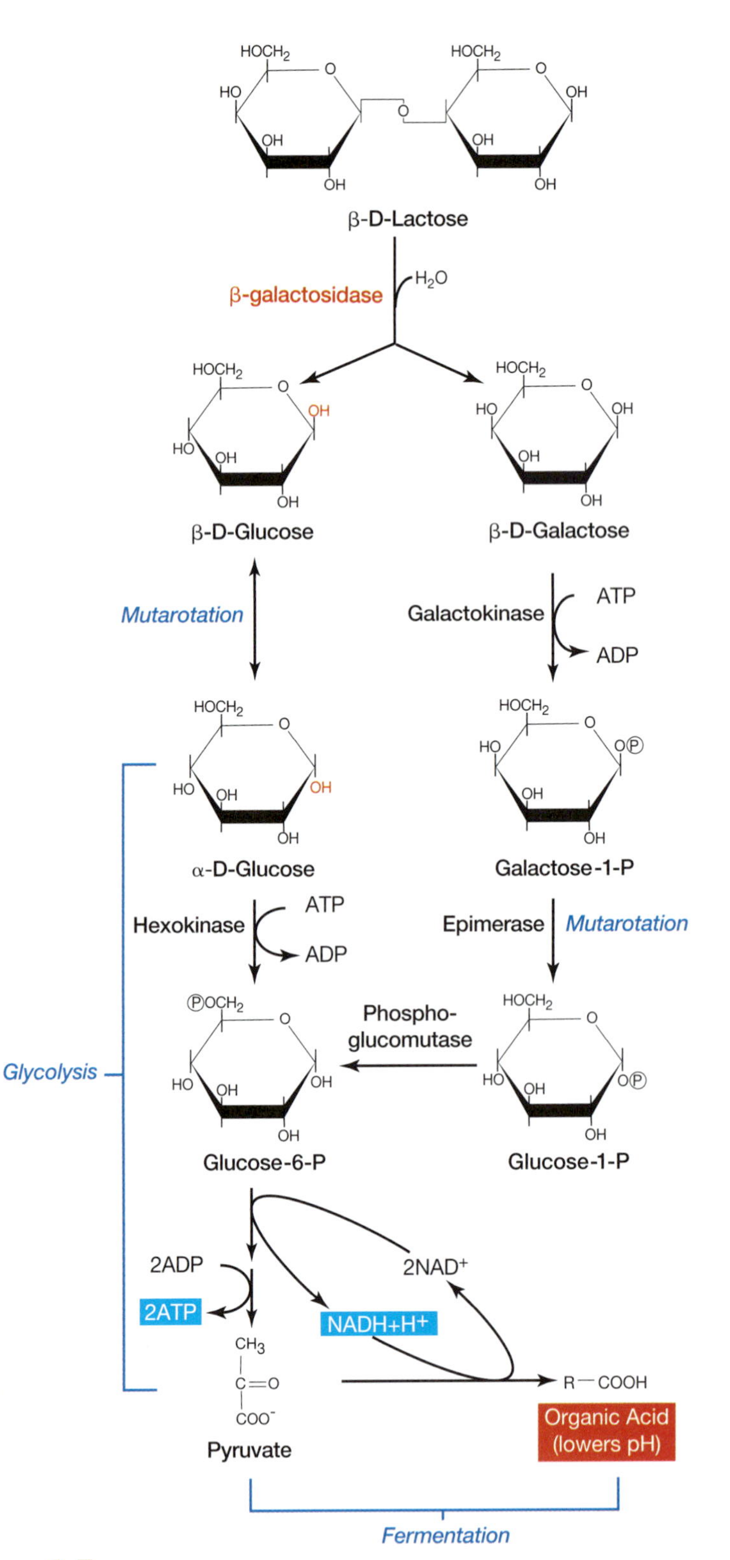


**4.5 Lactose Fermentation with Acid End-Products** ■ There are two important points to note here: one, that the fermentation provides a mechanism for oxidizing NADH back to $NAD^+$ so it can be used again in glycolysis; and two, that the end product of the fermentation is an acid—the acid detected by the neutral red pH indicator in MacConkey agar. For simplicity, only the relevant parts of glycolysis are shown. See Appendix A, Figure A.1 for details.

### Lab Two

1. Examine and compare the plates for color and for quality of growth.
2. Refer to Table 4-4 when recording your results on the data sheet, page 251.
3. Dispose of all plates in the appropriate autoclave container when finished.

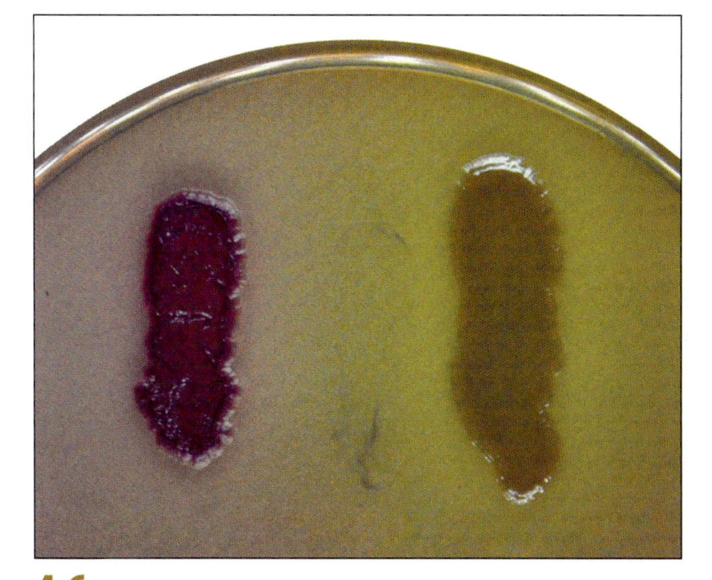

**4.6 MacConkey Agar (MAC)** ■ This MacConkey agar plate was inoculated with a Gram-positive coccus (center) and two members of the *Enterobacteriaceae* (Gram-negative rods—left and right). Obviously, the Gram positive was inhibited by the bile salts and crystal violet. The Gram negative on the left is a coliform and fermented lactose to acid end-products and is pinkish. Sometimes a pinkish bile precipitate forms around the growth, but it didn't with this specimen. The Gram negative on the right, a noncoliform, did not ferment lactose. It appears close to its natural color and has yellowed the medium. MacConkey agar would normally be streaked for isolation, but you will spot inoculate pure cultures for the sole purpose of seeing what the different results look like.

TABLE **4-4** **MacConkey Agar Results and Interpretations**

| Result | Interpretation | Presumptive ID |
|---|---|---|
| Poor growth (P) or no growth (N) | Organism is inhibited by crystal violet and/or bile | Gram positive |
| Good growth (G) | Organism is not inhibited by crystal violet or bile | Gram negative |
| Pink to red growth with or without bile precipitate (R) | Organism produces acid from lactose fermentation | Probable coliform |
| Growth is "colorless" (not red or pink) (C) | Organism does not ferment lactose | Noncoliform |

4

## References

Tille, Patricia M. Pages 86–87 in *Bailey & Scott's Diagnostic Microbiology*, 13th ed. St. Louis, MO: Mosby, 2014.

Winn, Washington C. et al. *Koneman's Color Atlas and Textbook of Diagnostic Microbiology*, 6th ed. Baltimore: Lippincott Williams & Wilkins, 2006.

Zimbro, Mary Jo and David A. Power. Page 334 in *Difco™ & BBL™ Manual–Manual of Microbiological Culture Media*. Sparks, MD: Becton, Dickinson and Co., 2003.

4

Name ______________________________

Date ______________________________

Lab Section ______________________________

I was present and performed this exercise (initials) ______________________________

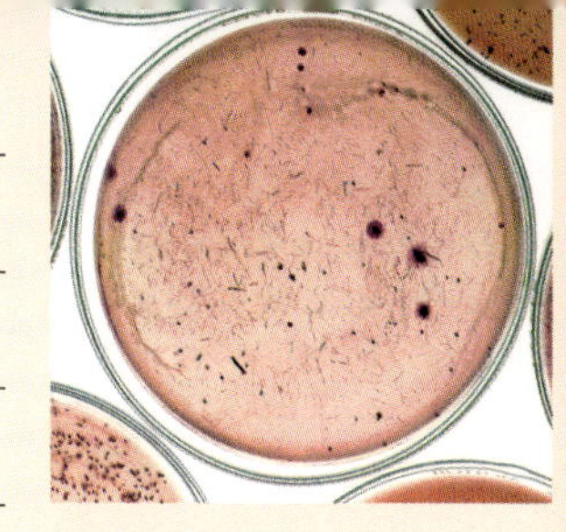

# DATA SHEET 4-4

## MacConkey Agar

### OBSERVATIONS AND INTERPRETATIONS

1 Refer to Table 4-4, page 249, when recording your results and interpretations in the table below. Use abbreviations or symbols as needed.

| Organism | Growth (N/P/G) | | MAC Growth Color (R/C) | Interpretation and Presumptive ID |
|---|---|---|---|---|
| | MAC | NA | | |
| | | | | |
| | | | | |
| | | | | |

### QUESTIONS

**1** *In your own words, what is the application (purpose) of MacConkey agar?*

**2** *Which ingredient(s) in MacConkey agar supply(ies)*

**a.** *Carbon?*

**b.** *Nitrogen?*

**3** *Is MacConkey agar a defined or an undefined medium? Provide the reasoning behind your choice and explain why this formulation is desirable.*

**4** *In your own words, what are the roles of crystal violet and bile salts in MacConkey agar?*

4

**5** *In your own words, what are the roles of neutral red and lactose in MacConkey agar?*

**6** *Think about the results you recorded.*

**a.** *Growth on the MacConkey agar and NA plates was recorded as "good growth," "poor growth," or "no growth." These are qualitative and, at least for the first two, subjective terms. What did you use to establish what constituted "good growth?"*

**b.** *Why wouldn't it be advisable to compare growth of the organisms on each plate to each other? There are at least two answers to this question!*

Name ______________________________

Date ______________________________

Lab Section ______________________________

I was present and performed this exercise (initials) ______________________________

# DATA SHEET 4-4

*(continued)*

**7** *Would removing bile salts and/or crystal violet from MacConkey agar alter the medium's sensitivity or specificity? Explain your answer. (See pages 8 and 9, "A Word About Experimental Design," for assistance with these terms.)*

**8** *Suppose a mistake is made in preparing a batch of MacConkey agar and the starting pH is 7.6 instead of 6.9–7.3. Would that affect the medium's sensitivity or specificity? Explain your answer. (See pages 8 and 9, "A Word About Experimental Design," for assistance with these terms.)*

**9** *Compare the recipes of nutrient agar and MacConkey agar. If an organism can grow on both media, on which would you expect it to grow better? Explain your answer.*

# EXERCISE 4-5

# Eosin Methylene Blue Agar

## Theory

Eosin methylene blue (EMB) agar is a complex (chemically undefined), selective, and differential medium. It contains digest of gelatin, lactose, and the dyes eosin Y and methylene blue. The gelatin provides nitrogen and organic carbon. Lactose is fermented to acid end-products by coliforms such as *Escherichia coli* and *Enterobacter aerogenes*, whereas it is not by pathogens such as *Proteus*, *Shigella*, and *Salmonella* species. (See Fig. 4.5 for an introduction to lactose fermentation.)

The purpose of the dyes is twofold: (1) they inhibit the growth of most Gram-positive organisms (some *Enterococcus* and *Staphylococcus* species are exceptions), and (2) they react with vigorous lactose fermenters whose acid end-products turn the growth dark purple or black. This dark growth is typical of *Escherichia coli* and is usually accompanied by a green metallic sheen (Fig. 4.7). Other, less-aggressive lactose fermenters, such as *Enterobacter* or *Klebsiella* species, produce colonies that can range from pink to dark purple on the medium. Lactose non-fermenters typically retain their normal color or take on the coloration of the medium.

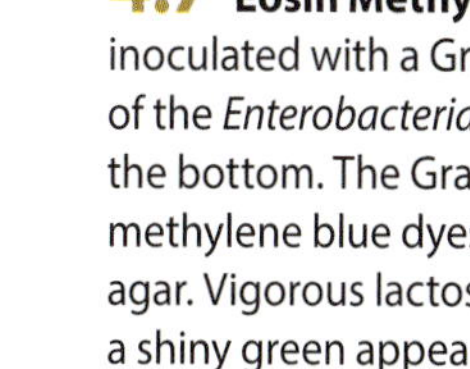

**4.7 Eosin Methylene Blue Agar (EMB)** ■ This EMB plate was inoculated with a Gram-positive coccus (center) and two members of the *Enterobacteriaceae*, a coliform at the top and a noncoliform at the bottom. The Gram-positive coccus was inhibited by eosin and methylene blue dyes but still grew enough to form a thin film on the agar. Vigorous lactose fermentation by the coliform gave its growth a shiny green appearance. The noncoliform was not inhibited by the dyes but also did not ferment the lactose, and so is a more natural color. EMB would normally be streaked for isolation, but you will spot inoculate pure cultures for the sole purpose of seeing what the different results look like.

## Application

EMB agar (Levine) is used for the isolation of fecal coliforms. It can be streaked for isolation or used in the Membrane Filter Technique as discussed in Exercise 7-5.

## In This Exercise

You will spot inoculate one EMB agar plate and one nutrient agar (NA) plate with four test organisms. The NA plate will serve as a comparison for growth quality on the EMB agar plate.

## Materials

### Per Student

- ☐ Lab coat
- ☐ Disposable gloves
- ☐ Chemical eye protection

### Per Student Group

- ☐ One EMB plate
- ☐ One NA plate
- ☐ Fresh broth cultures of these recommended organisms:
  - *Enterobacter aerogenes*
  - *Escherichia coli*
  - *Providencia stuartii*
  - *Staphylococcus epidermidis*

## Medium Recipes

### Eosin Methylene Blue Agar (Levine)

| | |
|---|---|
| ☐ Pancreatic digest of gelatin | 10.0 g |
| ☐ Lactose | 10.0 g |
| ☐ Dipotassium phosphate | 2.0 g |
| ☐ Methylene blue | 0.065 g |
| ☐ Eosin Y | 0.4 g |
| ☐ Agar | 15.0 g |
| ☐ Distilled or deionized water | 1.0 L |

*pH 6.9–7.3 at 25°C*

### Nutrient Agar

| | |
|---|---|
| ☐ Beef extract | 3.0 g |
| ☐ Peptone | 5.0 g |
| ☐ Agar | 15.0 g |
| ☐ Distilled or deionized water | 1.0 L |

*pH 6.6–7.0 at 25°C*

## PROCEDURE

### Lab One

1. Wear a lab coat, gloves, and chemical eye protection when performing this procedure.
2. Gently mix each culture according to your lab's standards.
3. Using a permanent marker, divide the bottom of each plate into four sectors.
4. Label the plates with the organisms' names, your name, and the date. Use the same positions for each specimen on both plates.
5. Spot inoculate (refer to Appendix B, "Spot Inoculation of an Agar Plate," p. 601) the sectors on the EMB plate with the test organisms.
6. Repeat step 5 with the nutrient agar plate. Be sure to inoculate the specimens in the sectors that correspond to their position on the EMB plate.
7. Invert and incubate the plate at 35 ± 2°C for 24 to 48 hours.
8. Save or dispose of the original cultures as directed by your instructor.

### Lab Two

1. Examine and compare the plates for color and quality of growth.
2. Record your results on the data sheet, page 257. Refer to Table 4-5 as needed.
3. Dispose of all plates in the appropriate autoclave container when finished.

### References


Tille, Patricia M. Page 84 in *Bailey & Scott's Diagnostic Microbiology*, 13th ed. St. Louis, MO: Mosby, 2014.

Winn, Washington C. et al. *Koneman's Color Atlas and Textbook of Diagnostic Microbiology*, 6th ed. Baltimore: Lippincott Williams & Wilkins, 2006.

Zimbro, Mary Jo and David A. Power. Page 218 in *Difco™ & BBL™ Manual–Manual of Microbiological Culture Media*. Sparks, MD: Becton, Dickinson and Co., 2003.[/REF]


TABLE **4-5** **EMB Results and Interpretations**

| Result | Interpretation | Presumptive ID |
|---|---|---|
| Poor growth (P) or no growth (N) | Organism is inhibited by eosin and/or methylene blue | Gram positive |
| Good growth (G) | Organism is not inhibited by eosin and methylene blue | Gram negative |
| Growth is pink and mucoid (Pi) | Organism ferments lactose with little acid production | Possible coliform |
| Growth is "dark" (purple to black, with or without green metallic sheen) (D) | Organism ferments lactose with acid production | Probable coliform |
| Growth is "colorless" (not purple, green, red, or pink) (C) | Organism does not ferment lactose | Noncoliform |

Name ______________________________

Date ______________________________

Lab Section ______________________________

I was present and performed this exercise (initials) ______________

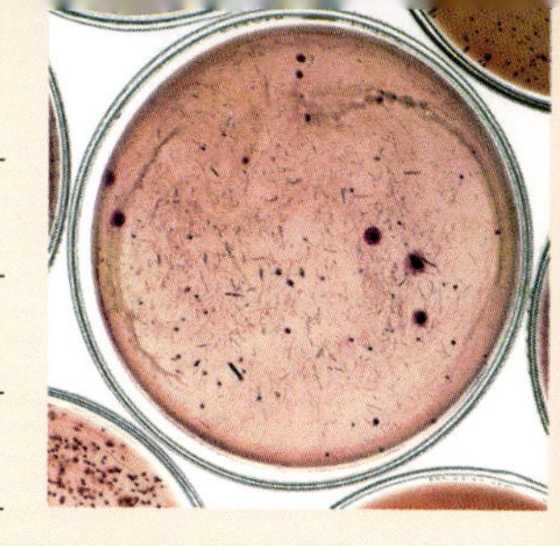

# DATA SHEET 4-5

## Eosin Methylene Blue Agar

### OBSERVATIONS AND INTERPRETATIONS

**1** Refer to Table 4-5, page 256, when recording your results and interpretations in the table below. Use abbreviations or symbols as needed.

| Organism | Growth (N/P/G) | | EMB Growth Color (Pi/D/C) | Interpretation and Presumptive ID |
|---|---|---|---|---|
| | EMB | NA | | |
| | | | | |
| | | | | |
| | | | | |

## QUESTIONS

**1** *In your own words, what is the application (purpose) of EMB agar?*

*Which ingredient(s) in EMB agar supply(ies)*

**a.** *Carbon?*

**b.** *Nitrogen?*

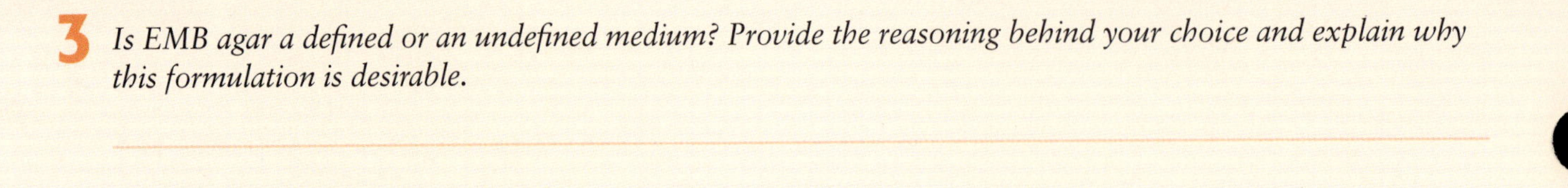

**3** *Is EMB agar a defined or an undefined medium? Provide the reasoning behind your choice and explain why this formulation is desirable.*

**4** *In your own words, what are the roles of eosin Y and methylene blue in EMB agar?*



**5** *In your own words, what is the role of lactose in EMB agar?*

**6** *Think about the results you recorded.*

**a.** *Growth on the EMB agar and NA plates was recorded as "good growth," "poor growth," or "no growth." These are qualitative and, at least for the first two, subjective terms. What did you use to establish what constituted "good growth?"*

**b.** *Why wouldn't it be advisable to compare growth of the organisms on each plate to each other? There are at least two answers to this question!*

Name ______________________________

Date ______________________________

Lab Section ______________________________

I was present and performed this exercise (initials) ______________________________



**7** *Would removing eosin Y and/or methylene blue from EMB agar alter the medium's sensitivity or specificity? Explain your answer. (See pages 8 and 9, "A Word About Experimental Design," for assistance with these terms.)*

**8** *Suppose a mistake is made in preparing a batch of EMB agar and the starting pH is 7.6 instead of 6.9–7.3. Would that affect the medium's sensitivity or specificity? Explain your answer. (See pages 8 and 9, "A Word About Experimental Design," for assistance with these terms.)*

**9** *Compare the recipes of nutrient agar and EMB agar. If an organism can grow on both media, on which would you expect it to grow better? Explain your answer.*

# Hektoen Enteric Agar

EXERCISE 4-6

## Theory

Hektoen enteric (HE) agar is a complex (chemically undefined), moderately selective, and differential medium designed to isolate *Salmonella* and *Shigella* species from other enterics. The test is based on the ability to ferment lactose, sucrose, or salicin to acid end-products, and to reduce sulfur (e.g., sodium thiosulfate) to hydrogen sulfide gas ($H_2S$).

Ferric ammonium citrate is included as a source of oxidized iron to react with any $H_2S$ produced to form the black precipitate ferrous sulfide (FeS). Bile salts are included to prevent or inhibit growth of Gram-positive organisms. The bile salts also have a moderate inhibitory effect on enterics, so relatively high concentrations of animal tissue and yeast extract are included to offset this situation. Bromothymol blue and acid fuchsin dyes are added to indicate pH changes.

Differentiation is possible as a result of the various colors produced in the colonies and in the agar. Enterics that produce acid from fermentation will produce yellow to salmon-pink colonies. Neither *Salmonella* nor *Shigella* species ferment any of the sugars; instead they break down the animal tissue, which raises the pH of the medium slightly and gives their colonies a blue-green color. Additionally, *Salmonella* species reduce sulfur to $H_2S$, so the colonies formed also contain FeS, which makes them partially or completely black. Refer to Figure 4.8.

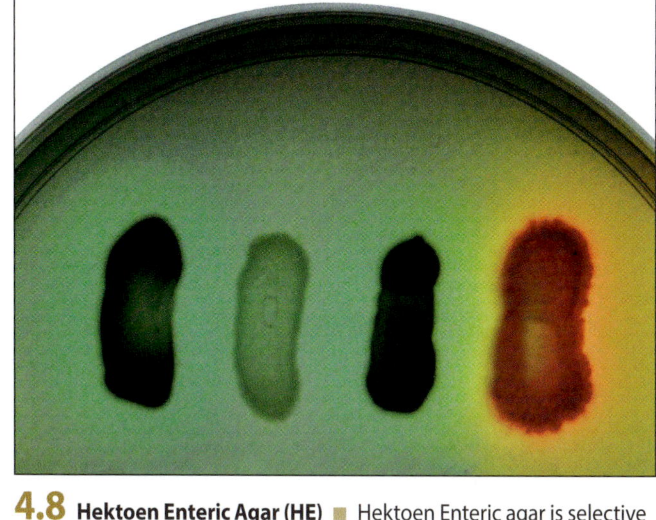

**4.8 Hektoen Enteric Agar (HE)** ■ Hektoen Enteric agar is selective for Gram negatives and is typically used to isolate members of the *Enterobacteriaceae*. It differentiates between coliforms, which produce acid from lactose fermentation, and noncoliforms, and also identifies sulfur reducers. This HE plate was inoculated with four Gram-negative rods. The organism on the right produced acid from lactose fermentation resulting in a yellow (or sometimes salmon) color of the growth and surrounding medium. The other three are noncoliforms and don't ferment lactose, resulting in a blue-green color best shown by the organism second from the left, which is negative for sulfur reduction. The black coloration largely hiding the blue-green color of the first and third organisms is due to sulfur reduction. HE plates would normally be streaked for isolation, but you will spot inoculate pure cultures for the sole purpose of seeing what the different results look like.

## Application

In a medical setting, HE agar is used to isolate and differentiate *Salmonella* and *Shigella* species from other Gram-negative enteric organisms in a patient's stool sample. It also is used in identifying fecal contamination of dairy and poultry products.

## In This Exercise

You will spot inoculate one Hektoen enteric agar plate and one nutrient agar plate with four test organisms. (In a clinical or agricultural application, the agar would be streaked for isolation.) The nutrient agar will serve as a comparison for growth quality on the Hektoen enteric agar plate.

## Materials

### Per Student

- ☐ Lab coat
- ☐ Disposable gloves
- ☐ Chemical eye protection

### Per Student Group

- ☐ One HE agar plate
- ☐ One NA plate
- ☐ Fresh broth cultures of these recommended organisms:
  - *Enterococcus faecalis* (BSL-2)
  - *Escherichia coli*
  - *Salmonella typhimurium* (BSL-2)
  - *Shigella flexneri* (BSL-2) or *Providencia stuartii*

## Medium Recipes

### Hektoen Enteric Agar

| | |
|---|---|
| □ Yeast extract | 3.0 g |
| □ Peptic digest of animal tissue | 12.0 g |
| □ Lactose | 12.0 g |
| □ Sucrose | 12.0 g |
| □ Salicin | 2.0 g |
| □ Bile salts | 9.0 g |
| □ Sodium chloride | 5.0 g |
| □ Sodium thiosulfate | 5.0 g |
| □ Ferric ammonium citrate | 1.5 g |
| □ Bromothymol blue | 0.064 g |
| □ Acid fuchsin | 0.1 g |
| □ Agar | 13.5 g |
| □ Distilled or deionized water | 1.0 L |

*pH 7.4–7.8 at 25°C*

4

### Nutrient Agar

| | |
|---|---|
| □ Beef extract | 3.0 g |
| □ Peptone | 5.0 g |
| □ Agar | 15.0 g |
| □ Distilled or deionized water | 1.0 L |

*pH 6.6–7.0 at 25°C*

## PROCEDURE

### Lab One

1. Wear a lab coat, gloves, and chemical eye protection when performing this procedure.
2. Gently mix each culture according to your lab's standards.
3. Using a permanent marker, divide the bottom of each plate into four sectors.
4. Label the plates with the organisms' names, your name, and the date. Use the same positions for each specimen on both plates.
5. Spot inoculate (refer to Appendix B, "Spot Inoculation of an Agar Plate," p. 601) the sectors on the HE plate with the test organisms. **Use BSL-2 precautions when transferring *E. faecalis, S. typhimurium,* and *S. flexneri* (if used).**
6. Repeat step 5 with the nutrient agar plate. Be sure to inoculate the specimens in the sectors that correspond to their position on the HE plate.
7. Invert and incubate the plate at 35 ± 2°C for 24 to 48 hours.
8. Save or dispose of the original cultures as directed by your instructor.

### Lab Two

1. Examine and compare the plates for color and quality of growth.
2. Record your results on the data sheet, page 263. Refer to Table 4-6 as needed.
3. Dispose of all plates in the appropriate autoclave container when finished.

### References


Tille, Patricia M. Page 86 in *Bailey & Scott's Diagnostic Microbiology,* 13th ed. St. Louis, MO: Mosby, 2014.

Winn, Washington C. et al. *Koneman's Color Atlas and Textbook of Diagnostic Microbiology*, 6th ed. Baltimore: Lippincott Williams & Wilkins, 2006.

Zimbro, Mary Jo and David A. Power. Page 265 in *Difco™ & BBL™ Manual–Manual of Microbiological Culture Media.* Sparks, MD: Becton, Dickinson and Co., 2003.


TABLE **4-6** **Hektoen Enteric Agar Results and Interpretations**

| Result | Interpretation | Presumptive ID |
|---|---|---|
| Poor growth (P) or no growth (N) | Organism is inhibited by bile and/or one of the dyes included | Gram positive |
| Good growth (G) | Organism is not inhibited by bile or any of the dyes included | Gram negative |
| Pink to orange growth (Pi) | Organism produces acid from lactose, sucrose, and/or salicin fermentation | Not *Shigella* or *Salmonella* |
| Blue-green growth with black precipitate (Bppt) | Organism does not ferment lactose, sucrose, or salicin, but reduces sulfur to hydrogen sulfide ($H_2S$) | Probable *Salmonella* |
| Blue-green growth without black precipitate (B) | Organism does not ferment lactose, sucrose, or salicin, or reduce sulfur. | Probable *Shigella* or rarely *Salmonella* |

Name ______________________

Date ______________________

Lab Section ______________________

I was present and performed this exercise (initials) ______________________

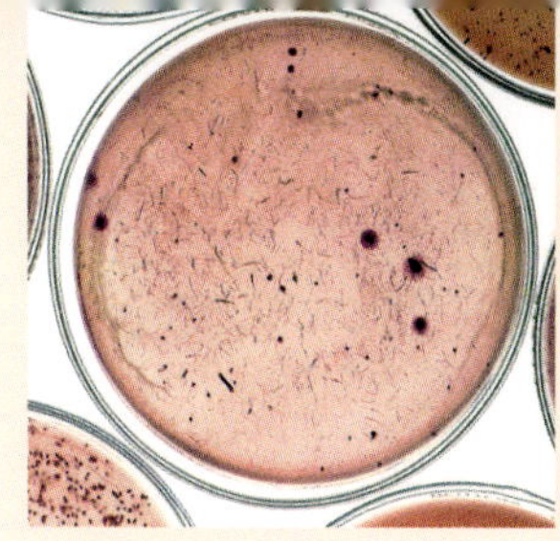

DATA SHEET 4-6

## Hektoen Enteric Agar

### OBSERVATIONS AND INTERPRETATIONS

**1** Refer to Table 4-6, page 262, when recording your results and interpretations in the table below. Use abbreviations or symbols as needed.

| Organism | Growth (N/P/G) | | HE Growth Color (Pi/Bppt/B) | Interpretation and Presumptive ID |
|---|---|---|---|---|
| | HE | NA | | |
| | | | | |
| | | | | |
| | | | | |

### QUESTIONS

**1** *In your own words, what is the application (purpose) of HE agar?*

**2** *Which ingredient(s) in HE agar supply(ies)*

**a.** *Carbon?*

**b.** *Nitrogen?*

**3** *Is HE agar a defined or an undefined medium? Provide the reasoning behind your choice and explain why this formulation is desirable.*

**4** *In your own words, what is the role of bile salts in HE agar?*

**5** *In your own words, what are the roles of lactose, sucrose, and salicin in HE agar?*

**6** *In your own words, what are the roles of acid fuchsin and bromothymol blue in HE agar?*

Name ______________________________

Date ______________________________

Lab Section ______________________________

I was present and performed this exercise (initials) ______________________________

# DATA SHEET 4-6
*(continued)*

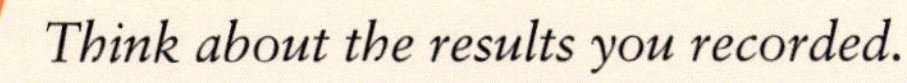

**7** *Think about the results you recorded.*

**a.** *Growth on the HE agar and NA plates was recorded as "good growth," "poor growth," or "no growth." These are qualitative and, at least for the first two, subjective terms. What did you use to establish what constituted "good growth?"*

**b.** *Why wouldn't it be advisable to compare growth of the organisms on each plate to each other? There are at least two answers to this question!*

**8** *Would removing bile salts from HE agar alter the medium's sensitivity or specificity? Explain your answer. (See pages 8 and 9, "A Word About Experimental Design," for assistance with these terms.)*

**9** *Suppose a mistake is made in preparing a batch of HE agar and the starting pH is 7.0 instead of 7.4–7.8. Would that affect the medium's sensitivity or specificity? Explain your answer. (See pages 8 and 9, "A Word About Experimental Design," for assistance with these terms.)*

**10** *Compare the recipes of nutrient agar and HE agar. If an organism can grow on both media, on which would you expect it to grow better? Explain your answer.*

**11** *All enterics ferment glucose. What would be some consequences of replacing the sugars in HE agar with glucose? What color combinations would you expect to see?*



**12** *Write a description of an organism whose colonies are green with black centers on HE agar.*

SECTION 5

# Differential Tests

In the last century, bacteria have been differentiated and characterized based on their enormous biochemical and physiological diversity using **differential tests**. Once characterized, those same tests—now acting in the role of **diagnostic tests**—can be used to identify unknown bacteria in a patient or environmental sample by matching the unknown's test results to the characterization of "known" bacteria that fits best. In this section, you will be introduced to a variety of differential tests. Then, in Section 9, you will have the opportunity to use many of these differential tests to identify one or more unknown bacterial species.

As you will see repeatedly in this section, bacterial physiological characteristics are brought about by enzymatic reactions that are part of specific and well-understood metabolic pathways. When microbiologists run differential tests, most of the time they are really looking for the ability of an organism to make a particular enzyme or group of enzymes. In many cases, a test identifies the organism's physiological ability, which in turn allows the microbiologist to infer the presence or absence of an enzyme or multiple enzymes (e.g., lactose fermentation). In other cases, the test more directly determines the presence or absence of an enzyme (e.g. the catalase test).

The tests included in this section were selected because they provide a sampling of metabolic diversity available for testing. They also are among the more routinely used tests by microbiologists in diagnostics. To help you place many of the tests into the overall workings of bacteria, Figure A.1, page 590, shows an integrated summary of metabolic pathways and in many instances indicates which exercise addresses that pathway.

Differential tests can be grouped into several basic categories, with individual tests in the category targeting a specific aspect of that category. The categories of differential tests in this section are:

- Energy metabolism, including fermentation of carbohydrates and aerobic and anaerobic respiration
- Utilization of a specified medium component
- Decarboxylation and deamination of amino acids
- Hydrolytic reactions requiring intracellular or extracellular enzymes
- Multiple reactions performed in a single combination medium
- Antimicrobial susceptibility
- Miscellaneous differential tests

## A Word about Biochemical Tests and Acid-Base Reactions

Differential test media provide limited specific growth conditions (usually including a single key substrate such as glucose or lactose in the case of fermentation media) whereby an organism is given the opportunity to carry out, if it can, a specified metabolic reaction or series of reactions of interest to the microbiologist. But however complex the chemistry may be, the evaluation of the test is often a simple check to see if an acid or base has been formed. Considering the diversity of physiological characteristics bacteria display, it is comforting to know that many of the biochemical tests used to differentiate them rely on a simple qualitative evaluation of pH! (For more about pH, see Exercise 2-9.)

An indicator dye (e.g., phenol red, bromocresol purple, or bromothymol blue) is usually included in the test medium (or may be added after incubation) that changes color when the pH changes past a certain point. Different dyes operate best in a specific pH range and allow media to be tailored to the type of organism or metabolic process being tested. Many oral bacteria, for example, prefer a slightly acidic environment (around pH 6) whereas enteric bacteria prefer a more neutral environment (around pH 7).

Because it is so easy to detect and many microbes produce acids and/or bases during their metabolism, pH change is an extremely useful tool in diagnostic microbiology. The principle is simple—if the pH changes (as evidenced by a color change) in one direction, the organism is said to have done one thing; if the pH (color) changes in the other direction, the organism is said to have done something else. If there is no change in color, it is determined that there was no reaction or *that the organism did not grow*. Naturally, the conditions of each biochemical test affect how the pH change is interpreted.

Obviously, there are more metabolic reactions than can be discussed in this manual, but it is helpful to know a little bit about why certain metabolic products are formed and how they affect their environment. As you will see throughout this section, one recurring theme is fermentation so we will use it as an example. Microbial fermentations produce three principal products: gas, alcohol (or other organic solvents), and/or acid.

Gas (frequently in the form of $CO_2$) can be an easy indicator of fermentation because it is easy to see in the medium. Typically, this is accomplished by the addition of a small inverted test tube, called a Durham tube, to broth media. When gas-producing fermentation occurs, microscopic gas bubbles are trapped inside the Durham tube, which coalesce to form a visible gas bubble. Figure A.6, page 597, illustrates several fermentation pathways that use reactions where carbon compounds are split and $CO_2$ is released.

Unfortunately, there are no simple qualitative tests to detect alcohols and other organic solvents (e.g., acetone) produced by fermentation. They are liquid and colorless, so they are not seen in the medium and because they do not ionize in water they have minimal effect (if any) on the medium's pH.

The weak acids (e.g., acetic acid, propionic acid, formic acid, succinic acid, and butyric acid) have a dramatic effect on the pH of any medium and can easily be detected by addition of one of the pH indicators mentioned above.

Acids, as defined by the Brønsted/Lowry theory of acids and bases, are substances that donate hydrogen ions ($H^+$) while bases are substances that accept hydrogen ions. All of the above-mentioned acids are called **carboxylic acids** because they contain the carboxyl functional group:

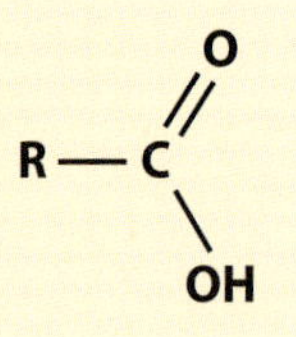


where "R" is the remainder of the molecule. (A carboxylic acid may also be represented as R–COOH because it fits on a line of text better!) Because carboxylic acids can exist as either R–COOH or $R\text{–}COO^-$ in the same solution, they form a **conjugate pair**, where R–COOH is the conjugate acid and $R\text{–}COO^-$ is the conjugate base. A typical reaction would look like this: $R\text{–}COOH + H_2O \rightleftharpoons R\text{–}COO^- + H_2O + H^+$ (also represented as $R\text{–}COO^- + H_3O^+$). As you can see, the reaction can run in either direction, but when the reaction goes to the right, hydrogen ions are released into the medium and this lowers the pH (Exercise 2-9).

Conversely, the removal of hydrogen ions or the addition of hydroxide ions ($OH^-$) to the water raises the pH. A typical way that an organism raises the pH in fermentation and other growth media is by breaking down proteins. Breaking down proteins raises the pH of media because the reaction releases ammonia (**deamination**), which further reacts with the water to form ammonium and hydroxide ions. The latter reaction looks like this: $NH_3 + H_2O \rightleftharpoons NH_4^+ + OH^-$.

It is important to note that a differential medium prepared in the laboratory is not intended to duplicate an organism's natural environment. Differential media are artificial mixtures of chemicals and organic substrates intended to "persuade" microbes to perform in ways that might never actually be expressed in nature.

So even though an organism may never radically lower the pH of its natural environment (because of low substrate availability or competing protein catabolism) we can manipulate it to do so in the lab by cultivating it in medium spiked with a specific sugar or other substrate. And when two or more species, otherwise indistinguishable by way of staining, microscopy, or colony characteristics, are thus compared, their differences frequently become clear. ■

## Introduction to Energy and Metabolism

Cells are chemical entities and the chemical reactions they perform are collectively referred to as **metabolism**. Some reactions result in a decrease in the reactant molecule's complexity. These are **catabolic** reactions. Conversely, **anabolic** reactions result in the product becoming more complex than the reactant. Inevitably, energy is involved in metabolic reactions, so to fully appreciate the role of the metabolic reactions you will encounter in this section we will need to introduce you to some characteristics of energy.

Energy behaves in consistently predictable ways that are summarized by the two laws of thermodynamics. In a nutshell, the First Law tells us that the total energy in a system remains constant; that is, energy cannot be created or destroyed. It also, and very importantly, tells us that energy can be transferred and change forms (e.g., electrical energy converted to light energy in a light bulb).

The Second Law is very complex and can be stated and recognized in many different ways. It tells us that energy transfers and conversions are never 100% efficient. It also tells us that in those transfers—or even if energy is "left alone"—high quality energy dissipates and becomes lower quality energy (less able to do "useful" work), with heat energy occupying the lowest level. Another way of looking at this is that order devolves to disorder. There is evidence of this everywhere—batteries run down, paint fades and peels, automobiles rust, fences fall down, and so on.

A measure of that disorder is called **entropy**, so entropy always increases. Organisms constantly confront this reality: They need a constant input of energy because the energy they have will eventually be dissipated as heat while they do the work of living and it needs to be replaced. Without a constant input of energy, they die and fall apart, just like an untended building.



The potential energy of the bonds within a biochemical is called **free energy** and is considered to be "useful" energy. Consistent with both laws, metabolic reactions can be coupled such that a portion of free energy released in one reaction can be captured in another. Thus, there is a loss of free energy in the first reaction and a gain of free energy in the coupled reaction, but with a net loss of free energy. A loss of free energy occurs in catabolic reactions (because a bond is broken), whereas anabolic reactions gain free energy (because a new bond is formed).

Another way to view this is that the potential energy contained in the bonds of a biochemical came from the energy released by anabolic reactions involved in its assembly from individual components. And again, obeying the law of entropy, *some* of this energy retained by the molecule can be released and stored in subsequent reactions or used to do other cellular work. One important use of released free energy is production of high-energy molecules such as ATP. Of course, some free energy is converted to heat at every step along the way.

Frequently, metabolic free energy changes are associated with **oxidation-reduction**, or "**redox**" reactions. Reduction occurs when a molecule (called an **oxidant**) gains one or more electrons, which requires oxidation of another molecule (called a **reductant**) to provide those electrons. Because electrons possess potential energy, a redox reaction results in an energy transfer.

Molecules either accept or donate electrons based on their relative **reduction potential**. Reduction potential, simply defined, is the tendency of a molecule to donate electrons, measured in volts. Without the input of energy, electron transfers will always go from a molecule with a higher reduction potential to a molecule with a lower reduction potential. The "electron tower" in Figure 5.1 illustrates the relationship of many important cellular oxidants and reductants.

Heterotrophic bacteria obtain their energy by performing either **respiration** or **fermentation** using ingested organic molecules. Both catabolic systems oxidize the organic molecules and transfer their chemical energy into high-energy bonds in **adenosine triphosphate (ATP)**. They differ principally in what **final electron acceptor** (**FEA**), which acts as an oxidant, is used; the presence or absence of an **electron transport chain** (**ETC**); and in their ATP yield.

Fermentations are characterized by the absence of an ETC and use an organic compound as the FEA, whereas respirations use an inorganic compound and possess an ETC. Oxygen serves as the FEA in **aerobic respiration**, whereas another inorganic compound (e.g., $NO_3$ or $SO_4$) acts in that role in **anaerobic respiration**.

Fermentation pathways are many and varied, but all use the process of **substrate-level phosphorylation** to make ATP, in which the energy for adding the phosphate comes directly from breaking a bond in a biochemical. Respiration accomplishes ATP synthesis by means of three well-understood pathways: **glycolysis**; the **citric acid cycle** (also known as **Krebs,** or **tricarboxylic acid cycle**); and an electron transport chain, which performs **oxidative phosphorylation** to synthesize ATP. ATP synthesis also occurs by substrate-level phosphorylation in the first two respiratory pathways.

In glycolysis and the citric acid cycle, certain reactions require **coenzymes** during the oxidation of carbon intermediates. Coenzymes are the part of the enzyme-coenzyme-substrate complex that directly takes part in the transfer of electrons with the substrate. Because they can exist in either the oxidized or reduced state, coenzymes are responsible for both oxidations and reductions.

The relevant coenzymes of fermentation and respiration are nicotinamide adenine dinucleotide ($NAD^+$) and flavin adenine dinucleotide (FAD). In their reduced form, these coenzymes reduce carriers in the respiratory electron transport chain and provide the energy for ATP synthesis (oxidative phosphorylation). For a more detailed description of these metabolic processes, see Appendix A and Figures A.2 through A.6. ■

## References


Madigan, Michael T., John M. Martinko, Kelly S. Bender, Daniel H. Buckley, and David A. Stahl. Chap. 3 in *Brock's Biology of Microorganisms*, 14th ed. San Francisco: Pearson/Benjamin Cummings, 2015.

Moat, Albert G., John W. Foster, and Michael P. Spector. Pages 373-376 in *Microbial Physiology*, 4th ed. New York, NY: Wiley–Liss, 2002.

White, David, James Drummond, and Clay Fuqua. Chap. 5 in *The Physiology and Biochemistry of Prokaryotes*, 4th ed. New York, NY: Oxford University Press, 2012.




| Redox Pair | Reduction Potential ($E_o'$ in V) | Number of Electrons |
|---|---|---|
| $CO_2$/glucose | −0.43 | $24e^-$ |
| $NAD^+$/NADH | −0.32 | $2e^-$ |
| $S^0/H_2S$ | −0.28 | $2e^-$ |
| Pyruvate/lactate | −0.19 | $2e^-$ |
| Cytochrome $b_{558\ ox/red}$ | −0.08 to −0.04 | $1e^-$ |
| Fumarate/succinate | +0.03 | $2e^-$ |
| Cytochrome $b_{562\ ox/red}$ | +0.13 to +0.26 | $1e^-$ |
| $Fe^{3+}/Fe^{2+}$ (pH 7) | +0.20 | $1e^-$ |
| $Ubiquinone_{ox/red}$ | +0.11 | $2e^-$ |
| Cytochrome $c_{ox/red}$ | +0.25 | $1e^-$ |
| Cytochrome $d_{ox/red}$ | +0.26 to +0.28 | $1e^-$ |
| Cytochrome $a_{ox/red}$ | +0.29 | $1e^-$ |
| Cytochrome $a_{3\ ox/red}$ | +0.39 | $1e^-$ |
| $NO_3^-/NO_2^-$ | +0.42 | $2e^-$ |
| $NO_3^-/½ N_2$ | +0.74 | $5e^-$ |
| $Fe^{3+}/Fe_2^+$ (pH 2) | +0.76 | $1e^-$ |
| $½ O_2/H_2O$ | +0.82 | $2e^-$ |

**5.1 Electron Tower** ■ Redox pairs are listed from top to bottom in order of decreasing reduction potential ($E_o'$), which means the more positive the charge is, the lower the reduction potential. (***Note***: Reduction potentials can be affected by physical factors, such as pH, so the values listed are not absolute. Also, the increments between redox pairs are not to scale.) The oxidant is on the left of each pair; the reductant is on the right. Electrons are spontaneously (i.e., without an external energy source) transferred downward from the reductant of one pair to the oxidant of a pair containing lower reduction potential. Therefore, all spontaneous transfers as shown in the tower are downward. Also included are the numbers of electrons transferred in any one reaction.

(After Madigan, et al., 2015 and White, et al., 2012.)

# Oxidation–Fermentation (O–F) Test

EXERCISE **5-1**

## Theory

The Oxidation–Fermentation (O–F) test is designed to differentiate bacteria on the basis of fermentative or oxidative metabolism of carbohydrates. In this medium, oxidative organisms oxidize the carbohydrate to $CO_2$, $H_2O$, and energy, using in the following order: glycolysis, the oxidation of pyruvate, the citric acid cycle, and finally the electron transport chain (ETC) with oxygen (the final electron acceptor, FEA) being reduced to $H_2O$.

Similarly, fermentative organisms convert the carbohydrate to pyruvate, but because either they *cannot* use $O_2$ as an FEA or it is *not available*, pyruvate, or one of its derivatives, acts as the FEA and it is reduced to the familiar acid, gas, and/or alcohol fermentation end products. Consequently, fermenters identified by this test acidify O–F medium to a greater extent than do oxidizers.

Hugh and Leifson's O–F medium includes a high sugar-to-peptone ratio to reduce the possibility that alkaline products from peptone utilization will neutralize weak acids produced by oxidation of the carbohydrate. Bromothymol blue dye, which is yellow at pH 6.0, green at pH 7.1, and blue at pH 7.6, is added as the indicator. A low agar concentration makes it a semisolid medium that also allows determination of motility.

The medium is prepared with glucose, lactose, sucrose, maltose, mannitol, or xylose and is not slanted. Two tubes of the specific sugar medium are stab inoculated several times with the test organism. After inoculation, one tube is sealed with a layer of sterile mineral oil to promote anaerobic growth and fermentation (Fig. 5.2). The mineral oil creates an environment unsuitable for oxidation because it prevents diffusion of oxygen from the air into the medium.

The other tube is left unsealed to allow aerobic growth and oxidation. (***Note***: Tubes of O–F medium are heated in boiling water and then cooled prior to inoculation. This removes free oxygen from the medium and ensures an anaerobic environment in all tubes. The tubes covered with oil will remain anaerobic, whereas the uncovered medium quickly will become aerobic as oxygen diffuses back in.) O–F Test results are summarized in Table 5-1 and shown in Figure 5.3.

Organisms that are able to ferment the carbohydrate or ferment *and* oxidize the carbohydrate will turn the sealed and unsealed media yellow throughout, as will organisms that are able to ferment only. Organisms that are able to oxidize only will turn the unsealed medium yellow (or partially yellow) and leave the sealed medium green or blue. Slow or weak fermenters will turn both tubes slightly yellow at the top. Organisms that are not able to metabolize the sugar will either produce no color change or turn the medium blue because of alkaline products from amino acid degradation (deamination).

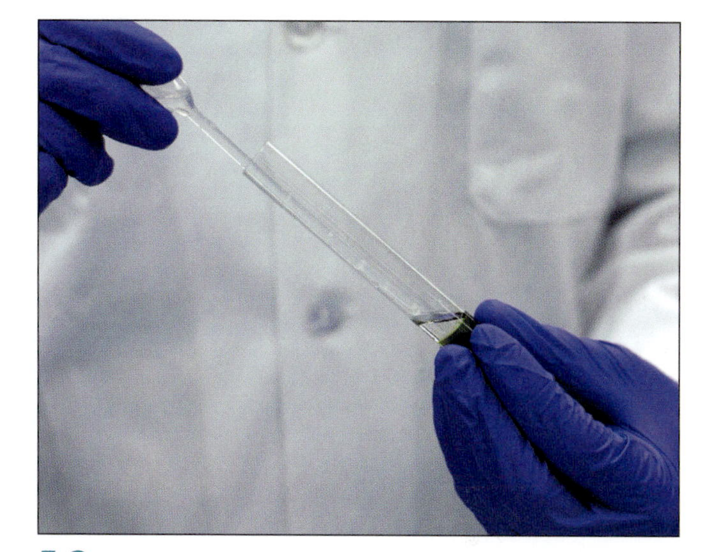

**5.2 Adding the Sterile Mineral Oil Layer** ■ Tip the tube slightly to one side, and gently add sterile mineral oil to a depth of 3 mm–4 mm. Be sure to use a sterile pipette for each tube.

## Application

The O–F test is used to differentiate bacteria based on their ability to oxidize or ferment specific sugars. It allows presumptive separation of the fermentative *Enterobacteriaceae* from the oxidative *Pseudomonas* and *Bordetella*, and the nonreactive *Alcaligenes* and *Moraxella*.

## In This Exercise

Eight O–F media tubes will be used for this exercise—six for inoculation (two for each organism) and two for controls. One of each pair will receive a mineral oil overlay to maintain an anaerobic environment and force fermentation if the organism is capable of doing so. You will not do a motility determination, because motility is covered in Exercise 5-24. For best results, use a heavy inoculum and several stabs for each tube. Use only sterile pipettes to add the mineral oil.

TABLE 5-1 **Oxidation–Fermentation (O–F) Test Results and Interpretations**

| Sealed (Anaerobic) | Unsealed (Aerobic) | Interpretation | Symbol |
|---|---|---|---|
| Green or blue | Any amount of yellow | Oxidation | O |
| Yellow throughout | Yellow throughout | Oxidation and fermentation, or fermentation only | O–F or F[1] |
| Slightly yellow at the top | Slightly yellow at the top | Slow oxidation and slow fermentation, or slow fermentation only | O–F or F[1] |
| Green or blue | Green or blue | No sugar metabolism; organism is nonsaccharolytic | N |

1 The results of a fermentative organism and one capable of both fermentative and oxidative metabolism (of that specific carbohydrate) look the same in this medium. Therefore, when both tubes are yellow, the result is recorded as (O-F) or (F).

**5.3 Oxidation–Fermentation (O–F ) Test Results** ■ These pairs of tubes illustrate three possible results in the O–F test. Reading from left to right, pair 1 was inoculated with an organism capable only of oxidative metabolism (O). Pair 2 was uninoculated for color comparison. Pair 3 was inoculated with an organism incapable of utilizing glucose. Note the blue color in the upper portion of the unsealed tube suggesting that the organism is both nonsaccharolytic (N) and strictly aerobic. Pair 4 was inoculated with an organism capable of both oxidative and fermentative, or just fermentative, utilization of glucose (O–F or F, because it is unclear if the organism is fermenting or oxidatively catabolizing the sugar in the unsealed tube).

## ▼ Materials

### Per Student

- □ Lab coat
- □ Gloves
- □ Chemical eye protection

### Per Student Group

- □ Eight O–F glucose tubes
- □ Sterile mineral oil
- □ Sterile transfer pipettes
- □ Fresh agar slant cultures of these recommended organisms:
  - *Alcaligenes faecalis*
  - *Escherichia coli*
  - *Kocuria rosea*

## ■ Medium Recipe

### Hugh and Leifson's O–F Medium with Glucose

- □ Pancreatic digest of casein 2.0 g
- □ Sodium chloride 5.0 g
- □ Dipotassium phosphate 0.3 g
- □ Agar 2.5 g
- □ Bromothymol blue 0.08 g
- □ Glucose 10.0 g
- □ Distilled or deionized water 1.0 L

*pH = 6.6–7.0 at 25°C*

## PROCEDURE

### Lab One

1. Wear a lab coat, gloves, and chemical eye protection when performing this procedure.
2. Working with your lab group, obtain eight O–F tubes. Label six (in three pairs) with the names of the organisms, your group name, and the date. Label the last pair "control."
3. Using an inoculating needle, stab inoculate each pair with the appropriate test organism. Stab several times to a depth of about 1 cm from the bottom of the agar to introduce a heavy inoculum. Do not inoculate the controls.
4. Using a different sterile transfer pipette for each organism overlay one of each pair of tubes (including the controls) with 3 mm–4 mm of sterile mineral oil (Fig. 5.2).
5. Incubate all tubes at 35 ± 2°C for 48 hours.
6. Save or dispose of the original culture tubes as directed by your instructor.

### Lab Two

1. Examine the tubes for color changes. Be sure to compare them to the controls before making a determination.
2. Record your results in the table provided on the data sheet, page 275.
3. Dispose of all tubes in the appropriate autoclave container when finished.

5

## *References*


Atlas, Ronald and James Snyder. Page 341 in *Manual of Clinical Microbiology,* 11th ed. James H. Jorgensen, Michael A. Pfaller, Karen C. Carroll, Guido Funke, Marie Louise Landry, Sandra S. Richter, and David W. Warnock, eds. Washington, DC: ASM Press, American Society for Microbiology, 2015.

Collins, C. H., Patricia M. Lyne, and J. M. Grange. Page 112 in *Collins and Lyne's Microbiological Methods*, 7th ed. Oxford, UK: Butterworth-Heinemann, 1995.

Delost, Maria Dannessa. Pages 218–219 in *Introduction to Diagnostic Microbiology.* St. Louis, MO: Mosby, 1997.

MacFaddin, Jean F. Page 379 in *Biochemical Tests for Identification of Medical Bacteria*, 3rd ed. Philadelphia: Lippincott Williams & Wilkins, 2000.

Smibert, Robert M. and Noel R. Krieg. Page 625 in *Methods for General and Molecular Bacteriology*. Philipp Gerhardt, R. G. E. Murray, Willis A. Wood, and Noel R. Krieg, eds. Washington, DC: American Society for Microbiology, 1994.

Tille, Patricia M. Pages 98 and 224 in *Bailey & Scott's Diagnostic Microbiology,* 13th ed. St. Louis: Mosby, 2014.

Zimbro, Mary Jo and David A. Power, eds. Page 410 in *Difco™ and BBL™ Manual—Manual of Microbiological Culture Media.* Sparks, MD: Becton Dickinson and Co., 2003.

Name ______________________________

Date ______________________________

Lab Section ______________________________

I was present and performed this exercise (initials) ______________

# DATA SHEET 5-1

## Oxidation–Fermentation (O–F) Test

### OBSERVATIONS AND INTERPRETATIONS

**1** Refer to Table 5-1, page 272, when recording your results and interpretations in the table below.

| Organism | Color Results | | Symbol | Interpretation |
|---|---|---|---|---|
| | Sealed | Unsealed | | |
| Uninoculated control | | | | |
| | | | | |
| | | | | |
| | | | | |

## QUESTIONS

**1** *Consider the controls.*

**a.** *What is the purpose of the uninoculated control tubes used in this test?*

**b.** *Are the controls positive or negative controls? Explain.*

**c.** *Why is it necessary to use two controls rather than just one? What are the specific purposes of each? Be thorough in your answer.*

5

**2** *Some microbiologists recommend inoculating a pair of O–F basal media (without carbohydrate) along with the carbohydrate media. Why do you think this is done?*

Name ______________________________

Date ______________________________

Lab Section ______________________________

I was present and performed this exercise (initials) ______________

# DATA SHEET 5-1

*(continued)*

**3** *All enterics* (Enterobacteriaceae) *are facultative anaerobes and are capable of both aerobic respiration and fermentation. What color results would you expect for organisms in O–F glucose media inoculated with an enteric? Remember to describe both sealed and unsealed tubes.*

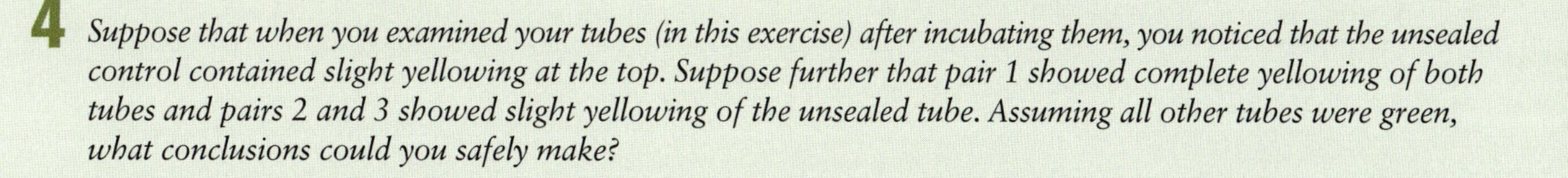

**4** *Suppose that when you examined your tubes (in this exercise) after incubating them, you noticed that the unsealed control contained slight yellowing at the top. Suppose further that pair 1 showed complete yellowing of both tubes and pairs 2 and 3 showed slight yellowing of the unsealed tube. Assuming all other tubes were green, what conclusions could you safely make?*

**a.** *Which results, if any, are reliable? Why?*

**b.** *Which results, if any, are not reliable? Why not?*

**5** *Uninoculated tubes and tubes inoculated with an organism that is (N) serve as negative controls. What information is provided by each?*

5

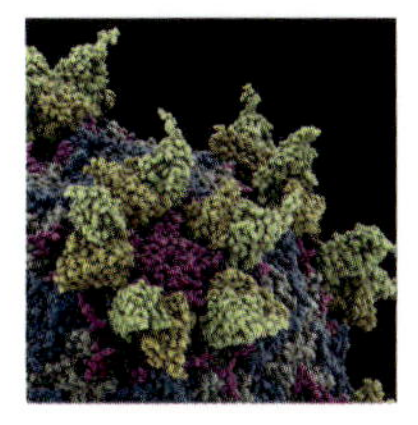

## Fermentation Tests

As defined at the beginning of this section, carbohydrate fermentation is the metabolic process by which an organic molecule acts as an electron donor (becoming oxidized in the process) and one or more of its organic products acts as the final electron acceptor (FEA) and becomes reduced. In actuality, the term **carbohydrate fermentation** is used rather broadly to include hydrolysis of disaccharides prior to the fermentation reaction.

Thus, a "lactose fermenter" is an organism that splits the disaccharide lactose into the monosaccharides glucose and galactose and then ferments the monosaccharides. In this section you will see the term **fermenter** frequently. Unless it is expressly used otherwise, this term should be assumed to include the initial hydrolysis and/or conversion reactions.

Fermentation of glucose begins with the production of pyruvate. Although some organisms use alternative pathways (e.g., Entner-Doudoroff pathway, Fig. A.3) most bacteria accomplish this by glycolysis. The end products of pyruvate fermentation include a variety of organic acids, alcohols, and hydrogen or carbon dioxide gas. The specific end products depend on the specific organism, its enzymes, and the substrate fermented. Refer to Figures 5.4 and A.6.

In the next two lab exercises, you will perform tests that identify different fermentations. In Exercise 5-2 the general-purpose fermentation medium phenol red broth is used. Typically, it contains any one of several carbohydrates (e.g., glucose, lactose, sucrose), a pH indicator, and a Durham tube to detect acid and gas formation, respectively.

MR-VP broth, used in Exercise 5-3, is a dual-purpose medium that tests an organism's ability to follow either (or both) of two specific fermentation pathways. The methyl red (MR) test detects what is called a **mixed acid fermentation**. The Voges-Proskauer (VP) test identifies bacteria that are able to produce acetoin as part of a **2,3-butanediol fermentation**. See Appendix A, "Oxidation of Pyruvate" (page 591), for more information about fermentation. ■



# Phenol Red Fermentation Broth

EXERCISE 5-2

## Theory

Phenol red (PR) broth is a differential fermentation medium composed of standard ingredients (the "base broth") to which a single carbohydrate is added. After inoculation and incubation, an organism's ability to ferment that particular carbohydrate can be determined, as can the end products of its fermentation. Because virtually any carbohydrate can be added to the base, PR broth is a versatile medium. Figure 5.4 shows fermentation pathways for glucose, lactose, and sucrose.

Included in the base medium are peptone and the pH indicator phenol red. Phenol red is yellow below pH 6.8, pink to magenta above pH 7.4, and red in between. During preparation the pH is adjusted to approximately 7.3 so it appears red. Finally, an inverted Durham tube is added to each tube as an indicator of gas production.

Acid production from fermentation of the carbohydrate lowers the pH below the neutral range of the indicator and turns the medium yellow (Fig. 5.5). Gas production, also from fermentation, is indicated by a bubble, or pocket, in the Durham tube where the broth has been displaced. Deamination of amino acids supplied by casein (milk protein) results in an alkaline reaction from the ammonia ($NH_3$) that is produced. Ammonia raises the pH and turns the broth pink if no acid is produced or if acid production has ceased due to consumption of the carbohydrate.

## Application

PR broth is used to differentiate members of *Enterobacteriaceae* and to distinguish them from other Gram-negative rods. It also is used to distinguish between Gram-positive fermenters, such as *Streptococcus* and *Lactobacillus* species.

## In This Exercise

Today you will inoculate PR glucose broths with four organisms to determine their fermentation characteristics. Use Table 5-2 as a guide when interpreting and recording your results. Be sure to use an uninoculated control of each medium for color comparison.

At your instructor's discretion, other PR broths (such as PR lactose and/or PR sucrose) can be used in addition to PR glucose.

## ▼ Materials

### Per Student

- □ Lab coat
- □ Gloves
- □ Chemical eye protection

### Per Student Group

- □ Four PR glucose broths with Durham tubes[1]
- □ (Optional) four PR base broths with Durham tubes
- □ Fresh agar slant cultures of these recommended organisms:
  - *Alcaligenes faecalis*
  - *Escherichia coli*
  - *Providencia stuartii*
  - (Optional) *Streptococcus gallolyticus*; recommended if additional PR broths are used

[1] This lab exercise can also be performed using Purple broths, which differ only slightly in ingredient composition, but use bromocresol purple as the pH indicator. Bromocresol purple turns yellow at pH 5.2 and is purple above pH 6.8.

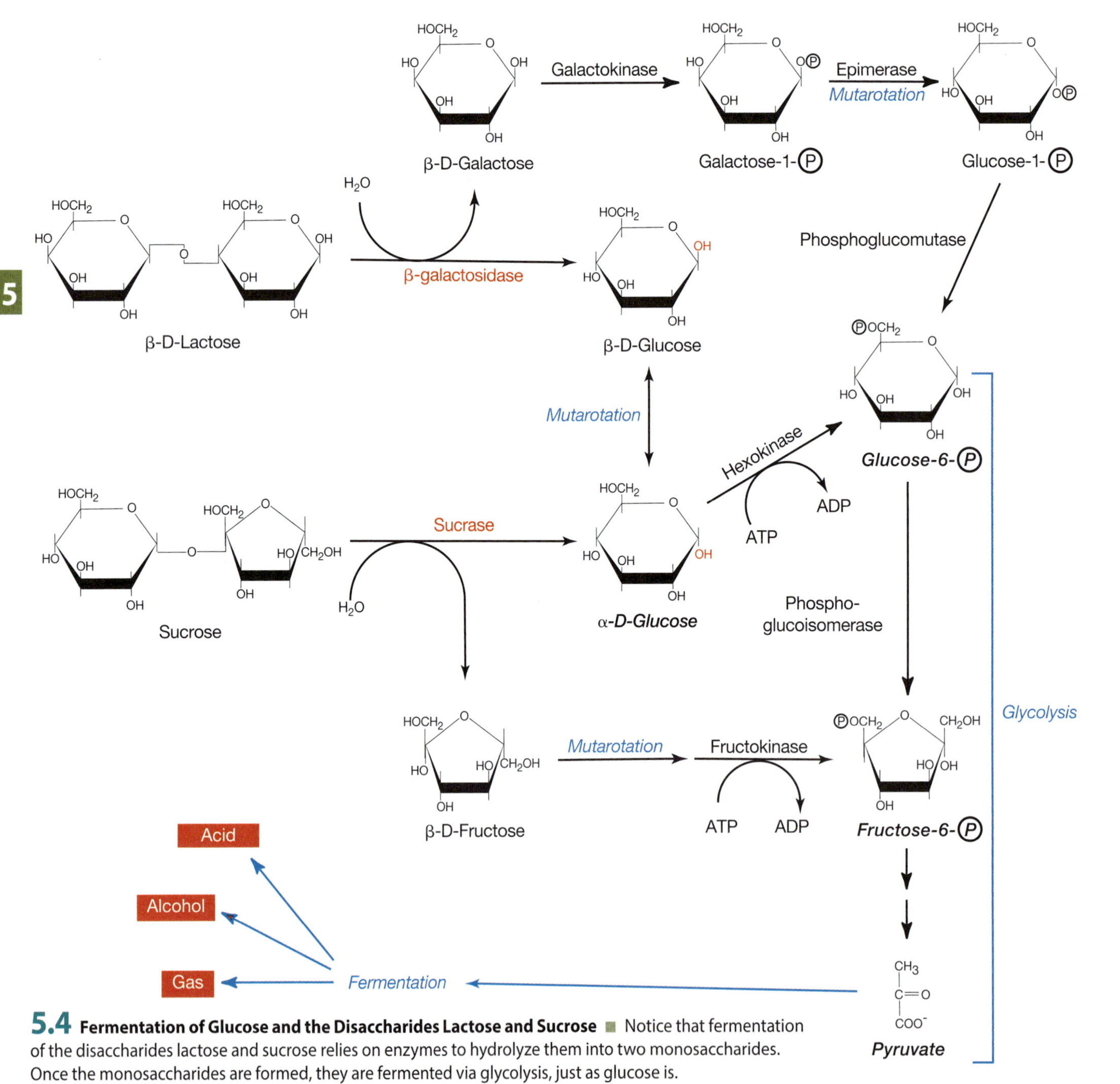


**5.4 Fermentation of Glucose and the Disaccharides Lactose and Sucrose** ■ Notice that fermentation of the disaccharides lactose and sucrose relies on enzymes to hydrolyze them into two monosaccharides. Once the monosaccharides are formed, they are fermented via glycolysis, just as glucose is.

TABLE 5-2 **Phenol Red Fermentation Broth Test Results and Interpretations**

| Result | Interpretation | Symbol |
|---|---|---|
| Yellow broth, bubble in tube | Fermentation of the specific carbohydrate with acid and gas end products | A/G |
| Yellow broth, no bubble in tube | Fermentation of the specific carbohydrate with acid end products; no gas produced, or gas produced in undetectable amounts | A/– |
| Red broth, no bubble in tube | No fermentation of the specific carbohydrate or products are not made in detectable amounts. | –/– |
| Pink broth, no bubble in tube | Degradation of peptone; alkaline end products | K |

**5.5 PR Glucose Fermentation Broth Results** ■ From left to right, these five PR glucose broths demonstrate acid and gas production (A/G), acid production without gas (A/–), uninoculated control tube for comparison, no reaction (–/–), and alkaline reaction as a result of peptone degradation (K).

## Medium Recipe

### PR (Carbohydrate) Broth[2]

- □ Pancreatic digest of casein — 10.0 g
- □ Sodium chloride — 5.0 g
- □ Carbohydrate (one only: glucose, lactose, sucrose, etc.) — 5.0 g
- □ Phenol red — 0.018 g
- □ Distilled or deionized water — 1.0 L

*pH = 7.1–7.5 at 25°C*

[2] This formulation without carbohydrate is PR base broth.

## PROCEDURE

### Lab One

1. Wear a lab coat, gloves, and chemical eye protection when performing this procedure.
2. Obtain four PR glucose broth tubes (and four PR base broths, if used). Label each with the name of the organism, your group name, the medium name, and the date. Label the fourth tube of each medium "control."
3. Inoculate one of each broth with each test organism.
4. Incubate all of the tubes at $35 \pm 2°C$ for 48 hours.
5. Save or dispose of the original culture tubes as directed by your instructor.

### Lab Two

1. Using the uninoculated controls for color comparison and Table 5-2 as a guide, examine all of the tubes and enter your results in the table provided on the data sheet, page 283. When indicating the various reactions, use the standard symbols as shown, with the acid reading first, followed by a slash and then the gas reading. K indicates alkalinity and –/– symbolizes no reaction. ***Note***: Do not try to read these results without the appropriate control tubes for comparison.
2. Dispose of all tubes in the appropriate autoclave container when finished.

## References


Atlas, Ronald and James Snyder. Page 342 in *Manual of Clinical Microbiology,* 11th ed. James H. Jorgensen, Michael A. Pfaller, Karen C. Carroll, Guido Funke, Marie Louise Landry, Sandra S. Richter, and David W. Warnock, eds. Washington, DC: ASM Press, American Society for Microbiology, 2015.

Lányi, B. Page 44 in *Methods in Microbiology,* Vol. 19. R. R. Colwell and R. Grigorova, eds. New York: Academic Press, 1987.

MacFaddin, Jean F. Page 57 in *Biochemical Tests for Identification of Medical Bacteria*, 3rd ed. Philadelphia: Lippincott Williams & Wilkins, 2000.

Zimbro, Mary Jo and David A. Power, eds. Page 440 in *Difco™ and BBL™ Manual—Manual of Microbiological Culture Media.* Sparks, MD: Becton Dickinson and Co., 2003.


5

Name ______________________

Date ______________________

Lab Section ______________________

I was present and performed this exercise (initials) ______________________

# DATA SHEET 5-2

## Phenol Red Fermentation Broth

### OBSERVATIONS AND INTERPRETATIONS

**1** Enter your results (color and bubble/no bubble) in the table below, using the symbols shown in Table 5-2, page 281. If you inoculated other PR broths in addition to PR glucose, be sure to include the relevant enzyme's name in your interpretation of that carbohydrate's fermentation.

| Organism | **Results** (color and bubble) | | | | Interpretation |
|---|---|---|---|---|---|
| | PR Base Broth | PR Glucose | PR Lactose (optional) | PR Sucrose (optional) | |
| Uninoculated control | | | | | BASE BROTH |
| | | | | | GLU |
| | | | | | OTHER |
| | | | | | BASE BROTH |
| | | | | | GLU |
| | | | | | OTHER |
| | | | | | BASE BROTH |
| | | | | | GLU |
| | | | | | OTHER |
| | | | | | BASE BROTH |
| | | | | | GLU |
| | | | | | OTHER |
| | | | | | BASE BROTH |
| | | | | | GLU |
| | | | | | OTHER |

## QUESTIONS

**1** *Consider the controls.*

**a.** *Were the uninoculated controls positive or negative controls, and what purpose did they serve?*

**b.** *What purpose did the PR base broths serve?*

**2** *Early formulations of this medium used a smaller amount of carbohydrate and occasionally produced false (pink/alkaline) results after 48 hours. This phenomenon is called a reversion.*

**a.** *Why do you think this happened?*

5

**b.** *List at least two steps that you as a microbiologist could take to prevent the problem.*

**3** *Suppose you inoculate a PR broth with an organism known to be a slow-growing fermenter. After 48 hours, you see slight turbidity but score it as (–/–).*

**a.** *Is this result a false positive or a false negative?*

**b.** *Is this false result caused by poor specificity or poor sensitivity of the test system?*

Name ______________________________

Date ______________________________

Lab Section ______________________________

I was present and performed this exercise (initials) ______________________________

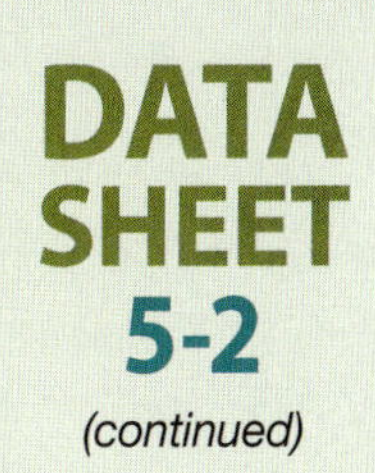

# DATA SHEET 5-2

*(continued)*

**4** *Assuming you tested an organism using the three carbohydrate broths and a base broth, which of the combinations of results in the following table would be reliable? Interpret each of the combinations and explain why the results are reliable or not. Assume all tubes in a series were inoculated with the same amount of organism and were incubated for 24 hours.*

| PR Base | PR Glucose | PR Lactose | PR Sucrose | Interpretation | Reliable? Y/N (explain why, if not) |
|---|---|---|---|---|---|
| –/– | A/G | A/G | A/G | | |
| –/– | A/– | A/– | –/– | | |
| A/G | A/G | A/G | A/G | | |
| –/– | A/G | A/G | K | | |

# Methyl Red and Voges-Proskauer Tests

EXERCISE **5-3**

## Theory

MR-VP broth is a combination medium used for both methyl red (MR) *and* Voges-Proskauer (VP) tests. It is a simple solution containing only peptone, glucose, and a phosphate buffer. The peptone and glucose provide protein (with its nitrogen) and fermentable carbohydrate, respectively, and the potassium phosphate resists pH changes in the medium.

The MR test is designed to detect organisms capable of performing a **mixed acid fermentation,** which overcomes the phosphate buffer in the medium and lowers the pH (Figs. 5.6, 5.7, and A.6). Succinate is produced from the addition of $CO_2$ to phosphoenolpyruvate, whereas the other end products are derived from the reduction of pyruvate to lactate or its oxidation to acetyl-CoA and formate.

Conversion of acetyl-CoA to acetate results in the formation of one ATP or it can be reduced to ethanol. Formate can be further broken down into $H_2$ and $CO_2$ gas. The acids produced by these organisms tend to be stable, whereas acids produced by other organisms may be converted to more neutral products or their fermentation shifts to produce more neutral products in response to the lowered pH.

Mixed acid fermentation is verified by the addition of methyl red indicator dye following incubation. Methyl red is red at pH 4.4 and yellow at pH 6.2. Between these two pH values, it is various shades of orange. Red color is the only true indication of a positive result. Orange is negative or inconclusive. Yellow is negative (Fig. 5.8).

The Voges-Proskauer test was designed to identify organisms that are able to ferment glucose, with the production of acetoin and 2,3-butanediol, which have a

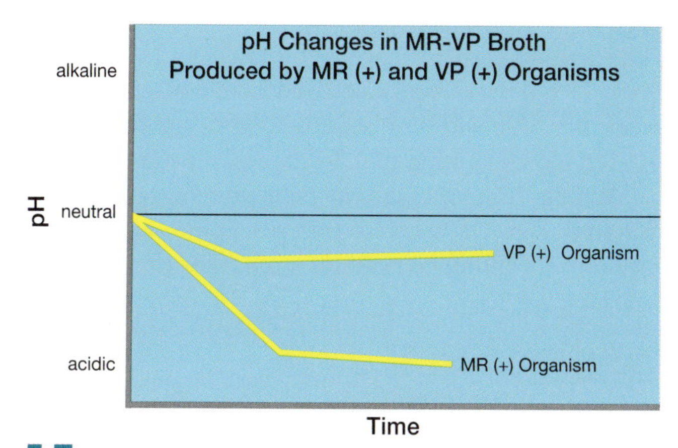


**5.7 pH Changes in MR-VP Broth** ■ The MR test identifies organisms that perform a mixed acid fermentation and produce stable acid end products. MR-positive organisms lower the broth's pH permanently. The VP test is used to identify organisms that perform a 2,3-butanediol fermentation. VP-positive organisms initially produce lactic acid and temporarily lower the pH, but the lowered pH activates the 2,3-butanediol fermentation pathway. Accumulation of its neutral end products raises the pH by completion of the test.

**5.8 Methyl Red Test Results** ■ The tube on the left is MR positive and the tube on the right is MR negative.

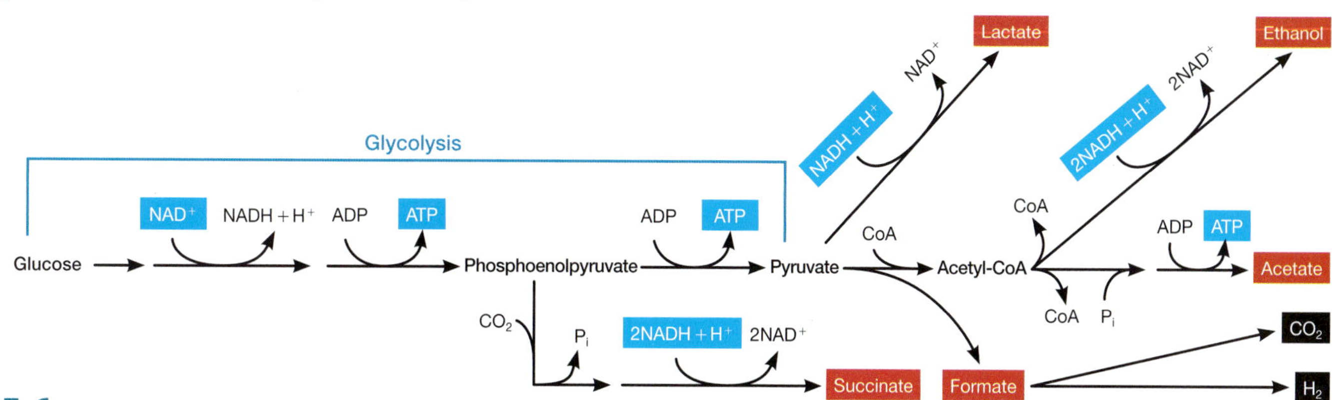


**5.6 Mixed Acid Fermentation of *E. coli*** ■ *E. coli* is a representative methyl red positive organism and is recommended as a positive control for the test. Its mixed acid fermentation produces (in order of abundance) lactate, carbon dioxide, hydrogen gas, ethanol, acetate, succinate, and formate. Most of the formate is converted to $H_2$ and $CO_2$ gases. All of the end products are derived from pyruvate with the exception of succinate, which is produced from phosphoenolpyruvate. *Salmonella* and *Shigella* also are methyl red positive, though some of the specific end products and amounts differ.

neutral pH (Figs. 5.7, 5.9, and A.6). VP-positive organisms can produce other fermentation end products in addition to 2,3-butanediol. They can reduce pyruvate to lactate (which lowers the pH), and they can make formate (which is broken down into $H_2$ and $CO_2$ gases) and acetyl-CoA (which is reduced to ethanol). The acetyl-CoA path results in no pH change, but formate (at least until its conversion to $H_2$ and $CO_2$) and lactate production lower pH. VP-positive organisms compensate for the lowered pH by shifting to 2,3-butanediol production.

Adding VP reagents to the medium oxidizes the **acetoin** to **diacetyl**, which in turn reacts with **guanidine nuclei** from peptone to produce a red color (Fig. 5.10). A positive VP result, therefore, is red. No color change (or development of copper color) after the addition of reagents is negative. The copper color is a result of interactions between the reagents and should not be confused with the true red color of a positive result (Fig. 5.11). Use of positive and negative controls for comparison is usually recommended.

To avoid conflicting results of the two tests, two 1 mL volumes are transferred from the MR-VP broth to separate test tubes after incubation. Methyl red indicator reagent is added to one tube, and VP reagents are added to the other. Color changes then are observed and documented. Refer to the procedural diagram in Figure 5.12.

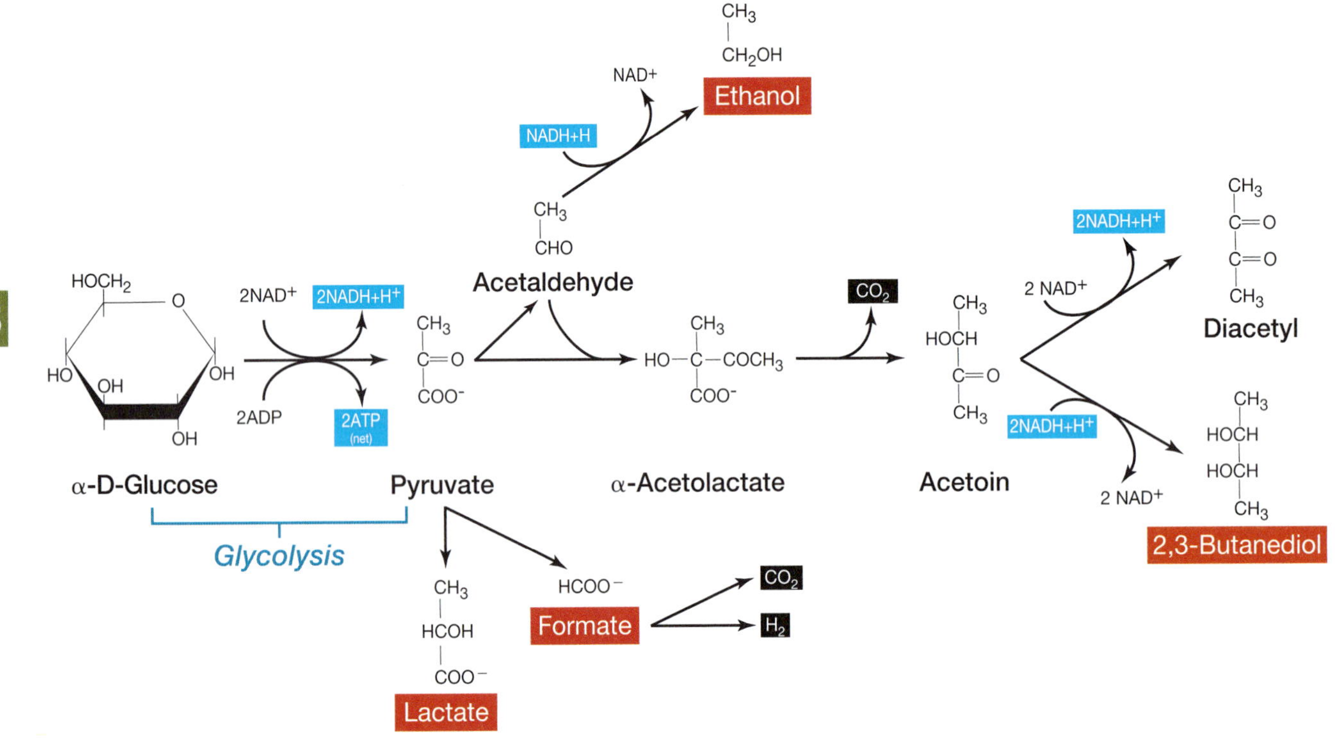


**5.9 2,3-Butanediol Fermentation** ■ VP-positive organisms have the potential to perform fermentations with ethanol, lactate, formate, 2,3-butanediol, carbon dioxide, and hydrogen gas as end products. As the pH drops due to accumulation of acid end products, VP-positive organisms switch to a 2,3 butanediol fermentation, which raises the intracellular pH. Acetoin is the last intermediate in this fermentation and its reduction by NADH produces the end product 2,3-butanediol. Acetoin can also be oxidized to diacetyl with the production of reducing power in the form of NADH+H$^+$. The VP indicator reaction artificially oxidizes acetoin to diacetyl (see Fig. 5.10).

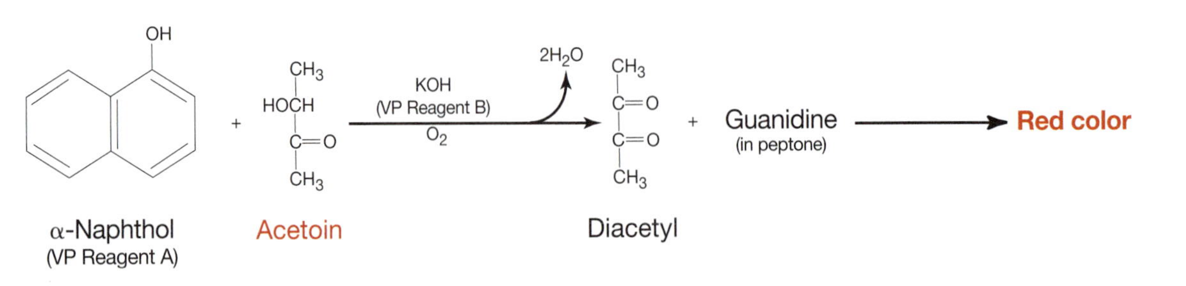


**5.10 Indicator Reaction of Voges-Proskauer Test** ■ Reagents A and B are added to VP broth after 48 hours of incubation. These reagents react with acetoin and oxidize it to diacetyl, which in turn reacts with guanidine (from the peptone in the medium) to produce a red color.

**5.11 Voges-Proskauer Test Results** ■ The tube on the left is VP negative and the tube on the right is VP positive.

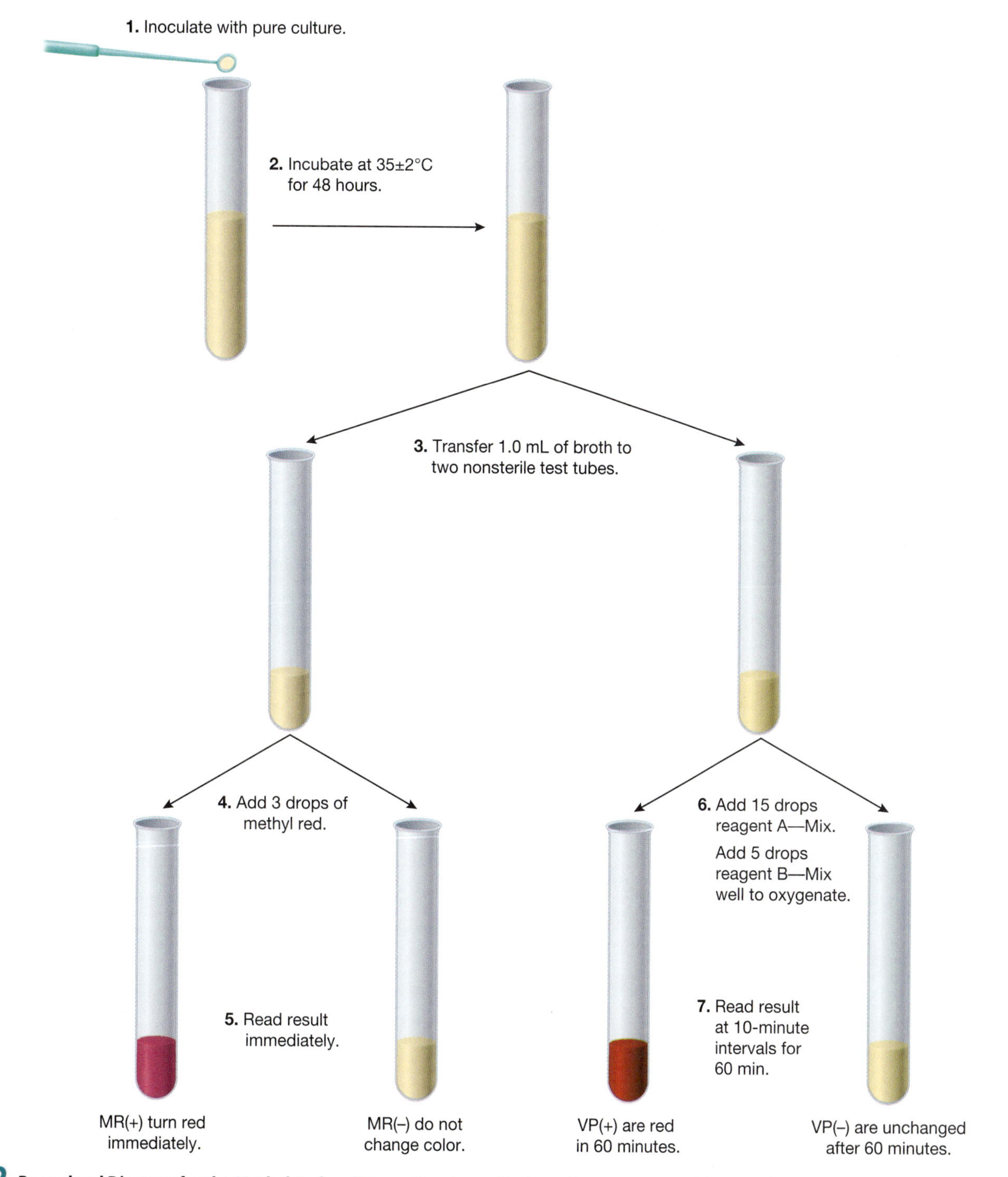


**5.12 Procedural Diagram for the Methyl Red and Voges-Proskauer Tests** ■ Be sure to wear a lab coat, gloves and chemical eye protection when inoculating the broths and after incubation when adding reagents, mixing, and reading results.

## Application

The methyl red and Voges-Proskauer tests are components of the *IMViC* battery of tests (Indole, Methyl red, Voges-Proskauer, and Citrate) used to distinguish between members of the family *Enterobacteriaceae* and differentiate them from other Gram-negative rods. Species in the genera *Escherichia*, *Shigella*, and *Salmonella* are MR positive, whereas *Enterobacter*, *Serratia*, and *Erwinia* species are VP positive. For more information about these fermentations, refer to Figure A.6 on page 597.

## In This Exercise

Today you will perform methyl red and Voges-Proskauer tests on *Escherichia coli* and *Enterobacter aerogenes*. After a 48-hour incubation period, you will observe the results of a mixed acid fermentation and a 2,3-butanediol fermentation. The exercise involves transfer of a portion of broth containing living organisms and the addition of measured reagents to complete the observable reactions. Please use care in your transfers and in your measurements. For help with the procedure, refer to Figure 5.12. For help interpreting your results, refer to Figures 5.8 and 5.11 and Tables 5-3 and 5-4.

5

TABLE 5-3 **Methyl Red Test Results and Interpretations**

| Result | Interpretation | Symbol |
|---|---|---|
| Red | Organism performs a mixed acid fermentation | + |
| No color change | Organism does not perform a mixed acid fermentation or the end products are not made in detectable amounts | − |

TABLE 5-4 **Voges-Proskauer Test Results and Interpretations**

| Result | Interpretation | Symbol |
|---|---|---|
| Red | Organism performs a 2,3-butanediol fermentation (acetoin produced) | + |
| No color change | Organism does not perform a 2,3-butanediol fermentation (acetoin is not produced or is not detectable) | − |

## Materials

### Per Student

- ☐ Lab coat
- ☐ Gloves
- ☐ Chemical eye protection

### Per Student Group

- ☐ Three MR-VP broths
- ☐ Methyl red reagent
- ☐ VP reagents A and B
- ☐ Six nonsterile 12 mm × 75 mm test tubes
- ☐ Three nonsterile 1 mL pipettes
- ☐ Fresh agar slant cultures of these recommended organisms:
  - *Escherichia coli*
  - *Enterobacter aerogenes*

## Medium and Reagent Recipes

### MR-VP Broth

- ☐ Buffered peptone 7.0 g
- ☐ Dipotassium phosphate 5.0 g
- ☐ Dextrose (glucose) 5.0 g
- ☐ Distilled or deionized water 1.0 L

*pH 6.7–7.1 at 25°C*

### Methyl Red Reagent

- ☐ Methyl red dye 0.1 g
- ☐ Ethanol 300.0 mL
- ☐ Distilled water to bring volume to 500.0 mL

### VP Reagent A

- ☐ α-naphthol 5.0 g
- ☐ Absolute alcohol to bring volume to 100.0 mL

### VP Reagent B

- ☐ Potassium hydroxide 40.0 g
- ☐ Distilled water to bring volume to 100.0 mL

## PROCEDURE

### Lab One

1. Wear a lab coat, gloves, and chemical eye protection when performing this procedure.
2. Obtain three MR-VP broths. Label two of them with the names of the organisms, your group name, and the date. Label the third tube "control."
3. Inoculate two broths with the test cultures. Do not inoculate the control.
4. Incubate all tubes at 35 ± 2°C for 48 hours.
5. Save or dispose of the original culture tubes as directed by your instructor.

## Lab Two

1. Wear a lab coat, gloves, and chemical eye protection when performing this part of the procedure.
2. Mix each broth well without splashing into the cap. Transfer 1.0 mL from each broth to two nonsterile tubes. The pipettes should be sterile, but the tubes need not be. Refer to the procedural diagram in Figure 5.12.
3. For the three pairs of tubes, add the reagents as follows.

### Tube #1: VP Test

a. Add 15 drops (0.6 mL) of VP reagent A. Mix well to oxygenate the medium, but be careful not to splash it.

b. Add 5 drops (0.2 mL) of VP reagent B. Carefully mix well again to oxygenate the medium.

c. Place the tubes in a test tube rack and observe for red color formation at regular intervals starting at 10 minutes for up to 1 hour. Mix again periodically to oxygenate the medium.

d. Using Table 5-4 as a guide, record your results in the table provided on the data sheet, page 293. Sometimes it is difficult to differentiate weak positive reactions from negative reactions. VP-negative reactions often produce a copper color. Weak VP-positive reactions produce a pink color. The strongest color change should be at the surface.

e. Dispose of the test tubes and culture tubes in the appropriate autoclave container.

### Tube #2: Methyl Red Test

a. Add three drops of methyl red reagent. Observe for red color formation immediately.

b. Using Table 5-3 as a guide, record your results in the table provided on the data sheet, page 293.

c. Dispose of the test tubes and culture tubes in the appropriate autoclave container.

## References


Atlas, Ronald and James Snyder. Pages 320 and 340 in *Manual of Clinical Microbiology,* 11th ed. James H. Jorgensen, Michael A. Pfaller, Karen C. Carroll, Guido Funke, Marie Louise Landry, Sandra S. Richter, and David W. Warnock, eds. Washington, DC: ASM Press, American Society for Microbiology, 2015.

Chapin, Kimberle C. and Tsai-Ling Lauderdale. Page 361 in *Manual of Clinical Microbiology,* 8th ed. Patrick R. Murray, Ellen Jo Baron, James H. Jorgensen, Michael A. Pfaller, and Robert H. Yolken, eds. Washington, DC: ASM Press, American Society for Microbiology, 2003.

Delost, Maria Dannessa. Pages 187–188 in *Introduction to Diagnostic Microbiology.* St. Louis, MO: Mosby, 1997.

MacFaddin, Jean F. Pages 321 and 439 in *Biochemical Tests for Identification of Medical Bacteria,* 3rd ed. Baltimore: Williams & Wilkins, 2000.

Moat, Albert G., John W. Foster, and Michael P. Spector. Pages 425-428 in *Microbial Physiology*, 4th ed. New York, NY: Wiley–Liss, 2002.

Tille, Patricia M. Page 214 in *Bailey & Scott's Diagnostic Microbiology*, 13th ed. St. Louis, MO: Mosby, 2014.

White, David, James Drummond, and Clay Fuqua. Pages 394-396 in *The Physiology and Biochemistry of Prokaryotes*, 4th ed. New York, NY: Oxford University Press, 2012.

Zimbro, Mary Jo and David A. Power, eds. Page 330 in *Difco™ and BBL™ Manual—Manual of Microbiological Culture Media.* Sparks, MD: Becton Dickinson and Co., 2003.


5

Name ______________________

Date ______________________

Lab Section ______________________

I was present and performed this exercise (initials) ______________________

# DATA SHEET 5-3

## Methyl Red and Voges-Proskauer Tests

### OBSERVATIONS AND INTERPRETATIONS

**1** Refer to Tables 5-3 and 5-4, page 290, as you record your results (color) and interpretations in the table below.

| Organism | MR Result | VP Result | Interpretation |
|---|---|---|---|
| Uninoculated control | | | |
| | | | |
| | | | |

## QUESTIONS

**1** *Some protocols call for a shorter incubation time for the MR and VP tests. Other protocols allow for up to 10 days incubation with virtually no risk of producing a false positive.*

**a.** *Which of the two tests would likely produce more false negatives with a shorter incubation time? Why?*

**b.** *Which test would likely benefit most from a longer incubation time? Why?*

**2** *Would a false negative result for the VP test more likely be attributable to poor sensitivity or to poor specificity of the test system?*

**3** *Why were you told to shake the VP tubes after the reagents were added?*

**5**

**4** *Why is the methyl red test read immediately after addition of methyl red reagent and the Voges-Proskauer read up to 60 minutes after addition of VP reagents A and B?*

**5** *Some microbiologists recommend reincubating organisms producing methyl red-negative results for an additional 2 to 3 days. Why do you think this is done?*

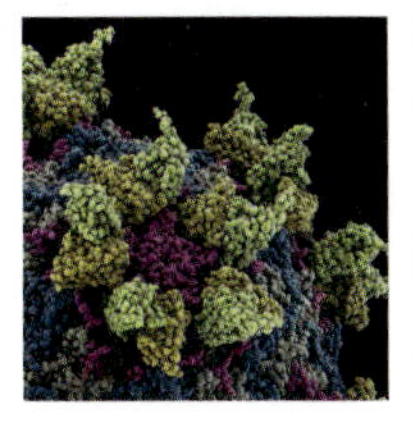

## Tests Identifying Microbial Ability to Respire

In the next three exercises we will examine techniques designed to differentiate bacteria based on their ability to respire. As mentioned earlier in this section, respiration transfers chemical potential energy in glucose into high-energy bonds of ATP using glycolysis, the citric acid cycle, and oxidative phosphorylation in an electron transport chain (ETC). The final electron acceptor (FEA) in all respiratory pathways is an inorganic molecule.

Tests that identify an organism as an aerobic or an anaerobic respirer generally are designed to detect specific products of (or constituent enzymes used in) the reduction of the final (or terminal) electron acceptor. Aerobic respirers reduce oxygen ($O_2$) to water. Anaerobic respirers reduce other inorganic molecules, such as nitrate ($NO_3^-$) or sulfate ($SO_4^{-2}$). Nitrate is reduced to nitrogen gas ($N_2$) or other nitrogenous compounds such as nitrite ($NO_2^-$), which also may be used as an FEA. Sulfate is reduced to hydrogen sulfide gas ($H_2S$). Regardless of the FEA, the carbon pathways for aerobic and anaerobic respiration are the same. These are described in more detail in Appendix A.

The following three exercises and organisms were chosen to directly or indirectly demonstrate the presence of an ETC. The catalase test (Exercise 5-4) detects organismal ability to produce catalase—an enzyme that detoxifies the cell by converting hydrogen peroxide ($H_2O_2$) produced in the ETC to $H_2O$ and $O_2$.

While the carbon pathways of respiration are the same, the specific components of the ETC are highly variable. The oxidase test (Exercise 5-5) identifies the presence of cytochrome c oxidase in the ETC of some aerobic respirers.

The nitrate reduction test (Exercise 5-6) examines bacterial ability to use nitrate as the FEA of anaerobic respiration. Exercises 5-18 and 5-19 demonstrate anaerobic respiration with $SO_4^{-2}$ as the FEA. ■

# Catalase Test

EXERCISE 5-4

5

## ■ Theory

The electron transport chains (ETC) of aerobic and facultatively anaerobic bacteria are composed of molecules capable of accepting and donating electrons as conditions dictate. As such, these molecules alternate between their oxidized and reduced forms, passing electrons down the chain to the final electron acceptor, $O_2$. Energy lost by electrons in this sequential transfer is used to perform oxidative phosphorylation (i.e., phosphorylate ADP to ATP).

In most cases, electrons in the aerobic ETC follow the stepwise path to oxygen, but other paths can be followed and these result in production of toxic forms of reduced oxygen[1]. For instance, one ETC carrier molecule called **flavoprotein** can bypass the next carrier in the chain and transfer electrons directly to oxygen (Fig. 5.13), which produces hydrogen peroxide ($H_2O_2$), a highly potent cellular toxin. Reduced flavin adenine dinucleotide ($FADH_2$) is capable of the same reaction (Fig. 5.13).

[1] As an analogy, a person can walk down stairs one tread at a time and reach the bottom safely. However, the option exists for jumping from the middle tread to the bottom and spraining an ankle. The person arrives at the same place, but damage is done! The potential energy is not released in a slow, controlled way, but in a larger chunk, which has damaging consequences.

Even if the electrons follow the complete ETC another toxin, the superoxide radical ($O_2^-$), can be produced in the final step because electrons reduce oxygen one at a time and sometimes it is released before it is completely reduced to $H_2O$ (Fig. 5.13).

Hydrogen peroxide and the superoxide radical are toxic because they oxidize biochemicals and make them

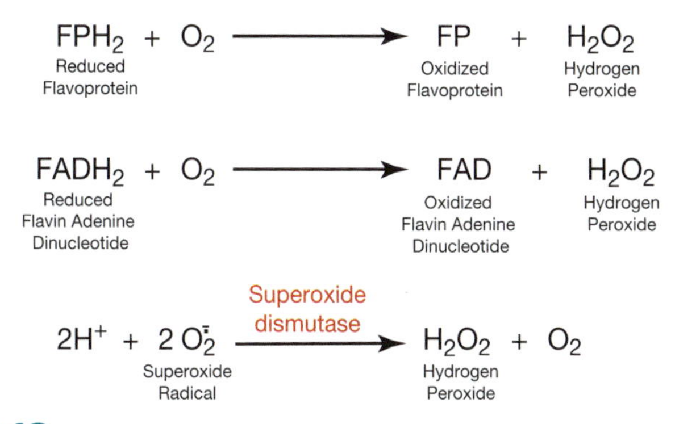


**5.13 Microbial Production of $H_2O_2$** ■ Hydrogen peroxide ($H_2O_2$) may be formed through the transfer of electrons from reduced flavoprotein or flavin adenine dinucleotide ($FADH_2$) to oxygen, or from the action of superoxide dismutase detoxifying the superoxide radical, another destructive form of reduced oxygen.

nonfunctional. However, organisms that produce them also produce enzymes capable of breaking them down. **Superoxide dismutase** catalyzes conversion of superoxide radicals (the more lethal of the two compounds) to hydrogen peroxide (Fig. 5.13). **Catalase** converts hydrogen peroxide into water and gaseous oxygen (Fig. 5.14). In large part (though exceptions exist), the ability to synthesize these protective enzymes accounts for an organism's ability to live in the presence of oxygen (Table 5-5).

Bacteria that produce catalase can be detected easily using typical store-grade hydrogen peroxide. When hydrogen peroxide is added to a catalase-positive culture, oxygen gas bubbles form immediately. If no bubbles appear, the organism is catalase-negative. This test can be performed on a microscope slide (Fig. 5.15) or by adding hydrogen peroxide directly to the bacterial growth (Fig. 5.16).

$$\underset{\text{Hydrogen Peroxide}}{2H_2O_2} \xrightarrow{\text{Catalase}} 2H_2O + O_{2(g)}$$

**5.14 Catalase Mediated Conversion of $H_2O_2$** ■ Catalase is an enzyme of aerobes, microaerophiles, and facultative anaerobes that converts hydrogen peroxide to water and oxygen gas and allows them to grow in the presence of oxygen.

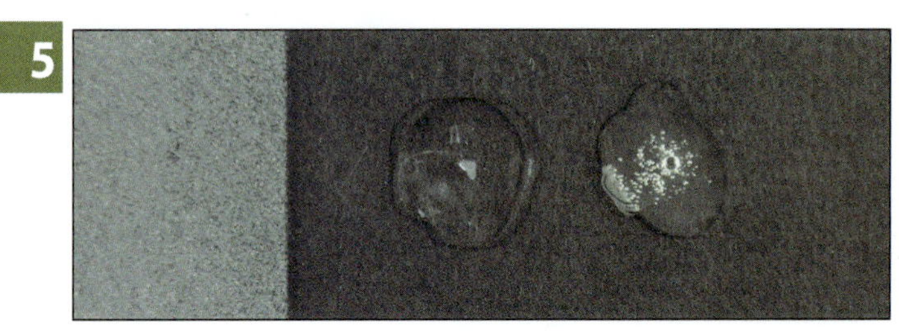

**5.15 Catalase Slide Test Results** ■ Visible bubble production indicates a positive result in the catalase slide test. A catalase-negative organism is on the left. A catalase-positive organism is on the right. For purposes of the photograph, no cover is shown, but one should be used to contain spattering from bubbles produced by a positive reaction (Fig. 5.17).

**5.16 Catalase Tube Test Results** ■ The catalase test also may be performed on an agar slant. A catalase-positive organism is on the left. A catalase-negative organism is on the right. Only perform this test when you are done with the culture, because adding hydrogen peroxide makes it unusable.

**TABLE 5-5 Protective Enzymes in Various Aerotolerance Groups**

| Aerotolerance Group | Superoxide Dismutase | Catalase |
|---|---|---|
| Obligate aerobe | Present | Present |
| Facultative anaerobe | Present | Present |
| Microaerophile | Present | Present in small amounts |
| Aerotolerant anaerobe | Present | Absent (alternative mechanism) |
| Obligate anaerobe | Absent | Absent |

## ■ Application

This test is used to identify organisms that produce the enzyme catalase. It is frequently used to differentiate catalase-positive *Micrococcus* and *Staphylococcus* from the catalase-negative *Streptococcus*, *Enterococcus*, and *Lactococcus*. Variations on this test also may be used in identification of *Mycobacterium* species.

## ■ In This Exercise

You will perform the catalase test. The materials provided will enable you to conduct both the slide test and the slant test. The slide test has to be done first because the tube test requires the addition of hydrogen peroxide directly to the slant, after which the culture will be unusable.

## ▼ Materials

### Per Student

- □ Lab coat
- □ Gloves
- □ Chemical eye protection

### Per Student Group

- □ Hydrogen peroxide (3% solution)
- □ Transfer pipettes
- □ Microscope slides
- □ Petri dish base or lid
- □ One nutrient agar slant
- □ 24-hour agar slant cultures of these recommended organisms:
  - *Staphylococcus epidermidis*
  - *Lactococcus lactis*

## PROCEDURE

### Slide Test

1. Wear a lab coat, gloves, and chemical eye protection when performing this procedure.
2. Transfer a visible amount of growth to a microscope slide with a sterile wooden stick. (If you use a loop, be sure to perform this test in the proper order. Placing the metal loop into $H_2O_2$ could catalyze a false-positive reaction.) Properly dispose of the stick. ***Note***: It is essential that this test be run with fresh (24-hour) cultures. Older cultures may not have enough active catalase to produce visible bubbles.
3. Aseptically place one or two drops of hydrogen peroxide directly onto the bacteria and immediately cover with the base or lid of a Petri dish (Fig. 5.17). This will contain any spattering due to bubbling of a catalase-positive result. Observe for the formation of bubbles. The cover can be removed when the bubbling has subsided and reused for further testing. Place it on a disinfectant-soaked towel between uses and dispose of it properly when done for the day. ***Note***: When running this test on an unknown, a positive control should be run simultaneously to verify the quality of the peroxide.

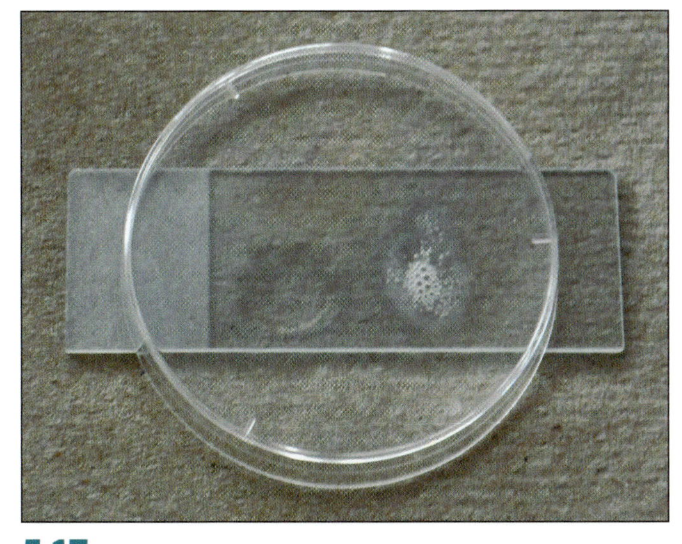

**5.17 Cover the Reaction** ■ As soon as you add the hydrogen peroxide to the slide, cover it with a Petri dish lid or base to prevent splattering from a positive result. When the reaction has subsided, the lid can be removed and used again for subsequent tests. Store it on a disinfectant-soaked paper towel between uses and dispose of it properly when finished for the day.

4. If you do not see obvious bubbling and your instructor permits it, you can observe the slide using the scanning objective on your microscope.
5. Record your results, in the table provided on the data sheet, page 299.
6. Save or dispose of the original culture tubes as directed by your instructor.

### Slant Test

1. Wear a lab coat, gloves, and chemical eye protection when performing this procedure.
2. Add approximately 1 mL of hydrogen peroxide to each of the three slants (one uninoculated control and two cultures) and replace the caps. Do this *one slant at a time*, observing and recording the results as you go. Refer to Table 5-6. This should be done when you are finished with the cultures because they will not be usable for anything else after hydrogen peroxide is added.
3. Record your results in the table provided on the data sheet, page 299.
4. Dispose of all tubes in an appropriate autoclave container when finished.

TABLE **5-6** **Catalase Test Results and Interpretations**

| Result | Interpretation | Symbol |
|---|---|---|
| Bubbles | Catalase is present | + |
| No bubbles | Catalase is absent or not detectable | − |

## References


Atlas, Ronald and James Snyder. Page 317 in *Manual of Clinical Microbiology,* 11th ed. James H. Jorgensen, Michael A. Pfaller, Karen C. Carroll, Guido Funke, Marie Louise Landry, Sandra S. Richter, and David W. Warnock, eds. Washington, DC: ASM Press, American Society for Microbiology, 2015.

Collins, C. H., Patricia M. Lyne, and J. M. Grange. Page 110 in *Collins and Lyne's Microbiological Methods,* 7th ed. Oxford, Boston: Butterworth-Heinemann, 1995.

Lányi, B. Page 20 in *Methods in Microbiology,* Vol. 19. R. R. Colwell and R. Grigorova, eds. New York: Academic Press, 1987.

MacFaddin, Jean F. Page 78 in *Biochemical Tests for Identification of Medical Bacteria,* 3rd ed. Baltimore: Lippincott Williams & Wilkins, 2000.

Smibert, Robert M. and Noel R. Krieg. Page 614 in *Methods for General and Molecular Bacteriology*. Philipp Gerhardt, R. G. E. Murray, Willis A. Wood, and Noel R. Krieg, eds. Washington, DC: American Society for Microbiology, 1994.

Stanier, Roger Y., John L. Ingraham, Mark K. Wheelis, and Page R. Painter. Pages 210–211 in *The Microbial World*, 5th ed. Englewood Cliffs, NJ: Prentice-Hall, 1986.

Tille, Patricia M. Pages 97 and 201 in *Bailey & Scott's Diagnostic Microbiology*, 13th ed. St. Louis, MO: Mosby, 2014.

White, David, James Drummond, and Clay Fuqua. Pages 383–384 in *The Physiology and Biochemistry of Prokaryotes*, 4th ed. New York, NY: Oxford University Press, 2012.

Willey, Joanne M., Linda M. Sherwood, and Christopher J. Woolverton. Pages 144–145 in *Prescott's Principles of Microbiology*. Boston: McGraw-Hill Higher Education, 2009.

Name ______________________________

Date ______________________________

Lab Section ______________________________

I was present and performed this exercise (initials) ______________________________

# DATA SHEET 5-4

## Catalase Test

### OBSERVATIONS AND INTERPRETATIONS

**1** Using Table 5-6, page 297, as a guide, record your results and interpretations in the table below.

| Organism | Bubbles? Y/N | +/– | Interpretation |
|---|---|---|---|
| Uninoculated control (slant test only) | | | |
| | | | |
| | | | |

### QUESTIONS

**1** *Think about the advice given in the procedure to test a known catalase-positive organism along with an unknown.*

**a.** *Is this a positive or a negative control?*

______________________________

**b.** *What information is provided by the results?*

______________________________

______________________________

______________________________

______________________________

______________________________

**2** *Consider the step in the tube test where hydrogen peroxide is added to the uninoculated tube.*

**a.** *Is this a positive or a negative control?*

**b.** *What information is provided by the results?*

**3** *When flavoprotein transfers electrons directly to oxygen, hydrogen peroxide is produced. What other consequences might result from electron carriers in the ETC being bypassed?*

5

**4** *Would a false positive from the reaction between the inoculating loop and hydrogen peroxide be caused by poor specificity or poor sensitivity of the test system? Explain.*

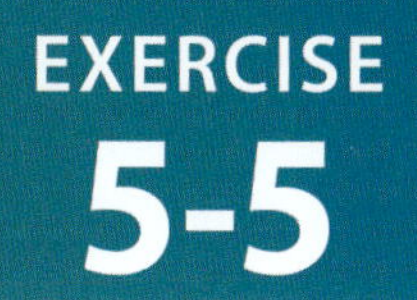


# Oxidase Test

## Theory

Consider the fate of a glucose molecule in a *respiring* cell[1]. If it is destined to be used for ATP production, it will become oxidized to 6 $CO_2$ with the concurrent reduction of the coenzymes 10 $NAD^+$ to 10 $NADH + H^+$ and 2 FAD to 2 $FADH_2$. In addition, 4 ATPs are made by substrate phosphorylation. These occur in reactions throughout glycolysis (Fig. A.1), the transition step, and the citric acid cycle (Fig. A.4).

As you can see, reduced coenzymes have the potential to accumulate rapidly. Therefore, in order to continue oxidizing glucose, these coenzymes must be converted back to their oxidized state. This is the job of the electron transport chain (Fig. 5.18).

Many aerobes, microaerophiles, facultative anaerobes, and even some anaerobes have ETCs. One function of the ETC is to pass electrons down a chain of membrane-bound molecules with increasingly positive reduction potentials (Fig. 5.1) to the terminal electron acceptor (e.g., $½O_2$, $NO_3^-$, $SO_4^{2-}$). A second function of some carriers in the ETC is to act as proton pumps, using the energy lost by the electrons to create a proton gradient across the cytoplasmic membrane that is higher on the outside.

Because of their charge, protons are incapable of diffusing back into the cell through the phospholipid bilayer of the membrane. However, the cytoplasmic membrane has channels through which protons *can* diffuse, and the kinetic energy of their diffusion is coupled to oxidative phosphorylation—ATP synthesis—catalyzed by membrane-bound ATPases. So, the energy possessed by electrons stripped away from glucose ultimately are used in ATP synthesis.

There are many different types of electron transport chains, but all share the characteristics listed above. Some organisms use more than one type of ETC, depending on the availability of oxygen or other final electron acceptor(s). *Escherichia coli*, for example, has two ETCs for respiring aerobically and at least one for respiring anaerobically (using nitrate as the FEA). Many bacteria have ETCs resembling mitochondrial ETCs in eukaryotes.

These chains contain a series of four large enzymes broadly named Complexes I, II, III, and IV, each of which contains several molecules jointly able to transfer electrons and use the free energy released in the reactions. The last enzyme in the chain, Complex IV, is called **cytochrome c oxidase** because it makes the final electron transfer of the chain from cytochrome c, residing in the periplasm or attached to the periplasmic side of the cytoplasmic membrane of Gram-negative cells, to oxygen inside the cell. In Gram-positive cells, it is membrane-bound.

The oxidase test is designed to identify the presence of cytochrome c oxidase. It is able to do this because cytochrome c oxidase has the unique ability to not only oxidize cytochrome c, but to catalyze the *reduction* of cytochrome c by a **chromogenic reducing agent** called tetramethyl-*p*-phenylenediamine. Chromogenic reducing agents are chemicals that develop color as they become oxidized (Fig. 5.19).

In the oxidase test, the reducing reagent is added directly to bacterial growth on solid media (Fig. 5.20A), or more conveniently, a bacterial colony is transferred to paper and reagent is added (Fig. 5.20B), or even *more* conveniently, bacteria are added to paper already containing the reagent (5.20C). A color change occurs within seconds if the reducing agent becomes oxidized, thus indicating that cytochrome c oxidase is present. Lack of color change within the allotted time means that cytochrome c oxidase is not present and signifies a negative result (Table 5-7).

TABLE 5-7 **Oxidase Test Results and Interpretations**

| Result | Interpretation | Symbol |
|---|---|---|
| Dark blue/purple within 20 seconds | Cytochrome c oxidase is present | + |
| No color change to blue/purple within 20 seconds | Cytochrome c oxidase is not present or not detectable | − |

## Application

This test is used to identify bacteria containing the respiratory enzyme cytochrome c oxidase. Among its many uses is the presumptive identification of the oxidase-positive *Neisseria* and *Moraxella*. It also can be useful in differentiating the oxidase-negative *Enterobacteriaceae* from the oxidase-positive *Pseudomonadaceae*.

[1] As you have seen in previous labs, glucose is not completely oxidized to 6 $CO_2$ in fermenting cells.

5

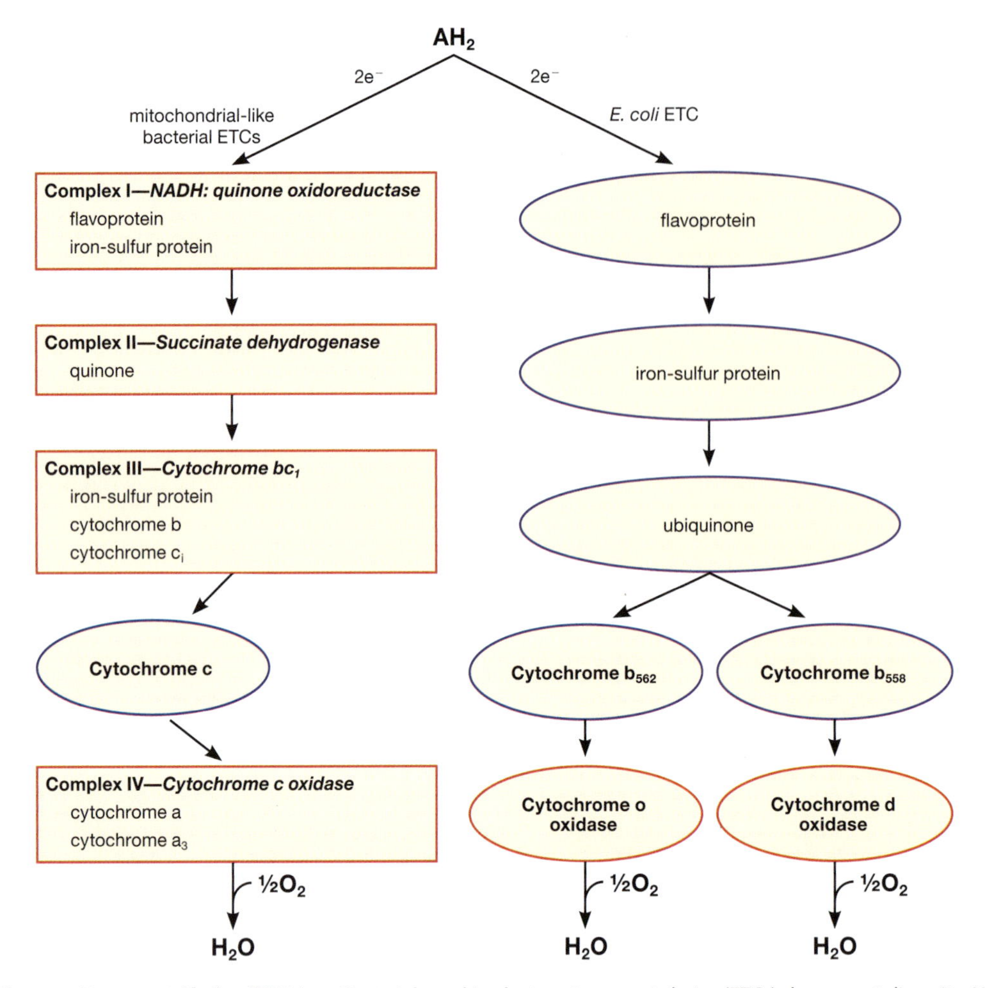


**5.18 Aerobic Electron Transport Chains (ETCs)** ■ Bacterial aerobic electron transport chains (ETCs) show great diversity. However, all begin with flavoproteins that receive electrons from reduced coenzymes, such as NADH or $FADH_2$ (represented by $AH_2$) and all have proton pumps that create the gradient for ATP synthesis. The part of the ETC relevant to the oxidase test, though, is at the end. Bacteria with chains that resemble the mitochondrial ETC in eukaryotes, shown on the left, contain cytochrome c oxidase (Complex IV), which transfers electrons from cytochrome c to oxygen. These organisms give a positive result for the oxidase test. Other bacteria, such as members of the *Enterobacteriaceae*, are capable of aerobic respiration, but have a different terminal oxidase system and give a negative result for the oxidase test. Both paths shown on the right are found in *E. coli*. The amount of available $O_2$ determines which pathway is more active. Enzymatic complexes and oxidases are outlined in red.

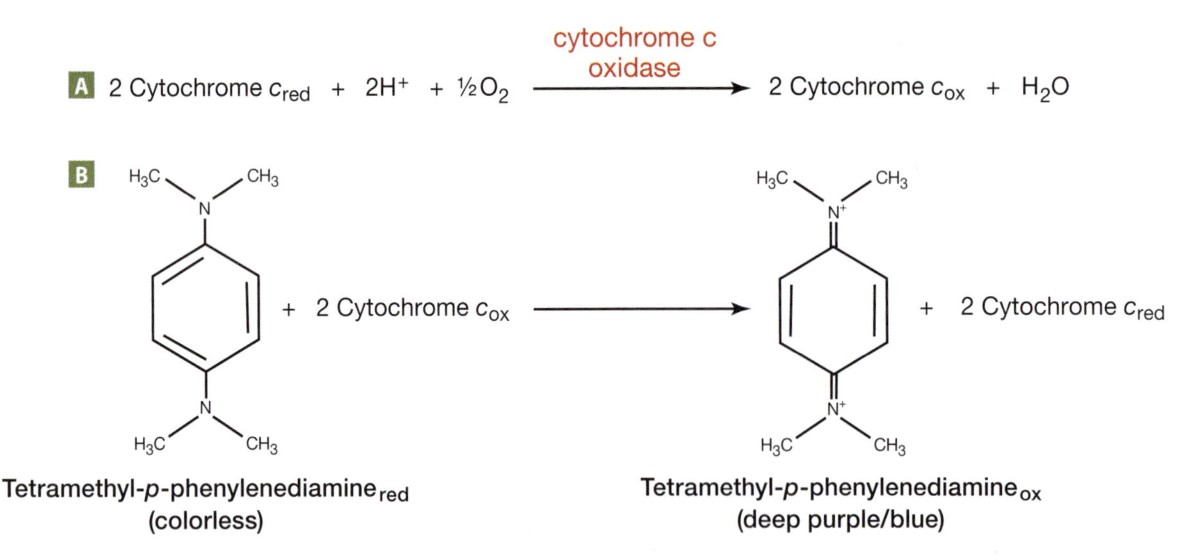


**5.19 Natural Oxidation of Cytochrome c and Reduction by the Oxidase Indicator Reaction** ■ (**A**) Under natural conditions, cytochrome c oxidase transfers electrons from cytochrome c to $O_2$, which produces water. (**B**) When in the oxidized state, cytochrome c is available to be reduced by the phenylenediamine reagent. The phenylenediamine reagent is colorless when reduced and blue/purple when oxidized, its condition in a positive oxidase test.

## In This Exercise

You will be performing the oxidase test either using one of several commercially available systems or the less expensive versions using "house" reagents (which we describe in the procedure—follow the instructions of the commercial test system if you use one). Your instructor will have chosen the system that is best for your laboratory. Regardless, most oxidase test systems produce a color change from colorless to blue/purple. Reagents used for this test are unstable and may oxidize independently shortly after they become moist. Therefore, it is important to take your readings within the recommended time (usually 20 seconds).

## Materials

### Per Student

- ☐ Lab coat
- ☐ Gloves
- ☐ Chemical eye protection

### Per Student Group

- ☐ Sterile wooden stick or disposable plastic loop
- ☐ Small pieces of filter paper (Fig. 5.20B)
- ☐ Sterile water
- ☐ Oxidase test reagent (Appendix E, p. 619)
- ☐ 24-hour agar slant cultures of these recommended organisms:
  - *Escherichia coli*
  - *Neisseria* spp. (e.g., *N. sicca* or *N. flava*)

## PROCEDURE

*Note*: We describe the oxidase test using filter paper and reagent. If you are using a commercial kit, follow its instructions.

1. Wear a lab coat, gloves, and chemical eye protection when performing this procedure.
2. Place the filter paper on a paper towel and transfer some of the culture to it using a sterile wooden stick or disposable loop. Properly dispose of the stick/loop. ***Note***: It is essential that this test be run with fresh (24-hour) cultures. Older cultures may not have enough active oxidase to produce a visible color change.
3. Add a few drops of oxidase reagent over the growth and observe for a color change within 20 seconds. ***Note***: If running an unknown, be sure to also test a known oxidase-positive organism.
4. Repeat steps one and two for the second organism. The same filter paper can be used if there is room. Properly dispose of the filter paper when finished.
5. Enter your results in the table provided on the data sheet, page 305.
6. Dispose of all tubes in an appropriate autoclave container when finished.

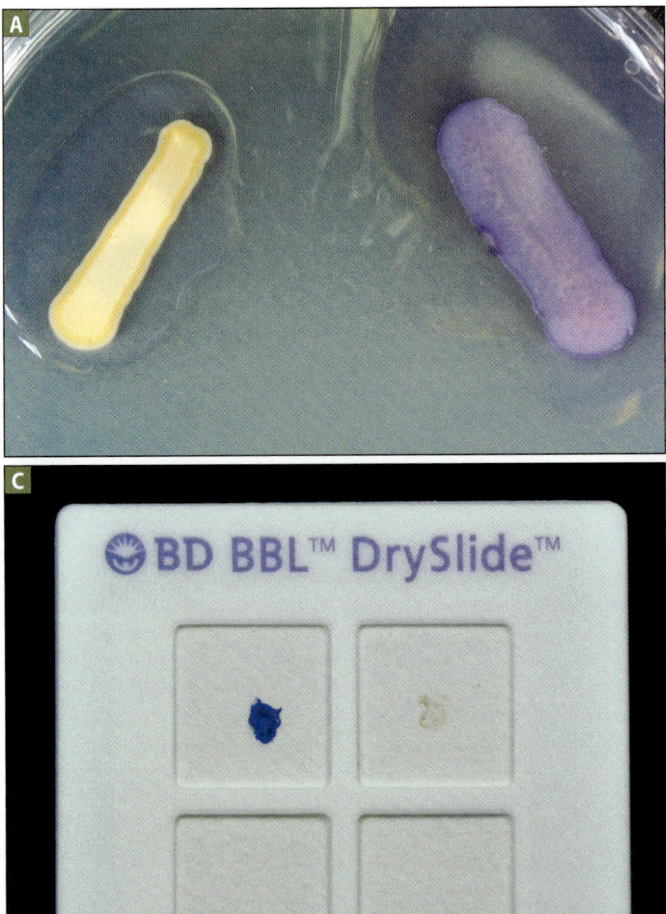

**5.20 Variations of the Oxidase Test** ■ Many variations of the oxidase test exist, but in all cases reaction with the phenylenediamine reagent produces a blue or purple color. Reading must be taken within 20 seconds because the reagent can react with moisture in the air and produce a false positive result. (**A**) Shown are oxidase-negative (left) and oxidase-positive (right) bacteria on an agar plate after addition of the phenylenediamine reagent. The organism on the left is its natural color; yellow is not a negative result. (**B**) Alternatively, a small amount of growth can be transferred to a piece of filter paper and reagent added. Shown is a positive result. (**C**) A commercial test system with reagent already in the slide is shown. All the microbiologist has to do is add the organism to a panel and wait 20 seconds for a color change. A positive result is shown in the upper left and a negative result is shown in the upper right panel.

(BBL™ DrySlide™ systems available from Becton Dickinson, Sparks, MD.)

5

## References

Atlas, Ronald and James Snyder. Page 319 in *Manual of Clinical Microbiology,* 11th ed. James H. Jorgensen, Michael A. Pfaller, Karen C. Carroll, Guido Funke, Marie Louise Landry, Sandra S. Richter, and David W. Warnock, eds. Washington, DC: ASM Press, American Society for Microbiology, 2015.

BBL™ *Dry*Slide™ package insert.

Collins, C. H., Patricia M. Lyne, and J. M. Grange. Page 116 in *Collins and Lyne's Microbiological Methods,* 7th ed. Oxford, Boston: Butterworth-Heinemann, 1995.

Lányi, B. Page 18 in *Methods in Microbiology,* Vol. 19. R. R. Colwell and R. Grigorova, eds. New York: Academic Press, 1987.

MacFaddin, Jean F. Page 368 in *Biochemical Tests for Identification of Medical Bacteria,* 3rd ed. Philadelphia: Lippincott Williams & Wilkins, 2000.

Pettigrew, Graham W. and Geoffrey R. Moore. Chap. 1 in *Cytochromes c, Biological Aspects.* Springer Series in Molecular Biology. Alexander Rich, ed. Berlin: Springer-Verlag, 1987.

Smibert, Robert M. and Noel R. Krieg. Page 625 in *Methods for General and Molecular Bacteriology*. Philipp Gerhardt, R. G. E. Murray, Willis A. Wood, and Noel R. Krieg, eds. Washington, DC: American Society for Microbiology, 1994.

Tille, Patricia M. Pages 97 and 223 in *Bailey & Scott's Diagnostic Microbiology*, 13th ed. St. Louis, MO: Mosby, 2014.

White, David, James Drummond, and Clay Fuqua. Pages 423–425 in *The Physiology and Biochemistry of Prokaryotes*, 4th ed. New York, NY: Oxford University Press, 2012.

Name ______________________________

Date ______________________________

Lab Section ______________________________

I was present and performed this exercise (initials) ______________

DATA SHEET

5-5

# Oxidase Test

## OBSERVATIONS AND INTERPRETATIONS

1 Using Table 5-7, page 301, as a guide, enter your results in the table below.

| Organism | Color Result | + / − | Interpretation |
|---|---|---|---|
| | | | |
| | | | |

## QUESTIONS

**1** *Think about the advice given in the procedure to test a known oxidase-positive organism along with an unknown.*

**a.** *Is this a positive or a negative control?*

______________________________

**b.** *What information is provided by the results?*

______________________________

______________________________

______________________________

______________________________

______________________________

**2** *Think about the 20-second time limit on the oxidase test.*

**a.** *What happens to the oxidase reagent after 20 seconds?*

______________________________

______________________________

______________________________

**b.** *Does this "reaction" only happen after 20 seconds? If not, why is a 20-second time limit set?*

______________________________

______________________________

______________________________

*Provide a* possible *explanation as to why the phenylenediamine reagent reduces cytochrome c (aa$_3$) oxidase and not other oxidases (such as cytochrome d of* E. coli *shown in Fig. 5.18). Figure 5.1 might also be helpful.*

# Nitrate Reduction Test

EXERCISE 5-6

## Theory

Anaerobic respiration involves the reduction of (transfer of electrons to) an inorganic molecule other than oxygen. Nitrate reduction is one such example. Many Gram-negative bacteria (including most *Enterobacteriaceae*) contain the enzyme **nitrate reductase** and perform a single-step reduction of nitrate to nitrite ($NO_3^- \rightarrow NO_2^-$). Other bacteria, in a multistep process known as **denitrification,** are capable of enzymatically converting nitrate to molecular nitrogen ($N_2$) or nitrous oxide ($N_2O$) via nitrite. Some products of nitrate reduction are shown in Figure 5.21.

Nitrate broth is an undefined medium of beef extract, peptone, and potassium nitrate ($KNO_3$). An inverted Durham tube is placed in each broth to trap a portion of any gas produced. In contrast to many differential media, no color indicators are included. The color reactions obtained in nitrate broth take place as a result of reactions between metabolic products and reagents added after incubation (Fig. 5.22).

Before a broth can be tested for nitrate reductase activity (nitrate reduction to nitrite), it must be examined for evidence of denitrification. This is simply a visual inspection for the presence of gas in the Durham tube (Fig. 5.23A). If the Durham tube contains gas the test is complete. Denitrification has taken place. Gas produced in a nitrate reduction test by an organism capable of

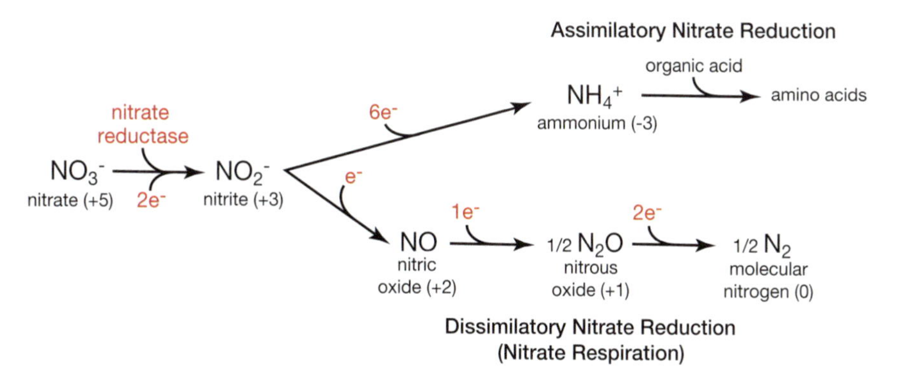


**5.21 Possible End Products of Nitrate Reduction** ■ Nitrate reduction is complex. Many different organisms under many different circumstances perform nitrate reduction with many different outcomes. Members of the *Enterobacteriaceae* simply reduce $NO_3$ to $NO_2$. Other bacteria, functionally known as "denitrifiers," reduce $NO_3$ all the way to $N_2$ via the intermediates shown, and are important ecologically in the nitrogen cycle. Both of these are anaerobic respiration pathways (also known as "nitrate respiration" and "dissimilatory nitrate reduction"). Other organisms are capable of assimilatory nitrate reduction, in which $NO_3^-$ or $NO_2^-$ is reduced to $NH_4^+$, which can be used in amino acid synthesis. The oxidation state of nitrogen in each compound is shown in parentheses.

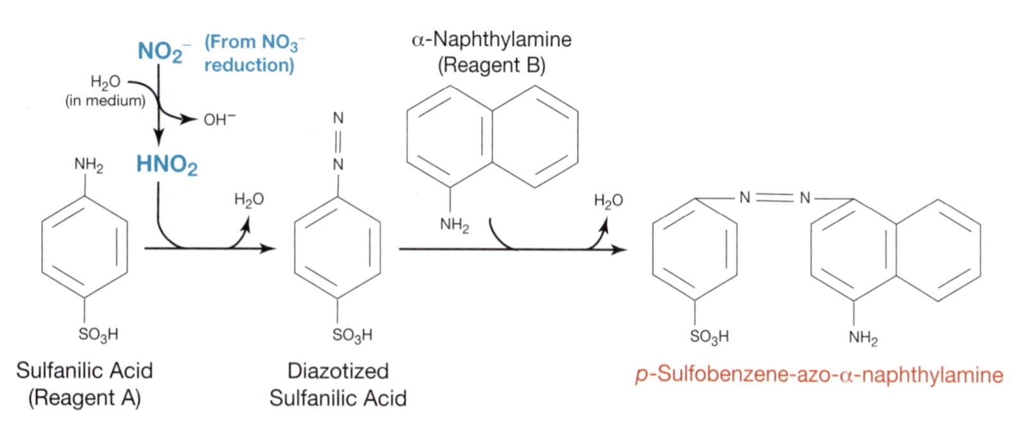


**5.22 Indicator Reaction for Nitrate Reduction to Nitrite** ■ If nitrate is reduced to nitrite, nitrous acid ($HNO_2$) will form in the medium. Nitrous acid then reacts with sulfanilic acid (reagent A) to form diazotized sulfanilic acid, which reacts with the α-naphthylamine (reagent B) to form *p*-sulfobenzene-azo-α-naphthylamine, which is red. Thus, a red color indicates the presence of nitrite and is considered a positive result for nitrate reduction to nitrite.

fermenting to gas end products is not determinative because the source of the gas is unknown. For more information on fermentation, see Exercise 5-2.

If there is no visual evidence of denitrification (or the test was not determinative), sulfanilic acid (nitrate reagent A) and α–naphthylamine (nitrate reagent B) are added to the medium to test for nitrate reduction to nitrite (Fig. 5.23B). If present, nitrite will form nitrous acid ($HNO_2$) in the aqueous medium. Nitrous acid reacts with the added reagents to produce a red, water-soluble compound.

Therefore, formation of red color after the addition of reagents indicates that the organism reduced nitrate to nitrite. If no color change takes place with the addition of reagents, the nitrate either was not reduced or was reduced to one of the other nitrogenous compounds shown in Figure 5.21. Because it is visually impossible to tell the difference between these two occurrences at this point, another test must be performed.

In this stage of the test, a small amount of powdered zinc is added to the broth to catalyze the reduction of any nitrate (which still may be present as $KNO_3$) to nitrite. If nitrate is present at the time zinc is added, it will be converted quickly to nitrite, and the above-described reaction between nitrous acid and reagents will follow and turn the medium red. In this instance, the red color indicates that nitrate was *not* reduced by the organism (Fig. 5.23C). No color change after the addition of zinc indicates that the organism reduced the nitrate to $NH_4^+$, NO, $N_2O$, or some other nongaseous nitrogenous compound other than nitrite.

## Application

Virtually all members of *Enterobacteriaceae* perform a one-step reduction of nitrate to nitrite. The nitrate test differentiates them from Gram-negative rods that either do not reduce nitrate or reduce it beyond nitrite to $N_2$ or other compounds.

## In This Exercise

After inoculation and incubation of the nitrate broths, you first will inspect the Durham tubes included in the media for bubbles—an indication of gas production. One or more of the organisms selected for this exercise are fermenters, and possible producers of gas other than molecular nitrogen; therefore, you will proceed with adding reagents to all tubes even if gas was produced. Record any gas observed in the Durham tubes and see if there is a correlation between that and the color results after the reagents and zinc dust have been added. Follow the procedure carefully, and use Table 5-8 as a guide when interpreting the results.

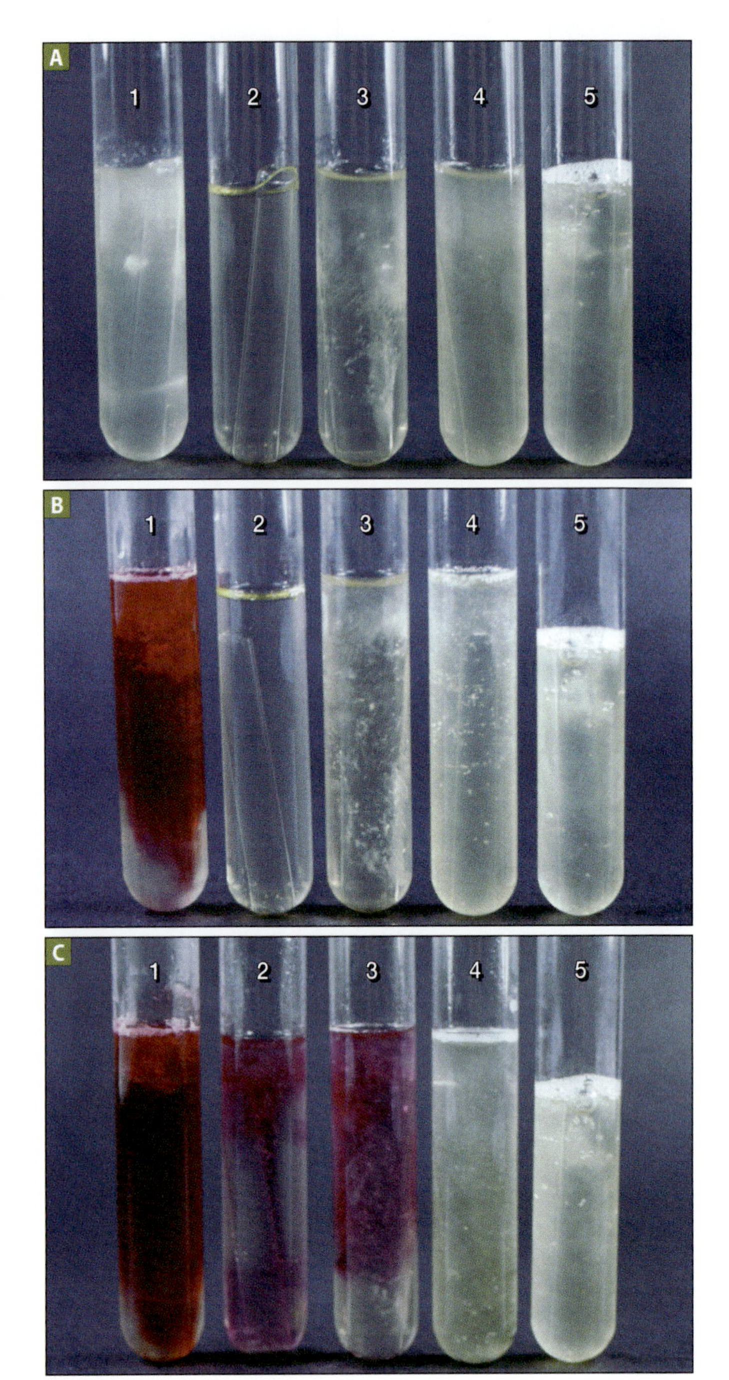


**5.23 Results of Incubated Nitrate Broth** ■ (**A**) These are nitrate broths immediately after incubation and prior to addition of reagents. Tube 2 is an uninoculated control used for color comparison. Note the gas produced by the organism in tube 5. (The bubble in the Durham tube is barely visible behind the layer of bubbles on the surface.) It is a known nonfermenter and, therefore, will receive no reagents. The gas produced is an indication of denitrification and a positive result for nitrate reduction. Tubes 1 through 4 will receive reagents. (**B**) After the addition of reagents, tube 1 shows a positive result. Tube 3 and tube 4 are inconclusive because they show no color change. Zinc dust must be added to tubes 2 (control), 3, and 4 to verify the presence or absence of nitrate. (**C**) Finally, a pinch of zinc was added to tubes 2, 3, and 4 because they have been colorless up to this point. Tube 2 (the control tube) and tube 3 turned red. This is a negative result because it indicates that nitrate is still present in the tube. Tube 4 did not change color, which indicates that the nitrate was reduced by the organism beyond nitrite to some other nitrogenous compound. This is a positive result.

## ▼ Materials

### Per Student

- ☐ Lab coat
- ☐ Gloves
- ☐ Chemical eye protection

### Per Student Group

- ☐ Four nitrate broths
- ☐ Sterile wooden sticks or disposable loops
- ☐ Nitrate test reagents A and B
- ☐ Zinc powder
- ☐ Fresh agar slant cultures of these recommended organisms:
  - *Alcaligenes faecalis*
  - *Escherichia coli*
  - (Optional) *Pseudomonas aeruginosa* (BSL-2)

## Medium and Reagent Recipes

### Nitrate Broth

| | |
|---|---|
| ☐ Beef extract | 3.0 g |
| ☐ Peptone | 5.0 g |
| ☐ Potassium nitrate ($KNO_3$) | 1.0 g |
| ☐ Distilled or deionized water | 1.0 L |

*pH 6.8–7.2 at 25°C*

### Reagent A

| | |
|---|---|
| ☐ Sulfanilic acid | 0.8 g |
| ☐ 5N acetic acid (30%) | 100.0 mL |

### Reagent B

| | |
|---|---|
| ☐ *N*,*N*-Dimethyl-α-naphthylamine | 0.6 g |
| ☐ 5N acetic acid (30%) | 100.0 mL |

## PROCEDURE

### Lab One

1. Wear a lab coat, gloves, and chemical eye protection when performing this procedure. **Use BSL-2 precautions with *P. aeruginosa*, if used.**
2. Obtain four nitrate broths. Label three of them with the names of the organisms, your group name, and the date. Label the fourth tube "control."
3. Inoculate three broths with the test organisms. Do not inoculate the control.
4. Incubate all tubes at 35 ± 2°C for 24 to 48 hours.
5. Save or dispose of the original culture tubes as directed by your instructor.

### Lab Two

1. Wear a lab coat, gloves, and chemical eye protection, then follow the procedural diagram in Figure 5.24.
2. Examine each tube for evidence of gas production. Record your results in the table provided on the data sheet, page 311. Refer to Table 5-8 when making your interpretations. (Be methodical when recording your results. Red color can have opposite interpretations depending on where you are in the procedure.)
3. Continue using Figure 5.24 as a guide and proceed to the addition of reagents to *all* tubes as follows:
   a. Add eight drops each (approximately 0.5 mL) of reagent A and reagent B to each tube. Mix well, but carefully, and let the tubes stand undisturbed for 10 minutes. Using the control as a comparison, record your test results in the table provided in the data sheet. Set aside any test broths that are positive.
   b. Where appropriate, add a pinch of zinc dust. (Only a small amount is needed. You can dip a wooden applicator into the zinc dust and transfer only the amount that clings to the wood.) Let the tubes stand for 10 minutes. Record your results in the table provided on the data sheet.
4. Dispose of all tubes in an appropriate autoclave container when finished.

**TABLE 5-8 Nitrate Reduction Test Results and Interpretations**

| Result | Interpretation | Symbol |
|---|---|---|
| Gas (non-fermenter) | Denitrification—production of nitrogen gas ($NO_3^- \rightarrow NO_2^- \rightarrow N_2$) | + |
| Gas (fermenter, or status is unknown) | Source of gas is unknown; requires addition of reagents | |
| Red color (after addition of reagents A and B) | Nitrate reduction to nitrite ($NO_3^- \rightarrow NO_2^-$); nitrate reductase is present | + |
| No color change (after the addition of reagents A and B) | Incomplete test; requires the addition of zinc dust | |
| No color change (after addition of zinc dust) | Nitrate reduction to nongaseous nitrogenous compounds ($NO_3^- \rightarrow NO_2^- \rightarrow NO$, ½ $N_2O$, or ½ $N_2$) | + |
| Red color (after addition of zinc dust) | No nitrate reduction | – |

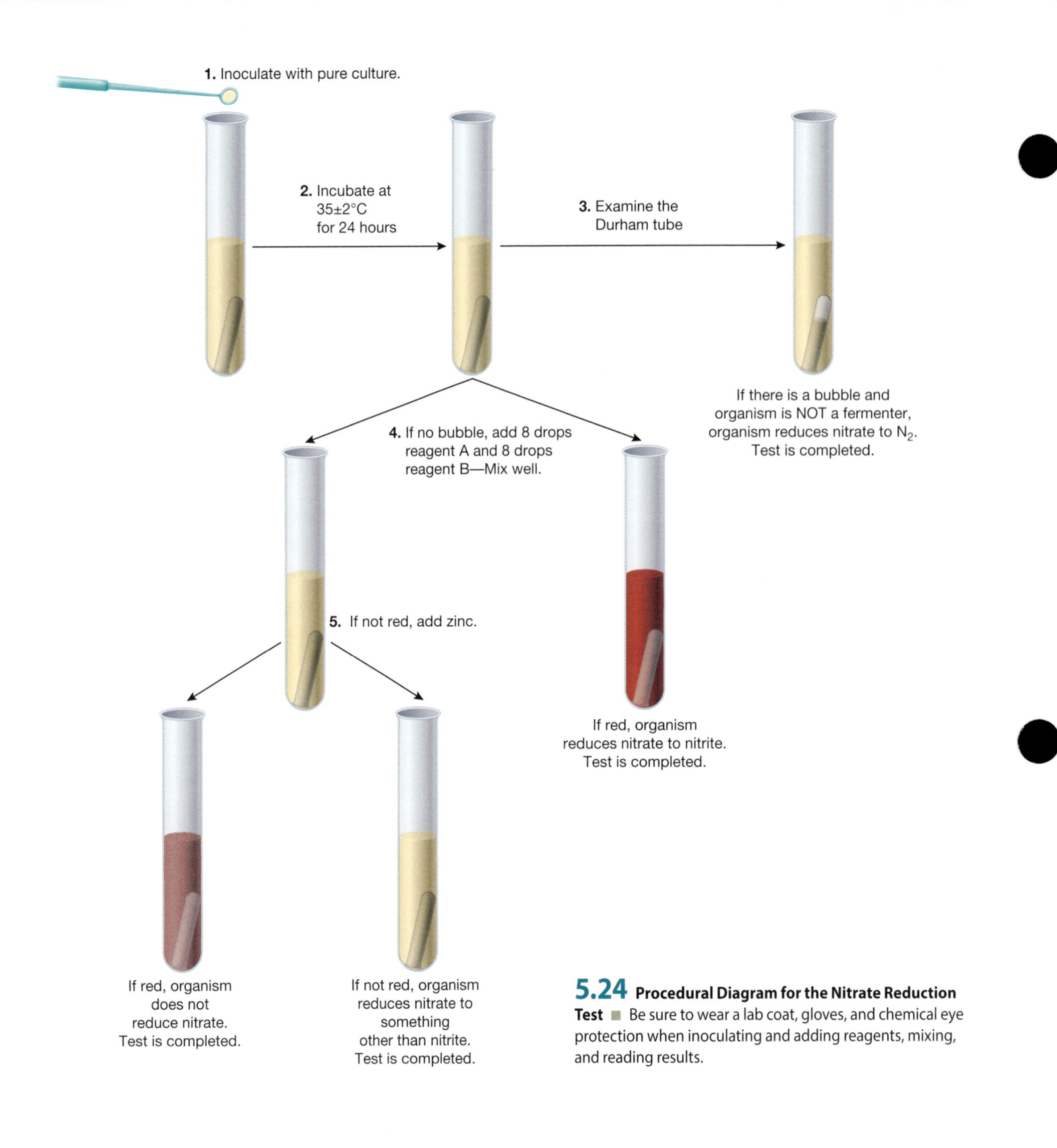


**5.24 Procedural Diagram for the Nitrate Reduction Test** ■ Be sure to wear a lab coat, gloves, and chemical eye protection when inoculating and adding reagents, mixing, and reading results.

## References


Atlas, Ronald M. and Richard Bartha. Pages 423–425 in *Microbial Ecology—Fundamentals and Applications,* 4th ed. Menlo Park, CA: Benjamin/Cummings Science Publishing, 1998.

Atlas, Ronald and James Snyder. Pages 319 and 341 in *Manual of Clinical Microbiology,* 11th ed. James H. Jorgensen, Michael A. Pfaller, Karen C. Carroll, Guido Funke, Marie Louise Landry, Sandra S. Richter, and David W. Warnock, eds. Washington, DC: ASM Press, American Society for Microbiology, 2015.

Lányi, B. Page 21 in *Methods in Microbiology,* Vol. 19. R. R. Colwell and R. Grigorova, eds. New York: Academic Press, 1987.

MacFaddin, Jean F. Page 348 in *Biochemical Tests for Identification of Medical Bacteria,* 3rd ed. Philadelphia: Lippincott Williams & Wilkins, 2000.

Moat, Albert G., John W. Foster, and Michael P. Spector. Chap. 14 in *Microbial Physiology,* 4th ed. New York: Wiley-Liss, 2002.

Tille, Patricia M. Page 219 in *Bailey & Scott's Diagnostic Microbiology,* 13th ed. St. Louis, MO: Mosby, 2014.

Zimbro, Mary Jo and David A. Power, eds. Page 400 in *Difco™ and BBL™ Manual—Manual of Microbiological Culture Media.* Sparks, MD: Becton Dickinson and Co., 2003.

Name ______________________________

Date ______________________________

Lab Section ______________________________

I was present and performed this exercise (initials) ______________

DATA SHEET 5-6

# Nitrate Reduction Test

## OBSERVATIONS AND INTERPRETATIONS

**1** Using Table 5-8, page 309, as a guide, enter your results and interpretations in the table below. If you have a positive for gas production, look up the fermentation characteristics of that organism in a standard reference, such as ***Bergey's Manual of Determinative Bacteriology,*** 9th ed. and include that result in your interpretation.

| Organism | Results | | | Interpretation (Be as specific as possible with respect to end products if positive) |
|---|---|---|---|---|
| | Gas | Color | | |
| | Y/N | After Reagents | After Zinc | |
| Uninoculated control | | | | |
| | | | | |
| | | | | |
| | | | | |

## QUESTIONS

**1** *Consider the uninoculated tube.*

**a.** *Is it a positive or a negative control?*

______________________________

**b.** *What information is provided by the uninoculated control?*

______________________________

______________________________

______________________________

**2** *Why is gas production not recognized as nitrate reduction when the organism is a known fermenter?*

**3** *Suppose you remove your test cultures from the incubator and notice that one of them—a known fermenter—has a gas bubble in the Durham tube. Knowing that fermenters frequently produce gas, you ignore the bubble and proceed to the next step. Adding reagents produces no change, and neither does adding zinc. Is this occurrence consistent with denitrification? Why?*



**4** *Would you change your answer to question 3 if the control broth did not change color after the addition of reagents? What if the control broth had changed color only after the addition of reagents and zinc?*

**5** *When testing microaerophiles, some microbiologists prefer to use a semisolid nitrate medium that contains a small amount of agar. Why do you think this is done?*

## Nutrient Utilization Tests

In the following exercise you will perform an example of a test using Simmons citrate medium, which belongs to a category referred to as "utilization media." Utilization media are highly defined formulations designed to differentiate organisms based on their ability to grow when an essential nutrient (e.g., carbon or nitrogen) is strictly limited. For example, citrate medium contains sodium citrate as the only carbon-containing compound and ammonium ion as the only nitrogen source. Utilization tests can be very useful in differentiating between species. ■

# Citrate Utilization Test

EXERCISE 5-7

## ■ Theory

The citrate utilization test was designed to differentiate members of the *Enterobacteriaceae*, all of which are facultative anaerobes. That is, they have the ability to ferment carbohydrates and they also have the ability to aerobically respire, which means they have a functional citric acid cycle. However, the citrate utilization test does not tell us about the citric acid cycle. Instead, it tells about the ability of organisms to use citrate as their sole carbon source and perform citrate fermentation.

Simmons citrate agar is a defined medium. That is, the amount and source of all ingredients are carefully controlled. Because sodium citrate is the only carbon source in the medium[1] it will not support a complex, high-energy yielding respiratory process like the citric acid cycle. It does, however, provide the means for bacterial species that possess the enzyme **citrate permease** (an **oxaloacetate decarboxylase $Na^+$ pump**) to transport citrate (actually, oxaloacetate—see below) into the cell and perform citrate fermentation. These organisms must also be able to survive with ammonium (in the form of ammonium phosphate) as the sole nitrogen source. Bacteria that do not possess citrate permease will not grow on this medium.

Citrate-positive bacteria hydrolyze citrate into oxaloacetate and acetate using the enzyme **citrate lyase** (Fig. 5.25). From there, oxaloacetate is decarboxylated

[1] Of course, there is $CO_2$ in the air above the agar, but the test is designed primarily to differentiate *Enterobacteriaceae*, none of which are autotrophic.

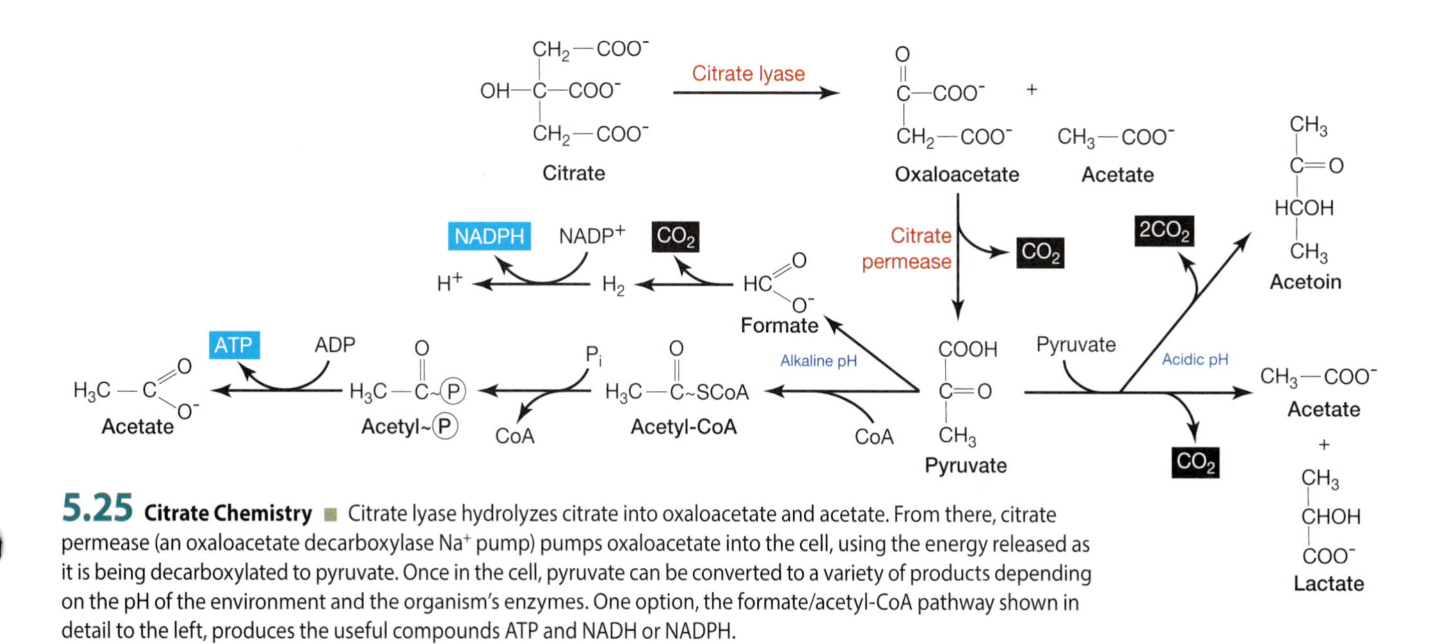


**5.25 Citrate Chemistry** ■ Citrate lyase hydrolyzes citrate into oxaloacetate and acetate. From there, citrate permease (an oxaloacetate decarboxylase $Na^+$ pump) pumps oxaloacetate into the cell, using the energy released as it is being decarboxylated to pyruvate. Once in the cell, pyruvate can be converted to a variety of products depending on the pH of the environment and the organism's enzymes. One option, the formate/acetyl-CoA pathway shown in detail to the left, produces the useful compounds ATP and NADH or NADPH.

5

to pyruvate, simultaneously using the energy released to pump oxaloacetate/pyruvate into the cell. A variety of products can be formed from pyruvate depending on the cell's pH.

Here is the secret: some bacterial fermentation pathways can make ATP and reducing power (NADH or NADPH). For instance, *Klebsiella* and *Enterobacter*, both of which are citrate positive, have the ability to make acetyl-CoA from pyruvate (nothing new) but then have the ability to convert it to acetyl phosphate, which can then phosphorylate ADP to ATP. In addition, the same bacteria are capable of using $H_2$ as a source of electrons to reduce NAD and/or NADP, the latter of which is used in synthesis reactions. Both of these are shown in Figure 5.25.

Bacteria that survive in the medium and utilize the citrate also convert the ammonium phosphate to ammonia ($NH_3$) and ammonium hydroxide ($NH_4OH$), both of which tend to alkalinize the agar. Bromothymol blue dye, which is green at pH 6.9 and blue at pH 7.6, is an indicator, and as the pH goes up the medium changes from green to blue (Fig. 5.26). Thus, conversion of the medium to blue is a positive citrate test result (Table 5-9).

Occasionally a citrate-positive organism will grow on a Simmons citrate slant without producing a change in color. In most cases, this is because of incomplete incubation. In the absence of color change, growth on the slant indicates that citrate is being utilized and is evidence of a positive reaction. To avoid confusion between actual growth and a heavy inoculum, which may be misinterpreted as growth, citrate slants typically are inoculated lightly with an inoculating needle rather than a loop.

**5.26 Citrate Utilization Test Results** ■ These Simmons citrate slants were inoculated with a citrate-negative organism on the left and a citrate-positive organism on the right. The slant in the center is an uninoculated control.

**TABLE 5-9 Citrate Utilization Test Results and Interpretations**

| Result | Interpretation | Symbol |
|---|---|---|
| Blue (even a small amount) | Citrate is utilized (Citrate permease is present) | + |
| No color change; growth | Citrate is utilized (Citrate permease is present) | + |
| No color change; no growth | Citrate is not utilized (Citrate permease is absent or not detectable) | − |

## ■ Application

The citrate utilization test is used to determine the ability of an organism to use citrate as its sole source of carbon. Citrate utilization is one part of a test series referred to as the IMViC (*I*ndole, *M*ethyl Red, *V*oges-Proskauer and *C*itrate tests) that distinguishes between members of the family *Enterobacteriaceae* and also from other Gram-negative rods.

## ■ In This Exercise

You will be inoculating Simmons citrate slants with a citrate-positive and a citrate-negative organism. To avoid confusing growth on the slant with a heavy inoculum, be sure to lightly inoculate using a needle instead of a loop.

## ▼ Materials

### Per Student

- □ Lab coat
- □ Gloves
- □ Chemical eye protection

### Per Student Group

- □ Three Simmons citrate slants
- □ Fresh agar slant cultures of these recommended organisms:
  - *Enterobacter aerogenes*
  - *Escherichia coli*

## ■ Medium Recipe

### Simmons Citrate Agar

- □ Ammonium dihydrogen phosphate 1.0 g
- □ Dipotassium phosphate 1.0 g
- □ Sodium chloride 5.0 g
- □ Sodium citrate 2.0 g
- □ Magnesium sulfate 0.2 g

- □ Agar 15.0 g
- □ Bromothymol blue 0.08 g
- □ Distilled or deionized water 1.0 L

*pH 6.7–7.1 at 25°C*

## PROCEDURE

### Lab One

1. Wear a lab coat, gloves, and chemical eye protection when performing this procedure.
2. Obtain three Simmons citrate tubes. Label two with the names of the organisms, your group name, and the date. Label the third tube "control."
3. Using an inoculating needle and *light inoculum*, streak the slants with the test organisms. Do not inoculate the control.
4. Incubate all tubes at 35 ± 2°C for 48 hours.
5. Save or dispose of the original culture tubes as directed by your instructor.

### Lab Two

1. Observe the tubes for color changes and/or growth.
2. Record your results in the table provided on the data sheet, page 317.
3. Dispose of all tubes in an appropriate autoclave container when finished.

## References

Atlas, Ronald and James Snyder. Page 344 in *Manual of Clinical Microbiology,* 11th ed. James H. Jorgensen, Michael A. Pfaller, Karen C. Carroll, Guido Funke, Marie Louise Landry, Sandra S. Richter, and David W. Warnock, eds. Washington, DC: ASM Press, American Society for Microbiology, 2015.

Collins, C. H., Patricia M. Lyne, and J. M. Grange. Page 111 in *Collins and Lyne's Microbiological Methods,* 7th ed. Oxford, Boston: Butterworth-Heinemann, 1995.

MacFaddin, Jean F. Page 98 in *Biochemical Tests for Identification of Medical Bacteria,* 3rd ed. Philadelphia: Lippincott Williams & Wilkins, 2000.

Moat, Albert G., John W. Foster, and Michael P. Spector. Pages 427–428 in *Microbial Physiology*, 4th ed. New York, NY: Wiley-Liss, 2002.

Smibert, Robert M. and Noel R. Krieg. Page 614 in *Methods for General and Molecular Bacteriology*. Philipp Gerhardt, R. G. E. Murray, Willis A. Wood, and Noel R. Krieg, eds. Washington, DC: American Society for Microbiology, 1994.

Steuber, Julia, Walter Krebs, Michael Bott, and Peter Dimroth. "A Membrane-Bound $NAD(P)^+$-Reducing Hydrogenase Provides Reduced Pyridine Nucleotides during Citrate Fermentation by *Klebsiella pneumoniae*." *J.Bacteriol.* 181, no. 1 (January 1999): 241–245.

Tille, Patricia M. Page 202 in *Bailey & Scott's Diagnostic Microbiology*, 13th ed. St. Louis, MO: Mosby, 2014.

Zimbro, Mary Jo and David A. Power, eds. Page 514 in *Difco™ and BBL™ Manual—Manual of Microbiological Culture Media.* Sparks, MD: Becton Dickinson and Co., 2003.

Name ______________________________

Date ______________________________

Lab Section ______________________________

I was present and performed this exercise (initials) ______________________________

# DATA SHEET 5-7

## Citrate Utilization Test

### OBSERVATIONS AND INTERPRETATIONS

**1** Using Table 5-9, page 314, as a guide, enter your results in the table below.

| Organism | Color Result | (+ or −) | Interpretation |
|---|---|---|---|
| Control | | | |
| | | | |
| | | | |

## QUESTIONS

**1** *Consider the uninoculated tube.*

**a.** *Is it a positive or a negative control?*

______________________________

**b.** *What information is provided by the uninoculated control?*

______________________________

______________________________

______________________________

**2** *Many bacteria that are able to metabolize citrate (as seen in the citric acid cycle) produce negative results in this test. Why? Be specific.*

**3** *Explain how an organism that possesses the citrate lyase enzyme might not test positively on Simmons citrate agar. Is this a false negative result? Why or why not?*

5

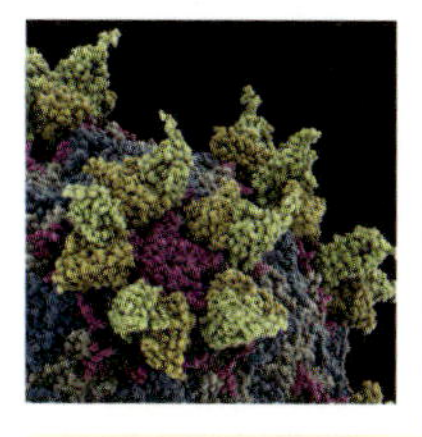

## Decarboxylation and Deamination Tests

Decarboxylation and deamination tests were designed to differentiate members of *Enterobacteriaceae* and to distinguish them from other Gram-negative rods. Most members of *Enterobacteriaceae* produce one or more enzymes that are necessary to break down amino acids. Enzymes that catalyze the removal of an amino acid's carboxyl group (COOH) are called **decarboxylases**. Enzymes that catalyze the removal of an amino acid's amine group ($NH_2$) are called **deaminases**.

Each decarboxylase and deaminase is specific to a particular substrate, which in these two exercises means a different amino acid. Decarboxylases catalyze reactions that produce alkaline products. Thus, we are able to identify the ability of an organism to produce a specific decarboxylase by preparing base medium, adding a known amino acid, and including a pH indicator to mark the shift to alkalinity. Differentiation of an organism in deamination media employs the same principle of substrate exclusivity by including a single known amino acid but requires the addition of a chemical reagent to produce a readable result.

In the next two exercises you will be introduced to the most common of both types of tests. In Exercise 5-8 you will test bacterial ability to decarboxylate the amino acids arginine, lysine, and ornithine. In Exercise 5-9 you will test bacterial ability to deaminate the amino acid phenylalanine. ■

# Amino Acid Decarboxylation (Decarboxylase Tests)

EXERCISE 5-8



## ■ Theory

Møller's decarboxylase base medium contains peptone, glucose, the pH indicator bromocresol purple, and the coenzyme pyridoxal phosphate. Bromocresol purple is purple at pH 6.8 and above, and yellow below pH 5.2. Base medium can be used with any one of a number of specific amino acid substrates, depending on the decarboxylase to be identified.

After inoculation, an overlay of mineral oil is used to seal the medium from external oxygen and promote fermentation (Fig. 5.27). Glucose fermentation in the anaerobic medium initially turns it yellow because of the accumulation of acid end products (all *Enterobacteriaceae* ferment glucose to acid end products). The low pH and presence of the specific amino acid induces **decarboxylase-positive** organisms to produce the enzyme. (That is, the specific decarboxylase gene is "switched on.")

Decarboxylation of the amino acid results in accumulation of amines, which are alkaline (Fig. 5.28) and turn the medium purple. If the organism is a glucose fermenter but does not produce the appropriate decarboxylase, the medium will turn yellow and remain so. If the organism does not ferment glucose, the medium will exhibit no color change. Purple color is the only positive result; all others are negative (Fig. 5.29). The three amino acids tested most frequently for decarboxylase activity are arginine, lysine, and ornithine (Figs. 5.30 through 5.32).

Arginine decarboxylase medium may, in fact, identify a second pathway for arginine catabolism. It is called the arginine dihydrolase system. In it, arginine is converted to citrulline and then into ornithine, ATP, $CO_2$, and two $NH_3$, which raises the pH and turns the medium purple. If the species is also positive for ornithine decarboxylase, it will produce putrescine and $CO_2$ from ornithine (Fig. 5.33).

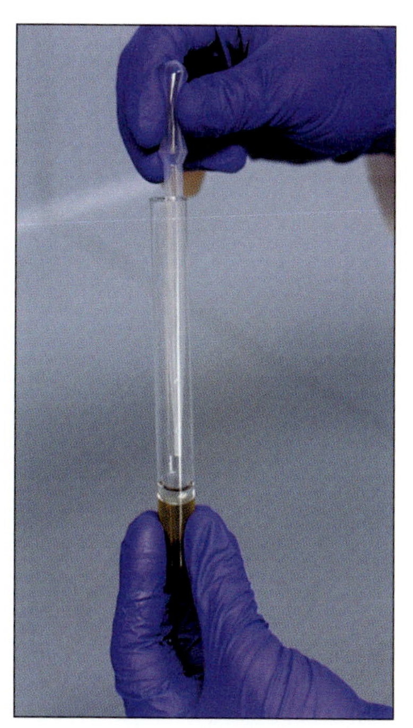

**5.27 Adding Sterile Mineral Oil to the Tubes** ■ Gently let the oil slide down the side of the test tube to a depth of approximately 4 mm (not mL). It is important to keep the mineral oil sterile to prevent contamination.

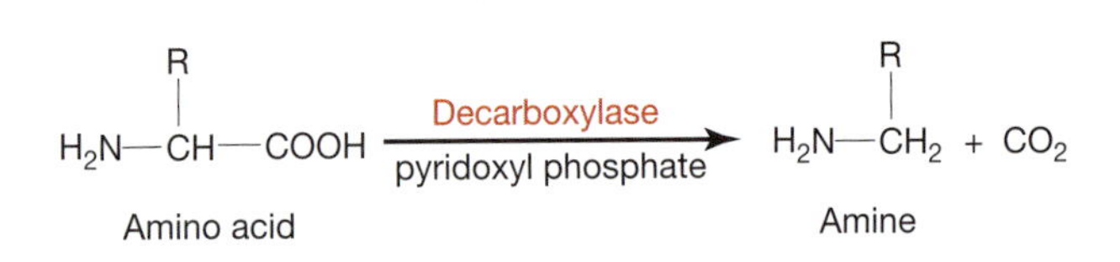


**5.28 Amino Acid Decarboxylation** ■ Removal of an amino acid's carboxyl group results in the formation of an amine and carbon dioxide.

**5.29 Decarboxylation Test Results** ■ A lysine decarboxylase-negative result is on the left, an uninoculated control is in the center, and a lysine decarboxylase-positive is on the right. The uninoculated control's color would match a decarboxylase-negative result without fermentation. The color results shown here are representative of arginine and ornithine decarboxylase media, as well.

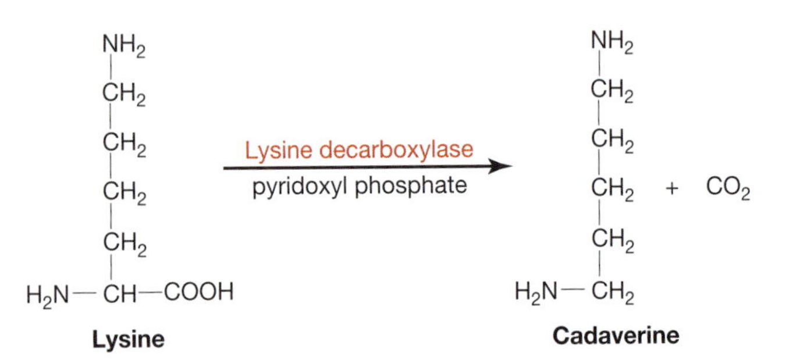

**5.30 Lysine Decarboxylation** ■ Decarboxylation of the amino acid lysine produces cadaverine and $CO_2$.

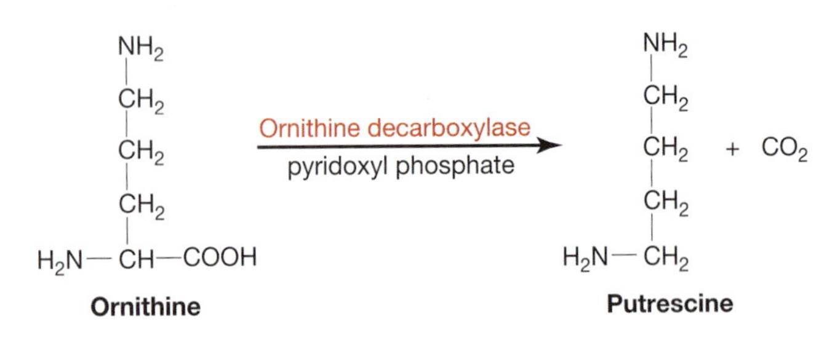


**5.31 Ornithine Decarboxylation** ■ Decarboxylation of the amino acid ornithine produces putrescine and $CO_2$.

## ■ Application

Decarboxylase media can include any one of several amino acids. Typically, these media are used to differentiate organisms in the family *Enterobacteriaceae* and to distinguish them from other Gram-negative rods. The most frequently used amino acids are arginine, lysine, and ornithine.

## ■ In This Exercise

You will perform the lysine decarboxylation test on three organisms in Møller's lysine and base decarboxylase media. This procedure requires the aseptic addition of mineral oil. Be careful not to introduce contaminants while adding the oil. Refer to Figure 5.29 and Table 5-10 when interpreting your results. (***Note***: positive and negative results—with and without acid production—look the same regardless of the amino acid in the decarboxylase medium. In Section 9 you may be asked to perform other decarboxylase tests, but you will read them the same way you read the results of lysine decarboxylase medium in this exercise.)

TABLE **5-10** **Decarboxylase Test Results and Interpretations**

| Result | Interpretation | Symbol |
|---|---|---|
| No color change | No decarboxylation (specific decarboxylase is absent or products are not detectable) | − |
| Yellow | Fermentation; no decarboxylation (specific decarboxylase is absent or products are not detectable) | − |
| Purple (may be slight) | Decarboxylation (organism produces the specific decarboxylase enzyme) | + |

## ▼ Materials[1]

### Per Student Group

- ☐ Three (or four) lysine decarboxylase broths
- ☐ Three decarboxylase base media
- ☐ Sterile mineral oil in microtubes
- ☐ Sixteen sterile transfer pipettes
- ☐ Fresh agar slant cultures of these recommended organisms:
  - *Enterobacter aerogenes*
  - *Providencia stuartii*
  - (Optional, see footnote) *Enterobacter cloacae*

[1] If desired, arginine and ornithine decarboxylase media may also be used in this exercise in preparation for unknowns in Section 9. If used, include *Enterobacter cloacae* as one of the test organisms and raise the number of each decarboxylase tube to four instead of three.

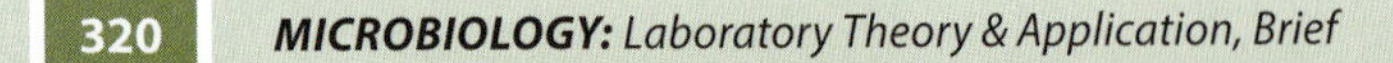

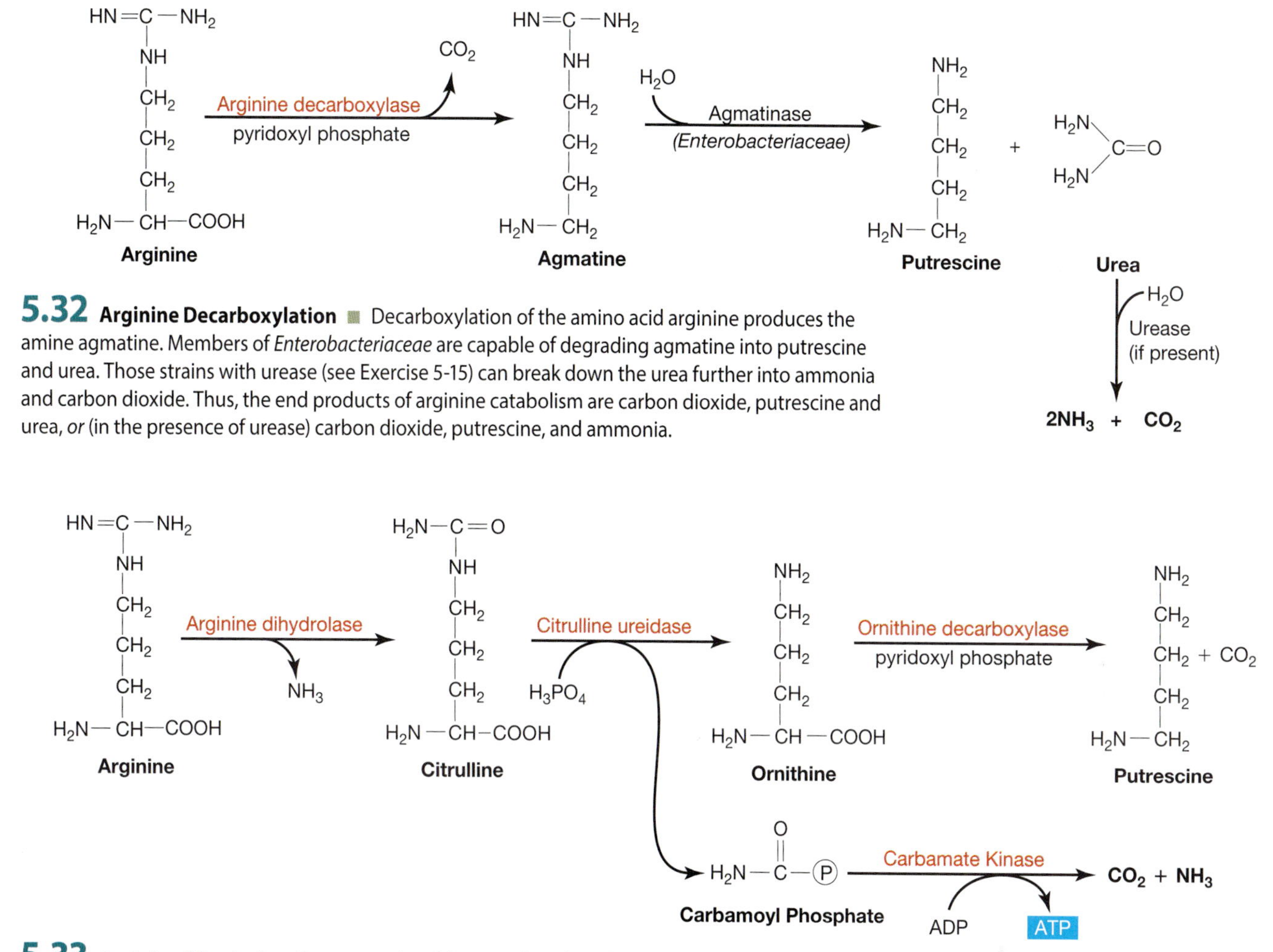


**5.32 Arginine Decarboxylation** ■ Decarboxylation of the amino acid arginine produces the amine agmatine. Members of *Enterobacteriaceae* are capable of degrading agmatine into putrescine and urea. Those strains with urease (see Exercise 5-15) can break down the urea further into ammonia and carbon dioxide. Thus, the end products of arginine catabolism are carbon dioxide, putrescine and urea, *or* (in the presence of urease) carbon dioxide, putrescine, and ammonia.

**5.33 Arginine Dihydrolase System** ■ In addition to decarboxylation, arginine may also be catabolized by a dihydrolase enzyme with the production of ornithine, $NH_3$, and $CO_2$. In ornithine decarboxylase-positive organisms, the ornithine is further degraded into putrescine and $CO_2$. The example shown here is a dihydrolase system of *Pseudomonas putida* in which ATP synthesis occurs. Regardless of the specifics, the arginine dihydrolase pathway is indistinguishable from the arginine decarboxylase pathway because both produce alkaline end products.

## Medium Recipe

### Lysine Decarboxylase Medium (Møller)

| | |
|---|---|
| □ Peptone | 5.0 g |
| □ Beef extract | 5.0 g |
| □ Glucose (dextrose) | 0.5 g |
| □ Bromocresol purple | 0.01 g |
| □ Cresol red | 0.005 g |
| □ Pyridoxal | 0.005 g |
| □ L-Lysine | 10.0 g |
| □ Distilled or deionized water | 1.0 L |

*pH 5.8–6.2 at 25°C*

## PROCEDURE

### Lab One

1. Wear a lab coat, gloves, and chemical eye protection when performing this procedure.
2. Obtain four of each decarboxylase broth. Label three of each with the name of the organism, your group name, and the date. Label the fourth of each broth "control."
3. Inoculate the media with the test organisms. Do not inoculate the control.
4. Overlay all tubes with 3 mm to 4 mm sterile mineral oil (Fig. 5.27).
5. Incubate all tubes aerobically at 35 ± 2°C for 96 hours (up to one week).
6. Save or dispose of the original culture tubes as directed by your instructor.

## Lab Two

1. Remove the tubes from the incubator and examine them for color changes.
2. Record your results in the table provided on the data sheet, page 323.
3. Dispose of all tubes in an appropriate autoclave container when finished.

## *References*


Atlas, Ronald and James Snyder. Page 316–317 in *Manual of Clinical Microbiology,* 11th ed. James H. Jorgensen, Michael A. Pfaller, Karen C. Carroll, Guido Funke, Marie Louise Landry, Sandra S. Richter, and David W. Warnock, eds. Washington, DC: ASM Press, American Society for Microbiology, 2015.

Collins, C. H., Patricia M. Lyne, and J. M. Grange. Page 111 in *Collins and Lyne's Microbiological Methods,* 7th ed. Oxford, Boston: Butterworth-Heinemann, 1995.

DIFCO Laboratories. Page 268 in *DIFCO Manual,* 10th ed. Detroit: DIFCO Laboratories, 1984.

Forbes, Betty A., Daniel F. Sahm, and Alice S. Weissfeld. Page 267 in *Bailey & Scott's Diagnostic Microbiology*, 11th ed. St. Louis, MO: Mosby, 2002.

Kakimoto, Toshio, Takeji Shibatani, Noriyuki Nishimura, and Ichio Chibata. "Enzymatic Production of L-Citrulline by *Pseudomonas putida. Applied Microbiology*. 22, no. 6 (December 1971): 992–999.

Lányi, B. Page 29 in *Methods in Microbiology*, Vol. 19. R. R. Colwell and R. Grigorova, eds. New York: Academic Press, 1987.

MacFaddin, Jean F. Page 120 in *Biochemical Tests for Identification of Medical Bacteria,* 3rd ed. Philadelphia: Lippincott Williams & Wilkins, 2000.

Tille, Patricia M. Page 204 in *Bailey & Scott's Diagnostic Microbiology*, 13th ed. St. Louis, MO: Mosby, 2014.

Name ______________________________

Date ______________________________

Lab Section ______________________________

I was present and performed this exercise (initials) ______________________________

# DATA SHEET 5-8

## Amino Acid Decarboxylation (Decarboxylase Tests)

### OBSERVATIONS AND INTERPRETATIONS

**1** Using Table 5-10, page 320, as a guide, enter your results in the table below.

| Organism | Results | | | | Interpretation |
|---|---|---|---|---|---|
| | Base | | Lysine | | |
| | Color | +/− | Color | +/− | |
| Control | | | | | |
| | | | | | |
| | | | | | |
| | | | | | |

## QUESTIONS

**1** *Consider the decarboxylase base broth.*

**a.** *Why isn't a different base broth required for each decarboxylase medium?*

______________________________

______________________________

______________________________

**b.** *Is the base broth a positive or a negative control?*

______________________________

**c.** *What information is provided by the base broth control?*

______________________________

______________________________

______________________________

______________________________

**2** *Consider the uninoculated lysine decarboxylase tube.*

**a.** *Is the uninoculated tube a positive or a negative control?*

**b.** *What information does the uninoculated tube provide?*

**c.** *Would it be acceptable to use uninoculated ornithine decarboxylase medium as a control for, say, lysine decarboxylase tests? Defend your answer.*

5

**3** *How can no color change in the broth and a conversion to a yellow color both be considered negative results?*

**4** *Incubation time for this medium is 96 hours.*

**a.** *Under what circumstances would an early reading be allowable? Explain.*

**b.** *Under what circumstances would an early reading not be allowable? Explain.*

Name ________________________________

Date ________________________________

Lab Section ________________________________

I was present and performed this exercise (initials) ________________

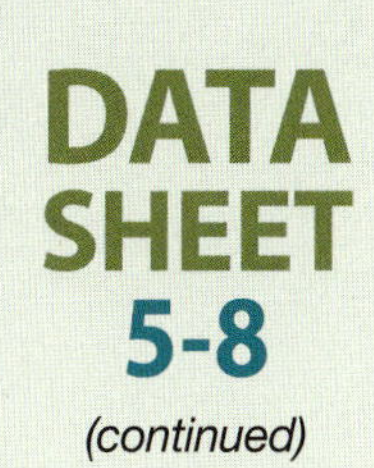

# DATA SHEET 5-8

*(continued)*

**5** *Think about the effect of the mineral oil on the test results.*

**a.** *Would omitting the sterile mineral oil affect the specificity of the test? Explain.*

**b.** *Would omitting the sterile mineral oil affect the sensitivity of the test? Explain.*

# Phenylalanine Deaminase Test

EXERCISE 5-9

## Theory

In Exercise 5-8 we saw one way amino acids can be catabolized—that is, by removal of the carboxyl group. In this lab we will examine a second way, and that is by the removal of the amine group ($NH_2$) in a process called **deamination** (Fig. 5.34).

Organisms that produce phenylalanine deaminase can be identified by their ability to deaminate the amino acid phenylalanine. The reaction, as shown in Figure 5.35, initially removes two hydrogen ions, which combine with oxygen to make water and produce an intermediate acid. The intermediate is then deaminated to produce ammonia ($NH_3$) and phenylpyruvic acid. Deaminase activity, therefore, is evidenced by the presence of phenylpyruvic acid.

Phenylalanine agar provides a rich source of phenylalanine. A reagent containing ferric chloride ($FeCl_3$) is added to the medium after incubation. The normally colorless phenylpyruvic acid reacts with the ferric chloride and turns a dark green color almost immediately (Fig. 5.36). Formation of green color indicates the presence of phenylpyruvic acid and, hence, the presence of phenylalanine deaminase. Yellow is negative (Fig. 5.37).

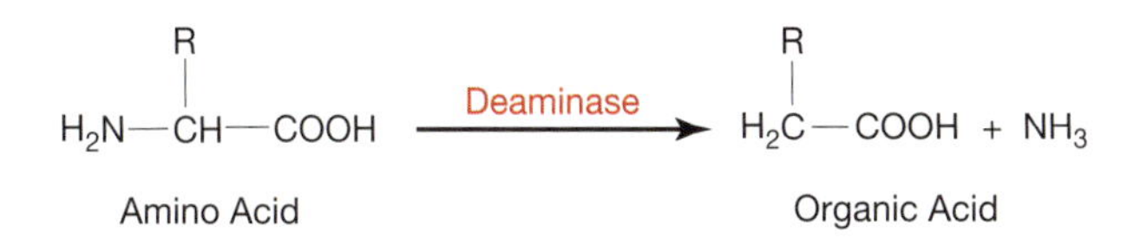


**5.34 Amino Acid Deamination** ■ Deamination is the removal of an amine ($NH_2$) from an amino acid.

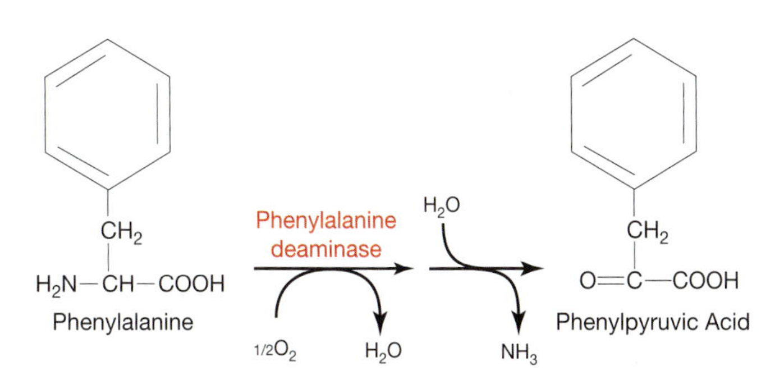


**5.35 Deamination of Phenylalanine** ■ Phenylalanine deamination occurs as a two-step process, but the net result is phenylpyruvic acid and ammonia.

$$\text{Phenylpyruvic Acid} + FeCl_3 \longrightarrow \text{Green Color}$$

**5.36 Phenylalanine Deaminase Test Indicator Reaction** ■ Phenylpyruvic acid produced by phenylalanine deamination reacts with $FeCl_3$ to produce a green color and indicates a positive phenylalanine deaminase test. Be sure to read the result immediately, as the color may fade.

**5.37 Phenylalanine Deaminase Test Results** ■ Note the color produced by the stream of ferric chloride in each tube. A phenylalanine deaminase-positive organism is on the left, an uninoculated control is in the middle, and a phenylalanine deaminase-negative organism is on the right.

## Application

This medium is used to differentiate the genera *Morganella*, *Proteus*, and *Providencia* from other members of the *Enterobacteriaceae*.

## In This Exercise

You will inoculate two phenylalanine agar slants with test organisms. After incubation you will add 12% $FeCl_3$ (phenylalanine deaminase test reagent) and watch for a color change. Refer to Figure 5.37 and Table 5-11 when making your interpretations and recording your results.

5

TABLE 5-11 **Phenylalanine Deaminase Test Results and Interpretations**

| Result | Interpretation | Symbol |
|---|---|---|
| Green color | Phenylalanine deaminase is present | + |
| No color change | Phenylalanine deaminase is absent or products are not detectable. | – |

## ▼ Materials

### Per Student

- ☐ Lab coat
- ☐ Gloves
- ☐ Chemical eye protection

### Per Student Group

- ☐ Three phenylalanine agar slants
- ☐ 12% ferric chloride solution
- ☐ Fresh agar slant cultures of these recommended organisms:
  - *Escherichia coli*
  - *Providencia stuartii*

5

## ■ Medium and Reagent Recipes

### Phenylalanine Agar

| | |
|---|---|
| ☐ DL-Phenylalanine | 2.0 g |
| ☐ Yeast extract | 3.0 g |
| ☐ Sodium chloride | 5.0 g |
| ☐ Sodium phosphate | 1.0 g |
| ☐ Agar | 12.0 g |
| ☐ Distilled or deionized water | 1.0 L |

*pH 7.1–7.5 at 25°C*

### Phenylalanine Deaminase Test Reagent

| | |
|---|---|
| ☐ Ferric chloride | 12.0 g |
| ☐ Concentrated hydrochloric acid | 2.5 mL |
| ☐ Deionized water (to bring the total to) | 100.0 mL |

## PROCEDURE

### Lab One

1. Wear a lab coat, gloves, and chemical eye protection when performing this procedure.
2. Obtain three phenylalanine agar slants. Label two slants with the name of the organism, your group name, and the date. Label the third slant "control."
3. Streak two slants with heavy inocula of the test organisms. Do not inoculate the control.
4. Incubate all slants aerobically at 35 ± 2°C for 18 to 24 hours.
5. Save or dispose of the original culture tubes as directed by your instructor.

### Lab Two

1. Wear a lab coat, gloves, and chemical eye protection when performing this procedure.
2. Add a few drops of 12% ferric chloride solution to each tube and observe for color change. (***Note***: This color may fade quickly, so read and record your results immediately.)
3. Record your results in the table provided on the data sheet, page 329.
4. Dispose of all tubes in an appropriate autoclave container when finished.

### *References*


Atlas, Ronald and James Snyder. Page 319 in *Manual of Clinical Microbiology,* 11th ed. James H. Jorgensen, Michael A. Pfaller, Karen C. Carroll, Guido Funke, Marie Louise Landry, Sandra S. Richter, and David W. Warnock, eds. Washington, DC: ASM Press, American Society for Microbiology, 2015.

Lányi, B. Page 28 in *Methods in Microbiology,* Vol. 19. R. R. Colwell and R. Grigorova, eds. New York: Academic Press, 1987.

MacFaddin, Jean F. Page 388 in *Biochemical Tests for Identification of Medical Bacteria,* 3rd ed. Philadelphia: Lippincott Williams & Wilkins, 2000.

Tille, Patricia M. Page 225 in *Bailey & Scott's Diagnostic Microbiology,* 13th ed. St. Louis, MO: Mosby, 2014.

Zimbro, Mary Jo and David A. Power, eds. Page 442 in *Difco™ and BBL™ Manual—Manual of Microbiological Culture Media.* Sparks, MD: Becton Dickinson and Co., 2003.

Name ___________________________________________

Date ___________________________________________

Lab Section ___________________________________________

I was present and performed this exercise (initials) ____________________

# DATA SHEET 5-9

## Phenylalanine Deaminase Test

### OBSERVATIONS AND INTERPRETATIONS

**1** Using Table 5-11, page 328, as a guide, enter your results in the table below.

| Organism | Color Result | (+ or −) | Interpretation |
|---|---|---|---|
| Control | | | |
| | | | |
| | | | |

### QUESTIONS

**1** *Consider the uninoculated phenylalanine agar tube.*

**a.** *Is this a positive or a negative control?*

___________________________________________

**b.** *What information is provided by this control?*

___________________________________________

___________________________________________

___________________________________________

___________________________________________

**2** *If you are performing this test on an unknown organism, why is it a good idea to run simultaneous tests on known phenylalanine-positive and phenylalanine-negative organisms?*

**3** *What do deamination and decarboxylation reactions have in common?*

5

**4** *In order to answer these questions you would need to have performed the decarboxylase test. Phenylalanine medium sometimes is prepared in broth form with a pH indicator similar to that used in decarboxylase medium so that increases in pH then can be detected by development of purple color.*

**a.** *When testing an organism for the ability to deaminate phenylalanine, would you expect to overlay this medium with mineral oil? Why or why not?*

**b.** *If you ran such a test, how would you know that an increase in pH was due to deamination and not to decarboxylation?*

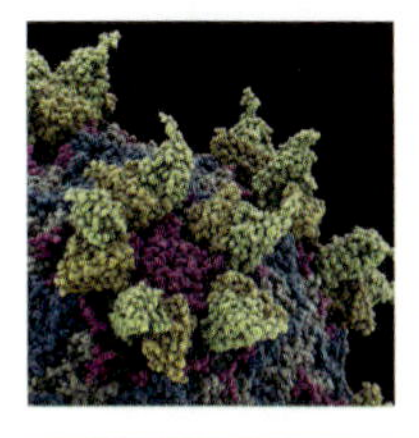

## Tests Detecting Hydrolytic Enzymes

Reactions that use water to split complex molecules are called **hydrolysis** (or **hydrolytic**) reactions. The enzymes required for these reactions are called hydrolytic enzymes. When the enzyme catalyzes its reaction inside the cell, it is referred to as **intracellular**. Enzymes secreted from the organism to catalyze reactions outside the cell are called **extracellular** enzymes, or **exoenzymes**. In this set of exercises you will be performing tests that identify the actions of both intracellular and extracellular types of enzymes.

Starch agar, DNase agar, tributyrin agar, and milk agar are plated media that identify extracellular enzymes capable of diffusing into the medium and producing a distinguishable halo of clearing around the bacterial growth. Nutrient gelatin detects gelatinase—an exoenzyme that liquefies the solid medium. The hydrolytic enzymes detected by the Urease and Bile Esculin tests are intracellular and produce identifiable color changes in the medium. Lastly, the PYR Test identifies a specific aminopeptidase that removes the amino acid on the amino end of a polypeptide. ■

# Starch Hydrolysis (Amylase Test)

EXERCISE 5-10

## Theory

Starch is a complex polysaccharide compound composed of α-glucose polymers in one of two forms—linear (amylose) and branched (amylopectin)—usually as a mixture with the branched configuration being predominant. The α-glucose molecules in both amylose and amylopectin are joined by 1,4-α-glycosidic (acetal) bonds (Fig. 5.38)[1]. The two forms differ in that amylopectin contains side chains covalently bonded to approximately every 30th glucose in the main chain by a 1,6-α-glycosidic linkage.

Starch is too large to pass through the bacterial cell membrane. Therefore, to be of metabolic value, it first must be digested extracellularly (or in the periplasmic space, in the case of Gram-negative bacteria) into smaller molecules (Fig. 5.38). There are three enzymes involved in starch digestion: **α-amylase, β-amylase**, and **α-1,6-glucosidase.** α-amylase breaks the starch polymer into the monosaccharide α-glucose and the disaccharide α-maltose, whereas β-amylase produces β-maltose. Neither amylase can disassemble the glucose subunits around branch points of amylopectin, so small, branched molecules called **limit dextran** remain after amylase activity. **α-(1,6)-glucosidase** breaks these down into short, linear oligosaccharides that α-amylase converts to α-glucose and α-maltose. Because the enzymes are secreted extracellularly, they are referred to as **exoenzymes.**

Starch agar is a simple plated medium of beef extract, soluble starch, and agar. When organisms that produce α-amylase, β-amylase, and oligo-1,6-glucosidase are cultivated on starch agar, they hydrolyze the starch in the area surrounding their growth. Because both starch and its sugar subunits are virtually invisible in the medium, the reagent iodine is used to detect the presence or absence of starch in the vicinity around the bacterial growth. Iodine reacts with starch and produces a blue or dark brown color; therefore, any microbial starch hydrolysis will be revealed as a clear zone surrounding the growth (Fig. 5.39 and Table 5-12).

TABLE 5-12 **Starch Hydrolysis Test Results and Interpretations**

| Result | Interpretation | Symbol |
|---|---|---|
| Clearing around growth | α-amylase, β-amylase, and/or α-1,6-glucosidase is present (or more simply, amylase is present) | + |
| No clearing around growth | α-amylase, β-amylase, and α-1,6-glucosidase are absent; (or more simply, amylase is absent or not detectable) | – |

[1] The carbons of organic molecules are numbered. A "1,4 linkage" is a covalent bond between the first carbon of one glucose and the fourth carbon of another glucose. A "1,6 linkage" is a covalent bond between the first and sixth carbons of two glucose molecules. The "α" and "β" refer to whether the two glucose molecules have the same orientation (α) or are "flipped" 180° relative to one another (β).

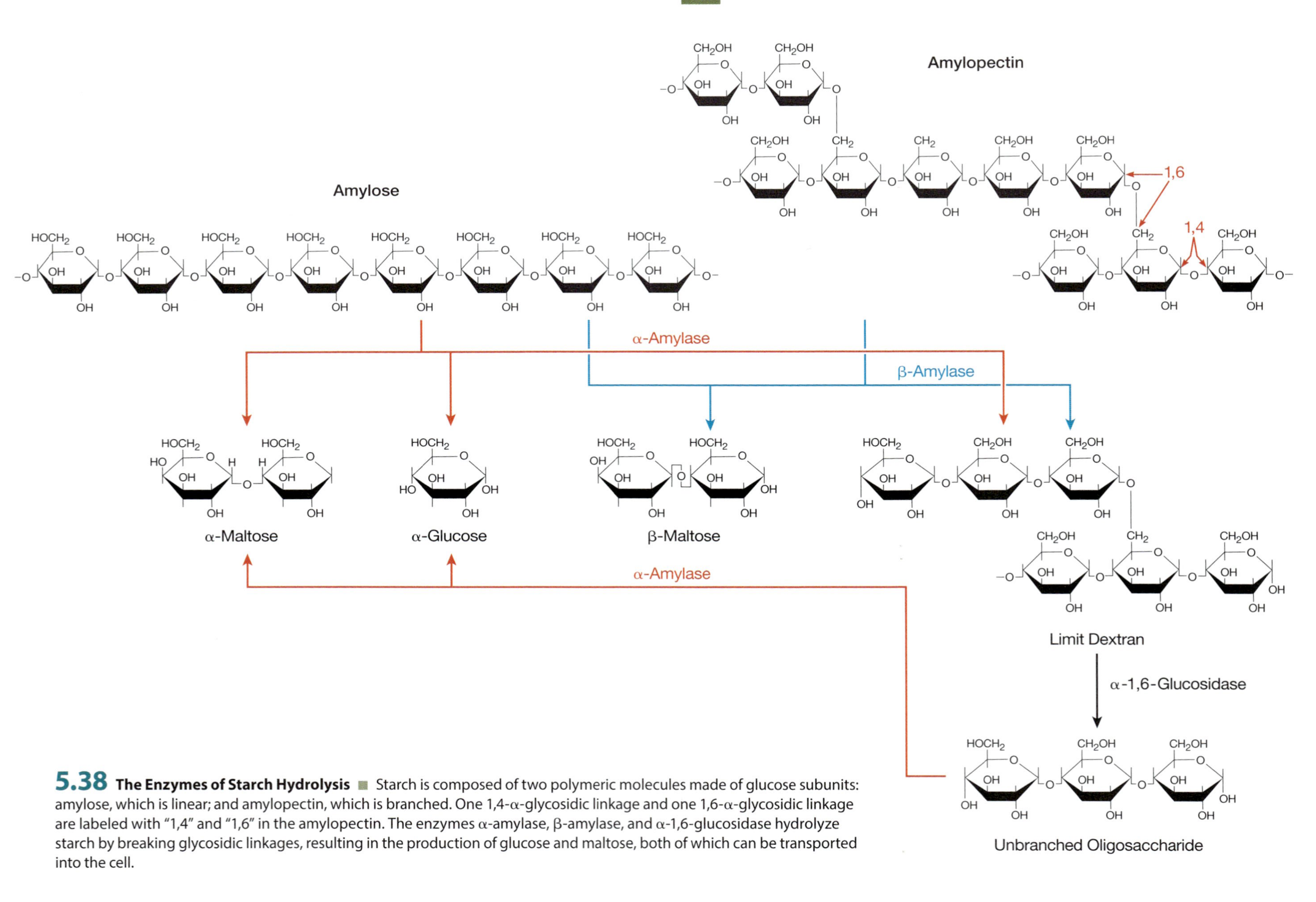


**5.38 The Enzymes of Starch Hydrolysis** ■ Starch is composed of two polymeric molecules made of glucose subunits: amylose, which is linear; and amylopectin, which is branched. One 1,4-α-glycosidic linkage and one 1,6-α-glycosidic linkage are labeled with "1,4" and "1,6" in the amylopectin. The enzymes α-amylase, β-amylase, and α-1,6-glucosidase hydrolyze starch by breaking glycosidic linkages, resulting in the production of glucose and maltose, both of which can be transported into the cell.

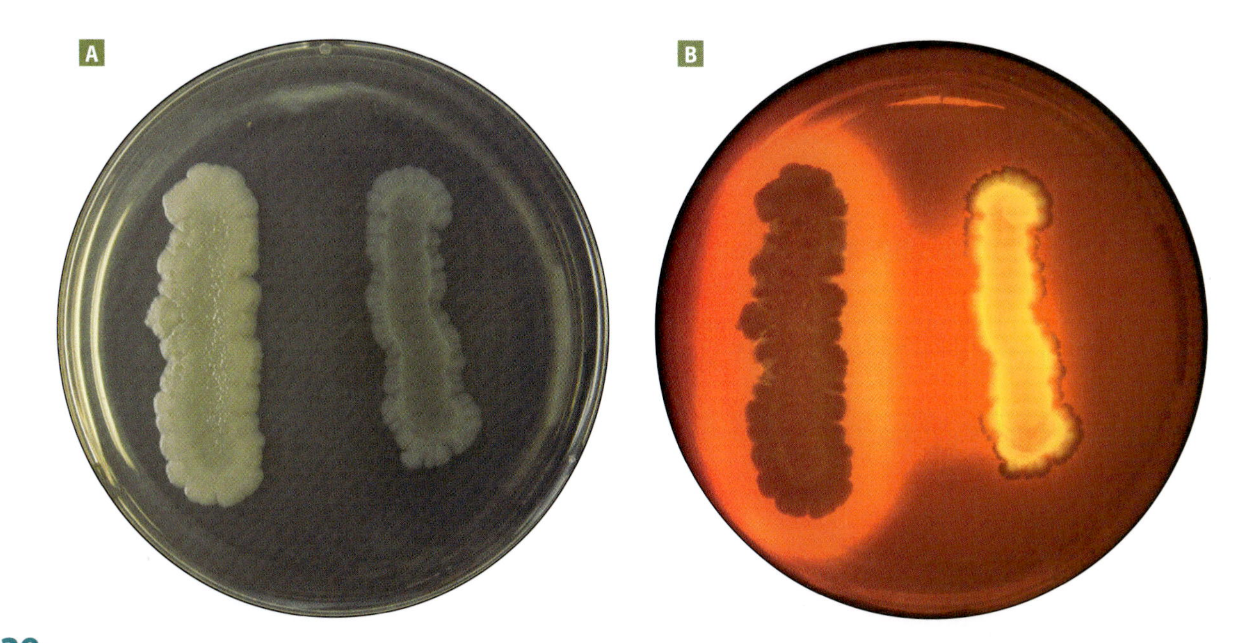


**5.39 Starch Hydrolysis Test Results** ■ (**A**) This is a starch agar plate inoculated with two organisms before iodine has been added. Notice the wavy margins of both organisms. (**B**) This is the same plate as in photo A after iodine has been added. The clearing in the medium around the organism on the left demonstrates a positive result for starch hydrolysis. The organism on the right, with no clearing, is negative. Note that the wavy margin of the negative organism produced a lighter region on the periphery of the growth that might be misinterpreted as clearing. To prevent recording a false positive, it is a good idea to establish the edges of growth of each organism by drawing a line on the plate's base prior to adding iodine.

## ■ Application

Starch agar originally was designed for cultivating *Neisseria*. It no longer is used for this, but with pH indicators, it is used to isolate and presumptively identify *Gardnerella vaginalis*. It aids in differentiating members of the genera *Corynebacterium, Clostridium, Bacillus, Bacteroides, Fusobacterium*, and *Enterococcus*, most of which have amylase-positive and amylase-negative species.

## ■ In This Exercise

You will inoculate a starch agar plate with two organisms. After incubation you will add a few drops of iodine to reveal clear areas surrounding the growth. Iodine will not color the agar where there is growth, only the area surrounding the growth. Before you add the iodine, make a line at the margins of growth on the plate so you do not confuse thinning growth at the edges with the halo produced by starch hydrolysis.

### ▼ Materials

#### Per Student

- □ Lab coat
- □ Gloves
- □ Chemical eye protection

#### Per Student Group

- □ One starch agar plate
- □ Gram iodine (from your Gram stain kit)
- □ Fresh broth cultures of these recommended organisms:
  - *Bacillus subtilis*
  - *Escherichia coli*

## ■ Medium Recipe

#### Starch Agar

| | |
|---|---|
| □ Beef extract | 3.0 g |
| □ Soluble starch | 10.0 g |
| □ Agar | 12.0 g |
| □ Distilled or deionized water | 1.0 L |

*pH 7.3–7.7 at 25°C*

## PROCEDURE

#### Lab One

1. Wear a lab coat, gloves, and chemical eye protection when performing this procedure.
2. Using a marking pen, divide the starch agar plate into three equal sectors. Be sure to mark on the bottom of the plate, not the lid.

3. Label the plate with the names of the organisms, your group name, and the date.
4. Spot inoculate two sectors with the test organisms near the edge of the plate.
5. Invert the plate and incubate it aerobically at $35 \pm 2°C$ for 24 hours. If incubated longer, you run the risk of one organism overgrowing the plate and obscuring the results of the other. Make arrangements to refrigerate your plate after 24 hours until your next lab meeting.
6. Save or dispose of the original culture tubes as directed by your instructor.

### Lab Two

1. Wear a lab coat, gloves, and chemical eye protection when performing this procedure.
2. Remove the plate from the incubator, and, before adding the iodine, note the location and appearance of the growth. Use a marking pen to indicate the margins of the growth.
3. Cover the growth and surrounding areas with Gram iodine. Do not flood the plate, but do put enough iodine on to colorize the agar completely. *Allow the iodine to soak in.* (Use a transfer pipette to remove any excess iodine.) Examine the areas surrounding the growth for clearing. ***Note:*** Usually the growth on the agar prevents contact between the starch and the iodine so no color reaction takes place at that point. Beginning students sometimes look at this lack of color change and judge it incorrectly as a positive result. Therefore, when examining the agar for clearing, look for a halo *around* the growth, not at the growth itself.
4. Record your results in the table provided on the data sheet, page 335.
5. Dispose of the plate in an appropriate autoclave container when finished.

5

## References


Collins, C. H., Patricia M. Lyne, and J. M. Grange. Page 117 in *Collins and Lyne's Microbiological Methods,* 7th ed. Oxford, Boston: Butterworth-Heinemann, 1995.

Lányi, B. Page 55 in *Methods in Microbiology*, Vol. 19. R. R. Colwell and R. Grigorova, eds. New York: Academic Press, 1987.

MacFaddin, Jean F. Page 412 in *Biochemical Tests for Identification of Medical Bacteria*, 2nd ed. Philadelphia: Lippincott Williams & Wilkins, 2000.

Moat, Albert G., John W. Foster, and Michael P. Spector. Pages 403–407 in *Microbial Physiology*, 4th ed. New York: Wiley-Liss, 2002.

Smibert, Robert M. and Noel R. Krieg. Page 630 in *Methods for General and Molecular Bacteriology*. Philipp Gerhardt, R. G. E. Murray, Willis A. Wood, and Noel R. Krieg, eds. Washington, DC: American Society for Microbiology, 1994.

Zimbro, Mary Jo and David A. Power, eds. *Difco™ and BBL™ Manual—Manual of Microbiological Culture Media.* Sparks, MD: Becton Dickinson and Co., 2003.

Name ______________________________

Date ______________________________

Lab Section ______________________________

I was present and performed this exercise (initials) ______________

DATA SHEET 5-10

## Starch Hydrolysis (Amylase Test)

**1** Using Table 5-12, page 331, as a guide, enter your results in the table below.

| Organism | Result | (+/−) | **Interpretation** (include enzyme) |
|---|---|---|---|
| Control sector | | | |
| | | | |
| | | | |

## QUESTIONS

**1** *Suppose you had poured iodine on your plate and noticed clearings in the uninoculated area, as well as around both of your transferred cultures.*

**a.** *What are some possible explanations for this occurrence?*

**b.** *Was integrity of the test compromised?*

**c.** *What measures could be taken to avoid this problem in the future?*

**2** *How would you expect both positive and negative results to be affected if you were to add glucose to the medium?*

**3** *In many tests it is acceptable to read a positive result before the incubation time is completed. Why is this not the case with starch agar?*



**4** *Suppose you could selectively prevent production of α-amylase or oligo-1,6-glucosidase in an organism that normally hydrolyzes starch. Which enzyme would the organism miss the most?*

EXERCISE 5-11

# DNA Hydrolysis (DNase Test)

## Theory

An enzyme that catalyzes the depolymerization of DNA into small fragments (oligonucleotides) or single nucleotides is called a **deoxyribonuclease**, or **DNase** (Fig. 5.40). DNase is an exoenzyme; that is, an enzyme that is secreted by a cell and acts on the substrate extracellularly. Extracellular digestion typically allows utilization of a macromolecule too large to be transported into the cell. Ability to produce this enzyme can be determined by culturing and observing an organism on a DNase test agar plate.

DNase test agar contains peptides derived from soybean and casein that serve as carbon and nitrogen sources, sodium chloride for osmotic balance, DNA as the substrate, and methyl green dye as an indicator of DNase activity. DNA also serves as an additional carbon and nitrogen source, or its subunits can be used for DNA synthesis by the growing organism. The dye forms a complex with polymerized DNA and gives the agar a blue-green color, but no complex is formed with nucleotides. Therefore, clearing that develops around bacterial growth on the medium is an indication of DNase activity (DNA hydrolysis) and is considered a positive result (Fig. 5.41).

## Application

DNase test agar is used to distinguish *Serratia* species (positive) from *Enterobacter* species (negative), *Moraxella catarrhalis* (positive) from *Neisseria* species (negative), and *Staphylococcus aureus* (positive) from other *Staphylococcus* species (negative).

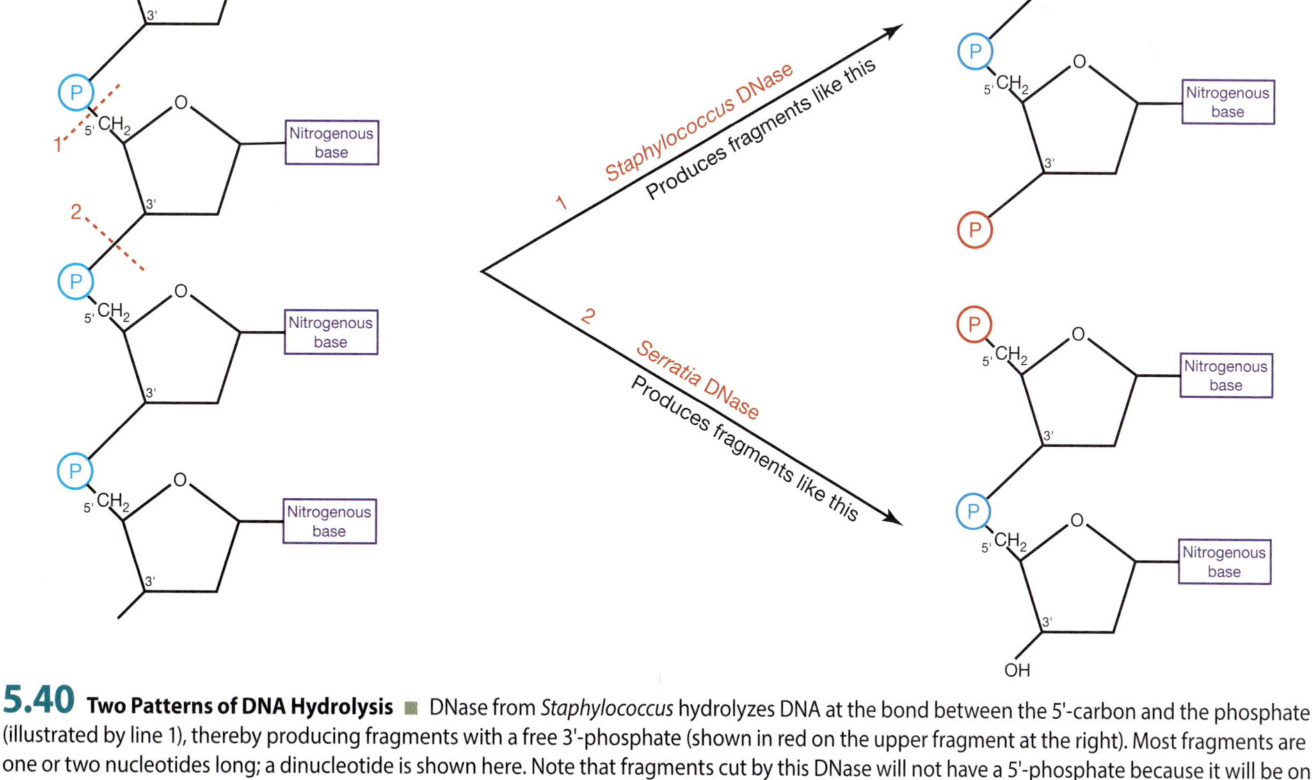


**5.40 Two Patterns of DNA Hydrolysis** ■ DNase from *Staphylococcus* hydrolyzes DNA at the bond between the 5'-carbon and the phosphate (illustrated by line 1), thereby producing fragments with a free 3'-phosphate (shown in red on the upper fragment at the right). Most fragments are one or two nucleotides long; a dinucleotide is shown here. Note that fragments cut by this DNase will not have a 5'-phosphate because it will be on the 3'-carbon of the next fragment. *Serratia* DNase cleaves the bond between the phosphate and the 3'-carbon (illustrated by line 2) and produces fragments with free 5'-phosphates (shown in red on the lower fragment at the right). Most fragments are two to four nucleotides in length; a dinucleotide is shown here. Note that the 3' end on each fragment cut this way would have a hydroxyl group, not a phosphate.

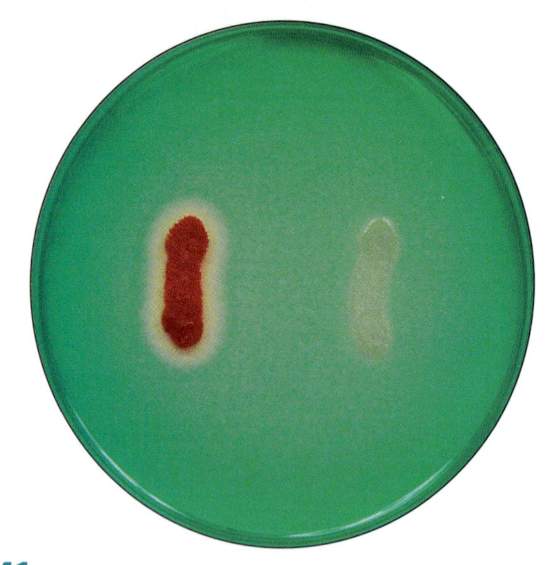

**5.41 DNA Hydrolysis Test Results on DNase Agar with Methyl Green** ■ This DNase test agar plate was inoculated with a DNase-positive organism (left—note the clearing of the agar) and a DNase-negative organism (right).

## In This Exercise

You will inoculate one DNase plate with two of the organisms the test was designed to differentiate. For best results, limit incubation time to 24 hours. Depending on the age of the plates and the length of incubation, the clearing around a DNase-positive organism may be subtle and difficult to see. When reading the plate, it might be helpful to look through it while holding it several inches above a white piece of paper. Use Table 5-13 to assist in interpretation.

### ▼ Materials

#### Per Student

- □ Lab coat
- □ Gloves
- □ Chemical eye protection

#### Per Student Group

- □ One DNase test agar plate
- □ Fresh broth cultures of these recommended organisms:
  - *Serratia marcescens*
  - *Enterobacter aerogenes*

## Medium Recipe

#### DNase Test Agar with Methyl Green

| | |
|---|---|
| □ Tryptose | 20.0 g |
| □ Deoxyribonucleic acid | 2.0 g |
| □ Sodium chloride | 5.0 g |
| □ Agar | 15.0 g |
| □ Methyl green | 0.05 g |
| □ Distilled or deionized water | 1.0 L |

*pH 7.1–7.5 at 25°C*

TABLE **5-13** **DNA Hydrolysis Test Results and Interpretations**

| Result | Interpretation | Symbol |
|---|---|---|
| Clearing in agar (loss of green color) around growth | DNase is present | + |
| No clearing in agar around growth | DNase is absent or not detectable | − |

## PROCEDURE

#### Lab One

1. Wear a lab coat, gloves, and chemical eye protection when performing this procedure.
2. Using a marking pen, divide the DNase test agar plate into three sectors. Be sure to mark on the bottom of the plate, not the lid.
3. Label the plate with the names of the organisms, your group name, and the date.
4. Spot inoculate two sectors with the test organisms and leave the third sector as a control.
5. Invert the plate and incubate it aerobically at 35 ± 2°C for 24 hours. If incubated longer, you run the risk of one organism overgrowing the plate and obscuring the results of the other. Make arrangements to refrigerate your plate after 24 hours until your next lab meeting.
6. Save or dispose of the original culture tubes as directed by your instructor.

#### Lab Two

1. Remove the plates from the incubator (or refrigerator) and examine the agar for clearing around the bacterial growth.
2. Record your results in the table provided on the data sheet, page 339.
3. Dispose of all plates in an appropriate autoclave container.

### References


Collins, C. H., Patricia M. Lyne, and J. M. Grange. Page 114 in *Collins and Lyne's Microbiological Methods*, 7th ed. Oxford, Boston: Butterworth-Heinemann, 1995.

Delost, Maria Dannessa. Page 111 in *Introduction to Diagnostic Microbiology*. St. Louis, MO: Mosby, 1997.

Lányi, B. Page 33 in *Methods in Microbiology*, Vol. 19. R. R. Colwell and R. Grigorova, eds. New York: Academic Press, 1987.

MacFaddin, Jean F. Page 136 in *Biochemical Tests for Identification of Medical Bacteria*, 3rd ed. Philadelphia: Lippincott Williams & Wilkins, 2000.

Tille, Patricia M. Page 205 in *Bailey & Scott's Diagnostic Microbiology*, 13th ed. St. Louis, MO: Mosby, 2014.

Zimbro, Mary Jo and David A. Power, eds. Page 170 in *Difco™ and BBL™ Manual—Manual of Microbiological Culture Media*. Sparks, MD: Becton Dickinson and Co., 2003.

Name ______________________________

Date ______________________________

Lab Section ______________________________

I was present and performed this exercise (initials) ______________

# DATA SHEET 5-11

## DNA Hydrolysis (DNase Test)

### OBSERVATIONS AND INTERPRETATIONS

**1** Using Table 5-13, page 338, as a guide, record your DNase results in the table below.

| Organism | Result | (+/−) | Interpretation (include enzyme) |
|---|---|---|---|
| Control sector | | | |
| | | | |
| | | | |

### QUESTIONS

**1** *In Theory on page 337 of this exercise, the disassembly of DNA was described as a "depolymerization." What other terms apply to the process? Choose one from each pair: anabolic/catabolic and dehydration/hydrolysis.*

______________________________

**2** *A positive result for the DNA hydrolysis test does not distinguish between* Staphylococcus *DNase and the DNase produced by* Serratia. *If you had the expertise to correct this weakness in the system, would you be improving the test's sensitivity or its specificity?*

______________________________

**3** *Suggest a reason why this test is read after only 24 hours while other tests (e.g., gelatinase test) may take a week. What would be the likely consequences of incubating DNase agar for a week?*

______________________________

 *Why is the uninoculated control sector relatively unnecessary in this test?*

**5** *Why is it advisable to use a positive control along with organisms that you are testing?*

5

**6** *In what ways can a DNase-positive organism use the products of DNA hydrolysis? (**Hint:** examine Figs. A.1 and A.5.)*

# Lipid Hydrolysis (Lipase Test)

EXERCISE 5-12

## Theory

The word **lipid** is generally used to describe all types of fats. Bacterial enzymes that hydrolyze lipids fall into a generic category called **lipases.** Many lipases are exoenzymes, that is, they are secreted by the organism to digest lipids outside the cell. Similar in function to other types of exoenzymes (Exercises 5-10, 5-11, 5-13, and 5-14) lipases are necessary because intact lipids are macromolecules and too large to be transported into the cell. One purpose of this exercise is to demonstrate that bacteria can be differentiated based on their ability to produce and secrete lipases.

Simple fats, known as **triglycerides** or **triacylglycerols,** are composed of glycerol and three long-chain fatty acids. After lipid hydrolysis (lipolysis), glycerol can be phosphorylated and oxidized in its conversion to dihydroxyacetone phosphate, which is an intermediate of glycolysis (Figs. 5.42 and A.5). The process costs one ATP, but reduces one $NAD^+$ to $NADH+H^+$. The fatty acids are catabolized by a process called **β-oxidation** in which two carbon fragments are sequentially removed and combined with Coenzyme A to produce acetyl-CoA, one NADH and one $FADH_2$. Acetyl-CoA can then enter the citric acid cycle. Alternatively, glycerol and fatty acids may be used in anabolic pathways.

A variety of simple fats can be used for the lipid hydrolysis (lipase) test, but tributyrin oil is a common choice because it is the simplest triglyceride found in natural fats. Tributyrin agar is prepared as an emulsion that makes the agar appear opaque. When the plate is inoculated with a lipase-positive organism, clear zones will appear around the growth as evidence of lipolytic activity (Fig. 5.43). If no clear zones appear, the organism is lipase-negative. Table 5-14 summarizes the results and interpretations.

TABLE 5-14 **Lipid Hydrolysis Test Results and Interpretations**

| Result | Interpretation | Symbol |
|---|---|---|
| Clearing in agar around growth | Lipase is present | + |
| No clearing in agar around growth | Lipase is absent or not detectable | − |

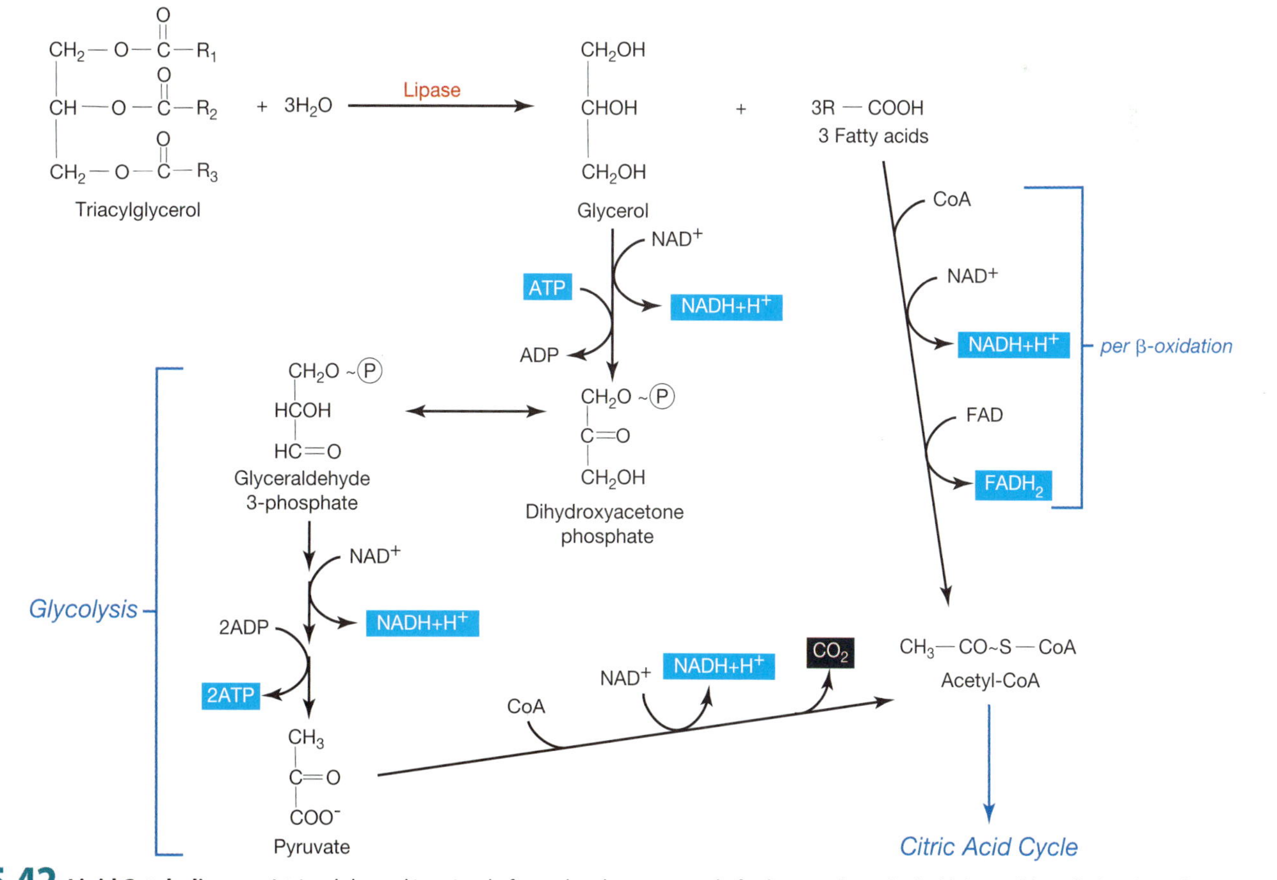


**5.42 Lipid Catabolism** ■ A triacylglycerol is a simple fat molecule, composed of a three-carbon alcohol (glycerol) bonded to three fatty acid chains (represented by $R_1$, $R_2$, and $R_3$ in this diagram). The products of lipid catabolism can be used in glycolysis (glycerol) and the citric acid cycle (acetyl-CoA).

**5.43 Lipid Hydrolysis Test Results on Tributyrin Agar** ■ This tributyrin agar plate was inoculated with a lipase-positive organism above (note the clearing in the agar) and a lipase-negative organism below.

## Application

The lipase test is used to detect and enumerate lipolytic bacteria, especially in high-fat dairy products. A variety of other lipid substrates, including corn oil, olive oil, and soybean oil, are used to detect differential characteristics among members of *Enterobacteriaceae*, *Clostridium*, *Staphylococcus*, and *Neisseria*. Several fungal species also demonstrate lipolytic ability.

5

## In This Exercise

You will inoculate a tributyrin agar plate with two organisms, one of which will produce a clearing in the medium surrounding the growth.

## Materials

### Per Student

- □ Lab coat
- □ Gloves
- □ Chemical eye protection

### Per Student Group

- □ One tributyrin agar plate
- □ Fresh broth cultures of these recommended organisms:
  - *Bacillus subtilis* or *B. cereus*
  - *Escherichia coli*

## Medium Recipe

### Tributyrin Agar

| | |
|---|---|
| □ Beef extract | 3.0 g |
| □ Peptone | 5.0 g |
| □ Agar | 15.0 g |
| □ Tributyrin oil | 10.0 mL |
| □ Distilled or deionized water | 1.0 L |

*pH 5.8–6.2 at 25°C*

## PROCEDURE

### Lab One

1. Wear a lab coat, gloves, and chemical eye protection when performing this procedure.
2. Using a marking pen, divide the plate into three equal sectors. Be sure to mark on the bottom of the plate, not on the lid.
3. Label the plate with the names of the organisms, your group name, and the date.
4. Spot inoculate two sectors with the test organisms, leaving the third sector uninoculated as a control.
5. Invert the plate and incubate it aerobically at 35 ± 2°C for 24 hours. If incubated longer, you run the risk of one organism overgrowing the plate and obscuring the results of the other. Make arrangements to refrigerate your plate after 24 hours until your next lab meeting.
6. Save or dispose of the original culture tubes as directed by your instructor.

### Lab Two

1. Examine the plates for clearing around the bacterial growth and record your results in the table provided on the data sheet, page 343. If no clearing has developed, reincubate for 24 to 48 hours. (This shouldn't happen with the organisms listed, but is something you should do if working on your unknowns.)
2. Dispose of all plates in an appropriate autoclave container when finished.

### *References*


Collins, C. H., Patricia M. Lyne, and J. M. Grange. Page 114 in *Collins and Lyne's Microbiological Methods,* 7th ed. Oxford, Boston: Butterworth-Heinemann, 1995.

Haas, Michael J. Chap. 15 in *Compendium of Methods for the Microbiological Examination of Foods,* 4th ed. Frances Pouch Downes and Keith Ito, eds. Washington, DC: American Public Health Association, 2001.

Knapp, Joan S. and Roselyn J. Rice. Page 335 in *Manual of Clinical Microbiology,* 6th ed. Patrick R. Murray, Ellen Jo Baron, Michael A. Pfaller, Fred C. Tenover, and Robert H. Yolken, eds. Washington, DC: ASM Press, 1995.

MacFaddin, Jean F. Page 286 in *Biochemical Tests for Identification of Medical Bacteria,* 3rd ed. Philadelphia: Lippincott Williams & Wilkins, 2000.

Zimbro, Mary Jo and David A. Power, eds. Page 521 in *Difco™ and BBL™ Manual—Manual of Microbiological Culture Media.* Sparks, MD: Becton Dickinson and Co., 2003.

Name ______________________________

Date ______________________________

Lab Section ______________________________

I was present and performed this exercise (initials) ______________________________

# DATA SHEET 5-12

## Lipid Hydrolysis (Lipase Test)

### OBSERVATIONS AND INTERPRETATIONS

**1** Using Table 5-14, page 341, as a guide, record your results and interpretations in the table below.

| Organism | Result | (+/−) | **Interpretation** (include enzyme) |
|---|---|---|---|
| Control sector | | | |
| | | | |
| | | | |

## QUESTIONS

**1** *Many organisms possessing many different lipases produce positive results on tributyrin agar. Is the inability of this medium to distinguish between these different enzymes a weakness in its specificity or its sensitivity? Explain your answer.*

**2** *Why do you think tributyrin agar is prepared in the blender as an emulsion rather than simply stirred as a solution?*

**3** *Tributyrin agar has a shelf life of only a few days before it loses its opacity.*

**a.** *With this in mind, explain the importance of positive and negative controls in this test.*

**b.** *How would expired tributyrin agar affect the results of lipase-positive and lipase-negative organisms?*

5

**4** *Imagine this situation: Species #1 and Species #2 are unrelated, but each produces an enzyme capable of hydrolyzing soybean oil. How can this be reconciled with the fact that enzymes are highly specific to their substrates?*

# Casein Hydrolysis (Casease Test)



## Theory

Many bacteria require proteins as a source of amino acids for synthesis of their own proteins and other compounds. The amino acids can also be used as an energy source. Some bacteria have the ability to produce and secrete enzymes into the environment that catalyze the hydrolysis of large proteins to smaller peptides or individual amino acids, thereby enabling their uptake across the membrane (Fig. 5.44).

**Casease** is a specific **proteolytic** enzyme that some bacteria produce to hydrolyze the milk protein **casein**, the molecule that gives milk its white color. When broken down into smaller fragments, the ordinarily white casein loses its opacity and becomes clear.

The presence of casease can be detected easily with the test medium milk agar (Fig. 5.45). Milk agar is an undefined medium containing pancreatic digest of casein, yeast extract, dextrose, and powdered milk. When plated milk agar is inoculated with a casease-positive organism, secreted casease will diffuse into the medium around the colonies and create a zone of clearing where the casein has been hydrolyzed. Casease-negative organisms do not secrete casease and do not produce clear zones around the growth.

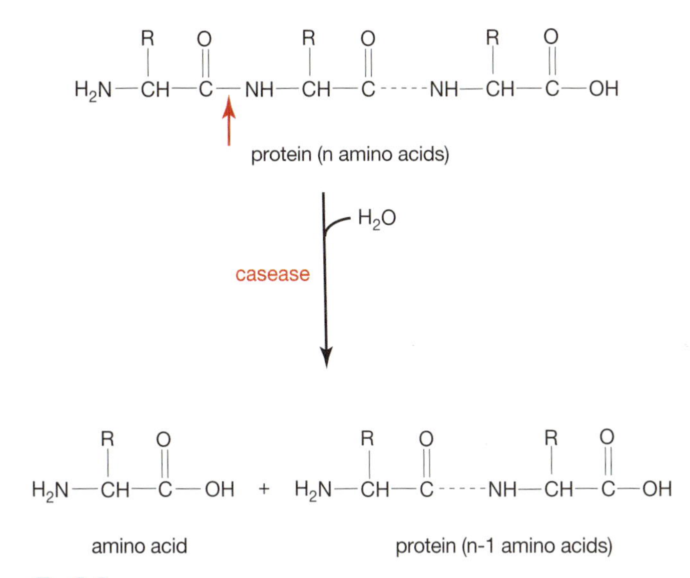


**5-44 Casein Hydrolysis** ■ Hydrolysis of any protein occurs by breaking peptide bonds (red arrow) between adjacent amino acids to produce short peptides or individual amino acids.

## Application

The casein hydrolysis test is used for the cultivation and differentiation of bacteria that produce the enzyme casease.

## In This Exercise

You will inoculate a single milk agar plate with a casease-positive and a casease-negative organism. Use Table 5-15 as a guide when making your interpretations.

**TABLE 5-15 Casein Hydrolysis Test Results and Interpretations**

| Result | Interpretation | Symbol |
|---|---|---|
| Clearing in agar | Casease is present | + |
| No clearing in agar | Casease is absent or not detectable | − |

**5.45 Casein Hydrolysis Test Results on Milk Agar** ■ This milk agar plate was inoculated with a casease-positive organism above and a casease-negative organism below. The clearing around the casease-positive organism is dark because of the background behind the plate, not because it should look that color.

## ▼ Materials

### Per Student

- □ Lab coat
- □ Gloves
- □ Chemical eye protection

### Per Student Group

- □ One milk agar plate
- □ Fresh broth cultures of these recommended organisms:
  - *Bacillus subtilis* or *B. cereus*
  - *Escherichia coli*

## ■ Medium Recipe

### Milk Agar

| | |
|---|---|
| □ Pancreatic digest of casein | 5.0 g |
| □ Yeast extract | 2.5 g |
| □ Powdered nonfat milk | 100.0 g |
| □ Glucose | 1.0 g |
| □ Agar | 15.0 g |
| □ Distilled or deionized water | 1.1 L |

*pH 6.9–7.1 at 25°C*

5

## PROCEDURE

### Lab One

1. Wear a lab coat, gloves, and chemical eye protection when performing this procedure.
2. Using a marking pen, divide the plate into three equal sectors. Be sure to mark on the bottom of the plate, not the lid.
3. Label the plate with the names of the organisms, your group name, and the date.
4. Spot inoculate two sectors with the test organisms and leave the third sector uninoculated as a control.
5. Invert the plate and incubate it aerobically at $35 \pm 2$°C for 24 hours. ***Note***: *Bacillus* species grow very well on milk agar. If you cannot read the plate in 24 hours, move it to the refrigerator or the *Bacillus* may have overgrown your entire plate by the time you get to it.
6. Save or dispose of the original culture tubes as directed by your instructor.

### Lab Two

1. Examine the plates for clearing around the bacterial growth.
2. Record your results in the table provided on the data sheet, page 347.
3. Dispose of all plates in an appropriate autoclave container when finished.

### *References*


Atlas, Ronald M. Page 1563 in *Handbook of Microbiological Media*, 3rd ed. Boca Raton, FL: CRC Press LLC, 2004.

Holt, John G., ed. *Bergey's Manual of Determinative Bacteriology*, 9th ed. Baltimore: Lippincott Williams & Wilkins, 1994.

Smibert, Robert M. and Noel R. Krieg. Page 613 in *Methods for General and Molecular Bacteriology*. Philipp Gerhardt, R. G. E. Murray, Willis A. Wood, and Noel R. Krieg, eds. Washington, DC: American Society for Microbiology, 1994.

Zimbro, Mary Jo and David A. Power, eds. Page 364 in *Difco™ and BBL™ Manual—Manual of Microbiological Culture Media*. Sparks, MD: Becton Dickinson and Co., 2003.

Name ______________________________

Date ______________________________

Lab Section ______________________________

I was present and performed this exercise (initials) ______________

# DATA SHEET 5-13

## Casein Hydrolysis (Casease Test)

**1** Using Table 5-15, page 345, as a guide, record your results and interpretations in the table below.

| Organism | Result | (+/−) | Interpretation (include enzyme) |
|---|---|---|---|
| Control sector | | | |
| | | | |
| | | | |

## QUESTIONS

**1** *What does the enzyme casease have in common with amylase (Exercise 5-10)? How are they different? (If you didn't do Exercise 5-10, you can answer the same question with respect to the enzymes in Exercises 5-11, 5-12, or 5-14 if you did those.)*

**2** *How do we know that casease is an exoenzyme and not a cytoplasmic enzyme?*

**3** *Consider incubation time for this test.*

**a.** *Is it acceptable to read a positive test before the incubation time is completed? Explain.*

**b.** *How about an early negative result? Explain.*

**4** *Why is the uninoculated control relatively unnecessary in this test?*

5

**5** *Why is it advisable to use a positive control along with organisms that you are testing?*

# Gelatin Hydrolysis (Gelatinase Test)

EXERCISE 5-14

## Theory

Gelatin is a protein derived from collagen—a component of vertebrate connective tissue. **Gelatinases** comprise a family of extracellular enzymes produced and secreted by some microorganisms to hydrolyze gelatin. Subsequently, the cell can absorb individual amino acids and use them for metabolic purposes. Bacterial hydrolysis of gelatin occurs in two sequential reactions, as shown in Figure 5.46.

The presence of gelatinases can be detected using nutrient gelatin, a simple test medium composed of gelatin, peptone, and beef extract. Nutrient gelatin differs from most other solid media in that the solidifying agent (gelatin) is also the substrate for enzymatic activity. Consequently, when a tube of nutrient gelatin is stab inoculated with a gelatinase-positive organism, secreted gelatinase (or gelatinases) will liquefy the medium. Gelatinase-negative organisms do not secrete the enzyme and do not liquefy the medium (Figs. 5.47 and 5.48). A 7-day incubation period is usually sufficient to see liquefaction of the medium. However, gelatinase activity is very slow in some organisms. All tubes still negative after 7 days should be incubated an additional 7 days.

A slight disadvantage of nutrient gelatin is that it melts at 28°C (82°F). Therefore, inoculated stabs are typically incubated at 25°C along with an uninoculated control to verify that any liquefaction is not temperature related.

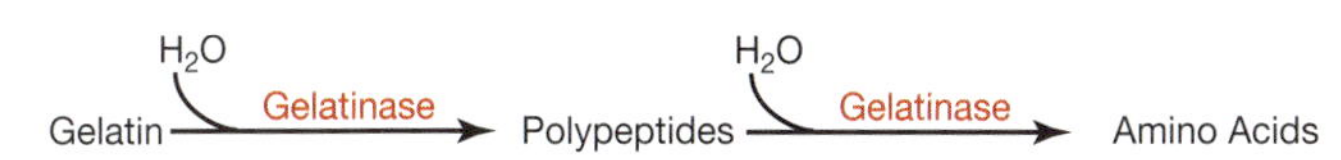


**5.46 Gelatin Hydrolysis** ■ Gelatin is hydrolyzed by the gelatinase family of enzymes.

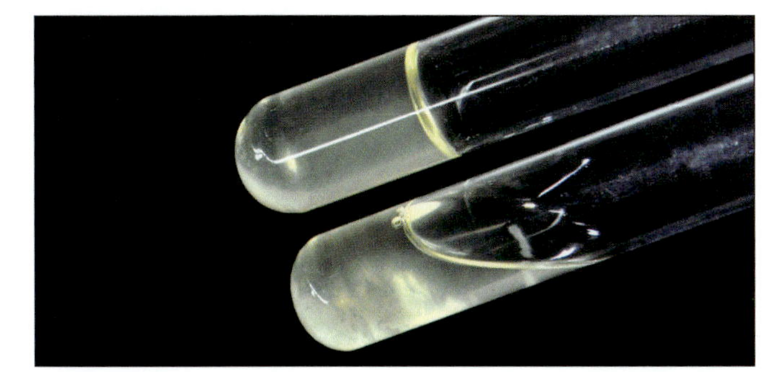

**5.47 Gelatin Hydrolysis Test Results in Nutrient Gelatin** ■ A gelatinase-negative organism was inoculated in the top tube; a gelatinase-positive organism was inoculated in the lower one. Make sure an uninoculated control (not shown) is solid before reading any inoculated tubes to ensure any liquefaction is due to gelatin hydrolysis and not to melting. It is not necessary to tip the tubes very much to see if the gelatin has been liquefied.

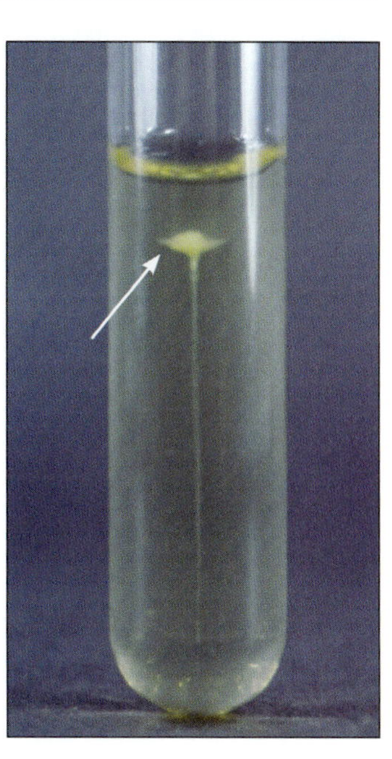

**5.48 Crateriform Liquefaction** ■ This form of gelatin liquefaction also may be of diagnostic use because not all gelatinase-positive microbes liquefy the gelatin completely. Shown here is an organism liquefying the gelatin in the shape of a crater (arrow).

## Application

This gelatin hydrolysis test is used to determine the ability of a microbe to produce gelatinases. *Staphylococcus aureus,* which is gelatinase-positive, can be differentiated from *S. epidermidis*. *Serratia* and *Proteus* species are gelatinase-positive members of *Enterobacteriaceae,* whereas most others in the family are negative. *Bacillus anthracis, B. cereus*, and several other *Bacillus* species are gelatinase-positive, as are *Clostridium tetani* and *C. perfringens*.

## In This Exercise

You will inoculate two nutrient gelatin tubes and incubate them for up to a week. Gelatin liquefaction is a positive result but is difficult (next to impossible) to differentiate from gelatin that has melted as a result of temperature. Therefore, be sure to incubate an uninoculated control tube along with the others. If the control melts as a result of temperature, refrigerate all tubes until it resolidifies. Use Table 5-16 as a guide to interpretation.

TABLE **5-16 Gelatin Hydrolysis Test Results and Interpretations**

| Result | Interpretation | Symbol |
|---|---|---|
| Gelatin is liquid (control is solid) | Gelatinase is present | + |
| Gelatin is solid | Gelatinase is absent or not detectable | – |

## ▼ Materials

### Per Student

- □ Lab coat
- □ Gloves
- □ Chemical eye protection

### Per Student Group

- □ Three nutrient gelatin stabs
- □ Ice bath or refrigerator
- □ Fresh agar slant cultures of these recommended organisms:
  - *Bacillus subtilis* or *B. cereus*
  - *Escherichia coli*

## ■ Medium Recipe

### Nutrient Gelatin

| | |
|---|---|
| □ Beef extract | 3.0 g |
| □ Peptone | 5.0 g |
| □ Gelatin | 120.0 g |
| □ Distilled or deionized water | 1.0 L |

*pH 6.6–7.0 at 25°C*

5

## PROCEDURE

### Lab One

1. Wear a lab coat, gloves, and chemical eye protection when performing this procedure.
2. Obtain three nutrient gelatin stabs. Label each of two tubes with the name of the organism, your group name, and the date. Label the third tube "control."
3. Stab inoculate two tubes with heavy inocula of the test organisms. Do not inoculate the control.
4. Incubate all tubes at 25°C for up to 1 week. A higher incubation temperature (e.g., 35°C) may be used, but an uninoculated control is an absolute necessity because all the tubes will melt.
5. Save or dispose of the original culture tubes as directed by your instructor.

### Lab Two

1. Examine the control tube. If the gelatin is solid, the test can be read. If it is liquefied, place all tubes in the ice bath or refrigerator until the control has resolidified.
2. When the control has solidified (it shouldn't take more than 15–20 minutes, but it must become solid before proceeding), examine the inoculated media for gelatin liquefaction. Typically, the tubes do not need to be tipped as much as is shown in Figure 5.47. Regardless, be careful not to spill liquefied gelatin out the top when tipping the tubes. If this happens, clean up the culture spill with disinfectant.
3. Record your results in the table provided on the data sheet, page 351.
4. Dispose of all tubes in an appropriate autoclave container when finished.

### References


Collins, C. H., Patricia M. Lyne, and J. M. Grange. Page 112 in *Collins and Lyne's Microbiological Methods*, 7th ed. Oxford, Boston: Butterworth-Heinemann, 1995.

Lányi, B. Page 44 in *Methods in Microbiology*, Vol. 19. R. R. Colwell and R. Grigorova, eds. New York: Academic Press, 1987.

MacFaddin, Jean F. Page 128 in *Biochemical Tests for Identification of Medical Bacteria*, 3rd ed. Philadelphia: Lippincott Williams & Wilkins, 2000.

Smibert, Robert M. and Noel R. Krieg. Page 617 in *Methods for General and Molecular Bacteriology*. Philipp Gerhardt, R. G. E. Murray, Willis A. Wood, and Noel R. Krieg, eds. Washington, DC: American Society for Microbiology, 1994.

Tille, Patricia M. Page 208 in *Bailey & Scott's Diagnostic Microbiology*, 13th ed. St. Louis, MO: Mosby, 2014.

Zimbro, Mary Jo and David A. Power, eds. Page 408 in *Difco™ and BBL™ Manual—Manual of Microbiological Culture Media*. Sparks, MD: Becton Dickinson and Co., 2003.

Name ______________________________

Date ______________________________

Lab Section ______________________________

I was present and performed this exercise (initials) ______________________________

DATA SHEET

5-14

# Gelatin Hydrolysis (Gelatinase Test)

**1** Using Table 5-16, page 349, as a guide, record your results and interpretations in the table below.

| Organism | Result | (+/−) | **Interpretation** (include enzyme) |
|---|---|---|---|
| Control sector | | | |
| | | | |
| | | | |

## QUESTIONS

**1** *Some microbiologists recommend incubating this medium at 37°C, along with an uninoculated control, and then transferring all tubes to the refrigerator prior to reading them. Why might this be the preferred technique in some situations? What potential problems can you see with this method?*

**2** *If the control is solid and an inoculated tube is liquid, is it acceptable to read the result before the complete incubation time (7 days) has elapsed? Explain.*

**3** *If the control is solid and an uninoculated tube is also solid, is it acceptable to read the result before the complete incubation time has elapsed? Explain.*



**4** *In the Theory, we mention that some organisms are slow gelatin liquefiers. What about an organism might make it a slow gelatin liquefier?*

**5** *Suppose that after 7 days a tube inoculated with a slow liquefier shows no evidence of liquefaction. Is this due to poor sensitivity or poor specificity of the medium? Explain.*

# Urea Hydrolysis (Urease Test)

EXERCISE **5-15**

## Theory

Decarboxylation of certain amino acids produces urea, which is the primary nitrogenous waste in the urine of many land animals (including mammals). It can be hydrolyzed to ammonia and carbon dioxide by bacteria possessing the enzyme **urease** (Fig. 5.49). Urea hydrolysis provides nitrogen in a usable form (ammonia) and it also acts as a virulence factor for some pathogens (e.g., *Helicobacter pylori*) by counteracting acid in the environment. *Streptococcus salivarius*, an oral commensal, also produces urease.

Many enteric bacteria possess the ability to metabolize urea, and the urease genes are activated when a usable nitrogen source is absent or urea is present. The original urease test used urea agar, which was formulated to differentiate enteric organisms capable of rapid urea hydrolysis ("rapid urease-positive bacteria," such as *Proteus*, *Morganella morganii*, and some *Providencia stuartii* strains) from slower urease-positive and urease-negative bacteria. A pH indicator was used to track the rising pH due to accumulation of ammonia. Rapid urease-positive organisms produce a positive result within a day; slower urease-positive organisms may take several days to produce a readable result.

In this exercise you will be using urea broth, which differs from urea agar in two important ways. First, its only nutrient source other than urea is a trace (0.0001%) of yeast extract. Second, it contains buffers strong enough to inhibit alkalinization of the medium by all but the rapid urease-positive organisms mentioned previously. Phenol red, which is yellow or orange below pH 8.4 and red or pink above, is included to expose any increase in pH. Pink color in the medium in less than 24 hours indicates a rapid urease-positive organism. Orange or yellow is negative (Fig. 5.50 and Table 5-17).

TABLE **5-17** **Urea Hydrolysis Test Results and Interpretations**

| Result | Interpretation | Symbol |
|---|---|---|
| Pink | Rapid urea hydrolysis; strong urease production | + |
| Orange or yellow | No urea hydrolysis; urease is absent or not produced in enough quantity to produce results in 24 hours; or the organism cannot live in urease broth | – |

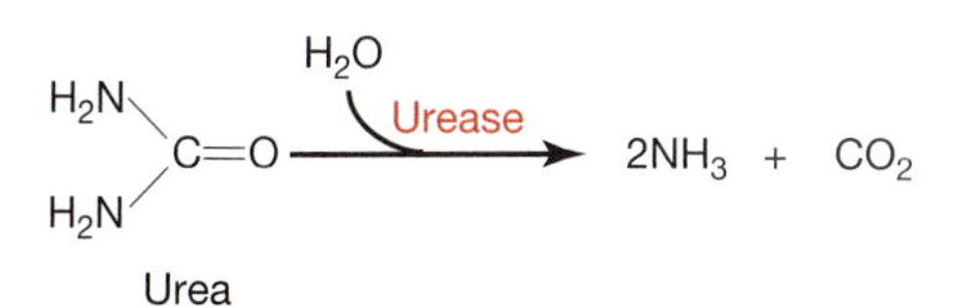


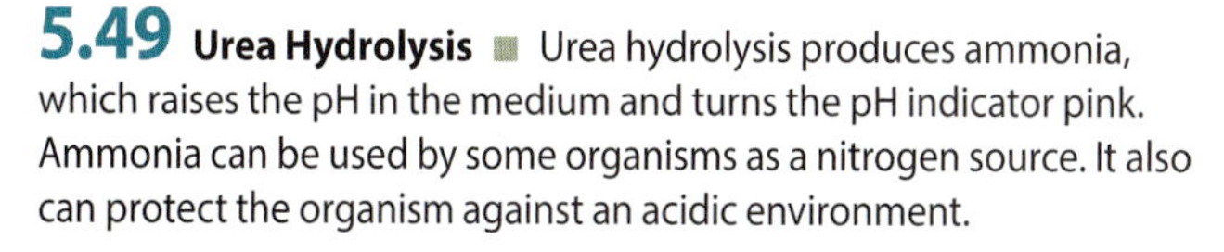
**5.49** **Urea Hydrolysis** ■ Urea hydrolysis produces ammonia, which raises the pH in the medium and turns the pH indicator pink. Ammonia can be used by some organisms as a nitrogen source. It also can protect the organism against an acidic environment.

**5.50** **Urea Hydrolysis Test Results** ■ These urea broth tubes after a 24-hour incubation illustrate a urease-negative organism on the left, an uninoculated control in the middle, and a urease-positive organism on the right.

## Application

This test is used to differentiate organisms based on their ability to hydrolyze urea with the enzyme urease. Urinary tract pathogens from the genus *Proteus* may be distinguished from other enteric bacteria by their rapid urease activity. It is also used in identifying *H. pylori*, which is associated with gastric and duodenal ulcers, as well as stomach cancer.

## In This Exercise

You will perform the urease test with urea broths, which should be read at 24 hours. If your labs are scheduled more than 24 hours apart, arrange to have someone place your broths in a refrigerator until you can examine them.

## ▼ Materials

### Per Student

- ☐ Lab coat
- ☐ Gloves
- ☐ Chemical eye protection

### Per Student Group

- ☐ Three urea broths
- ☐ Three sterile wooden sticks or sterile disposable loops
- ☐ Fresh agar slant cultures of these recommended organisms:
  - *Escherichia coli*
  - *Proteus vulgaris* (BSL-2)[1]

## ■ Medium Recipe

### Rustigian and Stuart's Urea Broth

| | |
|---|---|
| ☐ Yeast extract | 0.1 g |
| ☐ Potassium phosphate, monobasic | 9.1 g |
| ☐ Potassium phosphate, dibasic | 9.5 g |
| ☐ Urea | 20.0 g |
| ☐ Phenol red | 0.01 g |
| ☐ Distilled or deionized water | 1.0 L |

*pH 6.6–7.0 at 25°C*

5

## PROCEDURE

### Lab One

1. Wear a lab coat, gloves, and chemical eye protection when performing this procedure and **follow BSL-2 precautions if using *P. vulgaris*.**
2. Obtain three tubes of urease broth medium. Label two with the names of the organisms, your group name, and the date. Label the third tube of each medium "control."
3. Inoculate the two urea broths with heavy inocula from the test organisms and dispose of the sterile sticks/disposable loops according to your lab's practices. Do not inoculate the control.
4. Incubate all tubes aerobically at 35 ± 2°C for 24 hours.
5. Save or dispose of the original culture tubes as directed by your instructor.

### Lab Two

1. Remove all tubes from the incubator and examine them for color changes. Record all results in the table provided on the data sheet, page 355.
2. Dispose of all broth tubes in an appropriate autoclave container when finished.

[1] *Note to instructor:* The most consistently urease-positive organisms belong to the genus listed. If the goal of this exercise is simply to show a positive urease result and speed of reaction is not of concern, then *Staphylococcus epidermidis* and *Streptococcus salivarius* are BSL-1 organisms that are generally urease positive. Check your strain(s) and adjust the number of tubes accordingly.

### References


Atlas, Ronald and James Snyder. Page 319 in *Manual of Clinical Microbiology,* 11th ed. James H. Jorgensen, Michael A. Pfaller, Karen C. Carroll, Guido Funke, Marie Louise Landry, Sandra S. Richter, and David W. Warnock, eds. Washington, DC: ASM Press, American Society for Microbiology, 2015.

Collins, C. H., Patricia M. Lyne, and J. M. Grange. Page 117 in *Collins and Lyne's Microbiological Methods,* 7th ed. Oxford, Boston: Butterworth-Heinemann, 1995.

Delost, Maria Dannessa. Page 196 in *Introduction to Diagnostic Microbiology*. St. Louis, MO: Mosby, 1997.

DIFCO Laboratories. Page 1040 in *DIFCO Manual,* 10th ed. Detroit: DIFCO Laboratories, 1984.

Forbes, Betty A., Daniel F. Sahm, and Alice S. Weissfeld. Page 283 in *Bailey & Scott's Diagnostic Microbiology,* 11th ed. St. Louis, MO: Mosby, 2002.

Lányi, B. Page 24 in *Methods in Microbiology,* Vol. 19. R. R. Colwell and R. Grigorova, eds. New York: Academic Press, 1987.

MacFaddin, Jean F. Page 298 in *Biochemical Tests for Identification of Medical Bacteria,* 3rd ed. Philadelphia: Lippincott Williams & Wilkins, 2000.

Moat, Albert G., John W. Foster, and Michael P. Spector. Pages 491–492 in *Microbial Physiology*, 4th ed. New York: Wiley-Liss, 2002.

Smibert, Robert M. and Noel R. Krieg. Page 630 in *Methods for General and Molecular Bacteriology*. Philipp Gerhardt, R. G. E. Murray, Willis A. Wood, and Noel R. Krieg, eds. Washington, DC: American Society for Microbiology, 1994.

Tille, Patricia M. Page 229 in *Bailey & Scott's Diagnostic Microbiology,* 13th ed. St. Louis, MO: Mosby, 2014.

Winn, Jr., Washington, Stephen Allen, William Janda, Elmer Koneman, Gary Procop, Paul Schreckenberger, and Gail Woods. Pages 225 and 1458–1459 in *Koneman's Color Atlas and Textbook of Diagnostic Microbiology*, 6th ed. Baltimore: Lippincott Williams & Wilkins, 2006.

Zimbro, Mary Jo and David A. Power, eds. Page 606 in *Difco™ and BBL™ Manual—Manual of Microbiological Culture Media.* Sparks, MD: Becton Dickinson and Co., 2003.

Name ______________________________

Date ______________________________

Lab Section ______________________________

I was present and performed this exercise (initials) ______________________________

# DATA SHEET 5-15

## Urea Hydrolysis (Urease Test)

### OBSERVATIONS AND INTERPRETATIONS

**1** Using Table 5-17, page 353, as a guide, enter your results for the broth test after 24 hours.

| Organism | Color Result | (+/−) | **Interpretation** (include enzyme) |
|---|---|---|---|
| Control | | | |
| | | | |
| | | | |
| | | | |

## QUESTIONS

**1** *Suppose you ran this test with* Providencia stuartii *but it took 48 hours to turn pink.*

**a.** *Do you think this is a false result?*

______________________________

**b.** *If so, is it a false positive or a false negative? If not, why not? Give some possible reasons for this occurrence.*

______________________________

______________________________

______________________________

______________________________

______________________________

______________________________

**2** *Explain why it is acceptable to record positive tests before the suggested incubation time is completed but it is not acceptable to record a negative result early.*

**3** *Typically, urea broth is filter-sterilized because autoclaving breaks down the urea. However, even unsterilized broth rarely produces false-positive results. Why do you think this is true?*

# Bile Esculin Test

EXERCISE 5-16

## Theory

Bile esculin agar (BEA) is an undefined, selective, and differential medium containing beef extract, digest of gelatin, esculin, oxgall (bile), and ferric citrate. **Esculin** extracted from the bark of the horse chestnut tree is a glycoside composed of glucose and **esculetin** joined by a glycosidic bond (Fig. 5.51). Beef extract and gelatin provide nutrients and energy; bile (oxgall) is the selective agent added to separate Group D streptococci (the *Streptococcus bovis* group and enterococci) from other non-Group D streptococci. Ferric citrate is added as a source of oxidized iron to indicate a positive test.

Esculin hydrolysis results in the production of D-glucose and esculetin (Fig. 5.51). Hydrolysis can occur under acidic conditions, or a β-glucosidase enzyme generically referred to as **esculinase** can catalyze it. Many bacteria possess esculinase, but the number of bacteria that are able to hydrolyze esculin in the presence of bile is much more limited, making this a useful selective and differential medium.

BEA can be prepared as a plated or a slanted medium. If the task is to isolate a streptococcus or enterococcus from a mixed culture, then a BEA plate is streaked for isolation and the differential results are read from that. If testing a pure culture for esculinase activity in the presence of bile, then a slanted medium is used.

When esculin is hydrolyzed in BEA, the resulting esculetin reacts with ferric ions in the medium to produce a dark brown/black precipitate (Fig. 5.52), which darkens the medium surrounding the growth. When a BEA plate is used (as in this lab exercise), an organism that darkens the medium even slightly is bile esculin-positive. An organism that doesn't darken the medium is negative (Fig. 5.53). A positive result is recorded on BEA slants when more than half of the medium is blackened. No blackening to less than half-blackened is considered a negative result.

## Application

BEA is used for the isolation and presumptive identification of bile esculin-positive enterococci (typically *Enterococcus faecalis* and *E. faecium*) and the bovis group of streptococci (Group D *Streptococcus bovis, S. equinus, S. gallolyticus,*

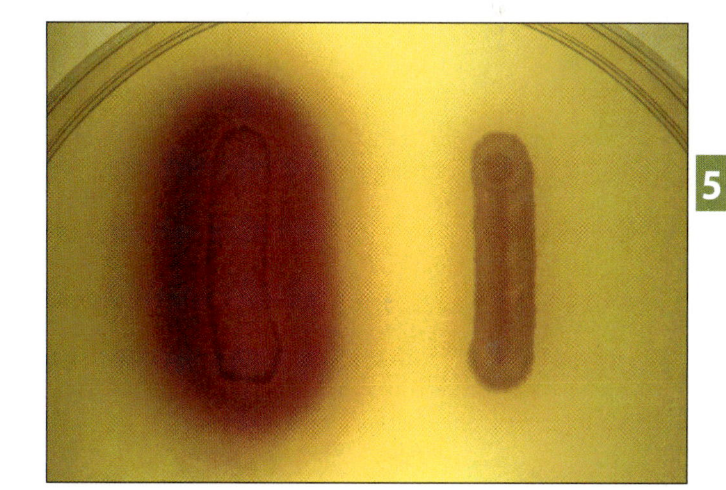

**5.53 Bile Esculin Test Results** ■ This plate was inoculated with two Gram-positive cocci. The one on the left is bile esculin positive and the one on the right is bile esculin negative.

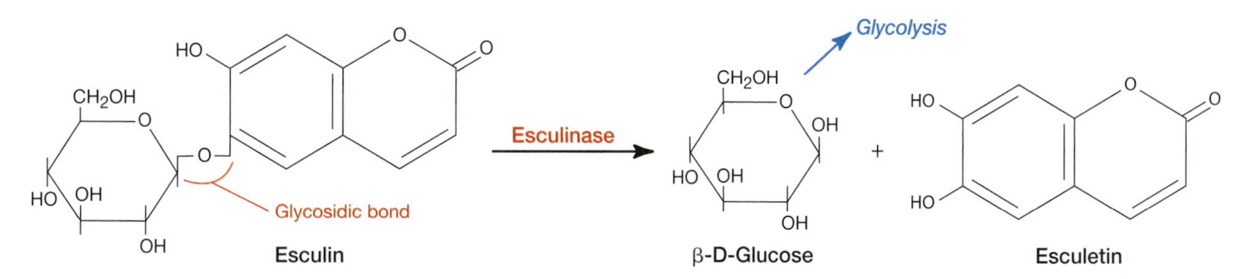


**5.51 Hydrolysis of Esculin** ■ Esculinase catalyzes the hydrolysis of esculin at the glycosidic bond joining glucose and esculetin. Many organisms produce esculinase, but the Group D streptococci and enterococci are unique in their ability to do this in the presence of bile salts, which is what this test detects.

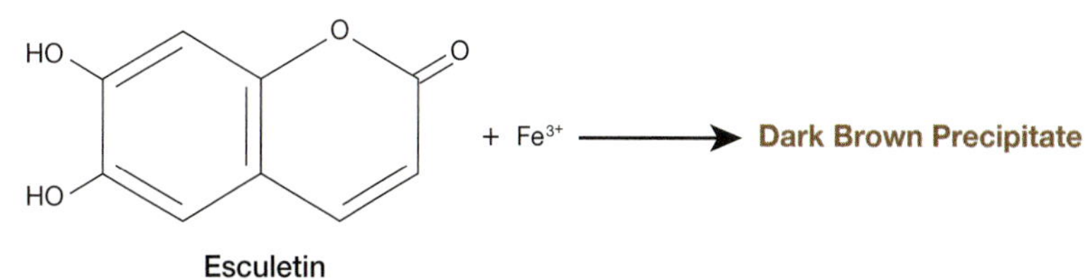


**5.52 Bile Esculin Test Indicator Reaction** ■ Esculetin, produced during the hydrolysis of esculin, reacts with $Fe^{3+}$ in the medium to produce a dark brown to black precipitate.

and other streptococci) from non-Group D streptococci. It also can be used to distinguish between the genera *Enterobacter, Klebsiella,* and *Serratia* (bile esculin-positive) from other genera in the *Enterobacteriaceae.*

## In This Exercise

Today you will inoculate one bile esculin agar plate with two test organisms and incubate it for up to 48 hours. (In a clinical application, BEA plates would be streaked for isolation.) Or, at the discretion of your instructor, you may inoculate two BE slants. Any blackening of the plated medium before the end of 48 hours can be recorded as positive. If using the slant, a positive result is recorded if the medium is more than half blackened within 48 hours. Negative results must incubate the full 48 hours before final determination is made. Use Table 5-18 as a guide.

## Materials

### Per Student

- ☐ Lab coat
- ☐ Gloves
- ☐ Chemical eye protection

### Per Student Group

- ☐ One bile esculin agar plate
- ☐ Fresh broth cultures of these recommended organisms:
  - *Lactococcus lactis* or *Streptococcus mutans*
  - *Enterococcus faecalis* (BSL-2) or *Streptococcus gallolyticus* subsp. *gallolyticus*

## Medium Recipe

### Bile Esculin Agar

| | |
|---|---|
| ☐ Pancreatic digest of gelatin | 5.0 g |
| ☐ Beef extract | 3.0 g |
| ☐ Oxgall (ox bile) | 20.0 g |
| ☐ Ferric citrate | 0.5 g |
| ☐ Esculin | 1.0 g |
| ☐ Agar | 14.0 g |
| ☐ Distilled or deionized water | 1.0 L |

*pH 6.6–7.0 at 25°C*

## PROCEDURE

### Lab One

1. Wear a lab coat, gloves, and chemical eye protection when performing this procedure. **Use BSL-2 precautions when transferring *E. faecalis*, if used.**
2. Gently mix each culture according to your lab's standards.
3. Obtain a bile esculin agar plate.
4. Using a permanent marker, divide the bottom into two halves.
5. Label the plate with the organisms' names, your group name, and the date.
6. Spot inoculate (refer to Appendix B, page 601) two of the sectors on the BEA plate with the test organisms.
7. Invert and incubate the plate at 35 ± 2°C for 24 to 48 hours.
8. Save or dispose of the original culture tubes as directed by your instructor.

### Lab Two

1. Examine the plate for any darkening of the medium.
2. Record your results on the data sheet, page 359.
3. Dispose of the plate in an appropriate autoclave container when finished.

### References


Atlas, Ronald and James Snyder. Pages 317 and 328 in *Manual of Clinical Microbiology,* 11th ed. James H. Jorgensen, Michael A. Pfaller, Karen C. Carroll, Guido Funke, Marie Louise Landry, Sandra S. Richter, and David W. Warnock, eds. Washington, DC: ASM Press, American Society for Microbiology, 2015.

Delost, Maria Dannessa. Page 132 in *Introduction to Diagnostic Microbiology.* St. Louis, MO: Mosby, 1997.

Lányi, B. Page 56 in *Methods in Microbiology,* Vol. 19. R. R. Colwell and R. Grigorova, eds. New York: Academic Press, 1987.

McFaddin, Jean F. Page 8 in *Biochemical Tests for Identification of Medical Bacteria,* 3rd ed. Philadelphia: Lippincott Williams & Wilkins, 2000.

Spellerberg, Barbara and Claudia Brandt. Page 342 in *Manual of Clinical Microbiology,* 10th ed. James Versalovic, Karen C. Carroll, Guido Funke, James H. Jorgensen, Marie Louise Landry, and David W. Warnock, eds. Washington, DC: ASM Press, 2011.

Tille, Patricia M. Page 199 in *Bailey & Scott's Diagnostic Microbiology,* 13th ed. St. Louis, MO: Mosby, 2014.

Zimbro, Mary Jo and David A. Power. Page 156 in *Difco™ & BBL™ Manual–Manual of Microbiological Culture Media.* Sparks, MD: Becton, Dickinson and Co., 2003.


TABLE **5-18** **Bile Esculin Test Results and Interpretations**

| Result | Interpretation | Symbol |
|---|---|---|
| Medium is darkened within 48 hours | Organism hydrolyzes esculin in the presence of bile; presumptive identification as a member of Group D *Streptococcus* or *Enterococcus* | + |
| No darkening of the medium after 48 hours | Organism doesn't hydrolyze esculin in the presence of bile or does but not in detectable amounts; presumptive determination as a non-Group D *Streptococcus* or *Enterococcus* | – |

Name ______________________________

Date ______________________________

Lab Section ______________________________

I was present and performed this exercise (initials) ______________________________

# DATA SHEET 5-16

## Bile Esculin Test

### OBSERVATIONS AND INTERPRETATIONS

**1** Refer to Table 5-18, page 358, when recording your results and interpretations in the table below.

| Organism | Color Result | (+/−) | Interpretation and Presumptive ID |
|---|---|---|---|
| | | | |
| | | | |

### QUESTIONS

**1** *In your own words, what is the application (purpose) of BEA as a plated medium?*

**2** *Which ingredient(s) in BEA supply(ies)*

**a.** *Carbon?*

**b.** *Nitrogen?*

**3** *Is BEA a defined or an undefined medium? Provide the reasoning behind your choice and explain why this formulation is desirable.*

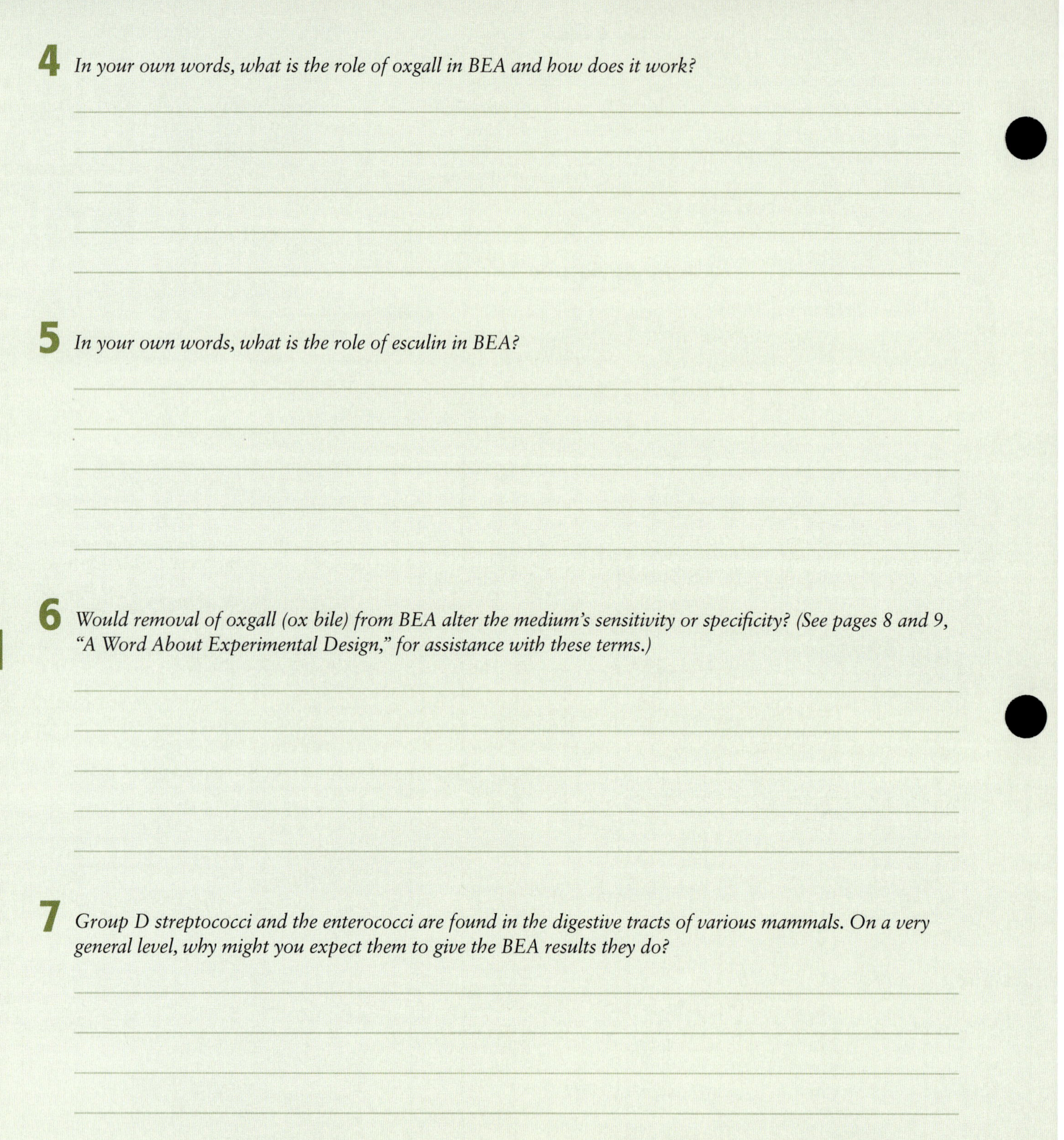

**4** *In your own words, what is the role of oxgall in BEA and how does it work?*

**5** *In your own words, what is the role of esculin in BEA?*



**6** *Would removal of oxgall (ox bile) from BEA alter the medium's sensitivity or specificity? (See pages 8 and 9, "A Word About Experimental Design," for assistance with these terms.)*

**7** *Group D streptococci and the enterococci are found in the digestive tracts of various mammals. On a very general level, why might you expect them to give the BEA results they do?*

# L-pyrrolidonyl–β–naphthylamide Hydrolysis (PYR Test)

EXERCISE 5-17

## Theory

**Aminopeptidases** are enzymes that remove amino acids from the amino end of a protein or peptide. A specific example is L-pyrrolidonyl arylamidase, which is produced by Group A streptococci (*Streptococcus pyogenes*) and enterococci. This enzyme hydrolyzes L-pyrrolidonyl–β–naphthylamide (PYR) to produce L-pyrrolidone and β-naphthylamine, all of which are colorless. In the indicator reaction, β-naphthylamine reacts with *p*-dimethylaminocinnamaldehyde to form a red precipitate (Fig. 5.54).

PYR may be performed as an 18-hour agar test, a 4-hour broth test or, as used in this example, a rapid disk test. In each case the medium (or disk) contains L-pyrrolidonyl–β–naphthylamide to which is added a heavy inoculum of the test organism. After the appropriate incubation or waiting period, a 0.01% *p*-dimethylaminocinnamaldehyde solution is added. Formation of a red color within a few minutes is interpreted as PYR-positive (Fig. 5.55). Yellow or orange is PYR-negative. Refer to Table 5-19 as a guide to interpretation.

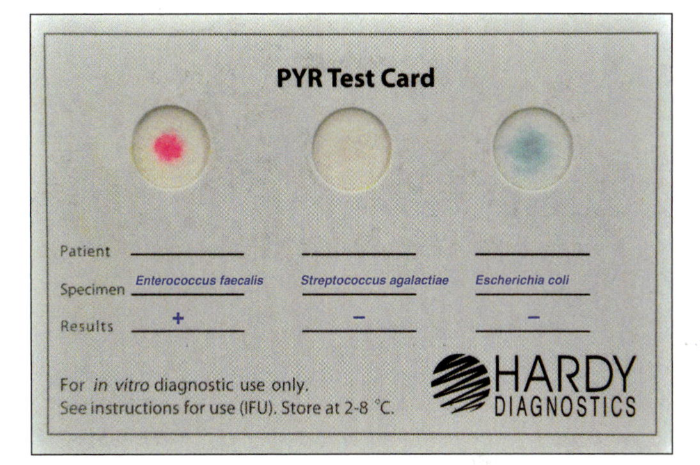


**5.55 PYR Disk Test Results** ■ Shown is a Hardy Diagnostics PYR Test card. The filter paper in each opening is impregnated with substrate. A pink/red color after addition of the PYR reagent (cinnamaldehyde) is considered positive. No color change (center opening) or an orange, salmon, or yellow color is considered negative. Certain *Enterobacteriaceae* (e.g., *E. coli*, as in this example) may produce the blue coloration seen on the right when grown on a high-tryptophan medium. This is also a negative result.

## Application

The PYR test is designed for presumptive identification of Group A streptococci (*S. pyogenes*) and enterococci by determining the presence of the enzyme L-pyrrolidonyl arylamidase.

## In This Exercise

You will inoculate two disks with test organisms to determine which one contains the enzyme L-pyrrolidonyl arylamidase.

TABLE 5-19 **PYR Disk Test Results and Interpretations**

| Result | Interpretation | Symbol |
|---|---|---|
| Red color formation | Organism produces L-pyrrolidonyl arylamidase; presumptive identification as Group A streptococcus (*S. pyogenes*) or enterococcus | + |
| Orange color or no color change | Organism does not produce L-pyrrolidonyl arylamidase or doesn't produce enough to be detectable | – |

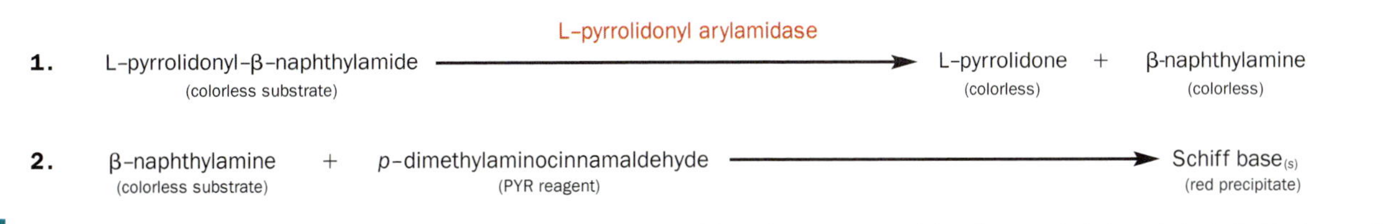


**5.54 PYR Test Chemistry** ■ Reaction 1 shows the hydrolytic splitting of the substrate L-pyrrolidonyl–β–naphthylamide included in the medium. In this exercise, the filter-paper disks were saturated with substrate. Both products of the reaction are colorless. Reaction 2 illustrates the reaction of the PYR reagent (*p*-dimethylaminocinnamaldehyde) and β–naphthylamine. In the reaction, the amino group ($H_2N$) of β–naphthylamine reacts with the aldehyde group ($-CH_4O$) of the reagent and produces a Schiff base ($-CH=N-$), which is red.

## ▼ Materials

### Per Student

- ☐ Lab coat
- ☐ Gloves
- ☐ Chemical eye protection

### Per Student Group

- ☐ One empty Petri dish or glass slide
- ☐ Three PYR disks
- ☐ 0.01% *p*-dimethylaminocinnamaldehyde
- ☐ Three sterile pipettes, unless the test reagent is in a dropper bottle
- ☐ Fresh agar slant cultures of these recommended organisms:
  - *Enterococcus faecalis* (BSL-2)
  - *Staphylococcus epidermidis*

## PROCEDURE

5

1. Wear a lab coat, gloves, and chemical eye protection when performing this procedure and **use BSL-2 precautions throughout while handling *E. faecalis*.**
2. Obtain a Petri dish or glass slide, three PYR disks, and three sterile pipettes (unless the PYR reagent is already in a dropper bottle).
3. Open the Petri dish and mark three spots where you will place the PYR disks. Label two spots with the name of its respective organism. Label the third spot "control."
4. Place three PYR disks inside on the appropriate spots.
5. With a disposable loop or sterile wooden stick, inoculate each disk with a large amount of its respective organism. Properly dispose of the loop/wooden stick immediately after use. **Use extra caution with *E. faecalis*.**
6. Save or dispose of the original culture tubes as directed by your instructor.
7. Using a different pipette for each, add a drop of the *p*-dimethylaminocinnamaldehyde solution to each disk.
8. If the disk turns red within 5 minutes (or within the time suggested by the disk manufacturer), it is positive. Orange or no color change is a negative result.
9. Record your results in the table on the data sheet, page 363.
10. Dispose of the original cultures, plate (or slide), and disks in an appropriate autoclave container(s) when finished.

### References


Atlas, Ronald and James Snyder. Page 319 in *Manual of Clinical Microbiology,* 11th ed. James H. Jorgensen, Michael A. Pfaller, Karen C. Carroll, Guido Funke, Marie Louise Landry, Sandra S. Richter, and David W. Warnock, eds. Washington, DC: ASM Press, American Society for Microbiology, 2015.

Claus, G. William. Chap. 14 in *Understanding Microbes—A Laboratory Textbook for Microbiology*. New York: W.H. Freeman and Company, 1989.

Collins, C. H., Patricia M. Lyne and J. M. Grange. Chap. 6 in *Collins and Lyne's Microbiological Methods*, 7th ed. Oxford, Boston: Butterworth-Heinemann, 1995.

Tille, Patricia M. Pages 97 and 225 in *Bailey & Scott's Diagnostic Microbiology*, 13th ed. St. Louis, MO: Mosby, 2014.

Winn, Jr., Washington, Stephen Allen, William Janda, Elmer Koneman, Gary Procop, Paul Schreckenberger, and Gail Woods. Page 1448 in *Koneman's Color Atlas and Textbook of Diagnostic Microbiology*, 6th ed. Baltimore: Lippincott Williams & Wilkins, 2006.

Name ______________________________

Date ______________________________

Lab Section ______________________________

I was present and performed this exercise (initials) ______________

# DATA SHEET 5-17

## L-pyrrolidonyl–β-naphthylamide Hydrolysis (PYR Test)

### OBSERVATIONS AND INTERPRETATIONS

**1** Using Table 5-19, page 361, as a guide, record your results and interpretations in the table below.

| Organism | Result | (+/−) | **Interpretation** (include enzyme) |
|---|---|---|---|
| Uninoculated control | | | |
| | | | |
| | | | |

### QUESTIONS

**1** *Consider the uninoculated disk.*

**a.** *Is it a positive or a negative control?*

**b.** *What purpose does it serve in the PYR test?*

**2** *How are the products of protein hydrolysis by an aminopeptidase used in metabolism?*

**3** *Deep red is the only color reaction that is considered positive for this test. Why do you think yellow or orange results are not positive (or weak positive)?*

**4** *Suppose you wanted to perform this test on an unknown organism but you were out of PYR reagent. Suppose further that a fellow lab employee said that you could substitute formaldehyde for PYR reagent and get good results. What is the chemical basis for your peer's suggestion?*

## Combination Differential Media

Combination differential media combine components of several compatible tests into one medium, thus saving critical diagnostic time and money. Typically, they include core tests to differentiate members of specific bacterial groups. SIM medium (Exercise 5-18), for example, tests for sulfur reduction, indole production, and motility—important characteristics of members of *Enterobacteriaceae*. Kligler Iron Agar (KIA; Exercise 5-19) tests for glucose and lactose fermentation, and for sulfur reduction, which again are diagnostically important tests for many Gram-negative organisms including *Enterobacteriaceae*. These two media will serve as examples of what can creatively be done by combining differential tests.

As with most biochemical tests, combination media can be used as a follow up to selective media. Selective media promote isolation of the organisms of interest and help determine the sequence of tests to follow. MacConkey agar (Exercise 4-4), for example, might be streaked initially to encourage and isolate Gram-negative organisms, followed by Kligler iron agar (KIA). The source of the sample and the results on MacConkey agar would dictate the direction followed by a medical laboratorian. ■

# SIM Medium: Determination of Sulfur Reduction, Indole Production, and Motility

EXERCISE 5-18

## ■ Theory

SIM medium is used for determination of three bacterial activities: sulfur reduction, indole production from tryptophan, and motility (from which the acronym "SIM" is derived). The semisolid medium includes casein and animal tissue as sources of amino acids, an iron-containing compound, and sulfur in the form of sodium thiosulfate.

Sulfur reduction to $H_2S$ can be accomplished by bacteria in two different ways, depending on the enzymes present.

1. The enzyme **cysteine desulfurase** catalyzes the hydrolysis of the sulfur-containing amino acid cysteine to pyruvate and $H_2S$ during putrefaction (Fig. 5.56).
2. In a totally unrelated way, the enzyme **thiosulfate reductase** catalyzes the reduction of sulfur (in the form of sulfate) to $H_2S$ at the end of an anaerobic respiratory electron transport chain (Fig. 5.57).

Both systems produce hydrogen sulfide ($H_2S$) gas. When either reaction occurs in SIM medium, the $H_2S$ produced combines with iron, in the form of ferrous ammonium sulfate, to form ferric sulfide (FeS), a black precipitate (Fig. 5.58). Any blackening of the medium is an indication of sulfur reduction and a positive test. No blackening of the medium indicates no sulfur reduction and a negative reaction (Fig. 5.59).

Indole production in the medium is made possible by the presence of the amino acid tryptophan (contained in casein and animal protein). Bacteria possessing the enzyme **tryptophanase** can hydrolyze tryptophan to pyruvate, ammonia (by deamination), and indole (Fig. 5.60).

$$3S_2O_3^{=} + 4H^+ + 4e^- \xrightarrow{\text{Thiosulfate reductase}} 2SO_3^{=} + 2H_2S\,(g)$$

**5.57 Sulfur Reduction: $H_2S$ Production from Anaerobic Respiration (Reduction of Thiosulfate)** ■ Anaerobic respiration with thiosulfate as the final electron acceptor also produces $H_2S$. Anaerobic respiration produces more ATP per glucose than fermentation, but less than aerobic respiration.

$$H_2S + FeSO_4 \longrightarrow H_2SO_4 + FeS\,(s)$$

**5.58 Indicator Reaction for $H_2S$ Production** ■ Hydrogen sulfide, a colorless gas, can be detected when it reacts with ferrous sulfate in the medium to produce the black precipitate ferric sulfide.

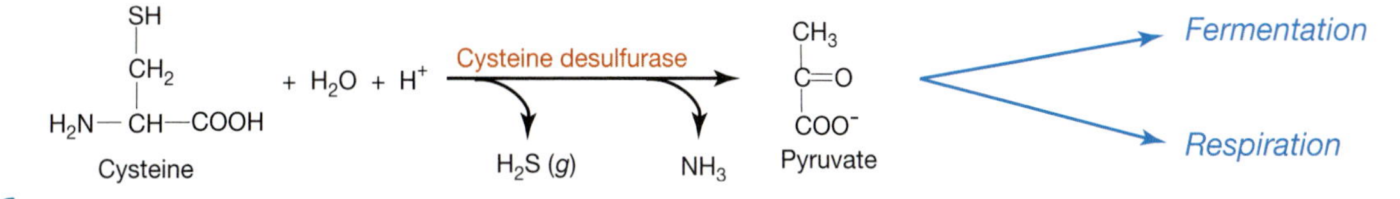


**5.56 Sulfur Reduction: $H_2S$ Production from Cysteine Hydrolysis (Putrefaction)** ■ Cysteine desulfurase catalyzes the hydrolysis of cysteine with the production of pyruvate, $NH_3$, and $H_2S$. This reaction is a mechanism for getting energy out of the amino acid cysteine.

Tryptophan hydrolysis in SIM medium can be detected by the addition of Kovac's reagent after the incubation period. Kovac's reagent contains ***p*-dimethylaminobenzaldehyde** (DMABA) and HCl dissolved in amyl alcohol. When a few drops of Kovac's reagent are added to the tube, it forms a layer over the agar. DMABA reacts with any indole present and produces a quinoidal compound that turns the reagent layer red (Figs. 5.61 and 5.62). The formation of red color in the reagent layer indicates a positive reaction and the presence of tryptophanase. No red color is indole negative.

Determination of motility in SIM medium is made possible by the reduced agar concentration (0.35% vs. 1.5% in nutrient agar) and the method of inoculation. The medium is inoculated with a single stab from an inoculating needle. Motile organisms are able to move about in the semisolid medium and can be detected by the radiating growth extending outward in all directions from the central stab line. Growth that radiates in *all* directions and appears slightly fuzzy is an indication of motility (Fig. 5.63). This should not be confused with the (seemingly) spreading growth produced by lateral movement of the inoculating needle when stabbing.

**5.59 Sulfur Reduction in SIM Medium** ■ The organism on the left is $H_2S$ negative. The organism on the right is $H_2S$ positive. Although without a control it is difficult to tell, the organism on the left is motile (note the haziness in the medium). However, when an organism is $H_2S$ positive it is nearly impossible to read motility in this medium.

**5.62 Indole Production Test Results in SIM Medium** ■ These SIM tubes were inoculated with an indole-negative organism on the left and an indole-positive organism on the right. Note how little Kovac's reagent is used. Both organisms are also motile, as evidenced by the haziness in the agar.

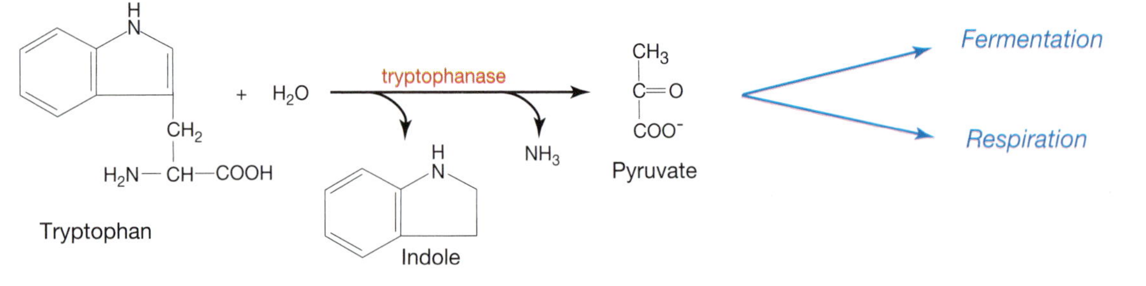


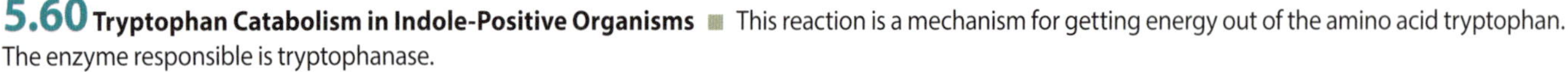

**5.60 Tryptophan Catabolism in Indole-Positive Organisms** ■ This reaction is a mechanism for getting energy out of the amino acid tryptophan. The enzyme responsible is tryptophanase.

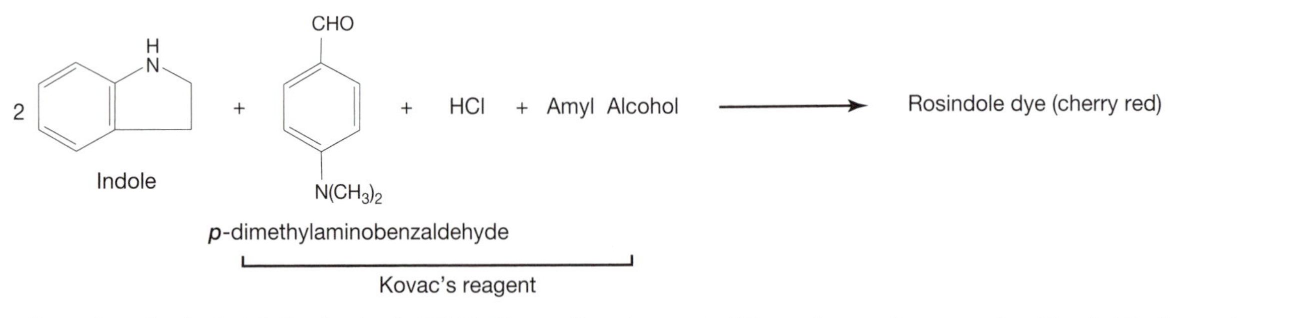


**5.61 Indicator Reaction for Indole Production in SIM Medium** ■ Kovac's reagent will form a layer on the agar surface. Then indole, if present, reacts with the DMABA in Kovac's reagent to produce a red color.

**5.63 Motility in SIM Medium** ■ These SIM tubes were inoculated with a nonmotile organism on the left and a motile organism on the right. With an adequate control, the haziness in the agar indicating motility is usually detectable. However, if results are equivocal, try motility agar (Exercise 5-24).

## Application

SIM medium is used to identify bacteria that are capable of producing indole using the enzyme tryptophanase. The indole test is one component of the IMViC battery of tests (Indole, Methyl red, Voges-Proskauer, and Citrate) used to differentiate the *Enterobacteriaceae*, especially *E. coli*-positive, from *Enterobacter*, *Klebsiella*, *Hafnia*, and *Serratia*-negative. SIM medium also is used to differentiate sulfur-reducing members of *Enterobacteriaceae*, especially members of the genera *Salmonella, Shigella,* and *Proteus*, from the negative *Morganella morganii* and *Providencia rettgeri*. In addition to the first two functions of SIM, motility is an important differential characteristic of *Enterobacteriaceae*.

## In This Exercise

You will stab inoculate SIM medium with three organisms demonstrating a variety of results. When reading your test results, make the motility and $H_2S$ determinations before adding the Kovac's (indole) reagent. Use Tables 5-20, 5-21 and 5-22, and Figures 5.59, 5.62, and 5.63 as guides when making your interpretations.

**TABLE 5-20 Sulfur Reduction ($H_2S$ Production) Test Results and Interpretations**

| Result | Interpretation | Symbol |
|---|---|---|
| Black in the medium | Sulfur reduction ($H_2S$ production); organism either respires anaerobically with sulfate as the final electron acceptor (thiosulfate reductase) or hydrolyzes the amino acid cysteine (cysteine desulfurase) | + |
| No black in the medium | Sulfur is not reduced, or is not reduced in detectable amounts | − |

**TABLE 5-21 Indole Production Test Results and Interpretations**

| Result | Interpretation | Symbol |
|---|---|---|
| Red in the alcohol layer of Kovac's reagent | Organism produces tryptophanase and hydrolyzes tryptophan into indole and pyruvate | + |
| Reagent color is unchanged | Tryptophanase is absent and indole is not produced, or is produced but not in detectable amounts | − |

**TABLE 5-22 Motility Test Results and Interpretations**

| Result | Interpretation | Symbol |
|---|---|---|
| Growth radiating outward from stab line | Motility | + |
| No radiating growth | Nonmotile | − |
| Blackened medium ($H_2S$ produced) | Motility not determined | ND |

## Materials

### Per Student

- □ Lab coat
- □ Gloves
- □ Chemical eye protection

### Per Student Group

- □ Four SIM tubes
- □ Kovac's reagent
- □ Microbial incinerator or disposable inoculating needle(s)
- □ Fresh agar slant cultures of these recommended organisms:
  - *Escherichia coli*
  - *Micrococcus luteus*
  - *Proteus mirabilis* (BSL-2)[1]

[1] *Note to instructor:* Approximately 75% of *Citrobacter freundii* strains are positive for sulfur reduction and it may serve as a suitable BSL-1 replacement for *P. mirabilis*, if desired. *Aeromonas hydrophila* may also work as a substitute. Check your lab's strains prior to assigning these.

## Medium and Reagent Recipes

### SIM (Sulfur-Indole-Motility) Medium

| | |
|---|---|
| □ Pancreatic digest of casein | 20.0 g |
| □ Peptic digest of animal tissue | 6.1 g |
| □ Ferrous ammonium sulfate | 0.2 g |
| □ Sodium thiosulfate | 0.2 g |
| □ Agar | 3.5 g |
| □ Distilled or deionized water | 1.0 L |
| *pH 7.1–7.5 at 25°C* | |

### Kovac's Reagent

| | |
|---|---|
| □ Amyl alcohol | 75.0 mL |
| □ Hydrochloric acid, concentrated | 25.0 mL |
| □ *p*-dimethylaminobenzaldehyde | 5.0 g |

## PROCEDURE

### Lab One

1. Wear a lab coat, gloves, and chemical eye protection when performing this procedure. ***Note*: If using *P. mirabilis*, BSL-2 precautions should be observed.**
2. Obtain four SIM tubes. Label three with the names of the organisms, your group name, and the date. Label one tube "control."
3. Stab inoculate three tubes with the test organisms. Insert the needle into the agar to within 1 cm of the bottom of the tube. Be careful to remove the needle along the original stab line. Do not inoculate the control. If using *P. mirabilis,* inoculation can be done with a wire needle if sterilized with a microbial incinerator. Otherwise, a disposable inoculating needle should be used.
4. Incubate all tubes aerobically at 35 ± 2°C for 24 to 48 hours.
5. Save or dispose of the original culture tubes as directed by your instructor.

### Lab Two

1. Wear a lab coat, gloves, and chemical eye protection when performing this procedure.
2. Examine the tubes for spreading from the stab line *and* formation of black precipitate in the medium. Record any $H_2S$ production and/or motility in the table provided on the data sheet, page 369.
3. Add Kovac's reagent to each tube to a depth of 2 mm–3 mm (*not mL*). After several minutes, observe for the formation of red color in the reagent layer.
4. Record your results in the table on the data sheet.
5. Dispose of the tubes in an appropriate autoclave container when finished.

### References


Atlas, Ronald and James Snyder. Pages 318 and 345 in *Manual of Clinical Microbiology,* 11th ed. James H. Jorgensen, Michael A. Pfaller, Karen C. Carroll, Guido Funke, Marie Louise Landry, Sandra S. Richter, and David W. Warnock, eds. Washington, DC: ASM Press, American Society for Microbiology, 2015.

Delost, Maria Dannessa. Page 186 in *Introduction to Diagnostic Microbiology*. St. Louis, MO: Mosby, 1997.

MacFaddin, Jean F. Page 162 in *Biochemical Tests for Identification of Medical Bacteria*, 2nd ed. Baltimore: Lippincott Williams & Wilkins, 1980.

Tille, Patricia M. Pages 97 and 210 in *Bailey & Scott's Diagnostic Microbiology*, 13th ed. St. Louis, MO: Mosby, 2014.

Winn, Jr., Washington, Stephen Allen, William Janda, Elmer Koneman, Gary Procop, Paul Schreckenberger, and Gail Woods. Pages 224, 226, and 1445–1446 in *Koneman's Color Atlas and Textbook of Diagnostic Microbiology*, 6th ed. Baltimore: Lippincott Williams & Wilkins, 2006.

Zimbro, Mary Jo and David A. Power, eds. Page 490 in *Difco™ and BBL™ Manual—Manual of Microbiological Culture Media.* Sparks, MD: Becton Dickinson and Co., 2003.

Name ______________________________

Date ______________________________

Lab Section ______________________________

I was present and performed this exercise (initials) ______________________________

# DATA SHEET 5-18

## SIM Medium: Determination of Sulfur Reduction, Indole Production, and Motility

### OBSERVATIONS AND INTERPRETATIONS

**1** Using Table 5-20, page 367, as a guide, record your results and interpretations in the table below.

| Sulfur Reduction ($H_2S$ Production) | | | |
|---|---|---|---|
| **Organism** | **Black PPT (Y/N)** | **(+/−)** | **Interpretation** (include enzyme) |
| Uninoculated control | | | |
| | | | |
| | | | |
| | | | |

**2** Using Table 5-21, page 367, as a guide, record your results and interpretations in the table below.

| Indole Production | | | |
|---|---|---|---|
| **Organism** | **Red Color? (Y/N)** | **(+/−)** | **Interpretation** (include enzyme) |
| Uninoculated control | | | |
| | | | |
| | | | |
| | | | |

**3** Using Table 5-22, page 367, as a guide, record your results and interpretations in the table below.

| Motility | | | |
|---|---|---|---|
| **Organism** | **Growth Location** | **(+/−)** | **Interpretation** |
| Uninoculated control | | | |
| | | | |
| | | | |
| | | | |

## QUESTIONS

**1** *Consider the uninoculated SIM tube.*

**a.** *Is it a positive or a negative control in each test?*

**b.** *What purpose does it serve in the sulfur reduction test?*

**c.** *What purpose does it serve in the indole test?*

**d.** *What purpose does it serve in the motility test?*

Name

Date

Lab Section

I was present and performed this exercise (initials)

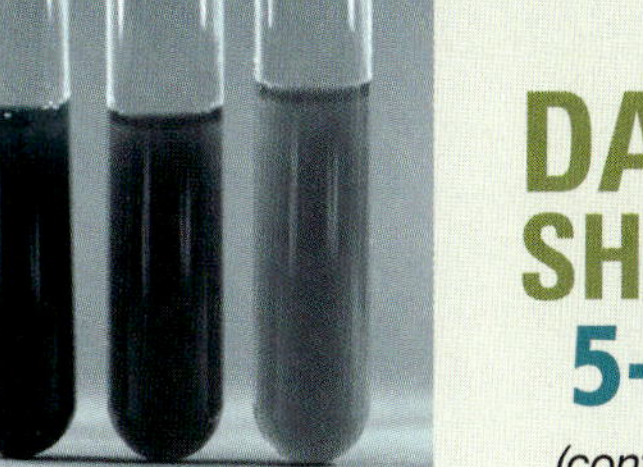

# DATA SHEET 5-18

*(continued)*

**2** *The sulfur reduction test is not able to differentiate $H_2S$ produced by anaerobic respiration and $H_2S$ produced by putrefaction. Is this inability the result of poor sensitivity or poor specificity of the test system? Explain.*

**3** *What factors dictate the choice of tests included in a combination medium?*

**4** *Which ingredient(s) could be eliminated if this medium were used strictly for testing indole production? Explain.*

# Triple Sugar Iron Agar / Kligler Iron Agar[1]

EXERCISE 5-19

## Theory

Triple sugar iron agar (TSIA) is a rich medium designed to differentiate bacteria (especially enterics–*Enterobacteriaceae*) on the basis of sulfur reduction and glucose, lactose, and sucrose fermentation. In addition to the three carbohydrates, it includes animal proteins as sources of carbon and nitrogen, and both ferrous sulfate and sodium thiosulfate as sources of oxidized sulfur. Phenol red is the pH indicator (yellow at pH less than 6.8 and reddish above pH 7.4), and the iron in the ferrous sulfate is the hydrogen sulfide indicator.

The medium is prepared as a shallow agar slant with a deep butt, thereby providing both aerobic and anaerobic growth environments. It is inoculated by a stab in the agar butt followed by a fishtail streak of the slant. The incubation period is 18 to 24 hours for carbohydrate fermentation and up to 48 hours for hydrogen sulfide reactions. Many reactions in various combinations are possible (Fig. 5.64 and Table 5-23).

When TSIA is inoculated with a glucose-only fermenter, acid products lower the pH and turn the entire medium yellow within a few hours. Because glucose is in short supply (0.1%), it will be exhausted within about 12 hours. As the glucose diminishes, ammonia produced from deamination of amino acids by organisms located in the aerobic region (slant) will begin to raise the pH and turn it red. This process, which takes 18 to 24 hours to complete, is called a **reversion.** It only occurs in the slant because the anaerobic conditions in the butt result in slower glucose consumption. Thus, a TSIA with a red slant and yellow butt after a 24-hour incubation period indicates that the organism ferments glucose but not lactose and/or sucrose.

Organisms that are able to ferment glucose *and* lactose *and/or* sucrose also turn the medium yellow throughout (see Fig. 5.4 for the biochemical relationship between these three fermentations). However, because the lactose and sucrose concentrations are 10 times higher than that of glucose, more acid is produced for a longer period of time, so both slant and butt will remain yellow after 24 hours. Therefore, a TSIA with a yellow slant and butt at 24 hours indicates that the organism ferments glucose and one or both of the other sugars. Gas produced by fermentation of any of the carbohydrates will appear as fissures in the medium or will lift the agar off the bottom of the tube (see tube on far right in Fig. 5.64).

Hydrogen sulfide ($H_2S$) may be produced by the reduction of thiosulfate in the medium or by the breakdown of cysteine in the peptone (Figs. 5.56 and 5.57). Ferrous sulfate reacts with the $H_2S$ to form a black precipitate, usually seen in the butt (Fig. 5.58). Acid conditions must exist for thiosulfate reduction; therefore, black precipitate in the medium is an indication of sulfur reduction *and* at the very least, glucose fermentation if the organism is suspected to be an enteric. If the black precipitate obscures the color of the butt, the color of the slant determines which carbohydrates have been fermented (i.e., red slant = glucose fermentation, yellow slant = glucose and lactose/sucrose fermentation).

An organism that does not ferment any of the carbohydrates but utilizes peptone and amino acids will alkalinize the medium and turn it red. If the organism can use the peptone aerobically and anaerobically, both

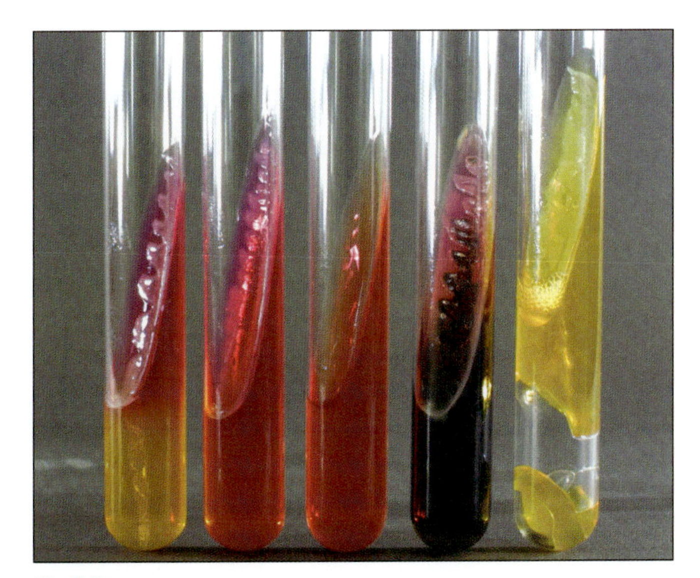

**5.64 Results in Kligler Iron Agar Slants** ■ These KIA slants from left to right illustrate: alkaline slant/acid butt (K/A); alkaline slant/no change in the butt (K/NC); uninoculated control; alkaline slant/acid butt, hydrogen sulfide present (K/A, $H_2S$); and acid slant/acid butt, gas (A/A, G). TSIA tubes with the same appearances would be scored the same. Refer to Table 5-23 for interpretations.

[1] This exercise is written for either triple sugar iron agar (TSIA) or Kligler iron agar (KIA). In form and function the two media are virtually identical except that KIA does not include sucrose. The results of both media look exactly the same. In results indicating fermentation of more than one carbohydrate, TSIA is reported as *glucose and lactose and/or sucrose fermentation*; KIA (because it contains no sucrose) is reported as *glucose and lactose fermentation.* To avoid the awkwardness of repeatedly including both names, we refer only to TSIA. If you are using KIA, simply substitute the name where applicable and do not expect results for sucrose fermentation.

the slant and butt will appear red. An obligate aerobe will turn only the slant red. (*Note*: The difference between a red butt and a butt unchanged by the organism may be subtle; therefore, comparison with an uninoculated control is always recommended.)

Not surprisingly, timing is critical when reading results. An early reading could reveal yellow throughout the medium, leading you to believe that the organism ferments more than one sugar when it simply has not yet exhausted the glucose. A reading after all of the sugars have been depleted could reveal a yellow butt and red slant, leading you to falsely believe that the organism is a glucose-only fermenter. The timing for interpreting sulfur reduction is not as critical, so tubes that have been interpreted for carbohydrate fermentation can be re-incubated for 24 hours before final $H_2S$ determination is made. Refer to Table 5-23 for information on the correct symbols and method of reporting the various reactions.

## Application

TSIA and KIA are primarily used to differentiate members of *Enterobacteriaceae* and to distinguish them from other Gram-negative rods such as *Pseudomonas aeruginosa*.

5

## In This Exercise

Today, you will inoculate four TSIA or KIA slants. Use a large inoculum and try not to introduce excessive air when stabbing the agar butt. Also be sure to remove all tubes from the incubator after no more than 24 hours to take fermentation readings.

## Materials

### Per Student

- □ Lab coat
- □ Gloves
- □ Chemical eye protection

### Per Student Group

- □ Five TSIA or KIA slants
- □ Microbial incinerator and wire inoculating needle or disposable inoculating needle if using BSL-2 organisms
- □ Fresh agar slant cultures of these recommended organisms[2]:
  - *Escherichia coli*
  - *Morganella morganii* (BSL–2)
  - *Pseudomonas aeruginosa* (BSL–2)
  - *Salmonella typhimurium* (BSL–2)

## Medium Recipes

### Triple Sugar Iron Agar

| | |
|---|---|
| □ Pancreatic digest of casein | 10.0 g |
| □ Peptic digest of animal tissue | 10.0 g |
| □ Dextrose (glucose) | 1.0 g |
| □ Lactose | 10.0 g |
| □ Sucrose | 10.0 g |

[2] *Note to instructor:* If your lab is not equipped to handle BSL-2 organisms, these will be adequate substitutes for the recommended BSL-2 organisms: *Alcaligenes faecalis, Citrobacter freundii* (Hardy Diagnostics #0315P), and *Citrobacter koseri* (Hardy Diagnostics #0106P).

TABLE **5-23** **TSIA and KIA Test Results and Interpretations**

| Result | Interpretation | Symbol |
|---|---|---|
| Yellow slant/yellow butt—KIA | Glucose and lactose fermentation with acid accumulation in slant and butt | A/A |
| Yellow slant/yellow butt—TSIA | Glucose and lactose and/or sucrose fermentation with acid accumulation in slant and butt | A/A |
| Red slant/yellow butt | Glucose fermentation with acid production; proteins catabolized aerobically (in the slant) with alkaline products (reversion) | K/A |
| Red slant/red butt | No fermentation; peptone catabolized aerobically and anaerobically with alkaline products; isolate is not from *Enterobacteriaceae* | K/K |
| Red slant/no change in the butt | No fermentation; peptone catabolized aerobically with alkaline products; isolate is not from *Enterobacteriaceae* | K/NC |
| No change in slant/no change in butt | Organism is growing slowly or not at all; isolate is not from *Enterobacteriaceae* | NC/NC |
| Black precipitate in the agar | Sulfur reduction (an acid condition from glucose fermentation of glucose, exists in the butt even if the yellow color is obscured by the black precipitate) | $H_2S$ |
| Cracks in or lifting of agar | Gas production | G |

| | |
|---|---|
| □ Ferrous ammonium sulfate | 0.2 g |
| □ Sodium chloride | 5.0 g |
| □ Sodium thiosulfate | 0.2 g |
| □ Agar | 13.0 g |
| □ Phenol red | 0.025 g |
| □ Distilled or deionized water | 1.0 L |

*pH 7.1–7.5 at 25°C*

### Kligler Iron Agar

| | |
|---|---|
| □ Pancreatic digest of casein | 10.0 g |
| □ Peptic digest of animal tissue | 10.0 g |
| □ Lactose | 10.0 g |
| □ Dextrose (glucose) | 1.0 g |
| □ Ferric ammonium citrate | 0.5 g |
| □ Sodium chloride | 5.0 g |
| □ Sodium thiosulfate | 0.5 g |
| □ Agar | 15.0 g |
| □ Phenol red | 0.025 g |
| □ Distilled or deionized water | 1.0 L |

*pH 7.2–7.6 at 25°C*

## PROCEDURE

### Lab One

1. Wear a lab coat, gloves, and chemical eye protection when performing this procedure and use **BSL–2 precautions throughout.**
2. Obtain five slants. Label four of the slants with the names of the organisms, your group name, and the date. Label the fifth slant "control."
3. Heavily inoculate four slants with the test organisms using an inoculating needle (not a loop). First stab the agar butt and then streak the slant. Do not inoculate the control. Be sure to properly dispose of the inoculating needle (if a disposable one is used). Otherwise, incinerate the inoculating needle.
4. Incubate all slants aerobically at 35 ± 2°C for 18–24 hours (no more or less). Be sure the caps are loose enough to allow gas exchange.
5. Save or dispose of the original culture tubes as directed by your instructor.

### Lab Two

1. Examine the tubes for characteristic color changes and gas production. Use Table 5-23 as a guide while recording your results on the data sheet, page 377. The proper format for recording results is: slant reaction/butt reaction, gas production, hydrogen sulfide production. For example, a tube showing yellow slant, yellow butt, fissures in or lifting of the agar, and black precipitate would be recorded as: A/A, G, $H_2S$.
2. Dispose of the tubes in an appropriate autoclave container when finished.

5

## References

Atlas, Ronald and James Snyder. Page 338 in *Manual of Clinical Microbiology,* 11th ed. James H. Jorgensen, Michael A. Pfaller, Karen C. Carroll, Guido Funke, Marie Louise Landry, Sandra S. Richter, and David W. Warnock, eds. Washington, DC: ASM Press, American Society for Microbiology, 2015.

Delost, Maria Dannessa. Pages 184–185 in *Introduction to Diagnostic Microbiology.* St. Louis, MO: Mosby, 1997.

MacFaddin, Jean F. Page 239 in *Biochemical Tests for Identification of Medical Bacteria,* 3rd ed. Philadelphia: Lippincott Williams & Wilkins, 2000.

Tille, Patricia M. Page 228 in *Bailey & Scott's Diagnostic Microbiology,* 13th ed. St. Louis, MO: Mosby, 2014.

Winn, Jr., Washington, Stephen Allen, William Janda, Elmer Koneman, Gary Procop, Paul Schreckenberger, and Gail Woods. Pages 216–218 in *Koneman's Color Atlas and Textbook of Diagnostic Microbiology,* 6th ed. Baltimore: Lippincott Williams & Wilkins, 2006.

Zimbro, Mary Jo and David A. Power, eds. Page 283 in *Difco™ and BBL™ Manual—Manual of Microbiological Culture Media.* Sparks, MD: Becton Dickinson and Co., 2003.

Name ______________________________

Date ______________________________

Lab Section ______________________________

I was present and performed this exercise (initials) ______________________________

# DATA SHEET 5-19

## Triple Sugar Iron Agar / Kligler Iron Agar

### OBSERVATIONS AND INTERPRETATIONS

**1** Refer to Table 5-23, page 374, when recording and interpreting your results.

| Organism | Color and Gas Result | Symbol | **Interpretation** (include enzyme when known) |
|---|---|---|---|
| Control | | | |
| | | | |
| | | | |
| | | | |
| | | | |

## QUESTIONS

**1** *As mentioned in Theory, the fermentation readings with TSIA and KIA must take place between 18 and 24 hours after inoculation.*

**a.** *Why is this true?*

**b.** *Is timing as critical with $H_2S$ readings? Why or why not?*

**2** *You learned in Theory that if the black precipitate obscures the color of the butt it must be acidic and scored as "A." Why do you think this is true?* ***Hints:*** *See Figure 5.57 and examine the starting pH of TSIA and KIA).*

**3** *TSIA and KIA are complex media with many ingredients. What would be the consequences of the following mistakes in preparing this medium? Consider each independently.*

**a.** *1% glucose is added rather than the amount specified in the recipe.*

**b.** *Ferrous ammonium sulfate (or ferric ammonium citrate in KIA) is omitted.*

**c.** *Casein and animal tissue are omitted.*

**d.** *Sodium thiosulfate is omitted.*

**e.** *Phenol red is omitted.*

**f.** *The initial pH is 8.2.*

**g.** *The agar butt is shallow rather than deep.*

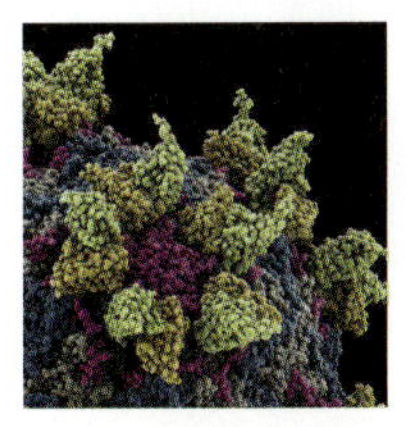

## Antimicrobial Susceptibility Testing

**Antimicrobials** are substances used to control or kill microorganisms. Some are produced naturally by microorganisms; others are synthetics produced in the laboratory. Most bacteria show susceptibility to one or more antimicrobial agents and, as illustrated in Exercise 7-2, may even react in predictable ways to the agent.

Antibiotics are antimicrobial chemicals produced by other microbes and susceptibility testing is frequently used in medicine to identify the one needed to treat an infection. But because of the predictable responses of some organisms to specific agents, susceptibility testing can also be helpful in microbial identification. All of the antibacterial susceptibility tests included in this unit apply the **disk diffusion method** and are used to aid in bacterial identification.

In the disk diffusion method, a commercially prepared filter paper disk impregnated with a specific concentration of antibiotic is placed on an agar plate immediately after it is inoculated to produce a "bacterial lawn." During incubation, two things happen simultaneously: (1) the antibacterial agent diffuses from the disk into the agar, becoming less concentrated the farther it travels; and (2) bacterial growth slowly develops on the agar surface.

In the area where antibiotic concentration in the agar is sufficient, growth of susceptible bacteria will be inhibited, which results in a clearing surrounding the disk called the **zone of inhibition**. The size of the zone relates to the **minimum inhibitory concentration** (**MIC**) of the antibacterial agent that is effective against the test organism.

The Clinical Laboratory Standards Institute (CLSI) periodically publishes a Performance Standards for Antimicrobial Testing, which includes Zone Diameter Interpretive Charts for use in testing and determining bacterial susceptibility or resistance to various antibacterial agents. The charts list the concentrations of all currently used antibacterial agents and their applicable zone diameters.

It is important to understand that even resistant organisms may be inhibited by heavy concentrations of an antimicrobial agent. Therefore, the *size* of the zone, not the mere presence of a zone, is what determines susceptibility or resistance. Additionally, specificity of disk diffusion tests is maintained largely by the amount of antibacterial agent added to the disk.

For example, many streptococci are susceptible to heavy concentrations of optochin, but resistant to small concentrations. Conversely, *Streptococcus pneumoniae* is susceptible to very small concentrations of optochin. By preparing disks with a scant 5 µg of optochin, the test system is able to accurately differentiate *S. pneumoniae* from other organisms in the genus. ■



# Bacitracin, Novobiocin, and Optochin Susceptibility Tests

EXERCISE 5-20

## Theory

**Bacitracin**, produced by *Bacillus licheniformis*, is a powerful peptide antibiotic that inhibits bacterial cell wall synthesis by interfering with transport of peptidoglycan subunits across the cytoplasmic membrane (Fig. 5.65). Thus, it is effective only on bacteria that have cell walls and are in the process of growing. A zone of inhibition 10 mm or greater around a 0.04 U bacitracin disk (called an "A" disk) is interpreted as bacitracin susceptibility (Fig. 5.66).

**Novobiocin** is an antibiotic, produced by *Streptomyces niveus*. It is related to coumarin and interferes with ATPase activity associated with DNA gyrase, an enzyme necessary during DNA replication. A zone of inhibition 16 mm or more around a 5 µg disk is interpreted as novobiocin susceptibility (Fig. 5.67).

**Optochin** is an antibiotic derived from quinine that disrupts ATP synthase activity, which results in reduced ATP production in susceptible bacteria. A zone of inhibition 14 mm or more around a 6 mm optochin disk (called a "P" disk) containing 5 µg of optochin is interpreted as optochin susceptibility (Fig. 5.68).

For more information on antimicrobial susceptibility, refer to Exercise 7-2, Antimicrobial Susceptibility Test.

$$CH_3\text{-}\overset{CH_3}{\overset{|}{C}}\text{=CH-}CH_2\text{-}[CH_2\text{-}\overset{CH_3}{\overset{|}{C}}\text{=CH-}CH_2]_9\text{-}CH_2\text{-}\overset{CH_3}{\overset{|}{C}}\text{=CH-}CH_2\text{-O-}\underset{O^-}{\overset{O}{\overset{\|}{P}}}\text{-}O^-$$

**5.65 Undecaprenyl Phosphate** ■ Undecaprenyl phosphate, a $C_{55}$ molecule derived from 11 isoprene subunits plus a phosphate, is involved in transporting peptidoglycan subunits across the cell membrane during cell wall synthesis. Bacitracin interferes with its release from the peptidoglycan subunit.

**5.66 Bacitracin Susceptibility Test on a Sheep Blood Agar Plate** ■ The zone diameter cut off for bacitracin susceptibility is 10 mm or more. The organism above has no clear zone and is resistant (R). The organism below has a zone diameter of 19 mm and is susceptible (S). The disks themselves are usually 6 mm and are referred to as "A" disks.

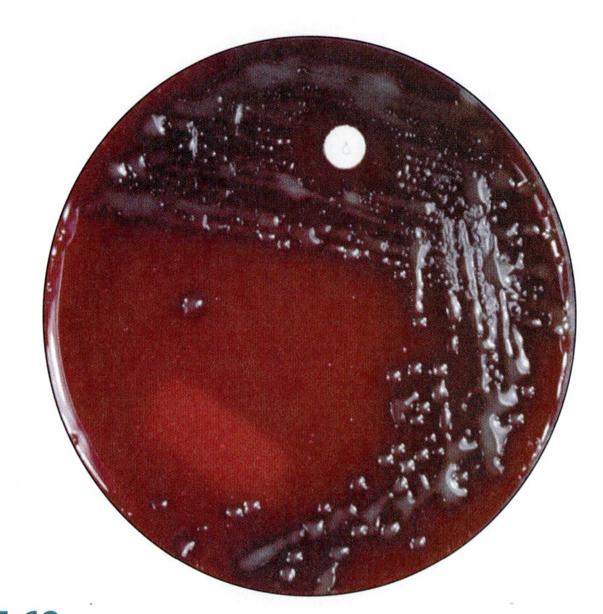

**5.68 Optochin Susceptibility Test on a Sheep Blood Agar Plate** ■ The zone diameter cut off for optochin susceptibility is 14 mm or more. The zone of inhibition surrounding the "P" disk is 19 mm, indicating susceptibility to optochin. The α-hemolysis coupled with optochin susceptibility leads to presumptive identification of *S. pneumoniae*.

**5.67 Novobiocin Susceptibility Test on a Sheep Blood Agar Plate** ■ The zone diameter cut off for novobiocin susceptibility is 16 mm or more. A novobiocin-resistant organism (R) with a zone diameter of 7 mm is on the left; a susceptible organism (S) with a zone diameter of 27 mm is on the right.

## Application

The bacitracin test is used to differentiate and presumptively identify β-hemolytic Group A streptococci (*Streptococcus pyogenes*–bacitracin susceptible) from other β-hemolytic streptococci (bacitracin resistant). It also differentiates the genus *Staphylococcus* (resistant) from the susceptible *Micrococcus*.

The novobiocin test is used to differentiate coagulase-negative staphylococci (Exercise 5-23). Most frequently it is used to presumptively identify the novobiocin-resistant *Staphylococcus saprophyticus*.

The optochin test is used to presumptively differentiate *Streptococcus pneumoniae* (susceptible) from other α-hemolytic streptococci.

## In This Exercise

You will inoculate from one to three blood agar plates (depending on the tests chosen by your instructor) to produce confluent growth of the test organisms. (***Note***: Blood agar is typically used in clinical settings for these tests, but TSA may be substituted.) You will then place specific antibiotic disks in the center of the inoculum. Following incubation, you will measure the clear zones surrounding the disks to determine susceptibility or resistance of the test organisms to the antibiotics.

## ▼ Materials

### Per Student

- □ Lab coat
- □ Gloves
- □ Chemical eye protection

### Per Student Group

- □ One blood agar plate per test (commercial preparation of TSA containing 5% sheep blood) (TSA may be substituted, if desired)
- □ Sterile cotton applicators (one per organism)
- □ Screw-cap jar of alcohol with forceps

### Bacitracin Test

- ☐ 0.04 unit bacitracin disks ("A" disks)
- ☐ Fresh broth cultures of these recommended organisms:
  - *Staphylococcus aureus* (BSL-2)
  - *Micrococcus luteus*

### Novobiocin Test

- ☐ 5 µg novobiocin disks
- ☐ Fresh broth cultures of these recommended organisms:
  - *Staphylococcus epidermidis*
  - *Staphylococcus saprophyticus*

### Optochin Test

- ☐ 5 µg (6 mm) optochin disks ("P" disks)
- ☐ Fresh broth cultures of these recommended organisms:
  - *Streptococcus pneumoniae* (BSL-2)
  - *Streptococcus agalactiae* (BSL-2)

## PROCEDURE (Per test)

### Lab One

1. Wear a lab coat, gloves, and chemical eye protection when performing this procedure. **Use BSL-2 precautions as appropriate** for the tests and organisms your class is assigned.
2. Obtain one blood agar plate for *each* test you are performing, and then label each plate with the antibiotic and your group name.
3. Next, draw a line on the bottom of each plate dividing it into two halves. Label one half with one of the organism's names and the other half with the other organism's name. Make sure you have the correct organisms for each antibiotic!
4. Using a sterile cotton applicator, inoculate half of one plate with one of its two test organisms. Make the inoculum as light as possible by wiping and twisting the wet cotton swab on the inside of the culture tube before removing it. Inoculate the plate by making a single streak nearly halfway across its diameter. Turn the plate 90° and spread the organism evenly to produce a bacterial lawn covering nearly half the agar surface. Refer to the photo in Figure 5.66.
5. Being careful not to mix the cultures, repeat the inoculation process in step 4 on the other half of the plate using the second test organism. Allow the broth to be absorbed by the agar for 5 minutes before proceeding to step 6.
6. Alcohol-flame forceps by placing them in the Bunsen burner flame long enough to ignite the alcohol, and then remove them. (***Note***: Do not hold the forceps in the flame and do not hold them near the alcohol jar. If the alcohol jar should ignite, smother the flame with the jar's lid.) Once the alcohol has burned off, use the forceps to place the appropriate antibiotic disk in the center of the half containing the first organism. Gently tap the disk into place to ensure that it makes full contact with the agar surface. Return the forceps to the alcohol.
7. Repeat step 6, placing a disk on the other half of the plate—the area containing the second organism. Tap the disk into place and return the forceps to the alcohol.
8. Repeat the inoculation and disk placement process (steps 4 through 7) for the other two plates.
9. Invert the plates and incubate them at 35 ± 2°C for 24 to 48 hours.
10. Save or dispose of the original culture tubes as directed by your instructor.

### Lab Two

1. Remove the plates from the incubator and examine them for clearing around the disks. Measure the zone diameters and record your results in the table on the data sheet, page 383. Refer to Tables 5-24, 5-25, and 5-26 to interpret your results.
2. Dispose of the plates in an appropriate autoclave container when finished.

TABLE 5-24 **Bacitracin Test Results and Interpretations**

| Result | Interpretation | Symbol |
|---|---|---|
| Zone of clearing 10 mm or greater | Organism is susceptible to bacitracin; probable *Streptococcus pyogenes* if a β-hemolytic streptococcus or *Staphylococcus* species if a Gram-positive coccus | S |
| Zone of clearing less than 10 mm | Organism is resistant to bacitracin; not *Streptococcus pyogenes*, or *Staphylococcus*; probable *Micrococcus* species if Gram-positive coccus | R |

TABLE 5-25 **Novobiocin Test Results and Interpretations**

| Result | Interpretation | Symbol |
|---|---|---|
| Zone of clearing 16 mm or greater | Organism is susceptible to novobiocin; not likely *Staphylococcus saprophyticus* if a coagulase-negative staphylococcus | S |
| Zone of clearing less than 16 mm | Organism is resistant to novobiocin; probable *Staphylococcus saprophyticus* if a coagulase-negative staphylococcus | R |

TABLE 5-26 **Optochin Test Results and Interpretations**

| Result | Interpretation | Symbol |
|---|---|---|
| Zone of clearing 14 mm or greater | Organism is susceptible to optochin; probable *Streptococcus pneumoniae* if an α-hemolytic streptococcus | S |
| Zone of clearing less than 14 mm | Organism is resistant to optochin; not *Streptococcus pneumoniae* if an α-hemolytic streptococcus | R |

5

## References


Delost, Maria Dannessa. Page 107 in *Introduction to Diagnostic Microbiology.* St. Louis, MO: Mosby, 1997.

DIFCO Laboratories. Page 292 in *DIFCO Manual*, 10th ed. Detroit: DIFCO Laboratories, 1984.

MacFaddin, Jean F. Page 3 in *Biochemical Tests for Identification of Medical Bacteria*, 3rd ed. Philadelphia: Lippincott Williams & Wilkins, 2000.

Spellerberg, Barbara and Claudia Brandt. Pages 391 and 394 in *Manual of Clinical Microbiology*, 11th ed. James H. Jorgensen, Michael A. Pfaller, Karen C. Carroll, Guido Funke, Marie Louise Landry, Sandra S. Richter, and David W. Warnock, eds. Washington, DC: ASM Press, American Society for Microbiology, 2015.

Tille, Patricia M. Page 198 and 222 in *Bailey & Scott's Diagnostic Microbiology*, 13th ed. St. Louis, MO: Mosby, 2014.

Winn, Jr., Washington, Stephen Allen, William Janda, Elmer Koneman, Gary Procop, Paul Schreckenberger, and Gail Woods. Pages 644–645, 1471–1472, and 1474 in *Koneman's Color Atlas and Textbook of Diagnostic Microbiology*, 6th ed. Baltimore: Lippincott Williams & Wilkins, 2006.

Name ____________________

Date ____________________

Lab Section ____________________

I was present and performed this exercise (initials) ____________________

# DATA SHEET 5-20

## Bacitracin, Novobiocin, and Optochin Susceptibility Tests

### OBSERVATIONS AND INTERPRETATIONS

**1** Refer to Tables 5-24, 5-25, and 5-26, page 382, when recording and interpreting your results in the table below.

| Antibacterial Agent | Organism | Zone Diameter (mm) | Susceptible (S) or Resistant (R) |
|---|---|---|---|
| Bacitracin | | | |
| | | | |
| Novobiocin | | | |
| | | | |
| Optochin | | | |
| | | | |

## QUESTIONS

**1** *Why should these tests be run with positive and negative controls?*

**2** *Why is it important to get a bacterial lawn rather than isolated colonies on the plate?*

**3** *Does the zone of inhibition's edge indicate the limit of the antibiotic's diffusion into the agar? Give reasons or evidence to support your answer.*

## Other Differential Tests

This last set of exercises includes tests that do not fit elsewhere but are important to consider. Blood agar (Exercise 5-21) detects hemolytic ability of Gram-positive cocci, and is especially useful in separating *Streptococcus* species. It also is used as a general-purpose growth medium appropriate for fastidious and nonfastidious microorganisms alike. Also targeting *Streptococcus* species (*Streptococcus agalactiae*) is the CAMP test (Exercise 5-22). The coagulase and clumping factor tests (Exercise 5-23) are commonly used to presumptively identify pathogenic *Staphylococcus aureus*. Motility test agar (Exercise 5-24) is used to detect bacterial motility, and is especially useful in differentiating *Enterobacteriaceae* and other Gram-negative rods. ■

# Blood Agar

EXERCISE 5-21

## ■ Theory

Several species of Gram-positive cocci produce exotoxins called **hemolysins**, which are able to destroy red blood cells (RBCs; erythrocytes) and hemoglobin. Blood agar includes 5% blood (frequently sheep blood) in a tryptic soy agar base. It allows differentiation of bacteria based on their ability to hemolyze RBCs.

The three major types of hemolysis are β-hemolysis, α-hemolysis, and γ-hemolysis. β-hemolysis, the complete destruction of RBCs and hemoglobin, results in a clearing of the medium around the colonies (Fig. 5.69). α-hemolysis is the partial destruction of RBCs and produces an olive-greenish discoloration of the agar around the colonies in reflected light (Fig. 5.70). Finally, γ-hemolysis is actually nonhemolysis and appears as simple growth with no change to the medium (Fig. 5.71).

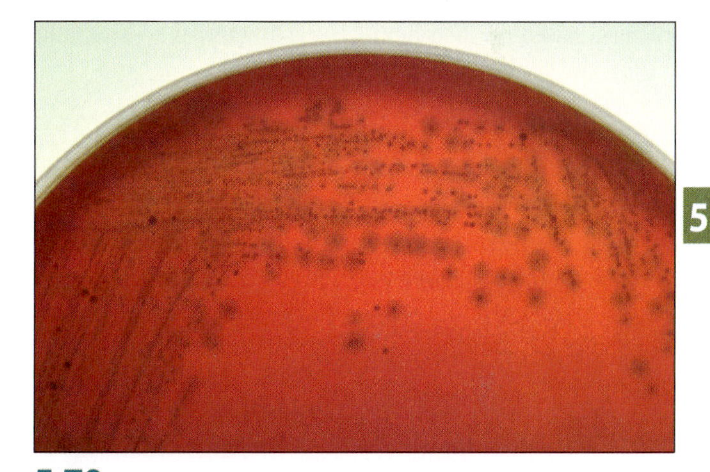

**5.70** α-**Hemolysis** ■ This is a streak plate of *Streptococcus pneumoniae* demonstrating α-hemolysis. The greenish zone around the colonies results from incomplete lysis of red blood cells. This photograph was taken with transmitted light, which may make the color appear more yellow-green than olive-green.

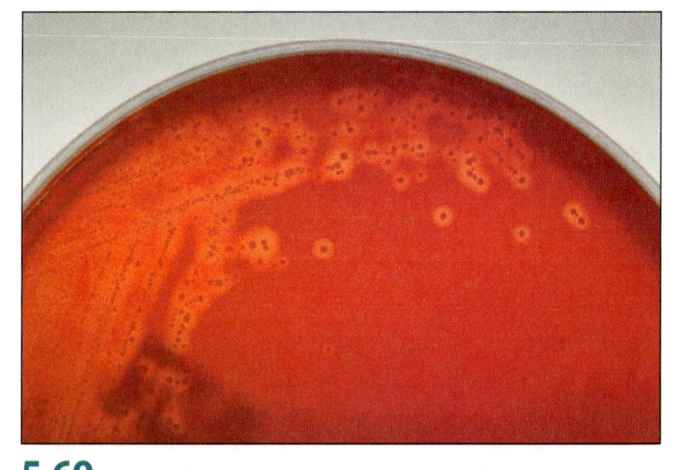

**5.69** β-**Hemolysis** ■ *Streptococcus pyogenes* demonstrates β-hemolysis. The clearing around the growth is a result of complete lysis of red blood cells. This photograph was taken with transmitted light.

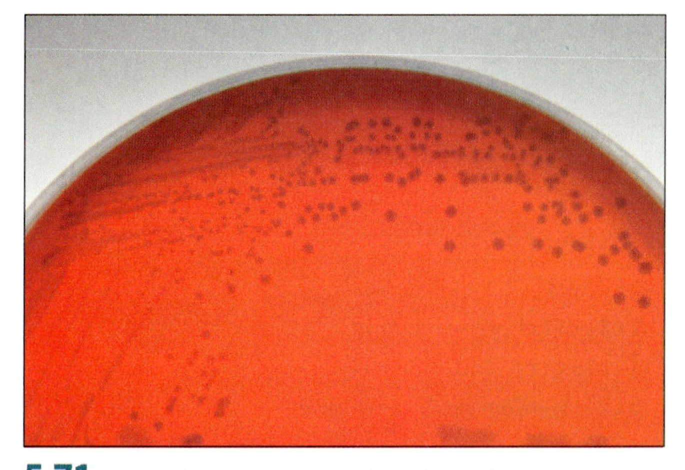

**5.71** γ-**Hemolysis** ■ This streak plate of *Staphylococcus epidermidis* on a sheep blood agar illustrates no hemolysis. This photograph was taken with transmitted light.

Hemolysins produced by streptococci are called **streptolysins.** They come in two forms—type O and type S. **Streptolysin O** is oxygen labile and expresses maximal activity under anaerobic conditions. **Streptolysin S** is oxygen stable but expresses itself optimally under anaerobic conditions as well. The easiest method of providing an environment favorable for streptolysins on blood agar is what is called the **streak-stab technique.** In this procedure the blood agar plate is streaked for isolation and then stabbed with a loop. The stab encourages streptolysin activity because of the reduced oxygen concentration of the subsurface environment. The result may be α-hemolysis on the surface, but β-hemolysis in the stab (Fig. 5.72).

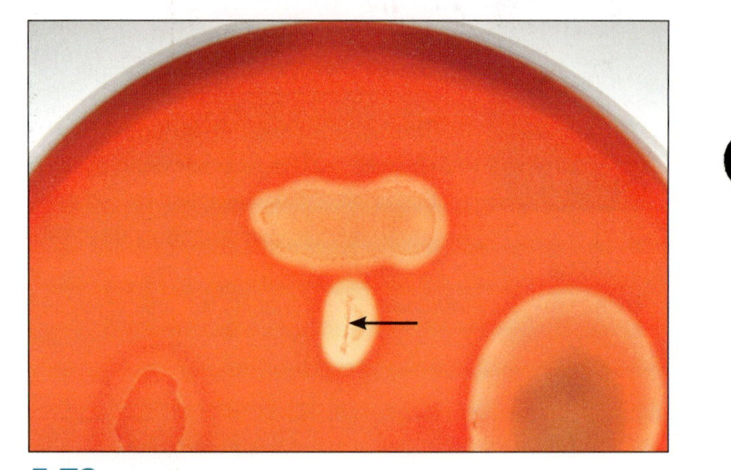

**5.72 Aerobic Versus Anaerobic Hemolysis** ■ An unidentified throat culture isolate demonstrates α-hemolysis when growing on the surface, but β-hemolysis beneath the surface surrounding the stab (arrow). This results from production of an oxygen-labile hemolysin.

## Application

Blood agar is used for isolation and cultivation of many types of fastidious bacteria. It also is used to differentiate bacteria based on their hemolytic characteristics, especially within the genera *Streptococcus*, *Enterococcus*, and *Aerococcus*.

5

## In This Exercise

You will not be provided with organisms to inoculate your plate. Instead, your partner will take a swab of your throat. Then you will inoculate the blood agar plate in one quadrant with the swab and complete a quadrant streak for isolation with your loop. Isolation of the different colonies is the only way to observe the different forms of hemolysis properly. (*Note*: When you streak your plate, do only the first streak with the swab, and do streaks two, three, and four with your inoculating loop. If you are not sure how to do this, refer to Exercise 1-5 (Figs. 1.33 through 1.35 and 1.39). Because of the potential for cultivating a pathogen from a throat swab, tape down the plate lid immediately after streaking (Fig. I.6). Upon removing your plate from the incubator, do not open it unless given permission by your instructor.

### Materials

**Per Student**

- □ Lab coat
- □ Gloves
- □ Chemical eye protection
- □ One sterile cotton swab
- □ One sterile tongue depressor
- □ Three disposable loops
- □ One sheep blood agar plate (commercially available—TSA containing 5% sheep blood)

## Medium Recipe

**5% Sheep Blood Agar (TSA ~ 5% Sheep Blood)**

| | |
|---|---|
| □ Infusion from beef heart (solids) | 2.0 g |
| □ Pancreatic digest of casein | 13.0 g |
| □ Sodium chloride | 5.0 g |
| □ Yeast extract | 5.0 g |
| □ Agar | 15.0 g |
| □ Defibrinated sheep blood | 50.0 mL |
| □ Distilled or deionized water | 1.0 L |

*pH 7.1–7.5 at 25°C*

## PROCEDURE

**Lab One**

1. Wear a lab coat, gloves, and chemical eye protection when performing this procedure and **use BSL-2 precautions throughout.**
2. Have your lab partner (also appropriately protected) obtain a culture from your throat. Follow the procedure in Appendix B, page 599.
3. Immediately transfer the specimen to a blood agar plate. Use the swab to begin a streak for isolation. Refer to Exercise 1-5, page 45 if necessary.
4. Dispose of the swab in a container designated for autoclaving.
5. Finish the isolation procedure by streaking quadrants 2, 3, and 4 with a different plastic disposable loop for each. Immediately dispose of each loop properly when done except for the last one (see step 6).
6. After completing the last streak, use your loop to stab the agar in two or three places in the first

streak pattern, and then in two or three places not previously inoculated. Properly dispose of the loop.

7. Label the plate with your name, the specimen source ("throat culture"), and the date.
8. Tape down the lid to prevent it from opening accidentally (Fig. I.6). Invert and incubate the plate aerobically at 35 ± 2°C for 24 hours. Remove the plate to the refrigerator if you cannot examine it at 24 hours.

### Lab Two

1. After incubation, do not open your plate until your instructor has seen it and given permission to do so.
2. Using transmitted and reflected light as necessary, and referring to Table 5-27, observe for color changes and clearing around the colonies. This can be done using a colony counter or by holding the plate up to a light. Record your results in the table on the data sheet, page 389.
3. Dispose of the plate in an appropriate autoclave container when finished.

TABLE **5-27** **Blood Agar Test Results and Interpretations**

| Result | Interpretation | Symbol |
|---|---|---|
| Clearing around growth | Organism hemolyzes RBCs completely | β-hemolysis |
| Greening around growth | Organism partially hemolyzes RBCs | α-hemolysis |
| No change in the medium | Organism does not hemolyze RBCs | no (γ) hemolysis |

5

## References

Atlas, Ronald and James Snyder. Page 317 in *Manual of Clinical Microbiology,* 11th ed. James H. Jorgensen, Michael A. Pfaller, Karen C. Carroll, Guido Funke, Marie Louise Landry, Sandra S. Richter, and David W. Warnock, eds. Washington, DC: ASM Press, American Society for Microbiology, 2015.

Delost, Maria Dannessa. Page 103 in *Introduction to Diagnostic Microbiology.* St. Louis, MO: Mosby, 1997.

Krieg, Noel R. Page 619 in *Methods for General and Molecular Bacteriology.* Philipp Gerhardt, R. G. E. Murray, Willis A. Wood, and Noel R. Krieg, eds. Washington, DC: American Society for Microbiology, 1994.

Power, David A. and Peggy J. McCuen. Page 115 in *Manual of BBL™ Products and Laboratory Procedures,* 6th ed. Cockeysville, MD: Becton Dickinson Microbiology Systems, 1988.

Tille, Patricia M. Page 64 in *Bailey & Scott's Diagnostic Microbiology,* 13th ed. St. Louis, MO: Mosby, 2014.

Winn, Jr., Washington, Stephen Allen, William Janda, Elmer Koneman, Gary Procop, Paul Schreckenberger, and Gail Woods. Chap. 13 in *Koneman's Color Atlas and Textbook of Diagnostic Microbiology,* 6th ed. Baltimore: Lippincott Williams & Wilkins, 2006.

Zimbro, Mary Jo and David A. Power, eds. *Difco™ and BBL™ Manual—Manual of Microbiological Culture Media.* Sparks, MD: Becton Dickinson and Co., 2003.

5

Name ______________________________

Date ______________________________

Lab Section ______________________________

I was present and performed this exercise (initials) ______________

DATA SHEET 5-21

# Blood Agar

## OBSERVATIONS AND INTERPRETATIONS

**1** Choose four different colonies (with a diversity of hemolysis reactions, including one stab) and fill in the table. Refer to Table 5-27, page 387, and Figure 2.4, page 68 when recording and interpreting your results.

| Source of Culture | Colony Morphology and Agar Appearance | Hemolysis Result ($\alpha$, $\beta$, $\gamma$) | Interpretation |
|---|---|---|---|
| | | | |
| | | | |
| | | | |
| | | | |

## QUESTIONS

**1** *The streak-stab technique, used to promote streptolysin activity, is preferred over incubating the plates anaerobically.*

**a.** *Why do you think this is so?*

______________________________

______________________________

______________________________

**b.** *Compare and contrast what you see as the advantages and disadvantages of each procedure.*

______________________________

______________________________

______________________________

______________________________

______________________________

______________________________

**2** *Assuming that all of the organisms cultivated in this exercise came from the throats of healthy students, why is it important to cover and tape the plates?*



**3** *Why is the streak plate preferred over the spot inoculations in this procedure?*

# CAMP Test

EXERCISE 5-22

## Theory

Group B *Streptococcus agalactiae* produces the CAMP factor—a hemolytic protein that acts synergistically with the β-hemolysin of *Staphylococcus aureus* subsp. *aureus*. When streaked perpendicularly to an *S. aureus* subsp. *aureus* streak on blood agar (Fig. 5.73), an arrowhead-shaped zone of hemolysis forms and is a positive result.

## Application

The CAMP test (an acronym of the developers of the test—Christie, Atkins, and Munch-Peterson) is used to differentiate Group B *Streptococcus agalactiae* (positive) from other *Streptococcus* species (negative).

## In This Exercise

You will inoculate one blood agar plate as shown in Figure 5.74 with *Streptococcus agalactiae* and *Staphylococcus aureus* subsp. *aureus*. You will do the same with *Streptococcus salivarius* and *S. aureus* subsp. *aureus* on a second plate. Following incubation, you will examine your plates for the characteristic arrowhead pattern of clearing.

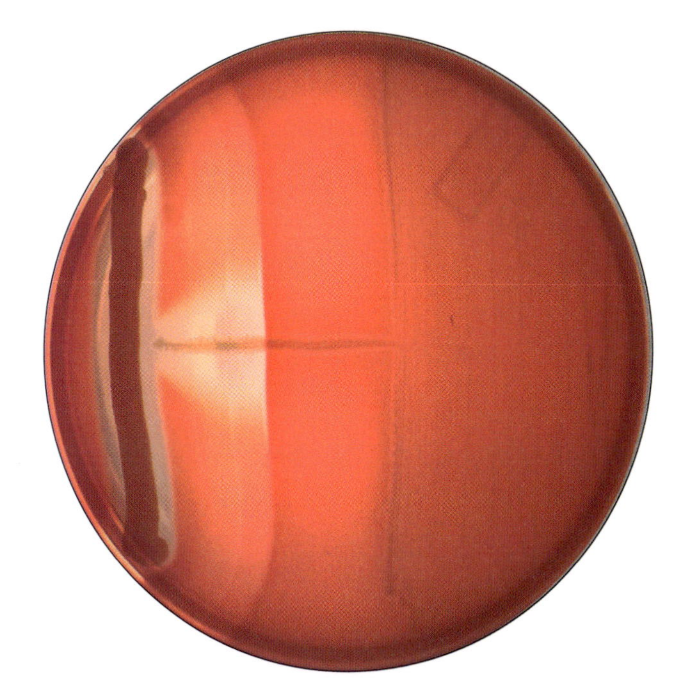

**5.73 Positive CAMP Test Results** ■ Note the arrowhead zone of clearing in the region where the CAMP factor of *Streptococcus agalactiae* acts synergistically with the β-hemolysin of *Staphylococcus aureus* subsp. *aureus*.

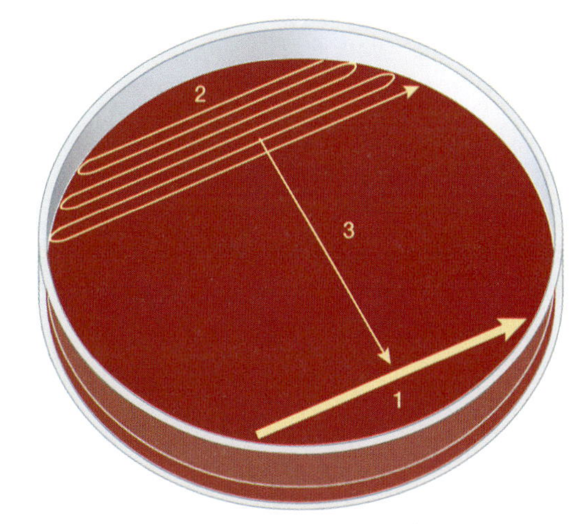


**5.74 CAMP Test Inoculation** ■ Two inoculations are made. First *Staphylococcus aureus* subsp. *aureus* is streaked along one edge of a fresh blood agar plate (1). Then the isolate (when testing an unknown organism) is inoculated densely in the other half of the plate opposite *S. aureus* (2). Finally, a single streak is made from inside streak II toward, but not touching, *S. aureus* (3).

5

## ▼ Materials

### Per Student

- ☐ Lab coat
- ☐ Gloves
- ☐ Chemical eye protection

### Per Student Group

- ☐ Two blood agar plates (commercial preparation of TSA containing 5% sheep blood)
- ☐ Three Sterile cotton applicators
- ☐ Fresh broth cultures of these recommended organisms:
  - *Staphylococcus aureus* subsp. *aureus* (BSL-2)
  - *Streptococcus agalactiae* (BSL-2)
  - *Streptococcus salivarius*

## PROCEDURE

### Lab One

1. Wear a lab coat, gloves, and chemical eye protection when performing this procedure and **use BSL-2 precautions throughout.**
2. Obtain two blood agar plates

3. Using a cotton swab, inoculate both plates as shown in Figure 5.74, each with a single streak of *S. aureus* subsp. *aureus* along one edge (streak 1). Properly dispose of the swab when finished streaking.
4. Inoculate one plate with a dense smear of *S. agalactiae* across from the *S. aureus* streak, as shown in streak 2. Then finish the plate with a single streak (as in streak 3) from the *S. agalactiae* species toward, but not touching, the *S. aureus*.
5. Repeat the procedure on the second plate with *S. salivarius*.
6. Label the plates with your group name, the date, and the names of the organisms. Tape them closed as in Figure I.6.
7. Invert the plates and incubate them at 35 ± 2°C for 24 hours.
8. Save or dispose of the original culture tubes as directed by your instructor.

### Lab Two

1. Do not open the plates.
2. Remove the plates from the incubator and observe for the characteristic arrowhead pattern.
3. Record your results in the table on the data sheet, page 393, referring to Table 5-28 to interpret your results.
4. Dispose of the plates in an appropriate autoclave container when finished.

**TABLE 5-28 CAMP Test Results and Interpretations**

| Result | Interpretation | Symbol |
|---|---|---|
| Arrowhead pattern | Organism produces hemolytic CAMP protein; presumptively identified as *Streptococcus agalactiae* | + |
| No arrowhead pattern | Organism does not produce hemolytic CAMP protein; presumptively identified as *not S. agalactiae* | – |

## References


Atlas, Ronald and James Snyder. Page 317 in *Manual of Clinical Microbiology*, 11th ed. James H. Jorgensen, Michael A. Pfaller, Karen C. Carroll, Guido Funke, Marie Louise Landry, Sandra S. Richter, and David W. Warnock, eds. Washington, DC: ASM Press, American Society for Microbiology, 2015.

Murray, Patrick R., Ken S. Rosenthal, and Michael A. Pfaller. Page 250 in *Medical Microbiology*, 5th ed. Philadelphia: Elsevier Mosby, 2005.

Spellerberg, Barbara and Claudia Brandt. Pages 391–392 in *Manual of Clinical Microbiology*, 11th ed. James H. Jorgensen, Michael A. Pfaller, Karen C. Carroll, Guido Funke, Marie Louise Landry, Sandra S. Richter, and David W. Warnock, eds. Washington, DC: ASM Press, American Society for Microbiology, 2015.

Tille, Patricia M. Pages 200 and 257 in *Bailey & Scott's Diagnostic Microbiology*, 13th ed. St. Louis, MO: Mosby, 2014.

Winn, Jr., Washington, Stephen Allen, William Janda, Elmer Koneman, Gary Procop, Paul Schreckenberger, and Gail Woods. Pages 717 and 1468 in *Koneman's Color Atlas and Textbook of Diagnostic Microbiology*, 6th ed. Baltimore: Lippincott Williams & Wilkins, 2006.

Name

Date

Lab Section

I was present and performed this exercise (initials)

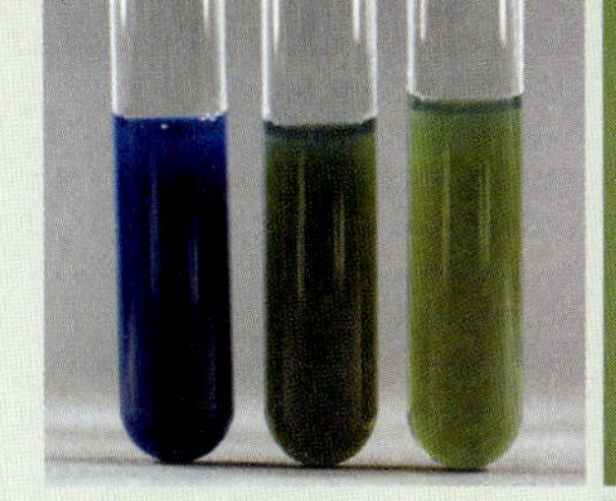

DATA SHEET

5-22

# CAMP Test

## OBSERVATIONS AND INTERPRETATIONS

**1** Refer to Table 5-28, page 392, when recording and interpreting your results in the table below.

| **Organism** (interacting with *S. aureus* subsp. *aureus*) | **Result** (+/−) | **Interpretation** |
|---|---|---|
| | | |
| | | |

## QUESTIONS

**1** *You were instructed to not allow streak 1 and streak 3 to touch. Why do you think this is important?*

**2** *Consider the arrowhead region of clearing.*

**a.** *Why do you think the clearing in a positive CAMP test is an arrowhead shape and not some other shape?*

**b.** *Would it likely make an important difference if you began streak 2 near* S. aureus *and streaked across the plate in the other direction? Why?*

5

**3** *What result would you expect to get if you accidentally reverse the organisms in the procedure?*

EXERCISE
5-23

# Coagulase and Clumping Factor Tests

## Theory

*Staphylococcus aureus* is an opportunistic pathogen that can be highly resistant to both the normal immune response and antimicrobial agents. Its resistance is due, in part, to the production of a coagulase enzyme. Plasma is the fluid portion of blood and includes, among other components, clotting factors. Coagulase works in conjunction with normal plasma components to form protective fibrin barriers around individual bacterial cells or groups of cells, shielding them from phagocytosis and other types of attack.

Coagulase enzymes occur in two forms—**free coagulase** and **bound coagulase**. Free coagulase is an extracellular enzyme (released from the cell) that reacts with a plasma component called coagulase-reacting factor (CRF). The resulting reaction is similar to the conversion of prothrombin and fibrinogen in the normal clotting mechanism. Bound coagulase (**clumping factor**) is part of the bacterial cell wall and when it reacts with plasma fibrinogen, the precipitated fibrin links the cells to produce visible clumps.

Two forms of the coagulase test have been devised to detect the enzymes: the tube test and the slide test. The tube test detects the presence of either bound or free coagulase and the slide test detects only bound coagulase. Both tests utilize rabbit plasma treated with anticoagulant (usually EDTA) to interrupt the normal clotting mechanisms.

The tube test is performed by adding the test organism to rabbit plasma in a test tube. Coagulation of the plasma (including any thickening or formation of fibrin threads) within 24 hours indicates a positive reaction (Fig. 5.75). The plasma typically is examined for clotting (without shaking) periodically for about 4 hours. After 4 hours coagulase-negative tubes can be incubated overnight, but no more than a total of 24 hours, because coagulation can take place early and revert to liquid within 24 hours due to activity of staphylococcal fibrinolytic enzymes.

In the slide test, bacteria are transferred to a slide containing a small amount of plasma. Agglutination of the cells on the slide within 1 to 2 minutes indicates the presence of bound coagulase (Fig. 5.76). Equivocal or negative slide test results typically are given the tube test for confirmation.

## Application

The coagulase test differentiates *Staphylococcus aureus* from other Gram-positive cocci.

**5.75 Coagulase Tube Test Results in Rabbit Plasma** ■ These coagulase tubes illustrate a coagulase-negative organism (below) and a coagulase-positive organism (above). The tube test identifies both bound and free coagulase enzymes. Coagulase increases bacterial resistance to phagocytosis and antibodies by surrounding infecting organisms with a clot. Tests must be read within 24 hours to avoid reversion of positive tests by staphylococcal fibrinolytic enzyme activity. **Use BSL-2 precautions when performing this test** with an unknown organism or *S. aureus*.

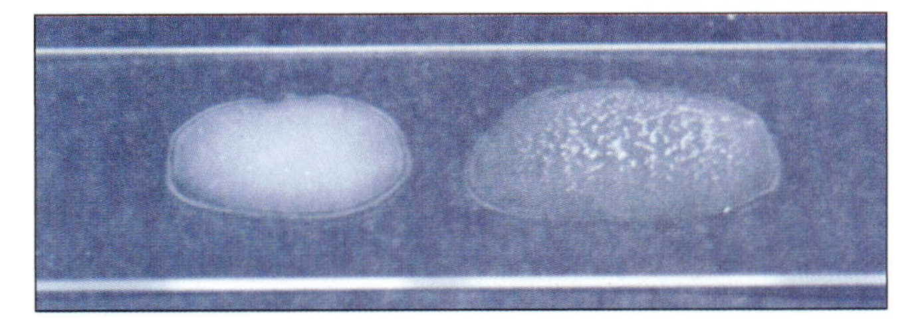

**5.76 Coagulase Slide Test Results (Clumping Factor)** ■ This slide illustrates a coagulase-negative organism on the left and a coagulase-positive organism on the right. Agglutination of the coagulase plasma is indicative of a positive result for bound coagulase. **Use BSL-2 precautions when performing this test** with an unknown organism or *S. aureus*.

## In This Exercise

You will perform both tube and slide coagulase tests. Immediately enter your slide test results on the data sheet, page 399. Read the tube test the following day. It would be most efficient if you inoculate the tube tests as early in your lab period as possible, with the hope that a positive result will be obtained before you leave.

### Materials

**Per Student**

- ☐ Lab coat
- ☐ Gloves
- ☐ Chemical eye protection

### Per Student Group

- □ Four sterile rabbit plasma tubes with EDTA (0.5 mL in 12 mm × 75 mm test tubes)
- □ Two sterile 1 mL transfer pipettes
- □ Six sterile wooden sticks or disposable plastic loops
- □ Sterile saline (0.9% NaCl)
- □ Two microscope slides
- □ Fresh agar slant cultures of these recommended organisms:
  - *Staphylococcus aureus* (BSL-2)
  - *Staphylococcus epidermidis*

## PROCEDURE

### Lab One: Tube Test

1. Wear a lab coat, gloves, and chemical eye protection when performing this procedure and **use BSL-2 precautions throughout.**
2. Obtain three coagulase tubes. Label two tubes with the names of the organisms, your group name, and the date. Label the third tube "control."
3. Using a sterile wooden stick or disposable loop, inoculate two tubes with the test organisms. There should be slightly visible turbidity. Immediately dispose of the stick/loop properly. Then mix the contents by gently rolling the tube between your hands. Do not inoculate the control.
4. Incubate all tubes at 35 ± 2°C for up to 24 hours, checking for coagulation periodically for the first 4 hours or remainder of your lab.
5. Do not exceed 24 hours of incubation. You will need to move the tubes to a refrigerator until your next lab period if you don't meet daily.

### Lab One: Slide Test (Clumping Factor)

1. Wear a lab coat, gloves, and chemical eye protection when performing this procedure and **use BSL-2 precautions throughout.**
2. Obtain two microscope slides and divide them in half with a marking pen. Label the sides A and B. Label one slide *S. aureus* and the other *S. epidermidis.*
3. Place a drop of sterile saline on side A and a drop of coagulase plasma (from the fourth coagulase tube) on side B of both slides. Use a different sterile transfer pipette for each solution.
4. Using a different sterile wooden stick or a disposable loop for each, transfer a loopful of *S. aureus* to the saline and plasma drops of one slide, making sure to gently and uniformly emulsify the bacteria in the solutions. Immediately dispose of the stick/loop properly. Gently rock the slide for 5 to 10 seconds, and then observe for agglutination within 2 minutes. Clumping after 2 minutes is not a positive result.
5. Repeat step 4 using the other slide and *S. epidermidis.*
6. Record your results in the table on the data sheet. Refer to Figure 5.76 and Table 5-29 when making your interpretations. Confirm any negative results by comparing with the corresponding completed tube test in 24 hours.
7. Dispose of the slides in an appropriate autoclave container when finished.

**TABLE 5-29 Clumping Factor (Slide Test) Results and Interpretations**

| Result | Interpretation | Symbol |
|---|---|---|
| Clumping of cells | Plasma has been coagulated by bound coagulase (clumping factor) | + |
| No clumping of cells | Plasma has not been coagulated or is undetectable | – |

### Lab Two

1. Remove all tubes from the incubator no later than 24 hours after inoculation. Examine for clotting of the plasma.
2. Record your results on the data sheet. Refer to Figure 5.75 and Table 5-30 when making your interpretations.
3. Dispose of the tubes in an appropriate autoclave container when finished.

TABLE **5-30** **Coagulase Tube Test Results and Interpretations**

| Result | Interpretation | Symbol |
|---|---|---|
| Medium is solid within 24 hours | Plasma has been coagulated by free or bound coagulase; probable *Staphylococcus aureus* | + |
| Medium remains liquid after 24 hours | Plasma has not been coagulated or is undetectable; not *S. aureus* | – |

5

## References


Atlas, Ronald and James Snyder. Page 317 in *Manual of Clinical Microbiology,* 11th ed. James H. Jorgensen, Michael A. Pfaller, Karen C. Carroll, Guido Funke, Marie Louise Landry, Sandra S. Richter, and David W. Warnock, eds. Washington, DC: ASM Press, American Society for Microbiology, 2015.

Becker, Karsten, Robert L. Skov, and Christof von Eiff. Pages 362–363 in *Manual of Clinical Microbiology,* 11th ed. James H. Jorgensen, Michael A. Pfaller, Karen C. Carroll, Guido Funke, Marie Louise Landry, Sandra S. Richter, and David W. Warnock, eds. Washington, DC: ASM Press, American Society for Microbiology, 2015.

Collins, C. H., Patricia M. Lyne, and J. M. Grange. Page 111 in *Collins and Lyne's Microbiological Methods,* 7th ed. Oxford, Boston: Butterworth-Heinemann, 1995.

Delost, Maria Dannessa. Pages 98–99 in *Introduction to Diagnostic Microbiology*. St. Louis, MO: Mosby, 1997.

DIFCO Laboratories. Page 232 in *DIFCO Manual,* 10th ed. Detroit: DIFCO Laboratories, 1984.

Holt, John G., ed. *Bergey's Manual of Determinative Bacteriology,* 9th ed. Baltimore: Lippincott Williams and Wilkins, 1994.

Lányi, B. Page 62 in *Methods in Microbiology,* Vol. 19. R. R. Colwell and R. Grigorova, eds. New York: Academic Press, 1987.

MacFaddin, Jean F. Page 105 in *Biochemical Tests for Identification of Medical Bacteria,* 3rd ed. Philadelphia: Lippincott Williams & Wilkins, 2000.

Tille, Patricia M. Page 203 in *Bailey & Scott's Diagnostic Microbiology,* 13th ed. St. Louis, MO: Mosby, 2014.

Winn, Jr., Washington, Stephen Allen, William Janda, Elmer Koneman, Gary Procop, Paul Schreckenberger, and Gail Woods. Pages 645–646 in *Koneman's Color Atlas and Textbook of Diagnostic Microbiology,* 6th ed. Baltimore: Lippincott Williams & Wilkins, 2006.

Name ______________________________

Date ______________________________

Lab Section ______________________________

I was present and performed this exercise (initials) ______________

DATA SHEET 5-23

# Coagulase and Clumping Factor Tests

## OBSERVATIONS AND INTERPRETATIONS

**1** Refer to Tables 5-29 and 5-30, pages 396–397, when recording and interpreting your results in the tables below.

**Slide Test Results**

| Organism | Result (include solution tested) | Interpretation (include coagulase type and provisional identification) |
|---|---|---|
| | Side A | |
| | Side B | |
| | Side A | |
| | Side B | |

**Tube Test Results**

| Organism | Result | Interpretation (include coagulase type and provisional identification) |
|---|---|---|
| Uninoculated control | | |
| | | |
| | | |

## QUESTIONS

**1** *Consider the controls.*

**a.** *Are the uninoculated tube and the saline on the slide positive or negative controls?*

**b.** *What information is provided by the uninoculated tube in the tube test?*

**c.** *What information is provided by the saline solution in the slide test?*

**d.** *Why is it a good idea to run a positive control in both the slide and tube tests?*

5

**2** *Why is it more important to use fresh cultures in the coagulase test than in a test medium such as milk agar or phenol red broth?*

**3** *How would you interpret a negative slide test and a positive tube test using the same organism?*

**4** *List possible reasons why the slide test is not appropriate for detecting free coagulase.*

# Motility Agar

EXERCISE 5-24

## Theory

Motility agar is a semisolid medium designed to detect bacterial motility. Its agar concentration is reduced from the typical 1.5% to 0.4%—just enough to maintain its form while allowing movement of motile bacteria. It is inoculated by stabbing with a straight transfer needle. Motility is detectable as diffuse growth radiating from the central stab line.

A tetrazolium salt (TTC) sometimes is added to the medium to make interpretation easier. TTC is used by the bacteria as an electron acceptor. In its oxidized form, TTC is colorless and soluble; when reduced it is red and insoluble (Fig. 5.77). A positive result for motility is indicated when the red (reduced) TTC is seen radiating outward from the central stab. A negative result shows red only along the stab line (Fig. 5.78).

## Application

This test is used to detect bacterial motility. Motility is an important differential characteristic of *Enterobacteriaceae* and other groups.

## In This Exercise

You will inoculate two motility stabs with an inoculating needle. Straighten the needle before you stab the medium. It is important to stab straight into the medium and remove the needle along the same line. Lateral movement of the needle will make interpretation more difficult.

## Materials

### Per Student

- ☐ Lab coat
- ☐ Gloves
- ☐ Chemical eye protection

### Per Student Group

- ☐ Three motility test agar tubes
- ☐ Fresh agar slant cultures of these recommended organisms:
  - *Enterobacter aerogenes*
  - *Streptococcus* spp. or *Staphylococcus epidermidis*

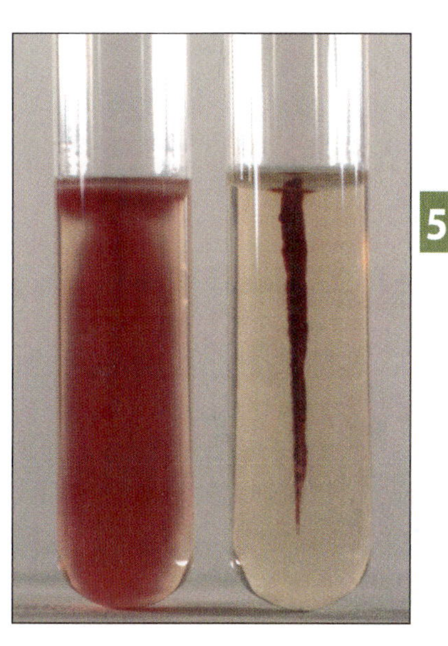

**5.78 Results in Motility Agar with TTC** ■ These motility test media were inoculated with a motile organism on the left and a nonmotile organism on the right.

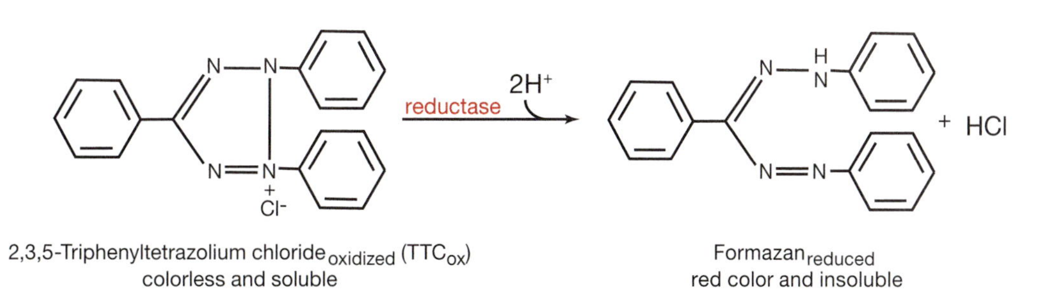


**5.77 Motility Agar Indicator Reaction: Reduction of TTC** ■ Reduction of 2,3,5-triphenyltetrazolium chloride (TTC) by metabolizing bacteria results in its conversion from colorless and soluble to the red and insoluble compound formazan. The location of growing bacteria can be determined easily by the location of the formazan in the medium.

## Medium Recipe

### Motility Test Medium

| | |
|---|---|
| □ Beef extract | 3.0 g |
| □ Pancreatic digest of gelatin | 10.0 g |
| □ Sodium chloride | 5.0 g |
| □ Agar | 4.0 g |
| □ Triphenyltetrazolium chloride (TTC) | 0.05 g |
| □ Distilled or deionized water | 1.0 L |

*pH 7.1–7.5 at 25°C*

### Lab One

1. Wear a lab coat, gloves, and chemical eye protection when performing this procedure.
2. Obtain three motility stab tubes. Label two tubes, each with the name of the organism, your group name, and the date. Label the third tube "control."
3. Using a straight inoculating needle, stab inoculate two tubes with the test organisms. (Motility can be obscured by careless stabbing technique. Try to avoid lateral movement when performing this stab.) Flame the inoculating needle after each stab.
4. Stab the control with a sterile inoculating needle. This really is unnecessary, but it provides one more opportunity to practice stabbing the motility agar in a straight line.
5. Incubate the tubes aerobically at 35 ± 2°C for 24 to 48 hours.
6. Save or dispose of the original culture tubes as directed by your instructor.

### Lab Two

1. Examine the tubes. Wherever the organism has grown the medium will appear red because the TTC has been reduced. Your interpretation will be based on growth only along the stab line or spreading from the stab line. Refer to Figure 5.78 and Table 5-31 when interpreting your results.
2. Record your results in the chart provided on the data sheet, page 403.
3. Dispose of the tubes in an appropriate autoclave container when finished.

TABLE **5-31** **Motility Agar Test Results and Interpretations**

| Result | Interpretation | Symbol |
|---|---|---|
| Red diffuse growth radiating outward from the stab line | The organism is motile | + |
| Red growth only along the stab line | The organism is nonmotile | − |

### References


MacFaddin, Jean F. Page 327 in *Biochemical Tests for Identification of Medical Bacteria,* 3rd ed. Philadelphia: Lippincott Williams & Wilkins, 2000.

Tille, Patricia M. Page 216 in *Bailey & Scott's Diagnostic Microbiology,* 13th ed. St. Louis, MO: Mosby, 2014.

Winn, Jr., Washington, Stephen Allen, William Janda, Elmer Koneman, Gary Procop, Paul Schreckenberger, and Gail Woods. Pages 227–228 in *Koneman's Color Atlas and Textbook of Diagnostic Microbiology,* 6th ed. Baltimore: Lippincott Williams & Wilkins, 2006.

Zimbro, Mary Jo and David A. Power, eds. Page 374 in *Difco™ and BBL™ Manual—Manual of Microbiological Culture Media.* Sparks, MD: Becton Dickinson and Co., 2003.

Name ______________________________

Date ______________________________

Lab Section ______________________________

I was present and performed this exercise (initials) ______________________________

# DATA SHEET 5-24

## Motility Agar

### OBSERVATIONS AND INTERPRETATIONS

**1** Refer to Table 5-31, page 402, when recording and interpreting your results in the table below.

| Organism | Result (location of red TTC) | Motility (+/−) | Interpretation |
|---|---|---|---|
| | | | |
| | | | |

### QUESTIONS

**1** *Consider the tube stabbed with the sterile inoculating needle.*

**a.** *Is this a positive or a negative control?*

______________________________

**b.** *What information is provided by the sterile-stabbed tube?*

______________________________

**2** *Why is it important to carefully insert and remove the needle along the same stab line?*

______________________________

**3** *Consider the TTC indicator.*

**a.** *Why is it essential that the reduced TTC be insoluble?*

**b.** *Why is there less concern about the solubility of the oxidized form of TTC?*

5

SECTION 6

# Quantitative Techniques

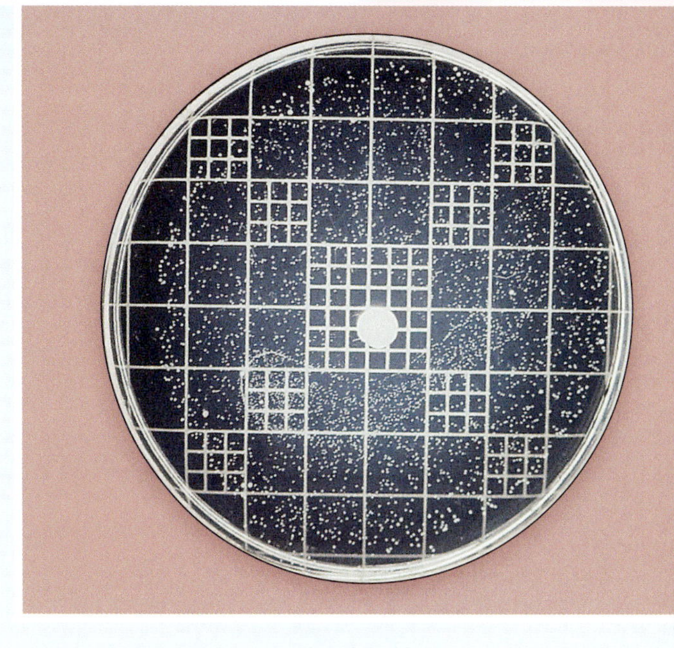

Knowledge of microbial population density is important in many areas of microbiology. Food and environmental microbiologists use population density for food production as well as the detection of food or water contamination. Medical microbiologists manipulate population density for use in standardized tests. Many researchers use population density to measure the effect of varying nutritional or environmental conditions. Industrial microbiologists maintain microbial populations at optimum levels in large-scale fermenters for the manufacture of many products, including enzymes, antibiotics, beer, and wine.

The most direct method of determining population density is to microscopically count cells in a known sample volume, which then can be converted to a value with the units of cells/mL. More frequently, however, density estimates are made indirectly by inoculating an agar plate with a known sample volume and counting the colonies that result.

In most instances actual *cell* density is not determined by this method, though, because that would require an assumption that each colony came from a single cell, which oftentimes is not the case. Recall that some bacteria grow in arrangements other than single cells, so a diplococcus, for instance, could start a single colony but the colony would have originated from *two* cells. Unless we can be sure every colony came from single cells, the more truthful estimate of colony forming unit density (CFU/mL) is reported.

The first three lab exercises in this section show different bacterial counting techniques. Exercise 6-1 uses a RODAC™ plate to sample environmental surfaces and estimate microbial density per square centimeter. It is the only method in this book where we will estimate density in an area rather than a volume.

Exercise 6-2 is a standard plate count (also known as a viable count), which is very versatile and is used in various food and industrial settings, among others. In it you will estimate the density of *E. coli* in a broth culture grown overnight, but the sample you use could just as easily be a dairy product being subjected to quality control.

Exercise 6-3 is a medical application and illustrates how microbial density in a urine sample can be determined. The procedure in Exercise 6-4 has a lot in common with the standard plate count, except that you will be determining the density of a virus in a sample instead of bacteria. There are some interesting differences, though, due to the different demands growing viruses place on a microbiologist.

Lastly, you will do a differential blood cell count in Exercise 6-5. While a differential blood count doesn't provide information about a microbial population density, it employs a direct microscopic counting technique that is useful as a clinical diagnostic tool. It is the only direct counting procedure in this section and determines the percentage of each white blood cell type.

As you will see, most of the techniques involve dilutions and all require calculations to determine the *original* population density. (You thought that in a section entitled "Quantitative Techniques" you would avoid doing calculations?) Fear not. The calculations are not terribly difficult and many are repetitive, so they become second nature quickly.

Microbiology is the science of extraordinarily small organisms in extraordinarily large numbers, which makes counting them a challenge. To avoid counting extraordinarily large microbial numbers in a sample, we count them in a fraction of a milliliter and then extrapolate from that number back to one milliliter.

We can do this in one of two ways: We can get the estimate by using a smaller volume, or we can dilute the original (which is actually a way of working with a fraction of the original). As long as we know the volumes used, we can successfully work backward to the original density.

Working with a smaller volume has its limitations. It is hard to work with volumes less than 0.001 mL (1/1000 mL, or $10^{-3}$ mL), so if a sample has $10^8$ cells/mL and we try to count the cells in 0.001 mL, we would still be counting: $10^8$ cells/mL $\times$ $10^{-3}$ mL $=$ $10^5$ cells! That's too many and we've got better things to do with our time!

The alternative is to dilute the sample, and sometimes a single dilution of a sample is sufficient to make counting manageable. However, more frequently we need to perform a serial dilution, in which a sample is diluted, then that first dilution is diluted again, then it is diluted a third time, and so on until the density becomes small enough to be used in an appropriate counting procedure (Fig. 6.3). As long as the amount of each dilution is known, then the original density can be calculated, because each dilution contains a known volume of the undiluted original.

Most of the quantitative techniques in this section were originally designed for measurements and calculations in milliliters. Many school laboratories now are equipped with digital micropipettors that have the ability to deliver volumes as small as 1.0 microliter. As long as you know that 1.0 µL = 0.001 mL and 1.0 mL = 1,000 µL, you will be fine!

6

EXERCISE 6-1

# Environmental Sampling: The RODAC™ Plate

## Theory

Monitoring of microbial surface contamination is an important practice in medical, veterinary, pharmaceutical, and food preparation settings. Often, the RODAC™ (Replicate Organism Detection and Counting) plate is used. It is a specially designed agar plate into which the medium is poured to produce a convex surface extending above the edge of the plate (Fig. 6.1). The special design allows support of the lid above the agar without touching it.

As a result, the sterile plate may be opened and pressed on a surface to be sampled. Most are 65 mm in diameter, which is smaller than standard-sized 100 mm Petri dishes. The smaller size makes it easier to apply uniform pressure across the whole plate when taking the sample. In addition, the base is marked in 16 1 cm squares, allowing an estimate of cell density on the surface (Fig. 6.2).

The plate can be filled with a variety of media, the choice of which depends on the surface being sampled and the microbes to be recovered. For instance, in this lab you will be using plates with Trypticase™ soy agar, a good, general-purpose growth medium. The medium can be supplemented with 5% sheep blood to improve recovery of fastidious bacteria. Fecal coliform density (always a concern) can be checked using m-FC agar or MacConkey agar. Monitoring yeast and mold contamination often employs Sabouraud dextrose agar. Because the surface sampled may have been recently disinfected, polysorbate 80 and lecithin are added to counteract the effect of residual disinfectant.

The acceptable amount of growth on a RODAC™ plate is determined by the surface being sampled. It stands to reason that a surgical area would have a lower limit of acceptability than a food preparation area. Table 6-1 provides some guidelines.

**6.1 RODAC™ Plate** ■ This plate is viewed from the side with the lid removed. Notice that the agar extends above the edges of the plate to allow contact with the surface to be sampled.

**6.2 Grid on the RODAC™ Plate** ■ Typically, RODAC™ plates are 65 mm in diameter. There is a grid of 16 squares molded into the base, each with an area of 1 $cm^2$ (seen here through the agar). Colonies growing in the grid can be counted and an average number of CFU per $cm^2$ of surface can be determined.

TABLE **6-1 Interpretation of RODAC™ Plate Colony Counts (Colonies Per Plate)**[1]

| Interpretation | Critical Surfaces[2] | Floors |
|---|---|---|
| Good | 0-5 | 0-25 |
| Fair | 6-15 | 26-50 |
| Poor | >16 | >50 |

[1] Adapted from BBL™ Trypticase™ Soy Agar with Lecithin and Polysorbate 80 package insert.

[2] Critical surfaces include those in operating rooms, nurseries, table tops, toilet seats, and other nonporous surfaces.

## Application

The RODAC™ plate is used to monitor surface contamination in food preparation, veterinary, pharmaceutical, and medical settings. The plates can also be used to assess the efficiency of decontamination of a surface by taking a sample with different plates before and after decontamination.

## In This Exercise

You will sample your work surface before and after decontamination to evaluate your technique.

## ▼ Materials

### Per Student

- □ Lab coat
- □ Disposable gloves
- □ Chemical eye protection

### Per Student Group

- □ Two Trypticase™ soy agar with lecithin and polysorbate 80 RODAC™ plates (available from BD Diagnostic Systems, http://www.bd.com/us/, Catalog number 221288). Alternatively, sterile RODAC™ plates may be purchased and filled with 16.5 mL to 17.5 mL TSA plus Lecithin and Polysorbate 80.
- □ Disinfectant solution used in your laboratory

## ■ Medium Recipe

### Trypticase™ Soy Agar with Lecithin and Polysorbate 80

|  |  |
|---|---|
| □ Pancreatic digest of casein | 15.0 g |
| □ Papaic digest of soybean meal | 5.0 g |
| □ Sodium chloride | 5.0 g |
| □ Lecithin | 0.7 g |
| □ Polysorbate 80 | 5.0 g |
| □ Agar | 15.0 g |
| □ Distilled or deionized water | 1.0 L |

*pH 7.1–7.5 at 25°C*

6

## PROCEDURE

### Lab One

1. IMPORTANT: For today's lab only, DO NOT disinfect your lab bench prior to beginning work.
2. Wear a lab coat, gloves, and chemical eye protection when performing this procedure.
3. Obtain two RODAC™ plates. Label the base of one plate "Before" and the base of the other "After." Also label the plates with your group name and date.
4. Remove the lid of the "Before" plate and gently press the agar onto the surface of your lab bench. Apply uniform vertical pressure and do not slide the plate. Replace the lid.
5. Wipe down the bench surface with your lab's disinfectant. Allow it to dry on its own before proceeding.
6. Remove the lid of the "After" plate and gently press the agar onto the surface of your lab bench as before. Apply uniform vertical pressure and do not slide the plate. Replace the lid.
7. Incubate both plates for 48–72 hours at 35 ± 2°C.

### Lab Two

1. Examine the plates for growth. Count the number of colonies in the grid to get an estimate of CFU density on the lab bench before and after decontamination.
2. Record your results on the data sheet, page 409.
3. Dispose of the plate in an appropriate autoclave container when finished.

### References


American Public Health Association (Laboratory Section), Committee on Microbial Contamination of Surfaces. 1970. "A Cooperative Microbiological Evaluation of Floor Cleaning Procedures in Hospital Patient Rooms." *Health Lab. Sci.* 7, no. 4 (1970): 256–264.

BBL™ Trypticase™ Soy Agar with Lecithin and Polysorbate 80 package insert. BD Diagnostic Systems, http://www.bd.com/us/.

Pepper, Ian L., Charles P. Gerba, and Terry J. Gentry. Page 174 in *Environmental Microbiology*, 3rd ed. San Diego, CA: Academic Press, 2015.

Name ______________________________

Date ______________________________

Lab Section ______________________________

I was present and performed this exercise (initials) ______________________________

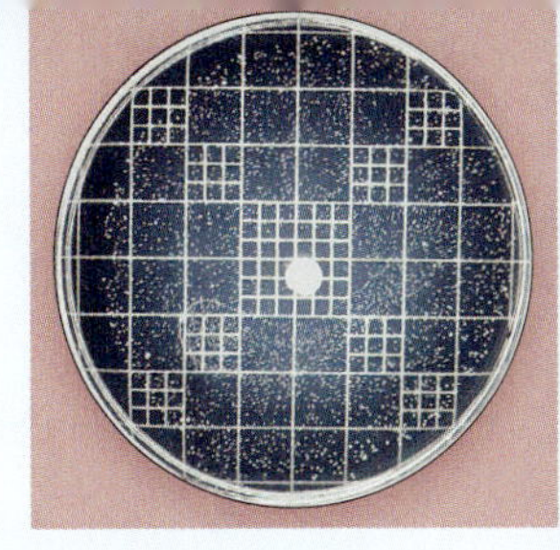

# DATA SHEET 6-1

## Environmental Sampling: The RODAC™ Plate

### OBSERVATIONS AND INTERPRETATIONS

**1** Record the results of each lab group and interpret them in the table below. To assess decontamination, compare the number of colonies on the "Before" and "After" plates. To interpret final colony count, apply the standards in Table 6-1 to the "After" plate.

**TABLE OF RESULTS**

| Lab Group | Source of Sample | Colonies Per Plate *Before* Decontaminating Lab Bench | Colonies Per Plate *After* Decontaminating Lab Bench | Assessment of Decontamination | Interpretation of Final Colony Count |
|---|---|---|---|---|---|
| | | | | | |
| | | | | | |
| | | | | | |
| | | | | | |
| | | | | | |
| | | | | | |
| | | | | | |
| | | | | | |

## QUESTIONS

**1** *What information is provided by sampling the surface immediately after treatment with disinfectant?*

**2** *What information would be provided by sampling the surface several hours after treatment with disinfectant?*

**3** *Why is it important to apply even pressure to the plate when sampling the surface?*

**4** *Why is it important not to slide the plate when sampling the surface?*

**5** *Why is it important that the plate is constructed so the lid is held above the agar (and not resting on it) during incubation?*

6

**6** *Why is it important to counteract the effects of residual disinfectant with Polysorbate 80 and Lecithin during incubation?*

**7** *Suppose a RODAC™ plate has no colonies on it after incubation. Can you assume that the surface is sterile? Explain your answer.*

# Standard Plate Count (Viable Count)

## Theory

The standard plate count is a procedure that allows microbiologists to estimate the population density in a liquid sample by plating a very dilute portion of that sample and counting the number of colonies it produces. The inoculum that is transferred to the plate contains a *known* proportion of the original sample because it is the product of a serial dilution.

As shown in Figure 6.3, a serial dilution is simply a series of controlled transfers down a line of dilution blanks (tubes containing a known volume of sterile diluent—water, saline, or buffer). The series begins with a sample containing an unknown concentration (density) of cells and ends with a very dilute mixture containing only a few—or no—cells. Each dilution blank in the series

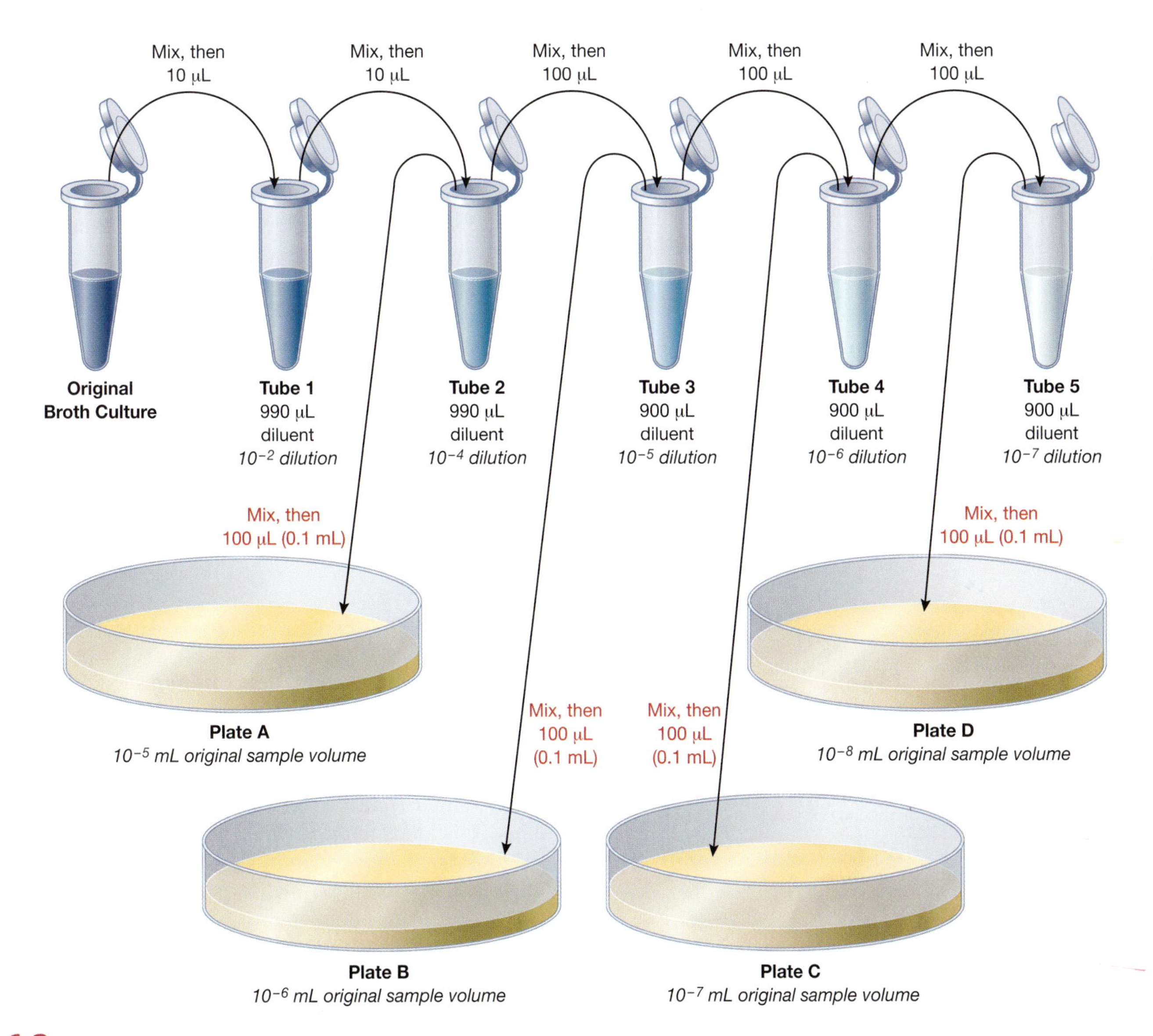


**6.3 Serial Dilution Procedural Diagram** This is an illustration of the dilution scheme outlined in the Procedure. The dilution assigned to each tube (written below the tube) represents the proportion of original sample inside that tube. For example, if the dilution is $10^{-2}$, the proportion of original sample inside the tube is 1/100th of the total volume. Be aware that the original cell density will be expressed in CFU/mL and the easiest way to do that is to have the original sample volumes on the plates also expressed in milliliters. However, the volume of diluted inoculum going to the plates in the diagram is in microliters (100 μL). Without showing it, we converted the 100 μL to 0.1 mL to calculate the original sample volumes shown below each plate. (Recall that original sample volume on the plate = volume plated × dilution factor.) Making this simple conversion at this point prevents a more complicated conversion later. It is convenient to measure in microliters, but try to think in milliliters.

receives a known volume from the mixture in the previous tube and delivers a known volume to the next, typically reducing the cell density to 1/10 or 1/100 at each step. (Greater dilutions in single steps are generally avoided, because they can be accomplished more conveniently and with greater accuracy by simply combining 1/10 or 1/100 dilutions.)

For example, if the original sample in Figure 6.3 contains 1,000,000 cells/mL, following the first transfer the 1/100 dilution (10 μL into 990 μL of diluent) in dilution tube 1 would contain 10,000 cells/mL (1,000,000 cells/mL × 1/100 = 10,000 cells/mL). In the second dilution (tube 2) the 1/100 dilution would reduce it further to 100 cells/mL (10,000 cells/mL × 1/100 = 100 cells/mL).

Because the cell density of the original sample is not known at this time, only the dilutions (without mL units) are recorded on the dilution tubes. By convention, dilutions are expressed in scientific notation. Therefore, a 1/10 dilution is written as $10^{-1}$ and a 1/100 dilution is written as $10^{-2}$.

A small portion of appropriate dilutions (depending on the *estimated* cell density of the original sample) is then spread onto agar plates to produce at least one countable plate. A countable plate is one that contains between 30 and 300 colonies (Fig. 6.4). A count lower than 30 colonies is considered statistically unreliable and greater than 300 is typically too many to be viewed as individual colonies on a standard 100 mm petri dish.

In examining the procedural diagram (Fig. 6.3), you can see that the first transfer in the series is a simple dilution, but that all successive transfers are compound

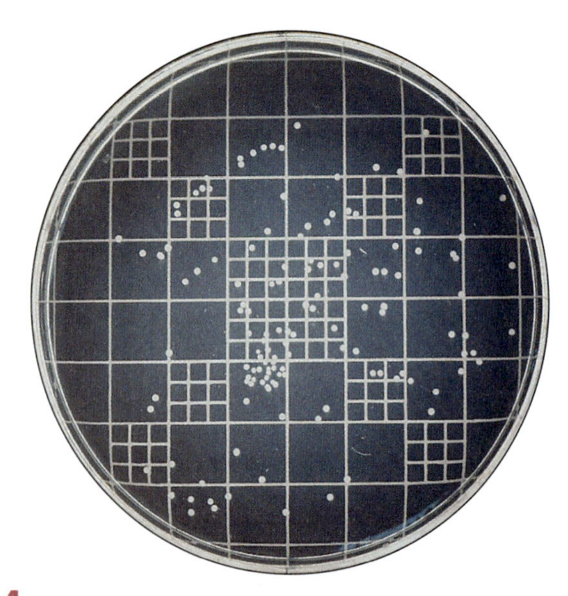

**6.4 Countable Plate** ■ A countable plate has between 30 and 300 colonies. Therefore, this plate (shown on a colony counter) with approximately 130 colonies is countable and can be used to calculate cell density in the original sample. Plates with fewer than 30 colonies are TFTC ("too few to count"). Plates with more than 300 colonies are TNTC ("too numerous to count").

dilutions. Both types of dilutions can be calculated using the following formula:

$$V_1D_1 = V_2D_2$$

where $V_1$ and $D_1$ are the volume and dilution of the concentrated broth, respectively, while $V_2$ and $D_2$ are the volume and dilution of the completed dilution. Undiluted samples are always expressed as 1. Therefore, to calculate the dilution of a 1 mL sample transferred to 9 mL of diluent, the permuted formula would be used as follows:

$$D_2 = \frac{V_1D_1}{V_2} = \frac{1.0 \text{ mL} \times 1}{10 \text{ mL}} = \frac{1}{10} = 10^{-1}$$

Notice that the dilution has no units because the milliliters cancel.

As mentioned above, compound dilutions are calculated using the same formula. However, because $D_1$ in compound dilutions no longer represents undiluted sample, but rather a fraction of the original density, it must be represented as something less than 1 (e.g., $10^{-1}$, $10^{-2}$, etc.) For example, if 1 mL of the $10^{-1}$ dilution from the last example were transferred to 9 mL of diluent, it would become a $10^{-2}$ dilution as follows:[1]

$$D_2 = \frac{V_1D_1}{V_2} = \frac{1.0 \text{ mL} \times 10^{-1}}{10 \text{ mL}} = 10^{-1} \times 10^{-1} = 10^{-2}$$

Spreading a known volume of this dilution onto an agar plate and counting the colonies that develop would give you all the information you need to calculate the original cell density (OCD). Below is the basic formula for this calculation:

$$OCD = \frac{CFU}{DxV}$$

CFU (colony forming units) is actually the number of colonies that develop on the plate. CFU is the preferred term because colonies could develop from single cells or from groups of cells, depending on the typical cellular arrangement of the organism. D is the dilution as written on the dilution tube from which the inoculum comes. V is the volume transferred to the plate. (***Note***: The volume is included in the formula because densities are expressed in CFU/mL, therefore a 0.1 mL inoculation [which would contain 1/10th as many cells as 1 mL] must be accounted for.)

[1] Permutations of this formula work with all necessary dilution calculations. For calculations involving unconventional volumes or dilutions, the formula is essential, but for simple tenfold or hundredfold dilutions like the ones described in this exercise and throughout this book, the final compounded dilution in a series can be calculated simply by multiplying each of the simple dilutions by each other. For example, a series of three $10^{-1}$ dilutions would yield a final dilution of $10^{-3}$ ($10^{-1} \times 10^{-1} \times 10^{-1} = 10^{-3}$). Three $10^{-2}$ dilutions would yield a final dilution of $10^{-6}$ ($10^{-2} \times 10^{-2} \times 10^{-2} = 10^{-6}$). We encourage you to use whatever method is best for you. In time you will be doing the calculations in your head.

As you can see in the formula, the volume of *original sample* being transferred to a plate is the product of the *volume transferred* and the *dilution* of the tube from which it came. Therefore 0.1 mL transferred from a $10^{-2}$ dilution contains only $10^{-3}$ mL of the original sample ($0.1 \text{ mL} \times 10^{-2} = 10^{-3} \text{ mL}$). The convention among microbiologists is to condense D and V in the formula into "Original sample volume"[2] (expressed in mL). The formula thus becomes,

$$\text{OCD} = \frac{\text{CFU}}{\text{Original sample volume}}$$

The sample volume is written on the plate at the time of inoculation. Following a period of incubation, the plates are examined, colonies are counted on the countable plates, and calculation is a simple division problem. Suppose, for example, you counted 37 colonies on a plate inoculated with 0.1 mL of a $10^{-5}$ dilution. Knowing that this plate now contains $10^{-6}$ mL of original sample, calculation would be as follows.

$$\text{OCD} = \frac{\text{CFU}}{\text{Sample volume}} = \frac{37 \text{ CFU}}{10^{-6} \text{ mL}} = 3.7 \times 10^{7} \text{ CFU/mL}$$

## Application

The viable count is one method of determining the density of a microbial population. It provides an estimate of actual *living* cells in the sample.

## In This Exercise

You will perform a dilution series and determine the population density of an *Escherichia coli* broth culture. You will inoculate the plates using the spread plate technique, as illustrated in Exercise 1-6. As described in Figure 1.43, the inocula from the dilution tubes will be evenly dispersed over the agar surface with a bent glass rod. You will be sterilizing the glass rod between inoculations by immersing it in alcohol and igniting it. Be careful to organize your work area properly and at all times keep the flame away from the alcohol jar. Should a fire start in the jar, extinguish it by replacing the lid on the jar.

## Materials

### Per Student

- □ Lab coat
- □ Disposable gloves
- □ Chemical eye protection

### Per Student Group

- □ Micropipettes (10–100 µL and 100–1,000 µL) with sterile tips
- □ Five sterile microtubes
- □ 50 mL flask with about 10 mL of sterile water or sterile saline
- □ Eight nutrient agar plates
- □ Jar (and its lid) containing ethanol, cotton in the bottom, and a bent glass rod
- □ Hand tally counter
- □ Colony counter
- □ 24-hour broth culture of *Escherichia coli* (this culture will have between $2 \times 10^{7}$ and $2 \times 10^{10}$ CFU/mL)

## PROCEDURE

Refer to the procedural diagram in Figure 6.3 as needed.

### Lab One

1. Wear a lab coat, gloves, and chemical eye protection when performing this procedure.
2. Obtain eight plates, organize them into four pairs, and label them $A_1$, $A_2$, $B_1$, $B_2$, etc.
3. Obtain the *E. coli* sample and the flask of sterile saline or water.
4. Obtain five microtubes and label them 1–5. These are your dilution tubes. Make sure they remain covered until needed.
5. Aseptically add 990 µL sterile water/saline to dilution tubes 1 and 2. Close the caps when finished. Aseptically add 900 µL sterile water/saline to dilution tubes 3, 4, and 5. Close the caps when finished.
6. Carefully, but thoroughly, mix the broth culture and aseptically transfer 10 µL to dilution tube 1; mix well. This is a $10^{-2}$ dilution.
7. Aseptically transfer 10 µL from dilution tube 1 to dilution tube 2; mix well. This is a $10^{-4}$ dilution.
8. Aseptically transfer 100 µL from tube 2 to dilution tube 3; mix well. This is a $10^{-5}$ dilution.
9. Aseptically transfer 100 µL from dilution tube 3 to dilution tube 4; mix well. This is a $10^{-6}$ dilution.
10. Aseptically transfer 100 µL from dilution tube 4 to dilution tube 5; mix well. This is a $10^{-7}$ dilution.
11. After mixing, aseptically transfer 100 µL from dilution tube 2 to plate $A_1$. Using the spread plate technique (Exercise 1-6), disperse the diluent evenly over the entire surface of the agar. Repeat the procedure with plate $A_2$, and label both plates "$10^{-5}$ mL original sample," or simply "$10^{-5}$."

[2] Some microbiologists refer to this as the "plate dilution." We prefer to use OSV in this introduction because it emphasizes what is really happening; that is, OSV is the volume of the original sample deposited on the plate.

12 Following the same procedure, transfer 100 μL volumes from dilution tubes 3, 4, and 5 to plates $B_1$ and $B_2$, $C_1$ and $C_2$, and $D_1$ and $D_2$, respectively. Label the plates with their appropriate original sample volumes.

13 Allow the inocula to soak in for a few minutes, and then invert the plates and incubate them at $35 \pm 2$ for 24 to 48 hours.

14 Save or dispose of the original *E. coli* tube as directed by your instructor. Tightly cap the dilution tubes and dispose of them in the appropriate autoclave container.

### Lab Two

1 After incubation, examine the plates and determine the countable pair—plates with 30 to 300 colonies. Only one pair of plates should be countable. The remainder should be "too few to count" (TFTC) or "too numerous to count" (TNTC).

2 Count the colonies on both plates and calculate the average (Fig. 6.5). Various methods are available for counting. A simple way is to use a hand tally counter and a felt-tip marker. Mark each colony on the plastic base and simultaneously click the counter.[3] Record these in the table provided on the data sheet, page 415. (***Note***: You may have more than one pair of plates that is countable. For the practice, count all plates that have between 30–300 colonies, and try to identify which pair you have the most confidence in. If no plates are in the 30–300 colony range, count the pair that is closest.)

3 Using the formula provided on the data sheet, calculate the density of the original sample and record it in the space provided.

4 Dispose of the plates in an appropriate autoclave container when finished.

[3] Other, more sophisticated counting methods are available. One uses an electronic pen that, when touched to the plastic Petri dish below a colony, records a tally. There are also software systems that capture an image of the plate and then colonies are counted by the computer.

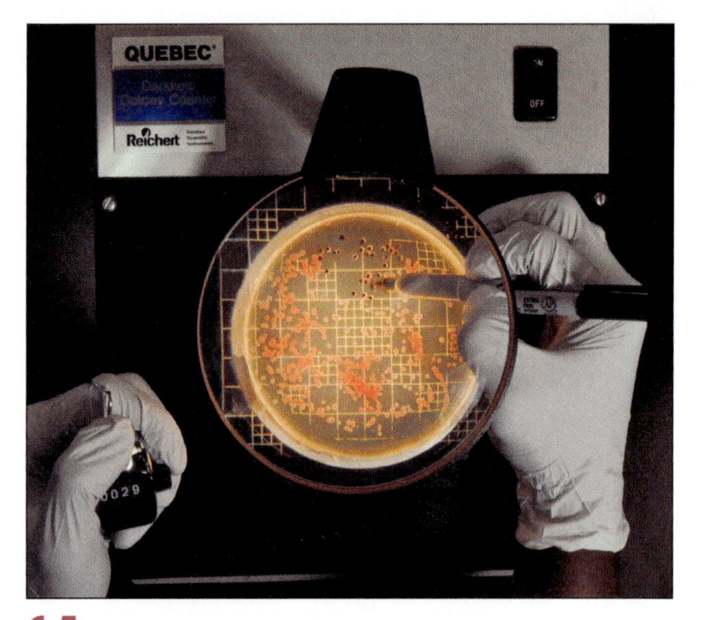


**6.5 Counting Bacterial Colonies** ■ Place the plate upside down on the colony counter. Turn on the light and adjust the magnifying glass until all the colonies are visible. Using the grid in the background as a guide, count colonies one section at a time. Mark each colony with a felt-tip marker as you record with a hand tally counter.

### References


Collins, C. H., Patricia M. Lyne, and J. M. Grange. Page 149 in *Collins and Lyne's Microbiological Methods*, 7th ed. Oxford, Boston: Butterworth-Heinemann, 1995.

Koch, Arthur L. Page 254 in *Methods for General and Molecular Bacteriology*. Philipp Gerhardt, R. G. E. Murray, Willis A. Wood, and Noel R. Krieg, eds. Washington, DC: American Society for Microbiology, 1994.

Postgate, J. R. Page 611 in *Methods in Microbiology*, Vol. 1. J. R. Norris and D. W. Ribbons, eds. New York: Academic Press, 1969.

Name ______________________

Date ______________________

Lab Section ______________________

I was present and performed this exercise (initials) ______________________

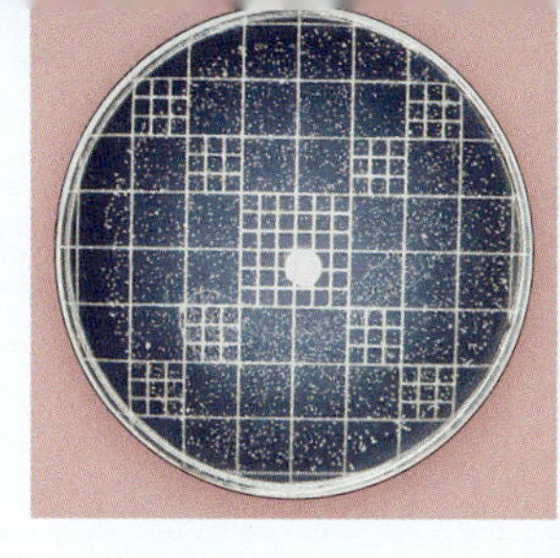

DATA SHEET 6-2

# Standard Plate Count (Viable Count)

## OBSERVATIONS AND INTERPRETATIONS

**1** Enter the number of colonies on each countable plate. Only one pair of plates should be countable, but for practice, record all countable plates anyway. For all plates containing more than 300 colonies, enter TNTC ("too numerous to count"). For plates containing fewer than 30, enter TFTC ("too few to count"). If no plates are countable, use the TFTC plate closest to 30 colonies for practice with the calculations. Make a note of this in step 3 below.

**2** Take the average number of colonies from the two (or more) countable plates and record it below.

| Plate | $A_1$ | $A_2$ | $B_1$ | $B_2$ | $C_1$ | $C_2$ | $D_1$ | $D_2$ |
|---|---|---|---|---|---|---|---|---|
| Colonies counted | | | | | | | | |
| Average # colonies | | | | | | | | |

**3** Calculate the original density in CFU/mL using the following formula:

$$\text{OCD} = \frac{\text{CFU}}{\text{Original sample volume}}$$

| Original density of *E. coli* in the broth | |
|---|---|

## QUESTIONS

When answering the following questions, assume all dilutions are in even powers of ten, unless told otherwise, and that 0.1 mL (100 μL) or 1.0 mL (1,000 μL) volumes were plated. Also, *cell density* and *cell concentration* are used interchangeably, as are *CFU* and *colonies*, depending on how the term is being used.

**1** *Suppose your professor handed you a test tube with 2.0 mL of an* E. coli *broth culture in it and told you to make a $10^{-1}$ dilution of the entire culture. Explain how you would do this. Show your calculations.*

**2** *Suppose your professor handed you a test tube with 2.0 mL of an* E. coli *broth culture in it and told you to make a $10^{-2}$ dilution of the entire culture. Explain how you would do this. Show your calculations.*

**3** *How would you produce a $10^{-1}$ dilution of a 3 mL bacterial sample using the entire 3 mL volume?*

**4** *How would you produce a $10^{-2}$ dilution of a 5 mL bacterial sample using the entire 5 mL volume?*

**5** *You have 0.05 mL of an undiluted culture at a density of $3.6 \times 10^6$ CFU/mL. You then add it to 4.95 mL sterile diluent. What is the dilution and what is the final density of cells?*

**6** *You have 0.3 mL of an undiluted culture at a density of $4.2 \times 10^7$ CFU/mL. You then add it to 2.7 mL sterile diluent. What is the dilution and what is the final density of cells?*

**7** *What is the dilution if 96 mL of diluent is added to 4 mL of a bacterial suspension?*

6

**8** *What is the dilution if 75 mL of diluent is added to 25 mL of a bacterial suspension?*

**9** *You were instructed to add 1.0 mL out of 5.0 mL of an undiluted sample to 99 mL of sterile diluent. Instead, you added all 5.0 mL to the 99 mL. What was the intended dilution and what was the actual dilution?*

**10** *Suppose you were instructed to add 0.2 mL of sample to 9.8 mL of diluent, but instead added 2.0 mL of sample. What was the intended dilution and what was the actual dilution?*

Name ______

Date ______

Lab Section ______

I was present and performed this exercise (initials) ______

# DATA SHEET 6-2

*(continued)*

**11** *Plating 1.0 mL of a sample diluted by a factor of $10^{-3}$ produced 43 colonies. What was the original cell density in the sample?*

______

**12** *Plating 0.1 mL of a sample diluted by a factor of $10^{-3}$ produced 43 colonies. What was the original cell density in the sample?*

______

**13** *A plate with a sample volume of $10^{-7}$ mL produced 72 colonies.*

**a.** *What was the original cell density?*

______

**b.** *How many colonies should be on the plate inoculated with a sample volume of $10^{-6}$ mL?*

______

**c.** *How many colonies should be on the plate inoculated with a sample volume of $10^{-8}$ mL?*

______

**14** *A plate with a sample volume of $10^{-6}$ mL produced 259 colonies.*

**a.** *What was the original cell density?*

______

**b.** *How many colonies should be on the plate inoculated with a sample volume of $10^{-5}$ mL?*

______

**c.** *How many colonies should be on the plate inoculated with a sample volume of $10^{-7}$ mL?*

______

**15** *A nutrient agar plate that received 1,000 μL of a bacterial sample diluted by a factor of $10^{-6}$ had 298 colonies on it after incubation. What was the original cell density?*

______

**16** *You inoculated a nutrient agar plate with 100 μL of a sample diluted by a factor of $10^{-3}$. After incubation, you count 58 colonies. What was the original cell density?*

______

**17** *A nutrient agar plate labeled $10^{-5}$ mL produced 154 colonies after incubation.*

**a.** *What was the cell density in the original sample?*

**b.** *What combination(s) of volumes and dilution factors could have been used to inoculate this plate?*

**18** *A nutrient agar plate labeled $10^{-7}$ mL produced 62 colonies after incubation.*

**a.** *What was the cell density in the original sample?*

**b.** *What combination(s) of volumes and dilution factors could have been used to inoculate this plate?*

**19** *The original cell density in a sample is $2.79 \times 10^{6}$ CFU/mL. Which sample volume should yield a countable plate? (Express your answer as $10^{\times}$ mL.)*

**20** *The original cell density in a sample is $5.1 \times 10^{9}$ CFU/mL. Which sample volume should yield a countable plate? (Express your answer as $10^{\times}$ mL.)*

6

**21** *A sample has a density of $1.37 \times 10^{5}$ CFU/mL.*

**a.** *What sample volume should yield a countable plate?*

**b.** *Which two dilution tubes could be used to produce this sample volume? How?*

**22** *A sample has a density of $7.9 \times 10^{9}$ CFU/mL.*

**a.** *What sample volume should yield a countable plate?*

**b.** *Which two dilution tubes could be used to produce this sample volume? How?*

Name ____________________

Date ____________________

Lab Section ____________________

I was present and performed this exercise (initials) ____________________

# DATA SHEET 6-2

*(continued)*

**23** *You are told that a sample has between $2.5 \times 10^3$ and $2.5 \times 10^5$ cells/mL. Devise a complete but efficient (that is, no extra plates!) dilution scheme that will ensure getting a countable plate.*

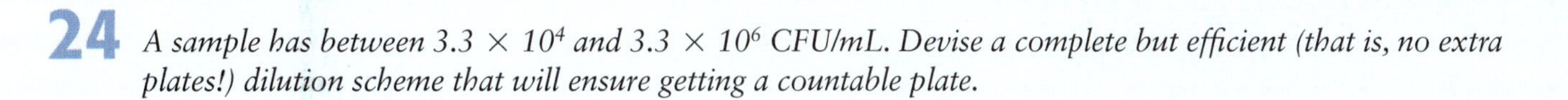

**24** *A sample has between $3.3 \times 10^4$ and $3.3 \times 10^6$ CFU/mL. Devise a complete but efficient (that is, no extra plates!) dilution scheme that will ensure getting a countable plate.*

**25** *Two plates received 100 µL from the same dilution tube. The first plate had 293 colonies, whereas the second had 158 colonies. Suggest reasonable sources of error.*

**26** *Two parallel dilution series were made from the same original sample. The plates with sample volumes of $10^{-5}$ mL from each dilution series yielded 144 and 93 colonies. Suggest reasonable sources of error.*

6

# Urine Culture

EXERCISE 6-3

## Theory

Urine culture is a semiquantitative CFU counting method that quickly produces countable plates without a serial dilution. The instrument used in this procedure is a volumetric loop, calibrated to hold 0.001 mL (1 μL) or 0.01 mL (10 μL) of sample. Urine culture procedures using volumetric loops are useful in situations where a rapid diagnosis is essential and approximations ($\pm 10^2$ CFU/mL) are sufficient to choose a course of action. Volumetric loops are useful in situations where population density is not likely to exceed $10^5$ CFU/mL.

In this standard procedure, a loopful of urine (generally from a clean catch midstream sample or a bladder catheter sample) is carefully transferred to a blood agar plate. The initial inoculation is a single streak across the diameter of the agar plate. Then the plate is turned 90° and (without flaming the loop) streaked again, this time across the original line in a zigzag pattern to evenly disperse the bacteria over the entire plate (Figs. 6.6 and 6.7). Following a period of incubation, the resulting colonies are counted and population density, usually referred to as "original cell density," or OCD, is calculated.

OCD is recorded in "colony forming units" per milliliter (CFU/mL), as described in the introduction to this section. CFU/mL is determined by dividing the number of colonies on the plate by the volume of the loop. For example, if 75 colonies are counted on a plate inoculated with a 0.001 mL loop, the calculation would be as follows:

**6.6 Semiquantitative Streak Method** ■ Streak 1 is a simple streak line across the diameter of the plate. Streak 2 is a tight streak across Streak 1 to cover the entire plate.

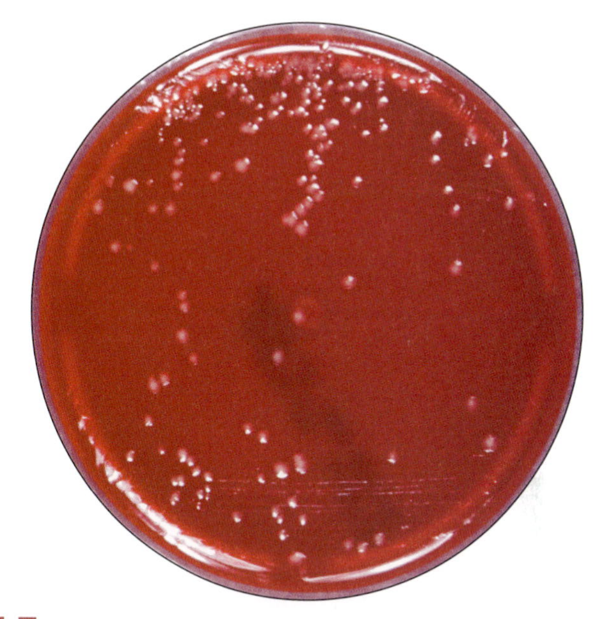

**6.7 Urine Streak on Sheep Blood Agar** ■ This plate was inoculated with a 0.01 mL volumetric loop. The cell density can be determined by dividing the number of colonies by 0.01.

$$\text{OCD} = \frac{\text{CFU}}{\text{Loop volume}}$$

$$\text{OCD} = \frac{75\ \text{CFU}}{0.001\ \text{mL}}$$

$$\text{OCD} = 7.5 \times 10^4\ \text{CFU/mL}$$

There is no universal minimum number of colonies recovered from a urine sample for it to be classified as a positive because samples are frequently contaminated with normal microbiota. In addition, the patient's sex and current symptoms, method of collection, and number of different potential pathogens recovered affect the interpretation. However, recovery of $10^2$ CFU/mL or more is typically regarded as a positive culture and further lab work is done to identify the organism(s) and their antibiotic susceptibilities.

## Application

Urine culture is a common method of detecting and quantifying urinary tract infections (UTIs). It frequently is combined with media for specific identification of members of *Enterobacteriaceae* (e.g., MacConkey agar) or *Streptococcus* (sheep blood agar).

## In This Exercise

You will estimate cell density in a "urine" sample using a volumetric loop and the above formula. Be sure to hold the loop vertically and transfer slowly.

## Materials

### Per Student

- ☐ Lab coat
- ☐ Disposable gloves
- ☐ Chemical eye protection

### Per Student Group

- ☐ One blood agar plate (TSA with 5% sheep blood)
- ☐ One sterile volumetric inoculating loop (either 0.01 mL or 0.001 mL)
- ☐ One 12 mm × 75 mm test tube with 2 mL (more or less) of artificial urine spiked with *Staphylococcus epidermidis*. ***Note to instructor***: To prepare these, make a $10^{-4}$ dilution of an overnight *S. epidermidis* broth culture (which should be about $10^8$ CFU/mL) in tryptic soy broth or synthetic urine (alarmingly, available online and at pharmacies!). Then dispense 2 mL aliquots into sterile, capped 12 mm × 75 mm test tubes and label each with a different sample number. (Be sure that your final volume after dilution is sufficient to fill the number of "urine" samples you need to make.) If you want different densities for each sample, Use a volumetric loop to add more cells from the diluted *S. epidermidis*. (Each 100 µL loopful will add $10^3$ cells to the 2 mL sample.) Or, add a drop or two of TSB/synthetic urine.
- ☐ Felt-tip pen
- ☐ Hand tally counter

### Lab One

1. Wear a lab coat, gloves, and chemical eye protection when performing this procedure.
2. Obtain a blood agar plate and "urine" sample. Label the plate with the "urine" sample number, your group name, and the date.
3. Thoroughly, but gently, mix the "urine" with a sterile wooden stick or suitable substitute. Dispose of the stick according to your lab practices (e.g., in a sharps container).
4. Volumetric loops are generally disposable and are in sterile packages. Remove the loop from its package and do not incinerate it.
5. Hold the loop vertically and gently immerse it in the "urine" sample. Then carefully withdraw it to obtain the correct volume of "urine." This loop is designed to fill to capacity in the vertical position. Do not tilt it until you get it in position over the plate.
6. Inoculate the blood agar by making a single streak across the diameter of the plate.
7. Turn the plate 90° and streak the "urine" across the entire surface of the agar, as shown in Figure 6.6. Dispose of the loop without incineration according to your lab practices.
8. Invert the plate and incubate it for 24 hours at 35 ± 2°C.
9. Save or dispose of the original "urine" tube as directed by your instructor.

### Lab Two

1. Remove the plate from the incubator and count the colonies using the felt-tip pen and hand tally counter. Touch the plate's base where there is a colony and simultaneously click the counter. Enter the data in the table on the data sheet, page 423.
2. Calculate the original cell density of the sample using the formula on the data sheet.
3. Dispose of the plate in an appropriate autoclave container when finished.

### References


Brooks, Geo. F., Karen C. Carroll, Janet S. Butel, Stephen A. Morse, and Timothy A. Mietzner. Pages 708 and 717–719 in *Jawetz, Melnick & Adelberg's Medical Microbiology*, 25th ed. New York, NY: McGraw Hill Medical Publishers, 2010.

Tille, Patricia M. Pages 926–927 in *Bailey & Scott's Diagnostic Microbiology*, 13th ed. St. Louis, MO: Mosby, 2014.

Winn, Washington C., et al. Page 85 in *Koneman's Color Atlas and Textbook of Diagnostic Microbiology*, 6th ed. Baltimore: Lippincott Williams & Wilkins, 2006.

Name ______________________________

Date ______________________________

Lab Section ______________________________

I was present and performed this exercise (initials) ______________________________

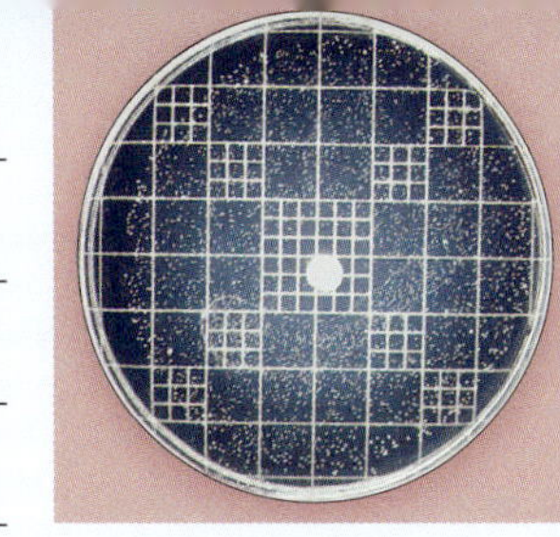

DATA SHEET 6-3

# Urine Culture

## OBSERVATIONS AND INTERPRETATIONS

**1** Enter your colony count and loop volume data in the table below. Then calculate the original cell density using the following formula:

$$OCD = \frac{CFU}{\text{Loop volume}}$$

Use the extra rows to calculate cell densities of "urine" samples of other groups. Label these appropriately.

| Urine Sample | Colonies Counted | Loop Volume (0.01 mL or 0.001 mL) | Original Cell Density (CFU/mL) |
|---|---|---|---|
| | | | |
| | | | |
| | | | |
| | | | |

## QUESTIONS

**1** *The plate pictured in Figure 6.7 was inoculated with a 0.01 mL volumetric loop and contains approximately 75 colonies. What was the original cell density?*

**2** *The equation shown in the Theory explanation is used for calculating cell density in urine when using a 0.001 mL calibrated loop. The urine transferred in the loop is not literally diluted, yet its volume is equivalent to a dilution factor.*

**a.** *What is the dilution factor (based on loop volume) expressed as a fraction?*

**b.** *What is the dilution factor in scientific notation?*

**3** *Calculation of original density in this exercise differs slightly from that offered in Exercise 6-2.*

**a.** *Compare and contrast the formula used today with that used in Exercise 6-2.*

**b.** *Could you have used the formula in Exercise 6-2 for today's calculations? Explain.*

**4** *Using a volumetric loop is a semiquantitative technique.*

**a.** *Why is it not quantitative?*

6

**b.** *Design a procedure that would make it quantitative. (**Hint**: Refer to Exercise 6-2 if necessary.)*

# Plaque Assay of Virus Titer

EXERCISE 6-4

## Theory

Viruses that attack bacteria are called bacteriophages, or simply phages (Fig. 6.8). Some viruses attach to the bacterial cell wall and inject viral DNA into the bacterial cytoplasm. The viral genome then commands the host cell to produce more viral DNA and viral proteins, which are used for the assembly of more phages. Once assembly is complete, the cell lyses and releases the phages, which then attack other bacterial cells and begin the replicative cycle all over again. This process, called the lytic cycle, is shown in Figure 6.9.

Lysis of bacterial cells growing in a lawn on an agar plate produces a clearing that can be viewed with the naked eye. These clearings are called plaques. The plaque assay uses this phenomenon as a means of calculating the phage concentration in a given sample. When a sample of bacteriophage (generally diluted by means of a serial dilution) is added to a plate inoculated with enough bacterial host to produce a lawn of growth, the number of plaques formed can be used to calculate the original phage titer, or density. Refer to the procedural diagram in Figure 6.10.

The plaque assay technique is similar to the standard plate count, in that it employs a serial dilution to produce countable plates needed for later calculations. (***Note***: Refer to Exercise 6-2, Standard Plate Count, as needed for a description of serial dilutions, dilution factors, and calculations.) One key difference is that the plaque assay is done using the pour-plate technique, in which bacterial cells and viruses are first added to molten agar and then poured into the plate.

In this procedure, diluted phage is added directly to a small amount of *E. coli* culture and allowed a 10-minute ($\pm 5$ minutes) adsorption period to attach to the bacterial cells. Then this phage–host mixture is added to a tube of soft agar, mixed, and poured onto prepared nutrient agar plates as an agar overlay. The consistency of the solidified soft agar is sufficient to immobilize the bacteria while allowing the smaller bacteriophages to diffuse short distances and infect surrounding cells. During incubation, the phage host produces a lawn of growth on the plate in which plaques appear where contiguous cells have been lysed by the virus (Fig. 6.11).

The procedure for counting plaques is the same as that for the standard plate count. To be statistically reliable, countable plates must have between 30 and 300 plaques. Calculating phage titer (original phage density) uses the same formula as other plate counts except that PFU (plaque forming units) instead of CFU (colony forming units) becomes the numerator in the equation. Phage titer, therefore, is expressed in PFU/mL and the formula is written as follows:

$$\text{Phage titer} = \frac{\text{PFU}}{\text{Volume plated} \times \text{Dilution}}$$

As with the standard plate count, it is customary to condense *volume plated* and *dilution* into *original sample volume*. The formula then becomes:

$$\text{Phage titer} = \frac{\text{PFU}}{\text{Original sample volume}}$$

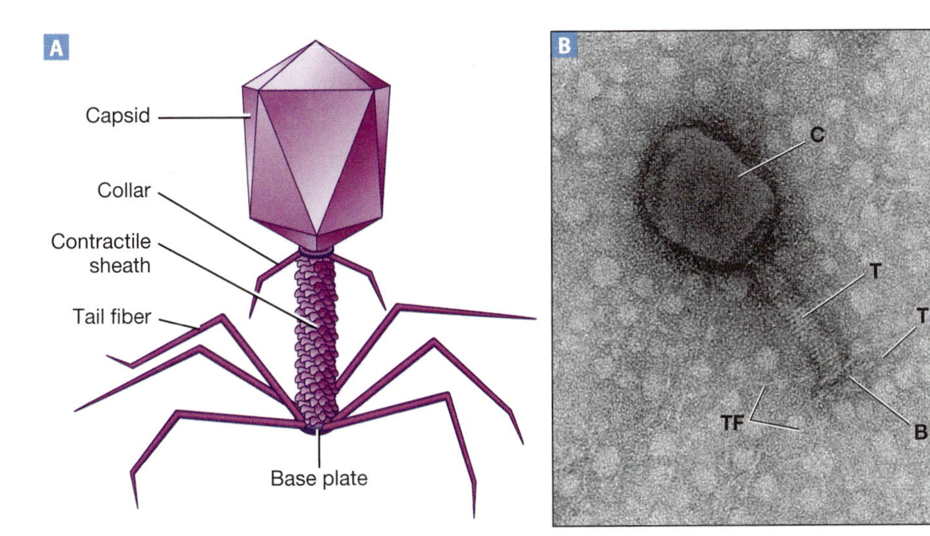


**6.8 T4 Coliphage** ■ (**A**) As shown in this artist's rendition, bacteriophages frequently have a complex structure that includes the protein capsid containing the nucleic acid genome and a protein tail with many parts. T4 phage has an icosahedral (20 triangular faces) capsid. (**B**) This is a negative stain of one T4 phage particle. Shown are the capsid (C), tail (T), base plate (B), and tail fibers (TF). The length of this phage from base plate to tip of capsid is approximately 180 nm (0.18 μm).

(Photo taken by author at the San Diego State University Electron Microscope Facility.)

The original sample volume is written on the plate at the time of inoculation. Following a period of incubation, the plates are examined, plaques are counted on the countable plates, and calculation is a simple division problem.

Suppose, for example, you inoculated a plate with 0.1 mL of a $10^{-4}$ dilution. (Remember, you are calculating the *phage density;* the *E. coli* has nothing to do with the calculations.) This plate now contains $10^{-5}$ mL of original phage sample. If you subsequently counted 115 plaques on the plate, calculation would be as follows:

$$\text{Original phage density} = \frac{1.15 \times 10^{2}\ \text{PFU}}{10^{-5}\ \text{mL}} = 1.15 \times 10^{7}\ \text{PFU/mL}$$

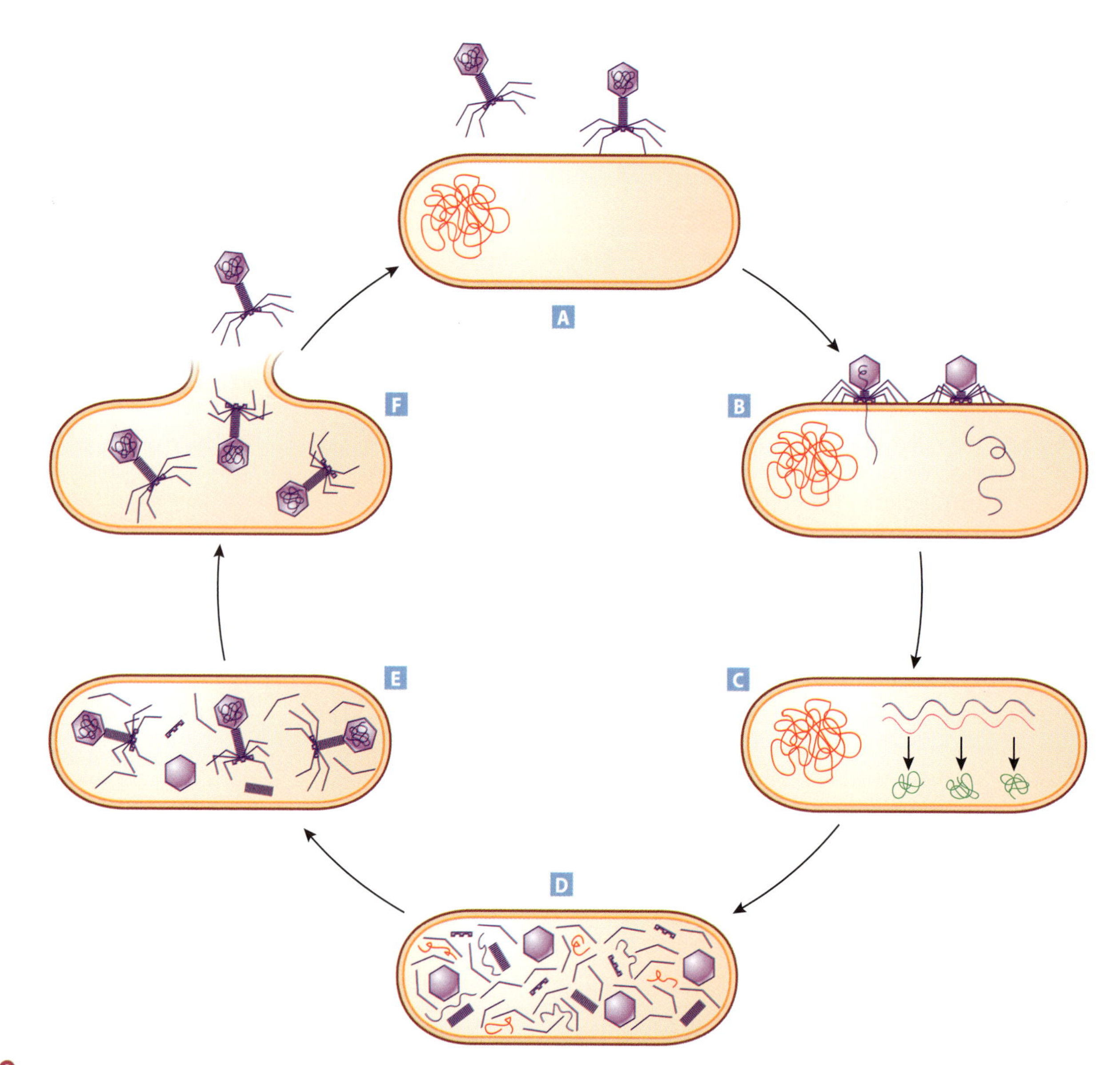


**6.9 T4 Phage Lytic Cycle** ■ Shown is a simplified diagram of the T4 coliphage's replicative cycle. In (A), an infective T4 phage (left) is approaching its host, *Escherichia coli*. On the right, the same phage is shown attached by its tail fibers to specific receptors on *E. coli*. This is the **attachment** phase. The *E. coli* chromosome is the red tangled line at the left of the cell. On the left of (B), T4 is in the process of transferring its DNA (purple line) into *E. coli*. This is the **penetration** phase. On the right, the entire phage genome is in the host. Removal of the viral genome from its capsid is called **uncoating**. In (C), phage DNA is in the process of being transcribed into mRNA (parallel red line), which is then translated into phage proteins (squiggly green lines). This is the **synthesis** phase and is performed by *E. coli* under the direction of the phage DNA. Simultaneously, phage DNA is being replicated, but this is not shown. Synthesis leads to **assembly** of the phage progeny, shown in (D) and (E), where the capsid subunits come together to form the capsid into which the genome is inserted, and the tail with all its diverse parts comes together and attaches to the capsid. Note that host DNA has been degraded, but remains as short, red lines. (This is important because occasionally *E. coli* DNA is put into a viral capsid and is transferred to a new *E. coli* host, providing it with new genes in a process called **generalized transduction**). In (F), the fully assembled phages are **released** as the cell bursts. At 37°C, this entire cycle can take as little as 25 minutes and release a few hundred phage progeny, each of which can repeat the process!

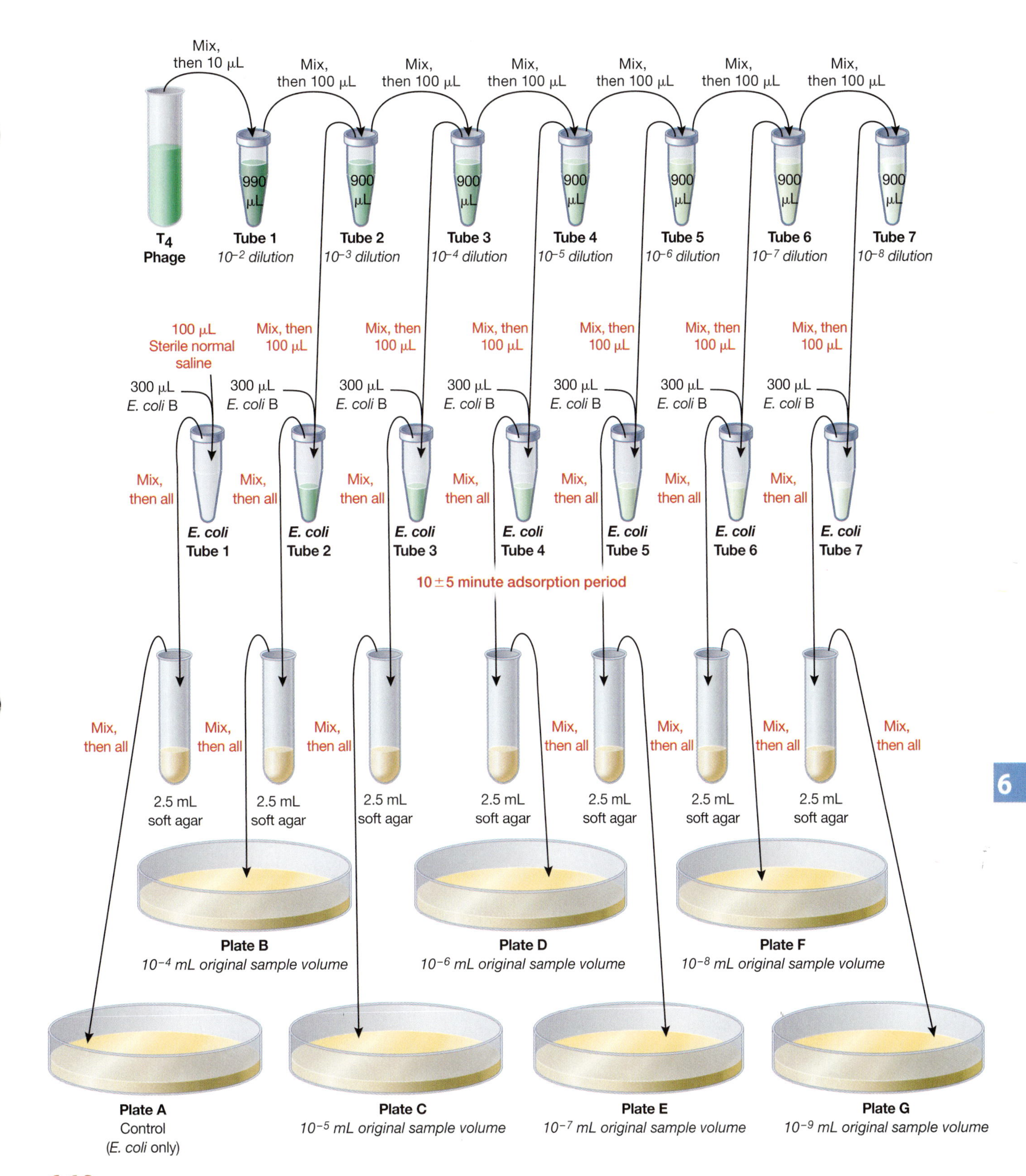


**6.10 Plaque Assay Procedural Diagram** ■ This is an illustration of the dilution scheme outlined in the Procedure. The dilutions assigned to the dilution tubes in the first row indicate the proportion of original phage sample present in the tube. For example, if the dilution is $10^{-4}$, the proportion of original sample inside the tube is 1/10,000th of the total volume. Be aware that the original phage density will be expressed in PFU/mL and the easiest way to do that is to have the original sample volumes on the plates also expressed in milliliters. However, the volume of diluted phage going to the plates (via the *E. coli*/soft agar tube mixture) is in microliters (100 μL). Without showing it, we converted the 100 μL to 0.1 mL to calculate the original sample volumes. (Recall that original sample volume on the plate = volume plated × dilution factor.) Making this simple conversion at this point prevents a more complicated conversion later. It is convenient to measure in microliters, but try to think in milliliters. (***Note:*** Do not be confused by the addition of *E. coli*; it does not affect the phage volume going to the plate.)

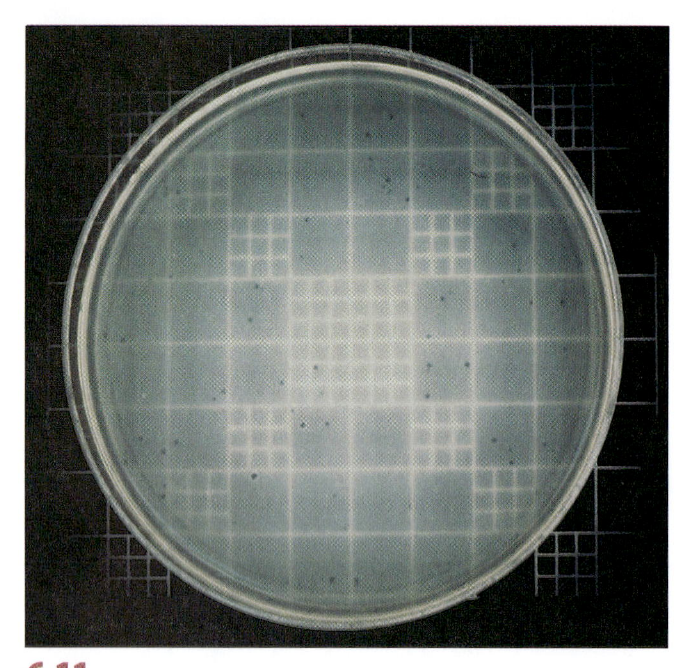

**6.11 Countable Plate** ■ This plaque assay plate has between 30 and 300 plaques; therefore, it is countable. They look dark because the photograph was shot against the dark background of a colony counter. Without the colony counter and viewed with transmitted light, the plaques will look like clear spots in the hazy *E. coli* B lawn.

## Application

This technique is used to determine the concentration of viral particles in a sample. Samples taken over a period of time can be used to construct a viral growth curve.

6

## In This Exercise

You will be estimating the density (titer) of a T4 coliphage sample using a strain of *Escherichia coli* (*E. coli* B) as the host organism.

## Materials

### Per Class

- ☐ 50°C hot-water bath containing tubes of 2.5 mL liquid soft agar (7 tubes per group)

### Per Student

- ☐ Lab coat
- ☐ Disposable gloves
- ☐ Chemical eye protection

### Per Student Group

- ☐ Micropipettes (10–100 μL and 100–1,000 μL) with sterile tips
- ☐ 14 sterile capped microtubes (1.5 mL or larger) in a microtube holder (rack)
- ☐ Seven tryptic soy agar or nutrient agar plates
- ☐ Seven tubes containing 2.5 mL soft agar
- ☐ 10 mL of sterile normal saline in a small flask
- ☐ Seven sterile transfer pipettes
- ☐ T4 coliphage
- ☐ 24-hour broth culture of *Escherichia coli* B (T-series phage host)

## Medium Recipe

### Soft Agar

| | |
|---|---|
| ☐ Beef extract | 3.0 g |
| ☐ Peptone | 5.0 g |
| ☐ Sodium chloride | 5.0 g |
| ☐ Tryptone | 2.5 g |
| ☐ Yeast extract | 2.5 g |
| ☐ Agar | 7.0 g |
| ☐ Distilled or deionized water | 1.0 L |

## PROCEDURE

Refer to the procedural diagram in Figure 6.10 as needed.

### Lab One

1. Wear a lab coat, gloves, and chemical eye protection when performing this procedure.
2. Obtain all materials except for the soft agar tubes. To keep the agar tubes liquefied, leave them in the water bath and take them out one at a time as needed.
3. Label seven microtubes 1 through 7. Label the other seven microtubes *E. coli* 1–7. Place all tubes in a microtube holder, pairing like-numbered tubes.
4. Label the TSA (or nutrient agar) plates A through G at the edge of the base along with your group name. Place them in the 35°C incubator to warm them. Take them out one at a time as needed. This will keep the soft agar (added at step 15) from solidifying too quickly and result in a smoother agar surface.
5. Aseptically transfer 990 μL sterile normal saline to dilution tube 1.
6. Aseptically transfer 900 μL sterile normal saline to dilution tubes 2–7.
7. Mix the *E. coli* culture and aseptically transfer 300 μL into each of the *E. coli* microtubes.
8. Mix the T4 suspension and aseptically transfer 10 μL into dilution tube 1. Mix well. This is a $10^{-2}$ dilution.

9. Aseptically transfer 100 μL from dilution tube 1 to dilution tube 2. Mix well. This is a $10^{-3}$ dilution.
10. Aseptically transfer 100 μL from dilution tube 2 to dilution tube 3. Mix well. This is a $10^{-4}$ dilution.
11. Continue in this manner through dilution tube 7. The dilution in tube 7 should be $10^{-8}$.
12. Aseptically transfer 100 μL sterile normal saline to *E. coli* tube 1. This will be mixed with 2.5 mL soft agar and used to inoculate a control plate.
13. Aseptically transfer 100 μL from dilution tube 2 to its companion *E. coli* tube. Repeat this procedure with the remaining five tubes.
14. This is the beginning of the adsorption period. Let all seven tubes stand undisturbed for $10 \pm 5$ minutes.
15. Remove one soft agar tube from the water bath and add the entire contents of *E. coli* tube 1 with a sterile transfer pipette. Mix well with the pipette and immediately transfer all of it onto plate A. Gently tilt the plate back and forth until the soft agar mixture is spread evenly across the solid medium.
16. Remove a second soft agar tube from the water bath and add the entire contents of *E. coli* tube 2 with a sterile transfer pipette. Mix well with the pipette and immediately transfer all of it onto plate B. Tilt back and forth to cover the agar and label it $10^{-4}$ mL (this is the phage sample volume).
17. Repeat this procedure with dilutions $10^{-4}$ through $10^{-8}$ and plates C through G. Label the plates with the appropriate original phage sample volumes.
18. Allow the agar to solidify completely.
19. When the agar has solidified, label plate A "control." Then label plates B through G with their appropriate sample volumes (e.g., plate B is $10^{-4}$ and plate G is $10^{-9}$).
20. Invert the plates and incubate aerobically at $35 \pm 2$°C for 24–48 hours.
21. Save or dispose of the original *E. coli* tube as directed by your instructor. Tightly cap the dilution tubes and dispose of them in the appropriate autoclave container.

### Lab Two

1. After incubation, examine plate A for *E. coli* B growth and the absence of plaques.
2. Examine the remainder of your plates and determine which one is countable (30 to 300 plaques). Count the plaques using the hand tally counter, marking each with the marking pen on the plate's base as you count. Record the plaque count in the table provided on the data sheet, page 431. Record all others as either TNTC (too numerous to count) or TFTC (too few to count). ***Note***: if you have more than one countable plate, count both and use each to calculate the phage titer. Then, try to determine which one you have more confidence in.
3. Using the sample volume on the countable plate and the formula provided on the data sheet, calculate the original phage titer. Record your results on the data sheet.
4. Dispose of the plates in an appropriate autoclave container when finished.

## References


Collins, C. H., Patricia M. Lyne, J. M. Grange. Page 149 in *Collins and Lyne's Microbiological Methods*, 7th ed. Oxford, Boston: Butterworth-Heinemann, 1995.

DIFCO Laboratories. Page 619 in *DIFCO Manual*, 10th ed. Detroit: DIFCO Laboratories, 1984.

Province, David L. and Roy Curtiss III. Page 328 in *Methods for General and Molecular Bacteriology*. Philipp Gerhardt, R. G. E. Murray, Willis A. Wood, and Noel R. Krieg, eds. Washington, DC: American Society for Microbiology, 1994.

Name ________________________________

Date ________________________________

Lab Section ________________________________

I was present and performed this exercise (initials) ____________________

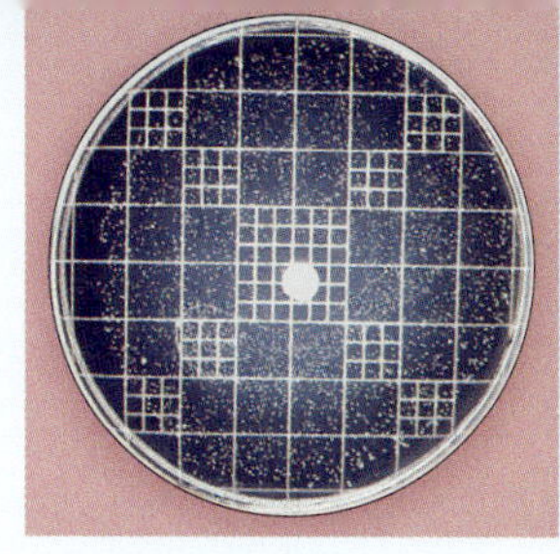

# Plaque Assay of Virus Titer

## OBSERVATIONS AND INTERPRETATIONS

**1** Enter the number of plaques counted on the countable plate. Only one plate should be countable, but record the numbers of any that have between 30 and 300 plaques. If there are more than 300 plaques, enter TNTC. If fewer than 30 plaques, enter TFTC.

| Plate | A | B | C | D | E | F | G |
|---|---|---|---|---|---|---|---|
| Plaques counted | | | | | | | |
| Sample volume | | | | | | | |

**2** Calculate the original density using the following formula, and enter the result below. Record your answer in PFU/mL.

$$\text{Phage titer} = \frac{\text{PFU}}{\text{Original sample volume}}$$

If you don't have any countable plates, use the TFTC plate closest to 30 in your calculation for practice. Make a note of this in your results.

| Phage titer | |
|---|---|

## QUESTIONS

**1** *In this exercise, there must be enough bacteria inoculated to produce a lawn of growth. Why is that important?*

**2** *Consider the control plate (Plate A) inoculated only with* E. coli *B.*

**a.** *Is it an example of a positive or a negative control?*

**b.** *What purpose did it serve? Be descriptive in your answer.*

**3** *Consider the adsorption step.*

**a.** *How might the results be altered if you had skipped the adsorption phase?*

**b.** *How might the results be altered if you extended it to 35 minutes?*



**4** *Consider the water bath.*

**a.** *Why was it set at 50°C?*

**b.** *What might be some consequences of raising its temperature? Of lowering its temperature?*

Name ______________________________

Date ______________________________

Lab Section ______________________________

I was present and performed this exercise (initials) ______________________________

# DATA SHEET 6-4

*(continued)*

**5** *Consider the soft agar.*

**a.** *Why was soft agar used for the agar overlay?*

**b.** *What would you expect to see if standard nutrient agar or trypticase soy agar had been used instead? (**Hint:** It might be helpful to look at the recipes for these in Appendix E.)*

**6** *For each data set, calculate the original phage titer. In each case, 100 μL of diluted phage was used.*

a. *Phage dilution tube:* $10^{-4}$; E. coli *B: 300 μL; soft agar: 2.5 mL; countable plate: 37 plaques.*

b. *Phage dilution tube:* $10^{-8}$; E. coli *B: 300 μL; soft agar: 2.5 mL; countable plate: 216 plaques.*

c. *Phage dilution tube:* $10^{-5}$; E. coli *B: 300 μL; soft agar: 2.5 mL; countable plate: 37 plaques.*

d. *Phage dilution tube:* $10^{-5}$; E. coli *B: 500 μL; soft agar: 2.5 mL; countable plate: 37 plaques.*

6

e. *Phage dilution tube:* $10^{-5}$; E. coli *B: 300 μL; soft agar: 2.0 mL; countable plate: 37 plaques.*

**7** *Examine Figure 6.11. Suppose the OSV on this plate was* $10^{-6}$ *mL. What was the original phage titer?*

## Differential Blood Cell Count

EXERCISE 6-5

### Theory

Leukocytes (white blood cells, or WBCs) are divided into two groups: granulocytes, which have prominent cytoplasmic granules, and agranulocytes, which lack them. There are three basic types of granulocytes: neutrophils, basophils, and eosinophils. The two types of agranulocytes are monocytes and lymphocytes. All leukocytes develop from bone marrow stem cells.

Neutrophils (Fig. 6.12A) are 12 µm–15 µm in diameter—about twice the size of an erythrocyte (RBC)[1] and are the most abundant WBCs in blood. Their most prominent cytoplasmic granules are neutral-staining and thus do not have the intense color of other granulocytes when prepared with Wright's or Giemsa stain. When needed, they leave the blood and enter tissues to phagocytize foreign material, whose recognition requires binding to a variety of different membrane receptors.

Once internalized by phagocytosis, bacteria are killed by several mechanisms acting in concert, including oxidation of cell structures by oxygen radicals, enzymatic destruction of the cell wall, and leakage through the cell membrane caused by cationic peptides. Neutrophils typically live only a few days before undergoing apoptosis ("programmed cell death"). Host cell debris, dead neutrophils, and dead bacteria form pus.

Mature neutrophils sometimes are referred to as segs because their nucleus usually is segmented into two to five lobes. Because of the variation in nuclear appearance, they also are called polymorphonuclear neutrophils (PMNs). Immature neutrophils lack this nuclear segmentation and are referred to as bands (Fig. 6.12B). This distinction is useful because a patient with an active infection increases neutrophil production, which leads to a higher percentage of immature band cells in the blood. The inactive X-chromosome is frequently seen as a "drumstick" extending from the neutrophil nucleus of females.

Eosinophils (Fig. 6.13) are 12 µm–15 µm in diameter (about twice the size of an RBC), generally have two lobes in their nucleus, and have red staining cytoplasmic granules. Most reside in connective tissues and are sometimes difficult to find in blood smears, where they represent less than 3% of all leukocytes. Their functions are varied and their numbers increase during allergic reactions, chronic inflammation, and parasitic infections. They phagocytose and digest antigen-antibody complexes, and their secretions include chemicals that control the inflammatory response (e.g., histaminase), and cytotoxic and neurotoxic chemicals that attack parasites.

Basophils (Fig. 6.14) are the least abundant WBCs in normal blood and are difficult to find. They are structurally and functionally similar to tissue mast cells and produce some of the same chemicals (histamine and heparin) as well as bind immunoglobin E (IgE), but are

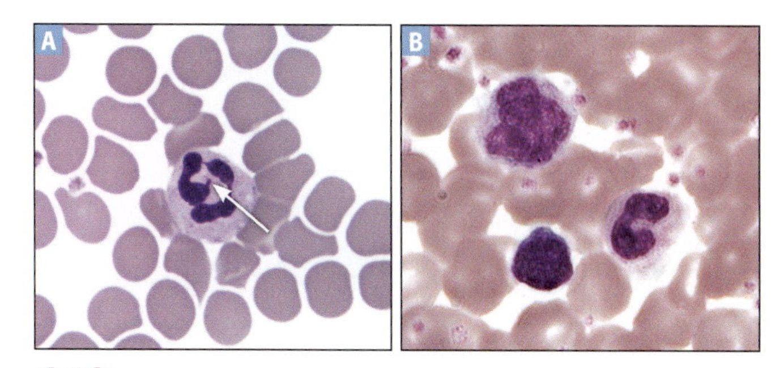


**6.12 Neutrophils** ■ (A) The segmented nucleus of this cell identifies it as a mature neutrophil (seg). About 30% of neutrophils in blood samples from females demonstrate a "drumstick" protruding from the nucleus, as in this specimen (arrow). This is the region of the inactive X chromosome. (B) At the right is an immature band neutrophil (or slightly beyond the band stage, but still immature) with an unsegmented nucleus. Also shown are a monocyte (above left) and a lymphocyte (lower left).

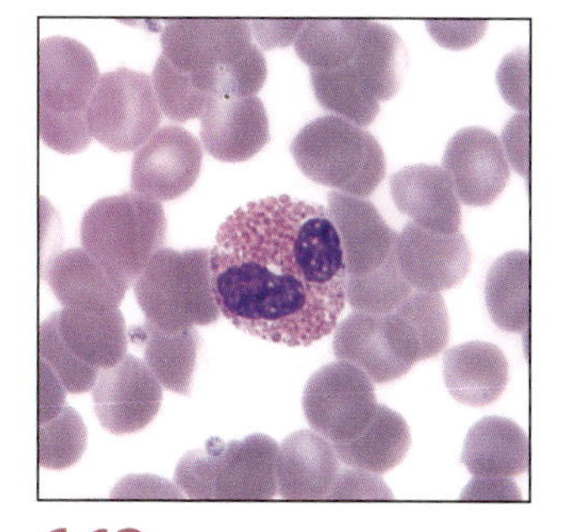

**6.13 Eosinophil** ■ These granulocytes are relatively rare and are about twice the size of red blood cells. Their cytoplasmic granules stain red, and their nucleus usually has two lobes.

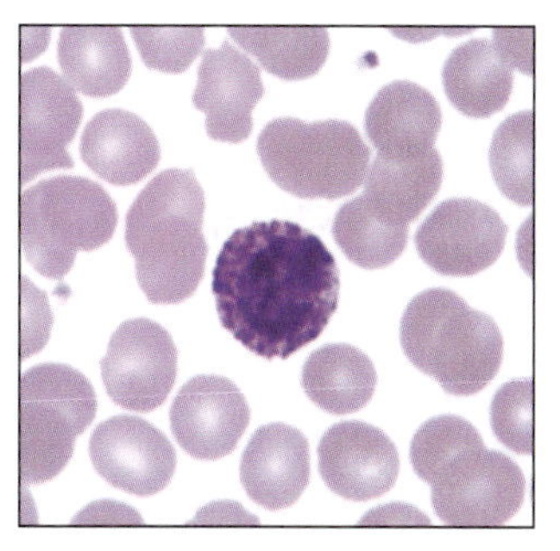

**6.14 Basophil** ■ Basophils comprise less than 0.2% of all white blood cells and require patience to find, because they are rare. (At 0.2%, you would theoretically have to look at 500 WBC just to see one.) They are slightly larger than red blood cells and have dark-purple cytoplasmic granules that obscure the nucleus. They release histamine and other chemicals that enhance the inflammatory response.

[1] It is convenient to discuss leukocyte size in terms of erythrocyte size because RBCs are so uniform in diameter. In an isotonic solution, erythrocytes are 7.5 µm in diameter.

derived from different stem cells in bone marrow. They are 12 μm–15 μm in diameter and their dark-staining cytoplasmic granules usually obscure the one- or two-lobed nucleus.

Agranulocytes include monocytes and lymphocytes. Monocytes (Fig. 6.15) are the blood form of macrophages, which phagocytize foreign cells, cell debris, and also act as antigen-presenting cells during many immune responses. They are the largest of the leukocytes, being two to three times the size of RBCs (12 μm–20 μm). Their nucleus is horseshoe-shaped (or at least indented), and the cytoplasm lacks prominent granules, but may appear finely granular.

Lymphocytes (Fig. 6.16) are cells of specific acquired immunity. They are approximately the same size as RBCs, or up to twice their size, and have very little cytoplasm visible around their spherical nucleus. Two functional types of lymphocytes are the T-cell, involved in cell-mediated immunity, and the B-cell, which converts to a plasma cell (Fig. 6.16B) when activated, and produces antibodies. Natural killer (NK) cells comprise a third class of lymphocyte that kills foreign or virally-infected cells without antigen-antibody interaction (Fig. 6.16C). They are larger than B- and T-lymphocytes, being up to 16 μm and have large cytoplasmic granules, which accounts for their other name: large granular lymphocytes.

In a differential white cell count, a sample of blood is observed under the microscope, and at least 100 WBCs are counted and tallied (this task is automated now). Approximate normal percentages for each leukocyte are as follows and as summarized in Table 6-2.

neutrophils (mostly segs) 55%–65%

lymphocytes 25%–33%

monocytes 3%–7%

eosinophils 1%–3%

basophils <0.2%.

TABLE **6-2** **Typical Features of Human Leukocytes in Blood**

| Cell | Abundance in Blood (%) | Diameter (μm) | Nucleus | Cytoplasmic Granules (Wright's or Giemsa Stain) | Functions |
|---|---|---|---|---|---|
| **Granulocyte** | | | | | |
| Neutrophil | 55–65 | 12–15 | 2–5 lobes | Present, but stain poorly; contain antimicrobial chemicals | Phagocytosis and digestion of (usually) bacteria |
| Eosinophil | 1–3 | 12–15 | 2 lobes | Present and stain red; contain antiparasitic chemicals and inflammatory moderators (e.g., histaminase) | Present in inflammatory reactions (removal of antigen-antibody complexes) and destruction of multicellular parasites (worms) with cytotoxins and neurotoxins |
| Basophil | <0.2 | 12–15 | Unlobed or 2 lobes | Present and stain dark purple; contain histamine and other chemicals | Sustain inflammatory response by releasing chemicals, such as heparin (an anticoagulant) and histamine (as vasodilator) |
| **Agranulocyte** | | | | | |
| Lymphocyte | 25–33 | 7–18 (rare) | Spherical (leaving little visible cytoplasm) | Absent | Active in specific acquired immunity (as T- and B-cells). Also active as NK cells. |
| Monocyte | 3–7 | 12–20 | Horseshoe-shaped (cytoplasm is prominent) | Absent | Phagocytosis and presentation of antigens to immune cells (as macrophages) |

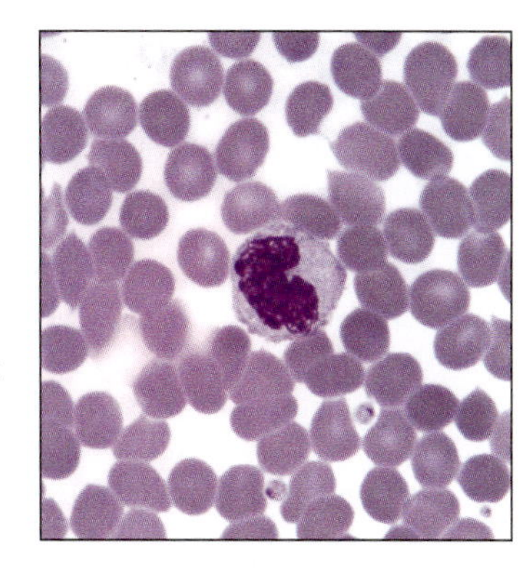

**6.15 Monocyte** Monocytes are the blood form of macrophages. They are about twice the size of red blood cells and have a round or indented nucleus.

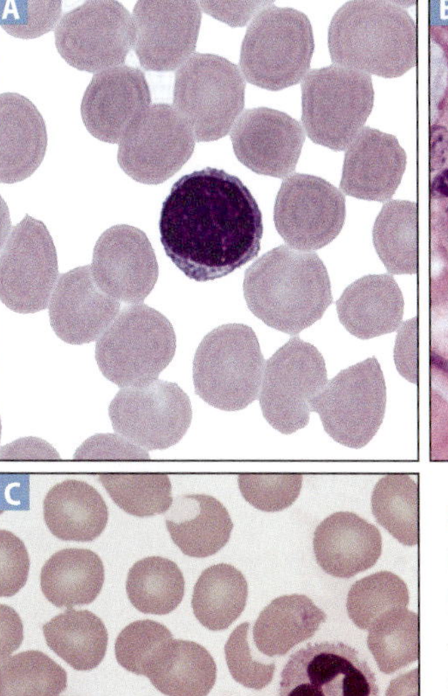

**6.16 Lymphocytes** Lymphocytes are common in the blood, comprising up to 33% of all WBCs. (A) Small- and medium-sized lymphocytes are about the size of red blood cells and have only a thin halo of cytoplasm encircling their round nucleus. They belong to functional groups called B-cells and T-cells (which are morphologically indistinguishable). (B) Although you won't see plasma cells in the blood smears, it doesn't hurt for you to see a micrograph of one in anticipation of studying immunology. B-cells differentiate into plasma cells when stimulated by the appropriate antigen. Once formed, plasma cells secrete protective antibodies against that antigen. This plasma cell was found in a slide of a thin section of the colon. (C) About 15% of blood lymphocytes are large granular lymphocytes, most of which are natural killer (NK) cells (left). Also visible at the right in this micrograph is a neutrophil.

## Application

A differential blood cell count is done to determine approximate numbers of the various leukocytes in blood. Excess or deficiency of all or a specific group is indicative of certain disease states. Even though differential counts are automated now, it is good training to perform one "the old-fashioned way" using a blood smear and a microscope to get an idea of the principle behind the technique. It will also put a face on many of the cells you will study during the immunology section of your microbiology course.

## In This Exercise

You will be examining prepared blood smears and doing a differential count of white blood cells. As an optional activity, you may look at smears of abnormal blood and compare their differential count to normal blood.

## Materials

### Per Student Group

- □ Commercially prepared human blood smear slides (Wright's or Giemsa stain)
- □ (Optional) commercially prepared abnormal human blood smear slides (e.g., infectious mononucleosis, eosinophilia, or neutrophilia)

## PROCEDURE

1. Obtain a blood smear slide and locate a field where the cells are spaced far enough apart to allow easy counting. (The cells should be fairly dense on the slide, but not overlapping.)
2. Using the oil-immersion lens, scan the slide using the pattern shown in Figure 6.17. Be careful not to overlap fields when scanning the specimen. Choose a "landmark" blood cell at the right side of the field, and move the slide horizontally until that cell disappears off the left side. Then, choose a "landmark" cell in at the bottom of the field and move the stage until it disappears off the top. When it disappears, choose a "landmark" cell at the *left* side of the field and scan horizontally *but in the opposite direction*. Repeat until you have counted the required number of cells. Avoid diagonal movement of the slide. As you scan, use the mechanical stage knobs separately to move the slide up and back or to the right and left in straight lines.

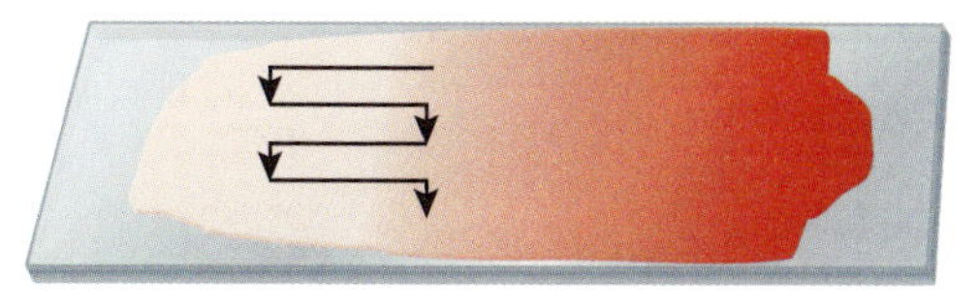

**6.17 Following a Systematic Path** A systematic scanning path is used to avoid wandering around the slide and perhaps counting some cells more than once. Remember that a microscope image is inverted. If you want the image to move left, you must move the slide to the right.

3. Make a tally mark in the appropriate box in the table on the data sheet, page 439, for the first 100 leukocytes you see. (Your instructor may have you count more in hope of you locating the less common leukocytes.)
4. Calculate percentages and compare your results with the accepted normal values.
5. Repeat with a pathological blood smear (if available).

## References


Brown, Barbara A. *Hematology—Principles and Procedures,* 6th ed. Philadelphia: Lea and Febiger, 1993.

Diggs, L. W., Dorothy Sturm, and Ann Bell. *The Morphology of Human Blood Cells,* 4th ed. North Chicago, IL: Abbott Laboratories, 1978.

Mak, Tak W., Mary E. Saunders, and Bradley D. Jett. Chap. 2 in *Primer to the Immune Response*, 2nd ed. Burlington, MA: AP Cell, Elsevier, 2014.

Male, David, Jonathan Brostoff, David B. Roth, and Ivan M. Roitt. Chap. 2 in *Immunology*, 8th ed. Waltham, MA: Elsevier, 2013.

Mescher, Anthony L. Chap. 12 in *Junqueira's Basic Histology, Text and Atlas,* 12th ed. New York: McGraw Hill, Lange, 2010.

Ross, Michael H. and Wojciech Pawlina. Chap. 10 in *Histology A Text and Atlas*, 6th ed. Philadelphia: Wolters Kluwer, Lippincott Williams & Wilkins., 2011.

Name ______

Date ______

Lab Section ______

I was present and performed this exercise (initials) ______

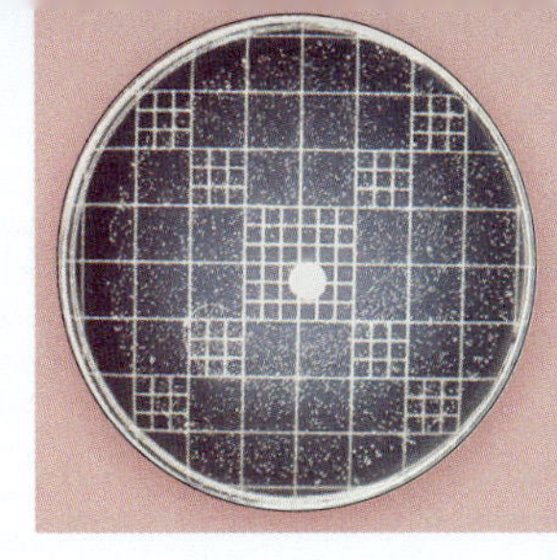

# DATA SHEET 6-5

## Differential Blood Cell Count

### OBSERVATIONS AND INTERPRETATIONS

**1** Record your data from the differential blood cell count in the table below. As you count the 100 white blood cells, make tally marks in the appropriate boxes. Then calculate the percentages of each type and compare them to the expected values.

| Normal Blood | | | | | | |
|---|---|---|---|---|---|---|
| | Monocytes | Lymphocytes | Segmented Neutrophils | Band Neutrophils | Eosinophils | Basophils |
| Number | | | | | | |
| Percentage | | | | | | |
| Expected percentage | 3–7% | 25–33% | 55–65% | — | 1–3% | <0.2% |

**2** If available, perform a differential count as in step 1 above with an abnormal blood smear.

| Abnormal Blood (Condition: ______ ) | | | | | | |
|---|---|---|---|---|---|---|
| | Monocytes | Lymphocytes | Segmented Neutrophils | Band Neutrophils | Eosinophils | Basophils |
| Number | | | | | | |
| Percentage | | | | | | |
| Expected percentage | 3–7% | 25–33% | 55–65% | — | 1–3% | <0.2% |

## QUESTIONS

**1** *How do the percentages of each WBC compare to the published values? What might account for any differences you noted?*

**2** *If you did differential counts on abnormal blood, how did the percentages compare to normal blood? How can any differences you noted be explained in the context of the disease and/or defense process? Be sure to name the blood condition in your answer.*

**3** *What is the purpose of scanning the slide in a systematic pattern (as described in the text)?*

6

**4** *Most leukocytes function outside of the blood stream. Given this fact, why are they found in blood smears?*

SECTION 7

# Medical, Environmental, and Food Microbiology

This section is a potpourri of lab exercises that span the range of traditional microbiology. Microbiology has its roots as a formal discipline of study in medical applications. However, over the years its scope broadened to include studying the important—and often irreplaceable—roles played by microbes in ecosystems of all types, their impact on public health, and how their activities can be exploited in food production.

## Medical Microbiology

Microbiology as a formal discipline of study has its roots in medicine (though some might make the argument that it really has its roots in food production, because *human* cultures worldwide always seemed to learn very early on what was available to them to ferment!). In fact, so much of microbiology has to do with its medical applications that a strong case could be made that most of the exercises you have already completed fall under this heading (though they may have additional applications in other microbiological fields). The superficial treatment of medical microbiology that follows in this section is not meant to represent a balanced body of information. Rather, the topics are intended to augment exercises you have already done related to medical microbiology.

In Exercise 7-1 you will perform the Snyder test, which is designed to detect susceptibility to dental decay in a patient. It is a simple procedure to perform and will give you an opportunity to know your oral microbiota a little better!

The procedure in Exercise 7-2 is performed routinely in medical laboratories worldwide on a daily basis because it addresses a current medical crisis: bacterial antibiotic resistance. The procedure is called the Kirby-Bauer, or Disk Diffusion, test and is used to determine susceptibility and resistance of a pathogen isolated from a patient to a variety of antibiotics. In this way, treatment can be tailored to the particular pathogenic strain that has infected the patient.

The last two exercises allow you to explore a whole different domain of microbiology—epidemiology. Literally, epidemiology means "the study of (disease) upon the population," and when applied to infectious disease it falls within the umbrella of medical microbiology. Epidemiologists don't spend a lot of time worrying about how to cultivate or identify a particular pathogen in the lab or looking at the interaction of pathogens and their host on an individual basis. Mostly, epidemiologists try to figure out who in the population is susceptible to the pathogen, how they got infected, and what can be done to minimize occurrence of that disease.

In the first (Exercise 7-3), you will use a standard resource—*Morbidity and Mortality Weekly Report* available from the Centers for Disease Control and Prevention (CDC) website—to follow the incidence of a disease of your choice over the course of one complete year. In Exercise 7-4, you will simulate an epidemic outbreak in your class, determine the source of the outbreak, and perform epidemiological analyses of your class population.

## Environmental Microbiology

The topic of environmental microbiology encompasses exploring the microbial world around us to discover its diversity in water, soil, and many other habitats, as well as testing the environment (e.g., water) for public health reasons. It is this latter area you will explore in Exercises 7-5 and 7-6. Both exercises demonstrate different approaches to testing water quality for fecal contamination.

## Food Microbiology

Food microbiology is devoted to the study and utilization of beneficial microbes, as well as to the control of many common, and in some cases, potentially deadly, contaminants. You will employ a simple test in Exercise 7-7 to check the quality of milk. Then in Exercise 7-8, you will have the opportunity to exploit the ability of certain bacteria to ferment by having them make yogurt for you!

# Snyder Test

EXERCISE 7-1

## Theory

The oral cavity presents a variety of habitats for microbial growth, including the tongue, floor of the mouth, and buccal (cheek) and palatal surfaces. These surfaces are covered with stratified squamous epithelium that sheds on a regular basis, thus making heavy microbial colonization difficult. Then there are the teeth, the only organs in the body that present hard surfaces to the outside world. Microbial communities attach firmly to the teeth in **biofilms**, more commonly known as **dental plaque**. This firm attachment to teeth makes colonization more permanent and more abundant than on the oral epithelial surfaces.

Tooth biofilms are inhabited by a diverse assemblage of bacteria that become attached to a proteinaceous film covering teeth called a **pellicle**. Initial attachment of bacteria to the tooth surface involves many factors, but a variety of bacterial **adhesins** bind specifically to host protein receptors, which accounts in large part for the presence of microorganisms characteristic of dental plaque. Many of these microbes are fermenters of sugars, such as glucose and sucrose, and produce acid as an end product. It is the acid that demineralizes tooth enamel and results in **caries** (cavities).

The worst offenders are *Streptococcus mutans*, other streptococci, *Actinomyces* species, and *Lactobacillus* species. Not surprisingly, the people who are most at risk are those with a high dietary intake of sugars and poor dental hygiene. To make matters worse, these organisms have evolved mechanisms to tolerate low pH conditions (pH of ~5.5) that inhibit growth of other less harmful bacteria.

Snyder test medium is formulated to favor the growth of oral bacteria (Fig. 7.1) and discourage the growth of other bacteria. This is accomplished by lowering the pH of the medium to 4.8. Glucose is added as a fermentable carbohydrate, and bromocresol green is the pH indicator. Lactobacilli and oral streptococci survive these harsh conditions, ferment the glucose, and lower the pH even further. The pH indicator, which is green at or above pH 4.8 and yellow below, turns yellow in the process. Development of yellow color in this medium, therefore, is evidence of fermentation and, further, is highly suggestive of the presence of dental decay-causing bacteria (Fig. 7.2).

The medium is autoclaved for sterilization, cooled to just over 45°C, and maintained in a warm-water bath until needed. The molten agar then is inoculated with a small amount of saliva, mixed well, and incubated for up to 72 hours. The agar tubes are checked at 24-hour intervals for any change in color. High susceptibility to dental caries is indicated if the medium turns yellow within 24 hours. Moderate and slight susceptibility are indicated by a change within 48 and 72 hours, respectively. No change by 72 hours is considered a negative result. These results are summarized in Table 7-1.

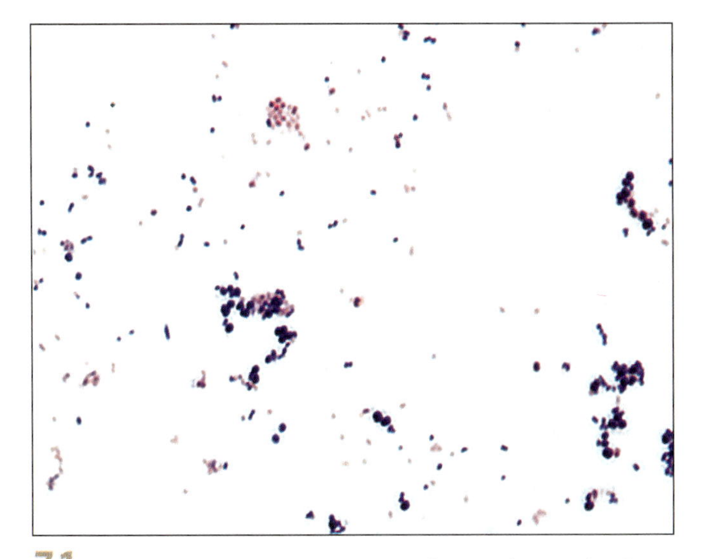

**7.1 Gram Stain of a Tooth Scraping** ■ This sample was taken from the surface of the second molar with a sterile wooden stick several hours after brushing. Note the cell diversity comprising the bacterial community on the tooth surface.

**7.2 Snyder Test Results** ■ A positive result is on the left and a negative result is on the right. Acid from glucose fermentation has lowered the pH in the positive tube, changing the pH indicator from green to yellow. To be of full value, the time it took for the agar to turn positive must be recorded. The test should be read every 24 hours and is complete at 72 hours. See Table 7-1 for an explanation of how time figures into the interpretation.

TABLE 7-1 Snyder Test Results and Interpretations

| Result | Interpretation |
|---|---|
| Yellow at 24 hours | High susceptibility to dental caries |
| Yellow at 48 hours | Moderate susceptibility to dental caries |
| Yellow at 72 hours | Slight susceptibility to dental caries |
| Yellow at > 72 hours | Negative |

## Application

The Snyder test is designed to measure susceptibility to dental caries (tooth decay), caused primarily by lactobacilli and oral streptococci.

## In This Exercise

You will check your own oral microflora for its potential to produce dental caries.

### Materials

#### Per Student

- ☐ Lab coat
- ☐ Disposable gloves
- ☐ Chemical eye protection
- ☐ 10 mL sterile beaker

#### Per Student Group

- ☐ One broth culture of *Lactobacillus delbrueckii*
- ☐ Sterile 1 mL pipettes with mechanical pipettor
- ☐ Canister for pipette disposal
- ☐ Three Snyder agar tubes

7

## Medium Recipe

#### Snyder Test Medium

| | |
|---|---|
| ☐ Pancreatic digest of casein | 13.5 g |
| ☐ Yeast extract | 6.5 g |
| ☐ Dextrose (glucose) | 20.0 g |
| ☐ Sodium chloride | 5.0 g |
| ☐ Agar | 16.0 g |
| ☐ Bromocresol green | 0.02 g |
| ☐ Distilled or deionized water | 1.0 L |

*pH 4.6–5.0 at 25°C*

## PROCEDURE

#### Lab One

1. Wear a lab coat, gloves, and chemical eye protection when performing this procedure.
2. Collect a small sample of saliva (about 0.5 mL) in the sterile beaker.
3. Aseptically add 0.2 mL of the sample to a molten Snyder agar tube (from the water bath), and roll it between your hands until the saliva is distributed uniformly throughout the agar. Label the culture with your name, the medium, and date and time of inoculation.
4. Inoculate one Snyder test medium with a few drops of *L. delbrueckii* and roll it between your hands as in step 3. Label the culture with the organism's name, the medium, and date and time of inoculation.
5. Remove a third Snyder agar tube and label it uninoculated.
6. Allow the agar in all three tubes to cool to room temperature. Do not slant.
7. Incubate the tubes at 35 ± 2°C.
8. Examine the tubes for color changes at 24-hour intervals for up to 72 hours.

#### Lab Two (Or whenever the tube shows a positive result)

1. Record your results in the table provided on the data sheet, page 445.

### References


Quivey, Robert G. Chap. 11, Caries, in *Oral Microbiology and Immunology.* Richard J. Lamont, Robert A. Burne, Marilyn S. Lantz, and Donald J. LeBlanc, eds. Washington, DC: ASM Press, 2006.

Scannapieco, Frank A. Chap. 3, The Oral Environment, in *Oral Microbiology and Immunology.* Richard J. Lamont, Robert A. Burne, Marilyn S. Lantz, and Donald J. LeBlanc, eds. Washington, DC: ASM Press, 2006.

Zimbro, Mary Jo and David A. Power, eds. Page 517 in *Difco™* and *BBL™ Manual—Manual of Microbiological Culture Media.* Sparks, MD: Becton Dickinson and Co., 2003.

Name ______________________________

Date ______________________________

Lab Section ______________________________

I was present and performed this exercise (initials) ______________

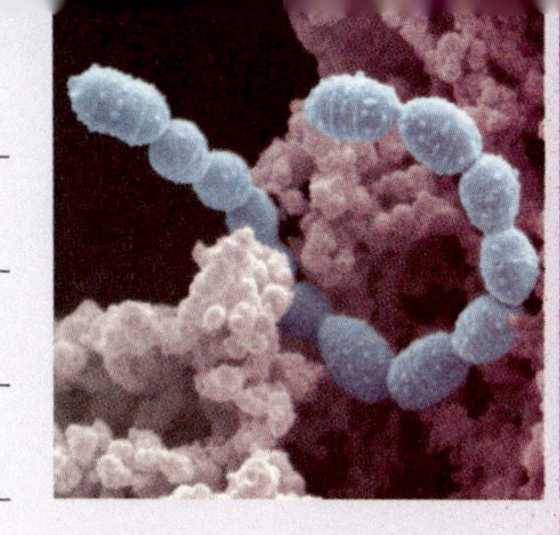

# DATA SHEET 7-1

## Snyder Test

### OBSERVATIONS AND INTERPRETATIONS

1. Record the color of the *Lactobacillus delbrueckii* culture at each time interval in the table below.
2. Record the color of your saliva culture at each time interval in the table below. Refer to Table 7-1, page 444, when recording and interpreting your results below.
3. Record the color of the uninoculated control at each time interval in the table below.

| | Time | | |
|---|---|---|---|
| | 24 Hrs. | 48 Hrs. | 72 Hrs. |
| *L. delbrueckii* culture | | | |
| Saliva sample | | | |
| Uninoculated control | | | |

4. Write your interpretation of the Snyder test using your saliva:

Based on the results of this test, I have a ______________ susceptibility to dental caries.

## QUESTIONS

1. *As demonstrated by this exercise,* Lactobacillus *species ferment glucose (dextrose). Considering their role in fermentation of milk, what other ingredient in Snyder test medium would you expect* Lactobacillus *species to digest?*

______________________________

______________________________

2. *Using Snyder test medium as a guide, what kinds of dietary items would likely increase the number of lactobacilli and streptococci in saliva (and in turn, on tooth biofilms)?*

______________________________

______________________________

______________________________

**3** *Why isn't the molten Snyder test agar allowed to solidify as a slant?*

**4** *What purpose does the uninoculated control serve?*

**5** *What purpose does the* L. delbrueckii *culture serve?*

**6** *How did the Snyder test results using your saliva compare to the results of the* L. delbrueckii *culture?*

7

# Antimicrobial Susceptibility Test: Disk Diffusion (Kirby-Bauer) Method



## Theory

**Antibiotics** are natural antimicrobial agents produced by microorganisms. One type of penicillin, for example, is produced by the mold *Penicillium notatum*. Today, because many agents used to treat bacterial infections are synthetic, the terms **antimicrobials** or **antimicrobics** are used to describe all substances used for this purpose.

In 1966 Alfred Bauer, William Kirby, and associates first published a standardized disk diffusion method of testing the effectiveness of over 20 antimicrobial chemicals. Their methodology, which bears their names, has become a valuable tool for establishing the effectiveness of antimicrobics against pathogenic microorganisms in clinical laboratories.

In the test, antimicrobic-impregnated paper disks are placed on an agar plate inoculated to form a bacterial lawn. The plates are incubated to allow growth of the bacteria and time for the agent to diffuse into the agar. As the drug moves through the agar, it establishes a concentration gradient. If the organism is susceptible to it, a clear zone—**zone of inhibition**—will appear around the disk where the concentration is high enough to stop growth (Fig. 7.3).

The junction of the zone of inhibition with growth is critical to interpretation of the test. It is at this junction where the concentration of antimicrobic has become too low to effectively stop growth. This junction represents the **minimum inhibitory concentration** (**MIC**) for that particular strain.

The basic disk diffusion test does not quantify the MIC (though variations are available that do this), but rather utilizes the zone of inhibition's size, which depends upon the sensitivity of the organism to the specific antimicrobial agent (among other factors). Interestingly, the clear zone does not necessarily mean that the bacteria have been killed. Drugs that kill the organism are said to be **bactericidal**, but other drugs are **bacteriostatic**; that is, they stop the bacteria from dividing, but do not kill them. Some mechanisms of antibiotic action and resistance are given in Table 7-2.

All aspects of the Kirby-Bauer procedure are standardized to ensure reliable results. Therefore, care must be taken to adhere to these standards. Mueller-Hinton agar, which has a pH between 7.2 and 7.4, is poured to a depth of 4 mm in either 150 mm or 100 mm Petri dishes. The depth is important because of its effect upon the diffusion. Thick agar slows lateral diffusion and thus produces smaller zones than plates held to the 4 mm standard.

Inoculation is made with a broth culture, generally incubated 4 to 6 hours to bring it into exponential phase, and subsequently diluted to match a **0.5 McFarland turbidity standard**, which corresponds to a cell density between 1 and $2 \times 10^8$ CFU/mL (Fig. 7.4). Alternatively, a spectrophotometer can be used to bring the broth culture to an absorbance of 0.08 to 0.10 at 625 nm in a 1 cm cuvette, which corresponds to the 0.5 McFarland standard.

The U.S. Food and Drug Administration establishes the disk concentration of each agent, which is printed on the disk along with a code indicating the chemical agent. Disks are then dispensed onto the inoculated plate (Fig. 7.5), which are incubated at $35 \pm 2$°C for 16 to 18 hours. After incubation, the plates are removed and the zone diameters are measured in millimeters (Fig. 7.6) and compared to a standardized table of results (Table 7-3).

The Clinical Laboratory Standards Institute (CLSI) in Wayne, Pennsylvania, has performed extensive research (and continues to do so) to establish the zone diameter interpretive standards in Table 7-3—and there are many more antimicrobial agent/pathogen combinations we have

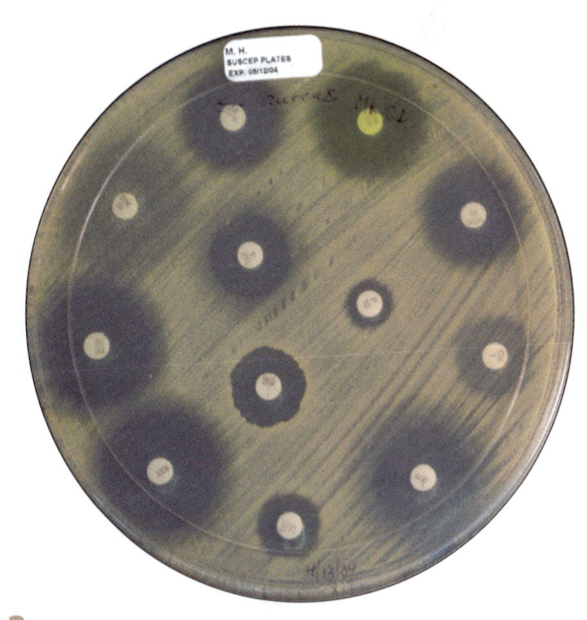


**7.3 Disk Diffusion Test of Methicillin-Resistant *Staphylococcus aureus* (MRSA)** ■ This plate illustrates the effect of (clockwise from top outer right) Nitrofurantoin (F/M300), Norfloxacin (NOR 10), Oxacillin (OX 1), Sulfisoxazole (G 0.25), Ticarcillin (TIC 75), Trimethoprim-Sulfamethoxazole (SXT), Tetracycline (TE 30), Ceftizoxime (ZOX 30), Ciprofloxacin (CIP 5), and (inner circle from right) Penicillin (P 10), Vancomycin (VA 30), and Trimethoprim (TMP 5) on Methicillin-resistant *Staphylococcus aureus* isolated from a patient.

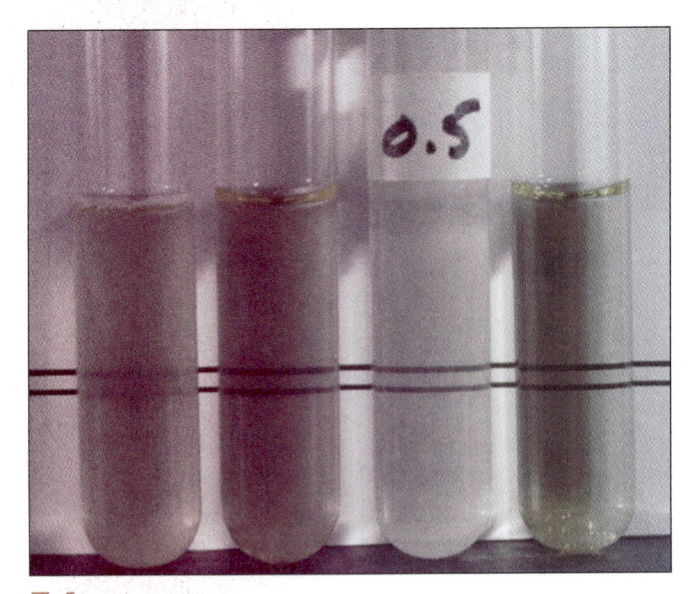


**7.4 McFarland Standards** ■ This is a comparison of a McFarland turbidity standard to three broths having varying degrees of turbidity. Each of the 11 McFarland standards (0.5 to 10) contains a specific percentage of precipitated barium sulfate to produce turbidity. In the Kirby-Bauer procedure, the test culture is diluted to match the 0.5 McFarland standard (roughly equivalent to $1.5 \times 10^8$ CFU/mL) before inoculating the plate. Comparison is made visually by placing a card with sharp black lines behind the tubes. Tube 3 is the 0.5 McFarland standard. Notice that the turbidity of Tube 2 matches the McFarland standard exactly, whereas Tubes 1 and 4 are too turbid and too clear, respectively. Alternatively, cultures may be standardized with a spectrophotometer. They should have an absorbance of 0.08 to 0.10 at 625 nm in a 1 cm cuvette.

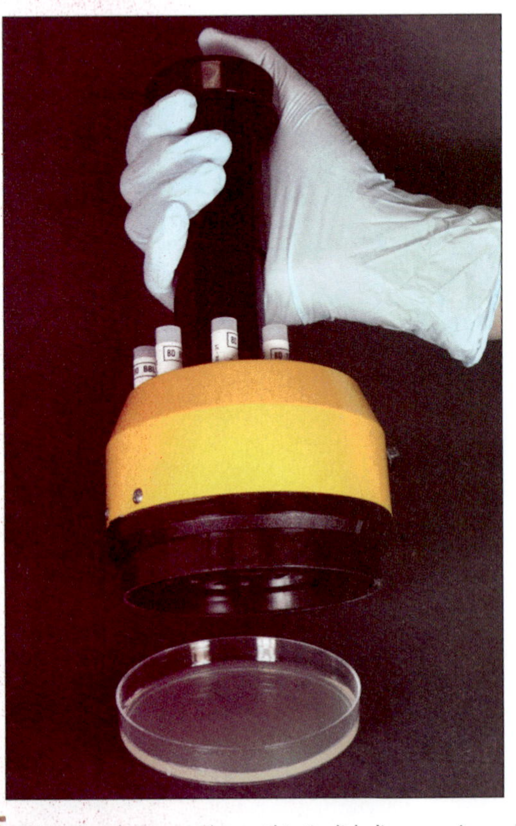

**7.5 Disk Dispenser** ■ This antibiotic disk dispenser is used to deposit disks uniformly on a Mueller-Hinton agar plate.

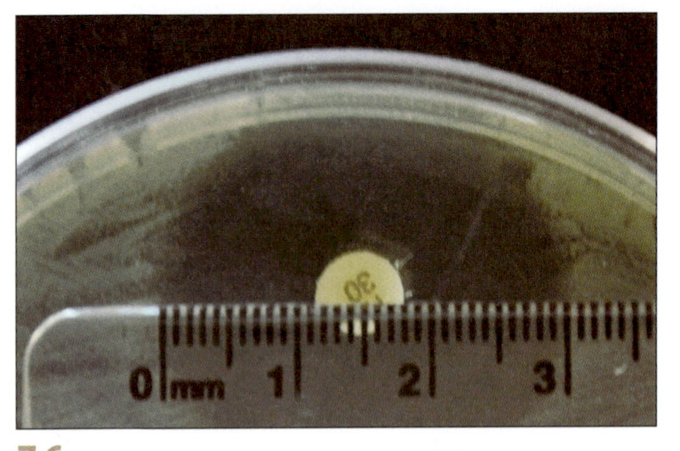


**7.6 Measuring the Antimicrobial Susceptibility Zones** ■ Using a metric ruler and a dark, nonreflective background, measure the diameter of each clearing. Standards for comparison are given in millimeters (mm). This zone is 29 mm in diameter.

not reproduced! These are published in a document titled, "M02–A11: Performance Standards for Antimicrobial Disk Susceptibility Tests; Approved Standard–Eleventh Edition (2012)" and updated in "M100–S24: Performance Standards for Antimicrobial Susceptibility Testing: Twenty-Fourth Informational Supplement (2014)."

Simply put, zone diameter breakpoints for resistance and susceptibility to a particular antimicrobial agent are determined after running a large number of specimens in which both the MIC and the zone diameter is established. The zone diameter below which all resistant strains fall is the **resistance breakpoint**. Likewise, the zone diameter above which all susceptible strains fall is the **susceptibility breakpoint**. In some cases, there may be an intermediate range of zone diameters in which some resistant and susceptible strains fall.

## ■ Application

The disk diffusion test is a standardized method used to measure the effectiveness of antibiotics and other chemotherapeutic agents on pathogenic microorganisms. In many cases, it is an essential tool in prescribing the appropriate treatment for a patient.

## ■ In This Exercise

You will test the susceptibility of *Escherichia coli* and *Staphylococcus aureus* strains to penicillin, chloramphenicol, tetracycline (optional), trimethoprim, streptomycin (optional), and ciprofloxacin. These antibiotics were chosen because they exhibit different modes of action on bacterial cells (Table 7-2).

TABLE **7-2** **Antibiotic Targets and Resistance Mechanisms** ■ Not all antibiotics affect cells in the same way. Some attack the bacterial cell wall, and others interfere with biosynthesis reactions. Resistance mechanisms can be broken down into four main categories: (a) altered target such that the antibiotic no longer can interact with the cellular process, (b) an alteration in how the drug is taken into the cell, (c) enzymatic destruction of the drug, and (d) development or increased activity of an efflux mechanism.

| Antibiotic | Cellular Target | Resistance Mechanism |
|---|---|---|
| **Chloramphenicol** | Prevents peptide bond formation during translation | 1. Poor uptake of drug<br>2. Inactivation of drug |
| **Ciprofloxacin** | Interferes with DNA replication | 1. Altered target<br>2. Poor uptake of drug |
| **Penicillin** | Inhibits cross-linking of the cell wall's peptidoglycan | One or more of:<br>1. Altered target<br>2. Poor uptake of drug<br>3. Production of β-lactamases |
| **Streptomycin** | Blocks initiation complex formation in protein synthesis | 1. Altered target |
| **Tetracycline** | Blocks attachment of aminoacyl tRNA to A site on ribosome | 1. Efflux mechanism |
| **Trimethoprim** | Inhibits purine and pyrimidine synthesis | 1. Altered target |

TABLE **7-3** **Zone Diameter Interpretive Chart**

| Antibiotic | Organism(s) | Code | Disk Potency | Zone Diameter Interpretive Standards (mm) | | |
|---|---|---|---|---|---|---|
| | | | | Resistant | Intermediate | Susceptible |
| **Chloramphenicol** | *Enterobacteriaceae* and *Staphylococcus* | C 30 | 30 µg | ≤ 12 | 13–17 | ≥ 18 |
| **Ciprofloxacin** | *Enterobacteriaceae* and *Staphylococcus* | CIP 5 | 5 µg | ≤ 15 | 16–20 | ≥ 21 |
| **Penicillin** | *Staphylococcus* | P 10 | 10 U | ≤ 28 | | ≥ 29 |
| **Streptomycin** | *Enterobacteriaceae* | S 10 | 10 µg | ≤ 11 | 12–14 | ≥ 15 |
| **Tetracycline** | *Enterobacteriaceae*<br>*Staphylococcus* | TE 30<br>TE 30 | 30 µg<br>30 µg | ≤ 11<br>≤ 14 | 12–14<br>15–18 | ≥ 15<br>≥ 19 |
| **Trimethoprim** | *Enterobacteriaceae* and *Staphylococcus* | TMP 5 | 5 µg | ≤ 10 | 11–15 | ≥ 16 |

This table includes the antibiotics used in this exercise and contains data provided by the Clinical and Laboratory Standards Institute (CLSI). Permission to use portions (specifically Tables 2A and 2C) of M100-S24 (Performance Standards for Antimicrobial Susceptibility Testing; Twenty-Fourth Informational Supplement) has been granted by CLSI. The interpretive data are valid only if the methodology in M02-A11 (Performance Standards for Antimicrobial Disk Susceptibility Tests—11th edition; Approved Standard) is followed. CLSI frequently updates the interpretive tables through new editions of the standard and supplements. Users should refer to the most recent editions. The current standard may be obtained from CLSI, 950 West Valley Road, Suite 2500, Wayne, PA 19087. Contact also may be made via the website (www.CLSI.org), email (customerservice@clsi.org) and by phone (1-877-447-1888).



## ▼ Materials

### Per Student

- ☐ Lab coat
- ☐ Disposable gloves
- ☐ Chemical eye protection

### Per Student Group

- ☐ Two Mueller-Hinton agar plates
- ☐ Penicillin, chloramphenicol, tetracycline (optional), trimethoprim, streptomycin (optional), and ciprofloxacin antibiotic disks (and/or other disks as available)
- ☐ Antibiotic disk dispenser or forceps for placement of disks
- ☐ Screw-capped jar of alcohol (for sterilizing forceps, if used)
- ☐ Two sterile cotton swabs
- ☐ One metric ruler
- ☐ Sterile saline (0.85%)
- ☐ Two sterile transfer pipettes
- ☐ One McFarland 0.5 standard with card
- ☐ Black, nonreflective poster board (8.5" × 11")
- ☐ Spectrophotometer (optional, in place of McFarland standard)
- ☐ Three matched, sterile, 1 cm cuvettes (optional, in place of McFarland standard)
- ☐ Three sterile transfer pipettes (optional, in place of McFarland standard)

- Two capped, sterile 12 mm × 75 mm test tubes (optional, in place of McFarland standard)
- Two tryptic soy agar (or nutrient agar) plates (for optional procedure)
- Fresh broth cultures of these recommended organisms:
  - *Escherichia coli*
  - *Staphylococcus aureus* (BSL-2)
  - *Staphylococcus epidermidis* (an optional substitute for *S. aureus*)

## Medium Recipe

### Mueller-Hinton II Agar

| | |
|---|---|
| Beef extract | 2.0 g |
| Acid hydrolysate of casein | 17.5 g |
| Starch | 1.5 g |
| Agar | 17.0 g |
| Distilled or deionized water | 1.0 L |

*pH 7.2–7.4 at 25°C*

## PROCEDURE

### Lab One

1. Wear a lab coat, gloves, and chemical eye protection when performing this procedure.
2. Gently mix the *E. coli* culture and the McFarland standard until they reach their maximum turbidity.
3. *Note*: If using a spectrophotometer, skip to step 5. Holding the culture and McFarland standard upright in front of you, place the card behind them so you can see the black line(s) through the liquid in the tubes. As you can see in Figure 7.4, the line becomes distorted by turbidity in the tubes. Use the black line to compare the turbidity level of the two tubes. Dilute the broth with sterile saline until it appears to have the same level of turbidity as the standard. If it is not turbid enough, incubate it until it reaches that level.
4. Repeat steps 2 and 3 with the *S. aureus* culture. Use **BSL-2 precautions when working with *S. aureus*.** Continue with step 8.
5. If preferred, a spectrophotometer may be used to bring the samples to the proper density. Transfer uninoculated broth (the same as used to grow the *E. coli* and *S. aureus* cultures) into one of the cuvettes using a sterile transfer pipette. Use this as a blank.
6. Mix the *E. coli* culture and aseptically transfer some broth into the second cuvette. Properly dispose of the pipette and the *E. coli* culture tube, and then determine absorbance of the *E. coli* broth at 625 nm. Either dilute the culture or let it grow in order to bring its absorbance to 0.08 to 0.10. Use a sterile transfer pipette to return the sample to one of the sterile 12 mm × 75 mm test tubes. Label it *E. coli.*
7. Repeat step 6 with the *S. aureus* culture. **Use BSL-2 precautions in transferring it.** When the two broths have the correct turbidity, continue with step 8.
8. Dip a sterile swab into the *E. coli* broth. As you remove it, press and rotate the cotton tip against the side of the tube to remove excess broth.
9. Inoculate a Mueller-Hinton plate with *E. coli* by streaking the entire surface of the agar three times with the swab. Your goal is confluent growth, so make the streaks right next to each other. When you have covered the surface, rotate the plate ⅓ turn and repeat the streaking of the inoculum already on the plate, using the same technique to produce confluent growth. Then rotate the plate another ⅓ turn and repeat. Properly dispose of the swab and replace the plate's lid.
10. Using a fresh sterile swab, inoculate the other plate with *S. aureus* in the same fashion to produce confluent growth. **Use BSL-2 precautions and properly dispose of the swab.** Replace the plate's lid.
11. Label the plates with the organisms' names, your group name, and the date.
12. Apply the penicillin, chloramphenicol, ciprofloxacin, and trimethoprim disks to the agar surface of each plate. You can apply the disks singly using alcohol-flamed forceps[1]. Be sure to space the disks sufficiently (4 cm to 5 cm) to prevent overlapping zones of inhibition and keep them away from the edge. Gently press the disks onto the agar with the forceps so they make good contact. ***Caution***: Keep the flaming forceps away from the alcohol jar. If the alcohol in the jar catches fire, extinguish it by replacing its lid. Continue with step 14.

[1] Keep the forceps in the alcohol jar until ready to use them. Then, close them as you remove them from the alcohol. This will trap some alcohol between the tips. Pass the forceps *through* the flame and allow the alcohol to burn off *away from the alcohol jar and the flame.* When the flame burns out, give the forceps a few seconds to cool, and then place the disk on the plate. Return the forceps to the alcohol jar until needed.

13 Alternatively, you can use a disk dispenser to deliver the disks (Fig. 7.5). If the dispenser is self-tamping, continue with step 14. If not, use the alcohol-flamed forceps to gently press each disk so it makes full contact with the agar. Continue with step 14.

14 Tape the plates closed around the edges. Invert them and incubate aerobically at 35 ± 2°C for 16 to 18 hours. Have a volunteer in the group remove and refrigerate the plates at the appropriate time if this does not match your lab schedule.

15 Dispose of the culture tubes as directed by your instructor.

### Lab Two

1 Remove the plates from the incubator (or refrigerator). Hold each plate over the black, nonreflective poster board and examine the plate with reflected light. The edge of a zone is where no growth is visible to the naked eye. Measure the diameter of each zone of inhibition in millimeters (Fig. 7.6).

2 Using Table 7-3 and those provided with your antibiotic disks (if you used additional antibiotics), record your results in the table provided on the data sheet, page 453.

### Optional Procedure Beginning with Lab Two

1 Wear a lab coat, gloves, and chemical eye protection when performing this procedure.

2 Obtain two TSA or NA plates. With your marking pen, divide the plates into four sectors (or more if you used more antibiotics). Label both plates with your group name.

3 Label one plate *E. coli* and the other *S. aureus*.

4 Label each sector with an antibiotic. It is easiest if you label in the same order as they are found on the MH plates.

5 Using a sterile loop for each transfer, obtain a sample from each antibiotic's zone of inhibition on the *E. coli* plate, and spot inoculate the corresponding sector on the TSA (or NA) plate. If there is no zone for a particular antibiotic, no transfer is necessary.

6 Repeat with the *S. aureus* MH plate. Use a sterile wooden stick or disposable loop for these transfers and dispose of them promptly and properly after each. **Use BSL-2 precautions when working with *S. aureus*.**

7 Incubate the plates aerobically at 35 ± 2° for 24–48 hours.

### Lab Three

1 Remove the plates from the incubator and examine each sector for growth.

2 Record your observations and answer the questions on the data sheet.

## References

Clinical Laboratory Standards Institute (CLSI). *Performance Standards for Antimicrobial Disk Susceptibility Tests; Approved Standard*, 11th ed. CLSI document M02–A11. Wayne, PA: CLSI, 2012.

Clinical Laboratory Standards Institute (CLSI). *Performance Standards for Antimicrobial Susceptibility Testing: Twenty-Fourth Informational Supplement*. CLSI document M100–S24. Wayne, PA: CLSI, 2014.

Collins, C. H., Patricia M. Lyne, and J. M. Grange. Page 128 in *Collins and Lyne's Microbiological Methods*, 7th ed. Oxford, Boston: Butterworth-Heinemann, 1995.

Ferraro, Mary Jane, and James H. Jorgensen. Chap. 15 in *Manual of Clinical Microbiology*, 8th ed. Patrick R. Murray, Ellen Jo Baron, James. H. Jorgensen, Michael A. Pfaller, and Robert H. Yolken, eds. Washington, DC: ASM Press, 2003.

Mims, Cedric, Hazel M. Dockrell, Richard V. Goering, Ivan Roitt, Derek Wakelin, and Mark Zuckerman. Chap. 33 in *Medical Microbiology*, 3rd ed. Philadelphia: Mosby, 2004.

Tille, Patricia M. 2014. Pages 168–170 and 174–178 in *Bailey & Scott's Diagnostic Microbiology*, 13th ed. St. Louis, MO: Mosby, 2014.

Winn, Jr., Washington, Stephen Allen, William Janda, Elmer Koneman, Gary Procop, Paul Schreckenberger, and Gail Woods. Pages 975 and 983–989 in *Koneman's Color Atlas and Textbook of Diagnostic Microbiology*, 6th ed. Philadelphia: Lippincott Williams & Wilkins, 2006.

Zimbro, Mary Jo, and David A. Power, eds. Page 376 in *Difco™ and BBL™ Manual—Manual of Microbiological Culture Media*. Sparks, MD: Becton Dickinson and Company, 2003.

7

Name ______________________________

Date ______________________________

Lab Section ______________________________

I was present and performed this exercise (initials) ______________________________

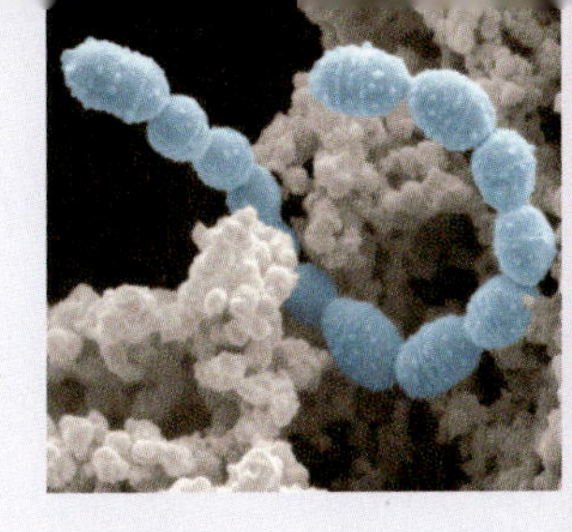

**DATA SHEET 7-2**

# Antimicrobial Susceptibility Test: Disk Diffusion (Kirby-Bauer) Method

## OBSERVATIONS AND INTERPRETATIONS

**1** Record the zone diameters in mm in Table 1 below. Then, use Table 7-3, page 449, as a guide and enter "S" if the organism is susceptible and "R" if it is resistant to the antibiotic. (Not all combinations of antibiotics and organisms are available.)

**Table 1: Interpretation of Zone Diameters**

| Organism | Chloramphenicol | | Ciprofloxican | | Trimethoprim | | Penicillin | | Streptomycin (optional) | | Tetracycline (optional) | |
|---|---|---|---|---|---|---|---|---|---|---|---|---|
| | Zone Diameter | S/R | Zone Diameter | S/R | Zone Diameter | S/R | Zone Diameter | S/R | Zone Diameter | S/R | Zone Diameter | S/R |
| *E. coli* | | | | | | | | | | | | |
| *S. aureus* | | | | | | | | | | | | |

**2** If you did the optional exercise, record "G" for growth and "NG" for no growth by cells recovered from the zone of inhibition on the TSA/NA plates in Table 2 below. Then, decide and record if the antibiotic is bactericidal or bacteriostatic.

**Table 2: Degree of Antibiotic Effect on Test Organisms**

| Organism | Chloramphenicol | | Ciprofloxican | | Trimethoprim | | Penicillin | | Streptomycin (optional) | | Tetracycline (optional) | |
|---|---|---|---|---|---|---|---|---|---|---|---|---|
| | NG/G | -static or -cidal | NG/G | -static or -cidal | NG/G | -static or -cidal | NG/G | -static or -cidal | NG/G | -static or -cidal | NG/G | -static or -cidal |
| *E. coli* | | | | | | | | | | | | |
| *S. aureus* | | | | | | | | | | | | |

## QUESTIONS

**1** *All aspects of the Kirby-Bauer test are standardized to assure reliability.*

**a.** *What might be the consequence of pouring the plates 2 mm deep instead of 4 mm deep?*

______________________________

______________________________

**b.** *The Mueller-Hinton II plates are supposed to be used within a specific time after their preparation and should be free of visible moisture. What negative effect(s) might moisture have on the test?*

______________________________

______________________________

**2** *In clinical applications of the Kirby-Bauer test, diluted cultures (for the McFarland standard comparison) must be used within 30 minutes. Why is this important?*

**3** E. coli *and* S. aureus *were chosen to represent Gram-negative and Gram-positive bacteria, respectively.*

**a.** *For a given antibiotic, is there a difference in susceptibility between the Gram-positive and Gram-negative bacteria?*

**b.** *If so, what difference(s) do you see?*

**c.** *Do these agree with the data in Table 7-3?*

**4** *How does the antibiotic get from the disk into the agar?*

**5** *After incubation, does the antibiotic extend into the agar beyond the zone of inhibition? How does your answer tie in with MIC?*

7

**6** *Suppose you do this test on a hypothetical* Staphylococcus *species with the antibiotics penicillin (P 10) and chloramphenicol (C 30). You record zone diameters of 25 mm for the chloramphenicol and penicillin disks. Which antibiotic would be more effective against this organism? What does this tell you about comparing zone diameters to each other and the importance of the zone diameter interpretive table?*

**7** *In the optional exercise, what was the purpose of inoculating TSA (or NA) plates with samples taken from the zones of inhibition?*

# Morbidity and Mortality Weekly Report (MMWR) Assignment

EXERCISE **7-3**

## Theory

**Epidemiology** is the study of the causes, occurrence, transmission, distribution, and prevention of diseases in a population. The Centers for Disease Control and Prevention (CDC) in Atlanta, Georgia, is the national clearinghouse for epidemiological data (Fig. 7.7). The CDC receives reports related to the occurrence of nearly 100 notifiable diseases (Table 7-4) from the United States and its territories, and compiles the data into tabular form, available in the publication *Morbidity and Mortality Weekly Report* (MMWR).

Two important disease measures that **epidemiologists** collect are related to **morbidity** (sickness) and **mortality** (death). Morbidity relative to a specific disease is the number of susceptible people who have the disease within a defined population during a specific time period. It usually is expressed as a rate and is calculated as follows:

$$\text{Morbidity Rate} = \frac{\text{Number of existing cases in a time period}}{\text{Size of at-risk population at midpoint of time period}} \times 10^n$$

Because population size constantly changes, it is conventional to use the population size at the midpoint of the study period, which would be July 1 if the study lasted a calendar year.

The resulting units for the rate fraction are "cases per person" in the time period and usually are small decimal fractions. To make the calculated rate more "user-friendly," it is multiplied by some power of 10 ($10^n$) to achieve a value that is a whole number, or at least a number greater than 1. Thus, a morbidity rate of 0.00002 is multiplied by 100,000 ($10^5$) so it can be reported as 2 cases per 100,000 people per time period rather than 0.00002 cases per person per time period.

The value chosen for $10^n$ is determined by the resulting decimal fraction and is somewhat arbitrary. For instance, if the fraction was 0.000024 cases per person, it could be reported in three ways. One, you could use $10^5$ as the multiplier and report 2.4 cases per 100,000 people. Or, you could use $10^6$ as the multiplier and report 24 cases per 1,000,000 people. Or, you could round off using $10^5$ as the multiplier and report 2 cases per 100,000 people.

Mortality can also be expressed as a rate. One form, called **cause-specific mortality rate**, is the number of people who die from a specific disease out of the total population afflicted with that disease in a specified time period. It, too, is multiplied by $10^n$ so the rate can be reported as a whole number of cases. The equation is:

$$\text{Mortality Rate} = \frac{\text{Number of deaths due to a disease in a time period}}{\text{Number of people with that disease in the time period}} \times 10^n$$

Minimally, an epidemiological study evaluates morbidity or mortality data in terms of **person** (age, sex, race, etc.), **place**, and **time**. Sophisticated analyses require training in biostatistics, but the simple epidemiological calculation you will be doing—**incidence rate**—can be performed with little mathematical background. Incidence rate is a measure of morbidity and is the occurrence of new cases of a disease within a defined population during a specific period of time. As before, $10^n$ is some power of 10 so the rate can be reported as a whole number of cases, using the following equation:

$$\text{Incidence Rate} = \frac{\text{Number of new cases in a time period}}{\text{Size of at-risk population at midpoint of time period}} \times 10^n$$

Because our focus is microbiology, we will deal only with **infectious diseases**—those caused by biological agents such as bacteria and viruses. **Noninfectious diseases**, such as stroke, heart disease, and emphysema, also are studied by epidemiologists but are not within the scope of microbiology.

**7.7 Centers for Disease Control and Prevention** ■ Shown is the entrance to the CDC campus in Atlanta, Georgia.

## Application

An understanding of the causes and distribution of diseases in a population begins with collecting relevant data and performing statistical analyses, such as morbidity, mortality, and incidence rates. These metrics (among many others) are useful to health-care providers in a couple of ways. First, they provide an awareness of what diseases are present in a population during a certain period of time, which aids in diagnosis. Second, they can lead to a better understanding of a disease, its causes, and transmission, which can be useful in implementing strategies for preventing it.

## In This Exercise

First you will choose a disease from the list of notifiable diseases in Table 7-4. Then you will collect incidence data for the United States over the past complete year. Once you have collected data, you will construct graphs illustrating the cumulative totals and incidence values for that year.

## Materials

- ☐ A computer with Internet access to *Morbidity and Mortality Weekly Report*

## PROCEDURE

1. Choose a disease from Table 7-4 and record its name on the data sheet, page 460.
2. Go to the CDC website (http://www.cdc.gov). Then follow these links. (***Note***: Websites often are revised, so if the current layout of the CDC site does not match this description exactly, it will probably be close. Improvise, and you will find what you need. These links are current as of July 2015):
   - Click on Morbidity and Mortality Weekly Report (MMWR) near the bottom left of the home page under "About CDC" and "Science."
   - Click on "Publications" in the menu bar at the left. This will drop down a new menu.

TABLE **7-4** **Notifiable Diseases in the United States and Its Territories Posted in Table II of MMWR (as of July 2015)** ■ Not all diseases have a high incidence, but do not be misled. Smaller numbers won't necessarily make your work any easier! The only diseases unsuitable for this assignment are **Invasive Pneumococcal Disease** (<5) and **Rabies, animal**, because you won't have a population size to use for your calculations.

| Selected Notifiable Diseases | |
|---|---|
| Babesiosis | Lyme Disease |
| Campylobacteriosis | Malaria |
| *Chlamydia trachomatis* infection | Meningococcal Disease, invasive, all serogroups |
| Coccidioidomycosis | Mumps |
| Cryptosporidiosis | Pertussis |
| Dengue Fever | **Rabies, animal** |
| *Ehrlichiosis chaffeensis* | Rubella |
| *Anaplasma phagocytophilum* | Salmonellosis |
| Girardiasis | Shiga toxin-producing *E. coli* (STEC) |
| Gonorrhea | Shigellosis |
| *Haemophilus influenzae*, invasive, all ages and serotypes | Spotted Fever Rickettsiosis (including RMSF)–confirmed |
| Hepatitis A | Syphilis, primary and secondary |
| Hepatitis B | Tetanus |
| Hepatitis C | Varicella (chickenpox) |
| Invasive Pneumococcal Disease (all ages) | Vibriosis |
| **Invasive Pneumococcal Disease** (<5) | West Nile Virus—neuroinvasive |
| Legionellosis | West Nile Virus—non-neuroinvasive |

- Click on "Weekly Report."
- Click on "Past Volumes (1982–201X)."
- Click on the link to the volume for the most recent complete year. For example, if the current year is 2016, then the most recent complete year would be 2015. It might look something like this: "Volume 64 (2015) Weekly *MMWR* Issues."[1]
- Scroll to the bottom of this new page and you will find the first issue for that volume. Then click on the link to "Notifiable Diseases and Mortality Tables."
- On this new page, click on the link to "Table II" and you are where you need to be![2]
- Table II in 2015 was divided into 12 parts, but that often changes from year-to-year, so be flexible. The diseases are listed alphabetically. Find the disease you have chosen and the *cumulative* number of cases in the United States for Week 1 of the year you are studying. For instance, if you are studying 2015, you would take the number from the column titled *Cum 2015*[3]. Record this in the data table for Week 1.
- Return to the page with the links to all the issues from the latest year you are pulling information from. Scroll down to Week 2 and click on the link to "Notifiable Diseases and Mortality Tables" and get the data for the year you are studying as before. Repeat this process for each week through *MMWR* Week 52. (Alternatively, you can just change the week number in the URL and press return.[4]) As you go, record the cumulative totals on the data sheet.

3. Answer the questions and complete the activities on the data sheet.
4. At some point before the assignment is due, go to the National Notifiable Diseases Surveillance System (NNDSS) page (http://wwwn.cdc.gov/nndss/) and read the information on the home page. Then, click the link to "Data Collection and Reporting" to see how the data you are using have been collected.

---

[1] Volumes for the next few years are: Volume 65, 2016; Volume 66, 2017; Volume 67, 2018; and Volume 68, 2019.

[2] Table I lists the occurrence of infrequently reported notifiable diseases (< 1,000 cases in the previous year). Your instructor may allow you to choose selected diseases from this table in addition to the ones listed in Table II.

[3] Be careful. Cumulative totals from the current and previous years are listed next to each other. And, using the years 2015 and 2014 as examples, you will find that the 2014 reported number of cases for a particular week in the 2015 table might differ from the reported number in the same week of the 2014 table. This is a result of corrections to the 2014 reported cases that have been incorporated into the 2015 table. Don't fret. This is out of your control! Just record the numbers as given.

[4] The URL looks like this: http://www.cdc.gov/mmwr/preview/mmwrhtml/mm**6427**md.htm?s_cid=mm6427md_w. The number highlighted in bold is the week number, preceded by the volume number. So, 6427 means Volume 64, Week 27. Change the 27 to 28, press return and you should end up on Table II, Week 28. Do not change the second set of "6427" numbers because that doesn't work. Thanks to former student Joe Montes, Jr. for this shortcut tip!

## References

CDC. "An Introduction to Applied Epidemiology and Biostatistics" in *Principles of Epidemiology in Public Health Practice*, 3rd ed. Atlanta, GA: Centers for Disease Control and Prevention, 2012. Available online. URL: http://www.cdc.gov/ophss/csels/dsepd/ss1978/lesson3/index.html.

Essex-Sorlie, Diane. Pages 48–50 in *Medical Biostatistics & Epidemiology*. Norwalk, CT: Appleton & Lange, 1995.

Lilienfeld, David E. and Paul D. Stolley. Page 109–110 in *Foundations of Epidemiology*, 3rd ed. New York: Oxford University Press, 1994.

Mausner, Judith S. and Shira Kramer. Pages 44–49 in *Epidemiology: An Introductory Text*, 2nd ed. Philadelphia: W.B. Saunders Company, 1985.

Name ______________________________

Date ______________________________

Lab Section ______________________________

I was present and performed this exercise (initials) ______________

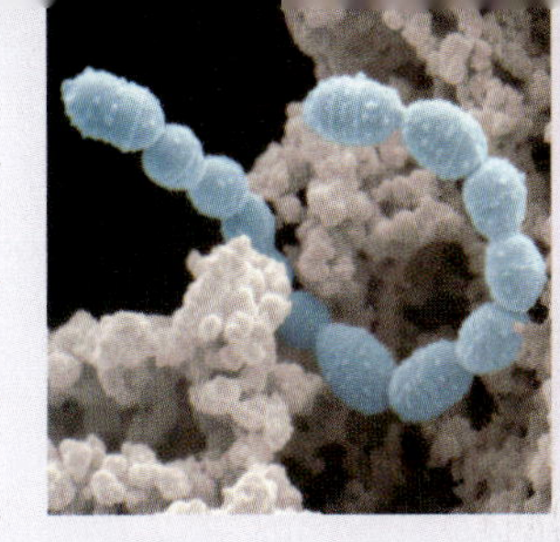

DATA SHEET 7-3

# Morbidity and Mortality Weekly Report (MMWR) Assignment

## DATA AND CALCULATIONS

1 Go to the MMWR Morbidity Tables at the CDC website (www.cdc.gov). Record the cumulative weekly totals for your selected disease for the last complete year in the data table on the following pages. No calculation is necessary regarding these numbers at this point. Because they are cumulative totals, they should be greater than or equal to the preceding week. Nevertheless, because the reported numbers are provisional and subject to change, this may not always be the case. Simply record the numbers from the CDC website as they are reported.

2 Calculate the number of new cases during each 4-week period (weeks 1–4, weeks 5–8, etc.) for the entire year by subtracting the total at the end of a 4-week period from the total at the end of the next 4-week period. That is, for the second 4 weeks subtract the total at the end of week 4 from the total at the end of week 8. Record these in the table on page 460 in the column labeled "4-Week Totals." Show a sample calculation for the *third 4-week total* in the space below.

3 Calculate national incidence values of each 4-week period during the entire year using the calculated 4-week totals (weeks 1–4, weeks 5–8, etc.) in the numerator. Use the U.S. resident population size on July 1 in the denominator. This value can be obtained from the U.S. Census Bureau website at http://www.census.gov/popest/, and then clicking on "Current Estimates Data" under "Quick Links." On the new page, you will see a table with four columns. The first row is for the Nation, Total Population. In the right column on this row there is a link (e.g., V2014). Click on it and on the new page you will see a link for "Monthly Population Estimates…" Click on this and find the estimated resident population for July 1 of your study year and record it below. Be sure to choose an appropriate value for $10^n$, and do not abuse significant figures! Show a sample calculation for *the third 4-week period* of your study year in the space below.

Estimated population size for the year 20 _____ :

$$\text{Incidence Rate} = \frac{\text{Number of new cases in a time period}}{\text{Size of at-risk population at midpoint of time period}} \times 10^n$$

Morbidity Data for ______________________ during the year 20 ______

| Week | Cumulative Totals by Week Year (20 ) | 4-Week Totals Year (20 ) | Incidence Values Year (20 ) | Week | Cumulative Totals by Week Year (20 ) | 4-Week Totals Year (20 ) | Incidence Values Year (20 ) |
|---|---|---|---|---|---|---|---|
| 1 | | | | 29 | | | |
| 2 | | | | 30 | | | |
| 3 | | | | 31 | | | |
| 4 | | | | 32 | | | |
| 5 | | | | 33 | | | |
| 6 | | | | 34 | | | |
| 7 | | | | 35 | | | |
| 8 | | | | 36 | | | |
| 9 | | | | 37 | | | |
| 10 | | | | 38 | | | |
| 11 | | | | 39 | | | |
| 12 | | | | 40 | | | |
| 13 | | | | 41 | | | |
| 14 | | | | 42 | | | |
| 15 | | | | 43 | | | |
| 16 | | | | 44 | | | |
| 17 | | | | 45 | | | |
| 18 | | | | 46 | | | |
| 19 | | | | 47 | | | |
| 20 | | | | 48 | | | |
| 21 | | | | 49 | | | |
| 22 | | | | 50 | | | |
| 23 | | | | 51 | | | |
| 24 | | | | 52 | | | |
| 25 | | | | | | | |
| 26 | | | | | | | |
| 27 | | | | | | | |
| 28 | | | | | | | |

Name ______________________________

Date ______________________________

Lab Section ______________________________

I was present and performed this exercise (initials) ______________________________

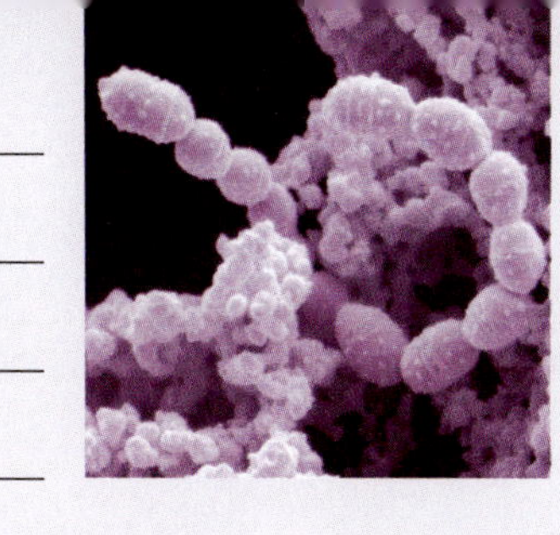

# DATA SHEET 7-3

*(continued)*

## QUESTIONS

**1** *Characterize the disease you have chosen. Be sure to include causative agent (by scientific name), symptoms, mode of transmission, relevant diagnostic tests, and treatment. Cite your references.*

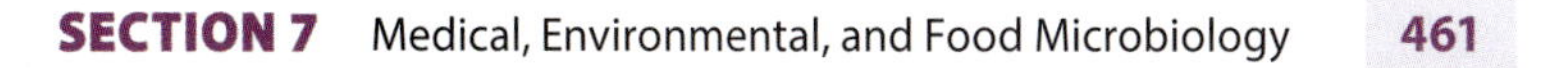

**2** *Prepare a graph that illustrates the cumulative data for the year studied. Be sure to include a title and axis labels with appropriate units in your graph.*

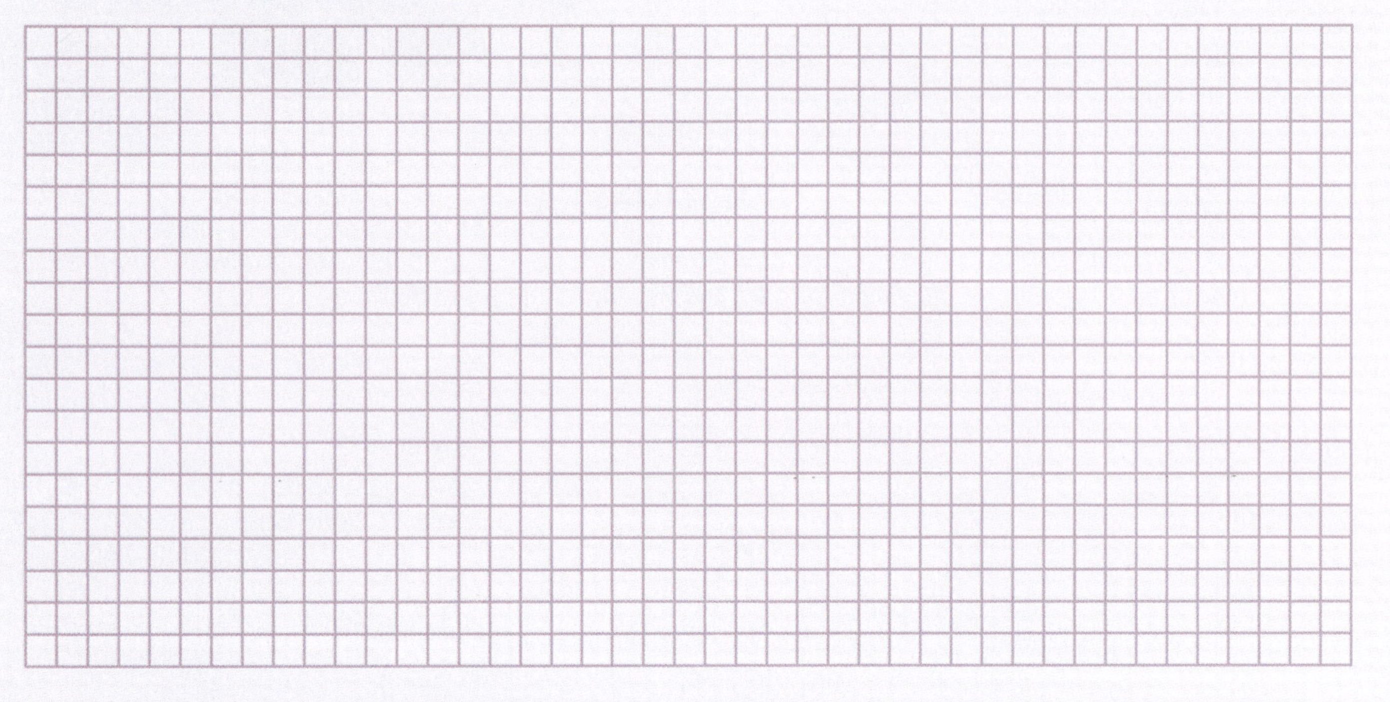

**3** *Graph the calculated national incidence values over the year studied. In your graph, be sure to include a title and axis labels with appropriate units.*

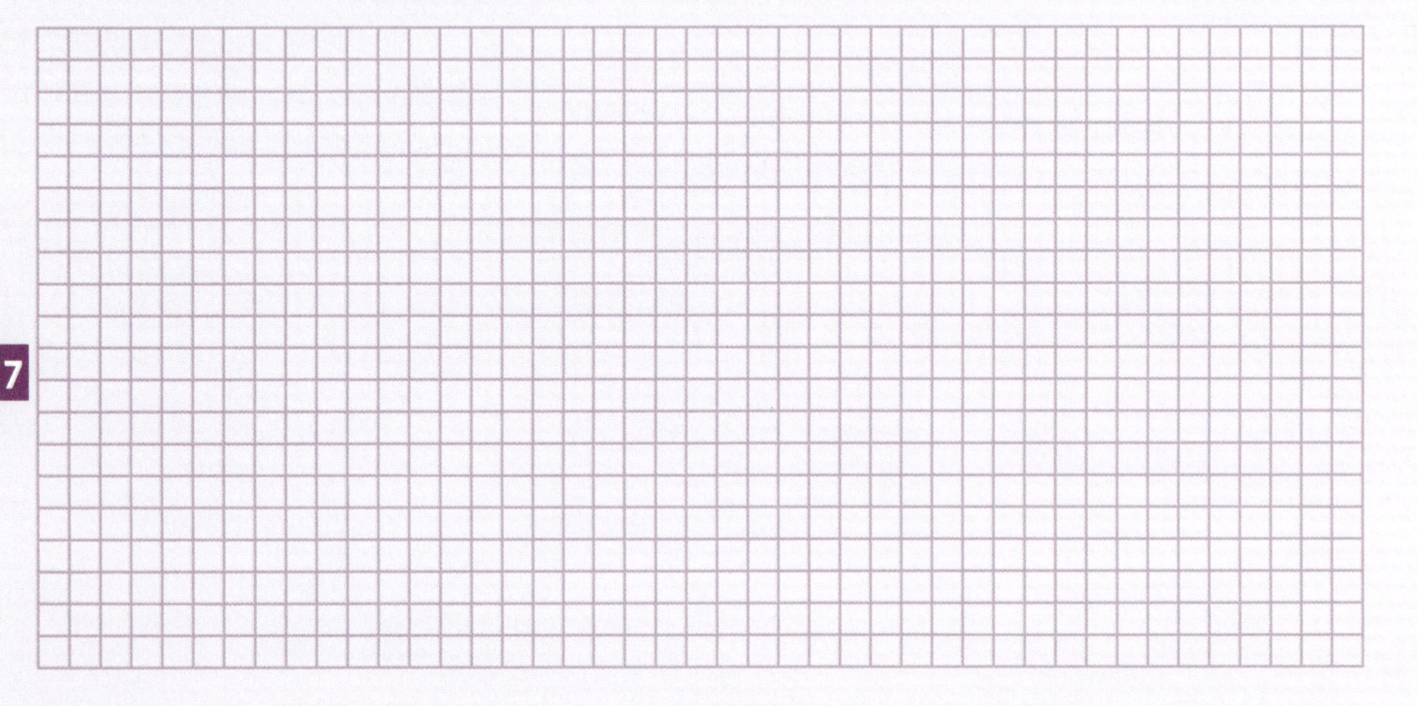

Name ______________________________

Date ______________________________

Lab Section ______________________________

I was present and performed this exercise (initials) ______________________________

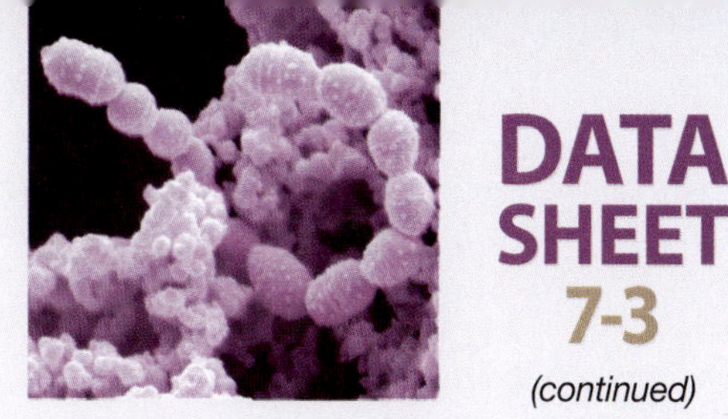

# DATA SHEET 7-3

*(continued)*

**4** *Briefly describe the national trend of your chosen disease for the year. Does the incidence appear to be seasonal? Support your answer by referencing portions of the appropriate graph.*

EXERCISE
# 7-4
# Epidemic Simulation

## Theory

As mentioned in Exercise 7-3, epidemiology is the study of the causes, occurrence, transmission, distribution, and prevention of diseases in a population. While there is no single approach to finding answers to these epidemiological questions, identification of the pathogen frequently happens first, and then is followed by discovering its mode of transmission.

Introduction of a pathogen into a new host can occur through ingestion, inhalation, direct skin contact, open wounds or lesions in the skin, animal bites, direct blood-to-blood contact as in blood transfusions, and sexual contact. The exact route is referred to as a **portal of entry** (e.g., through the GI tract or respiratory tract).

Infectious diseases can be transmitted to the portal of entry by way of sick people or animals, healthy people or animals carrying the infectious organism or virus, water contaminated with human or animal feces, contaminated objects (**fomites**), aerosols, or biting insects (**vectors**).

When a disease is transmitted from an area such as the heating or cooling system of a building or from contaminated water that infects many people at once, it is called a **common source epidemic. Propagated transmission** is where the pathogen is passed from person to person. The first case of such a disease is called the **index case.** Determining the index case in a simulated epidemic involving propagated transmission is the object of today's lab exercise.

A second objective of today's lab is to gather data to be used in performing simple epidemiological calculations. You will be calculating **incidence** and **prevalence rates** for the "disease" spreading through your class.

Incidence rate is the number of new cases of a disease reported in a defined population during a specific time period. Incidence rate is calculated using the following equation:

$$\text{Incidence rate} = \frac{\text{Number of new cases in a time period}}{\text{Size of at-risk population at midpoint of time period}} \times 10^n$$

Because population size constantly changes, it is conventional to use the population size at the midpoint of the study period. Also, the usually small decimal fraction obtained for the rate is multiplied by some power of 10 ($10^n$) to get its value up to a number bigger than one. That is, an incidence rate of 0.00003 is multiplied by 100,000 ($10^5$) so it can be reported as three cases per 100,000 people per time period, rather than 0.00003 cases per person per time period.

You also will be calculating one form of prevalence rate called **period prevalence**, the number of cases of a disease at a specific point in time in a defined population. It is calculated using the following equation:

$$\text{Period prevalence} = \frac{\text{Number of existing cases over a period of time}}{\text{At-risk population size at midpoint of time period}} \times 10^n$$

As with incidence rate, the calculated prevalence is multiplied by some power of 10 ($10^n$) to bring the number up to one or more.

Alternatively, if the average duration (D) of a disease is known, then prevalence can be estimated using the incidence rate (I), as follows:

$$P = I \times D$$

## Application

Epidemiologists have the task of identifying the source of a disease and establishing the mode of its transmission. Further, epidemiologists characterize diseases quantitatively, using measures such as incidence and prevalence. Incidence rate indicates the probability an individual will contract the disease in the stated time period, whereas prevalence is a useful indicator of the disease's presence in a population. Based on these and other information, recommendations for prevention can be made.

## In This Exercise

One of you will become the "index case" in a simulation of an epidemic, and many of you who contact this person directly and indirectly will become "infected." Your job, besides having a little fun, will be to collect the data and determine which one of you is the index case. You will also use the data to calculate incidence and prevalence rates for your class over a hypothetical time period.

### Materials

#### Per Student

- ☐ Lab coat
- ☐ Disposable gloves (latex or synthetic)
- ☐ Chemical eye protection

- ☐ Numbered Petri dish containing a piece of hard candy (one will be contaminated with *Serratia marcescens*[1] or other microbe as chosen by your instructor; the others will be moistened with water)[2]
- ☐ One nutrient agar or tryptic soy agar plate
- ☐ Two sterile cotton applicators and sterile saline
- ☐ Marking pen

## PROCEDURE

### Lab One

1. Wear a lab coat, gloves, and chemical eye protection when performing this procedure.
2. You must follow the instructions precisely. We recommend that the instructor orchestrate the entire lab so that the class does each step at the same time so specific procedures can be emphasized. Do not get ahead. This is being done for safety. Containers for swab and glove disposal should be located around the room so all students have easy access. You should also look at Figure 7.8 to see how to separate gloves safely.
3. Obtain a Petri dish containing candy, two sterile cotton applicators, a tube of sterile saline, and a nutrient agar plate. Circle the number of your candy dish in the "case" column of the data sheet, page 469. Then, put away all your papers and books for the duration.
4. Mark a line on the bottom of the agar plate to divide it into two halves. Label the sides "1" and "2." Also write your name and the number of your candy dish on your NA plate.
5. Open a package of sterile, cotton-tipped applicators so they can be easily removed. They are usually packaged in pairs, and that is all you need.
6. Put a glove on your nondominant hand.
7. Using your ungloved hand, remove one cotton applicator, moisten it in the sterile saline, and sample the palm of your gloved hand. Be sure not to touch anything else with the palm of the glove. When finished, zigzag inoculate side 1 of the plate with the cotton applicator (Fig. 1.38). Discard the applicator in an appropriate autoclave container.
8. Pick up the candy with your gloved hand and roll it around until a good amount of it is transferred to the glove. Drop the candy back into the Petri dish, close it with your ungloved hand, and touch nothing with your gloved hand. Just sit with your elbow on the table and your gloved hand in the air. All but one student in the class just has messy candy on their glove at this point, but *you* might be the one with *S. marcescens. Be mindful of this.*
9. **Read this entire step before doing anything!** Candy dish numbers identify each student as a case number. Using gloved hands only, case #1 rubs hands with case #2, making sure to transfer anything that may be on the palm of the glove to the palm of the other student's glove. Both students must hold on to the cuff of their glove throughout this step so it doesn't come off. After a few seconds of rubbing, they should *gently slide their hands apart, as shown in Figure 7.8.* They must not "snap" the gloves as they separate because this will produce aerosols. Case #1 should go back to his or her desk and sit with the elbow of their gloved hand on the table and their hand in the air.
10. Then case #2 rubs gloved hands with #3, and #3 with #4, and so on until all students have contacted gloved hands. Continue to be careful not to snap the gloves when you separate. Hold on to the cuff when transferring and separating to prevent the glove from coming off your hand.
11. When all students have rubbed hands, the last (highest numbered) case should rub hands with case #1.
12. With a second sterile applicator, sample the palm of your gloved hand as before and inoculate side 2 of the plate. Be sure to sample the area of the glove that contacted other students' gloves. Dispose of the cotton applicator in the nearest appropriate autoclave container. Do not walk across the room with the swab. If necessary, put it into a can, tip down until the exercise is completed.
13. Remove your glove by inserting the thumb of your other hand under the cuff and rolling it off your hand, inverting it in the process (Fig. I.4). Dispose of it properly.

[1] *S. marcescens* was chosen because of its obvious color. Other, less-distinctive organisms can be substituted if desired.

[2] Petri dishes should be numbered according to the number of students in the class. There should be no gaps in the sequence.

14 Wash your hands thoroughly with antiseptic soap, if available.

15 Discard all materials (except your inoculated plate) in appropriate containers. Be sure to tape the lid on your candy dish before you dispose of it.

16 Wipe down your work area with disinfectant.

17 Tape the lid on your inoculated plate. You may find it convenient to tape all plates from your group in a single stack so they can be retrieved more easily.

18 Invert the inoculated plate and incubate it at 25°C for 24 to 48 hours.

### Lab Two

1 Remove the plates from the incubator, and examine them for characteristic reddish growth of *Serratia marcescens*.

2 If your plate shows *S. marcescens* growth on side 2, enter a " + " next to your case number on the data sheet, page 469. If there is no evidence of *S. marcescens* growth, enter a " – " instead.

3 All class members will need to report whether or not they had *S. marcescens* growth on side 2 of their plate. Your instructor may do a roll call, have you post results on the board, or use some other mechanism for dispensing this information. Enter class results on your data sheet and follow the instructions for establishing the index case, incidence rate, and period prevalence.

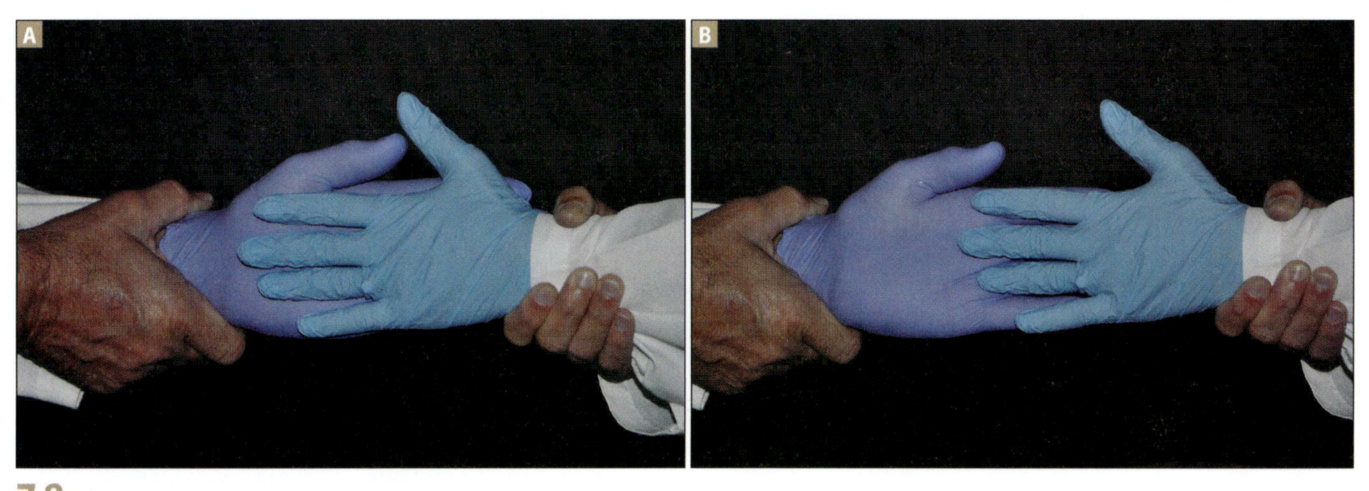


**7.8 Sliding the Gloves Apart** ■ After rubbing palms, students must separate by gently sliding their gloved hands apart (as in A and B). Be sure to hold on to the glove's cuff to prevent it from coming off. These safety precautions must be followed and are designed to prevent the gloves from snapping apart (they will be sticky) and producing aerosols.

## References

Essex-Sorlie, Diane. Pages 47–48 in *Medical Biostatistics & Epidemiology*. Norwalk, CT: Appleton & Lange, 1995.

Lilienfeld, David E. and Paul D. Stolley. Page 109–110 in *Foundations of Epidemiology*, 3rd ed. New York: Oxford University Press, 1994.

Mausner, Judith S. and Shira Kramer. Pages 44–50 in *Epidemiology: An Introductory Text*, 2nd ed. Philadelphia: W.B. Saunders Company, 1985.

Name ______________________________

Date ______________________________

Lab Section ______________________________

I was present and performed this exercise (initials) ______________________________

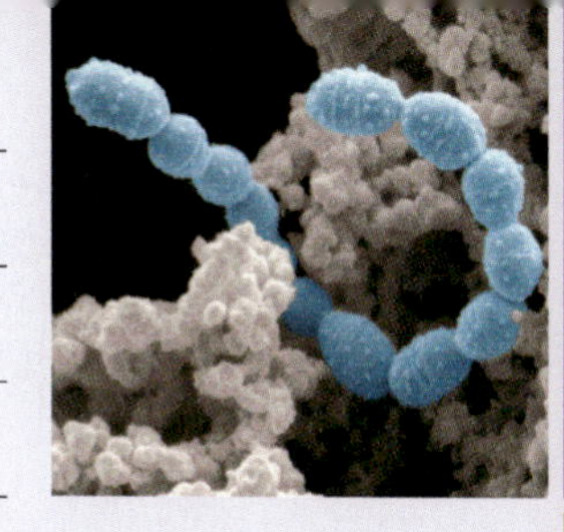

# DATA SHEET 7-4

## Epidemic Simulation

### DATA AND CALCULATIONS

1 On the following table, circle the case number that matches your candy dish number.

2 After the plates have been incubated, compile data for your class population (each case is the student's candy number in the exercise). The organism spread during contact between student pairs was *Serratia marcescens*, which produces reddish-orange colonies. You will likely see growth on both sides of the plate, but red colonies are the only sign of "disease." If there is reddish-orange growth on side 2 of the plate, consider the person to have the "disease" and record a "+" in the appropriate box. If there is no red growth on the plate or there is no difference between the two sides, consider the person to be healthy and record a "–" in the appropriate box.

| Case | Week | Disease (+/−) |
|---|---|---|
| 1 | 1 | |
| 2 | 1 | |
| 3 | 1 | |
| 4 | 1 | |
| 5 | 2 | |
| 6 | 2 | |
| 7 | 2 | |
| 8 | 2 | |
| 9 | 2 | |
| 10 | 3 | |

| Case | Week | Disease (+/−) |
|---|---|---|
| 11 | 3 | |
| 12 | 3 | |
| 13 | 3 | |
| 14 | 3 | |
| 15 | 3 | |
| 16 | 4 | |
| 17 | 4 | |
| 18 | 4 | |
| 19 | 4 | |
| 20 | 4 | |

| Case | Week | Disease (+/−) |
|---|---|---|
| 21 | 4 | |
| 22 | 4 | |
| 23 | 5 | |
| 24 | 5 | |
| 25 | 5 | |
| 26 | 5 | |
| 27 | 5 | |
| 28 | 6 | |
| 29 | 6 | |
| 30 | 6 | |

| Case | Week | Disease (+/−) |
|---|---|---|
| 31 | 7 | |
| 32 | 7 | |
| 33 | 7 | |
| 34 | 7 | |
| 35 | 8 | |
| 36 | 8 | |
| 37 | 8 | |
| 38 | 9 | |
| 39 | 9 | |
| 40 | 9 | |

3 Identify the case number of the index patient by writing "index patient" next to the case number.

4 Use the data obtained to calculate incidence and period prevalence rates in your class population. For this purpose, assume that the cases have been identified over a 9-week period (or a shorter period, depending on how many students participated in the simulation). Also assume that the duration of the disease is 7 days, so new cases in 1 week are still diseased in the next week but are healthy by the following week.

Record the population size: ____________

Record incidence and point prevalence rates for each week in the table.
The incidence and period prevalence rates should be reported as "cases per $10^n$ per week."

| Week | Incidence Rate | Period Prevalence Rate |
|---|---|---|
| 1 | | |
| 2 | | |
| 3 | | |
| 4 | | |
| 5 | | |
| 6 | | |
| 7 | | |
| 8 | | |
| 9 | | |

## QUESTIONS

**1** *Explain how you determined the index case.*

____________

____________

**2** *What was the purpose of side 1 on each plate?*

____________

____________

Name ______________________________

Date ______________________________

Lab Section ______________________________

I was present and performed this exercise (initials) ______________________________

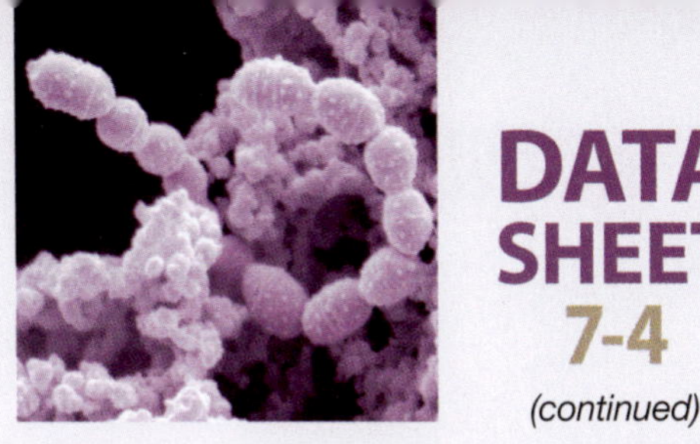

# DATA SHEET 7-4

*(continued)*

**3** *Suggest possible reasons (based on the execution of the simulation) for cases showing no* S. marcescens *growth between the cases that exhibited growth.*

______________________________

______________________________

**4** *Suggest how the gaps could represent "subclinical infections" or "carriers" in this simulation.*

______________________________

______________________________

**5** *Within the context of this simulation, how could you characterize any "non-*Serratia marcescens *growth on side 1 or 2 of your plate? (**Hint:** "Contamination" is not an adequate answer!)*

______________________________

______________________________

**6** *Examine the chart below. It shows the number of cases of a disease over a 4-month period in a population of 2,563. Each line's length indicates the duration of that case. Calculate the incidence and prevalence for each month in the table on page 472.* **Note:** *Case #1 started in December and cases 16 and 17 extended into May.*

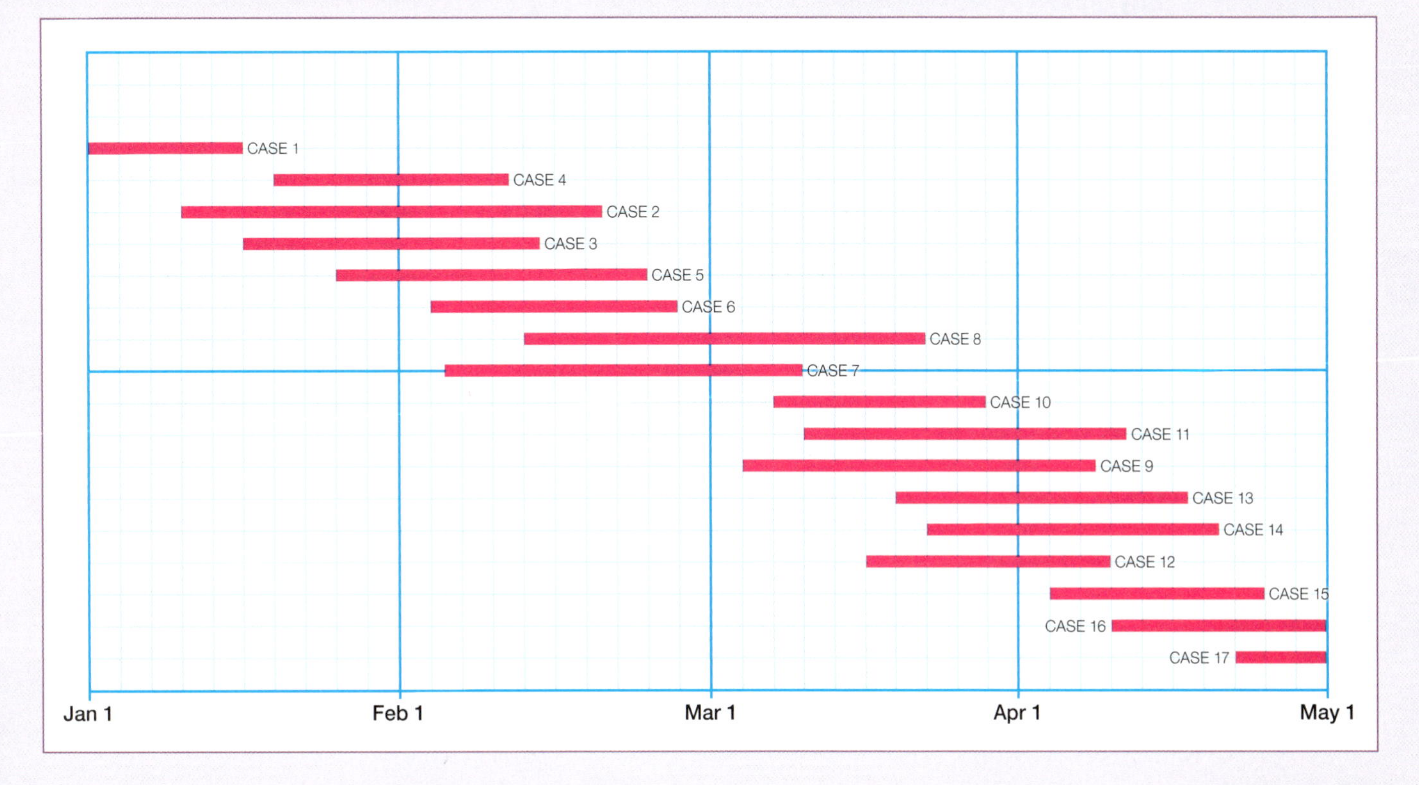

| Month | Incidence Rate | Period Prevalence Rate |
|---|---|---|
| January | | |
| February | | |
| March | | |
| April | | |

**7** *Use the graph paper below to plot a case chart for your class in the style of the one in question 6.*
***Note:*** *You will have to estimate the starting points of cases because more precise information is not known.*

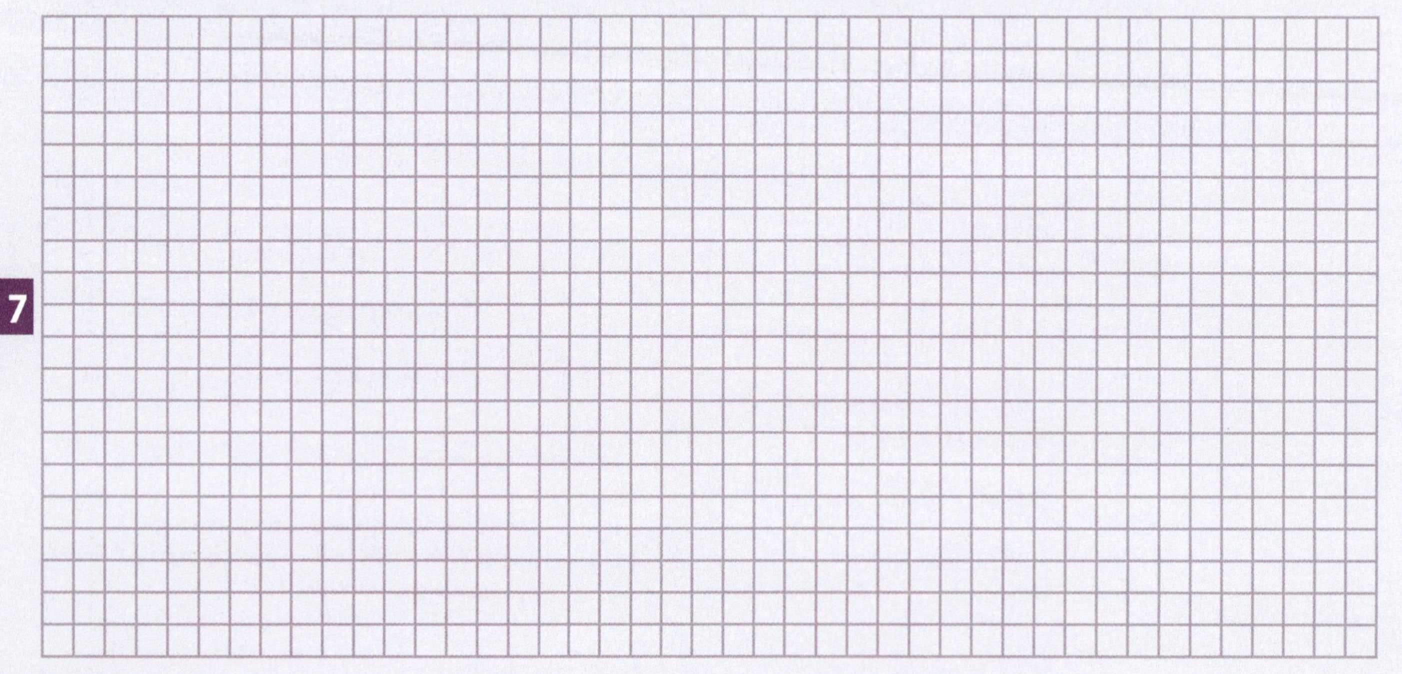

7

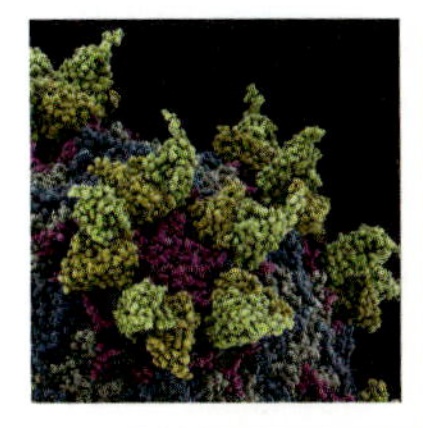

## Water Quality Testing

Water quality is essential for maintaining public health. Appropriate community agencies routinely test drinking water and recreational bodies of water for microbial and chemical contamination. Obviously, it is the former with which we are concerned in the following exercises.

Many pathogens are transmitted through water contaminated with feces. Examples of bacterial pathogens include *Vibrio cholerae* (cholera), *Salmonella typhi* (typhoid fever), and *Shigella dysenteriae* (dysentery), among others. Protozoans, such as *Entamoeba histolytica* (amebiasis) and *Giardia lamblia* (giardiasis), and viruses (Hepatitis A) are also waterborne pathogens found in feces.

It is not practical to screen water for every possible pathogen it might contain. Instead, water can be tested for fecal contamination using common intestinal flora as indicator species. For instance, *Escherichia coli* makes a good indicator species because it is abundant in feces, it is not fastidious and can survive with relative ease outside the body, and it is easily grown in the laboratory.

*E. coli* is a member of the *Enterobacteriaceae*, which is informally divided into coliform and noncoliform bacteria. The difference between the two groups is simply the ability of coliforms to ferment lactose to acid and gas end products. It is this ability that, within the context of the test system being used, identifies the presence of coliforms in general and sometimes *E. coli* in particular, both of which are indicative of fecal contamination.

Evidence of fecal contamination based on the presence of coliforms or *E. coli* does not mean the water is definitely carrying pathogens, or if it is, that the number of pathogens is likely to be high enough to cause infection. Standards have been established that set acceptable levels of coliform contamination in water, and these differ based on the intended use of it.

The following two exercises provide you with the opportunity to collect water samples and test them for fecal contamination. Because it is unknown if they are contaminated, or not, you will treat them as if they are, so please pay attention to the precautions in each exercise. ■

# Membrane Filter Technique

EXERCISE 7-5

## Theory

Fecal contamination is a common pollutant in open water and a potential source of serious disease-causing organisms. Certain members of *Enterobacteriaceae*, (Gram-negative facultative anaerobes) such as *Escherichia coli*, *Klebsiella pneumoniae*, and *Enterobacter aerogenes*, are able to ferment lactose rapidly and produce large amounts of acid and gas.

These organisms, called **coliforms**, are used as the indicator species when testing water for fecal contamination because they are relatively abundant in feces and easy to detect (Fig. 7.9). Once fecal contamination is confirmed by the presence of coliforms, any noncoliforms also present in the sample can be tested and identified as pathogenic or otherwise.

In the membrane filter technique, a water sample is drawn through a special porous membrane designed to trap microorganisms larger than 0.45 µm. After filtering the water sample, the membrane (filter) is applied to the surface of plated Endo agar and incubated for 24 hours (Fig. 7.10).

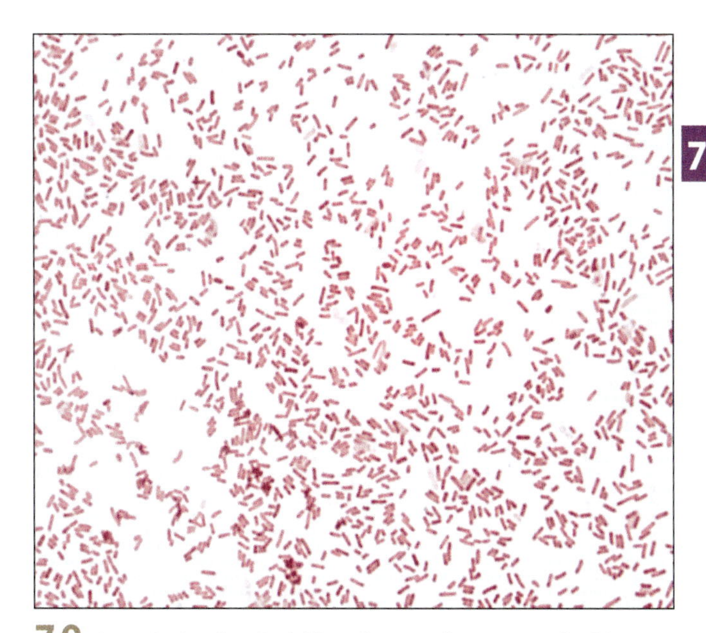

**7.9 Gram Stain of *Escherichia coli*** ■ *E. coli* is a principal coliform detected by the membrane filter technique. These cells were grown in culture.

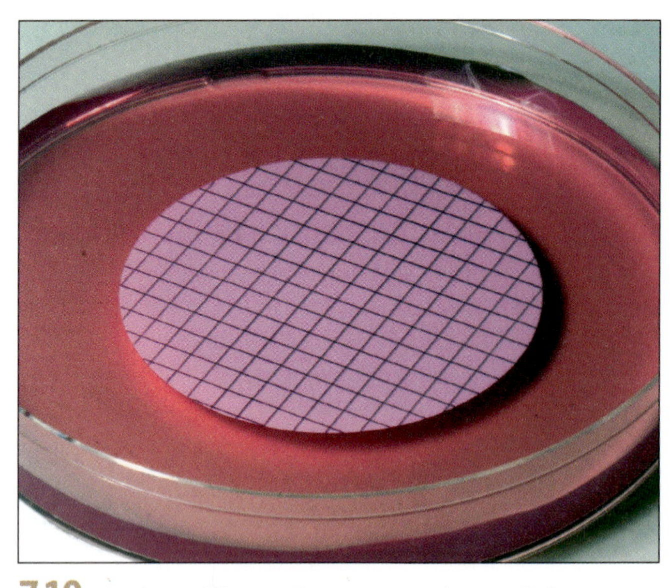

**7.10 Membrane Filter** ■ This porous membrane will allow water to pass through but will trap bacteria and particles larger than 0.45 µm. It is shown in position on an m-Endo agar LES plate.

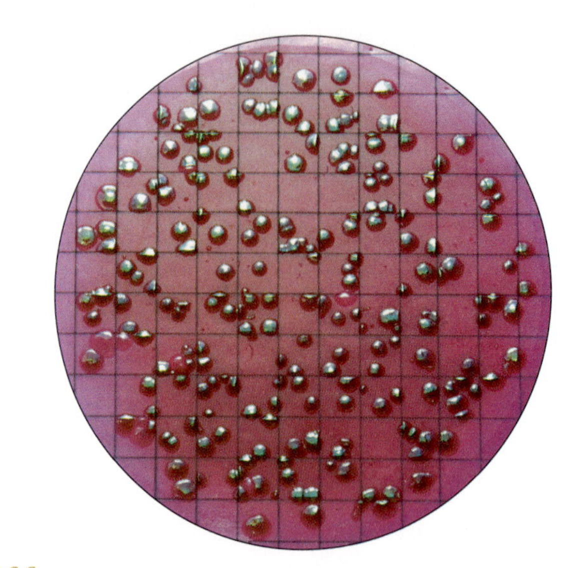

**7.11 Coliform Colonies on a Membrane Filter** ■ Note the characteristic dark colonies with a metallic sheen, indicating that this water sample is contaminated with fecal coliforms. Potable water has less than 1 coliform per 100 mL of sample tested.

m-Endo agar LES is a selective medium that encourages Gram-negative bacterial growth and inhibits Gram-positive growth. It contains lactose for fermentation and basic fuchsin to indicate pH changes. Coliforms produce acetaldehyde, which causes colonies to appear red. Subsequent rapid lactose fermentation with acid end products causes the colonies to become mucoid with a metallic sheen. Noncoliform bacteria (including several dangerous pathogens) tend to be pale pink, colorless, or the color of the medium.

After incubation, all red or metallic colonies are counted and are used to calculate "coliform colonies/100 mL" using the following formula:

$$\frac{\text{Total coliforms}}{100 \text{ mL}} = \frac{\text{Coliform colonies counted} \times 100}{\text{Volume of original sample in mL}}$$



A "countable" plate contains between 20 and 80 coliform colonies with a total colony count no larger than 200 (Fig. 7.11). To assure that the number of colonies will fall within this range, it is customary to dilute heavily polluted samples, thereby reducing the number of cells collecting on the membrane. When dilution is necessary, it is important to record only the volume of *original sample* passed through the membrane, not any added water. Potable water contains less than 1 coliform per 100 milliliters of sample.

## Application

The membrane filter technique is commonly used to identify the presence of fecal coliforms in water. While other media could be used, m-Endo agar LES is the standard in most applications.

## In This Exercise

To determine total coliform population density, you will be using collected water samples and performing a membrane filter technique. Your results will be recorded in coliforms per 100 mL of sample.

### ▼ Materials

**Per Student**

- ☐ Lab coat
- ☐ Disposable gloves
- ☐ Chemical eye protection

**Per Student Group**

- ☐ 100 mL water dilution bottle (to be distributed in the preceding lab)
- ☐ 100 mL water sample (obtained by student)
- ☐ Household disinfectant and paper towels
- ☐ Hand sanitizer
- ☐ Ziploc® bag large enough to hold the collection bottle
- ☐ One m-Endo agar LES plate
- ☐ One sterile membrane filter (pore size 0.45 µm)
- ☐ Small screw-capped jar containing alcohol and forceps
- ☐ 100 mL sterile water (for rinsing the apparatus)

**Per Class**

- ☐ One broth culture of *Escherichia coli*
- ☐ One broth culture of *Staphylococcus epidermidis*
- ☐ Two sterile cotton swabs
- ☐ Vacuum source (pump or aspirator)
- ☐ Sterile membrane filter suction apparatus (Fig. 7.12)

**7.12 Membrane Filter Apparatus** ■ Assemble the membrane filter apparatus as shown in this photograph. Use two suction flasks (as shown) to avoid getting water into the vacuum source. Secure the flasks on the table, because the tubing may make them top-heavy.

## Medium Recipe

### m-Endo Agar LES

| | |
|---|---|
| ☐ Yeast extract | 1.2 g |
| ☐ Casitone | 3.7 g |
| ☐ Thiopeptone | 3.7 g |
| ☐ Tryptose | 7.5 g |
| ☐ Lactose | 9.4 g |
| ☐ Dipotassium phosphate | 3.3 g |
| ☐ Monopotassium phosphate | 1.0 g |
| ☐ Sodium chloride | 3.7 g |
| ☐ Sodium desoxycholate | 0.1 g |
| ☐ Sodium lauryl sulfate | 0.05 g |
| ☐ Sodium sulfite | 1.6 g |
| ☐ Basic fuchsin | 0.8 g |
| ☐ Agar | 15.0 g |
| ☐ Distilled or deionized water | 1.0 L |

*pH 7.3–7.7 at 25°C*

## PROCEDURE

### Prelab

1. Obtain a 100 mL water dilution bottle from your instructor. These have a white line inscribed on them that indicates the 100 mL level. (If you use another collection bottle or jar, it would be wise to mark the 100 mL level in lab with a permanent marking pen.)
2. Choose an environmental source to sample. You will be treating the site as potentially contaminated with feces. (Your instructor may decide to approve your choice to avoid duplication among lab groups.)
3. Visit the environmental site as close to your lab period as possible. Bring your water dilution bottle, a pair of gloves, some household disinfectant, hand sanitizer, and paper towels. While wearing gloves, fill the bottle to the white line (100 mL) and replace the cap.
4. Wipe the outside of the bottle with disinfectant-soaked towels. Put the bottle in the Ziploc® bag and close it.
5. Dispose of the towels and gloves in the trash. Cleanse your hands thoroughly with hand sanitizer.
6. Store the sample in the laboratory refrigerator until your lab period. If the sample must sit for a while before your lab, leave the cap slightly loose to allow some aeration.

### Lab One

1. Wear a lab coat, gloves, and chemical eye protection when performing this procedure.
2. Pinch the forceps in the alcohol to capture some alcohol between the tips. Then, pass the forceps through the flame; don't hold them *in* the flame (Fig. 7.13). Use care when flaming the forceps. Hold them away from the flame and the alcohol jar until the alcohol burns off of them. If the alcohol jar catches on fire, smother the flame with the jar's lid. Then, use the forceps to place the membrane filter (grid facing up) on the lower half of the filter housing (Fig. 7.14). Do not touch the filter with your fingers.
3. Clamp the two halves of the filter housing together and insert the filter housing into the suction flask as shown in Fig. 7.15. (This assembly can be a little top-heavy, so have someone hold it or otherwise secure it to prevent tipping.)
4. Pour the appropriate volume of water sample into the funnel. (Refer to Table 7-5 for suggested sample volumes. If the sample size is smaller than 10.0 mL, add 10 to 20 mL of sterile water before filtering, to help distribute the cells evenly on the surface of the membrane filter.)
5. Turn on the suction pump (or aspirator), and filter the sample into the flask.
6. Before removing the membrane, and with the suction pump still running, rinse the sides of the funnel two or three times with 20 mL or 30 mL of sterile water. This will wash off any cells adhering to the funnel and reduce the likelihood of contaminating the next sample.
7. Alcohol-flame the forceps again and carefully transfer the filter to the Endo agar plate by gently rolling it onto the surface grid side up. Be careful not to fold it or create air pockets (Fig. 7.16).
8. Wait a few minutes to allow the filter to adhere to the agar, then invert the plate and incubate it aerobically at 35 ± 2°C for 22 to 24 hours.

9 One group should spot inoculate an Endo agar plate with *E. coli* and *S. epidermidis* and incubate it aerobically at 35 ± 2°C for 22 to 24 hours. Then, save or dispose of the culture tubes as directed by your instructor.

### Lab Two

1 Remove the plate and count the colonies on the membrane filter that are dark red, purple, have a black center, or produce a green or gold metallic sheen.

2 Record your data in the table on the data sheet, page 479.

3 Calculate the coliform CFU per 100 milliliters using the following formula:

$$\frac{\text{Total coliforms}}{100\text{ mL}} = \frac{\text{Coliform colonies counted} \times 100}{\text{Volume of original sample in mL}}$$

4 Record your coliform count and the class coliform counts in the table on the data sheet.

**7.13 Alcohol-Flaming the Forceps** ■ Dip the forceps into the beaker of alcohol. Remove the forceps and quickly pass them through the Bunsen burner flame—just long enough to ignite the alcohol. **Be sure to keep the flame away from the jar of alcohol and have no stray papers or flammables on the table. If the jar catches fire, smother the flame with the jar's lid. If a drop of flaming alcohol falls on the table top, it will probably quickly burn out on its own. If not, smother it.**

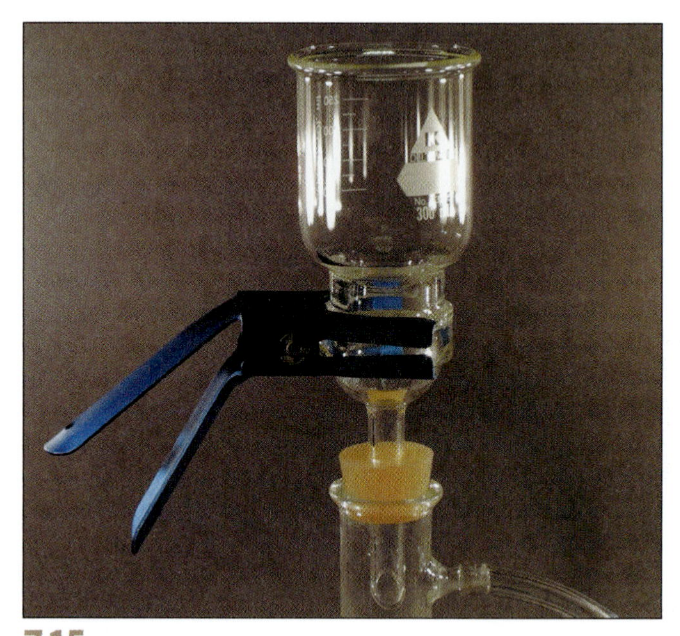

**7.15 Membrane Filter Assembly** ■ The membrane filter assembly is made up of a two-piece funnel and clamp. The membrane filter is inserted between the two funnel halves, and the whole assembly is clamped together.

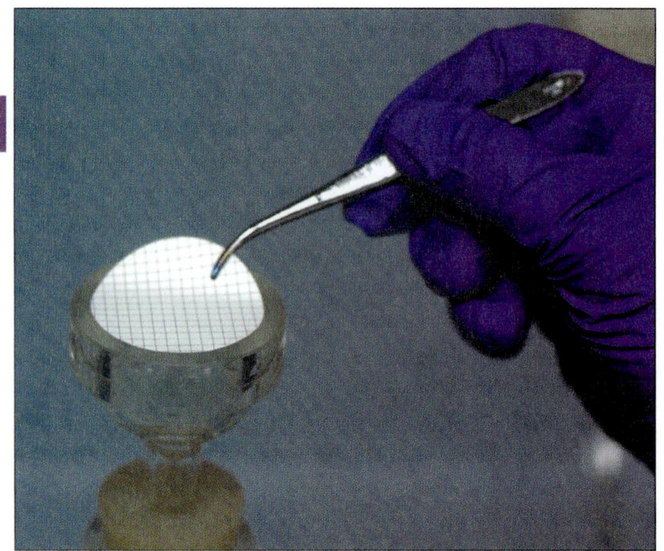

**7.14 Placing the Filter on the Filter Housing** ■ Carefully place the filter, grid up, on the bottom half of the filter housing. Clamp the filter funnel over the filter.

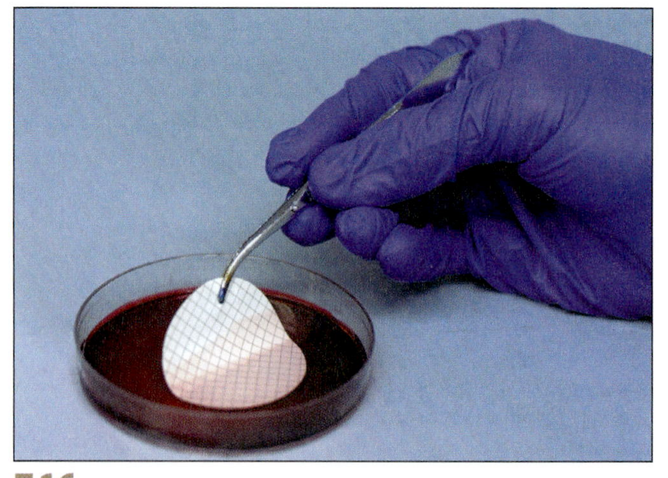

**7.16 Placing the Filter on the Agar Plate** ■ Using alcohol-flamed forceps, carefully place the filter onto the agar surface with the grid facing up. Try not to allow any air pockets under the filter as you roll it onto the agar, because contact with the agar surface is essential for growth of bacteria. If air pockets form, try again. Allow a few minutes for the filter to adhere to the agar before inverting the plate.

TABLE **7-5** **Suggested Sample Volumes for Membrane Filter Total Coliform Test**

| Water Source | Volume (•) To Be Filtered (mL) | | | | | | | |
|---|---|---|---|---|---|---|---|---|
| | 100 | 50 | 10 | 1.0 | 0.1 | 0.01 | 0.001 | 0.0001 |
| Drinking water | • | | | | | | | |
| Swimming pools | • | | | | | | | |
| Wells, springs | • | • | • | | | | | |
| Lakes, reservoirs | • | • | • | | | | | |
| Water supply intake | | | • | • | • | | | |
| Bathing beaches | | | • | • | • | | | |
| River water | | | | • | • | • | • | |
| Chlorinated sewage | | | | • | • | • | | |
| Raw sewage | | | | | • | • | • | • |

(Reprinted with permission from the American Public Health Association)

7

## References

Collins, C. H., Patricia M. Lyne, and J. M. Grange. Page 270 in *Collins and Lyne's Microbiological Methods,* 7th ed. Oxford, Boston: Butterworth-Heinemann, 1995.

Eaton, Andrew D. (American Water Works Association), Lenore S. Clesceri (Water Environment Federation), Eugene W. Rice and Arnold E. Greenberg (American Public Health Association), and Mary Ann H. Franson. Chap. 9 in *Standard Methods for the Examination of Water and Wastewater*, 21st ed. Washington, DC: APHA Publication Office, 2005.

Zimbro, Mary Jo, and David A. Power, eds. Page 208 in *Difco™ and BBL™ Manual—Manual of Microbiological Culture Media.* Sparks, MD: Becton Dickinson and Co., 2003.

Name ______________________

Date ______________________

Lab Section ______________________

I was present and performed this exercise (initials) ______________________

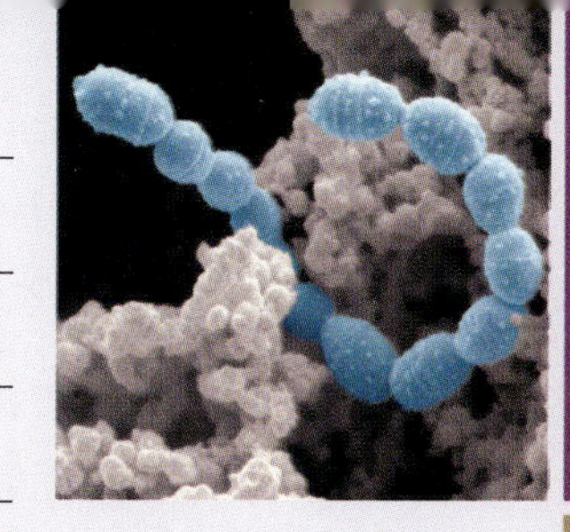

DATA SHEET

7-5

# Membrane Filter Technique

## OBSERVATIONS AND INTERPRETATIONS

1 Enter the number of colonies counted in each group's sample and calculate the total coliforms per 100 milliliters of water using the following formula:

$$\frac{\text{Total coliforms}}{100\text{ mL}} = \frac{\text{Coliform colonies counted} \times 100\text{ mL}}{\text{Volume of original sample in mL}}$$

2 Enter the results in the table below.

| Sample | Number of Colonies | Total Coliforms per 100 mL | Potable? Y/N |
|---|---|---|---|
| | | | |
| | | | |
| | | | |
| | | | |
| | | | |
| | | | |

## QUESTIONS

1 *Why is Endo agar used for this test instead of nutrient agar or tryptic soy agar? (The reason is the same for both.)*

**2** *Consider the composition of m-Endo agar LES.*

**a.** *How would adding glucose to this medium affect the results?*

**b.** *Would glucose affect the medium's sensitivity or specificity? Explain.*

**3** *Consider the Endo agar plate inoculated with* E. coli *and* S. epidermidis.

**a.** *What purpose did inoculating Endo agar with* E. coli *serve?*

**b.** *What purpose did inoculating Endo agar with* S. epidermidis *serve?*



**4** *Suppose you were to count one coliform colony produced from a 10 mL water sample. What is the coliform density of the water in cells per 100 mL? Is the water potable? Explain.*

# Multiple Tube Fermentation Method for Total Coliform Determination

EXERCISE **7-6**

## Theory

The **multiple tube fermentation method**, also called **most probable number**, or **MPN**, is a common means of estimating the number of coliforms present in 100 mL of a sample using a statistical approach. It is not an actual counting of cells. The same procedure can be used to obtain the MPN for both **total coliforms** and *E. coli*.

The three media used in the procedure are: lauryl tryptose broth (LTB), brilliant green lactose bile (BGLB) broth, and EC (*E. coli*) broth. All three media contain lactose that coliforms ferment to acid and gas, the latter of which is used to indicate coliform presence. They also contain an ingredient that inhibits the growth of noncoliforms. In LTB the inhibitory agent is sodium lauryl sulfate. In BGLB and EC broth, the inhibitory agent is bile (oxgall).

Because LTB does not screen out all noncoliforms, it is used to *presumptively* determine the presence or absence of coliforms. BGLB broth is used to *confirm* the presence of coliforms because of its greater inhibitory effect on noncoliforms. EC broth, which includes lactose and bile salts, is selective for *E. coli* when incubated at 45.5°C.

All broths are prepared in 10 mL volumes and contain an inverted Durham tube to trap any gas produced by lactose fermentation. The LTB tubes are arranged in up to 10 groups of five (Fig. 7.17, Procedural Diagram). Each tube in the first set of five receives 1.0 mL of the original sample. Each tube in the second group receives 1.0 mL of a $10^{-1}$ dilution. Each tube in group three receives 1.0 mL of $10^{-2}$, etc.[1]

**Note**

The volume of LT broth in the tubes is not part of the calculation of dilution factor. Dilutions of the water sample are made using sterile water prior to inoculating the broths. One milliliter of diluted sample is added to each tube in its designated group. (Refer to Exercise 6-2 for help with serial dilutions.)

After inoculation, the LTB tubes are incubated at 35 ± 2°C for up to 48 hours, and then examined for gas production. (This is why the volume of LTB is not considered in the dilution. At this point, you are only concerned with the ability of a particular diluted water sample's ability to produce gas or not.). Any positive LTB tubes (Fig. 7.18) then are used to inoculate BGLB tubes. Each BGLB receives one or two loopfuls from its respective positive LTB tube (again, volumes are not critical).

The cultures are incubated 48 hours at 35 ± 2°C and examined for gas production (Fig. 7.19). Positive BGLB cultures then are transferred to EC broth and incubated at 45.5°C for 48 hours (Fig. 7.20). After incubation the EC tubes with gas are counted. The same formula is used to calculate total coliform (BGLB) MPN and *E. coli* (EC) MPN, but they are calculated separately only using data from the appropriate tubes. The formula is:

$$\text{MPN per 100 mL} = \frac{100P}{\sqrt{V_n V_a}}$$

Where

P = total number of positive results (BGLB or EC)

$V_n$ = combined volume of sample in LTB tubes that produced negative results in BGLB or EC

$V_a$ = combined volume of sample in all LTB tubes

It is customary to calculate and report *both* total coliform and *E. coli* densities. Total coliform MPN is calculated using BGLB broth results, and *E. coli* MPN is based on EC broth results.

Using the data from Table 7-6 and the formula above, the calculation for total coliform MPN would be as follows:

$$\text{MPN per 100 mL} = \frac{100P}{\sqrt{V_n V_a}}$$

$$\text{MPN per 100 mL} = \frac{100 \times 9}{\sqrt{0.24 \times 5.55}}$$

$$\text{MPN per 100 mL} = 780$$

## Application

This standardized test is used to measure coliform density (cells/100 mL) in water. It may be used to calculate the density of all coliforms present (total coliforms) or to calculate the density of *Escherichia coli* specifically.

[1] This exercise has been simplified for ease of instruction. The number of groups, number of tubes in each group, dilutions necessary, volume of broth in each tube, and volumes of sample transferred vary significantly, depending on the source and expected use of the water being tested. As a time-saver, we also have you inoculating BGLB and EC tubes from positive LTB, when the convention is to only inoculate EC tubes from positive BGLB tubes.

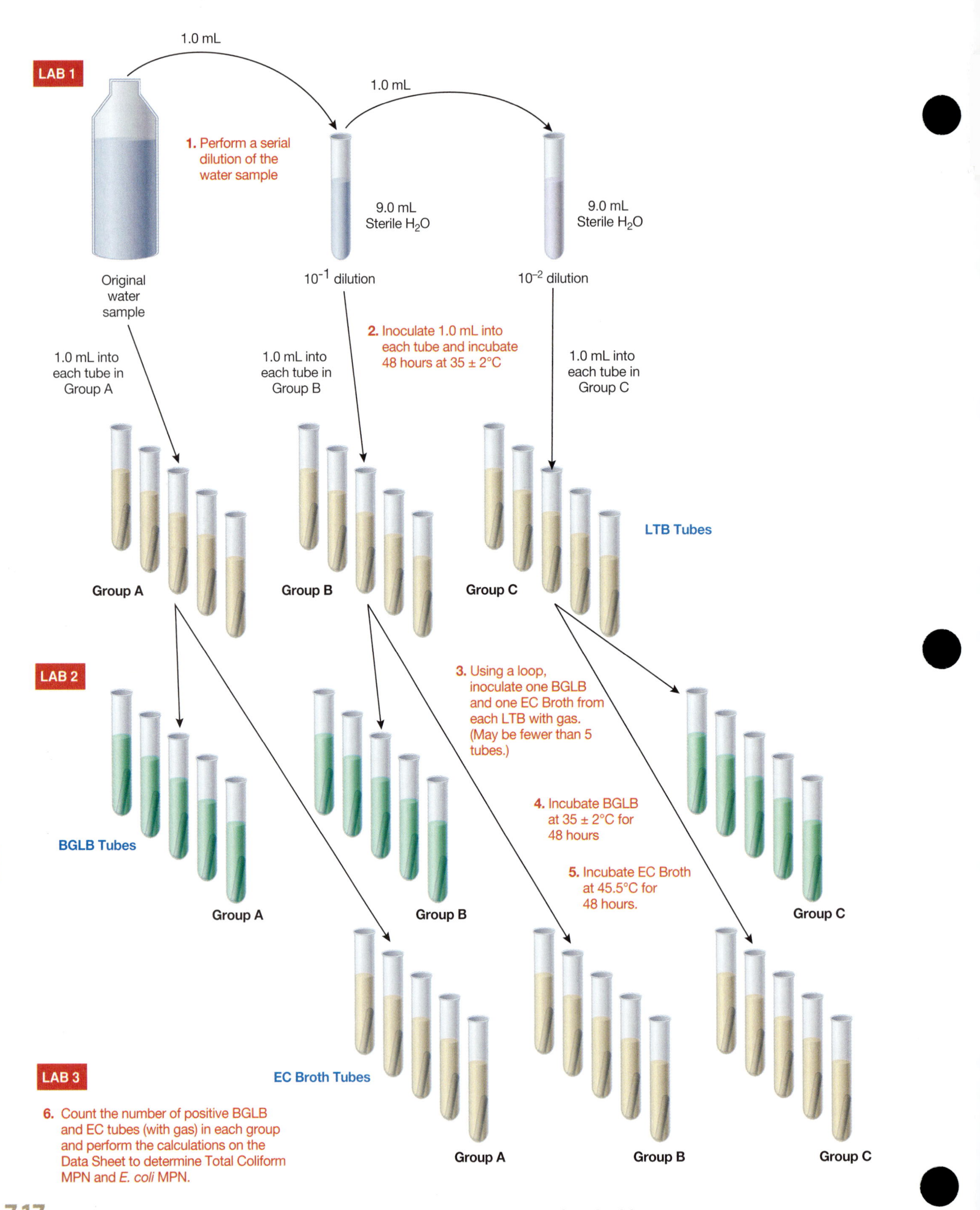


7

**7.17 Procedural Diagram of Multiple Tube Fermentation Method for Determination of Total Coliform**

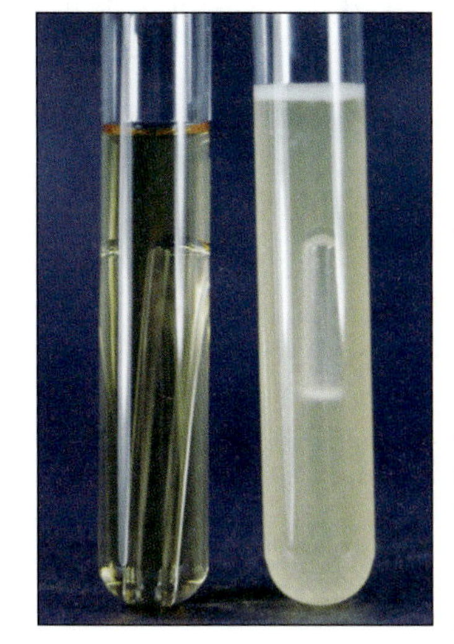

**7.18 LTB Results** ■ The bubble in the Durham tube on the right is *presumptive* evidence of coliform contamination. The tube on the left is negative. All LTB tubes showing a presumptive positive result are used to inoculate BGLB and EC tubes. At this point, treat any tube with turbidity as a presumptive positive.

**7.19 BGLB Broth Results** ■ The bubble in the Durham tube on the right is seen as *confirmation* of coliform contamination. The tube on the left is negative. At this point, only gas production is considered a positive result.

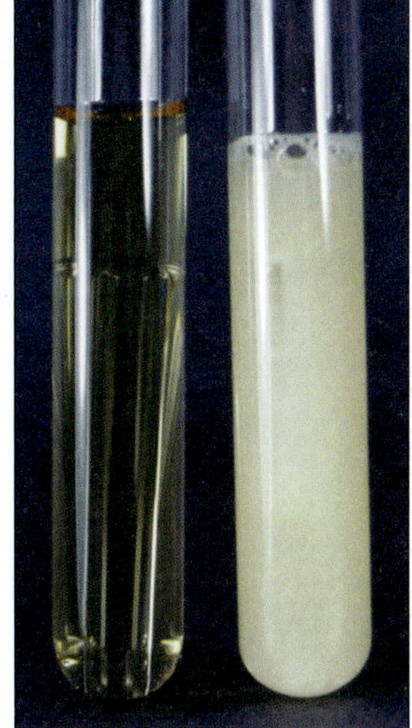

**7.20 EC Broth Results** ■ The bubble in the Durham tube on the right is seen as *confirmation* of *E. coli* contamination. The tube on the left is negative.

## In This Exercise

You will be using collected water samples to perform a multiple tube fermentation. The data collected then will be used to calculate a total coliform count and an *E. coli* count. Counts will be recorded in coliforms per 100 mL of sample and *E. coli* per 100 mL of sample, respectively.

## ▼ Materials

### Per Student

- ☐ Lab coat
- ☐ Disposable gloves
- ☐ Chemical eye protection

### Per Student Group

- ☐ 100 mL water dilution bottle (to be distributed in the preceding lab)
- ☐ 100 mL water sample (obtained by student)
- ☐ Household disinfectant and paper towels
- ☐ Hand sanitizer
- ☐ Ziploc® bag large enough to hold the collection bottle
- ☐ Test tube rack
- ☐ Mechanical pipettor
- ☐ One sterile 10.0 mL pipette
- ☐ Two sterile test tubes (minimum volume of 15 mL)
- ☐ Flask with 20 mL sterilized water
- ☐ A minimum of 5 sterile 1.0 mL pipettes
- ☐ 15 lauryl tryptose broth (LTB) tubes (containing 10 mL broth and an inverted Durham tube)
- ☐ Up to 15 brilliant green lactose bile (BGLB) broth tubes. (These tubes are needed for Lab Two. The number of tubes required will be determined by the results of the LTB test.)
- ☐ Up to 15 EC broth tubes. (These tubes are needed for Lab Two. The number of tubes required will be determined by the results of the LTB test.)
- ☐ Water bath set at 45.5°C (Lab Two)
- ☐ Labeling tape
- ☐ Marking pen

## Medium Recipes

### Brilliant Green Lactose Bile Broth

|  |  |
|---|---|
| ☐ Peptone | 10.0 g |
| ☐ Lactose | 10.0 g |
| ☐ Oxgall | 20.0 g |
| ☐ Brilliant green dye | 0.0133 g |
| ☐ Distilled or deionized water | 1.0 L |

*pH 7.0–7.4 at 25°C*

### Lauryl Tryptose Broth

|  |  |
|---|---|
| ☐ Tryptose | 20.0 g |
| ☐ Lactose | 5.0 g |
| ☐ Dipotassium phosphate | 2.75 g |
| ☐ Monopotassium phosphate | 2.75 g |
| ☐ Sodium chloride | 5.0 g |
| ☐ Sodium lauryl sulfate | 0.1 g |
| ☐ Distilled or deionized water | 1.0 L |

*pH 6.6–7.0 at 25°C*

### EC Broth

- □ Tryptose 20.0 g
- □ Lactose 5.0 g
- □ Bile Salts No. 3 1.5 g
- □ Dipotassium phosphate 4.0 g
- □ Monopotassium phosphate 1.5 g
- □ Sodium chloride 5.0 g
- □ Distilled or deionized water 1.0 L

*pH 6.7–7.1 at 25°C*

## PROCEDURE

### Prelab

1. Obtain a 100 mL water dilution bottle from your instructor. These have a white line inscribed on them that indicates the 100 mL level. (If you use another collection bottle or jar, it would be wise to mark the 100 mL level in lab with a permanent marking pen.)
2. Choose an environmental source to sample. You will be treating the site as potentially contaminated with feces. (Your instructor may decide to approve your choice to avoid duplication among lab groups.)
3. Visit the environmental site as close to your lab period as possible. Bring your water dilution bottle, a pair of gloves, some household disinfectant, hand sanitizer, and paper towels. While wearing gloves, fill the bottle to the white line (100 mL) and replace the cap.
4. Wipe the outside of the bottle with disinfectant-soaked towels. Put the bottle in the Ziploc® bag and close it.
5. Dispose of the towels and gloves in the trash. Cleanse your hands thoroughly with hand sanitizer.
6. Store the sample in the laboratory refrigerator until your lab period. If the sample must sit for a while before your lab, leave the cap slightly loose to allow some aeration.

### Lab One

Refer to the procedural diagram in Figure 7.17 as needed to complete the steps in Labs One and Two.

1. Wear a lab coat, gloves, and chemical eye protection when performing this procedure.
2. Using the sterile pipette, dispense 9 mL of sterile water into each of the test tubes. These are the dilution blanks.
3. Mix the water sample. Then use a sterile 1 mL pipette to aseptically make a dilution by adding 1.0 mL of the water sample to one of the dilution blanks. Mix well. The 10 mL in this tube now consists of only 1/10 of the original sample; therefore, the dilution is 1/10 or 0.1 or $10^{-1}$ (preferred). Label the tube $10^{-1}$. For help with dilutions, refer to Exercise 6-2.
4. Mix the $10^{-1}$ dilution, then use a 1 mL pipette to aseptically make a second dilution by adding 1.0 mL of the $10^{-1}$ dilution to the second dilution blank. Mix well. This 10 mL dilution contains 1/10 of 1/10 = 1/100 original sample and is $10^{-2}$. Label the tube $10^{-2}$.
5. Arrange the 15 LTB tubes into 3 groups of 5 in a test tube rack. Label the groups A, B, and C, respectively.
6. Mix the water sample once again. Then, using a 1 mL pipette aseptically transfer 1.0 mL of the undiluted water sample to each LTB tube in the group labeled A. You can use the same pipette for these transfers as long as you do not contaminate it. Mix well. This is undiluted sample; therefore it is a $10^0$ dilution.
7. Using a fresh, sterile 1 mL pipette, aseptically add 1.0 mL of the $10^{-1}$ dilution to each of the LTB tubes in group B. Again, you may use the same pipette for each transfer as long as you do not contaminate it. Mix well. (***Remember:*** The volume of LTB in this step and step 8 is not part of the dilution calculation even though the original sample is more dilute.)
8. Using a fresh, sterile 1 mL pipette, aseptically add 1.0 mL of the $10^{-2}$ dilution to each of the LTB tubes in the C group.
9. Incubate the LTB tubes at 35°C to 37°C for 48 hours.
10. Dispose of your original water sample in the appropriate autoclave container when finished.

### Lab Two

1. Remove the LTB tubes from the incubator and examine the Durham tubes for accumulation of gas one group at a time. Gas production is a positive result; absence of gas is negative if the broth is clear. If the LTB broth is turbid but contains no gas, the presence of coliforms is not yet ruled out. Therefore, record any turbidity as a positive result and include these tubes with those containing gas. Record positive and negative results in Table 1 on the data sheet, page 487, and calculate $V_a$.
2. Using a sterile loop, inoculate one BGLB broth with each positive LTB tube. Make sure that each BGLB tube is *clearly labeled* A, B, or C according to the LTB tube from which it is inoculated.

3. Inoculate EC broths with the positive LTB tubes in the same manner as the BGLB above. Again, be sure to clearly label all EC tubes appropriately.
4. Incubate the BGLB at 35°C to 37°C for 48 hours. Incubate the EC tubes in the 45.5°C water bath for 48 hours.
5. Dispose of all LTB tubes in the appropriate autoclave container when finished.

### Lab Three

1. Remove all tubes from the incubator and water bath and examine the Durham tubes for gas accumulation. At this point, turbidity does *not* count as positive; the only positive result is the presence of gas in the Durham tube. Count the positive BGLB tubes and enter your results in Table 2 on the data sheet, page 488.
2. Using Table 7-6 as a guide, complete Table 2 on the data sheet.
3. Using the data in Table 2, calculate the total coliform MPN using the formula on page 488. (This formula is used to calculate both Total Coliform MPN and *E. coli* MPN.)
4. Count the positive EC broth results in the same manner as the BGLB test. Record the results in Table 3 on the data sheet, page 488. Then, perform the calculations to complete Table 3 using the formula on page 481.
5. Determine the *E. coli* MPN in the same manner as described for total coliform count, and record your results in the chart on the data sheet.
6. Dispose of all BGLB and EC tubes in the appropriate autoclave container when finished.

TABLE **7-6 Example of BGLB Test Results** ■ The results shown here are of a hypothetical BGLB test using three groups of five tubes (A, B, and C). (***Note:*** On your data sheet, you are given three similar tables—one for LTB results, one for BGLB results, and one for EC broth results. Use this table as a sample to help you with completing Tables B and C using your own data.) In this example, the original water sample was diluted to $10^0$, $10^{-1}$, and $10^{-2}$. The three dilutions were used to inoculate the broths in groups A, B, and C, respectively. **Row 1:** The dilution of the inoculum used per group. **Row 2:** The actual amount of original sample that went into each LTB tube. **Row 3:** The number of tubes in each group. **Row 4:** The number of tubes from each group of five that showed evidence of gas production. This total (in red) inserts into the equation as P. **Row 5:** The number of tubes from each group that did *not* show evidence of gas production. **Row 6:** Is used for calculating the "combined volume of sample in negative tubes" and refers to the inoculum that went into the LTB tubes that produced a negative result. This total (in red) inserts into the equation as $V_n$. **Row 7:** Used for calculating the "combined volume of sample in all tubes" and refers to the total volume of inoculum that went into all LTB tubes. This total (in red) inserts into the equation as $V_a$. As you can see, the undiluted inoculation (Group A) produced five positive results and zero negative results; the $10^{-1}$ dilution (Group B) produced three positive results and two negative; and the $10^{-2}$ dilution (Group C) produced one positive and four negative results. The total volume of original sample that went into LTB tubes was 5.55 mL, 0.24 mL of which produced no gas (shown in red in rows 7 and 6, respectively).

| | Group | Group A | Group B | Group C | Totals (A + B + C) |
|---|---|---|---|---|---|
| 1 | Dilution (D) | $10^0$ | $10^{-1}$ | $10^{-2}$ | NA |
| 2 | Volume of dilution added to each LTB tube that is original sample (1.0 mL × D) | 1.0 mL | 0.1 mL | 0.01 mL | NA |
| 3 | # LTB tubes in group | 5 | 5 | 5 | NA |
| 4 | # BGLB positive results | 5 | 3 | 1 | 9 |
| 5 | # BGLB negative results | 0 | 2 | 4 | NA |
| 6 | Total volume of *original sample* in all LTB tubes that produced negative BGLB results (D × 1.0 mL × # negative tubes) | 0 mL | 0.2 mL | 0.04 mL | 0.24 mL |
| 7 | Volume of *original sample* in all LTB tubes inoculated (D × 1.0 mL × # tubes) | 5.0 mL | 0.5 mL | 0.05 mL | 5.55 mL |

## Reference

Eaton, Andrew D. (American Water Works Association), Lenore S. Clesceri (Water Environment Federation), Eugene W. Rice and Arnold E. Greenberg (American Public Health Association), and Mary Ann H. Franson. Chap. 9 in *Standard Methods for the Examination of Water and Wastewater*, 21st ed. Washington, DC: APHA Publication Office, 2005.

7

Name ______________________________

Date ______________________________

Lab Section ______________________________

I was present and performed this exercise (initials) ______________________________

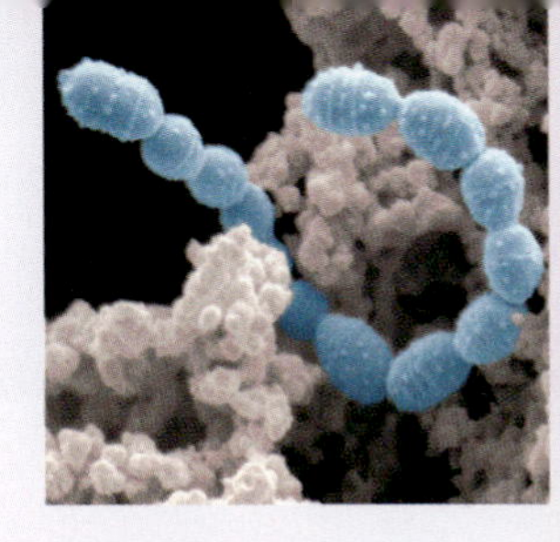

# DATA SHEET 7-6

# Multiple Tube Fermentation Method for Total Coliform Determination

## DATA AND CALCULATIONS

1 Enter your LTB data here. Perform the calculations to complete the table. No MPN calculations are required.

**Table 1** LTB Data—Presumptive Test

| Group | Group A | Group B | Group C | Totals (A + B + C) |
|---|---|---|---|---|
| Dilution (D) | $10^{0}$ | $10^{-1}$ | $10^{-2}$ | NA |
| Volume of dilution added to each LTB tube that is *original sample* (1.0 mL × D) | 1.0 mL | 0.1 mL | 0.01 mL | NA |
| # LTB tubes in group | 5 | 5 | 5 | NA |
| # Positive results (Gas or turbidity) | | | | P = |
| # Negative results (No gas or turbidity) | | | | NA |
| Volume of *original sample* in all LTB tubes inoculated (D × 1.0 mL × # tubes) | | | | $V_a$ = |

**2** Enter your BGLB data here. Perform the calculations to complete the table.

**Table 2** BGLB Data—Confirmation of Coliforms

| Group | Group A | Group B | Group C | Totals (A + B + C) |
|---|---|---|---|---|
| Dilution (D) | $10^0$ | $10^{-1}$ | $10^{-2}$ | NA |
| Volume of dilution added to each LTB tube that is *original sample* (1.0 mL × D) | 1.0 mL | 0.1 mL | 0.01 mL | NA |
| # LTB tubes in group | 5 | 5 | 5 | NA |
| # BGLB positive results (Gas only) | | | | P = |
| # BGLB negative results (No gas) | | | | NA |
| Total volume of *original sample* in all LTB tubes that produced negative results in BGLB tubes (D × 1.0 mL × # negative tubes) | | | | $V_n$ = |
| Volume of *original sample* in all LTB tubes inoculated (From Table 1) | | | | $V_a$ = |

Calculate the total coliform MPN per 100 mL using the BGLB data.

$$\text{Total coliform MPN/100 mL} = \frac{100P}{\sqrt{V_n V_a}}$$

**3** Enter your EC broth data here. Perform the calculations to complete the table.

**Table 3** EC Broth Data—Confirmation of *E. coli*

| Group | Group A | Group B | Group C | Totals (A + B + C) |
|---|---|---|---|---|
| Dilution (D) | $10^0$ | $10^{-1}$ | $10^{-2}$ | NA |
| Volume of dilution added to each LTB tube that is *original sample* (1.0 mL × D) | 1.0 mL | 0.1 mL | 0.01 mL | NA |
| # LTB tubes in group | 5 | 5 | 5 | NA |
| # EC positive results (Gas) | | | | P = |
| # EC negative results (No gas) | | | | NA |
| Total volume of *original sample* in all LTB tubes that produced negative results in EC broth tubes (D × 1.0 mL × # tubes) | | | | $V_n$ = |
| Volume of *original sample* in all LTB tubes inoculated (from Table 1) | | | | $V_a$ = |

Calculate the *E. coli* MPN per 100 mL using the EC broth data.

$$E.\ coli\ \text{MPN/100 mL} = \frac{100P}{\sqrt{V_n V_a}}$$

Name

Date

Lab Section

I was present and performed this exercise (initials)

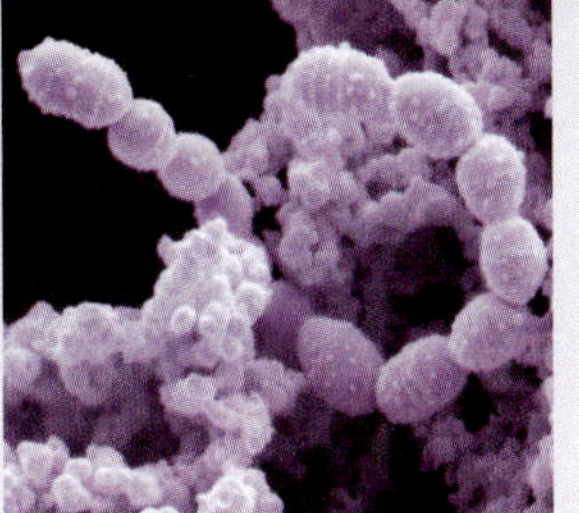

# DATA SHEET 7-6

*(continued)*

## QUESTIONS

**1** *In the following formula, $V_n$ and $V_a$ symbolize critical volumes in calculating the most probable number of coliforms in 100 milliliters of a sample. Do the symbols represent the amount of dilution added to tubes of LTB or the amount of original sample added to the tubes? Defend your answer in full and explain what would happen if you were to choose the wrong volumes in your calculation.*

$$\text{MPN/100 mL} = \frac{100\text{P}}{\sqrt{V_n V_a}}$$

**2** *Suppose you were to run this test on a water sample and, after 48 hours incubation of the LTB tubes, found turbidity but no gas bubbles. What should you do next?*

**3** *What if you found gas in the LTB tubes but none in the BGLB? How does this affect the EC test?*

**4** *All coliforms ferment glucose, but none of the media used for this test includes glucose. Why do you think glucose is not used?*

**5** *Would adding glucose increase or decrease the sensitivity? Specificity?*

**6** *Shown are data from row 4 of the BGLB and EC tables obtained after testing a hypothetical water sample.*

| | Group | Group A | Group B | Group C | Totals (A + B + C) |
|---|---|---|---|---|---|
| 1 | Dilution (D) | $10^0$ | $10^0$ | $10^0$ | NA |
| 4 | # positive BGLB results | 5 | 3 | 2 | 10 |
| 4 | # positive EC results | 3 | 1 | 0 | 4 |

*Explain the discrepancy in the number of positive tubes.*

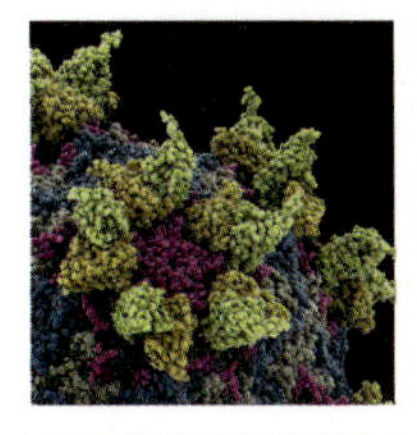

## Food Microbiology

Food microbiology is the study, utilization, and control of microorganisms in food. In Exercise 2-1 you learned that microorganisms exist virtually everywhere life exists. Left undisturbed in their own habitat, most are beneficial in some way. Some microorganisms have been used successfully to produce our favorite foods, such as yogurt, wine, beer, sauerkraut, buttermilk, vinegar, bread, and cheeses.

Generally speaking, however, the unintended introduction of microorganisms to our food (including otherwise beneficial microorganisms) can be a serious health hazard. For example, some strains of *Escherichia coli*, one of the most common enterics in humans and other mammals, can cause mild to severe illness or even death if ingested or introduced into other parts of the body.

It is not unusual to find unwanted microorganisms in unprocessed food. It is unavoidable and expected. This is why there are agencies such as the Food and Drug Administration (FDA) and the Centers for Disease Control and Prevention (CDC). These agencies were designed, in part, to protect consumers from the foodborne illnesses caused by improper management of food items. But these agencies exercise control only in the public arena; their influence over the practices of people in their homes typically is advisory only. This is why it is important to follow their recommendations and not only practice good personal hygiene, but also to wash and cook food properly.

In the remaining two exercises of this section, you will be performing a simple test of milk quality and then you will have a little fun by making yogurt. Don't forget to wash your hands before you start! ■

# Methylene Blue Reductase Test (MBRT)

EXERCISE 7-7

## ■ Theory

Milk is a good source of protein (casein) and carbohydrate (lactose) not only for mammals, but also for microbes—just think back at how many of the media used in earlier exercises had casein and/or lactose in them! Because it is such a good nutrient source and it has the potential of becoming contaminated with pathogenic microbes from the cow's skin or udder, or even unsterile machinery, milk quality is always a public health concern. The reduction of methylene blue dye may be used as an indicator of milk quality.

Methylene blue is blue when oxidized and colorless when reduced. It can be reduced enzymatically either aerobically or anaerobically. In the aerobic electron transport system, methylene blue is reduced by cytochromes but immediately is returned to the oxidized state when it subsequently reduces oxygen. Anaerobically, the dye is in the reduced form, and in the absence of an oxidizing substance, remains colorless.

In the methylene blue reductase test, a small quantity of a dilute methylene blue solution is added to a sterilized test tube containing raw milk. The tube then is sealed tightly and incubated in a 35°C water bath. The time it takes the milk to turn from blue to white (because of methylene blue reduction) is a qualitative indicator of the number of microorganisms living in the milk (Fig. 7.21).

Here is the basic idea: The higher the microorganism concentration, the faster the oxygen consumption in the sealed tube, and the faster methylene blue becomes reduced. Good-quality milk takes longer than 6 hours to convert the methylene blue (Table 7-7). At the other extreme, very poor quality milk will convert in less than 30 minutes.

**7.21 Methylene Blue Reductase Test** ■ The tube on the left is a negative control to illustrate the original color of the oxidized medium. The tube on the right took 20 hours before bacterial reduction of methylene blue occurred. The speed of reduction is related to the concentration of microorganisms present in the milk.

TABLE 7-7 **Milk Quality Standards for the Methylene Blue Reductase Test (MBRT)**

| Reduction Time ($t_E$) | Milk Quality |
|---|---|
| Longer than 8 hours | Excellent |
| Between 6 hours and 8 hours | Good |
| Between 2.0 hours and 6 hours | Fair |
| Between 30 minutes and 2.0 hours | Poor |
| Less than 30 minutes | Very poor |

## Application

This procedure is a qualitative method of determining milk quality. It also gives an indication of the contaminants present, with rapid reduction being associated with contamination by high levels of enteric bacteria and *Streptococcus lactis*. It has largely been replaced by quantitative methods, such as the standard plate count (Exercise 6-2) and the Multiple Tube Fermentation test (Exercise 7-6). Still, it's a fun and easy way to assess milk quality.

## In This Exercise

You will test milk quality by measuring how long the indicator dye, methylene blue, takes to become reduced by any bacterial contaminants present.

### Materials

**Per Student**

- ☐ Lab coat
- ☐ Disposable gloves
- ☐ Chemical eye protection

7

**Per Student Group**

- ☐ Milk samples (raw or processed; a variety is best)
- ☐ Sterile screw-capped test tubes
- ☐ Sterile 1 mL and 10 mL pipettes with mechanical pipettors
- ☐ Hot-water bath set at 35°C
- ☐ Methylene blue reductase reagent
- ☐ Overnight broth culture of *Escherichia coli*
- ☐ Clock or wristwatch

## Reagent

**Methylene Blue Reductase Reagent**

- ☐ Methylene blue dye 8.8 mg
- ☐ Distilled or deionized water 200.0 mL

## PROCEDURE

1. Wear a lab coat, gloves, and chemical eye protection when performing this procedure.
2. Obtain sterile tubes for as many samples as you are testing, plus two more, to be used as positive and negative controls. Label them appropriately.
3. Using a sterile 10 mL pipette, aseptically add 10 mL of milk to each tube.
4. Inoculate the tube marked "positive control" with 1 mL of *E. coli* culture.
5. Aseptically add 1.0 mL methylene blue solution to each test tube. Cap the tubes tightly and invert them three times to mix thoroughly.
6. Place the tube marked "negative control" in the refrigerator to prevent it from changing color.
7. Place all other tubes in the hot-water bath and note the time.
8. After 5 minutes, remove the tubes, invert them once to mix again, and then return them to the water bath. Record the time in the table in the data sheet, page 493, under "Starting Time" ($t_S$).
9. Using the control tubes for color comparison, check the tubes at 30 minutes and every hour thereafter (e.g., 1½ hours, 2½ hours, etc.). Record the time when each becomes white ("Completion Time"–$t_C$). By convention, a positive is recorded when 80% or more of the milk in a tube has changed. Record the $t_C$ as follows: If positive at 30 minutes, record "<30 minutes." When the milk has turned on any ½ hour, record the $t_C$ as the last complete hour (i.e., if the milk has turned at the 2½ hour reading, record $t_C$ as 2 hours.) Refer to Table 7-7 for milk quality classifications based on the MBRT.
10. Using the table provided in the data sheet, calculate the "Elapsed Time" ($t_E$) it takes for each milk sample to become white.
11. Dispose of all tubes in the appropriate autoclave container when finished.

### References


Atherton, H.V. and J.A. Newlander. Pages 276–277 in *Chemistry and Testing of Dairy Products*, 4th ed. Westport, CT: AVI Publishing Co., 1977.

Bailey, R. W. and E. G. Scott. Pages 114 and 306 in *Diagnostic Microbiology*, 2nd ed. St. Louis, MO: Mosby, 1966.

Benathen, Isaiah. Page 132 in *Microbiology with Health Care Applications*. Belmont, CA: Star Publishing, 1993.

Power, David A. and Peggy J. McCuen. Page 62 in *Manual of BBL™ Products and Laboratory Procedures*, 6th ed. Cockeysville, MD: Becton Dickinson Microbiology Systems, 1988.

Richardson, Gary H., Ed. *Standard Methods for the Examination of Dairy Products*, 15th ed. Washington DC: American Public Health Association, 1985.

Name ______________________________

Date ______________________________

Lab Section ______________________________

I was present and performed this exercise (initials) ______________________________

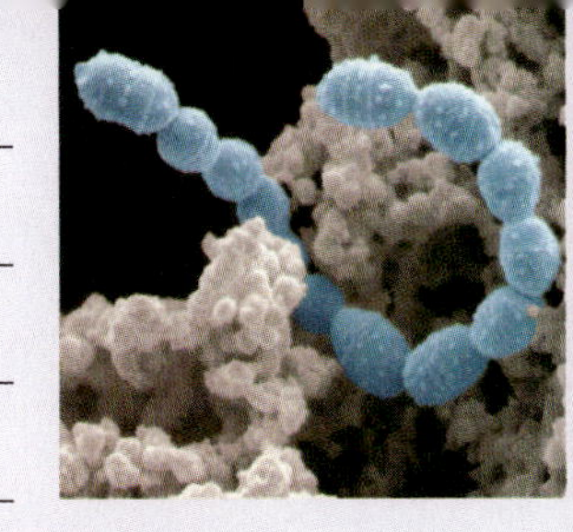

DATA SHEET

7-7

# Methylene Blue Reductase Test (MBRT)

## OBSERVATIONS AND INTERPRETATIONS

**1** Record your results and interpretations in the table below.

| **Milk Sample** (brand and raw or processed) | **Starting Time ($t_S$)** (milk is blue) | **Completion Time ($t_C$)** (milk is white) | **Elapsed Time** ($t_E = t_C - t_S$) | **Milk Quality** (acceptable/unacceptable) |
|---|---|---|---|---|
| | | | | |
| | | | | |
| | | | | |
| | | | | |

## QUESTIONS

**1** *Why were you told to cap the tubes tightly? Explain.*

______________________________

______________________________

______________________________

______________________________

______________________________

______________________________

______________________________

______________________________

______________________________

______________________________

**2** *What results would you expect if the tubes were inoculated with a strict aerobe? A strict anaerobe?*

7

# Making Yogurt

EXERCISE **7-8**

## Theory

Several species of bacteria are used in the commercial production of yogurt. Most formulations include combinations of two or more species to synergistically enhance growth and to produce the optimum balance of flavor and acidity. One common pairing of organisms in commercial yogurt is that of *Lactobacillus delbrueckii* subsp. *bulgaricus*, and *Streptococcus thermophilus*.

Yogurt gets its unique flavor from acetaldehyde, diacetyl, and acetate produced from fermentation of the milk sugar lactose. The proportions of products, and ultimately the flavor, in the yogurt depend upon the types of enzyme systems possessed by the species used. Both species mentioned above contain constitutive β-galactosidase systems that break down lactose and convert the glucose to lactate, formate, and acetate via pyruvate in the glycolytic pathway. (See Appendix A, Figure A.6.)

As you may remember, lactose is a disaccharide composed of glucose and galactose. *S. thermophilus* does not possess the enzymes needed to metabolize galactose, and *L. delbrueckii* preferentially metabolizes glucose. This results in an accumulation of galactose, which adds sweetness to the yogurt. Acetaldehyde is produced directly from pyruvate by *S. thermophilus* and through the conversion of proteolysis products threonine and glycine by *L. delbrueckii*. Some strains of *S. thermophilus* also produce glucose polymers, which give the yogurt a viscous consistency.

## Application

This exercise is designed to keep you away from the Internet for a few minutes.

## In This Exercise

You will produce yogurt using a simple home recipe with a commercial yogurt as a starter. Read the label to see which microorganisms are included. We hope you enjoy it.

### Materials

**Per Student Group**

- ☐ Whole, low-fat, or skim milk
- ☐ Plain yogurt with active cultures (bring from home or supermarket)
- ☐ Medium-size saucepan
- ☐ Medium-size bowl
- ☐ Wire whisk
- ☐ Hot plate
- ☐ Cooking thermometer
- ☐ Measuring cup
- ☐ Plastic wrap
- ☐ Fresh fruit
- ☐ (Optional) sugar
- ☐ pH meter or pH paper

## PROCEDURE[1]

**Lab One**

1. Obtain all materials, and set them up in a clean work area.
2. While stirring, slowly heat 5 cups of milk in the saucepan to 185°F. Remove the milk from the heat, and let it cool to 110°F.
3. Place ¼ cup starter yogurt in the bowl. Slowly, stir in the cooled milk, about ⅓ cup to ½ cup at a time, mixing until smooth after each addition.
4. Cover the bowl with plastic wrap, and puncture several times to allow gases and excess moisture to escape.
5. Label the bowl with your name, the date, and the cultures present in your yogurt starter.
6. Incubate 5–6 hours at 30°C–35°C. Remove the bowl from the incubator at the correct time, and place it in the refrigerator.

**Lab Two**

1. Remove your yogurt from the refrigerator.
2. Perform a simple stain on a smear from your yogurt and examine with the microscope.
3. Compare flavor, consistency, and starter cultures with other groups in the lab. With a pH meter or pH paper, measure the pH of your yogurt. Record your results in the table provided on the data sheet, page 497.
4. Serve with fresh fruit and enjoy!

[1] Because food and beverages are not allowed in microbiology labs, this exercise is best performed in an alternate location.

## References

Downes, Frances Pouch and Keith Ito. Chapter 47 in *Compendium of Methods for the Microbiological Examination of Foods*, 4th ed. Washington, DC: American Public Health Association, 2001.

Ray, Bibek. Chapter 13 in *Fundamental Food Microbiology*, 2nd ed. Boca Raton, FL: CRC Press LLC, 2001.

Name ______________________________

Date ______________________________

Lab Section ______________________________

I was present and performed this exercise (initials) ______________________________

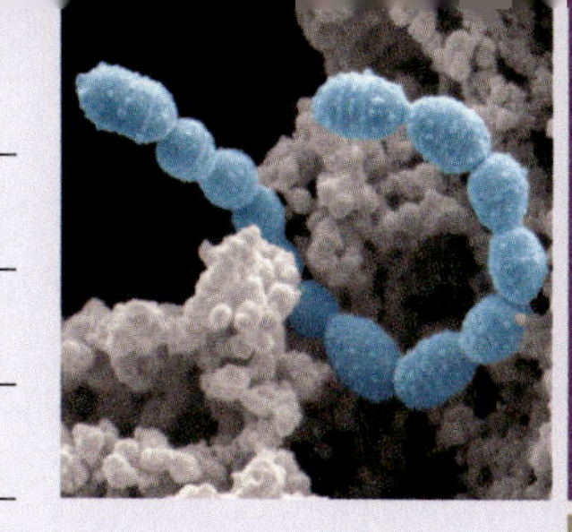

# DATA SHEET 7-8

## Making Yogurt

### OBSERVATIONS AND INTERPRETATIONS

1 Record yogurt made by student groups below.

| Culture Organisms | Cell Morphology and Arrangement | Flavor | Consistency | pH |
|---|---|---|---|---|
| | | | | |
| | | | | |
| | | | | |
| | | | | |
| | | | | |
| | | | | |
| | | | | |

SECTION 8

# Microbial Genetics and Serology

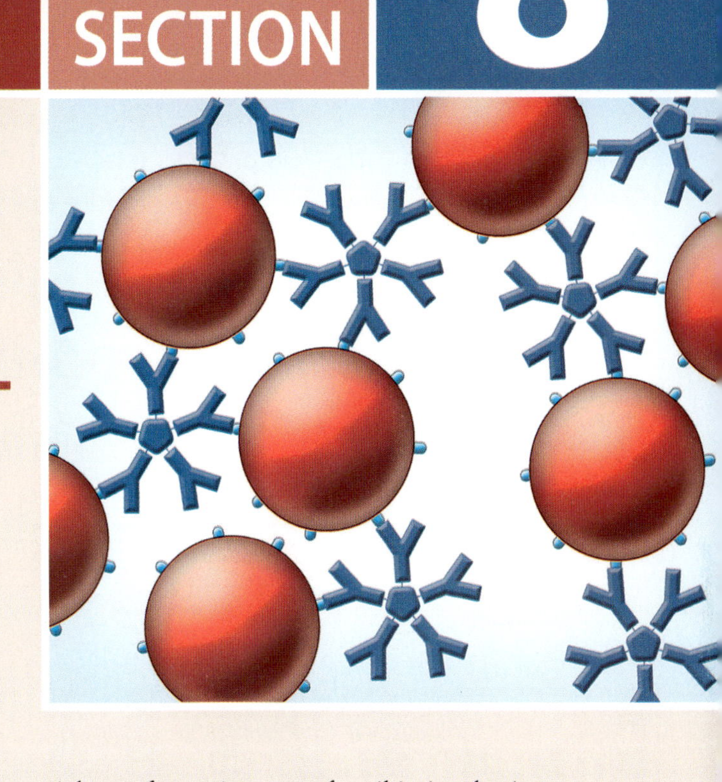

The unifying theme of this section's exercises is molecular biology. In the first three exercises, the focus is on DNA. First you will perform a simple extraction of *E. coli* DNA in Exercise 8-1. DNA extraction is often the first step in biotechnology procedures.

Exercise 8-2 allows you to do some biotechnology. You will be transferring a jellyfish gene into *E. coli* cells, selecting for only those cells that have been successfully transformed, and then manipulating the environment so the organisms produce the gene product only when you want them to. This exercise incorporates into one genetic-engineering experiment bacterial transformation, use of antibiotic selective media, and regulation of gene expression—all of which are techniques that can be applied to many other biotechnological procedures.

Then, in Exercise 8-3, you will examine the mutagenic effects of ultraviolet (UV) radiation on bacterial DNA and how bacteria are able, to some extent, to repair that damage.

The remainder of this section deals with practical applications of the immune system's antibodies, a discipline called serology. Serological tests rely on the specificity with which an antigen and an antibody react. It is this specificity, which is at the heart of all serological tests, coupled with the flexibility to construct tests that identify either antigen or antibody in a sample (whichever is appropriate) that makes serology such an important diagnostic tool.

Many different styles of serological tests have been developed, but they all rely on the reaction between a specific antigen and a specific antibody. Where they differ is in the indicator reaction used to tell if antigen-antibody reaction has occurred, which requires different processes to set up that indicator reaction. A very useful and relatively simple indicator involves the reaction between insoluble antigens and antibodies. When they react a visible clump is formed—an agglutinate. Exercises 8-4 and 8-5 demonstrate two agglutination tests.

In Exercise 8-6 you will conduct an ELISA test that detects antibody in a sample. ELISAs are more sensitive than agglutination tests and the procedure is a bit more complex, but fundamentally they still simply indicate the presence of antigen or antibody in a sample.

# Extraction of DNA from *Escherichia coli* Cells[1]

EXERCISE 8-1

## Theory

DNA extraction from cells is surprisingly easy and occurs in three basic stages.

1. A detergent (e.g., sodium dodecyl sulfate—SDS) is used to lyse cells and release cellular contents, including DNA.
2. This is followed by a heating step (at approximately 65°C–70°C) that denatures protein (including DNases that would destroy the extracted DNA) and other cell components. Temperatures higher than 80°C will denature DNA, and this is undesirable. A protease also may be added to remove proteins. (Other techniques for purification may be used, but these will not be included in this exercise.)
3. Finally, the water-soluble DNA is precipitated in cold alcohol as a whitish, mucoid mass (Fig. 8.1).

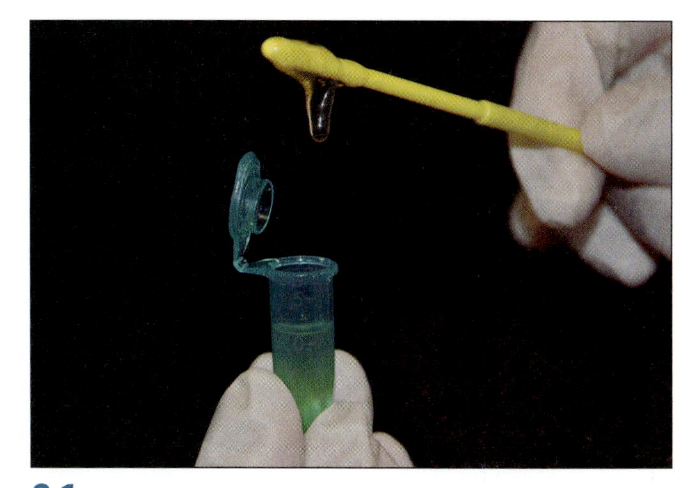

**8.1 Precipitated DNA** ■ This *E. coli* DNA has been spun onto the handle of a disposable inoculating loop.

As an optional follow up to extraction, an ultraviolet spectrophotometer (Fig. 8.2) will be used to estimate DNA concentration in the sample by measuring absorbance at 260 nm, the optimum wavelength for absorption by DNA. An absorbance of $A_{260nm}$ of 1 corresponds to 50 µg/mL of double-stranded DNA (dsDNA). To calculate the DNA concentration (**X**), use an equation of proportions:

$$\frac{\mathbf{X}\ \mu g/mL\ dsDNA}{A_{260}} = \frac{50\ \mu g/mL\ dsDNA}{A_{260}1.0}$$

$$\mathbf{X}\ \mu g/mL\ dsDNA = 50\ \mu g/mL\ dsDNA \times A_{260}$$

Reading absorbance at 280 nm and calculating the following ratio provides an estimate of the sample's purity:

$$\text{Sample Purity} = \frac{A_{260\ nm}}{A_{280\ nm}}$$

If the sample is reasonably pure nucleic acid, the ratio will be about 1.8 (between 1.65 and 1.85). Because protein absorbs maximally at 280 nm, a ratio of less than 1.6 is likely because of protein contamination. If purity is crucial, the DNA extraction can be repeated. If the ratio is greater than 2.0, the sample is diluted and read again.

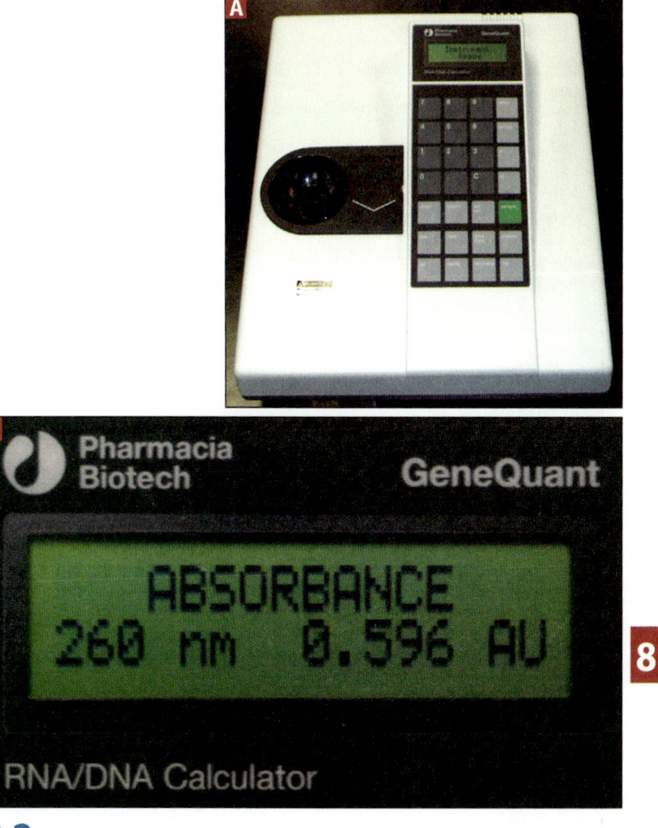


**8.2 Ultraviolet Spectrophotometer** ■ (**A**) A UV spectrophotometer can be used to determine DNA concentration. A quartz cuvette is shown in the sample port. (**B**) This specimen has an $A_{260\ nm}$ of 0.596. Because an $A_{260\ nm}$ of 1.0 is equal to 50 µg/mL of dsDNA, this specimen has a concentration of 29.8 µg dsDNA/mL. Absorbance also can be used to determine purity of the sample. A relatively pure DNA sample will have an $A_{260\ nm}/A_{280\ nm}$ value of approximately 1.8.

[1] Thanks to the following individuals who offered helpful suggestions for this protocol: Dr. Melissa Scott of San Diego City College; Allison Shearer of Grossmont College; Donna Mapston and Dr. Ellen Potter of Scripps Institute for Biological Studies; and Dr. Sandra Slivka of Miramar College.

## Application

Extraction of DNA is a starting point for many lab procedures, including DNA sequencing and cloning.

## In This Exercise

You will extract DNA from *E. coli*. To improve yield, you first will concentrate an *E. coli* broth culture. The actual extraction involves cell lysis, denaturation of protein, and precipitation of the DNA in alcohol. Following extraction, an optional procedure may be used to determine DNA yield and the purity of your extract.

## Materials

### Per Student

- □ Lab coat
- □ Disposable gloves
- □ Chemical eye protection

### Per Pair of Students

- □ Overnight culture of *Escherichia coli* in Luria-Bertani broth (young cultures work best)
- □ Water bath set to 65°C
- □ Microtube floats
- □ Ice bath (small cups with crushed ice work well)
- □ 300 µL 10% sodium dodecyl sulfate (SDS)
- □ 50 µL 20 mg/mL Proteinase K solution (stored in freezer between uses) (***Note:*** Meat tenderizer is an inexpensive substitute, though it may result in DNA hydrolysis if too much is used.)
- □ 300 µL 1.0 M sodium acetate solution (pH = 5.2)
- □ 3 mL 1× tris-acetate-EDTA (TAE) buffer (dilute 10× TAE 9 + 1)
- □ 2 mL 90% isopropanol (stored in a freezer or an ice bath)
- □ Two calibrated disposable transfer pipettes
- □ 100–1,000 µL digital pipettor and tips
- □ Container for tip disposal
- □ 25 mL centrifuge tube
- □ Tabletop centrifuge
- □ Minicentrifuge
- □ Disposable inoculating loop
- □ Two microtubes (at least 1.5 mL in volume)

### Additional Materials for Optional Procedure

- □ Vortex mixer
- □ Ultraviolet spectrophotometer
- □ Two quartz cuvettes
- □ Four transfer pipettes
- □ Container for liquid disposal

## Medium Recipe

### Luria-Bertani Broth

| | |
|---|---|
| □ Tryptone | 10.0 g |
| □ Yeast extract | 5.0 g |
| □ NaCl | 10.0 g |
| □ Distilled or deionized water | 1.0 L |

*pH 7.4 at 25°C*

## PROCEDURE

Refer to the procedural diagram in Figure 8.3 as you read and follow this protocol.

1. Wear a lab coat, gloves, and chemical eye protection when performing this procedure.
2. Obtain the *E. coli* culture. Mix the suspension until uniform turbidity is seen, then transfer 5 mL to a clean, nonsterile centrifuge tube.
3. Spin the 5 mL sample in the tabletop centrifuge slowly for 10 minutes to produce a cell pellet.
4. After spinning and without disturbing the pellet, remove 4.5 mL of the supernatant (the liquid above the pellet) with a transfer pipette. Dispose of the supernatant in the original *E. coli* culture tube.
5. Transfer the remaining 0.5 mL to a nonsterile microtube and dispose of the pipette in an appropriate autoclave container.
6. Add 200 µL of 10% SDS to the *E. coli*.
7. Add 30 µL of 20 mg/mL Proteinase K solution, or "half a pinch" of meat tenderizer.
8. Close the cap and gently mix the tube for 5 minutes by tipping it upside down every few seconds.
9. Place the tube in a float and incubate in a 65°C water bath for 5 minutes.
10. Add 200 µL 1 M sodium acetate solution and mix gently.
11. Place the tube in the ice bath until it is at or below room temperature.
12. When cooled, squirt 400 µL of cold 95% isopropanol into the preparation. (If there are bubbles on the surface, remove them with a nonsterile transfer pipette prior to adding the isopropanol.)
13. Use either end of a disposable loop to mix the preparation, moving in and out and turning occasionally. A glob of mucoid DNA will begin to appear as you mix and will adhere to the loop.
14. Properly dispose of the original *E. coli* tube in the appropriate autoclave container when finished.
15. At this point you have two options: If instructed to do so, continue with the optional procedure. Or,

you can admire your DNA extract, contemplate its awesomeness, then clean up and go home!

**Optional Procedure** (if your lab has an ultraviolet spectrophotometer) Refer to the procedural diagram in Figure 8.4 as you read and complete the following protocol.

1. Remove the DNA from the original microtube and transfer it to a second microtube, using the loop.
2. Resuspend the DNA in 1,000 μL isopropanol using a nonsterile transfer pipette, and then spin it in a minicentrifuge for a few seconds. The DNA should be at the bottom of the tube.
3. With a nonsterile transfer pipette, remove the supernatant and allow the DNA pellet to air-dry.
4. Resuspend the dried DNA in 1,000 μL of TAE buffer. Mix vigorously to dissolve the DNA in the TAE. This mixing may be done by hand, or a vortex mixer may be used.
5. Transfer the suspended DNA solution into a quartz cuvette.
6. Prepare a second cuvette containing 1,000 μL of TAE as a blank.

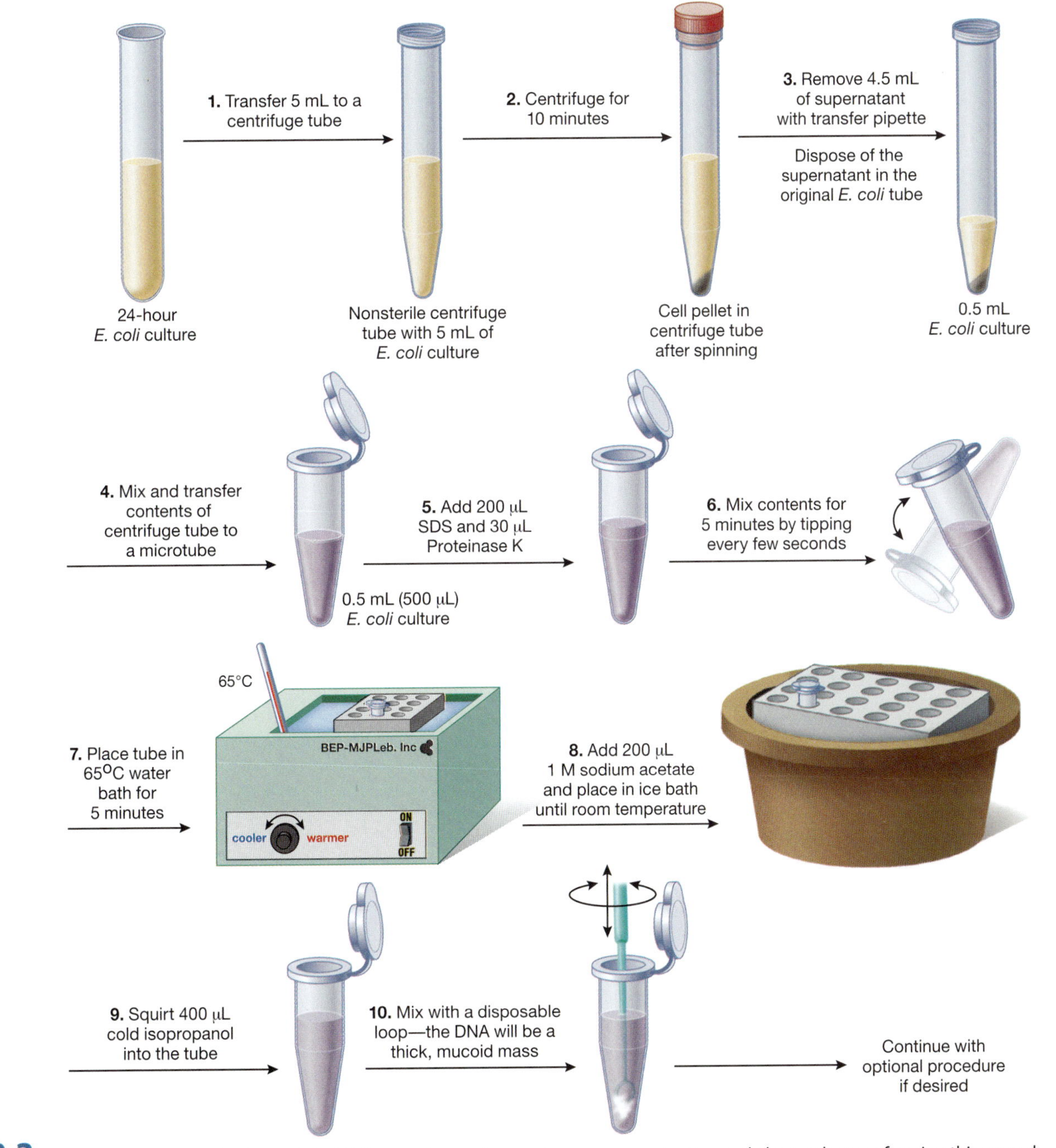


**8.3 Procedural Diagram for Bacterial DNA Extraction** ■ Be sure to wear eye protection and gloves when performing this procedure and to dispose of pipettes in appropriate autoclave container.

**7** Set the spectrophotometer to 260 nm wavelength. Follow the instructions for your UV spectrophotometer to check the absorbance of the extracted DNA, and record on the data sheet, page 505.

**8** Set the spectrophotometer to 280 nm wavelength. Follow the instructions for your UV spectrophotometer to check the absorbance of the extracted DNA, and record on the data sheet. If not continuing with additional absorption readings (step 9), properly clean and dry the cuvettes, and return them to their storage container. Then, go to step 10.

**9** If desired, absorbencies at other wavelengths may be taken to produce an absorption spectrum for DNA. Suggested wavelengths are: 220 nm, 240 nm, 300 nm, and 320 nm, in addition to the measurements for 260 nm and 280 nm taken above. Record these on the data sheet.

**10** Calculate the probable purity of your extracted DNA sample using the formula provided, and record on the data sheet.

**11** If step 9 was done, plot the absorption spectrum (Absorption versus Wavelength) of the DNA sample on the data sheet.

## References


Bost, Rod. *Down and Dirty DNA Extraction.* Research Triangle Park, NC: Carolina Genes, North Carolina Biotechnology, 1989.

Davis, Leonard G., Mark D. Dibner, and James F. Battey. *Basic Methods in Molecular Biology*. New York: Elsevier Science Publishing, 1986.

Freifelder, David. Pages 504–505 in *Physical Biochemistry,* 2nd ed. New York: W. H. Freeman and Co, 1982.

Kreuzer, Helen and Adrianne Massey. *Recombinant DNA and Biotechnology—A Guide For Teachers,* 2nd ed. Washington, DC: ASM Press, 2001.

Zyskind, Judith W. and Sanford I. Bernstein. *Recombinant DNA Laboratory Manual.* San Diego, CA: Academic Press, 1992.


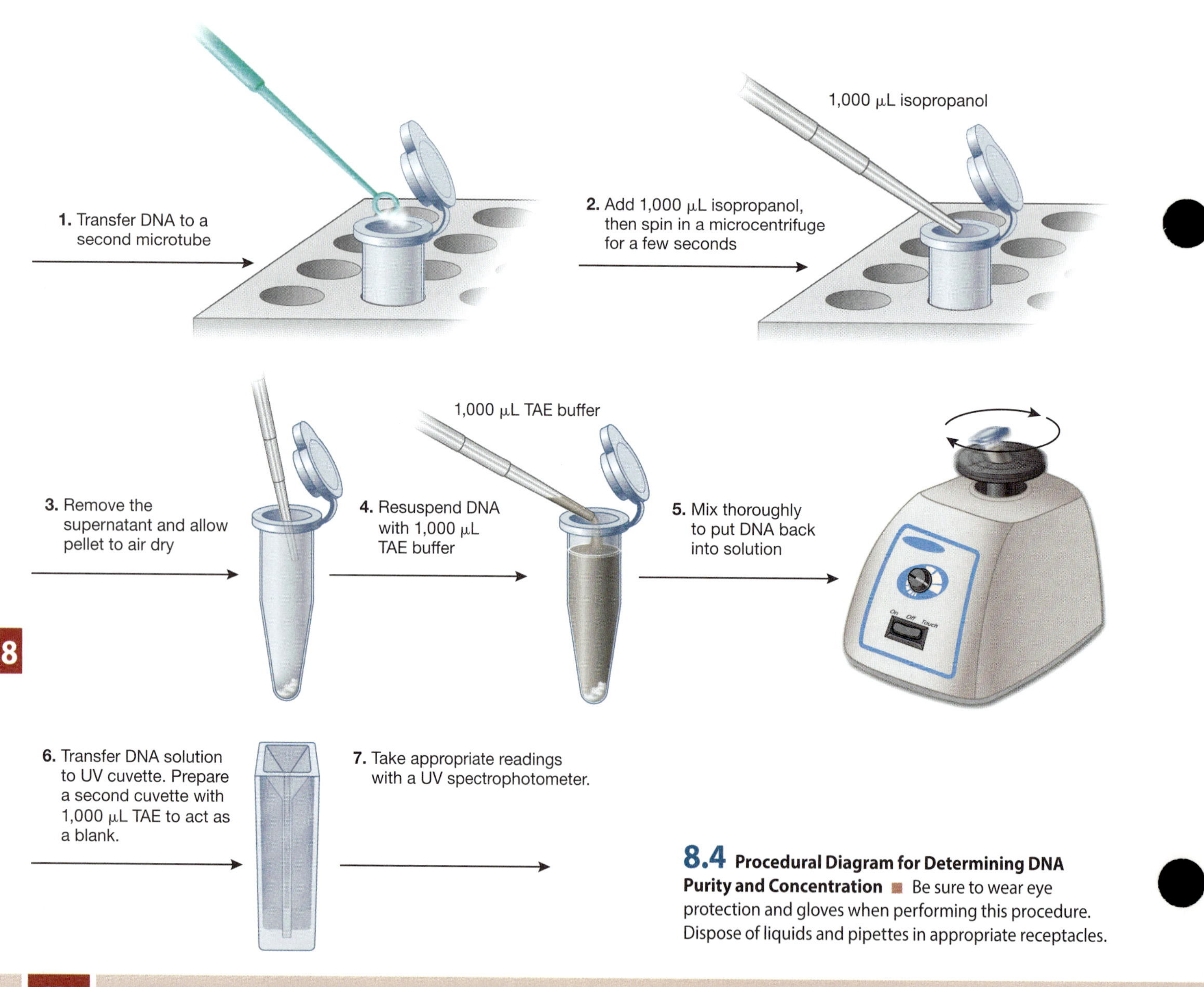


**8.4 Procedural Diagram for Determining DNA Purity and Concentration** ■ Be sure to wear eye protection and gloves when performing this procedure. Dispose of liquids and pipettes in appropriate receptacles.

Name ____________________

Date ____________________

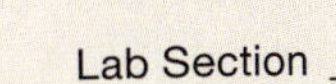

Lab Section ____________________

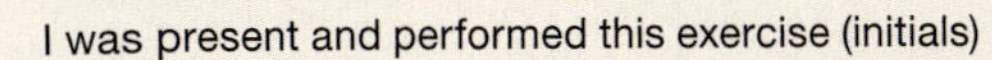

I was present and performed this exercise (initials) ____________________

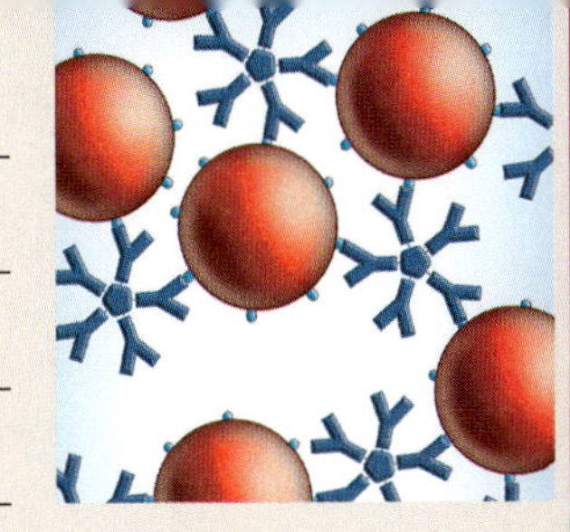

# DATA SHEET 8-1

## Extraction of DNA from *Escherichia coli* Cells

### OBSERVATIONS AND INTERPRETATIONS

**1** In the following table, record the absorbance values for the wavelengths used.

| **Wavelength** (nm) | **Absorbance** |
|---|---|
| 220 (optional) | |
| 240 (optional) | |
| 260 | |
| 280 | |
| 300 (optional) | |
| 320 (optional) | |

**2** Calculate the DNA concentration in your sample. Be sure to take any dilutions into account. (You suspended the DNA in 1 mL [1,000 µL]) TAE. Because the units of concentration are µg/mL, that isn't considered a dilution. The only dilution would have been made if the absorbance readings were over 2.)

**3** Calculate the purity of your sample. Is your extract pure or not? Defend your answer.

**4** Plot the absorption spectrum (Absorption versus Wavelength) of the DNA sample if absorbance at additional wavelengths was determined.

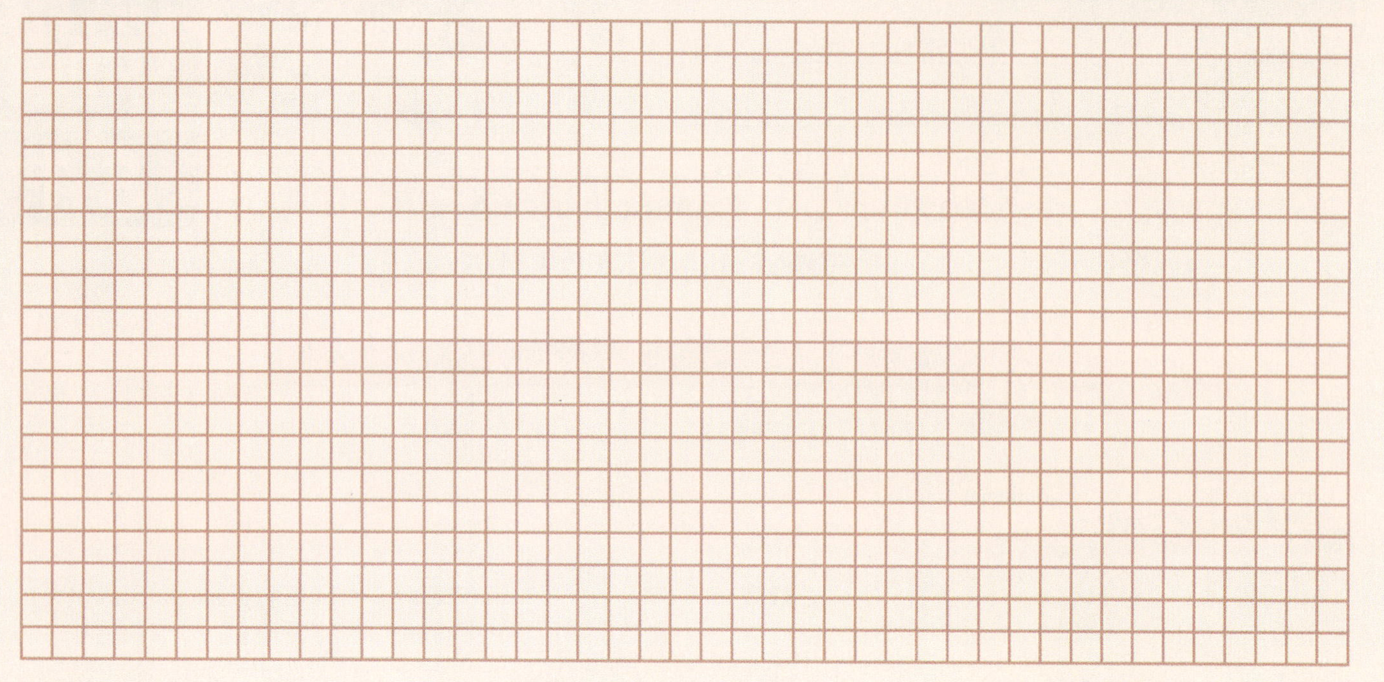

## QUESTIONS

**1** *What is the importance of heating the cell lysate?*

**2** *Why does the extraction work better with cold 95% isopropanol than with room-temperature 95% isopropanol?*

**3** *Based on your results, what wavelength gives maximum absorption of DNA? Does this match the accepted wavelength for maximum absorption? If not, suggest reasons for the discrepancy.*

8

# Bacterial Transformation: The pGLO™ System

EXERCISE **8-2**

## Theory

This exercise utilizes a kit produced by Bio-Rad Laboratories[1] that efficiently illustrates the following principles of microbial genetics:

- bacterial transformation
- use of an antibiotic selective medium to identify transformed cells
- the operon as a mechanism of regulating microbial gene expression

Bacterial **transformation** is the process by which **competent** bacterial cells pick up DNA from the environment and make use of the genes it carries. It was first demonstrated in a strain of pneumococcus in 1928 by English bacteriologist Fred Griffith and since has been found to occur naturally in only certain genera. Modern techniques, however, have allowed biologists to make most cells (including eukaryotic cells) artificially competent, and this has made transformation a useful tool in genetic engineering. You will be using a $CaCl_2$ transforming solution and heat shock to make *Escherichia coli* cells competent.

Green fluorescent protein (GFP) is responsible for **bioluminescence** in the jellyfish *Aequorea victoria* (Fig. 8.5). In developing this kit, Bio-Rad isolated the jellyfish GFP gene and altered it so the GFP fluoresces more than the natural version. You will be introducing the GFP gene into competent *E. coli* cells and will be using the cell's newly acquired ability to fluoresce as visual evidence of successful transformation and subsequent gene expression.

**Operons** are structural and functional genetic units of prokaryotes. Each operon minimally includes a **promoter site** (for binding **RNA polymerase**) and two or more **structural genes** coding for enzymes in the same metabolic pathway. In this exercise you will be using part of the arabinose operon.

The complete arabinose operon (Fig. 8.6) consists of the promoter ($P_{BAD}$) and three structural genes (*ara*B, *ara*A, and *ara*D) that code for enzymes used in arabinose digestion. In vivo, the arabinose enzymes are needed only when arabinose is present. After all, there is no point in the cell expending a lot of energy making the enzymes if the substrate is not there.

[1] Catalog Number 166-0003-EDU. Bio-Rad Laboratories Main Office 2000 Alfred Nobel Drive, Hercules, CA 94547; www.bio-rad.com, 1-800-424-6723.

A DNA binding protein called *ara*C attaches to the promoter of the arabinose operon and acts like a switch. When arabinose is not present, *ara*C prevents RNA polymerase from binding to the promoter, so transcription *cannot* occur—the switch is "off." When arabinose is present, it binds to *ara*C and changes its shape so RNA polymerase *can* bind to the promoter and transcribe the genes—the switch is "on."

Then this sequence of events occurs: The genes are transcribed, the enzymes are synthesized, they catalyze their reactions, and eventually the arabinose is consumed. Now, without arabinose *ara*C returns to its "off" shape and the genes are no longer transcribed. In this exercise you will be using the regulatory portion of the arabinose operon (the arabinose promoter and the *ara*C gene), but the structural genes have been replaced with the GFP gene.

All that remains is a means of carrying the GFP gene into the cell and replicating it—a **vector**. In this exercise you will be using a **plasmid** as a vector: the pGLO™ plasmid (Fig. 8.7). Plasmids are small, naturally occurring, circular DNA molecules that possess only a few genes and replicate independent of the chromosome (because they have their own replication origin). Although they are nonessential, they often carry genes that are beneficial to the bacterium, such as antibiotic resistance. The antibiotic resistance gene (*bla*) used in this experiment produces an enzyme called β-lactamase (hence, "*bla*"), which hydrolyzes the β-lactam ring within the ampicillin molecule (and many other antibiotics, including penicillin).

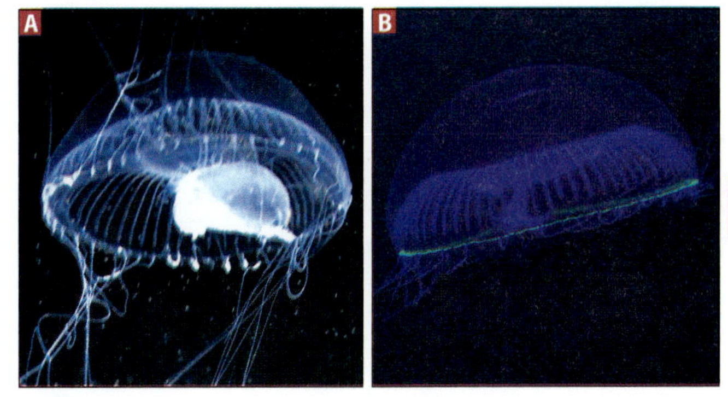


**8.5 Jellyfish *Aequorea victoria*** ■ (**A**) *A. victoria* produces GFP, but it is not visible to our eyes when illuminated with "natural" light. (**B**) When viewed with UV light, the ring of GFP around the edge of *A. victoria's* bell fluoresces.

(Photos taken at Cabrillo Marine Aquarium, San Pedro, CA.)

8

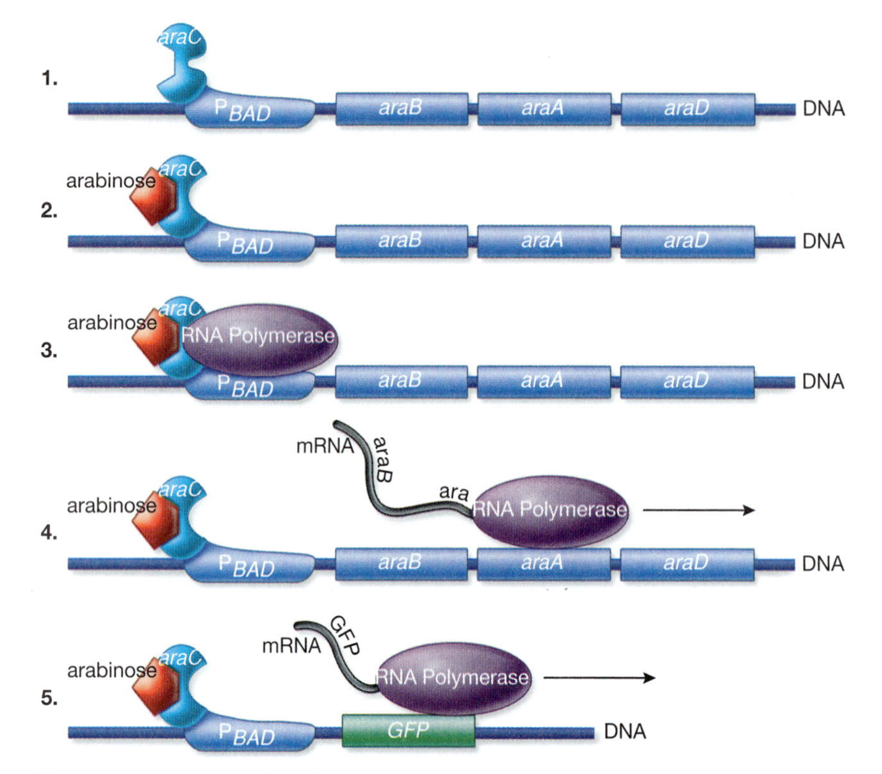


**8.6 The Functioning Arabinose Operon—Normal and Genetically Engineered** ■ Operons are prokaryotic structural and functional genetic units: They carry genes for enzymes in the same metabolic pathway, and they are regulated together. Operons with genes for catabolic enzymes are transcribed only when the specific substrate is present. In this case, the substrate is the pentose sugar arabinose. (**1**) The arabinose operon consists of a promoter ($P_{BAD}$) and three structural genes (*ara*B, *ara*A, and *ara*D). The DNA binding protein (*ara*C) attaches to the promoter and acts like a switch. Without arabinose present, RNA polymerase is unable to bind to the promoter and begin transcription of the genes. (**2**) and (**3**) Arabinose binds to a receptor on *ara*C and causes it to change to a shape that allows binding of RNA Polymerase to the promoter. (**4**) Transcription of the structural genes occurs, the enzymes are produced, and arabinose is catabolized. (**5**) The pGLO™ plasmid has been engineered to carry the arabinose promoter and the gene for green fluorescent protein (GFP) instead of the genes for arabinose catabolism. If arabinose is present, the GFP gene will be transcribed.

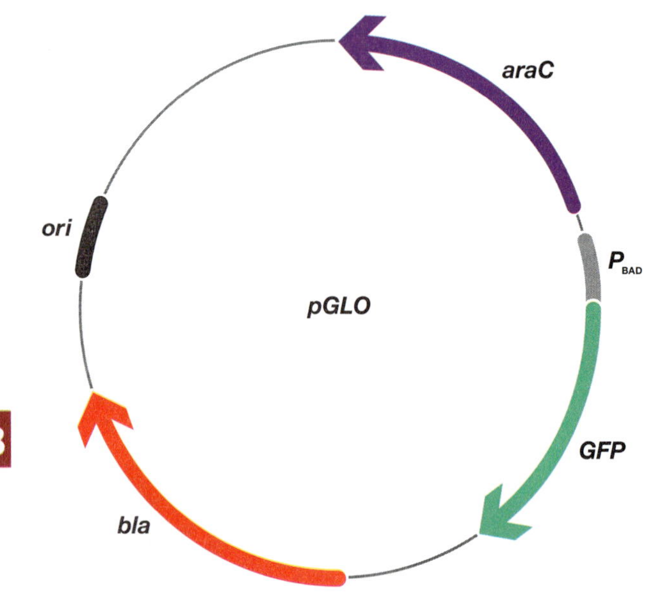


**8.7 pGLO™ Plasmid Map** ■ The pGLO™ plasmid has been genetically engineered to carry a replication origin (*ori*), a bacterial promoter from the arabinose operon ($P_{BAD}$), the GFP gene (*GFP*) from the jellyfish *Aequorea victoria*, an antibiotic resistance gene (*bla*) coding for the enzyme β-lactamase, and the gene for the DNA binding protein from the arabinose operon (*araC*). Arrows indicate the direction of transcription.

The bottom line is this: The pGLO™ plasmid used in this experiment has been genetically engineered to contain the arabinose promoter ($P_{BAD}$), the gene for *ara*C, an antibiotic-resistance gene (*bla*), the gene for green fluorescent protein (GFP), and a replication origin (*ori*).

As you interpret the results of your experiment, you will see how these components come together and provide you with information about what is happening at the molecular level in your *E. coli* culture. Their functions are summarized in Table 8-1.

## Application

Introduction of foreign DNA into a cell, identification of transformed cells, and regulation of an introduced gene's expression are skills used in genetic engineering.

## In This Exercise

You will first make *E. coli* cells competent. Then, you will transform the competent *E. coli* cells with a plasmid carrying the gene for green fluorescent protein. This technique of introducing a plasmid into cells is done

TABLE **8-1** **Cast of Characters and a Legend of Abbreviations**

| Name | Symbol | Function in This Experiment |
|---|---|---|
| Green fluorescent protein | GFP | It serves as an indicator of successful transformation and gene transcription (expression) in this experiment. |
| Plasmid | pGLO™ | It is the vector used to introduce the GFP gene into recipient *E. coli* cells. |
| Arabinose promoter | $P_{BAD}$ | Any promoter is the attachment site for RNA polymerase during transcription. *E. coli* RNA polymerase requires a prokaryotic promoter in order to transcribe genes, so the eukaryotic GFP gene must be spliced in next to one on the pGLO™ plasmid. (In *E. coli*, $P_{BAD}$ is the attachment site for RNA polymerase during transcription of the *ara*B, *ara*A, and *ara*D genes.) |
| DNA binding protein | *ara*C | It binds to and regulates the arabinose promoter. In the presence of arabinose, *ara*C has a shape that allows RNA polymerase to bind to the promoter. Without arabinose, *ara*C's shape prevents RNA polymerase from binding to the promoter. |
| Antibiotic resistance gene | *bla* | It is the gene for β-lactamase production, an enzyme that hydrolyzes antibiotics with a β-lactam structure, including ampicillin. Its gene product provides a means of differentiating cells that were transformed and those that were not. |
| Replication origin | *ori* | DNA replication begins at a replication origin. The pGLO™ plasmid is capable of replicating inside a transformed cell because it has this. As a result of replication, copies of the plasmid can be distributed to the descendants of the original, transformed *E. coli* cell(s). |

routinely in genetic engineering protocols. You also will use arabinose as a genetic switch to regulate expression of the GFP gene by *E. coli*.

## ▼ Materials

### Per Student

- ☐ Lab coat
- ☐ Disposable gloves
- ☐ Chemical eye protection

### Per Student Group

One Bio-Rad pGLO™ Transformation kit contains enough material for eight student workstations. Each workstation requires the following:

- ☐ One Luria-Bertani (LB) agar plate with *E. coli*[2] colonies
- ☐ One sterile LB plate
- ☐ Two sterile LB + ampicillin plates (LB/amp)
- ☐ One sterile LB + ampicillin + arabinose plate (LB/amp/ara)
- ☐ One microtube of $CaCl_2$ transformation solution
- ☐ Two sterile microtubes
- ☐ One microtube of LB broth
- ☐ Seven disposable inoculating loops
- ☐ Five disposable calibrated transfer pipettes
- ☐ One foam microtube holder/float
- ☐ One container of crushed ice (big enough to accommodate the foam microtube holder)
- ☐ One marking pen

### Per Class

In addition, a class supply of the following is required:

- ☐ Hydrated pGLO™ plasmid
- ☐ 42°C water bath and thermometer
- ☐ Long-wave UV lamp

## PROCEDURE

### Lab One

Refer to the procedural diagram in Figure 8.8 as you read and perform the following procedure.

1. Wear a lab coat, gloves, and chemical eye protection when performing this procedure.
2. Obtain two closed microtubes. Label them with your group's name, then label one "+DNA" and the other "−DNA." Put both tubes in the microtube holder/float.
3. Using a sterile calibrated transfer pipette, dispense 250 μL of $CaCl_2$ transformation solution into each tube. Close the caps. The pipette calibration marks are shown in Figures 8.8 and 8.9.
4. Return the two tubes to the microtube holder/float and place them in the ice bath.

[2] *Note to instructor:* The Bio-Rad kit comes with a particular *E. coli* strain that is ampicillin susceptible. We strongly urge use of their *E. coli* and not one out of your culture stock.

**5** With a sterile loop, transfer one entire *E. coli* colony into the +DNA tube. Agitate the loop until all the growth is off of it and the cells are dispersed uniformly in the transformation solution. Close the lid and properly dispose of the loop.

**6** Repeat step 5 with a sterile loop and the −DNA tube.

**7** Hold the UV light next to the vial of pGLO™ plasmid solution. Record your observation on the data sheet, page 513.

**8** Using a sterile loop, remove a loopful of pGLO™ plasmid DNA solution. Be sure there is a film across the loop. Then transfer the loopful of plasmid solution to the +DNA tube and mix.

**9** Leave the tubes on ice. Make sure they are far enough down in the microtube holder/float that they make good contact with the ice. Leave them on ice for 10 minutes.

**10** As the tubes are cooling on ice for 10 minutes, label the four LB agar plates.

- Label one LB/amp plate: +DNA
- Label the LB/amp/ara plate: +DNA
- Label the other LB/amp plate: −DNA
- Label the LB plate: −DNA

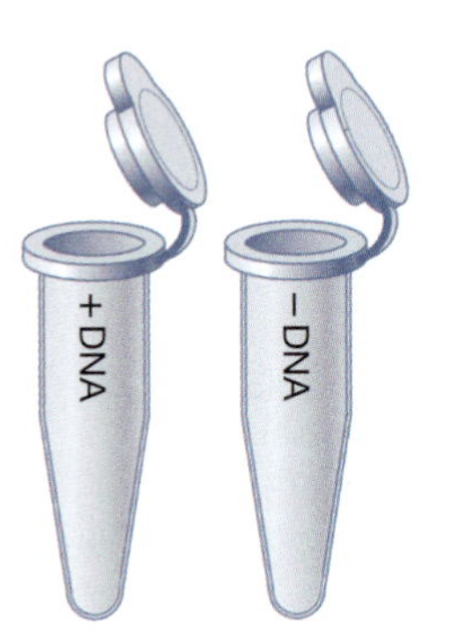


1. Label one microtube "+DNA" and the other "−DNA."

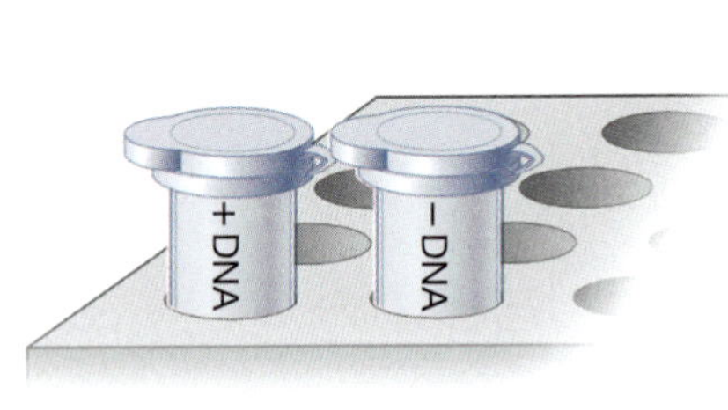


2. Place the tubes in the foam microtube rack.

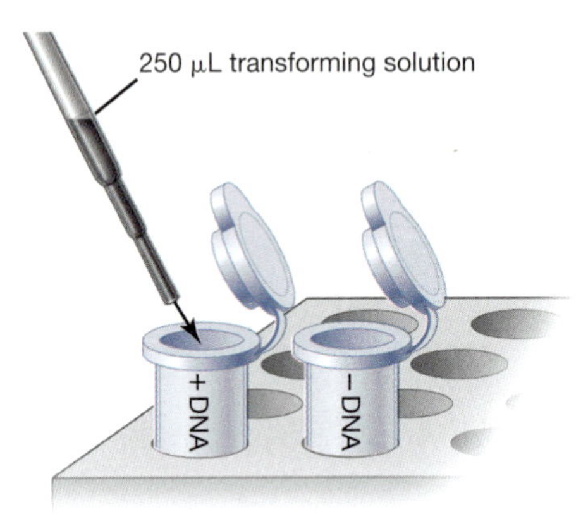


3. Use a sterile calibrated transfer pipette to dispense 250 μL transformation solution into each tube. Close the caps.

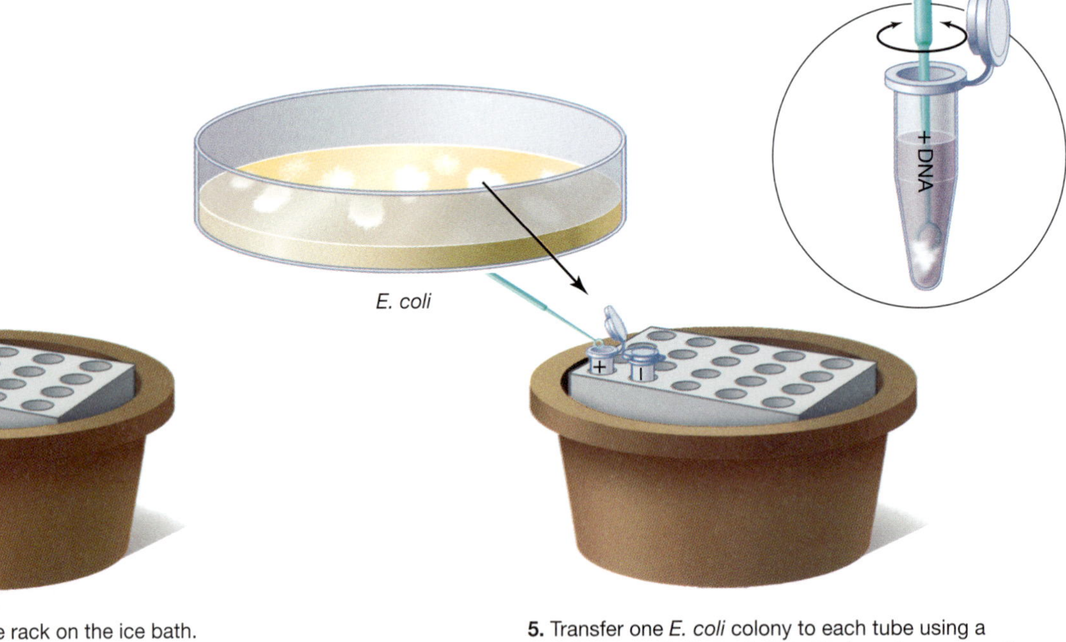


4. Place the microtube rack on the ice bath. Push the tubes down into the ice.

5. Transfer one *E. coli* colony to each tube using a different sterile loop. Agitate the loop to remove all growth (see detail). Close the caps and properly dispose of the loops.

**8.8 Procedural Diagram** ■ Wear a lab coat, gloves, and chemical eye protection when performing this procedure. Be sure to dispose of all pipettes and loops properly. *(Continues on next page)*

**11** After 10 minutes on ice, transfer the microtube holder/float (with both tubes in it) to the 42°C water bath for exactly 50 seconds. This transfer from ice to warm water must be done rapidly. Also, make sure the tubes make good contact with the water.

**12** After 50 seconds, quickly place both tubes back in the ice for 2 minutes. This process of heat shock makes the cell membranes more permeable to DNA. (Timing is critical. According to Bio-Rad Laboratories, 50 seconds is optimal. No heat shock results in a 90% *reduction* in transformants, whereas a 90-second heat shock yields about half the transformants.)

**13** After 2 minutes, remove the microtube holder/float from the ice and place it on the table.

**14** Using a sterile calibrated transfer pipette, add 250 μL of LB broth to the +DNA tube. Repeat with another sterile pipette and the −DNA tube. Properly dispose of both pipettes according to your lab's procedures.

**15** Incubate both tubes for 10 minutes at room temperature. Then mix the tubes by tapping them with your fingers.

**16** Using a different sterile calibrated transfer pipette for each, inoculate the LB/amp/+DNA plate and the LB/amp/ara/+DNA plate with 100 μL from the +DNA tube.

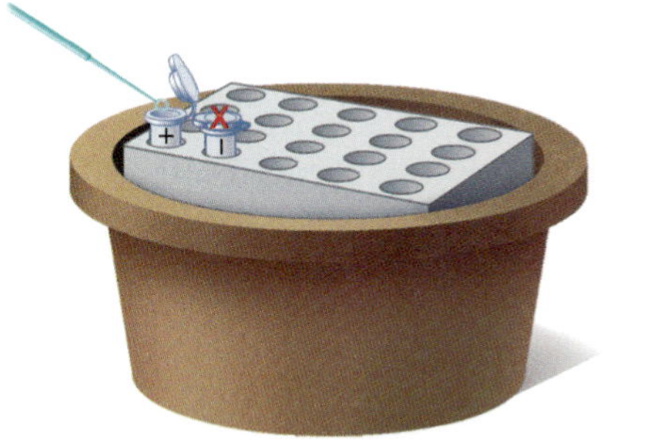

**6.** Transfer one loopful of pGLO™ plasmid DNA to the "+DNA" tube only. Continue icing both tubes for 10 minutes.

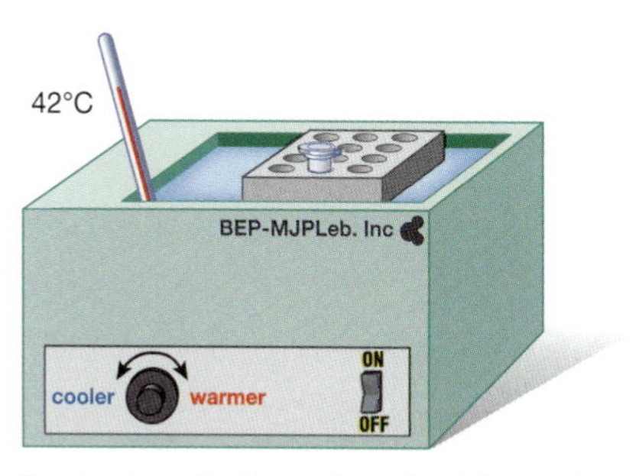


**7.** Quickly transfer the entire microtube rack to the 42°C water bath for exactly 50 seconds. Make sure the tubes contact the water.

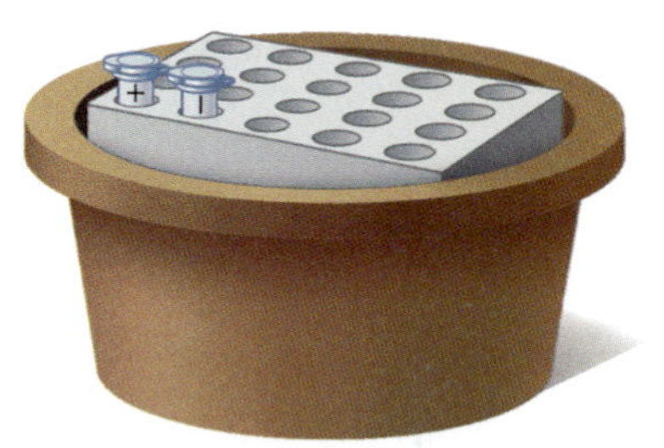

**8.** Quickly place the microtube rack back on ice for two minutes.

**9.** Place the microtube rack on the table.

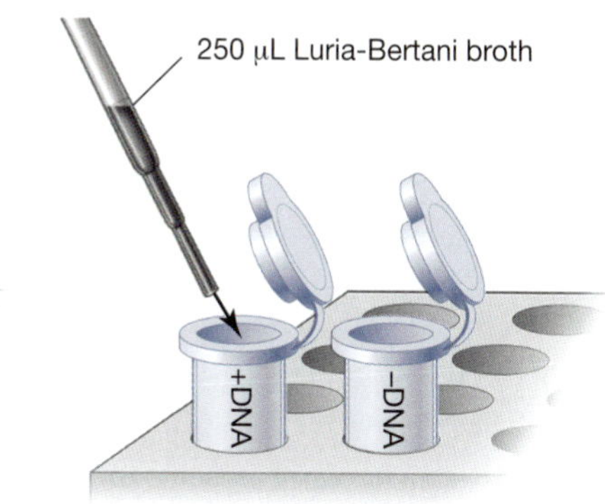


**10.** Add 250 μL LB broth to each tube with a different sterile transfer pipette.

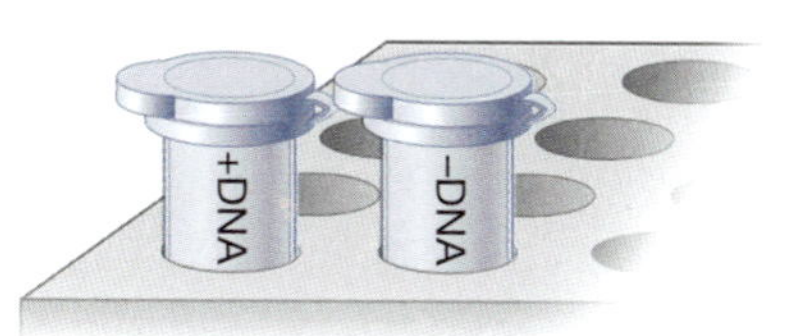


**11.** Incubate the tubes for 10 minutes at room temperature.

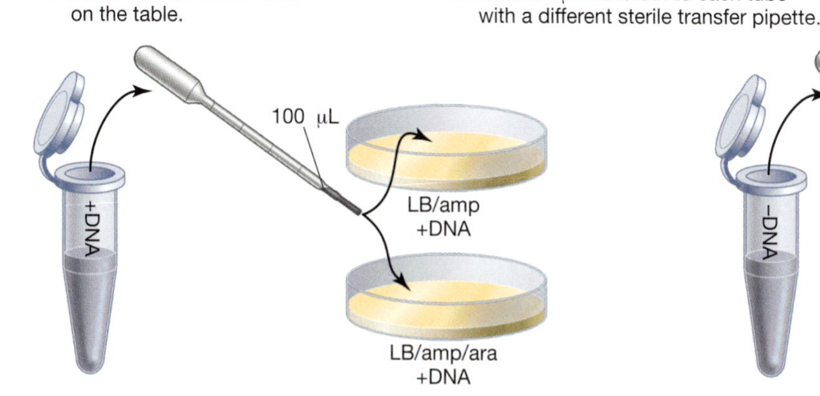


**12.** Using a different pipette for each, transfer 100 μL of +DNA to an LB/amp and an LB/amp/ara plate. Using a different sterile loop for each, spread the inoculum for confluent growth. Use the "face" of the loop for this, not its edge. Incubate for 24 hours at 37°C.

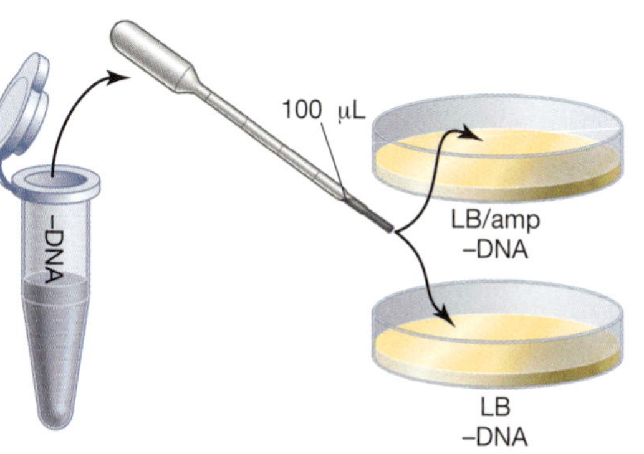


**13.** Using a different pipette for each, transfer 100 μL of –DNA to an LB/amp and an LB plate. Using a different sterile loop for each, spread the inoculum for confluent growth. Use the "face" of the loop for this, not its edge. Incubate for 24 hours at 37°C.

**8.8 Procedural Diagram** *(Continued)*

17 Using a different sterile loop for each, spread the inoculum over the surface of both plates to get confluent growth. Use the "face" of the loop for this, not its edge. Properly dispose of the loops.

18 Using a different sterile calibrated transfer pipette for each, inoculate the LB/amp/−DNA plate and the LB/−DNA plate with 100 µL from the −DNA tube.

19 As before, spread the inoculum over the surface of both plates using the "face" of the loops to get confluent growth. Properly dispose of the loops.

20 Allow a few minutes for the inocula to soak into the agar, then tape the plates together in a stack so they face the same direction. Then label them with your group name and the date, and incubate them for 24 hours at 37°C in an inverted position.

21 Save or dispose of the original *E. coli* plate as directed by your instructor.

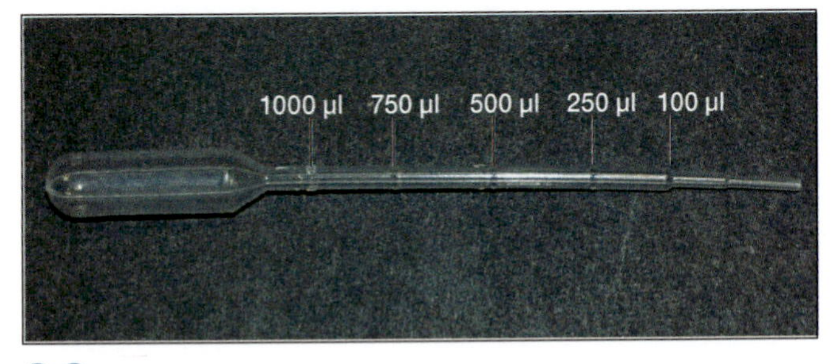


**8.9 Disposable Transfer Pipette Calibrations** ■ The transfer pipettes supplied with the pGLO™ Transformation kit are calibrated to deliver specific volumes. The volumes and their corresponding lines are shown here.

## Lab Two

1 Retrieve your plates. Observe them in ambient room light, then in the dark with UV illumination. Record your observations on the data sheet, page 513, and answer the questions.

2 When finished, properly dispose of all plates according to your lab's procedures.

## Reference

Bio-Rad Laboratories. Instruction pamphlet for the *Bacterial Transformation—The pGLO™ System* kit (Catalog Number 166-0003-EDU). Bio-Rad Laboratories, Hercules, CA.

8

Name

Date

Lab Section

I was present and performed this exercise (initials)

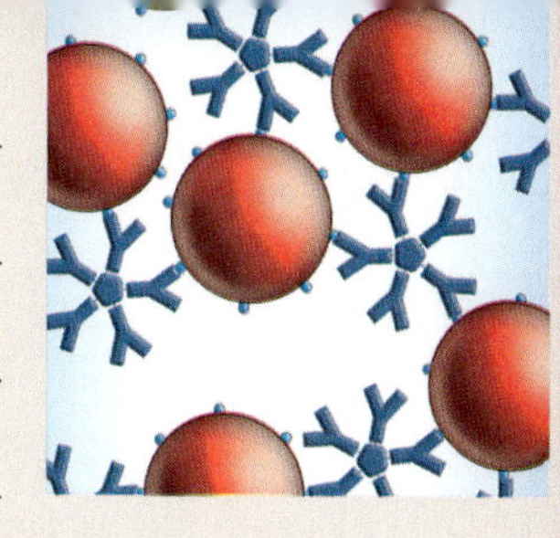

# DATA SHEET 8-2

## Bacterial Transformation: the pGLO™ System

### OBSERVATIONS AND INTERPRETATIONS

**1** Record the appearance (clear, violet, or fluorescent green) of the pGLO™ plasmid solution in natural light and in the dark with only UV illumination.

**2** Record your observations of the plates after incubation in the table below.

| Plate | Inoculum | Number of Colonies | Colony Appearance in Ambient Light | Colony Appearance with UV Light |
|---|---|---|---|---|
| LB | *E. coli* (−pGLO™ DNA) | | | |
| LB/amp | *E. coli* (−pGLO™ DNA) | | | |
| LB/amp | *E. coli* (+pGLO™ DNA) | | | |
| LB/amp/ara | *E. coli* (+pGLO™ DNA) | | | |

## QUESTIONS

**1** *Fill in the genotype of* E. coli *prior to transformation and after transformation for the GFP and the bla genes.*

| *E. coli* | GFP Gene (GFP$^+$ or GFP$^-$) | *Bla* (*bla*$^+$ or *bla*$^-$) |
|---|---|---|
| Before transformation | | |
| After transformation | | |

**2** *What was the purpose of examining the original pGLO™ solution with and without UV illumination?*

**3** *What was the purpose of transferring the +DNA and −DNA tubes from ice to hot water to ice again?*

**4** *Why were microtubes incubated for 10 minutes in LB broth rather than transferring their contents directly to the plates?*

**5** *What information is provided by the LB/–DNA plate?*

**6** *Obviously, transformation could occur only if the pGLO™ plasmid was introduced into the solution. Which plates exhibit transformation?*

**7** *What information is provided by the LB/amp/–DNA plate?*

**8** *Why does the LB/amp/ara/+DNA plate fluoresce when the LB/amp/+DNA plate does not?*

**9** *Use the following information to calculate transformation efficiency.*

**a.** *You put 10 μL (one loopful—the loop is calibrated for this volume) of a 0.03 μg/μL pGLO™ solution into the +DNA tube. Calculate the μg of DNA you put into the tube.*

**b.** *The +DNA tube contained 510 μL of solution (10 μL of plasmid, 250 μL of transforming solution, and 250 μL of LB broth) prior to plating, but you did not use all of it. Calculate the fraction of the +DNA solution you used on each plate. Round off your answer to one decimal place.*

**c.** *Use your answers to questions 9a and 9b to calculate the micrograms of DNA you plated.*

**d.** *Using the number of colonies on the LB/amp/ara +DNA plate, calculate the transformation efficiency. (**Hint:** The units are transformants/μg of pGLO™ DNA.)*

**e.** *According to the manual supplied by Bio-Rad Laboratories, this protocol should yield a transformation efficiency between $8.0 \times 10^2$ and $7.0 \times 10^3$ transformed cells per microgram of pGLO™ DNA. How does your transformation efficiency compare? Account for any discrepancy.*

# Ultraviolet Radiation Damage and Repair

EXERCISE 8-3

## Theory

Ultraviolet radiation is part of the electromagnetic spectrum, but with shorter, higher energy wavelengths than visible light. Prolonged exposure can be lethal to cells because when DNA absorbs UV radiation at 254 nm, the energy is used to form new covalent bonds between adjacent pyrimidines: cytosine-cytosine, cytosine-thymine, or thymine-thymine. Collectively, these are known as pyrimidine dimers, with **thymine dimers** being the most common. These dimers distort the DNA molecule and interfere with DNA replication and transcription (Fig. 8.10).

Many bacteria have mechanisms to repair such DNA damage. *Escherichia coli* performs **light repair**, or **photoreactivation**, in which the repair enzyme, **DNA photolyase**, is activated by visible light (300–500 nm) and simply monomerizes the dimer by reversing the original reaction.

A second *E. coli* repair mechanism, **excision repair**, or **dark repair**, involves a number of enzymes (Fig. 8.11). The thymine dimer distorts the sugar-phosphate backbone of the strand. This is detected by **UvrABC endonuclease** (also known as ABC exinuclease) that breaks two bonds—one is eight nucleotides in the 5' direction from the dimer, and the other is four nucleotides in the 3' direction. A **helicase** (UvrD, also known as helicase II) removes the 13-nucleotide fragment (including the dimer), leaving single-stranded DNA. **DNA polymerase I** inserts the appropriate complementary nucleotides in a 5' to 3' direction to make the molecule double-stranded again.

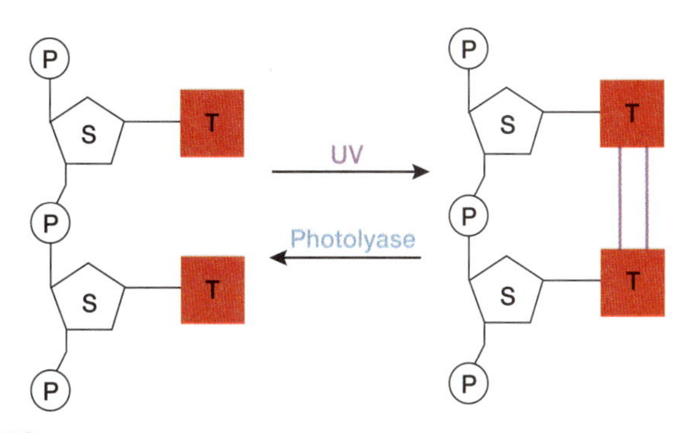


**8.10 A Thymine Dimer in One DNA Strand and its Photoreactivation Repair** ■ Thymine (or any pyrimidine) dimers form when DNA absorbs UV radiation with wavelengths in the neighborhood of 260 nm. The energy is used to form two new covalent bonds between adjacent thymines, resulting in distortion of the DNA strand. In photoreactivation repair, DNA photolyase can break this bond to return the DNA strand to its normal shape and function. If it doesn't and excision repair fails, the distortion interferes with DNA replication and transcription.

Finally, **DNA ligase** closes the gap between the last nucleotide of the new segment and the first nucleotide of the old DNA, and the repair is complete. Both mechanisms are capable of repairing a small amount of damage, but long and/or intense exposures to UV produce more damage than the cell can repair, making UV radiation lethal.

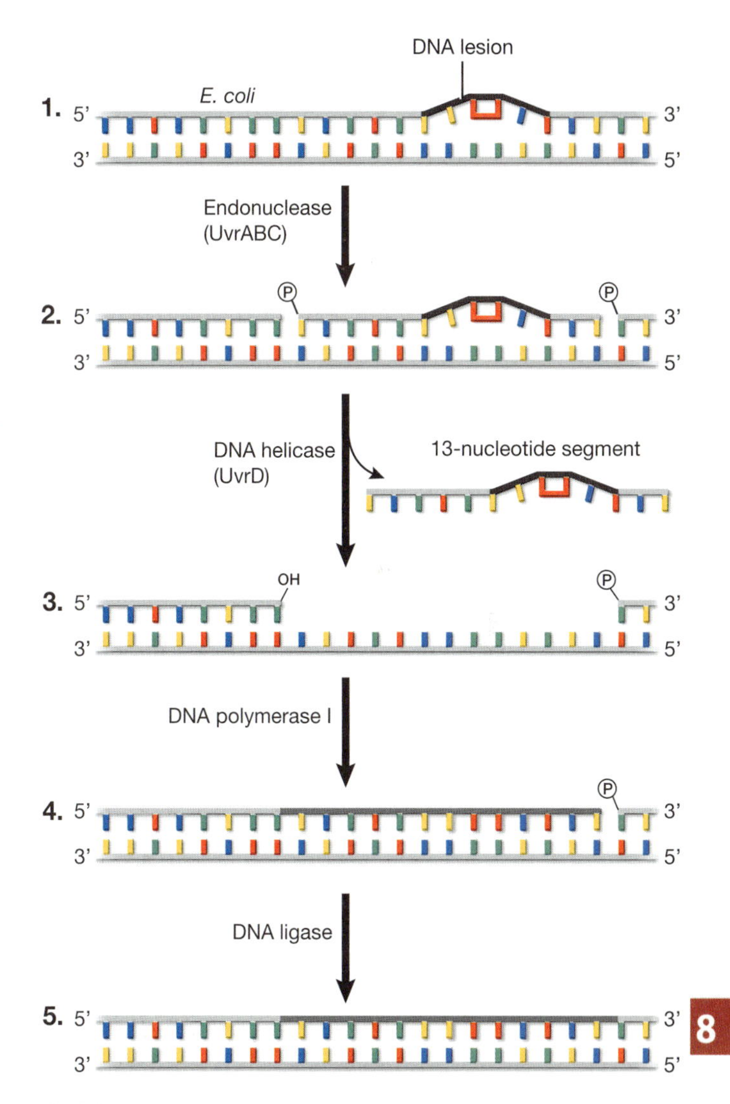


**8.11 Excision, or Dark Repair, in *E. coli*** ■ (**1**) Shown is a segment of DNA with a thymine dimer, which distorts the normally uniform double-helical structure of DNA. (**2**) Four enzymes are used in this repair process. An endonuclease (UvrABC) detects the distortion and breaks two covalent bonds in the sugar-phosphate backbone of the damaged strand to produce a segment 13 nucleotides in length. (**3**) A helicase (UvrD) removes the damaged segment. (**4**) DNA polymerase I synthesizes a new strand to replace the damaged segment. (**5**) DNA ligase forms a covalent bond between the new and the original strands. (After Nelson and Cox, 2013)

## Application

Because ultraviolet radiation has a lethal effect on bacterial cells, it can be used in decontamination. Its use is limited, however, because it penetrates materials such as glass and plastic, poorly. In addition, bacterial cells have natural mechanisms to repair UV damage.

## In This Exercise

The lethal effects of ultraviolet radiation, its ability to penetrate various objects, and the cells' ability to repair UV damage will all be demonstrated. Do not be misled by the apparent simplicity of the experimental design—there is a lot going on!

### Materials

#### Per Student

- ☐ Lab coat
- ☐ Disposable gloves
- ☐ Chemical eye protection

#### Per Student Group

- ☐ Seven nutrient agar plates
- ☐ (Optional) three TGYM plates
- ☐ Disinfectant
- ☐ Poster-board masks with 1" to 2" cutouts (Fig. 8.12)
- ☐ Short wavelength ultraviolet lamp (UV-C) with appropriate shielding and support
- ☐ Sterile cotton applicators
- ☐ 24-hour tryptic soy broth culture of *Serratia marcescens*
- ☐ (Optional) 24-hour TGYM culture of *Deinococcus radiodurans* or *D. radiophilus*

## PROCEDURE

#### Lab One

Refer to the procedural diagram in Figure 8.13 as you read and follow this procedure.[1]

1. Wear a lab coat, gloves, and chemical eye protection when performing this procedure.
2. Using a sterile cotton applicator, streak an agar plate to form a bacterial lawn over the entire surface by using a tight pattern of streaks. Rotate the plate one-third of a turn and repeat, then rotate it another one-third of a turn and streak one last time. Repeat this process for all remaining plates.

[1] Thanks to Roberta Pettriess of Wichita State University for her helpful suggestions to improve this exercise.

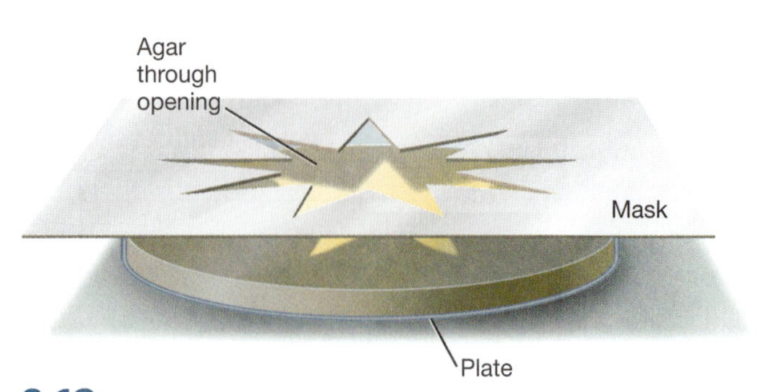


**8.12 Poster-board Mask** ■ This is an example of a mask placed over a Petri dish. The cutout may be any shape but should leave the outer 25% of the plate masked.

3. Number the plates 1, 2, 3, 4, 5, 6, and 7 and label them with your group name.
4. Remove the lid from plate 1 and set it on a disinfectant-soaked towel. Place the plate under the UV light and cover it with a mask.
5. Turn the UV lamp on, but do not look at it. After 30 seconds, turn off the UV lamp, remove the mask, and replace its lid. If space permits under the lamp, you may combine this step with step 6. ***Note:*** If you expose more than one plate at a time in this or subsequent steps, make sure all the plates are getting equal exposure to the UV beam.
6. Repeat the process for plate 2. If space permits, you may combine this step with plate 1 in step 5.
7. Repeat the process for plate 3, but leave the UV lamp on for 3 minutes. If space permits, you may combine this step with plates 4 and 5 in steps 8 and 9.
8. Irradiate plate 4 for 3 minutes, but leave the lid on and cover with the mask.
9. Repeat step 8 for plate 5.
10. Do not irradiate plates 6 and 7.
11. Incubate plates 1, 3, 4, and 6 for 24 to 48 hours at room temperature in an inverted position where they can receive natural light (e.g., a windowsill). Do not stack them. (Usually 24 hours is plenty of time. Longer incubation can lead to pigment loss by *S. marcescens*.)
12. Wrap plates 2, 5, and 7 in aluminum foil, invert them, and place with the others for 24 to 48 hours. Do not stack them unless space is limited.
13. (**Optional**) Inoculate three TGYM plates with *Deinococcus* spp. as in step 2. Then, expose them to 2, 5, and 8 minutes of UV irradiation with a mask but no lid. Label the plates 8, 9, and 10, and with their appropriate exposure times. Also label them with your group name. Incubate them in sunlight for 24 to 48 hours with the other plates. Do not stack them.

### Lab Two

1 Remove the plates and examine the growth patterns.

2 Record your results on the data sheet, page 519.

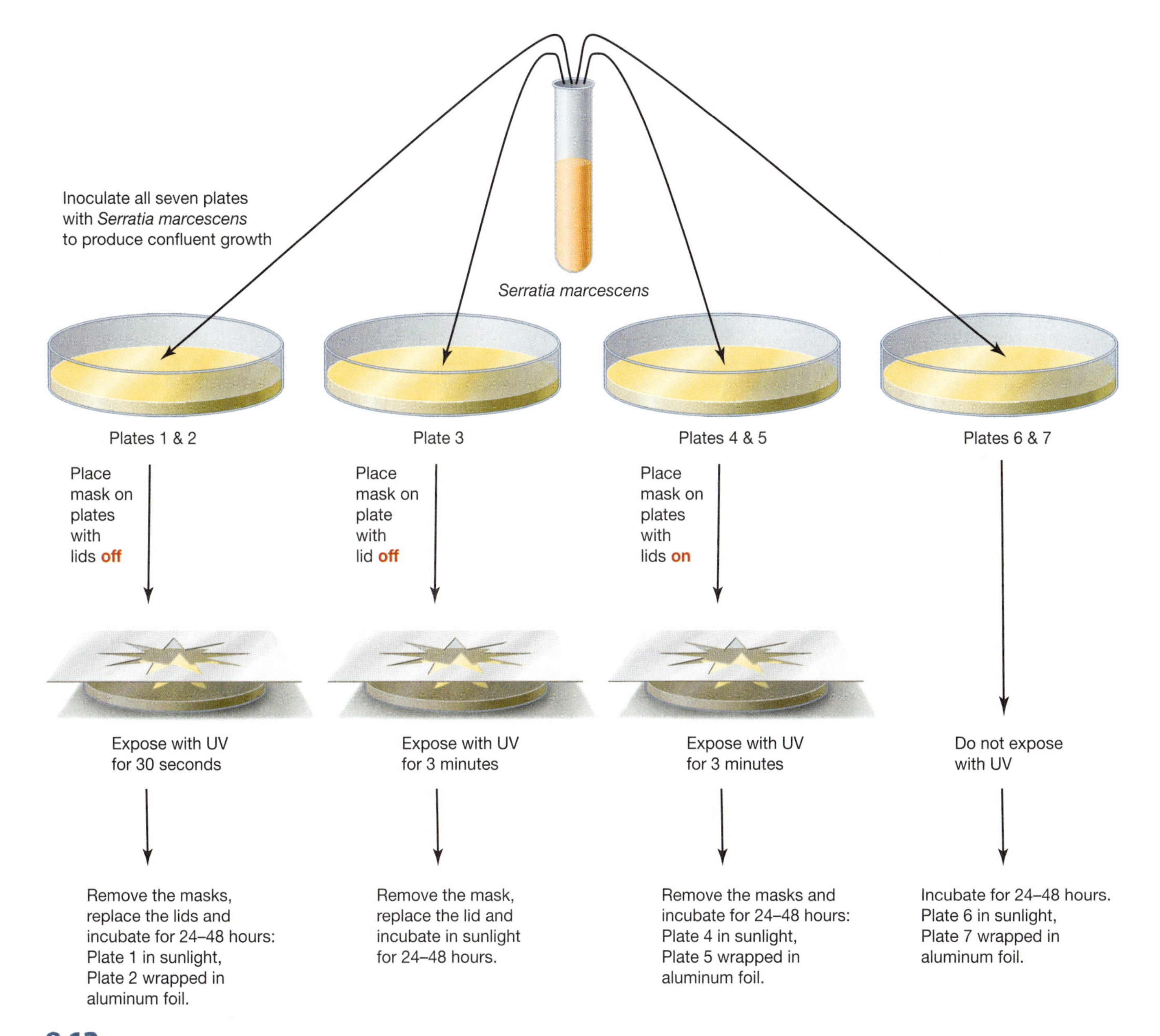


**8.13 Procedural Diagram** ■ Inoculate and expose the plates as directed. Be sure to shield the UV light source adequately. Do not look at the lamp.

## References


Alcorn, Joseph L. and Claud S. Rupert. *Journal of Bacteriology* 172, no. 12 (1990): 6885–6891.

Moat, Albert G., John W. Foster, and Michael P. Spector. Chap. 3 in *Microbial Physiology,* 4th ed. New York: Wiley-Liss, 2002.

Nelson, David L. and Michael M. Cox. Pages 1031–1034 in *Lehninger's Principles of Biochemistry,* 6th ed. New York: W. H. Freeman and Co., 2013.

White, David, James Drummond, and Clay Fuqua. Pages 415–417 in *The Physiology and Biochemistry of Prokaryotes*, 4th ed. New York, NY: Oxford University Press, 2012.

## Introduction to Antigens and Antibodies

The body's defenses against infectious agents can be divided into nonspecific defenses (also called innate immunity) and specific acquired immunity. The former include external structures, such as epithelial barriers; chemical secretions, such as stomach acid and mucus; and mechanical properties, such as peristalsis and ciliary action.

Nonspecific internal defenses include phagocytosis, the complement system, fever, inflammation, and interferon. All of these are categorized as nonspecific because their activation and performance is generic; for instance, phagocytic cells are not individually dedicated to a particular infectious agent, nor do interferon molecules each target a different virus.

On the other hand, the specific acquired immune system involves the production of **antibodies** or **immune cells** that each react with a single **antigen**. Because the following lab exercises utilize antibodies and not immune cells our focus will be on antibodies.

Using a simple definition, antigens are high molecular weight molecules with a definite three-dimensional shape that the body recognizes as being "foreign." More importantly, antigens stimulate the production of specific antibodies (or immune cells) and will react specifically with them once produced. In general, proteins make the best antigens, followed by polysaccharides. Lipids and nucleic acids do not make good antigens.

Antibodies are glycoproteins (**immunoglobins–Ig**) and each is composed of two identical long polypeptides called heavy chains and two identical shorter polypeptides called **light chains** (Fig. 8.14A). Functionally, each antibody molecule behaves as if it has the shape of the letter "Y." There are two antigen binding sites, one at the end of each arm of the "Y," formed by one heavy and one light chain in combination. This part of the antibody is called the **$F_{ab}$ portion**. The stem of the "Y" (the **$F_c$** portion) is formed by the remainder of the two heavy chains and performs **effector functions**, such as binding to phagocytic cells or activating the complement system.

Obviously, there must be great antibody diversity in order to protect against all the antigens the world presents. On a gross level, antibody diversity begins with the type of heavy chain present. The five basic immunoglobin classes (with heavy chains in parentheses) are: IgG ($\gamma$), IgM ($\mu$), IgA ($\alpha$), IgD ($\delta$), and IgE ($\epsilon$). Additional diversity occurs because light chains come in two forms: kappa ($\kappa$) and lambda ($\lambda$), which are able to combine with any of the heavy chains. Further, antibodies may be joined into more complex arrangements, such as in IgM, where five Y-shaped antibodies form a pentamer, and in IgA that exists as a dimer in mucus and other secretions (Fig. 8.14B).

The majority of antibody diversity occurs as a result of each lymphocyte's ability to produce its own version of its heavy and light chains. Each heavy and light chain has an amino acid sequence along most of its length that is more or less characteristic of its type. This is called the **constant region**. However, each heavy and light chain also contains a **variable region**, which differs in amino acid sequence from those produced by all other lymphocytes and from each other. When assembled, the variable regions combine to form a unique three-dimensional pocket that will bind an antigen with a complementary shape.

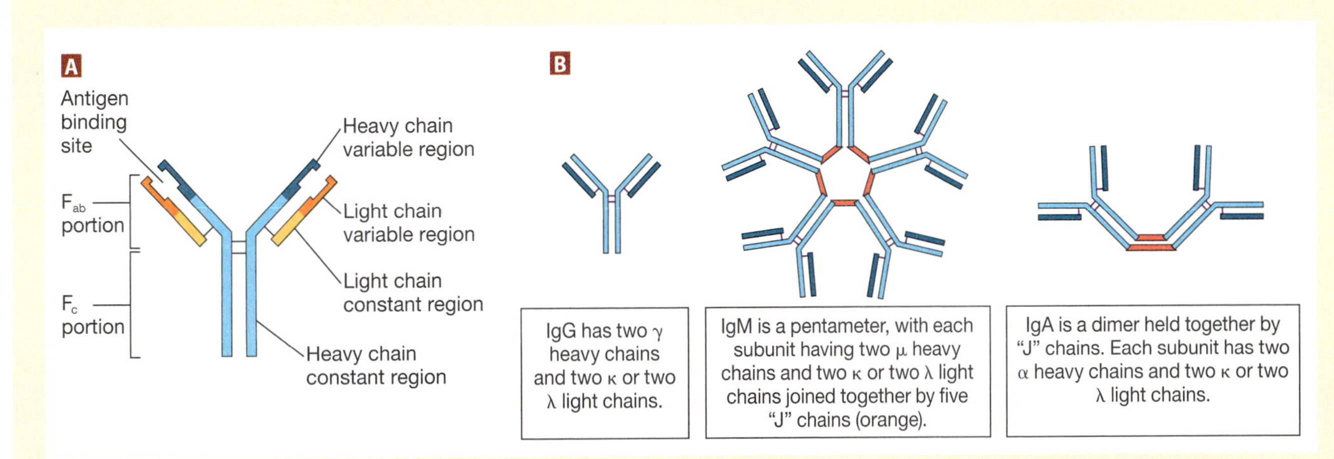


**8.14 Antibody Structure** ■ **(A)** The fundamental antibody structure is two identical heavy chains and two identical light chains held together by covalent disulfide bonds. Each chain has a constant region and a variable region composed of "standard" and unique amino acid sequences. The variable regions are unique to each B lymphocyte and join together to form the antigen binding site, thus accounting for the ability of each B-cell to produce antibodies that react with only a single antigen (epitope). **(B)** There are five antibody classes identified by the heavy chain in each. IgG, IgD, and IgE have the basic "Y" structure, but IgM and IgA combine to form more complex arrangements.

Molecules have a three-dimensional shape and a surface topography. Because antigens have a relatively unchanging three-dimensional shape with a varied surface topography, different parts of an antigen can stimulate production of different antibodies, each reacting with a different surface feature. These surface features are called **epitopes**, and their presence makes an immune response to any antigen molecule more complex than just one antibody for one antigen (Fig. 8.15). ■

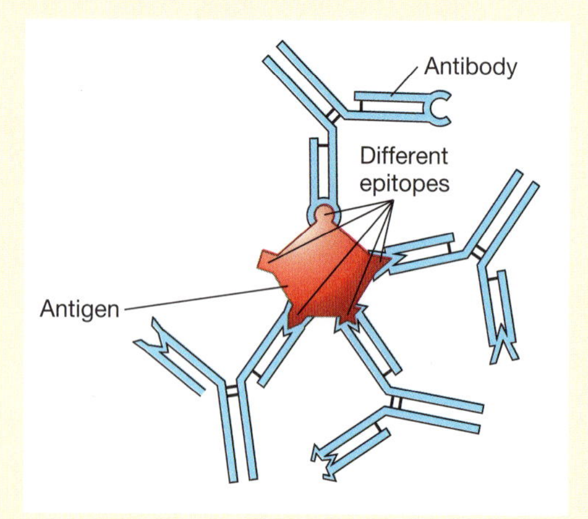


**8.15 Epitopes** ■ Shown is an artist's representation of an antigen molecule with multiple epitopes. Each epitope has the ability to stimulate lymphocytes to produce antibodies that will bind them because of the complementary shapes. Often biologists refer to the whole molecule (and sometimes even the whole cell or virus particle) as an "antigen," but in reality the immune system "sees" only epitopes. In this illustration, a single molecule has resulted in the production of four antibodies targeting it. The "square" surface feature on the left didn't stimulate antibody production and is therefore not an epitope. However, in another individual that same "square" might act as an epitope. Every individual makes their own arsenal of antibodies.

## Serological Reactions

Antigen–antibody reactions are highly specific and occur in vitro as well as in vivo. Serology is the discipline that exploits this specificity as an in vitro diagnostic tool. All serological tests can be designed to identify either antigen or antibody in a sample. The choice for any particular test largely depends on the circumstances of the disease process; that is, whether it is easier to identify antigens or antibodies in a patient sample. All serological tests also fundamentally rely on the reaction between antigen and **homologous antibody**, but they differ in the way a positive reaction is indicated. For instance, one simple serological reaction—agglutination—is used in the following three exercises because it results in the formation of complexes that can be viewed with the naked eye and without sophisticated equipment.

The slide agglutination test (Exercise 8-4) can be an important diagnostic (and highly specific) tool for the identification of organisms. It is especially useful for serotyping large genera such as *Salmonella*. Hemagglutination, a type of agglutination reaction, detects specific antigens on red blood cells and is the standard test for determining blood type (Exercise 8-5). The most sophisticated serological test covered in this section is in Exercise 8-6. It is an ELISA designed to detect antibody in a sample. Its indicator reaction is the production of a colored product from a colorless substrate by an enzyme attached to a secondary antibody that is present in the reaction mixture only if it has bound to the antibody being looked for in the patient sample. ■

# Slide Agglutination: The *Salmonella* O Antigen

EXERCISE **8-4**

## Theory

Particulate antigens (such as whole cells) may combine with homologous antibodies to form visible clumps called **agglutinates**. In vivo, these agglutinates immobilize the cells until they can be removed by phagocytic cells. In vitro, **agglutination** can serve as evidence of antigen–antibody reaction and is considered a positive result. Agglutination reactions are highly sensitive and may be used to detect either the presence of antigen or antibody in a sample.

There are many variations of agglutination tests (Fig. 8.16). **Direct agglutination** relies on the combination of antibodies and naturally particulate antigens. **Indirect (passive) agglutination** relies on artificially constructed systems in which agglutination will occur. These involve coating particles (such as RBCs or latex microspheres) with either antibody or antigen, depending on what is being looked for in the sample.

Addition of the homologous antigen or antibody then will result in clumping of the artificially constructed particles and turns what would have naturally been a less sensitive indicator into a more sensitive agglutination. Agglutination tests may be performed in test tubes, microtiter plates, or on microscope slides.

While agglutination results are frequently reported as simply "positive" or "negative," in some cases scoring is semiquantitative, in which degree of reaction is given on a scale of 1+ to 4+, with 2+ corresponding to an agglutinate barely visible to the naked eye; 1+ would require microscopic examination to see clumping. Still other agglutinating tests using diluted antibody samples provide a mechanism to estimate antibody titer in the patient sample.

The sample is diluted in twofold increments (e.g., ½, ¼, ⅛, etc.) and added to antigen samples of constant concentration. The most dilute sample that produces agglutination is used to calculate titer by taking its reciprocal. For example, if the highest dilution producing agglutination is 1/32, the titer would be reported as 32.

Clearly, agglutination tests are very versatile in the information they provide, but today you will perform a simple indirect agglutination slide test involving the genus *Salmonella*. *Salmonella* classification is complicated in large part because of conflicting nomenclatural systems (i.e., subspecies vs. serological varieties—also known as "serovars" or "serotypes") and a tradition of microbiologists equating serotypes with species. Currently, only two *Salmonella* species are recognized: *S. choleraesuis* with six subspecies and *S. bongori*. However, change is in the air and the name *S. choleraesuis* is in the process of being replaced (in order to bring more clarity to the confusion) with *S. enterica*. Further, *S. typhi* may be officially recognized as a third species. But we digress...

According to the CDC, there are over 2,500 *Salmonella* serotypes, of which fewer than 100 are responsible for the majority of human infection. Each serovar has a unique combination of three antigens: O, Vi, and H. The O antigen is the O polysaccharide

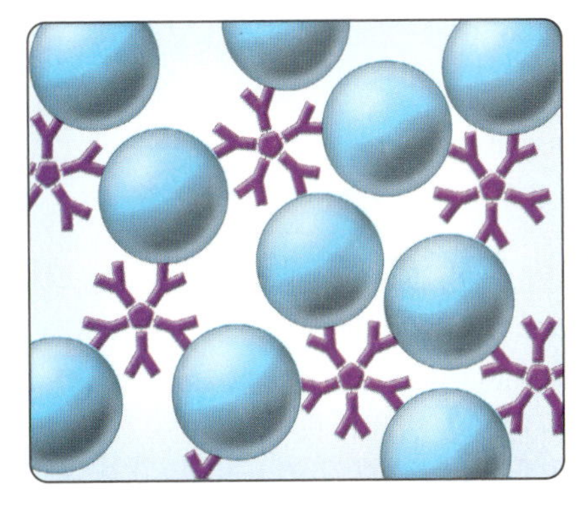

Direct agglutination occurs with naturally particulate antigens. Either antigen or antibody may be detected in a sample using this style of agglutination test.

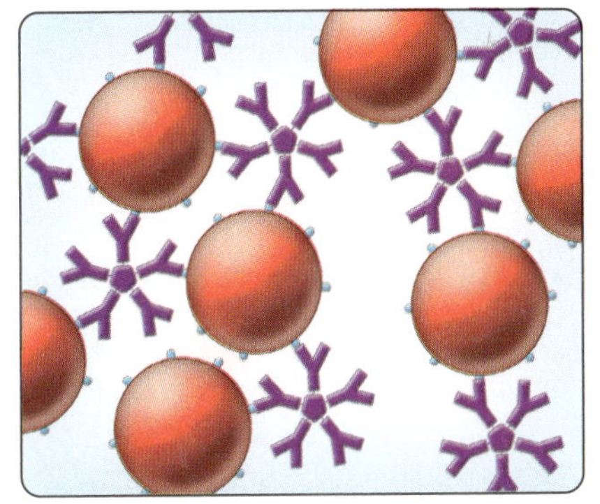

Detection of antibody in a sample can be done by indirect (passive) agglutination. A test solution is prepared by artificially attaching homologous antigen (blue) to a particle (red) such as red blood cells or latex beads and mixing with the sample suspected of containing the antibody (purple).

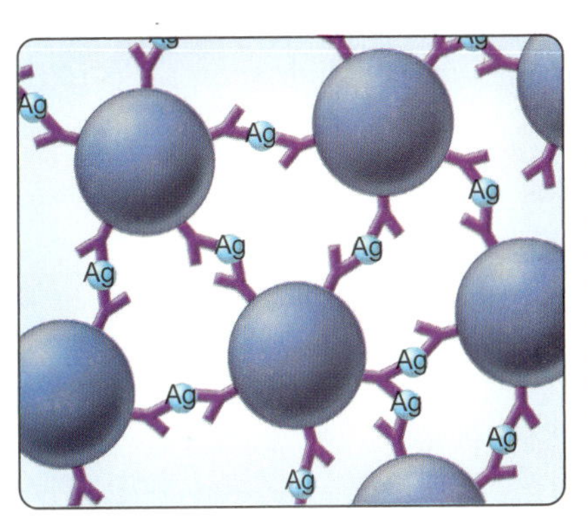


Reverse passive agglutination detects antigen (light blue) in a sample. Antibodies (purple) are artificially attached to particles (dark blue), which are then mixed with the sample suspected of containing the antigen.

**8.16 Direct and Indirect (Passive) Agglutination Tests** ■ Direct agglutinations involve naturally particulate antigens. Indirect (passive) agglutination relies on attaching either the antigen or antibody to a particle, such as a latex bead or red blood cell.

8

found in the outer membrane of Gram-negative bacteria (Fig. 3.49) and is the one you will be working with today. The Vi (standing for "virulence") antigen is found in the extracellular capsule and is mostly associated with two pathogenic serovars: Typhi (*S. typhi*) and Paratyphi C (*S. paratyphi* C). The H antigen is flagellin protein, which comprises the flagellar filament.

Philosophical taxonomic issues aside, *Salmonella* serovar identification has practical applications because it allows rapid preliminary identification of a patient isolate and can complement traditional biochemical/physiological identification methods. Clinically, a slide agglutination test determines the O antigen group (most pathogens belong to group A, B, $C_1$, $C_2$, D, or E) and the presence or absence of the Vi antigen. More specific antigenic determination is left to reference laboratories for surveillance purposes.

## Application

Agglutination reactions may be used to detect the presence of either antigen or antibody in a sample. Direct agglutination reactions are used to identify some pathogens (like *Salmonella* and *Streptococcus* sp.—Exercise 8-6), determine if a patient has been exposed to a certain pathogen based on the presence of antibodies (e.g., infectious mononucleosis), and are involved in blood typing (Exercise 8-5). Indirect agglutination is used in some pregnancy tests as well as in diagnosing disease either by detecting antibodies (e.g., syphilis) or antigens (e.g., *Crytococcus neoformans* and *Neisseria meningitidis*) in a patient sample.

## In This Exercise

You will perform an agglutination test to demonstrate the presence of *Salmonella* O antigen in a sample. Unlike a practicing clinical microbiologist, who would test a patient sample using the whole cell, you will be using a solution of just the O antigen and reacting it with O antiserum to demonstrate what agglutination looks like.

## Materials

### Per Student

- □ Lab coat
- □ Disposable gloves
- □ Chemical eye protection

### Per Student Group

- □ One clean microscope slide
- □ Two toothpicks
- □ Marking pen
- □ Dropper bottle of sterile normal saline
- □ *Salmonella* O Group A antigen (BD Difco Salmonella O QC antigen [Cat. No. 221301])
- □ *Salmonella* O antiserum, Poly A (BD Difco Salmonella O antiserum [Cat. No. 225341])

## PROCEDURE

1. Wear a lab coat, gloves, and chemical eye protection when performing this procedure.
2. Using a marking pen, draw two circles approximately the size of a dime on a microscope slide (Fig. 8.17). Label one "S" and the other "O."
3. Place a drop of *Salmonella* O antiserum in each circle.
4. Place a drop of *Salmonella* O antigen in the "O" circle and a drop of sterile saline in the "S" circle (Fig. 8.18). Be careful not to touch the dropper to the antiserum already on the slide.
5. Using a *different* toothpick for each circle, mix until each antigen is completely emulsified with the antiserum. Do not overmix. Discard the toothpicks in a sharps container.
6. Gently rock this slide for a few minutes and observe for agglutination. Record your observations on the data sheet, page 527. (When rocking, the fluid should not spread. If it does, you are rocking too hard.)
7. Properly dispose of the slides in a sharps container.

## Alternative Test Procedure

Your instructor will cover the labels of the *Salmonella* antigen and sterile saline bottles and label them "1" and "2." You will use this procedure to identify which one contains the *Salmonella* O antigen.

1. Wear a lab coat, gloves, and chemical eye protection when performing this procedure.
2. Using a marking pen, draw two circles approximately the size of a dime on a microscope slide (Fig. 8.17). Label them "1" and "2."
3. Place a drop of *Salmonella* O antiserum in each circle.
4. Place a drop from bottle 1 in circle "1" and a drop from bottle 2 in circle "2" (Fig. 8.18). Be careful not to touch the dropper to the antiserum solutions already on the slide.
5. Using a *different* toothpick for each circle, mix until each antigen is completely emulsified with the antiserum. Do not *over*-mix. Discard the toothpicks in a biohazard container.
6. Gently rock this slide for a few minutes, and then observe for agglutination. Record your observations on the data sheet, page 527.
7. Properly dispose of the slides in a sharps container.

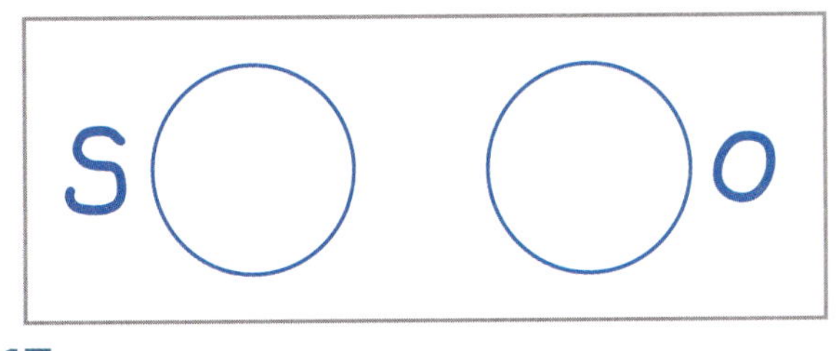


**8.17 Prepare the Slide** ■ With your marking pen, draw two dime-size circles on the slide. Label one circle "S" and the other "O." The circles are where you perform the agglutination tests.

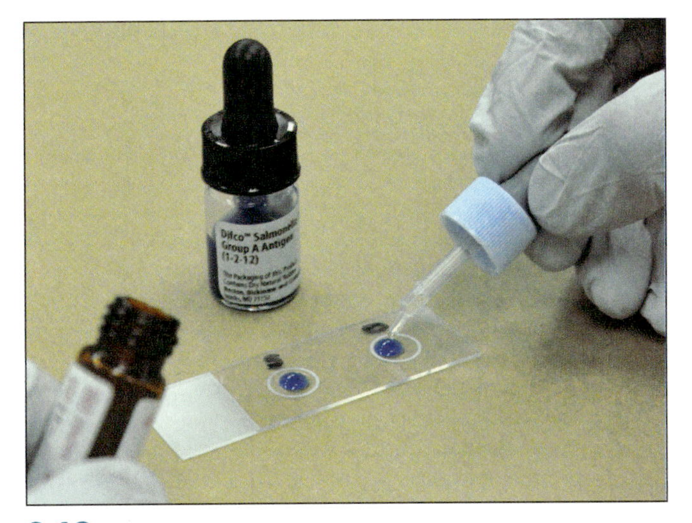

**8.18 Adding the Antigen** ■ A drop of *Salmonella* O antiserum has been added to the "S" and "O" circles. A drop of *Salmonella* O antigen is being added to the "O" circle. Next, a drop of saline will be added to the "S" circle. It is important not to touch the droppers to the antiserum on the slide when dispensing the antigen and saline.

## References

Brenner, Don J. and J.J. Farmer III. Page 606 in *Bergey's Manual of Systematic Bacteriology*, Vol. 2, *The Proteobacteria, Part B—The Gammaproteobacteria*. New York: Springer, 2005.

CDC. *National* Salmonella *Surveillance Overview*. Atlanta, GA: US Department of Health and Human Services, CDC, 2011. Available online. URL: http://www.cdc.gov/nationalsurveillance/PDFs/NationalSalmSurveillOverview_508.pdf.

CDC. Serotypes and the Importance of Serotyping *Salmonella*. Atlanta, GA: US Department of Health and Human Services, CDC, 2011. Available online. URL: http://www.cdc.gov/salmonella/reportspubs/salmonella-atlas/serotyping-importance.html.

Collins, C. H., Patricia M. Lyne, J. M. Grange. Page 118 in *Collins and Lyne's Microbiological Methods*, 7th ed. Oxford, Boston: Butterworth-Heinemann, 1995.

Khanna, Raj. Pages 363–366 in *Immunology*. New Delhi, India: Oxford University Press, 2011.

Lam, Joseph S. and Lucy M. Mutharia. Pages 118–119 in *Methods for General and Molecular Bacteriology*. Philipp Gerhardt, R. G. E. Murray, Willis A. Wood, and Noel R. Krieg, eds. Washington, DC: American Society for Microbiology, 1994.

Pier, Gerald B., Jeffrey B. Lyczak, and Lee M. Wetzler. Pages 220–224 in *Immunology, Infection, and Immunity*. Washington, DC: ASM Press, 2004.

Popoff, Michel Y. and Léon E. Le Minor. Genus XXXIII. "*Salmonella*" in *Bergey's Manual of Systematic Bacteriology*, Vol. 2, *The Proteobacteria, Part B—The Gammaproteobacteria*. New York: Springer, 2005.

Strockbine, Nancy A., Cheryl A. Bopp, Patricia J. Fields, James B. Kaper, and James P. Nataro. Pages 699–704 in *Manual of Clinical Microbiology*, Vol. 1, 11th ed. James H. Jorgensen, Michael A. Pfaller, Karen C. Carroll, Guido Funke, Marie Louise Landry, Sandra S. Richter, and David W. Warnock, eds. Washington, DC: ASM Press, American Society for Microbiology, 2015.

Theel, Elitza S., A. Betts Carpenter, and Matthew J. Binnicker. Pages 96–97 in *Manual of Clinical Microbiology*, Vol. 1, 11th ed. James H. Jorgensen, Michael A. Pfaller, Karen C. Carroll, Guido Funke, Marie Louise Landry, Sandra S. Richter, and David W. Warnock, eds. Washington, DC: ASM Press, American Society for Microbiology, 2015.

Tille, Patricia M. Pages 135–136 and 322 in *Bailey & Scott's Diagnostic Microbiology*, 13th ed. St. Louis, MO: Mosby, 2014.

Name ______________________________

Date ______________________________

Lab Section ______________________________

I was present and performed this exercise (initials) ______________________________

DATA SHEET

8-4

# Slide Agglutination: The *Salmonella* O Antigen

## OBSERVATIONS AND INTERPRETATIONS

**1** Sketch and label your results in the diagram below.

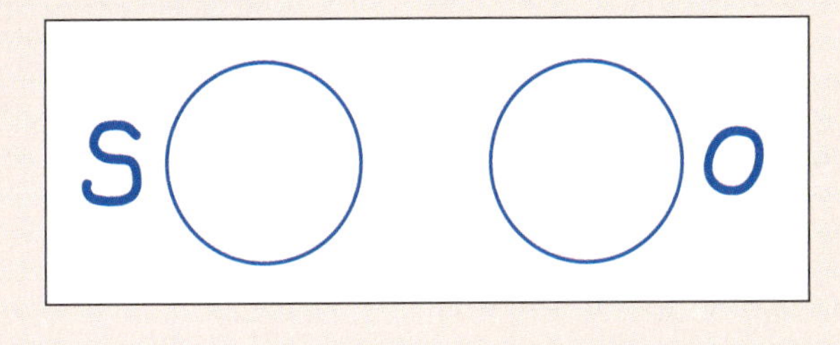


**2** If you performed the optional procedure, sketch and label your results in the diagram below.

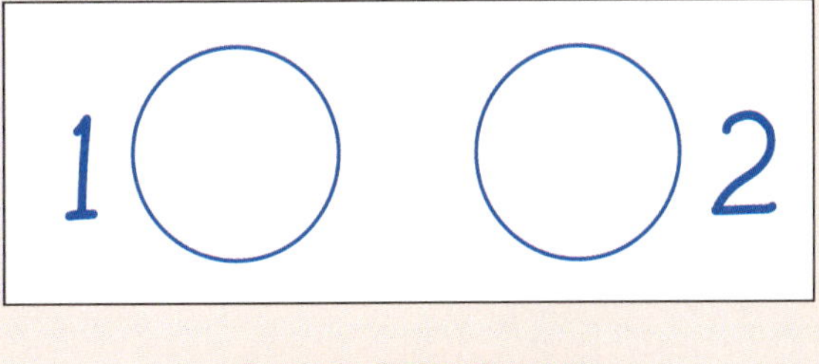


## QUESTIONS

***Note:*** *These questions apply to both versions of this exercise.*

**1** *Suppose you performed this test and got agglutination with O antigen and sterile saline. Eliminating contamination of the antigen and antiserum samples as a possibility, provide an explanation of this hypothetical result. Is this attributable to poor sensitivity or specificity of the test system?*

**2** *Suppose you performed this test and neither sample produced agglutination, even though you know the antigen and antibody should react. Would this be attributable to poor sensitivity or specificity of the test system?*

**3** *Higher vertebrates produce antibodies;* Salmonella *is a bacterium. Given these facts, how can* Salmonella *antiserum be produced?*

8

# Blood Typing

EXERCISE 8-5

## Theory

Hemagglutination is a general term applied to any agglutination test in which clumping of red blood cells indicates a positive reaction. Blood tests as well as a number of indirect diagnostic serological tests are hemagglutinations.

The most common form of blood typing detects the presence and absence of **A** and/or **B antigens** on the surface of red blood cells. An individual with type A blood has RBCs with the A antigen and produces anti-B antibodies. Conversely, an individual with type B blood has RBCs with the B antigen and produces anti-A antibodies. People with type AB blood have *both* A and B antigens on their RBCs and lack anti-A and anti-B antibodies. Type O individuals lack A and B antigens but produce *both* anti-A and anti-B antibodies.

While genetically determined, the A and B antigens are not direct gene products, but rather are produced by enzymatic reactions that modify the O antigen, a short (5-sugar) polysaccharide, found on all RBC membranes (Fig. 8.19). People with type A blood produce an enzyme that adds the sugar N-acetylgalactosamine to the end of the O antigen to make the A antigen. In like fashion, people with type B blood have a different enzyme that adds the sugar galactose to the end of the O antigen. People with type AB blood make both enzymes and modify the O antigen in both ways, and people with type O blood can only synthesize the O antigen, which remains unmodified.

ABO blood type is ascertained by adding a drop of a patient's blood separately to anti-A and anti-B antiserum and observing any signs of agglutination (Table 8-2 and Fig. 8.20). Agglutination with anti-A antiserum indicates the presence of the A antigen and type A blood. Agglutination with anti-B antiserum indicates the presence of the B antigen and type B blood. If both agglutinate, the individual has type AB blood; lack of agglutination occurs in individuals with type O blood.

TABLE **8-2** **Interpretation of Blood Typing Results**

| Reaction With | | | | | |
|---|---|---|---|---|---|
| **Anti-A Antiserum** | **Anti-B Antiserum** | **Anti-Rh Antiserum** | **Interpretation** | **Symbol** | **(Possible) Genotype** |
| Agglutination | No Agglutination | Agglutination | A antigen present<br>B antigen absent<br>Rh antigen present | A+ | $I^AI^A$ or $I^Ai$ and *DD* or *Dd* |
| Agglutination | No Agglutination | No Agglutination | A antigen present<br>B antigen absent<br>Rh antigen absent | A− | $I^AI^A$ or $I^Ai$ and *dd* |
| No Agglutination | Agglutination | Agglutination | A antigen absent<br>B antigen present<br>Rh antigen present | B+ | $I^BI^B$ or $I^Bi$ and *DD* or *Dd* |
| No Agglutination | Agglutination | No Agglutination | A antigen absent<br>B antigen present<br>Rh antigen absent | B− | $I^BI^B$ or $I^Bi$ and *dd* |
| Agglutination | Agglutination | Agglutination | A and B antigens present<br>Rh antigen present | AB+ | $I^AI^B$ and *DD* or *Dd* |
| Agglutination | Agglutination | No Agglutination | A and B antigens present<br>Rh antigen absent | AB− | $I^AI^B$ and *dd* |
| No Agglutination | No Agglutination | Agglutination | A and B antigens absent<br>Rh antigen present | O+ | *ii* and *DD* or *Dd* |
| No Agglutination | No Agglutination | No Agglutination | A and B antigens absent<br>Rh antigen absent | O− | *ii* and *dd* |

8

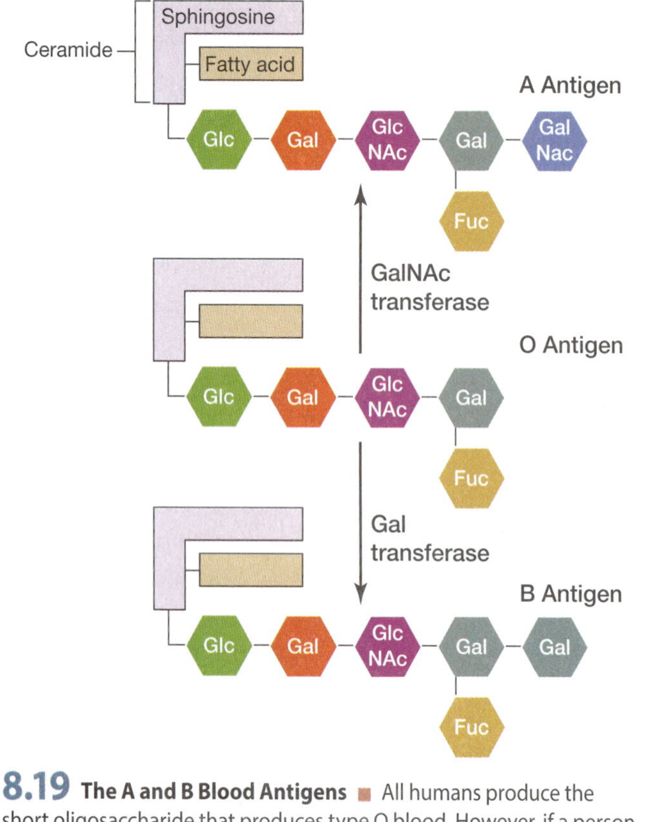


**8.19 The A and B Blood Antigens** ■ All humans produce the short oligosaccharide that produces type O blood. However, if a person has at least one gene (either genotype $I^AI^A$ or $I^Ai$) to produce the enzyme *N*-acetylgalactosamine transferase (GalNAc transferase) they add the sugar *N*-acetylgalactosamine to the O oligosaccharide and have type A blood. If a person has at least one allele (either genotype $I^BI^B$ or $I^Bi$) to produce the enzyme galactose transferase (Gal transferase) they add the sugar galactose to the O oligosaccharide and have type B blood. The presence of both enzymes (genotype $I^AI^B$) results in both reactions occurring and type AB blood. The absence of both enzymes (genotype ***ii***) leaves the O oligosaccharide unmodified and the person has type O blood. Because it is present in all humans, it is not detected as an antigen. Key: glucose (Glc), galactose (Gal), *N*-acetylglucosamine (GlcNAc) *N*-acetylgalactosamine (GlcNAc), and fucose (Fuc).

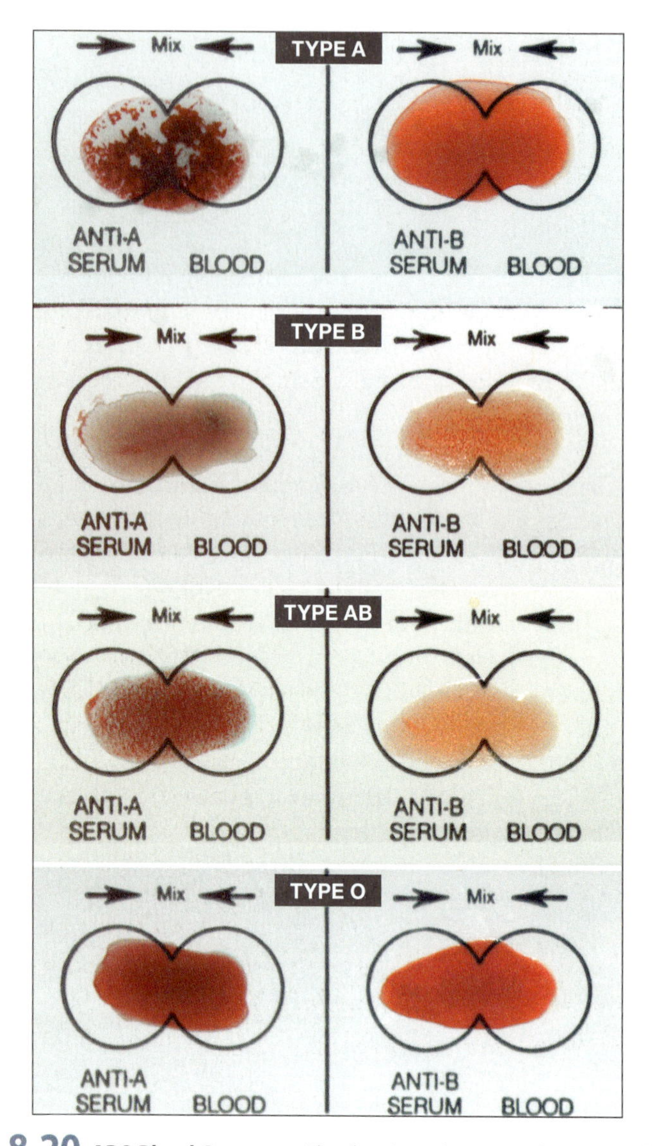


**8.20 ABO Blood Groups** ■ Blood typing relies on agglutination of RBCs by anti-A and/or anti-B antisera. The blood types are as shown. Visually, a positive Rh factor test looks the same.

A similar test is used to determine the presence or absence of the **Rh factor**, an erythrocyte membrane polypeptide. There are actually three Rh antigens (C, D, and E), and possession of any one of them makes the person Rh positive, but the one usually detected in blood testing is the D antigen. If clumping of the patient's blood occurs when mixed with anti-D antiserum, the patient is Rh positive.

The ABO and Rh blood factors are genetically unrelated, so any combination of the two is possible (e.g., "O–positive" or "O–negative"). The A and B blood factors are determined by the codominant $I^A$ and $I^B$ alleles, respectively. The recessive ***i*** allele, which all humans have, produces the molecule to which the $I^A$ and $I^B$ gene products (*N*-acetylgalactosamine and galactose, respectively) attach. The Rh factor is most simply viewed as being the product of a dominant allele ***D***, with the recessive ***d*** allele producing no Rh factor. See Table 8-2 for the possible genotypes for each blood type.

## Application

Blood typing is a simple example of an agglutination test using blood. Clinically, it is used for cross-matching donor and recipient blood prior to transfusion. It is also used to prophylactically treat Rh-negative mothers carrying Rh-positive fetuses to prevent erythroblastosis fetalis in that or any subsequent Rh-positive fetus.

## In This Exercise

You will determine your blood type with respect to two different markers: the ABO group and the Rh factor. Then, you will compile class data and compare it to blood type frequencies in the United States (Table 8-3).

TABLE **8-3** **Approximate Percentages of Blood Types in the United States**

| Blood Type | Caucasian | African American | Hispanic | Asian |
|---|---|---|---|---|
| O+ | 37 | 47 | 53 | 39 |
| O– | 8 | 4 | 4 | 1 |
| A+ | 33 | 24 | 29 | 27 |
| A– | 7 | 2 | 2 | 0.5 |
| B+ | 9 | 18 | 9 | 25 |
| B– | 2 | 1 | 1 | 0.4 |
| AB+ | 3 | 4 | 2 | 7 |
| AB– | 1 | 0.3 | 0.1 | 0.1 |

## ▼ Materials

### Per Student Group

- □ Blood typing anti-A antiserum
- □ Blood typing anti-B antiserum
- □ Blood typing anti-Rh (anti-D) antiserum
- □ Two microscope slides
- □ Three toothpicks
- □ Marking pen
- □ Sterile lancets
- □ Alcohol wipes
- □ Small adhesive bandages
- □ Sharps container (or other suitable receptacle for blood slide disposal) within arm's reach of each student's work station
- □ Disposable gloves

## PROCEDURE

1. Place two clean microscope slides on a paper towel at your work station. Make sure a sharps container is within easy reach.
2. With your marking pen, draw two circles on one microscope slide. Label one circle "A" and the other "B" (Fig. 8.21A).
3. Draw a single circle with your marking pen in the center of a second microscope slide. Label it "Rh" (Fig. 8.21B).
4. Place a drop of anti-A antiserum in the "A" circle.
5. Place a drop of anti-B antiserum in the "B" circle.
6. On the second microscope slide, place a drop of anti-D antiserum.
7. Clean the tip of your finger with an alcohol wipe. Let the alcohol dry.
8. Open a lancet package and remove the lancet. Hold it in your dominant hand, being careful not to touch the tip before you use it.
9. Shake your hand and "milk" blood down to the end of the finger you are going to prick. The tip of the ring finger is a good target.
10. Prick the end of your finger and immediately place a drop of blood beside each drop of antiserum. Do not touch the slide or antisera with your finger. It is okay to have someone else prick your finger, but make sure he or she wears protective gloves.
11. Discard the lancet in the sharps container.
12. Put an adhesive bandage on your wound.
13. Using a circular motion, mix each set of drops with a toothpick. Be sure to *use a different toothpick* for each antiserum. Dispose of the toothpicks in the sharps container.
14. Gently rock the slides back and forth for a few minutes or until agglutination occurs.
15. After the agglutination reaction is complete, record the results in the table provided on the data sheet, page 533. Once read, immediately dispose of the blood slides in a sharps container immediately. Do not spill blood on the opening.
16. Compare your results with the possible results in Table 8-2 to determine your blood type.
17. Collect class data and record these in the table provided on the data sheet. Compare class data with U.S. population data in Table 8-3.

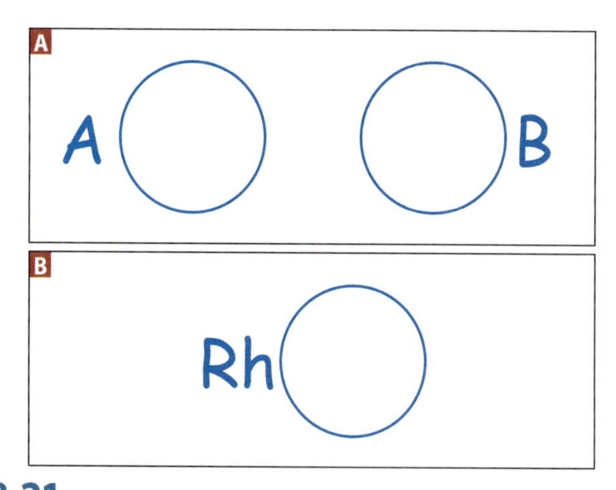


**8.21 Prepare the Slides** ■ (**A**) Draw two circles on a slide and label them "A" and "B" for the type of antiserum each will receive. (**B**) Draw one circle on a second slide and label it "Rh."

## References

American Red Cross. http://www.redcrossblood.org

Lodish, Harvey, Arnold Berk, Chris A. Kaiser, Monty Krieger, Anthony Bretscher, Hidde Ploegh, Angelika Amon, and Matthew P. Scott. Pages 461–462 in *Molecular Cell Biology*, 7th ed. New York: W. H. Freeman, 2013.

Nelson, David L. and Michael M. Cox. Pages 367–368 in *Lehninger's Principles of Biochemistry*, 6th ed. New York: W. H. Freeman, 2013.

Pier, Gerald B., Jeffrey B. Lyczak, and Lee M. Wetzler. Page 223 in *Immunology, Infection, and Immunity*. Washington, DC: ASM Press, 2004.

Ross, Michael H. and Wojciech Pawlina. Pages 273–274 in *Histology A Text and Atlas*, 6th ed. Philadelphia: Wolters Kluwer, 2011.

8

Name ______________________________

Date ______________________________

Lab Section ______________________________

I was present and performed this exercise (initials) ______________________________

DATA SHEET

8-5

# Blood Typing

## OBSERVATIONS AND INTERPRETATIONS

**1** Record your results below.

| Antiserum | Agglutination (+/−) |
|---|---|
| Anti-A | |
| Anti-B | |
| Anti-Rh | |

My blood type is: ______________

**2** Record class data in the table below.

| Blood Type | Percentage in U.S. Population | Number in Class | Percentage in Class | Deviation from National Values (±%) |
|---|---|---|---|---|
| O+ | 37.4 | | | |
| O− | 6.6 | | | |
| A+ | 35.7 | | | |
| A− | 6.3 | | | |
| B+ | 8.5 | | | |
| B− | 1.5 | | | |
| AB+ | 3.4 | | | |
| AB− | 0.6 | | | |

Source: Stanford School of Medicine. Copyright 2014.

## QUESTIONS

**1** *Examine the blood type data obtained from your class. Attempt to explain any deviations from the national values.*

**2** *People with type "O" are said to be "universal donors" in blood transfusions. Explain the reasoning behind this. Which blood type is designated as the "universal recipient"?*

**3** *What biological fact makes the concepts of "universal donor" and "universal recipient" misnomers in practice?*

**4** *Maternal-fetal Rh incompatibility (where the Rh− mother's anti-Rh antibodies destroy the fetus's Rh+ red blood cells) is a well-known phenomenon. Less well known are situations of maternal-fetal ABO incompatibility. Suggest combinations of maternal and fetal blood types that could lead to this situation.*



**5** *Why are red blood cells used in many indirect agglutination tests?*

# ELISA for Detecting Antibodies in a Patient's Sample[1] (The Antibody Capture Method)

EXERCISE **8-6**

## Theory

ELISA is an acronym for **Enzyme Linked Immunosorbant Assay**. As with other serological tests, ELISAs can be used to detect antigen or antibody in a sample. If antigen in a sample is to be identified, you must start with pure antibody to that antigen (homologous antibody). If the goal is to identify antibody in the sample, you must start with the homologous antigen. In either case, a secondary antibody with an attached, or **conjugated**, enzyme as an indicator of antigen-antibody reaction is used.

In this exercise, you will be looking for the presence of antibody in a sample made to simulate a patient's serum. Follow along with Figure 8.22 (pages 536–539) as the general process is described. In this version of ELISA, the **purified antigen** (PA) homologous to the antibody being tested for is added to the wells of a microtiter strip. The antigen will **adsorb** (attach) to the plastic wall of the wells.

After a wash with buffer to remove unbound antigen, a **blocking agent** (such as gelatin or other protein) may be used to coat the well where antigens have not attached. This prevents antibodies used in subsequent steps from attaching to the well. Following another wash step, the patient's "serum" is added. If the antibodies are present, they will bind specifically to the purified antigen coating the well. Another wash step removes unbound antibody and equally importantly, other components in the patient's serum.

At this point, the test is either positive or negative, but we cannot see it. What is required is an indicator, and that is where the enzyme-linked (**secondary** or **conjugate**) antibody comes in. This antibody is an **immunoglobin antibody**—its antigen is actually an antibody! The first step in the indicator reaction is adding secondary antibodies to the wells. They will bind to the patient's antibodies in the sample, if present. After allowing time for the secondary antibodies to react with their antigen (the patient's antibody), the well is washed again with buffer to remove any unbound secondary antibodies.

The second step in the indicator reaction is to use the enzyme on the secondary antibody to indicate its presence. The specific enzyme chosen for use in an ELISA is only important in that it catalyzes a reaction that can be detected easily. A commonly used enzyme is **horseradish peroxidase** (HRP). HRP catalyzes the conversion of hydrogen peroxide to water and oxygen ($2H_2O_2 \rightarrow O_2 + 2H_2O$).

In this exercise, the hydrogen peroxide substrate is associated with a **chromogenic cosubstrate** (3,3',5,5'–tetramethyl-benzidene–TMB). It turns a blue color when it becomes oxidized by the $O_2$ produced by peroxidase activity, indicating a positive result; a negative result shows no color change because the secondary antibody is not present and the peroxide is not broken down (Fig. 8.23).

## Application

An ELISA may be used to detect the presence and amount of either antigen or antibody in a sample. Screening of patients for the presence of HIV antibodies (4–8 weeks after exposure), rubella virus antibodies, West Nile virus antibodies, and others use the antibody capture ELISA. An antigen capture ELISA is used to detect hormones (such as HCG in some pregnancy tests and LH in ovulation tests), drugs, and viral (e.g., HIV p24 antigen within 4–8 weeks after exposure) and bacterial antigens.

## In This Exercise

You will perform an ELISA to determine presence of an antibody in a patient's "serum" sample.

### Materials

#### Per Student

- ☐ Lab coat
- ☐ Disposable gloves
- ☐ Chemical eye protection

#### Per Student Group

(The kit is designed for 12 groups of 4 students each)

- ☐ 4 "patient serum" test samples, each with 250 µL (one per student, with or without primary antibody in yellow microtubes, numbered from 1 up to 48, depending on class size)
- ☐ 1 microtube of purified antigen containing 1.5 mL (green microtube—labeled "PA")

[1] Bio-Rad ELISA Immuno Explorer™ Kit (Catalog #166-2400EDU), *Bio-technology Explorer™ Instruction Manual*, Rev. D. Bio-Rad Laboratories, Life Science Education. 1-800-4-BIORAD (800-424-6723), www.explorer.bio-rad.com.

8

- □ 1 microtube of positive control containing 500 μL (violet microtube—labeled " + ")
- □ 1 microtube of negative control containing 500 μL (blue microtube—labeled " – ")
- □ 1 microtube of secondary antibody containing 1.5 mL (orange microtube—labeled "SA")
- □ 1 microtube of substrate containing 1.5 mL (brown microtube—labeled "SUB")
- □ (Optional) 1 microtube of 1% blocking agent containing 1.5 mL—labeled "BA"[2]
- □ 2 microtiter plate strips (one per pair of students)
- □ 2 (better with 4) micropipettors capable of dispensing 50 μL and tips
- □ 4 disposable transfer pipettes
- □ 2 beakers of 33 mL wash buffer labeled "WB" (one per pair of students)[3]
- □ Large stack of paper towels
- □ 2 (up to 4) black marking pens
- □ Microtube rack
- □ 2 receptacles (plastic cup or something similar—some use a sharps container for this) for pipette tip disposal labeled "tip disposal"

### Preliminary Notes

- The materials are supplied for a group of 4 students, but you will work in pairs.
- Wear a lab coat, gloves, and chemical eye protection when performing this procedure.
- The written "steps" in the procedure correspond to steps in the procedural diagram in Figure 8.22.
- As a group (of four) obtain all materials listed and shown in Fig. 8.24. Each group member should choose a patient's "serum" sample to test (numbered yellow microtubes). Record its number in boxes 7–9 or 10–12 in the chart on the data sheet, page 541.

[2] Using blocking agent is an optional step in this protocol. Tubes for this are not supplied in the kit, so must come from instructor's stock. Blocking agent is 1% w/v gelatin, albumin, or nonfat dry milk.

[3] If the blocking step is used, 33 mL of wash buffer is just enough to carry out all the wash steps, but do not waste any! If it is not used, then another 5 mL may be added to the beaker. This presumes the prep is for the maximum 12 student groups of four.

### Step 1 Labeling the Microtiter Strip

Working in pairs, label the sides of one 12-well microtiter plate strip with a permanent marker as follows: The first three wells with " + ," the second three wells with " – ," the third three wells with one lab partner's sample number, and the fourth three wells with the other lab partner's sample number.

### Step 2 Adding Purified Antigen

Using a micropipettor with a fresh tip, dispense 50 μL of purified antigen (green tube labeled "PA") to each of the wells. Incubate at your desk for 5 minutes. Dispose of the pipette tip in the container provided.

### Step 3 Removing Unbound Purified Antigen

Invert the strip and place it on the stack of paper towels (or a paper towel folded several times). Gently tap out the liquids onto the paper towel to remove unbound purified antigen.

### Steps 4 and 5 Washing the Wells

Fill (but do not overflow) each well with wash buffer using a disposable pipette (which can be used for all subsequent wash steps). Remove the wash buffer onto fresh paper towels. Avoid contaminating wells with "drainage" from other wells on the towel. Replace the towel if it becomes so wet that there is the potential of contaminating wells with the "drainage." Wash and drain the wells a second time.

### Steps 6, 7, 8, and 9 Optional: Using a Blocking Agent

Using a new tip, add 50 μL of blocking agent to each well. Incubate at your lab desk for 5 minutes. Remove unbound blocking agent on fresh towels and wash twice as before.

+ + + – – – 1 1 1 2 2 2

**Step 1: Labeling the Microtiter Strip** Using a permanent marker, label the outside of each well as shown. Replace numbers 1 and 2 with the "Patient Sample" numbers you have been assigned.

**8.22 Procedural Diagram for the ELISA** ■ Pay special attention to volumes, solutions, and pipettes used in each step. Fresh pipette tips must be used for each solution and disposed of properly. The same disposable pipette may be used for transferring wash buffer throughout. Take care not to mix solutions from neighboring wells by overfilling or by emptying over a contaminated part of the paper towels. *(Continues)*

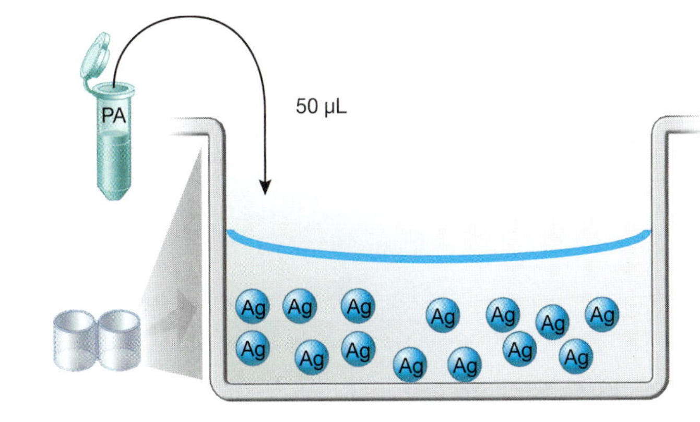


**Step 2: Adding Purified Antigen (Ag)** Add 50 μL of purified antigen (green microtube labeled "PA") to each of the 12 wells. Incubate at your lab desk for 5 minutes. Dispose of the pipette tip as directed by your instructor.

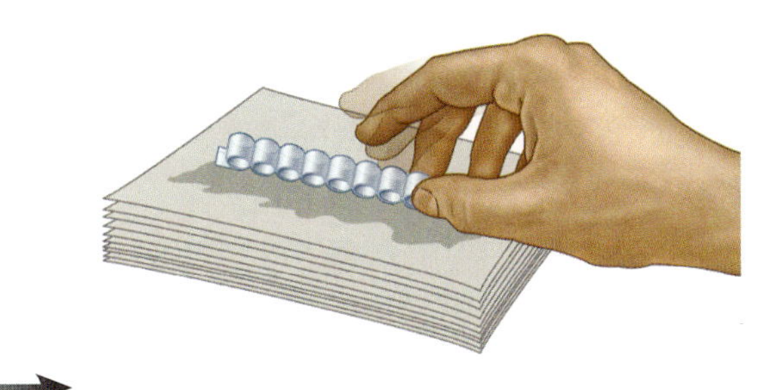

**Step 3: Remove Unbound Purified Antigen** Gently tap the strip on several thicknesses of fresh paper towels to remove unbound antigen. Dispose of wet towels as directed by your instructor.

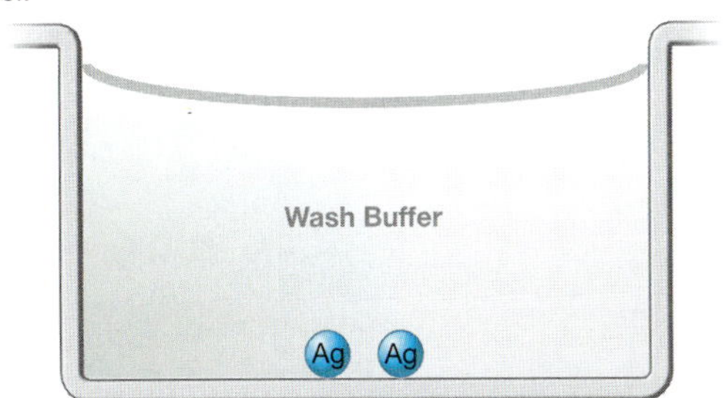


**Step 4: Washing the Wells** Using the disposable transfer pipette, fill (but don't overfill and mix neighboring well contents) all 12 wells with wash buffer. You can save the disposable pipette for subsequent wash steps. (***Note:*** In reality, at this point the entire lining of the well would be coated with Ag, but only two are shown to reduce cluttering in the art.)

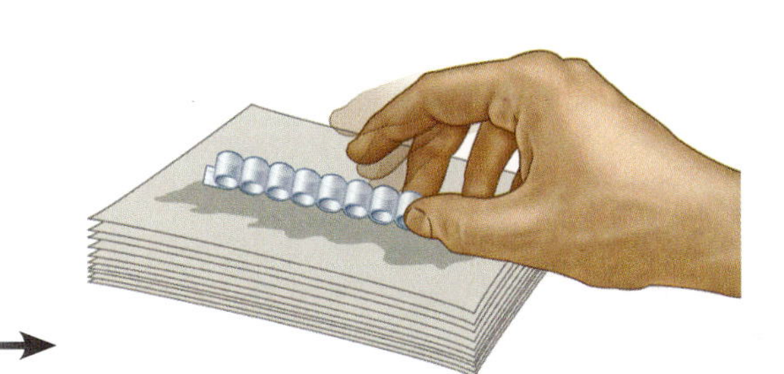

**Step 5: Removing the Wash Buffer** Drain the wells onto several thicknesses of fresh paper towels, and then gently tap out the remainder. Do not allow mixing between wells. Dispose of wet towels. **Repeat steps 4 and 5 to wash a second time**, then continue with step 6 (optional) or step 10.

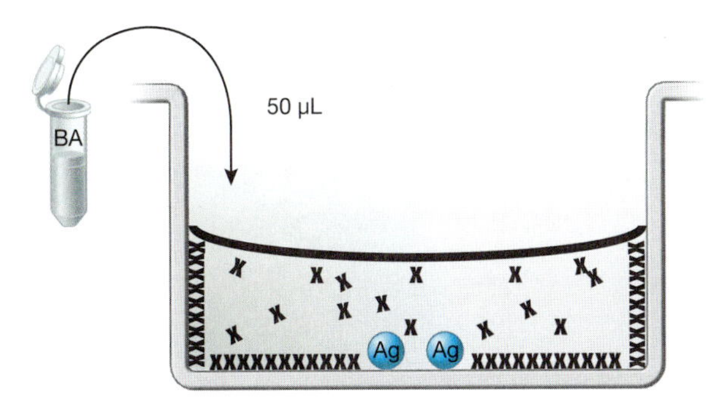


**Step 6: Adding Blocking Agent (Optional)** Add 50 μL of blocking agent (shown as x's in the drawing) to each well. Incubate at your lab desk for 5 minutes.

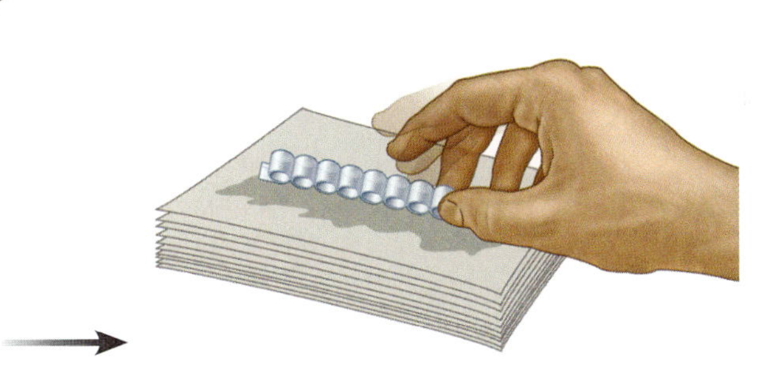

**Step 7: Removing Unbound Blocking Agent** Gently tap the strip on several thicknesses of fresh paper towels to remove unbound blocking agent.

**Step 8: Washing the Wells** Using the disposable transfer pipette, fill (but don't overfill and mix neighboring well contents) all 12 wells with wash buffer.

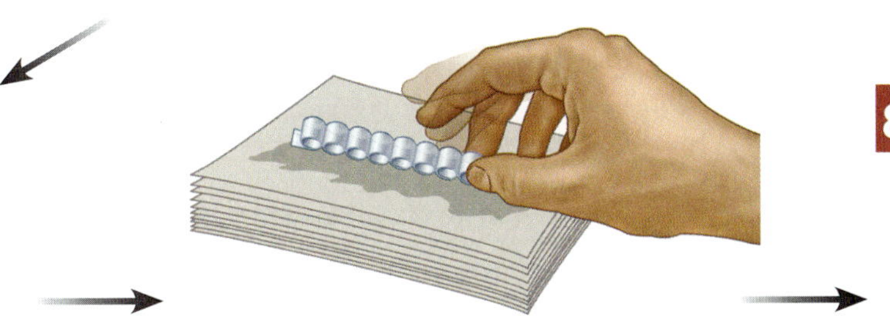

**Step 9: Removing the Wash Buffer** Drain the wells onto several thicknesses of fresh paper towels. Tap out the last remnants of wash buffer. Dispose of wet towels. **Repeat steps 8 and 9 to wash a second time, then continue with step 10.**

**8.22 Procedural Diagram for the ELISA** *(Continues)*

## Step 10 Adding the Positive Control

Using the micropipettor, add 50 µL of positive control (violet tube labeled " + ") to each of the wells labeled " + ." Dispose of the tip in the designated container. Continue immediately with step 11.

## Step 11 Adding the Negative Control

Using a new tip, add 50 µL of negative control (blue tube labeled " – ") to each of the wells labeled " – ." Dispose of the tip in the designated container. Continue immediately with step 12.

## Step 12 Adding the Patient "Serum" Samples

Using a new tip, add 50 µL of one sample (yellow microtube) to the wells marked with its number. Dispose of the tip in the designated container. Using a new tip, add 50 µL of the second sample (yellow microtube) to the wells marked with its number. Dispose of the tip in the designated container. Incubate at your lab desk for 5 minutes.

## Step 13 Removing the Control and Patient "Serum" Sample Solutions and Washing the Wells

Drain the controls and patient samples out of the wells onto a fresh paper towel and gently tap out any residual as in step 3. Be careful not to mix solutions between wells. Wash the wells twice as in steps 8 and 9.

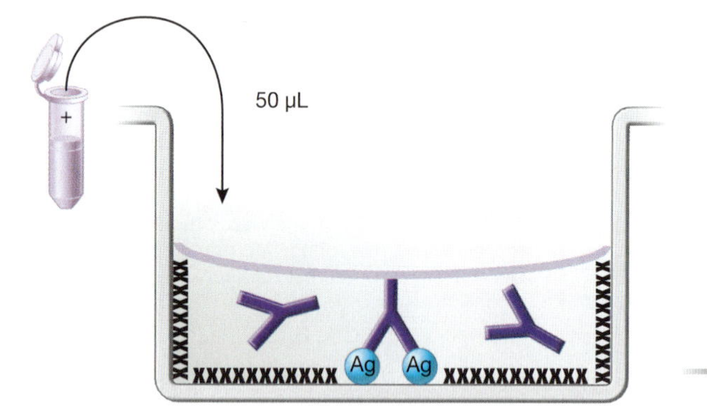


**Step 10:** Adding the Positive Control Add 50 µL of positive control (violet microtube labeled " + ") **to the three " + " wells. Continue immediately with step 11.**

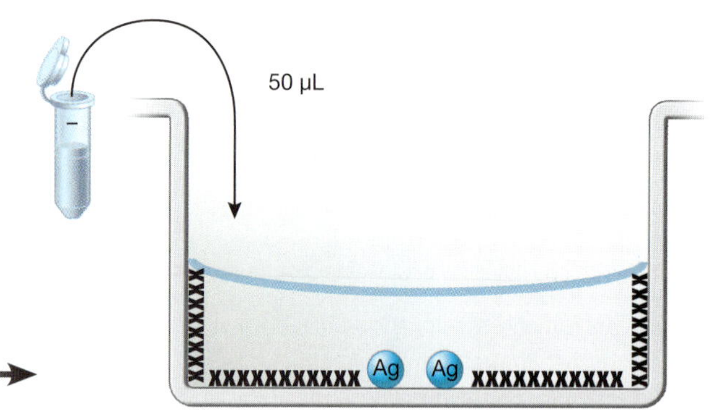


**Step 11:** Adding the Negative Control Add 50 µL of negative control (blue microtube labeled " – ") **to the three " – " wells. Continue immediately with step 12.**

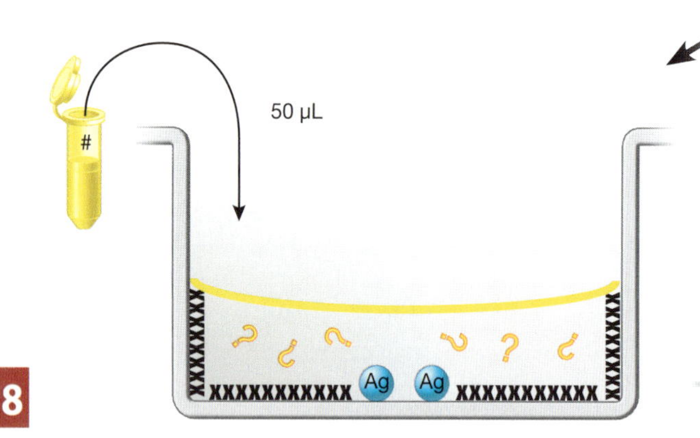


**Step 12:** Adding Patient "Serum" Samples Add 50 µL of patient "serum" (yellow microtubes labeled with numbers) to the appropriate remaining wells. Incubate for 5 minutes at your lab desk. The question marks indicate that the contents of the patient's sample are unknown.

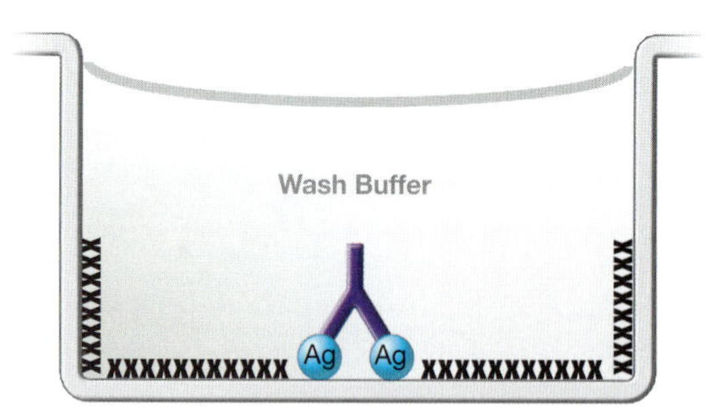


**Step 13:** Removing the Control and Patient "Serum" Sample Solutions and Washing the Wells Gently tap the strip on several thicknesses of fresh paper towels to remove the control and patient "serum" solutions. Be careful not to mix contents between wells. **Perform two washes with wash buffer as shown in steps 8 and 9.** The illustration shows what a positive result looks like at this point. A negative result would not have the antibody attached to the antigens and would look like the art in step 11.

**8.22 Procedural Diagram for the ELISA** *(Continues)*

### Step 14 Adding Secondary (Conjugate) Antibody

Using a new tip, dispense 50 μL of secondary antibody (orange microtube labeled "SA") to each of the wells. Incubate at your desk for 5 minutes.

### Step 15 Removing Unbound Secondary Antibody

Remove the secondary antibody by gently tapping the strip on fresh towels as in step 3, and then wash three times as in steps 8 and 9. Replace towels as necessary.

### Step 16 Adding Substrate

Using a new tip, dispense 50 μL of substrate (brown tube labeled "SUB") to all 12 wells. Incubate at your lab desk for 5 minutes. Observe for color changes and record your results on the data sheet, page 541. Take your reading within 5 minutes because there is the potential for a blue color developing (due to a reaction with light) if allowed to incubate longer.

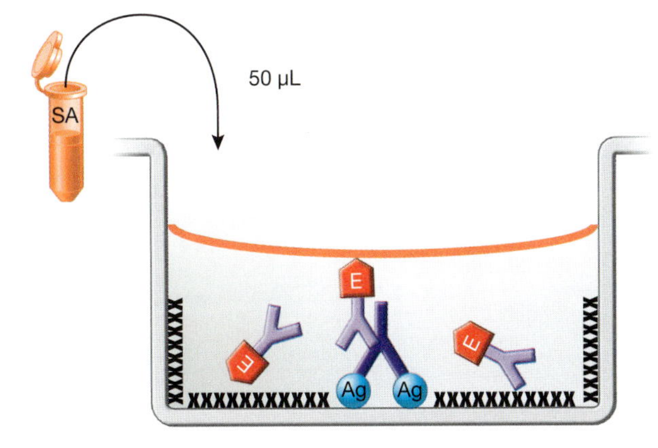


**Step 14: Adding Secondary (Conjugate) Antibody** Add 50 μL of secondary antibody (orange microtubes labeled with "SA") to the 12 wells. Incubate for 5 minutes at your lab desk.

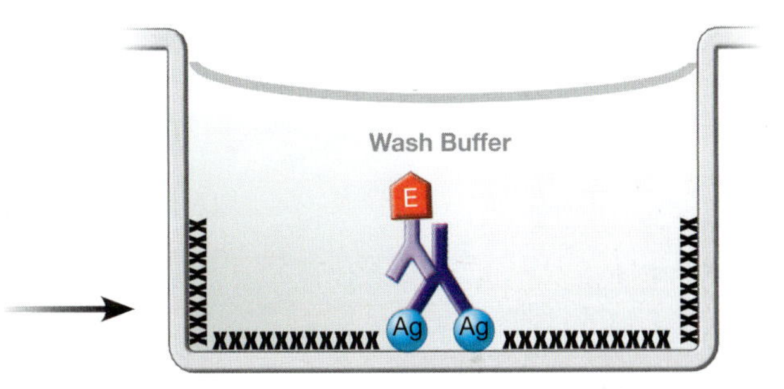


**Step 15: Removing Unbound Secondary Antibody** Gently tap the strip on several thicknesses of fresh paper towels to remove unbound secondary antibody solution. Be careful not to mix contents between wells. **Perform three washes with wash buffer as shown in steps 8 and 9.** The illustration shows what a positive result looks like at this point. A negative result would not have either antibody present and would look like the art in step 11.

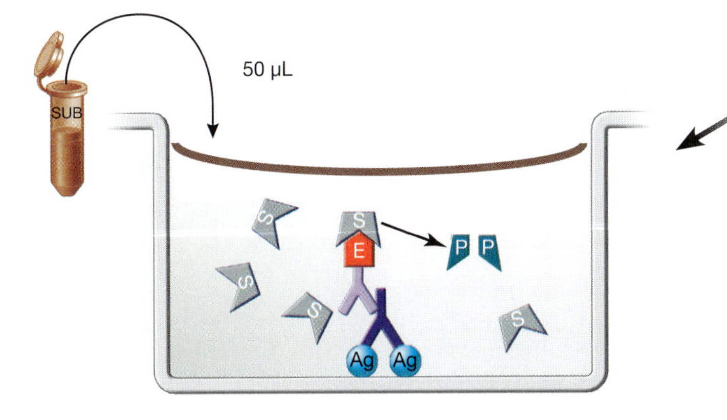


**Step 16: Adding Substrate** Add 50 μL of substrate (brown microtube labeled "SUB") to the 12 wells. Incubate at your lab desk and read results within 5 minutes. A blue color is a positive result. **Waiting longer than 5 minutes to read results could lead to the substrate turning blue through its exposure to light, not enzyme.**

**8.22 Procedural Diagram for the ELISA** *(Continued)* ■

**8.23 Positive and Negative Results** ■ A positive result in this ELISA produces a blue color within 5 minutes (left); no reaction is colorless (right). Waiting longer than 5 minutes may lead to a nonspecific (that is, not due to enzyme activity) color change.

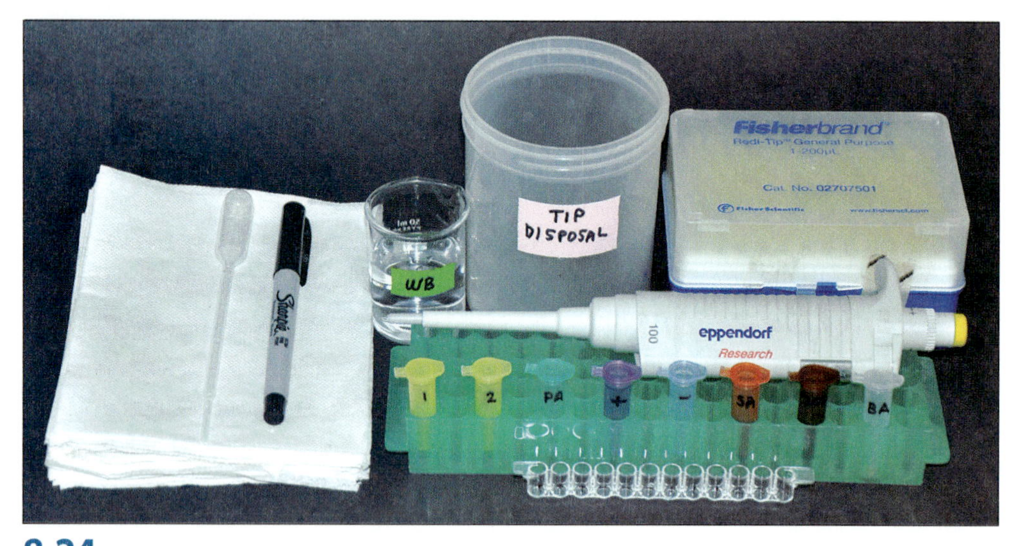


**8.24 Materials Needed** ■ Shown are the materials needed for this procedure, but not necessarily in the proper quantity (due to space limitations). The tubes in the rack are as follows: yellow are numbered patient "serum" samples, green is purified antigen ("PA"), violet is the positive control ("+"), blue is the negative control ("−"), orange is the secondary antibody ("SA"), brown is the substrate ("SUB"), and clear is the optional 1% blocking agent ("BA"). Towels, a disposable transfer pipette, a permanent marker, wash buffer (WB), micropipettor and tips, can for tip disposal, and microtiter strip (foreground) are also shown.

## References

Bio-Rad ELISA Immuno Explorer™ Kit (Catalog #166-2400EDU), *Bio-technology Explorer™ Instruction Manual*, Rev. D. Bio-Rad Laboratories, Life Science Education. 1-800-4-BIORAD (800-424-6723), www.explorer.bio-rad.com.

Lam, Joseph S. and Lucy M. Mutharia. Chap. 5 in *Methods for General and Molecular Bacteriology*. Philipp Gerhardt, R.G.E. Murray, Willis A. Wood, and Noel. R. Krieg, eds. Washington, DC: ASM Press, 1994.

Madigan, Michael T., John M. Martinko, Paul V. Dunlop, and David P. Clark. Pages 922–926 in *Brock's Biology of Microorganisms*, 12th ed. San Francisco: Pearson/Benjamin Cummings, 2009.

Shores, Teri. Pages 102–103 in *Understanding Viruses*. Sudbury, MA: Jones and Bartlett Publishers, 2009.

Name ______________________________

Date ______________________________

Lab Section ______________________________

I was present and performed this exercise (initials) ______________________________

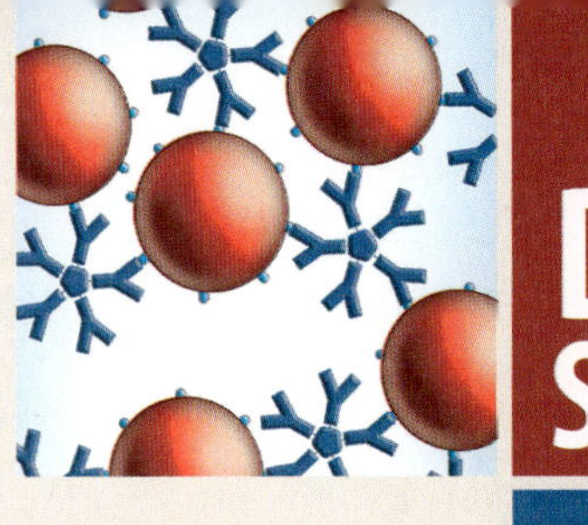

DATA SHEET 8-6

# ELISA for Detecting Antibodies in a Patient's Sample (The Antibody Capture Method)

Take your readings within 5 minutes because there is the potential for a blue color developing (due to a reaction with light) if allowed to incubate longer.

## OBSERVATIONS AND INTERPRETATIONS

**1** Write the number of yours and your partner's Patient Sample Numbers in the table below.

| | 1 | 2 | 3 | 4 | 5 | 6 | 7 | 8 | 9 | 10 | 11 | 12 |
|---|---|---|---|---|---|---|---|---|---|---|---|---|
| **Tube labels** | + | + | + | – | – | – | | | | | | |
| **Results** | | | | | | | | | | | | |

**2** Record the results in each well of the microtiter strip in the table above.

## QUESTIONS

**1** *Describe the expected consequences of the following procedural errors,*

**a.** *Mixing the positive control solution in a negative control well during step 13.*

**b.** *Mixing the negative control solution in a positive control well during step 13.*

**c.** *Mixing primary antibody in a negative control well after step 13.*

**d.** *Mixing secondary antibody with negative control after step 15.*

**2** *Consider the positive controls.*

**a.** *Given the design of this ELISA, what is the essential component in the positive control solution? Choose from antigen, primary antibody, secondary antibody, and substrate.*

**b.** *What purpose do the positive controls serve?*

**c.** *What are possible causes of negative results in all the positive controls?*

**d.** *How would negative results in all the positive controls affect your interpretation of results given by the patient samples?*

**3** *Consider the negative controls.*

**a.** *Given the design of this ELISA, what component must be absent in the negative control solution? Choose from antigen, primary antibody, secondary antibody, and substrate.*

**b.** *What purpose do the negative controls serve?*

**c.** *What are possible causes of positive results in all the negative controls?*



**d.** *How would positive results in all the negative controls affect your interpretation of results given by the patient samples?*

Name ______________________________

Date ______________________________

Lab Section ______________________________

I was present and performed this exercise (initials) ______________________________

# DATA SHEET 8-6

*(continued)*

**4** *Consider the samples tested.*

**a.** *Why are the samples run in triplicate?*

______________________________

______________________________

**b.** *What might explain inconsistent results within each triplicate set and how might they affect your interpretations of the patient samples?*

1) *Positive controls*

______________________________

______________________________

2) *Negative controls*

______________________________

______________________________

3) *Patient sample*

______________________________

______________________________

**5** *Refer to the procedural diagram on pages 536–539 for the following questions:*

**a.** *In step 4, what could be the consequence of washing too vigorously with wash buffer? Of not washing enough?*

**b.** *What might be the consequence of not adding blocking agent in step 6? (Omit this question if blocking agent was not used.)*

**c.** *In step 13, what would be the consequence of washing too vigorously with wash buffer? Of not washing enough?*

**d.** *In step 15, what would be the consequence of washing too vigorously with wash buffer? Of not washing enough?*

SECTION 9

# Identification of Unknowns

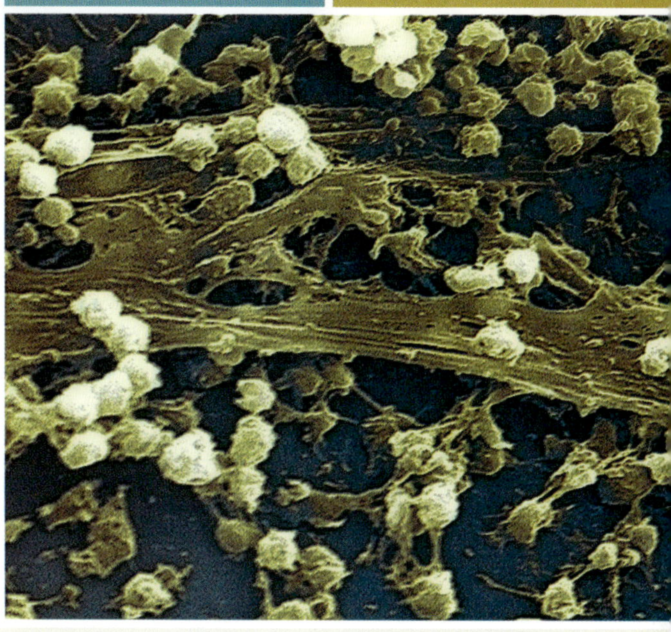

The exercises in this section are a culmination of all you have done up to this point. You will be using skills developed in earlier exercises and apply them as a practicing microbiologist would. That is, you will be identifying unknown microbes from samples supplied to you by your instructor.

The opportunity to apply what you have learned to a real problem with practical applications is both challenging and exciting for most students. A word of caution, though. Your instructor will know what your unknown is and can tell you if your identification is correct. But out of the classroom and in a professional laboratory, no one can tell the microbiologist if he or she is correct or not. In fact, "identification" is more about finding out what the unknown *isn't*, rather than finding out with *certainty* what it *is*.

There is an unspoken understanding that the unknown's identity is based on the best match between the unknown's test results and the accepted results for the same tests of the named species. But, the identification is provisional because other tests that were not run might lead to a different conclusion. Further, don't forget about those nasty false positives and false negatives. Now, with that in mind...

The first three exercises employ traditional methods of microbial identification. That is, you will use staining properties, cell morphology, cell arrangement, and biochemical test results to identify an unknown organism assigned to you. In each exercise, a flowchart will direct you in your choice of tests. Matching your unknown's results to the flowchart will lead you, by the process of elimination, to a provisional identification.

Although several multi-test systems that utilize computer databases are available (Exercises 9-4 and 9-5, for example), flowcharts are still a useful way to visualize the process of identification by elimination. It is striking how often just a few test results, when taken in combination, are necessary to identify an organism. This methodology will be applied to common species of Gram-negative rods (Exercise 9-1), Gram-positive cocci (Exercise 9-2), and Gram-positive rods (Exercise 9-3).

You probably have already done most tests in the flowcharts. Some, such as fermentation of sugars, use the same PR medium as Exercise 5-2, just with a different sugar, and results are read in the same way. The same is true of the various amino acid decarboxylases (Exercise 5-8).

In Exercise 9-4 and Exercise 9-5, you will have the opportunity to identify unknown enterics using multi-test systems. These take a shotgun approach but allow more rapid identification than traditional methods. Each has several biochemical tests that are inoculated simultaneously. These tests are based on the same principles as their tubed and plated forebears, but the combination of results produces a number that is compared to a database for identification.

# Identification of Selected *Enterobacteriaceae*

EXERCISE
9-1

## Theory

The *Enterobacteriaceae* (Fig. 9.1) comprise a diverse family of Gram-negative rods that were first isolated from the human intestinal tract. The family has grown from 12 genera and 36 species in *Bergey's Manual of Determinative Bacteriology*, 8th ed. (1974) to 44 genera and 176 species in *Bergey's Manual of Systematic Bacteriology*, 2nd ed., Volume 2, *Part B–The Gammaproteobacteria* (2005), with the expectation that these numbers will increase as more is learned.

This expansion has brought additional variability into the group and the recognition that the human intestines are not their main habitat. *Enterobacteriaceae* are also common in the environment (soil) and have hosts as diverse as plants and insects. Still, the following major characteristics are true for the majority of species:

- Growth on MacConkey agar
- Gram-negative rods
- Facultative anaerobes (respiratory and fermentative metabolism)
- Oxidase negative
- Acid production from glucose (with or without gas)
- Most are catalase positive
- Most reduce $NO_3$ to $NO_2$
- Peritrichous flagella, if motile

Many **enterics** are human pathogens (e.g., *Escherichia coli* 0157:H7—bloody diarrhea; *Klebsiella pneumoniae*—pneumonia of various types; *Proteus mirabilis*—urinary tract infections; *Salmonella typhi*—typhoid fever; *Shigella dysenteriae*—bacillary dysentery; and *Yersinia pestis*—plague) but just as many are harmless gut **commensals** or **opportunists.** The organisms for which the identification flowcharts in this exercise work are listed in Figure 9.2 and some are commonly associated with human infections. Your instructor will choose enteric unknowns appropriate to your microbiology course and facilities.

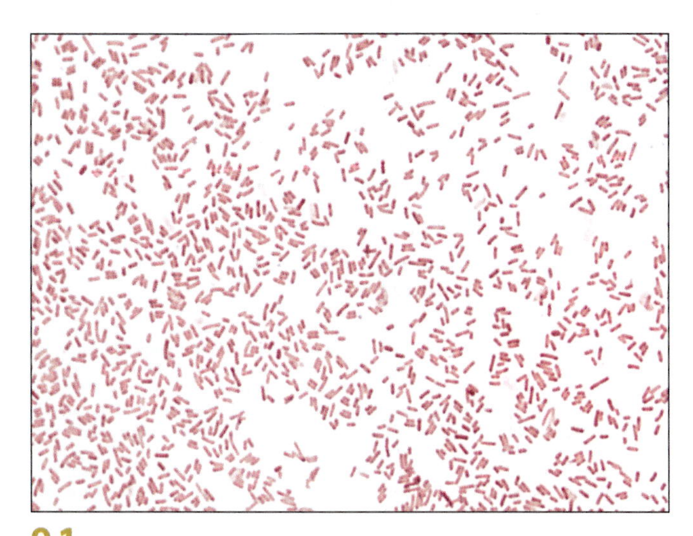

**9.1 *Escherichia coli*, a Common Species in the Family *Enterobacteriaceae*** ■ All *Enterobacteriaceae* are Gram-negative, oxidase-negative rods. They also ferment glucose to acid end products.

**List of *Enterobacteriaceae* in Identification Charts and IMViC Results**

*Citrobacter freundii* *
*Citrobacter koseri*
*Enterobacter aerogenes*
*Enterobacter cloacae*
*Escherichia coli*
*Hafnia alvei* *
*Klebsiella pneumoniae* *+ (BSL-2)
*Morganella morganii* (BSL-2)
*Proteus mirabilis* * (BSL-2)
*Proteus vulgaris* (BSL-2)
*Providencia alcalifaciens*
*Providencia stuartii*
*Salmonella enterica* ++ (BSL-2)
*Serratia liquefaciens*
*Serratia marcescens* *
*Shigella flexneri (Group B)* * (BSL-2)
*Shigella sonnei (Group D)* (BSL-2)

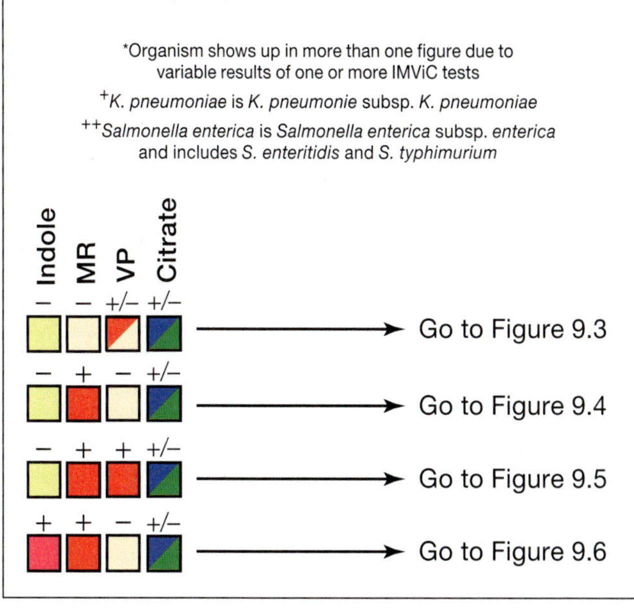


**9.2 List of Organisms and IMViC Results of *Enterobacteriaceae* Used in this Exercise** ■ The organisms listed are *Enterobacteriaceae* used in this exercise, though not all may be assigned as unknowns in your class. Match the IMViC results of your isolate with one of the four shown, and then proceed with the flowchart in the figure indicated. Some organisms are listed in more than one flowchart because of variable results for a particular test. A box divided diagonally with the symbol "+/–" means either result for that test is a match. (Most results after Brenner and Farmer, 2005.) ***Note:*** Each flowchart can be used to identify only the organisms listed.

9

## Application

Identification of enterics from human specimens using traditional methods requires a coordinated and integrated use of biochemical tests and stains. It is a process that is performed on a daily basis in medical labs worldwide.

## In This Exercise

This exercise will span several lab periods. You will be assigned an unknown enteric from the list in Figure 9.2. Then you will run biochemical tests as directed by Figure 9.2 and the appropriate flowchart in Figures 9.3 through 9.6 to identify your unknown. Table 9-1 provides a list of possible biochemical tests you will be using, with colored icons to remind you of what results look like.

## Materials

### Per Class

- □ Media (and appropriate reagents) listed in Table 9-1 for biochemical testing
- □ Pure cultures of organisms listed in Figure 9.2 to be used for positive controls and unknowns. (All organisms are available from either Ward's Science or Carolina Biological Supply companies. **BSL-2 organisms should not be chosen as unknowns by schools or classes not adequately prepared to handle them.**)

### Per Student

- □ Lab coat
- □ Disposable gloves
- □ Chemical eye protection
- □ Two MacConkey agar plates
- □ One unknown organism[1] from the list in Figure 9.2 (numbered to an instructor's key)
- □ TSA slants (enough to keep the culture fresh over the time required for identification)
- □ Gram-stain kit
- □ Microscope slides
- □ Compound microscope with oil-objective lens and ocular micrometer
- □ Immersion oil
- □ Lens paper
- □ Bibulous paper
- □ Sterile wooden sticks or disposable loops
- □ (Optional) bacterial incinerator

[1] *Note to instructor:* This exercise can be combined with Exercises 9-2 or 9-3 by passing out a mixture of an enteric and a Gram-positive coccus or rod as unknowns. The student must isolate the two organisms from mixed culture and then proceed to identify the two independently. If this option is chosen, grow the two unknowns separately and then mix them immediately prior to handing them out in lab.

TABLE **9-1** **Key to Tests and Result Icons Used in This Exercise** The list indicates tests you may be using to identify your unknown and refers you to the appropriate exercise for instructions on how to run them. Colors are to remind you what results look like. They should not be used to compare with your results. ***Note:*** For the fermentation tests, "A+" means acid is produced; "A–" means no acid is produced.

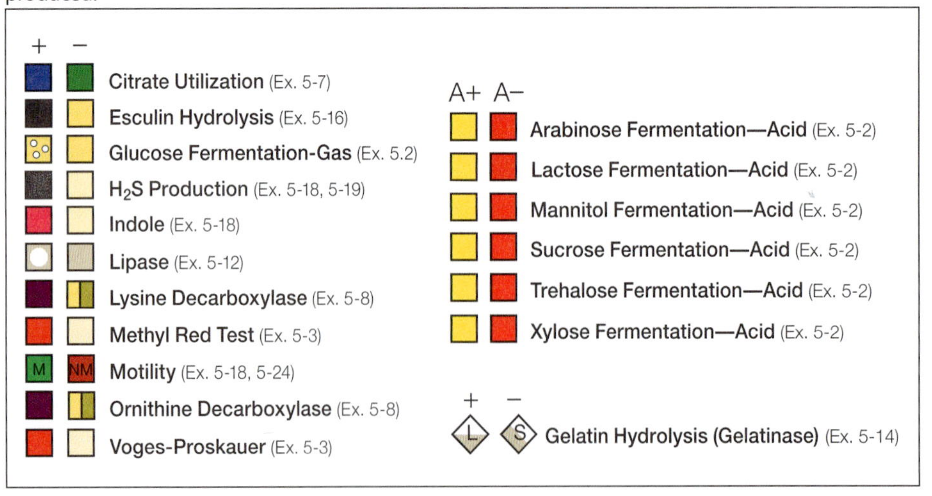

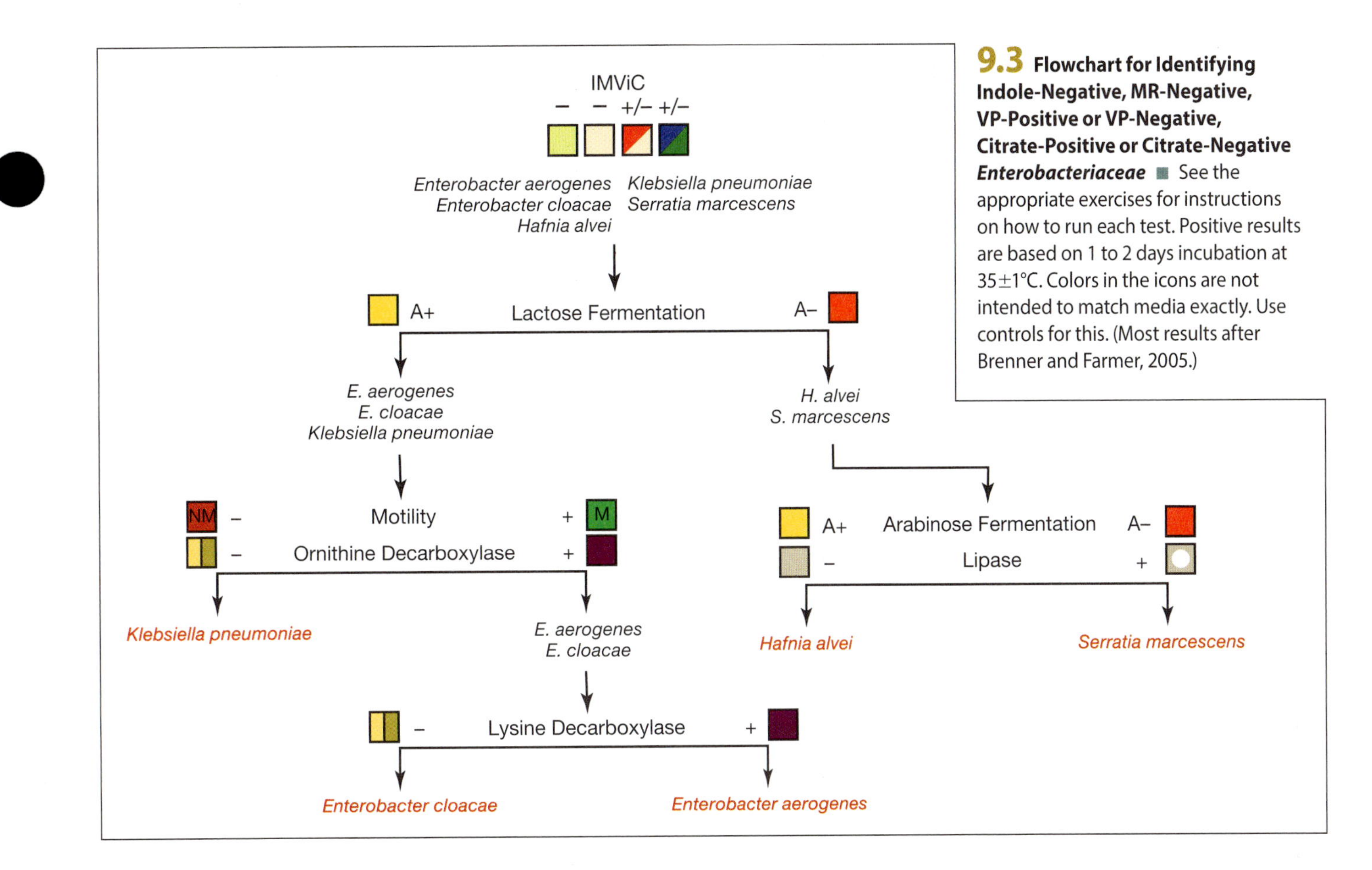


**9.3 Flowchart for Identifying Indole-Negative, MR-Negative, VP-Positive or VP-Negative, Citrate-Positive or Citrate-Negative *Enterobacteriaceae*** ■ See the appropriate exercises for instructions on how to run each test. Positive results are based on 1 to 2 days incubation at 35±1°C. Colors in the icons are not intended to match media exactly. Use controls for this. (Most results after Brenner and Farmer, 2005.)

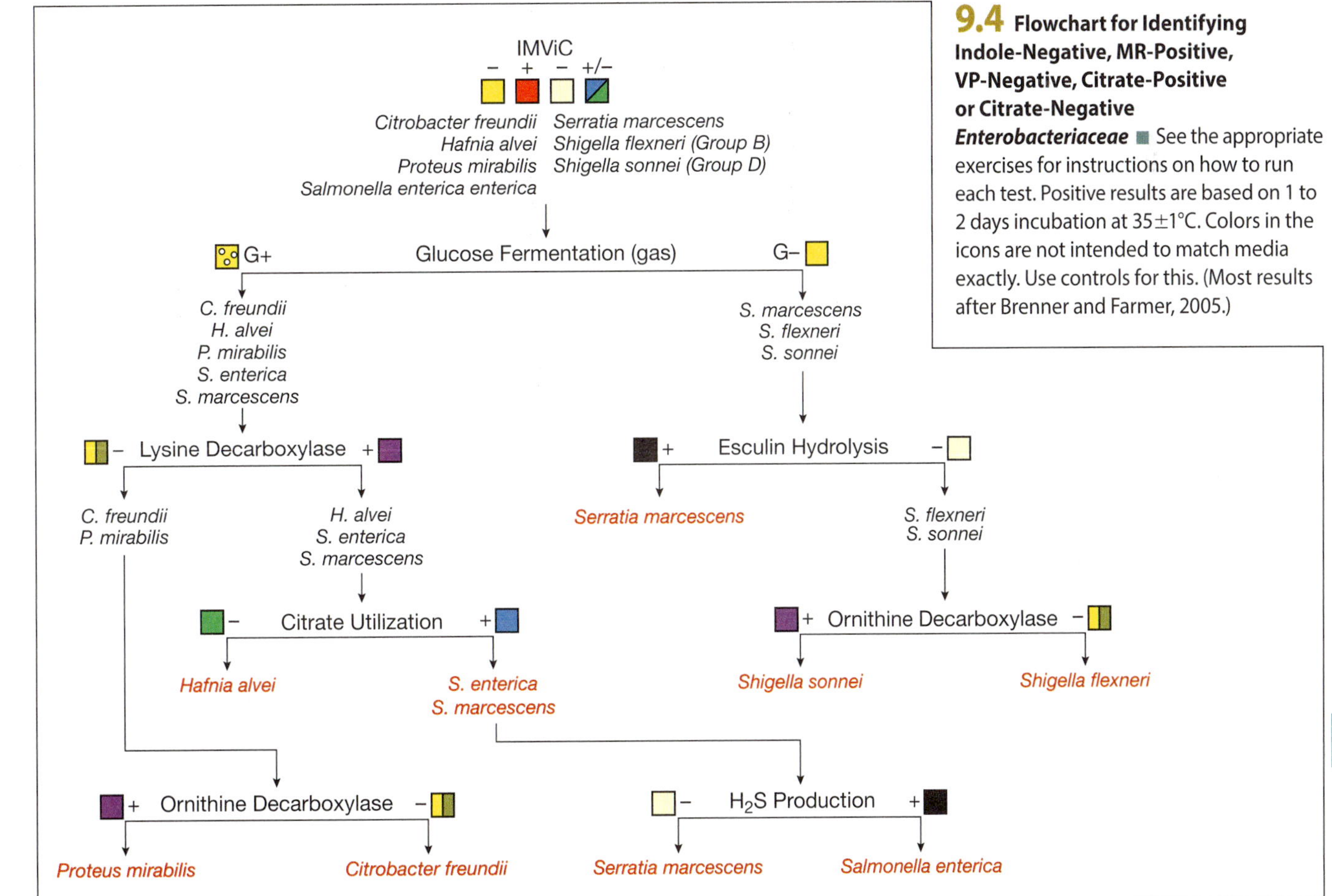


**9.4 Flowchart for Identifying Indole-Negative, MR-Positive, VP-Negative, Citrate-Positive or Citrate-Negative *Enterobacteriaceae*** ■ See the appropriate exercises for instructions on how to run each test. Positive results are based on 1 to 2 days incubation at 35±1°C. Colors in the icons are not intended to match media exactly. Use controls for this. (Most results after Brenner and Farmer, 2005.)

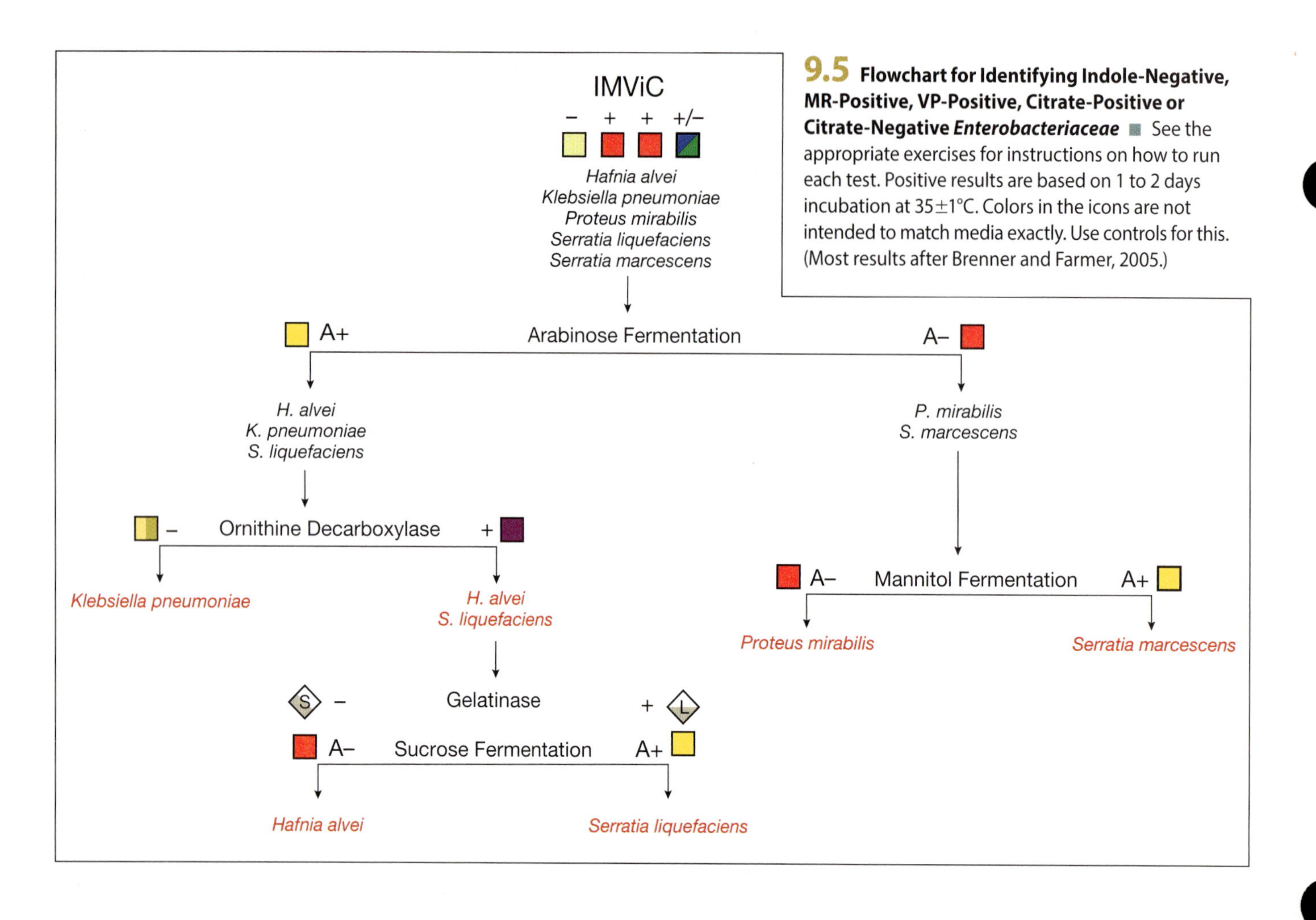


**9.5 Flowchart for Identifying Indole-Negative, MR-Positive, VP-Positive, Citrate-Positive or Citrate-Negative *Enterobacteriaceae*** ■ See the appropriate exercises for instructions on how to run each test. Positive results are based on 1 to 2 days incubation at 35±1°C. Colors in the icons are not intended to match media exactly. Use controls for this. (Most results after Brenner and Farmer, 2005.)

## PROCEDURE

1. Wear a lab coat, gloves, and chemical eye protection when performing this procedure. **In addition, all procedures should be run using BSL-2 precautions unless otherwise directed by your instructor.**
2. Obtain an unknown. Record its number on the data sheet, page 555. ***Note***: You are urged to keep accurate records of all your activities related to identification of this unknown as you proceed (and not after the fact). It is surprising how quickly and easily we (humans) forget details…
3. Streak the sample for isolation on two MacConkey agar plates. Incubate one at 25°C and the other 35±2°C for 24 hours or more. Record your activities on the data sheet under Isolation Procedure.
4. It is best to check your plate for isolation at 24 hours. If you have isolation, continue with step 5. If you don't have time to continue with step 5, refrigerate your plates until you do. If you do not see isolation, ask your instructor if you should re-streak for isolation, let the original plates continue to incubate, or do both. Record your activities under Isolation Procedure as directed on the data sheet.
5. Examine the plates and record optimum growth temperature under Preliminary Observations on the data sheet. Find a well-isolated colony on one of the plates and record its morphology (including medium and incubation temperature) on the data sheet. Also include the result of the differential component of MacConkey agar as Test #1 under Differential Tests on the back of the data sheet. Continue to record your isolation procedure on the data sheet through inoculation of your pure culture (step 6).
6. Perform a Gram stain on a portion of one well-isolated colony (preferably the one whose morphology you described). Record its Gram reaction and cell morphology, arrangement, and size on the data sheet under Preliminary Observations. If the isolate is a Gram-negative rod, transfer a portion of the *same* colony to a tryptic soy agar slant (or another suitable growth medium as available in your lab) and incubate it at its optimum temperature. This is your pure culture to be used as a source of organisms for further testing. (**Once it has grown, keep it in the refrigerator and make fresh pure cultures as necessary. Cultures will last roughly three weeks in refrigeration.**) Complete the record of your isolation procedure on the data sheet.

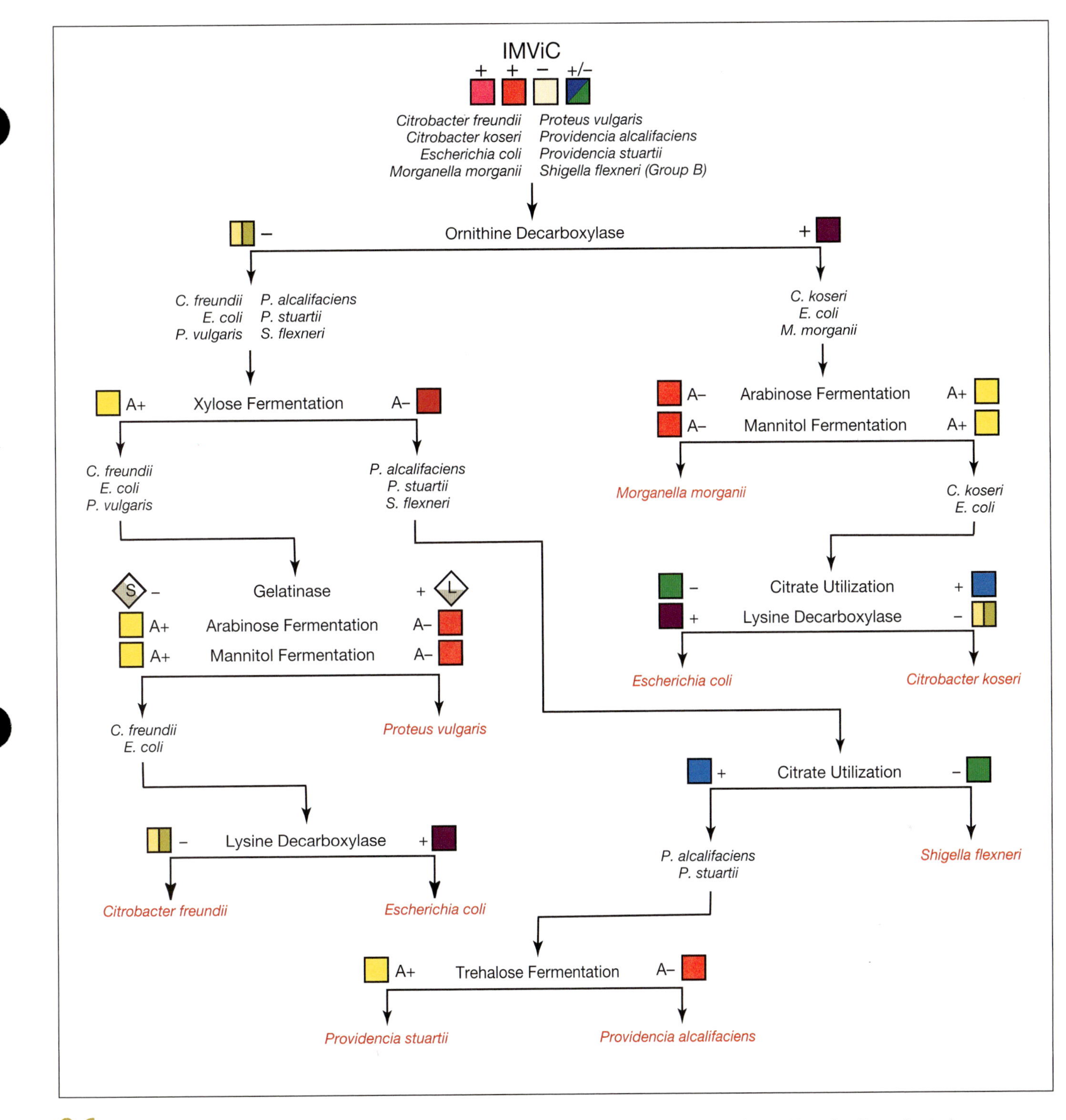


**9.6 Flowchart for Identifying Indole-Positive, MR-Positive, VP-Negative, Citrate-Positive or Citrate-Negative *Enterobacteriaceae***
See the appropriate exercises for instructions on how to run each test. Positive results are based on 1 to 2 days incubation at 35±1°C. Colors in the icons are not intended to match media exactly. Use controls for this. (Most results after Brenner and Farmer, 2005.)

**7** Perform an oxidase test on what remains of the isolated colony, or wait until your pure culture has grown and do it then. Record the result as Test #2 under Differential Tests on the data sheet. If the isolate grew on MacConkey agar and is an oxidase-negative, Gram-negative rod, it probably is a member of the *Enterobacteriaceae*. At the discretion of your instructor, further tests chosen from the characteristics listed under Theory may be run to assure that it is an organism from *Enterobacteriaceae*. (Unless this exercise has been combined with either Exercise 9-2 or 9-3, the oxidase test and any others assigned by your instructor are done for practice because you were given an unknown from the *Enterobacteriaceae*.)

8. You now will begin biochemical testing to identify your organism. The tests in the flowcharts were chosen because of their uniformity of results as published in standard microbiological references, so they give the best chance of correct identification. Be aware, however, that the symbol " + " indicates that 90% or more of the strains tested give a positive result. This means that as many as 10% of the strains tested give a negative result. The same is true of the symbol " – "; that is, as many as 10% of the strains give a positive result for that test. Bottom line: There are no guarantees that your particular lab strains will behave in the majority, and if not, you will misidentify your unknown. This issue, if relevant to your unknown, will be addressed in step 13.
9. Perform the IMViC series of tests (Indole Production, Methyl Red Test, Voges-Proskauer Test, and Citrate Utilization) at your isolate's optimum temperature. You may run these simultaneously.
   - Record the relevant inoculation information and results on the data sheet under Differential Tests. Use Test #3 for indole, Test #4 for MR and VP, and Test #5 for citrate.
   - After incubation, match your isolate's IMViC results with one shown in Figure 9.2, and proceed to the appropriate identification flowchart (Figs. 9.3 through 9.6). (***Note***: If a result is shown as +/– [as it is for all citrate tests], either result for that test is a match.) Record the figure number of the flowchart you will use on the data sheet.
10. Follow the tests in the appropriate flowchart to identify your unknown. Use these guidelines.
    - Do not run more than one test at a time (except for the IMViC) unless your instructor tells you to do so.
    - You will probably have to refer to the appropriate exercise for instructions on running and reading each test. We advise you to strictly follow the incubation time given—most are 24 to 48 hours—unless told otherwise by your instructor, but to incubate your unknown at *its* optimum temperature (which may differ from the one given in the exercise).
    - Where multiple tests are listed at a branch point, run only one, and then move to the next level in the flowchart as indicated by the result of that test. Choose the test whose incubation time best fits your lab schedule, or the easiest one, or the hardest one (for practice).
    - Continue to record the relevant inoculation information and results on the data sheet as you go. Keep accurate records of what you have done as you proceed. Do not enter all your data at the end of the project.
    - As you progress, either keep test cultures in the refrigerator or dispose of them as you go. Your instructor will advise you on your lab's convention.
11. When you have identified your organism, use *Bergey's Manual* or another standard reference to find one more test to run for confirmation. The confirmatory test does not have to separate the final organisms on the last branch of the flowchart. It only has to be a test for which you know the result and that you have not run already as part of the identification process. Record this test on the data sheet and identify it as the confirmatory test. Record the *expected* result on the Comments line. After incubation, note whether the organism's result matches (confirms) or differs from the expected result.

**12** After the confirmatory test, write the name of the organism on the back of the data sheet where it says, "My unknown is" and check with your instructor to see if you are correct. If you are, congratulations! You're finished! If you aren't correct, continue with step 13.

**13** Your instructor knows the identity of your unknown and can look at your results and the flowchart to find the test(s) responsible for the discrepancy between your provisional identification and your unknown's actual identity. After determining which test(s) gave an "incorrect" result, your instructor will write the test name(s) in the "rerun" space next to your identification.[2] These should be rerun with appropriate controls from your school's inventory of organisms. Record the test(s), control(s) used, and results for your unknown and controls on your data sheet as a continuation of differential tests. (That is, start with the space following your confirmatory test.) **This process is an important component of the unknown project. Checking the results with controls and your unknown will indicate if the "incorrect" test result was truly incorrect when previously run or if the strain of organism your school is using doesn't match the majority of strains for that test result.**

**14** Back to work for you, using your instructor's guidance! When you get a new provisional identification supported by a confirmatory test result or have evidence that one of the rerun tests gives the same the "incorrect" result, check with your instructor.

**15** Dispose of all cultures in the appropriate autoclave container.

[2] *Note to instructor:* Because of time and resource limitations, we look for tests the student has already run with results that match their actual unknown. We put a check mark next to those tests and tell students not to rerun them, but to use those results when they encounter the tests again and to move to the next branch in the flowchart as if they had rerun them.

9

## References


Brenner, Don J. and J. J. Farmer, III. Order XIII. "Enterobacteriales" in *Bergey's Manual of Systematic Bacteriology*, Vol. 2, *The Proteobacteria, Part B—The Gammaproteobacteria*. New York: Springer, 2005.

Farmer III, J. J., K. D. Boatwright, and J. Michael Janda. Chap. 42 in *Manual of Clinical Microbiology*, 9th ed. Patrick R. Murray, Ellen Jo Baron, James H. Jorgensen, Marie Louise Landry, and Michael A. Pfaller, eds. Washington, DC: ASM Press, 2007.

Forbes, Betty A., Daniel F. Sahm, and Alice S. Weissfield. Chap. 25 in *Bailey & Scott's Diagnostic Microbiology,* 11th ed. St. Louis, MO: Mosby, 2002.

Forsythe, Stephen H., Sharon L. Abbott, and Johann Pitout. Chap. 38 in *Manual of Clinical Microbiology,* Vol. 1, 11th ed. James H. Jorgensen, Michael A. Pfaller, Karen C. Carroll, Guido Funke, Marie Louise Landry, Sandra S. Richter, and David W. Warnock, eds. Washington, DC: ASM Press, American Society for Microbiology, 2015.

MacFaddin, Jean F. *Biochemical Tests for Identification of Medical Bacteria*, 3rd ed. Philadelphia: Lippincott Williams & Wilkins, 2000.

Strockbine, Nancy A., Cheryl A. Bopp, Patricia J. Fields, James B. Kaper, and James P. Nataro. Chap. 37 in *Manual of Clinical Microbiology,* Vol. 1, 11th ed. James H. Jorgensen, Michael A. Pfaller, Karen C. Carroll, Guido Funke, Marie Louise Landry, Sandra S. Richter, and David W. Warnock, eds. Washington, DC: ASM Press, American Society for Microbiology, 2015.

Winn Jr., Washington, Stephen Allen, William Janda, Elmer Koneman, Gary Procop, Paul Schreckberger, and Gail Woods. Chap. 6 in *Koneman's Color Atlas and Textbook of Diagnostic Microbiology,* 6th ed. Philadelphia: Lippincott Williams & Wilkins, 2006.

Name ______________________________

Date ______________________________

Lab Section ______________________________

I was present and performed this exercise (initials) ______________________________

# DATA SHEET 9-1

## Identification of Selected *Enterobacteriaceae*

**Unknown Number** ______________________________

### Isolation Procedure

Record all activities associated with isolation of your organisms—from mixed culture to pure culture. Always include the date, source of inoculum, destination, type of inoculation, incubation temperature, and any other relevant information. Also make note of transfers made to keep your pure culture fresh. (This log must be kept current.)

### Preliminary Observations

Colony Morphology (include medium and incubation temperature) ______________________________

Gram stain ______________ Cell dimensions ______________ Optimum temperature ______________

Cellular morphology and arrangement ______________________________

### Differential Tests

Begin recording with the differential MacConkey result, followed by the oxidase test, and then your IMViC results. This log must be kept current.

Record the figure number of the identification chart you are using. If you change charts, indicate the figure number of the new chart and the date of the change.

**Test #1:** ______ Date Begun: ______ Date Read: ______ Result: ______

Comments: ______

**Test #2:** ______ Date Begun: ______ Date Read: ______ Result: ______

Comments: ______

**Test #3:** ______ Date Begun: ______ Date Read: ______ Result: ______

Comments: ______

**Test #4:** ______ Date Begun: ______ Date Read: ______ Result: ______

Comments: ______

**Test #5:** ______ Date Begun: ______ Date Read: ______ Result: ______

Comments: ______

**Test #6:** ______ Date Begun: ______ Date Read: ______ Result: ______

Comments: ______

**Test #7:** ______ Date Begun: ______ Date Read: ______ Result: ______

Comments: ______

**Test #8:** ______ Date Begun: ______ Date Read: ______ Result: ______

Comments: ______

**Test #9:** ______ Date Begun: ______ Date Read: ______ Result: ______

Comments: ______

**Test #10:** ______ Date Begun: ______ Date Read: ______ Result: ______

Comments: ______

**Test #11:** ______ Date Begun: ______ Date Read: ______ Result: ______

Comments: ______

**Test #12:** ______ Date Begun: ______ Date Read: ______ Result: ______

Comments: ______



My unknown is: ______ Rerun[1]: ______

______ Rerun: ______

______ Rerun: ______

[1] Your instructor will write what tests to rerun in this space if you misidentify your unknown.

# Identification of Selected Gram-Positive Cocci

EXERCISE **9-2**

## Theory

Gram-positive cocci are frequent isolates in a clinical setting because they are common inhabitants of skin and mucous membranes. Five genera, representing very diverse groups, are briefly described below and will be used in this lab exercise.

### *Staphylococcus* (Fig. 9.7)

- Phylum *Firmicutes*, Order *Bacillales*
- Gram-positive cocci in singles, pairs, tetrads, or clusters (especially when grown in broth)
- Catalase positive
- With rare exception, facultatively anaerobic
- Most are oxidase negative
- G + C content within the genus ranges between 27% and 41%.
- Grow in 6.5% NaCl
- Most produce acid from glucose
- Most are resistant to bacitracin
- Key pathogen is *Staphylococcus aureus* (toxic shock syndrome and a variety of other skin and deep organ infections, including bacteremia)

### *Streptococcus* (Fig. 9.8)

- Phylum *Firmicutes*, order *Lactobacillales*
- Gram-positive cocci to ovoid cocci in singles, pairs, or short chains (especially when grown in broth)
- Gray to white, moist colonies are frequently observed
- Catalase negative
- Oxidase negative
- Facultatively anaerobic
- Nutritionally fastidious
- Some require 5% $CO_2$ for growth
- Ferment glucose and other carbohydrates, mostly to lactic acid
- G + C content within the genus ranges between 33% and 46%
- Many species produce hemolysins that either completely (β-hemolysins) or partially (α-hemolysins) destroy erythrocytes; some species are nonhemolytic
- Beta-hemolytic streptococci have traditionally been grouped by antigens first described by Rebecca Lancefield; important Lancefield groups include Group A (*S. pyogenes*) and Group B (*S. agalactiae*); *S. pneumoniae* has no Lancefield antigen
- Key pathogens are *Streptococcus pyogenes* (strep throat, necrotizing fasciitis, scarlet fever) and *S. pneumoniae* (bacterial pneumonia, otitis media, and bacteremia)

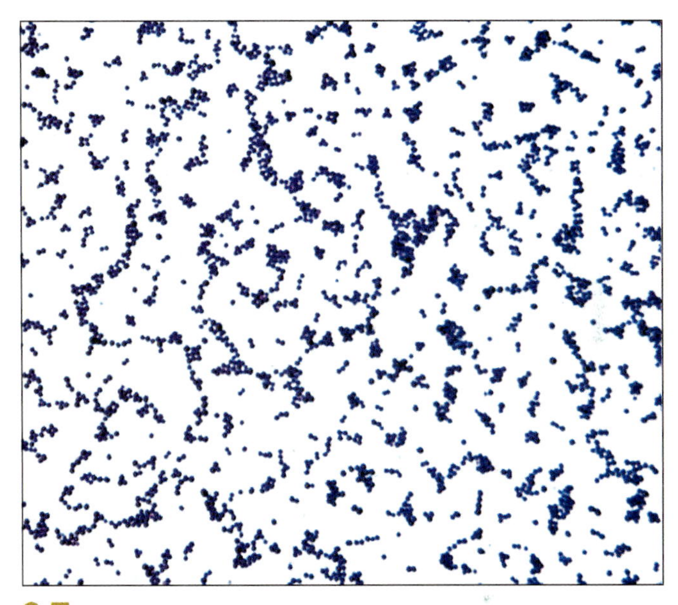

**9.7 Gram Stain of *Staphylococcus aureus*** ■ This specimen grown in broth illustrates the grapelike clusters of cells characteristic of the genus. Specimens grown on solid media may not show the clusters as clearly.

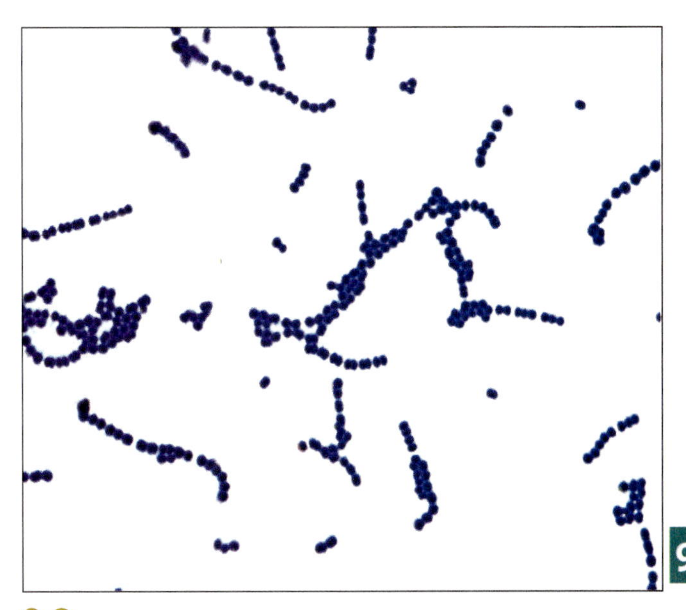

**9.8 Gram Stain of *Streptococcus salivarius*** ■ This specimen grown in broth illustrates the streptococcal arrangement of cells characteristic of the genus. Specimens grown on solid media may not show the chains as clearly.

### *Enterococcus*

- Phylum *Firmicutes*, order *Lactobacillales*
- Formerly members of the genus *Streptococcus*
- Gram-positive cocci to ovoid cocci in singles, pairs, or short chains (especially when grown in broth); may be more rod-shaped if grown on solid media
- Catalase negative
- Oxidase negative
- Facultatively anaerobic
- Lactic acid, but no gas, is the sole end product of fermentation
- Grow in 6.5% NaCl broth
- Grow in bile esculin
- Most are PYR positive
- G+C content within the genus ranges between 35% and 45%
- Most express the Lancefield Group D antigen
- Most species are commensals or opportunistic pathogens; the key opportunistic pathogen is *E. faecalis* (urinary tract infections, wound infections, and bacteremia in seriously ill elderly persons)

### *Micrococcus*

- Phylum *Actinobacteria*, order *Micrococcales*
- Gram-positive cocci in pairs and tetrads
- Obligately aerobic
- Oxidase positive
- Do not produce acid from glucose
- Bacitracin susceptible
- Catalase positive
- G+C content within the genus ranges between 69% and 76%.
- Grow in 6.5% NaCl
- Commensals or opportunistic pathogens. *M. luteus* is a common skin commensal

### *Kocuria*

- Phylum *Actinobacteria*, order *Micrococcales*
- A small genus; most species were formerly classified in the genus *Micrococcus*
- Gram-positive cocci in pairs and tetrads
- Obligately aerobic
- Do not produce acid from glucose
- Bacitracin susceptible
- Catalase positive
- G+C content within the genus ranges between 66% and 75%.
- Commensals or opportunistic pathogens, especially among immunocompromised patients

The organisms to be used as unknowns are listed in Figure 9.9, as are the tests to be used in their identification. (*Note*: Your instructor will choose Gram-positive coccus unknowns appropriate to your microbiology course and facilities.)

## Application

Identification of Gram-positive cocci from human specimens requires a coordinated and integrated use of biochemical tests and stains, something that is done daily in medical labs worldwide. Although several serological tests allow rapid identification, flowcharts are still a useful way to visualize the process of identification by elimination.

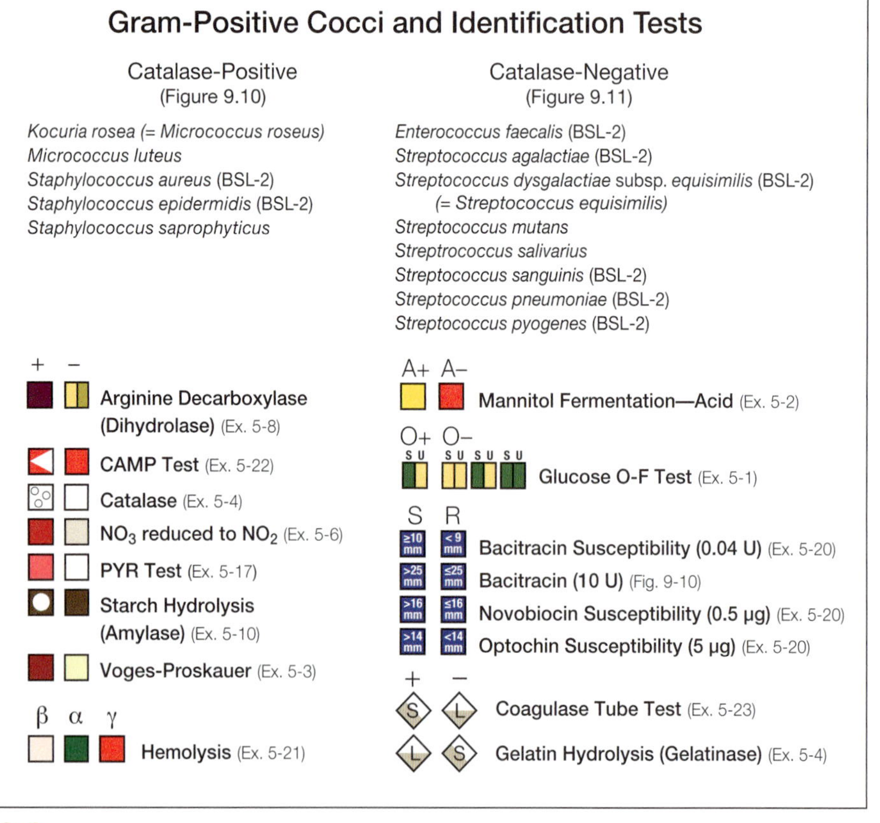


**9.9 List of Organisms and a Key to Test and Result Icons Used in This Exercise** The organisms listed are Gram-positive cocci used in this exercise, though not all may be assigned as unknowns in your class. The test list indicates ones you may be using to identify your unknown and refers you to the appropriate exercise for instructions on how to run them. Icon colors are to remind you what the results look like. They should not be used to compare with your results. ***Note:*** For the fermentation tests, "A+" means acid is produced; "A−" means no acid is produced.

## In This Exercise

This exercise will span several lab periods. You will be given an unknown Gram-positive coccus from the organisms listed in Figure 9.9, and then run biochemical tests as directed by the flowcharts in Figures 9.10 or 9.11 to identify it.

## Materials

### Per Class

- Media (and appropriate reagents) listed in Figure 9.9 for biochemical testing
- Pure cultures of organisms listed in Figure 9.9 to be used for positive controls and unknowns. All are available from either Ward's Science or Carolina Biological Supply companies. (**BSL-2 organisms should not be chosen as unknowns by schools or classes not adequately prepared to handle them.**)
- Todd-Hewitt broth, tryptic soy broth, or brain–heart infusion broth
- (Optional) candle jar setup
- (Optional) $CO_2$ incubator

### Per Student

- Lab coat
- Disposable gloves
- Chemical eye protection
- Two PEA, CNA, or 5% sheep blood agar plates
- One unknown organism[1] from the list in Figure 9.9 (numbered to an instructor's key)
- TSA slants (enough to keep the culture fresh over the time required for identification)
- Gram-stain kit
- Microscope slides
- 3% $H_2O_2$
- Compound microscope with oil-objective lens and ocular micrometer
- Immersion oil
- Lens paper
- Bibulous paper
- Sterile wooden sticks or disposable loops
- (Optional) bacterial incinerator

[1] *Note to instructor:* This exercise can be combined with Exercise 9-1 by passing out a mixture of a Gram-negative enteric and a Gram-positive coccus as unknowns. The student then must isolate the two organisms from a mixed culture, and then proceed to identify the two independently. If this option is chosen, grow the two unknowns separately, and then mix them immediately prior to handing them out in lab.

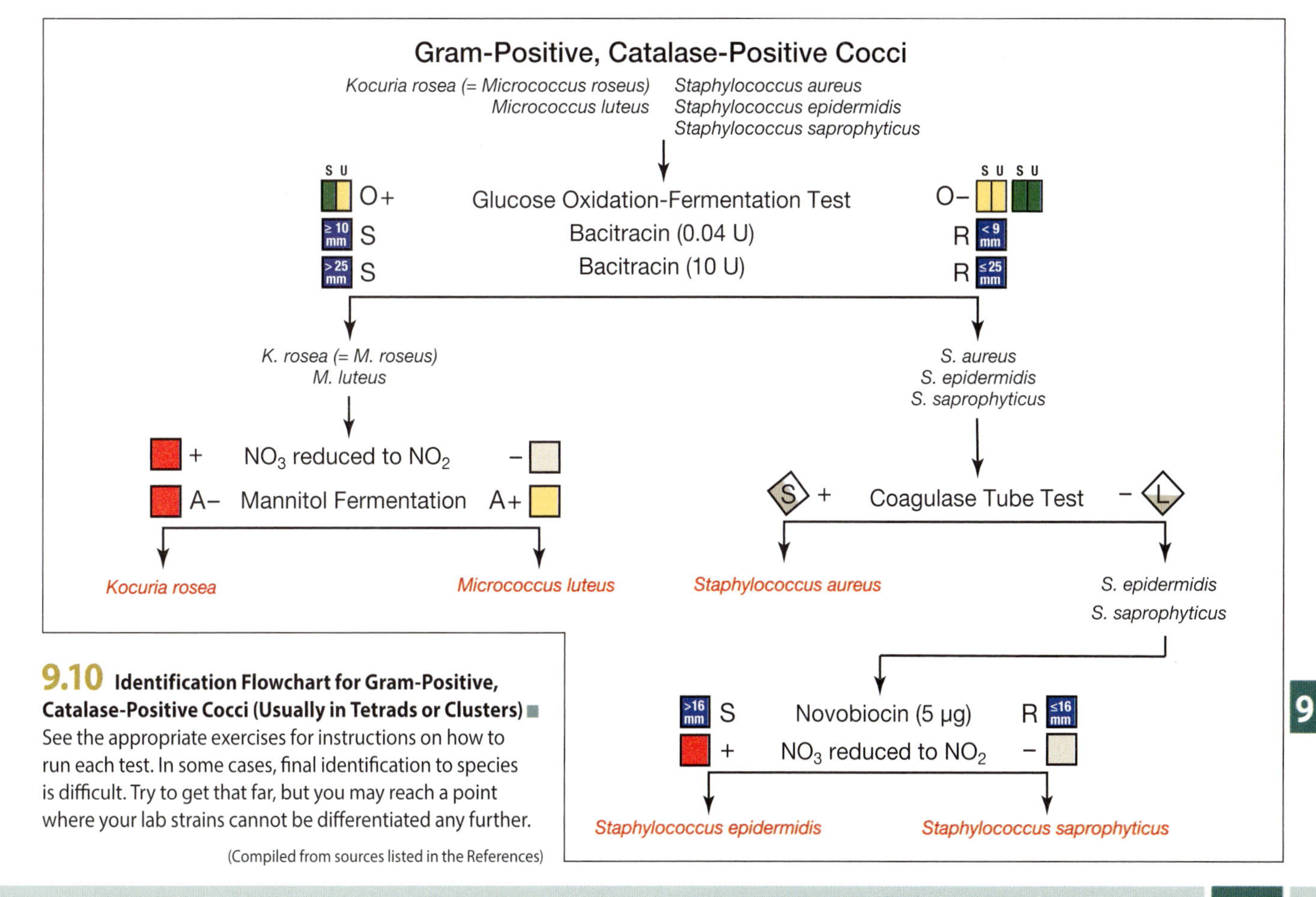


**9.10 Identification Flowchart for Gram-Positive, Catalase-Positive Cocci (Usually in Tetrads or Clusters)** See the appropriate exercises for instructions on how to run each test. In some cases, final identification to species is difficult. Try to get that far, but you may reach a point where your lab strains cannot be differentiated any further.

(Compiled from sources listed in the References)

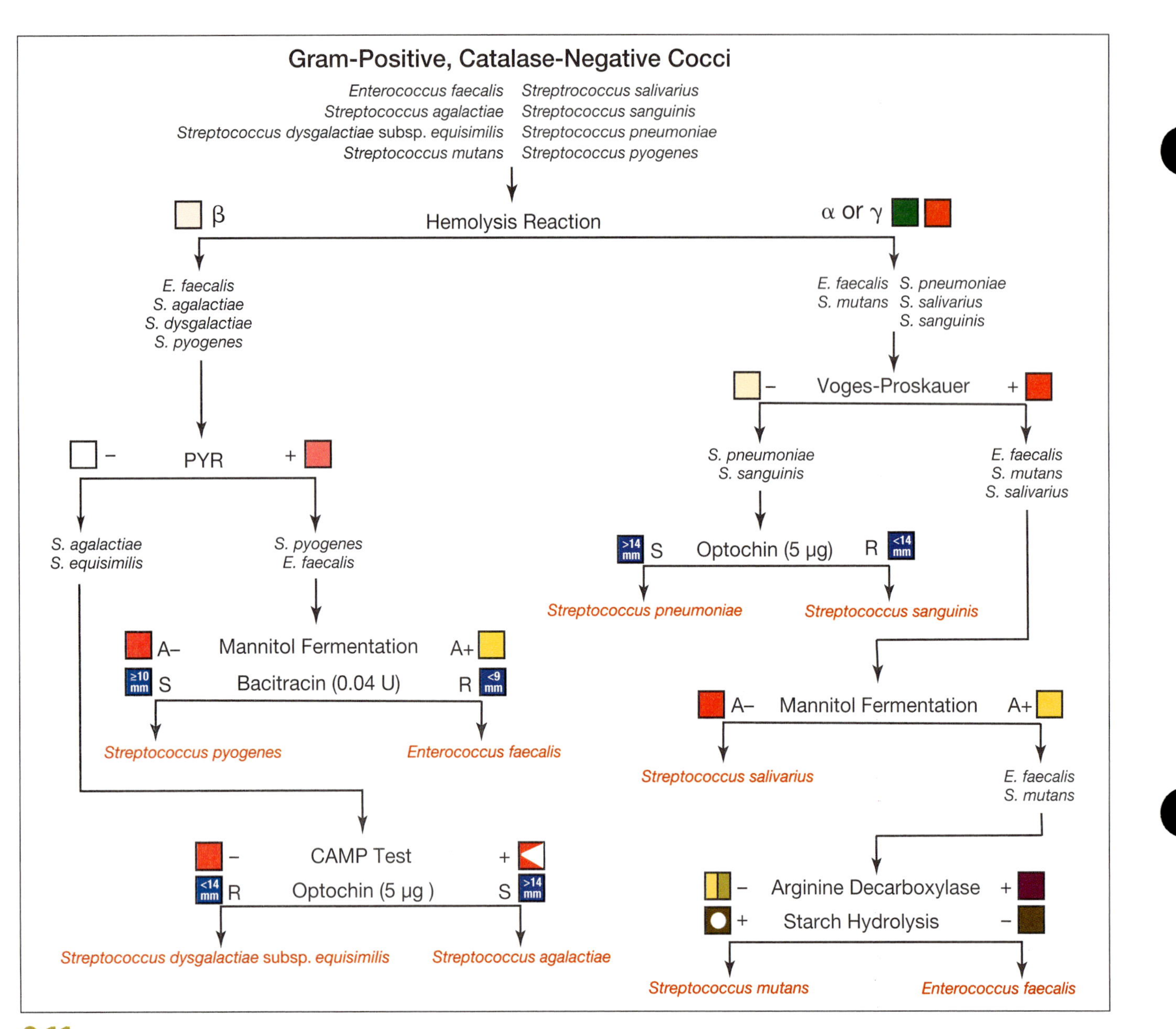


**9.11 Identification Flowchart for Gram-Positive, Catalase-Negative Cocci (Usually in Pairs or Chains)** ■ See the appropriate exercises for instructions on how to run each test. In some cases, final identification to species is difficult. Try to get that far, but you may reach a point where your lab strains cannot be differentiated any further. (Compiled from sources listed in the References)

## PROCEDURE

1. Wear a lab coat, gloves, and chemical eye protection when performing this procedure. **In addition, all procedures should be run using BSL-2 precautions unless otherwise directed by your instructor.**
2. Obtain an unknown. Record its number on the data sheet, page 563. ***Note***: You are urged to keep accurate records of all your activities related to identification of this unknown as you proceed (and not after the fact). It is surprising how quickly and easily we (humans) forget details…
3. Streak the sample for isolation on two sheep blood agar, PEA, or CNA plates. If instructed to do so, incubate one in the candle jar and the other aerobically. Incubate both at 30°C–35°C. Otherwise, incubate aerobically at 30°C–35°C. Record your activities on the data sheet under Isolation Procedure.
4. It is best to check your plate for isolation at 24 hours. If you have isolation, continue with step 5. If you don't have time to continue with step 5, refrigerate your plates until you do. If you don't see isolation, ask your instructor if you should re-streak for isolation, let the original plates continue to incubate, or do both. Record your activities under Isolation Procedure as directed on the data sheet.
5. Examine the plates and record optimum growth temperature and $CO_2$ requirement (if determined)

under Preliminary Observations on the data sheet. Find a well-isolated colony on one of the plates and record its morphology (including medium and incubation temperature) on the data sheet. Also record the result of the differential component of blood agar as Test #1 on the back of the data sheet if you used sheep blood agar for isolation. Continue to record your isolation procedure on the data sheet through inoculation of your pure culture (step 6).

6 Perform a Gram stain on a portion of the colony whose morphology you described. Continue testing colonies until you find one that is a Gram-positive coccus. ***Note:*** If the colonies are too small for a Gram stain and transferring to an agar slant, follow the second bullet below to start a pure culture, then Gram stain the pure culture. The pure culture takes priority; Gram staining can wait!
   - Record its Gram reaction and cell morphology, arrangement, and size under Preliminary Observations on the data sheet. If you have difficulty discerning cell arrangement, you have the option of starting your pure culture (next bullet) in a broth, which when Gram stained should be easier to read.
   - If the isolate is a Gram-positive coccus, transfer a portion of the *same* colony to an agar slant (as available in your lab) and incubate it at its optimum temperature. This is your pure culture to be used as a source of organisms for further testing. Complete the record of your isolation procedure on the data sheet.
   - Once your pure culture has grown, keep it in the refrigerator and make fresh pure cultures as necessary. They will last roughly three weeks in refrigeration.

7 Perform a catalase test on what remains of the isolated colony, or wait until your pure culture has grown and do it then. (If your pure culture is in a broth, you may need to transfer it to a slant and wait until it grows before running the catalase test.) Record the result as Test #1 (or #2 if hemolysis reaction is already recorded) under Differential Tests on the data sheet. (***Note***: Be sure to use growth from the top of the colony if isolation was performed on a blood agar plate. This minimizes the possibility of a false positive from the catalase-positive erythrocytes in the medium. Alternatively, you can postpone the catalase test until your pure culture has grown.)

8 Use the chart in Figure 9.10 for identification of catalase-positive cocci (usually in tetrads or clusters). Use the chart in Figure 9.11 if the isolate is catalase-negative (usually with cocci in pairs or chains). Record the figure number of the flowchart you will use on the data sheet. If your isolate is catalase-negative, ask your instructor if you would get better growth of your pure culture using Todd-Hewitt broth or brain–heart infusion broth and incubating in a candle jar or $CO_2$ incubator.

9 Now you will begin biochemical testing to identify your organism. The tests in the flowcharts were chosen because of their uniformity of results as published in standard microbiological references, so they give the best chance for correct identification. Be aware, however, that the symbol " + " indicates that 90% or more of the strains tested give a positive result. This means that as many as 10% of the strains tested give a negative result. The same is true of the symbol " – "; that is, as many as 10% of the strains give a positive result. Bottom line: There are no guarantees that your lab strains will behave in the majority, and if not, you will misidentify your unknown. This issue, if relevant to your unknown, will be addressed in step 13.

10 Follow the tests in the appropriate flowchart to identify your unknown.[2] Use these guidelines.
   - Do not run more than one test at a time unless your instructor tells you to do so.
   - You will probably have to refer to the appropriate exercise for instructions on running and reading each test. We advise you to strictly follow the incubation time given—most are 24 to 48 hours—unless told otherwise by your instructor, but to incubate your unknown at *its* optimum temperature (which may differ from the one given in the exercise).
   - Where multiple tests are listed at a branch point, run only one, and then move to the next level in the flowchart as indicated by the result of that test. Choose the test whose incubation time best fits your lab schedule, or the easiest one, or the hardest one (for practice).
   - Continue to record the relevant inoculation information and results on the data sheet as you go. Keep accurate records of what you have done as you proceed. *Do not* enter all your data at the end of the project.
   - As you progress, either keep test cultures in the refrigerator or dispose of them as you go. Your instructor will advise you on your lab's convention.

[2] These flowcharts can be used only for the organisms listed in Figure 9.9.

**11** When you have identified your organism, use *Bergey's Manual* or another standard reference to find one more test to run for confirmation. The confirmatory test does not have to separate the final organisms on the last branch of the flowchart. It only has to be a test for which you know the result and that you have not run already as part of the identification process. Record this test on the data sheet, and identify it as the confirmatory test. Record the *expected* result on the Comments line. After incubation, note whether the organism's result matches (confirms) or differs from the expected result.

**12** After the confirmatory test, write the organism's name on the back of the data sheet under where it says, "My unknown is" and check with your instructor to see if you're correct. If you are, congratulations! You're finished! If you aren't correct, continue with step 13.

**13** Your instructor knows the identity of your unknown and can look at your results and the flowchart to find the test(s) responsible for the discrepancy between your provisional identification and your unknown's actual identity. After determining which test(s) gave an "incorrect" result, your instructor will write the test name(s) in the "rerun" space next to your identification.[3] These should be rerun with appropriate controls from your school's inventory of organisms. Record the test(s), control(s) used, and results for your unknown and controls on your data sheet as a continuation of differential tests. (That is, start with the space following your confirmatory test.) **This process is an important component of the unknown project. Checking the results with controls and your unknown will indicate if the "incorrect" test result was truly incorrect when previously run or if the strain of organism your school is using doesn't match the majority of strains for that test result.**

**14** Back to work for you, using your instructor's guidance! When you get a new provisional identification supported by a confirmatory test result or have evidence that one of the rerun tests gives the same the "incorrect" result, check with your instructor.

**15** Dispose of all cultures in the appropriate autoclave container.

[3] *Note to instructor:* Because of time and resource limitations, we look for tests the student has already run with results that match their actual unknown. We put a check mark next to those tests and tell students not to rerun them, but to use those results when they encounter the tests again and to move to the next branch in the flowchart as if they had rerun them.

## References


Becker, Karsten, Robert L. Skov, and Christof von Eiff. Chap. 21 in *Manual of Clinical Microbiology*, Vol. 1, 11th ed. James H. Jorgensen, Michael A. Pfaller, Karen C. Carroll, Guido Funke, Marie Louise Landry, Sandra S. Richter, and David W. Warnock, eds. Washington, DC: ASM Press, American Society for Microbiology, 2015.

Busse, Hans-Jürgen. Genus I. *Micrococcus*, pages 571–576 in *Bergey's Manual of Systematic Bacteriology*, 2nd ed., Vol. 5A, The Actinobacteria. New York: Springer, 2012.

MacFaddin, Jean F. *Biochemical Tests for Identification of Medical Bacteria*, 3rd ed. Philadelphia: Lippincott Williams & Wilkins, 2000.

Ruoff, Kathryn L. Chap. 22 in *Manual of Clinical Microbiology*, 10th ed. James Versalovic, Karen C. Carroll, Guido Funke, James H. Jorgensen, Marie Louise Landry, and David W. Warnock, eds. Washington, DC: ASM Press, 2011.

Spellerberg, Barbara and Claudia Brandt. Chap. 22 in *Manual of Clinical Microbiology*, Vol. 1, 11th ed. James H. Jorgensen, Michael A. Pfaller, Karen C. Carroll, Guido Funke, Marie Louise Landry, Sandra S. Richter, and David W. Warnock, eds. Washington, DC: ASM Press, American Society for Microbiology, 2015.

Stackebrandt, Eriko and Peter Schumann. Genus V. *Kocuria*, pages 626–635 in *Bergey's Manual of Systematic Bacteriology*, 2nd ed., Vol. 5A, The Actinobacteria. New York: Springer, 2012.

Švec, Pavel and Luc A. Devriese. Pages 594–607 in *Bergey's Manual of Systematic Bacteriology*, 2nd ed. Vol. 3, The Firmicutes. New York: Springer, 2009.

Teixeira, Lúcia Martins, Maria da Glória Siqueira Carvalho, Richard R. Facklam, and Patricia Lynn Shewmaker. Chap. 23 in *Manual of Clinical Microbiology*, Vol. 1, 11th ed. James H. Jorgensen, Michael A. Pfaller, Karen C. Carroll, Guido Funke, Marie Louise Landry, Sandra S. Richter, and David W. Warnock, eds. Washington, DC: ASM Press, American Society for Microbiology, 2015.

Teuber, Michael. Pages 711–722 in *Bergey's Manual of Systematic Bacteriology*, 2nd ed., Vol. 3, The Firmicutes. New York: Springer, 2009.

Tille, Patricia M. Chapter 14 in *Bailey & Scott's Diagnostic Microbiology*, 13th ed. St. Louis: Mosby, 2014.

Whiley, Robert A. and Jeremy M. Hardie. Pages 655–711 in *Bergey's Manual of Systematic Bacteriology*, 2nd ed., Vol. 3, The Firmicutes. New York: Springer, 2009.

Name ______________________________

Date ______________________________

Lab Section ______________________________

I was present and performed this exercise (initials) ______________

DATA SHEET 9-2

# Identification of Selected Gram-Positive Cocci

**Unknown Number** ______________

### Isolation Procedure

Record all activities associated with isolation of your organisms—from mixed culture to pure culture. Always include the date, source of inoculum, destination, type of inoculation, incubation temperature, and any other relevant information. Also make note of transfers made to keep your pure culture fresh. (This log must be kept current.)

### Preliminary Observations

Colony Morphology (include medium and incubation temperature) ______________

Gram stain ______________ Cell dimensions ______________ $CO_2$ requirement ______________

Cellular morphology and arrangement ______________ Optimum temperature ______________

### Differential Tests

Begin recording with the hemolysis result (if determined), followed by the catalase test. This log must be kept current.

Record the figure number of the identification chart you are using. If you change charts, indicate the figure number of the new chart and the date of the change.

**Test #1:** ____________ Date Begun: ________ Date Read: ________ Result: ________

Comments: ______________________________

**Test #2:** ____________ Date Begun: ________ Date Read: ________ Result: ________

Comments: ______________________________

**Test #3:** ____________ Date Begun: ________ Date Read: ________ Result: ________

Comments: ______________________________

**Test #4:** ____________ Date Begun: ________ Date Read: ________ Result: ________

Comments: ______________________________

**Test #5:** ____________ Date Begun: ________ Date Read: ________ Result: ________

Comments: ______________________________

**Test #6:** ____________ Date Begun: ________ Date Read: ________ Result: ________

Comments: ______________________________

**Test #7:** ____________ Date Begun: ________ Date Read: ________ Result: ________

Comments: ______________________________

**Test #8:** ____________ Date Begun: ________ Date Read: ________ Result: ________

Comments: ______________________________

**Test #9:** ____________ Date Begun: ________ Date Read: ________ Result: ________

Comments: ______________________________

**Test #10:** ____________ Date Begun: ________ Date Read: ________ Result: ________

Comments: ______________________________

**Test #11:** ____________ Date Begun: ________ Date Read: ________ Result: ________

Comments: ______________________________

**Test #12:** ____________ Date Begun: ________ Date Read: ________ Result: ________

Comments: ______________________________

9

My unknown is: ____________________ Rerun[1]: ____________________

____________________ Rerun: ____________________

____________________ Rerun: ____________________

[1] Your instructor will write what tests to rerun in this space if you misidentify your unknown.

# Identification of Selected Gram-Positive Rods

EXERCISE 9-3

## Theory

Several genera of Gram-positive rods have environmental, industrial, and clinical importance. Among these are *Bacillus*, *Clostridium*, *Corynebacterium*, *Lactobacillus*, and *Mycobacterium*. Each is briefly characterized below.

### *Bacillus* (Fig. 9.12)

- Phylum *Firmicutes*, order *Bacillales*
- A large, heterogeneous group comprising 162 species (as of 2011)
- Gram-positive rods (at least early in growth—Fig. 9.12A), in singles or chains
- Aerobic or facultatively anaerobic
- Produce endospores aerobically; spore shape and position are variable, but never more than one per cell
- Most are catalase positive
- Most are motile
- Most are soil saprophytes and are chemoorganotrophic
- Key pathogens are *B. anthracis* (anthrax) and *B. cereus* (food poisoning and opportunistic infections)

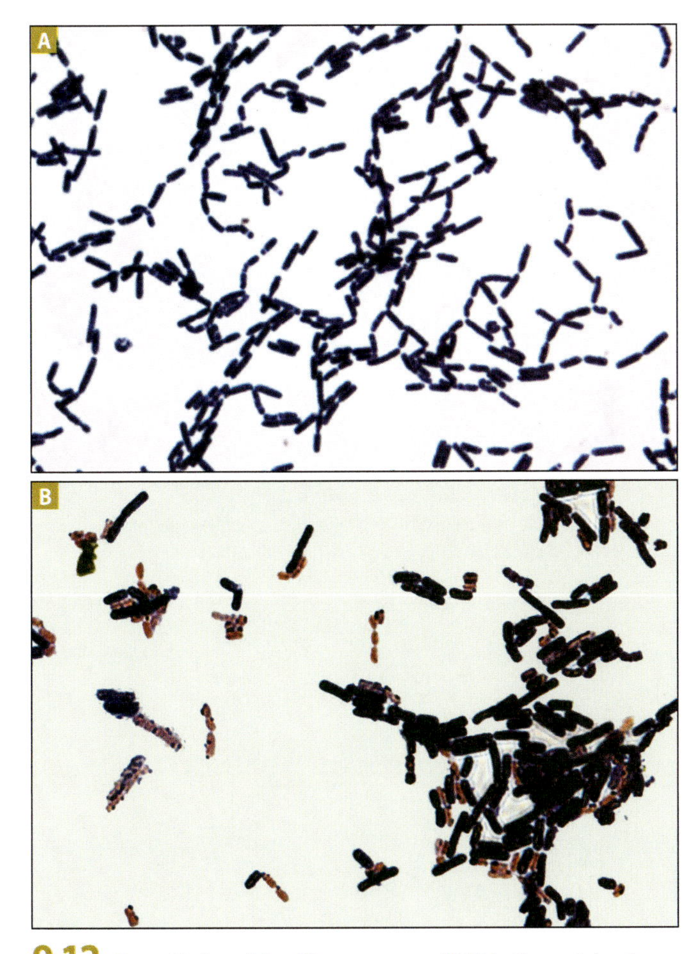


**9.12 Gram Stains of *Bacillus cereus*** ■ (**A**) This Gram stain of a young (<24 hours old) *B. cereus* culture stains uniformly Gram-positive. (**B**) With age, many *Bacillus* species lose their ability to retain the crystal violet during a Gram stain and appear pink, some with purple spots, as in this specimen.

### *Lactobacillus* (Fig. 9.13)

- Phylum *Firmicutes*, order *Lactobacillales*
- Gram-positive rods, sometimes in chains
- Facultatively anaerobic with fermentative metabolism; lactic acid is an abundant product
- Catalase negative
- Oxidase negative
- No endospores
- Nonmotile
- Isolated from a variety of foods (dairy, fish, grain, meat), and many are in normal flora (mouth, intestines, and vagina)
- No key pathogens and therefore not usually identified to species in clinical setting

### *Clostridium* (Fig. 9.14)

- Phylum *Firmicutes*, order *Clostridiales*
- A large, heterogeneous genus comprising over 160 species belonging to at least 12 different groups
- Gram-positive rods (at least early in growth), in singles, pairs, or chains

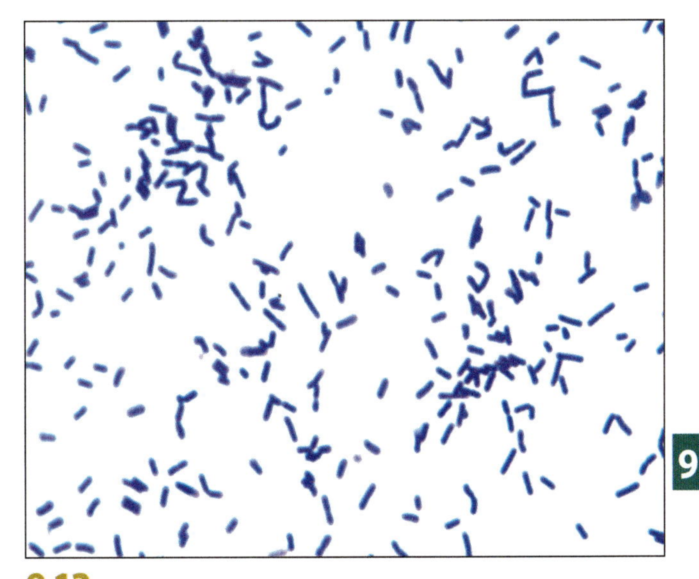

**9.13 Gram Stain of *Lactobacillus* spp.** ■ Cells in the genus *Lactobacillus* range from long, slender rods to coccobacilli.

- Most are obligate anaerobes, but some are microaerophiles
- Produce endospores, but not aerobically; spore shape and position are variable, but usually cause distention (bulging) of the cell
- Most are catalase negative
- Most are chemoorganotrophic and are isolated from soil, sewage, or marine sediments
- Key pathogens are *C. tetani* (tetanus), *C. botulinum* (botulism), *C. perfringens* (food poisoning and gas gangrene), and *C. difficile* (pseudomembranous colitis)

### *Corynebacterium* (Fig. 9.15)

- Phylum *Actinobacteria*, order *Corynebacteriales*
- Gram-positive rods, often club-shaped, in singles, pairs (V-forms), or arranged in stacks (palisades) or irregular clusters
- Metachromatic granules often present
- Facultatively anaerobic, but some are aerobes
- Catalase positive
- No endospores
- Nonmotile
- Some are found in the environment and many are in the normal flora of humans (skin and mucous membranes)
- Key pathogen is *C. diphtheriae* (diphtheria), although many species are opportunistic pathogens

### *Mycobacterium* (Fig. 9.16)

- Phylum *Actinobacteria*, order *Corynebacteriales*
- Over 130 recognized species divided into slow growers (>7 days to form colonies on Löwenstein-Jensen medium—about 70 species) and rapid growers (<7 days to form colonies on L-J medium—about 60 species)

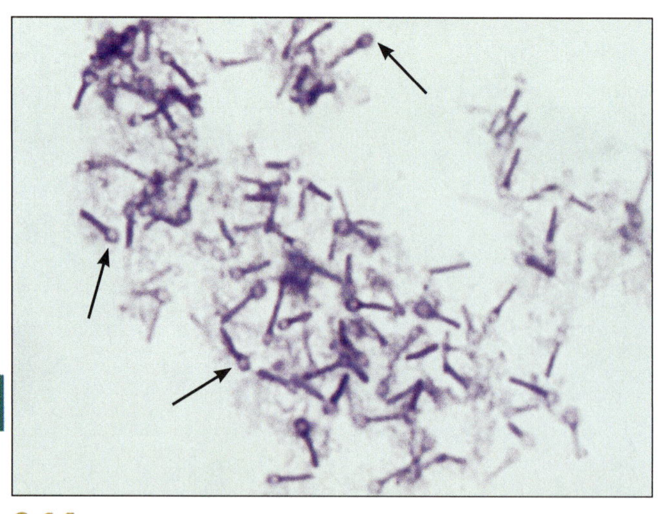

**9.14 Gram Stain of *Clostridium tetani*** ■ Note the round, terminal endospores (arrows) that distend the cells.

- Weakly Gram-positive curved or straight rods, sometimes branched or filamentous
- Acid-fast (generally in early stages of growth)
- Aerobic to microaerophilic
- Catalase positive
- Nonmotile
- No endospores
- Colony pigmentation with and without light is of taxonomic use
- Many are aquatic saprophytes
- Key pathogens are *M. tuberculosis* (tuberculosis) and *M. leprae* (leprosy or Hansen's disease). Many rapidly growing species are opportunists.

## ■ Application

Staining reactions and biochemical testing are used widely in bacterial identification.

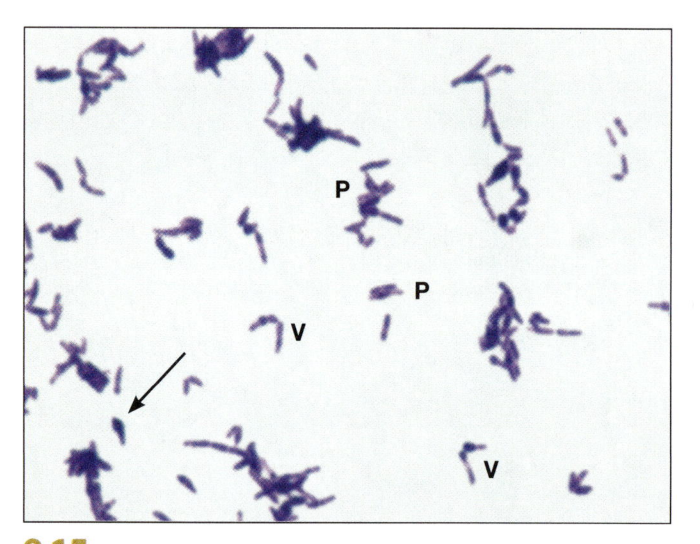


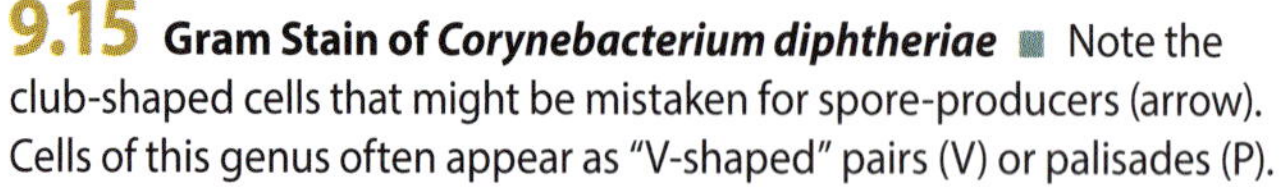

**9.15 Gram Stain of *Corynebacterium diphtheriae*** ■ Note the club-shaped cells that might be mistaken for spore-producers (arrow). Cells of this genus often appear as "V-shaped" pairs (V) or palisades (P).

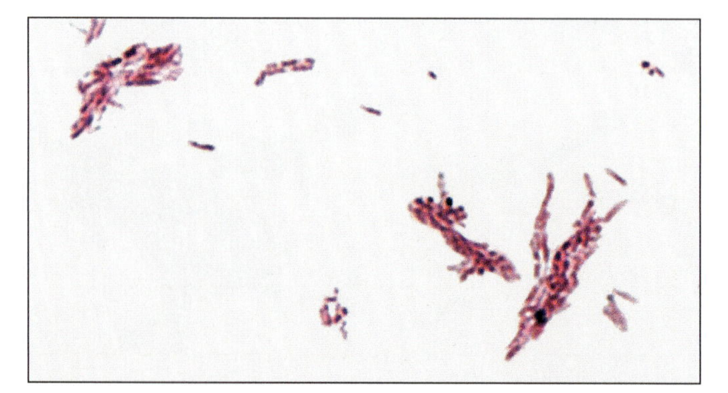

**9.16 Acid-Fast Stain of *Mycobacterium nonchromogenicum*** ■ *Mycobacterium* species are typically composed of slender rods with a "beaded" appearance (note the dark spots). Because of their waxy walls, they are difficult to emulsify and are frequently seen in clumps.

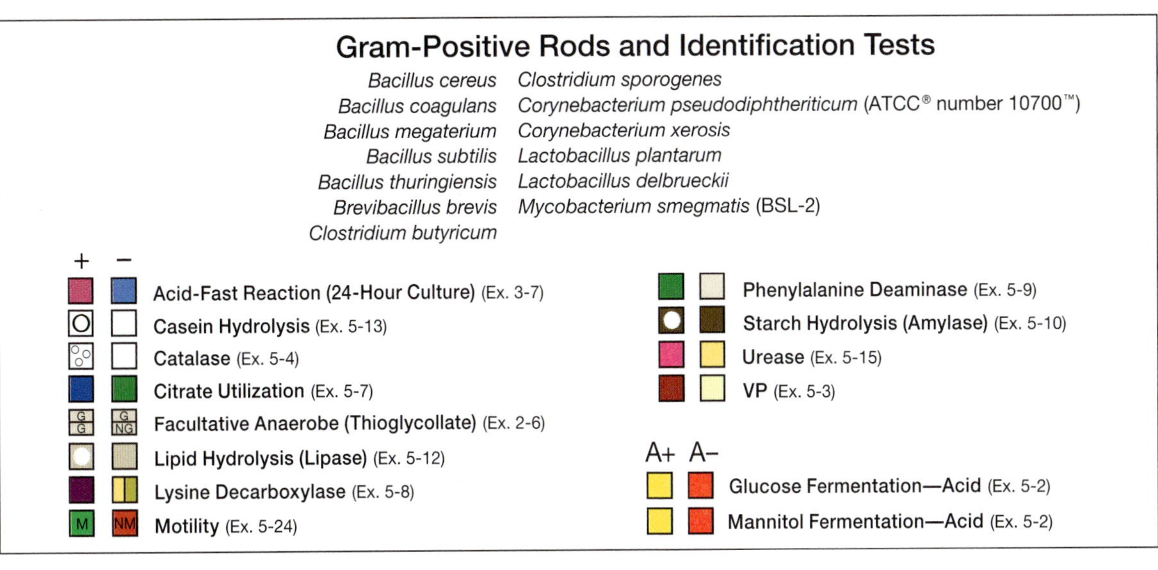


**9.17 List of Organisms and a Key to Test and Result Icons Used in This Exercise** ■ The organisms listed are Gram-positive rods available from either Ward's Science or Carolina Biological Supply companies. Icon colors are to remind you what the results look like. They should not be used to compare your results. ***Note:*** For the fermentation tests, "A+" means acid is produced; "A–" means no acid is produced.

## In This Exercise

This exercise will span several lab periods. You will be assigned an unknown Gram-positive rod from the list in Figure 9.17 and run biochemical tests as directed by the appropriate flowchart in Figure 9.18 to identify it. For obvious reasons, pathogens are not included, but the organisms chosen will give you an overview of the characteristics of each genus as well as practice in the identification process.

## Materials

### Per Class

- Media (and appropriate reagents) listed in Figure 9.17 for biochemical testing
- Pure cultures of organisms listed in Figure 9.17 to be used for positive controls and unknowns. All are available from either Ward's Science or Carolina Biological Supply companies. (**BSL-2 organisms should not be chosen as unknowns by schools or classes not adequately prepared to handle them.**)
- (Optional) anaerobic jars and GasPaks to accommodate one plate per student

### Per Student

- Lab coat
- Disposable gloves
- Chemical eye protection
- One unknown organism from the list in Figure 9.17 in thioglycollate broth[1] (numbered to an instructor's key)
- Two agar plates for streaking: instructor will choose from tryptic soy agar, TSA plus 5% sheep blood, PEA, or brain–heart infusion agar
- TSA slants or thioglycollate broths (enough to keep the culture fresh over the time required for identification)
- Gram-stain kit
- Microscope slides
- Compound microscope with ocular micrometer and oil objective
- Immersion oil
- Lens paper
- Bibulous paper
- Sterile wooden sticks or disposable loops
- Bacterial incinerator

## PROCEDURE

1. Wear a lab coat, gloves, and chemical eye protection when performing this procedure. **In addition, all procedures should be run using BSL-2 precautions unless otherwise directed by your instructor.**
2. Obtain an unknown and record its number on the data sheet, page 571. ***Note:*** You are urged to keep accurate records of all your activities related to identification of this unknown as you proceed (and not after the fact). It is surprising how quickly and easily we (humans) forget details...

[1] *Note to instructor:* This exercise can be combined with Exercise 9-1 by passing out a mixture of a Gram-positive rod and an enteric. The student must isolate the two organisms from mixed culture and then proceed to identify the two independently. If this option is chosen, grow the two unknowns separately and then mix them immediately prior to handing them out in lab.

3. Streak the sample for isolation on two agar plates (medium as provided by the instructor). If instructed to do so, incubate both at 30°C–35°C, one aerobically and the other in the anaerobic jar. Otherwise, incubate aerobically at 30°C–35°C. Record your activities on the data sheet under Isolation Procedure.
4. It is best to check your plate for isolation at 24 hours. If you have isolation, continue with step 5. If you don't have time to continue with step 5, refrigerate your plates until you do. If you do not see isolation, ask your instructor if you should re-streak for isolation, let the original plates continue to incubate, or do both. Record your activities under Isolation Procedure as directed on the data sheet.
5. After incubation examine the plates and find a well-isolated colony. Record colony morphology (including medium) under Preliminary Observations and note any differences between aerobic and anaerobic growth (if done). Also record the result of the differential component of blood agar as Test #1 on the back of the data sheet if you used TSA plus 5% sheep blood agar for isolation. Continue to record your isolation procedure on the data sheet through inoculation of your pure culture (step 6).
6. Perform a Gram stain on a portion of the colony whose morphology you described. Continue testing colonies until you find one that is a Gram-positive rod. *Note:* If the colonies are too small for a Gram stain and transferring to an agar slant, follow the second bullet below to start a pure culture, then Gram stain the pure culture. The pure culture takes priority; Gram staining can wait!
   - Record its Gram reaction and cell morphology, arrangement, and size under Preliminary Observations on the data sheet.
   - If the isolate is a Gram-positive rod, transfer a portion of the same colony to a tryptic soy agar slant (or another suitable growth medium as available in your lab) or thioglycollate broth (if anaerobic) and incubate it at its optimum temperature. This is your pure culture to be used as a source of organisms for further testing. Complete the record of your isolation procedure on the data sheet.
   - Once your pure culture has grown, keep it in the refrigerator and make fresh pure cultures as necessary. They will last roughly three weeks in refrigeration.
7. You now will begin biochemical testing to identify your organism. The tests in the flowchart (Fig. 9.18) were chosen because of their uniformity of results, as published in standard microbiological references, so they give the best chance of correct identification. Be aware, however, that the symbol " + " indicates that 90% or more of the strains tested give a positive result. This means that as many as 10% of the strains tested give a negative result. The same is true of the symbol " – "; that is, as many as 10% of the strains give a positive result for that test. Bottom line: There are no guarantees that your particular lab strains will behave in the majority and if not, you will misidentify your unknown. This issue, if relevant to your unknown, will be addressed in step 11.
8. Follow the tests in Figure 9.18 to identify your unknown.[2] Use these guidelines.
   - Do not run more than one test at a time unless your instructor tells you to do so.
   - You will probably have to refer to the appropriate exercise for instructions on running and reading each test. We advise you to strictly follow the incubation time given—most are 24 to 48 hours—unless told otherwise by your instructor, but to incubate your unknown at *its* optimum temperature (which may differ from the one given in the exercise).
   - If your unknown is an obligate anaerobe, you will have to incubate media in an anaerobic jar. Coordinate your incubation with others who need the anaerobic jar to conserve GasPaks and incubator space. The catalase-negative facultative anaerobes would also benefit from incubation in an anaerobic jar.
   - Where multiple tests are listed at a branch point, run only one, and then move to the next level in the flowchart as indicated by the result of that test. Choose the test whose incubation time best fits your lab schedule, or the easiest one, or the hardest one (for practice).
   - Continue to record the relevant information and results on the data sheet as you go. Keep accurate records of what you have done as you proceed. Do *not* enter all of your data after you complete the project.
   - As you progress, either keep test cultures in the refrigerator or dispose of them as you go. Your instructor will advise you on your lab's convention.

[2] This flowchart can be used only with the organisms listed in Figure 9.17.

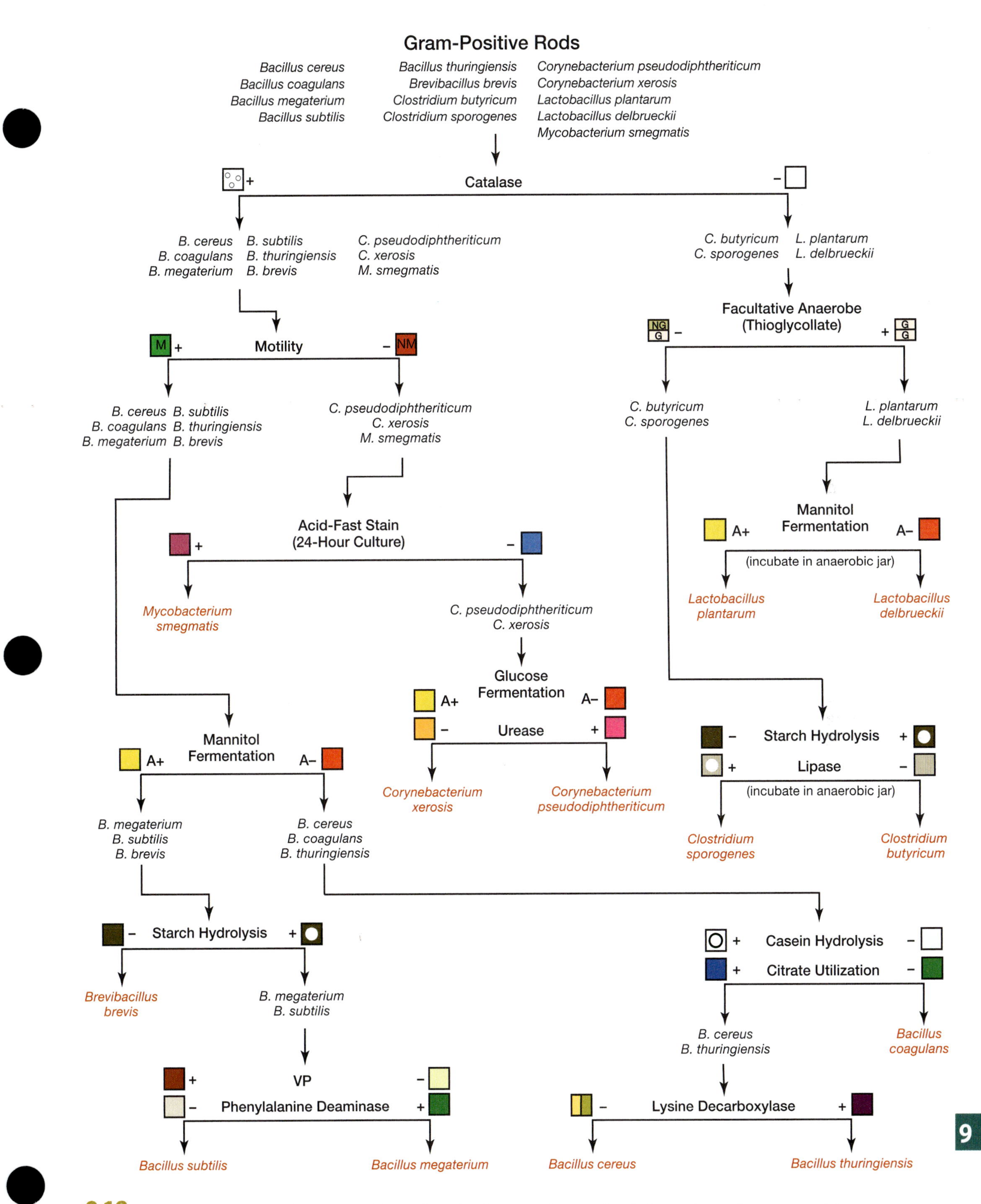


**9.18 Identification Flowchart for Gram-Positive Rods** ■ See the appropriate exercises for instructions on how to run each test.

(Compiled from sources listed in References)

9. When you have identified your organism, use *Bergey's Manual* or another standard reference to find one more test to run for confirmation. The confirmatory test does not have to separate the final organisms on the last branch of the flowchart. It only has to be a test for which you know the result and that you have not run already as part of the identification process. Record this test on the data sheet and identify it as the confirmatory test. Record the *expected* result on the Comments line. After incubation, note whether the organism's result matches (confirms) or differs from the expected result.
10. After the confirmation test, write the organism's name on the back of the data sheet where it says, "My unknown is" and check with your instructor to see if you're correct. If you are, congratulations! You're finished! If you aren't correct, continue with step 12.
11. Your instructor knows the identity of your unknown and can look at your results and the flowchart to find the test(s) responsible for the discrepancy between your provisional identification and your unknown's actual identity. After determining which test(s) gave an "incorrect" result, your instructor will write the test name(s) in the "rerun" space next to your identification.[3] These should be rerun with appropriate controls from your school's inventory of organisms. Record the test(s), control(s) used, and results for your unknown and controls on your data sheet as a continuation of differential tests. (That is, start with the space following your confirmatory test.) **This process is an important component of the unknown project. Checking the results with controls and your unknown will indicate if the "incorrect" test result was truly incorrect when previously run or if the strain of organism your school is using doesn't match the majority of strains for that test result.**
12. Back to work for you, using your instructor's guidance! When you get new a provisional identification supported by a confirmatory test result or have evidence that one of the rerun tests gives the same the "incorrect" result, check with your instructor.
13. Dispose of all cultures in the appropriate autoclave container.

[3] *Note to instructor:* Because of time and resource limitations, we look for tests the student has already run with results that match their actual unknown. We put a check mark next to those tests and tell students not to rerun them, but to use those results when they encounter the tests again and to move to the next branch in the flowchart as if they had rerun them.

## References


Bernard, Kathryn A. and Guido Funke. Pages 245–289 in *Bergey's Manual of Systematic Bacteriology*, 2nd ed., Vol. 5, The Actinobacteria. New York: Springer, 2012.

Collins, M. D. and C. S. Cummins. Section 15 in *Bergey's Manual of Systematic Bacteriology*, Vol. 2. John G. Holt, ed. Baltimore: Lippincott Williams & Wilkins, 1986.

Funke, Guido and Kathryn A. Bernard. Chap. 28 in *Manual of Clinical Microbiology*, Vol. 1, 11th ed. James H. Jorgensen, Michael A. Pfaller, Karen C. Carroll, Guido Funke, Marie Louise Landry, Sandra S. Richter, and David W. Warnock, eds. Washington, DC: ASM Press, American Society for Microbiology, 2015.

Hammes, Walter P. and Christian Hertel. Pages 465–511 in *Bergey's Manual of Systematic Bacteriology*, 2nd ed., Vol. 3, The Firmicutes. New York: Springer, 2009.

Logan, Niall A. and Paul De Vos. Pages 21–128 in *Bergey's Manual of Systematic Bacteriology*, 2nd ed., Vol. 3, The Firmicutes. New York: Springer, 2009.

Logan, Niall A. and Paul De Vos. Pages 305–316 in *Bergey's Manual of Systematic Bacteriology*, 2nd ed., Vol. 5, The Actinobacteria. New York: Springer, 2009.

Magee, John G. and Alan C. Ward. Pages 312–326 in *Bergey's Manual of Systematic Bacteriology*, 2nd ed., Vol. 5, The Actinobacteria. New York: Springer, 2012.

MacFaddin, Jean F. *Biochemical Tests for Identification of Medical Bacteria*, 3rd ed. Philadelphia: Lippincott Williams & Wilkins, 2000.

Pfyffer, Gaby E. Chap. 30 in *Manual of Clinical Microbiology*, Vol. 1, 11th ed. James H. Jorgensen, Michael A. Pfaller, Karen C. Carroll, Guido Funke, Marie Louise Landry, Sandra S. Richter, and David W. Warnock, eds. Washington, DC: ASM Press, American Society for Microbiology, 2015.

Rainey, Fred A., Becky Jo Hollen, and Alanna Small. Pages 738–828 in *Bergey's Manual of Systematic Bacteriology*, 2nd ed., Vol. 3, The Firmicutes. New York: Springer, 2009.

Stevens, Dennis L., Amy E. Bryant, and Karen C. Carroll. Chap. 53 in *Manual of Clinical Microbiology*, Vol. 1, 11th ed. James H. Jorgensen, Michael A. Pfaller, Karen C. Carroll, Guido Funke, Marie Louise Landry, Sandra S. Richter, and David W. Warnock, eds. Washington, DC: ASM Press, American Society for Microbiology, 2015.

Turenne, Christine Y., James W. Snyder, and David C. Alexander. Chap. 26 in *Manual of Clinical Microbiology*, Vol. 1, 11th ed. James H. Jorgensen, Michael A. Pfaller, Karen C. Carroll, Guido Funke, Marie Louise Landry, Sandra S. Richter, and David W. Warnock, eds. Washington, DC: ASM Press, American Society for Microbiology, 2015.

Wayne, Lawrence G., and George P. Kubrica. Section 15 in *Bergey's Manual of Systematic Bacteriology*, Vol. 2. John G. Holt, ed. Baltimore: Lippincott Williams & Wilkins, 1986.

Name ______________________________

Date ______________________________

Lab Section ______________________________

I was present and performed this exercise (initials) ______________

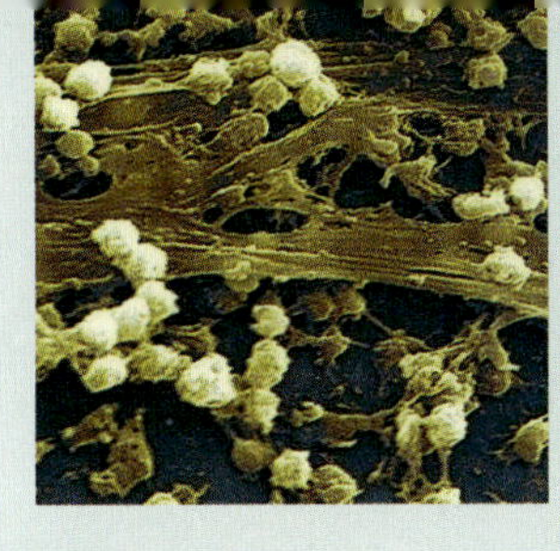

DATA SHEET 9-3

# Identification of Selected Gram-Positive Rods

**Unknown Number** ______________

## Isolation Procedure

Record all activities associated with isolation of your organisms—from mixed culture to pure culture. Always include the date, source of inoculum, destination, type of inoculation, incubation temperature, and any other relevant information. Also make note of transfers made to keep your pure culture fresh. (This log must be kept current.)

## Preliminary Observations

Colony Morphology (include medium, incubation temperature, and differences between aerobic and anaerobic growth, if applicable)

Optimum temperature ______________

Gram stain ______________ Cellular morphology, arrangement, and dimensions ______________

### Differential Tests

Begin recording with the catalase test. This log must be kept current.

Record the figure number of the identification chart you are using. If you change charts, indicate the figure number of the new chart and the date of the change.

**Test #1:** ________ Date Begun: ________ Date Read: ________ Result: ________

Comments: ________

**Test #2:** ________ Date Begun: ________ Date Read: ________ Result: ________

Comments: ________

**Test #3:** ________ Date Begun: ________ Date Read: ________ Result: ________

Comments: ________

**Test #4:** ________ Date Begun: ________ Date Read: ________ Result: ________

Comments: ________

**Test #5:** ________ Date Begun: ________ Date Read: ________ Result: ________

Comments: ________

**Test #6:** ________ Date Begun: ________ Date Read: ________ Result: ________

Comments: ________

**Test #7:** ________ Date Begun: ________ Date Read: ________ Result: ________

Comments: ________

**Test #8:** ________ Date Begun: ________ Date Read: ________ Result: ________

Comments: ________

**Test #9:** ________ Date Begun: ________ Date Read: ________ Result: ________

Comments: ________

**Test #10:** ________ Date Begun: ________ Date Read: ________ Result: ________

Comments: ________

**Test #11:** ________ Date Begun: ________ Date Read: ________ Result: ________

Comments: ________

**Test #12:** ________ Date Begun: ________ Date Read: ________ Result: ________

Comments: ________

My unknown is: ________ Rerun[1]: ________

________ Rerun: ________

________ Rerun: ________

[1] Your instructor will write what tests to rerun in this space if you misidentify your unknown.

## Multiple Test Systems

Multiple test systems are systems designed to run an entire battery of tests simultaneously. They employ the same biochemical principles discussed earlier in Section 5 and are read in essentially the same manner. All conditions possible with standard tubed media also can be achieved with multiple test media, including aerobic or anaerobic growth conditions and the addition of reagents for indicator reactions.

The savings in time and money alone make these systems enormously valuable. Even more important, what might take days or weeks to do with media preparation and individual tests, multiple test systems can do in as little as 18 hours. In addition, these multi-test systems come with databases for fast and easy identification.

These two exercises will give you an opportunity to have fun while using some of the skills you have learned. The organisms chosen for these tests will be provided as unknowns for you to identify. ■

# api® 20 E Identification System for *Enterobacteriaceae* and Other Gram-Negative Rods

EXERCISE **9-4**

## Theory

The api® 20 E system is designed to identify members of the *Enterobacteriaceae* and other Gram-negative rods. It is one of a series of similar, multi-test systems api manufactures that identify Gram-negative nonenteric bacteria, staphylococci and micrococci, streptococci and enterococci, coryneforms, anaerobes, and yeasts.

The api® 20 E is a plastic strip with 20 microtubes and cupules, partially filled with different dehydrated substrates. Bacterial suspension is added to the microtubes, rehydrating the media and inoculating them at the same time. As with the other biochemical tests, color changes take place in the tubes either during incubation or after addition of reagents. These color changes reveal the presence or absence of chemical action and, thus, a positive or negative result (Fig. 9.19).

After incubation, spontaneous reactions—those that do not require addition of reagents—are evaluated first. Then tests that require addition of reagents are performed and evaluated. Finally, the results are entered on the api® 20 E Result Sheet (Fig. 9.20). An oxidase test is performed separately and constitutes the 21st test.

As shown in Figure 9.20, the Result Sheet divides the tests into groups of three, with the members of a group having numerical values of 1, 2, or 4, respectively. These numbers are assigned for positive (+) results only. Negative (−) results are not counted. The values for positive results in each group are added together to produce a number from 0 to 7, which is entered in the oval below the three tests. The totals from each group are combined sequentially to produce a seven-digit code, which can then be interpreted in the Analytical Profile Index (Fig. 9.21) or via api*web*™ identification software.[1]

[1] BioMérieux-USA.com: https://apiweb.biomerieux.com

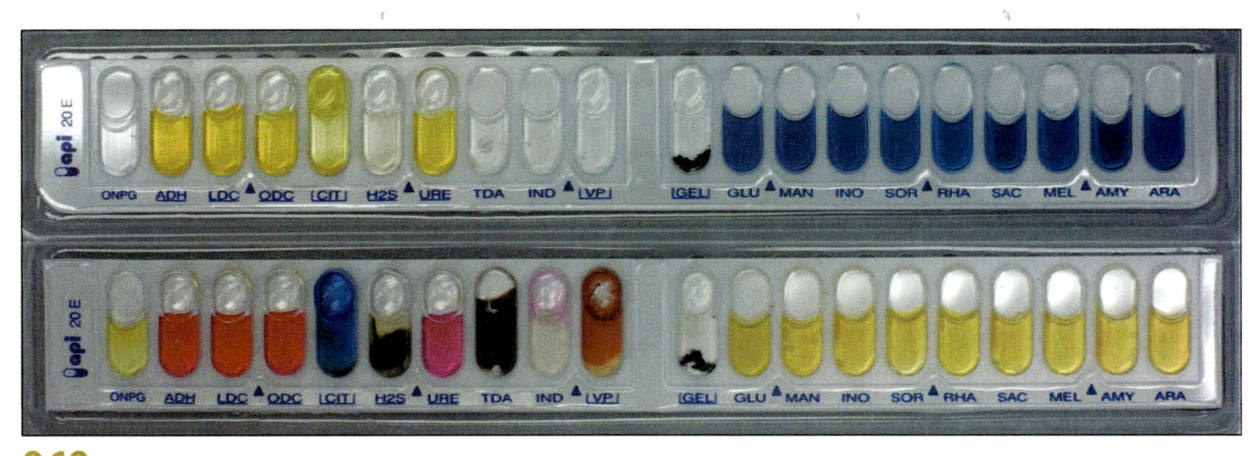


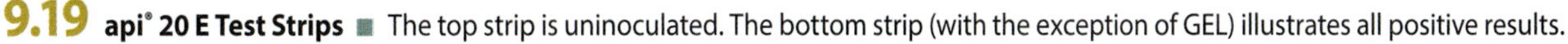

**9.19 api® 20 E Test Strips** ■ The top strip is uninoculated. The bottom strip (with the exception of GEL) illustrates all positive results.

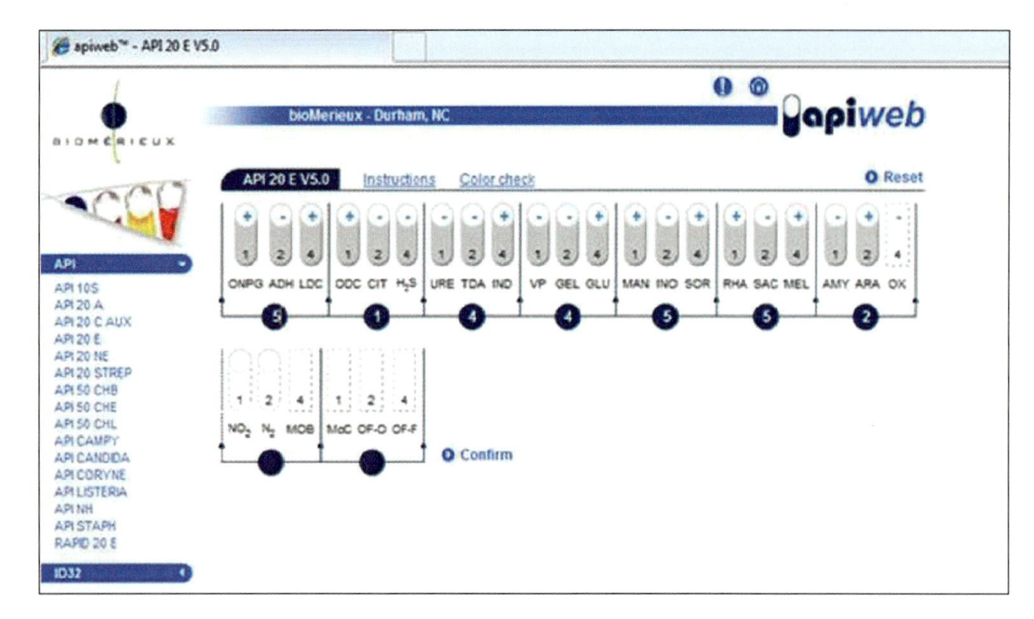


**9.20 Recording api® 20 E Results** ■ The 21 test results of the api® 20 E (oxidase is not part of the strip and is run separately) are divided into groups of three, with each positive result given a numerical value of 1, 2, or 4. In this screen shot from api*web*™, results from a sample have been recorded electronically. The isolate is positive for ONPG, LDC, ODC, IND, GLU, MAN, SOR, RHA, MEL, and ARA. The sum of the numbers assigned to those positive results within each group is shown in the blue circles below each group. For instance, group 1 was positive for ONPG (1) and LDC (4) and a 5 (1+4=5) is recorded below. Collectively, the blue circles produce a seven-digit number, 5144552, which is matched to the database in the printed Analytical Profile Index or on api*web*™ (Fig. 9.21). Additional tests, shown below the main tests, can be run if identification is equivocal. (Photo courtesy of bioMérieux)

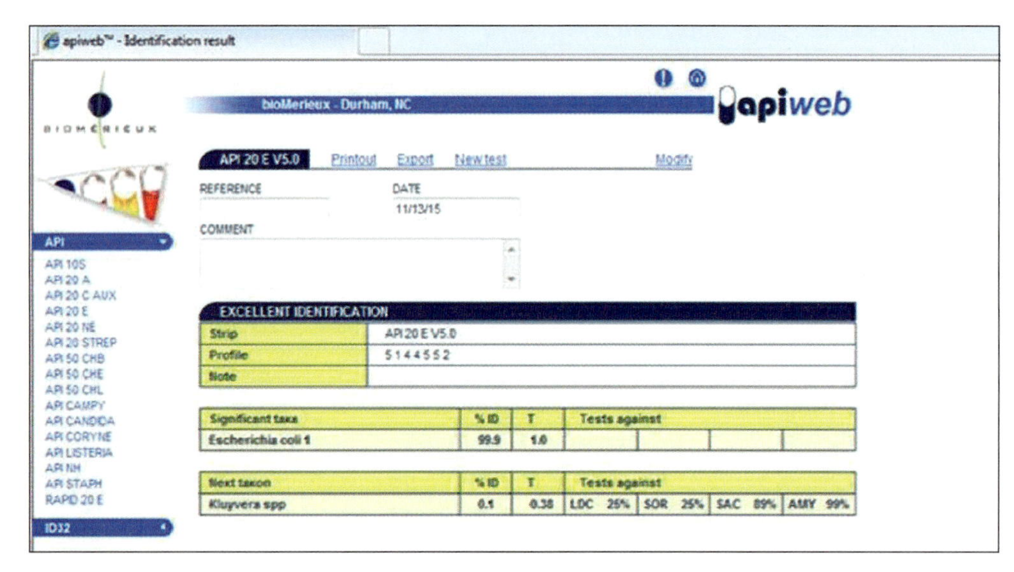


**9.21 Identification Using api*web*™** ■ The printed Analytical Profile Index and the website api*web*™ have an extensive database of test results for *Enterobacteriaceae* and other non-fastidious Gram-negative rods for which the api® 20 E was designed. Shown is a screen shot with the seven-digit number from Figure 9.20 entered. The resulting identification was *Escherichia coli* with 99.9% confidence. The next closest match (0.1% confidence) was the genus *Kluyvera*. Tests not conforming to *Kluyvera* are listed to the right. (Photo courtesy of bioMérieux)

In rare instances, information from the 21 tests (and the seven-digit code) is not discriminating enough to identify an organism. When this occurs, the organism is grown and examined on MacConkey agar and supplemental tests are performed for nitrate reduction, oxidation/reduction of glucose, and motility. The results are entered separately in the supplemental spaces on the Result Sheet and used for final identification.

## ■ Application

The api® 20 E multi-test system (available from BioMérieux, Inc.) is used clinically for the rapid identification of *Enterobacteriaceae* (more than 5,500 strains) and other Gram-negative rods (more than 2,300 strains).

## In This Exercise

Today you will use a multi-test system to run an entire battery of tests designed to identify a member of the *Enterobacteriaceae*.

## Materials

### Per Student

- □ Lab coat
- □ Disposable gloves
- □ Chemical eye protection

### Per Student Group[2]

- □ api® 20 E identification system for *Enterobacteriaceae* and other non-fastidious Gram-negative rods
- □ api® 20 E Results Sheet
- □ Sterile 0.85% saline solution
- □ Ferric chloride reagent (see Exercise 5-9)
- □ James reagent (available from bioMérieux)
- □ Potassium hydroxide reagent (see Exercise 5-3)
- □ α-naphthol reagent (see Exercise 5-3)
- □ Sulfanilic acid reagent (see Exercise 5-6)
- □ *N,N*-Dimethyl-1-naphthylamine reagent (see Exercise 5-6)
- □ Zinc powder (see Exercise 5-6)
- □ Oxidase reagent and filter paper (see Exercise 5-5)
- □ 3% hydrogen peroxide (see Exercise 5-4)
- □ Sterile mineral oil
- □ Sterile transfer pipettes
- □ Sterile wooden sticks or disposable loops
- □ One sterile 12 mm × 75 mm test tube
- □ Distilled water
- □ Microscope slides
- □ Gram-stain kit
- □ Bibulous paper
- □ Immersion oil
- □ Lens paper
- □ Compound microscope with oil-immersion lens
- □ Streak plates (numbered to an instructor's key), each containing oxidase-negative, Gram-negative unidentified bacterial colonies (**BSL-2 organisms should not be chosen as unknowns by schools or classes not adequately prepared to handle them.**)

### Per Class

- □ api*web*™ subscription (https://apiweb.biomerieux.com)
- □ One streak plate of a Gram-negative, oxidase-negative "unknown" per student

## PROCEDURE

### Lab One

1. Wear a lab coat, gloves, and chemical eye protection when performing this procedure.
2. Obtain an unknown and a results sheet. Write the unknown number on the results sheet. **If notified by your instructor that BSL-2 organisms have been distributed as unknowns, follow BSL-2 procedures throughout.**
3. Perform a Gram stain on your unknown organism to confirm that it is a Gram-negative rod.
4. Open an incubation box and distribute about 5 mL of distilled water into the honeycombed wells of the tray to create a humid atmosphere (Fig. 9.22).
5. Record your name and the unknown number on the elongated flap of the tray.
6. Remove the strip from its packaging and place it in the tray.
7. Perform the oxidase test on a colony identical to the colony that will be tested. If you need help with this, refer to Exercise 5-5. Record the result on the score sheet as the 21st test.
8. Using a sterile Pasteur pipette, remove a single, well-isolated colony from the plate (Fig. 9.23) and fully emulsify it in a tube of suspension medium or sterile 0.85% saline (Fig. 9.24). Try not to pick up any agar.

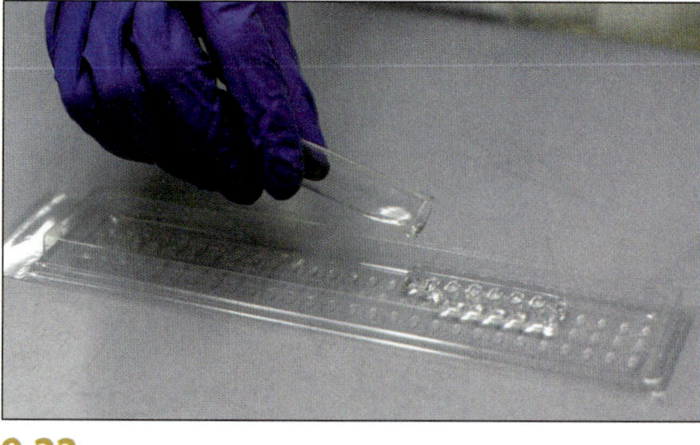

**9.22 Add Water to Honeycombed Wells** ■ The purpose of the water is to humidify the strip during incubation. Spread the water as uniformly as possible in the honeycomb, but it is not necessary to add identical amounts to each well.

[2] This lab can be run with unknowns being identified by each student or by conveniently sized groups. The workload can easily be distributed among several students so they all share in the experience.

**9.23 Pick a Colony** ■ Being careful not to disturb the agar surface, pick up one isolated colony with a sterile transfer pipette and immediately transfer it to sterile saline (Fig. 9.24).

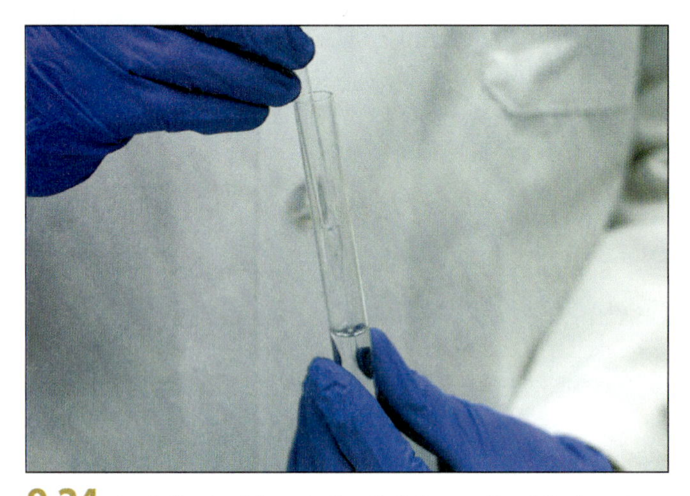

**9.24 Emulsify the Colony in Sterile Saline** ■ Emulsify the colony in 5 mL sterile 0.85% saline. Do this by filling and emptying the pipette several times until uniform turbidity is achieved in the solution.

9. Using the same pipette, fill both tube and cupule of test CIT, VP, and GEL with bacterial suspension (Fig. 9.25). Fill only the tubes (not the cupules) of all other tests.
10. Overlay the ADH, LDC, ODC, $H_2S$, and URE microtubes by filling the cupules with sterile mineral oil using a sterile transfer pipette.
11. Close the incubation box and incubate at 35 ± 2°C for 24 hours. (*Note*: If the tests cannot be read at 24 hours, have someone place the incubation box in a refrigerator until the tests can be read.)
12. Save or dispose of the original culture plate as directed by your instructor.

### Lab Two

1. Wear a lab coat, gloves, and chemical eye protection when performing this procedure.

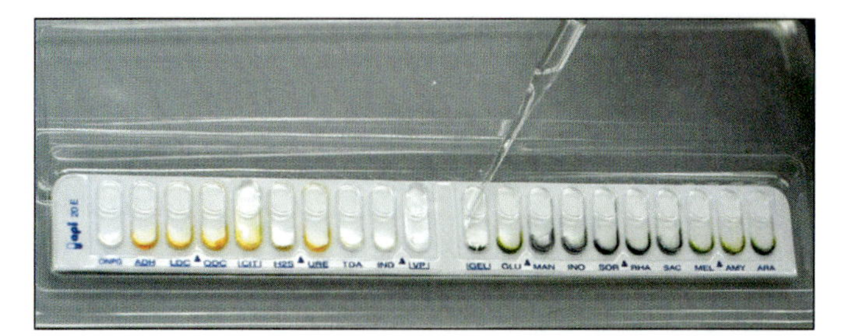

**9.25 Inoculate the Tubes with the Suspension** ■ Inoculate the tubes by placing the pipette at the side of the tube and gently filling each with the suspension. If necessary to avoid creating bubbles, tilt the strip slightly. Tap gently with your index finger if necessary to remove bubbles.

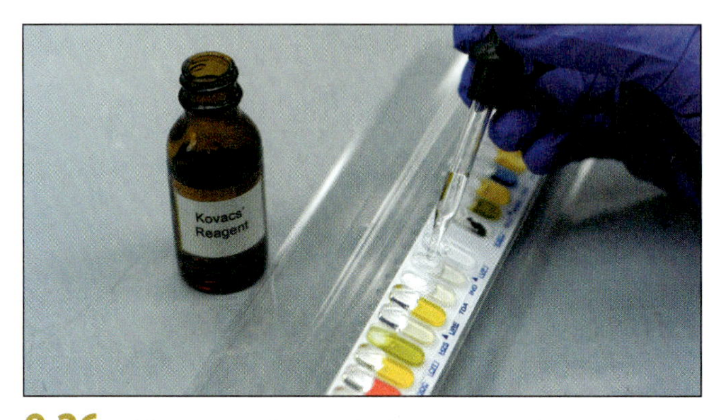


**9.26 Add Reagents after Incubation** ■ Follow the instructions for each in the text.

2. Refer to Table 9-2 (Table 2 on the package insert) as you examine the test results on the strip.
3. Examine your test strip. Record all spontaneous reactions on the Result Sheet. (Spontaneous reactions are reactions completed without the addition of reagents.)
4. **TDA (Tryptophan Deaminase):** Add 1 drop of ferric chloride to the TDA microtube (Fig. 9.26). A reddish-brown color indicates a positive reaction.
5. **VP (Voges–Proskauer):** Add 1 drop of potassium hydroxide reagent and 1 drop of α-naphthol reagent to the VP microtube. A pink or red color indicates a positive reaction to be recorded on the Result Sheet. A slightly pink color appearing in 10 minutes should be considered negative.
6. **NIT (Nitrate Reduction) done in GLU microtube:** Add 1 drop of sulfanilic acid reagent and 1 drop of N,N-Dimethyl-1-naphthylamine reagent to the GLU microtube. Wait 2 to 5 minutes. A red color is a positive reaction ($NO_3 \rightarrow NO_2$). Enter this on the Result Sheet as positive. A yellow color at this point is inconclusive because the nitrate may have been reduced to nitrogen gas ($NO_3 \rightarrow N_2$, sometimes evidenced by gas bubbles). If the substrate in the tube remains yellow after adding reagents, add 2 mg to 3 mg of zinc powder to the tube. A yellow

tube after 5 minutes is positive for $N_2$ and is recorded on the Result Sheet as positive for nitrate reduction. If the test turns pink-red after the addition of zinc, the nitrates remaining in the tube have been reduced by the zinc. This is recorded as a negative result for nitrate reduction.

7. **IND (Indole Production):** Add 1 drop of James reagent to the IND microtube. A pink color throughout the cupule indicates a positive reaction.
8. When you are satisfied that the results have been read correctly, properly dispose of the api® test strip as directed by your instructor.
9. Add the positive results within each group on the Result Sheet, using the values given, and enter the number in the circle below each group's results (Fig. 9.20).
10. Access the api*web*™ database: https://apiweb.biomerieux.com. Locate the seven-digit code and identify your organism. Enter the name on the Result Sheet and tape it on the data sheet, page 579.
11. Answer the questions on the data sheet.

TABLE **9-2** **api® 20 E Results and Interpretations**

| # | Test | Substrate/Activity | Result | Interpretation | Symbol |
|---|---|---|---|---|---|
| 1 | ONPG | 2-nitophenyl-β-D-galactopyranoside | Yellow[a] | Organism produces β-galactosidase (ONPG) | + |
| | | | Colorless | Organism does not produce β-galactosidase (ONPG) | – |
| 2 | ADH | L-arginine | Red/orange[b] | Organism produces arginine dihydrolase | + |
| | | | Yellow | Organism does not produce arginine dihydrolase | – |
| 3 | LDC | L-lysine | Red/orange[b] | Organism produces lysine decarboxylase | + |
| | | | Yellow | Organism does not produce lysine decarboxylase | – |
| 4 | ODC | L-ornithine | Red/orange[b] | Organism produces ornithine decarboxylase | + |
| | | | Yellow | Organism does not produce ornithine decarboxylase | – |
| 5 | CIT | Trisodium citrate | Blue-green/blue[c] | Organism utilizes citrate as sole carbon source | + |
| | | | Pale green/yellow | Organism does not utilize citrate | – |
| 6 | $H_2S$ | Sodium thiosulfate | Black deposit/thin line | Organism reduces sulfur | + |
| | | | Colorless/grayish | Organism does not reduce sulfur | – |
| 7 | URE | Urea | Red/orange[b] | Organism produces urease | + |
| | | | Yellow | Organism does not produce urease | – |
| 8 | TDA | L-tryptophan | Reddish-brown | Organism produces tryptophan deaminase | + |
| | | | Yellow | Organism does not produce tryptophan deaminase | – |
| 9 | IND | L-tryptophan | Pink throughout | Organism produces indole | + |
| | | | Colorless/pale green/yellow | Organism does not produce indole | – |
| 10 | VP | Sodium pyruvate | Pink/red[d] | Organism produces acetoin | + |
| | | | Colorless/pale pink | Organism does not produce acetoin | – |
| 11 | GEL | Gelatin (bovine origin) | Diffusion of black pigment | Organism produces gelatinase | + |
| | | | No black pigment diffusion | Organism does not produce gelatinase | – |
| 12 | GLU | D-glucose[e] | Yellow | Organism ferments glucose | + |
| | | | Blue/blue-green | Organism does not ferment glucose | – |
| 13 | MAN | D-mannitol[e] | Yellow | Organism ferments mannitol | + |
| | | | Blue/blue-green | Organism does not ferment mannitol | – |
| 14 | INO | Inositole[e] | Yellow | Organism ferments inositol | + |
| | | | Blue/blue-green | Organism does not ferment inositol | – |

*(continues)*

TABLE **9-2** **api® 20 E Results and Interpretations** *(continued)*

| # | Test | Substrate/Activity | Result | Interpretation | Symbol |
|---|---|---|---|---|---|
| **15** | SOR | D-sorbitol[e] | Yellow | Organism ferments sorbitol | + |
| | | | Blue/blue-green | Organism does not ferment sorbitol | – |
| **16** | RHA | L-rhamnose[e] | Yellow | Organism ferments rhamnose | + |
| | | | Blue/blue-green | Organism does not ferment rhamnose | – |
| **17** | SAC | D-sucrose[e] | Yellow | Organism ferments sucrose | + |
| | | | Blue/blue-green | Organism does not ferment sucrose | – |
| **18** | MEL | D-melibiose[e] | Yellow | Organism ferments melibiose | + |
| | | | Blue/blue-green | Organism does not ferment melibiose | – |
| **19** | AMY | Amygdalin[e] | Yellow | Organism ferments amygdalin | + |
| | | | Blue/blue-green | Organism does not ferment amygdalin | – |
| **20** | ARA | L-arabinose[e] | Yellow | Organism ferments arabinose | + |
| | | | Blue/blue-green | Organism does not ferment arabinose | – |
| **21** | OX | Separate test done on paper test strip | Violet | Organism produces cytochrome-oxidase | + |
| | | | Colorless | Organism does not produce cytochrome-oxidase | – |
| **22** | GLU | Potassium nitrate | Red after addition of reagents | Organism reduces nitrate to nitrite | + |
| | | | Yellow after addition of reagents | Organism does not reduce nitrate to nitrite | – |
| | | | Yellow after addition of zinc | Organism reduces nitrate to $N_2$ gas | + |
| | | | Orange-red after addition of zinc | Organism does not reduce nitrate | – |
| **23** | MOB | Motility medium or wet mount slide | Motility | Organism is motile | + |
| | | | Nonmotility | Organism is not motile | – |
| **24** | McC | MacConkey medium | Growth | Organism is probably *Enterobacteriaceae* | + |
| | | | No growth | Organism is not *Enterobacteriaceae* | – |
| **25** | OF | Glucose | Yellow under mineral oil | Organism ferments glucose | + |
| | | | Green under mineral oil | Organism does not ferment glucose (not *Enterobacteriaceae*) | – |
| | | | Yellow without mineral oil | Organism either ferments glucose or utilizes it oxidatively | + |
| | | | Green without mineral oil | Organism does not utilize glucose (not *Enterobacteriaceae*) | – |

**a** A very pale yellow is also positive.
**b** Orange after 36 hours is negative.
**c** Reading made in the cupule (aerobic).
**d** A slightly pink color after 10 minutes is negative.
**e** Fermentation begins in the lower portion of the tube; oxidation begins in the cupule.

## 9 *References*


api® 20 E Identification System for *Enterobacteriaceae* and Other Gram-Negative Rods package insert.

Name ______________________________

Date ______________________________

Lab Section ______________________________

I was present and performed this exercise (initials) ______________________________

DATA SHEET

9-4

# api® 20 E Identification System for *Enterobacteriaceae* and Other Gram-Negative Rods

## OBSERVATIONS AND INTERPRETATIONS

**1** Tape your api® 20 E Result Sheet here.

## QUESTIONS

**1** *Why is it important to perform the reagent tests last?*

**2** *In clinical applications of this test system, reagents are added only if the glucose (oxidation/fermentation) test result is yellow or at least three other tests are positive. If these conditions are not met, a MacConkey agar plate is streaked and additional tests are performed confirming glucose metabolism, nitrate reduction, and motility. Why do you think this is so? Be specific.*

**3** *Suppose, after 24 hours incubation, you notice no growth in the tubes containing mineral oil. Assuming that it is behaving properly under these conditions, what do you know about the organism and what predictions can you safely make about its performance in the decarboxylase tests, fermentation tests, and nitrate reduction test? Is it a member of* Enterobacteriaceae?

9

# EnteroPluri-*Test*

## Preface

For many years, the Enterotube® II was a popular multi-test system designed to identify enteric bacteria and other select oxidase-negative, Gram-negative bacteria. While there may still be unused Enterotubes® in college labs around the country, as far as we can determine it is no longer available for purchase. However, a very similar system called EnteroPluri-*Test* is manufactured by the Italian company Liofilchem® and sold by various vendors in the United States[1]. This lab exercise is written for the EnteroPluri-*Test*, but it will also work for the Enterotube® II system. There are some slight differences in inoculation and test reading, so check the package insert. For clarity during this transitional time, we refer only to EnteroPluri-*Test*.

[1] A quick web search located Hardy Diagnostics and Carolina Biological Supply companies as sources of the EnteroPluri-*Test* and EnteroPluri-*Test* Codebook.

## Theory

The EnteroPluri-*Test* is a multiple test system designed to identify enteric bacteria based on: glucose, adonitol, lactose, arabinose, sorbitol, and dulcitol fermentation; lysine and ornithine decarboxylation; sulfur reduction; indole production; acetoin production from glucose fermentation; phenylalanine deamination; urea hydrolysis; and citrate utilization.

The EnteroPluri-*Test*, as shown in Figure 9.27, is a tube containing 12 individual chambers with the capability of performing 15 tests (Table 9-3). Inside the tube, running lengthwise through its center, is a removable wire. After the end-caps are removed (aseptically), one end of the wire is touched to an isolated colony on a streak plate and drawn back through the tube to inoculate the media in each chamber (Figs. 9.28 through 9.31), air holes are opened for tests requiring aerobic conditions (Fig. 9.32), caps are replaced, and the tube is incubated for 18–24 hours.

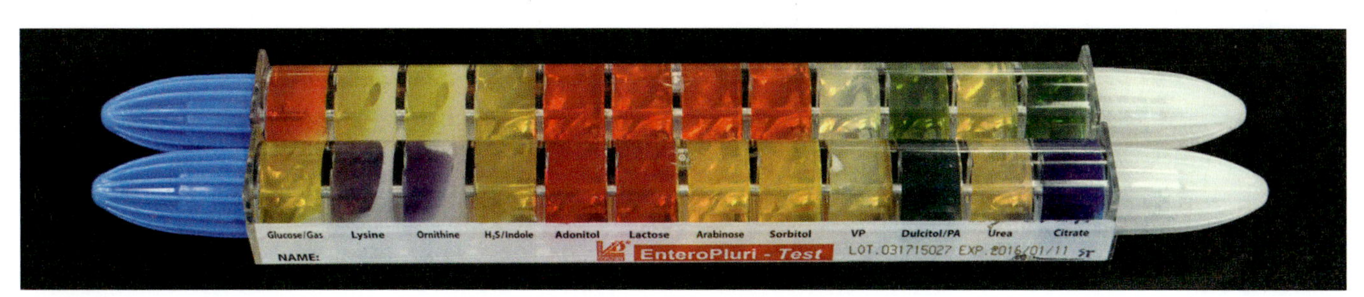


**9.27 EnteroPluri-*Test* System** ■ An inoculated tube is in front; an uninoculated control is behind. Each tube contains 12 compartments, with the ability to run 15 tests. From left to right: acid from glucose/gas from glucose; lysine decarboxylase; ornithine decarboxylase; sulfur reduction/indole production; acid from adonitol; acid from lactose; acid from arabinose; acid from sorbitol; Voges-Proskauer; acid from dulcitol/phenylalanine deaminase; urease; and citrate utilization. The scored data chart for this organism in shown in Figure 9.35.

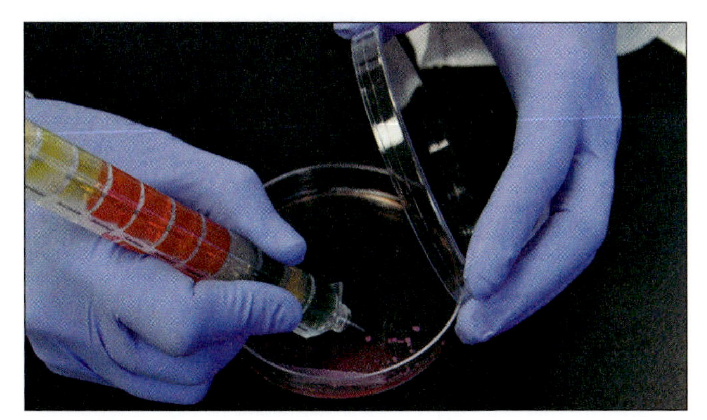

**9.28 Pick a Colony** ■ Unscrew both caps and place them on the paper towel. The wire's handle is under the blue cap. The tip is under the white cap. Using the sterile tip of the wire, aseptically remove growth from a colony on the agar surface. Do not dig into the agar. The inoculum should be large enough to be visible. Remember, there must be enough to inoculate all 12 compartments.

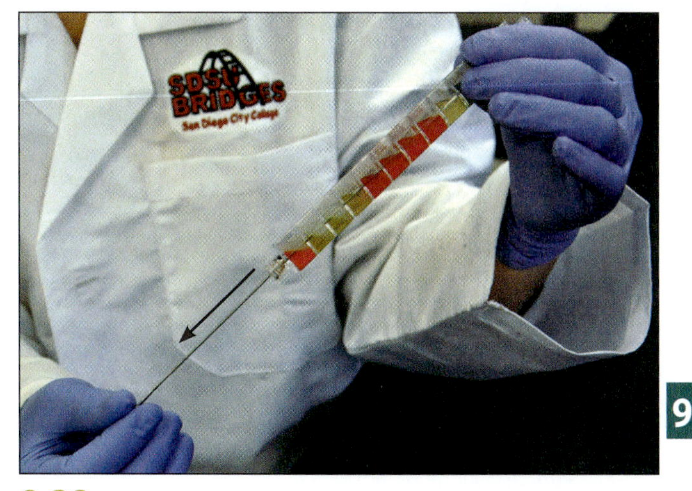


**9.29 Pull Wire Through** ■ Loosen the wire by turning it slightly. While continuing to rotate it withdraw the wire until its tip is inside the last compartment (glucose).

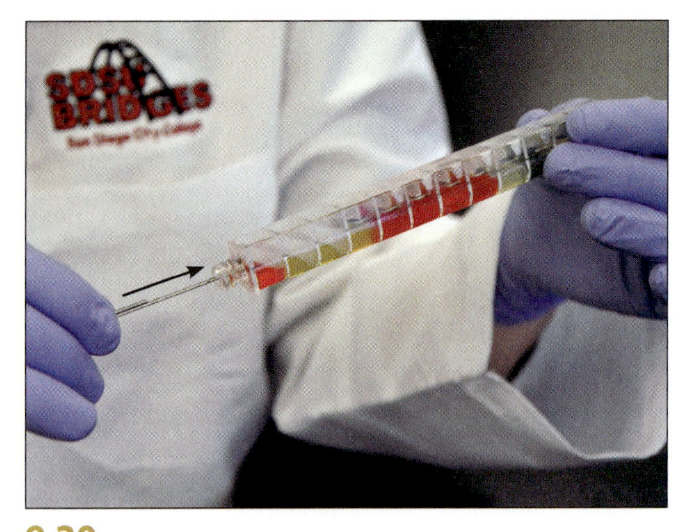


**9.30 Reinsert Wire** ■ Slide the wire back into the tube until you can see the tip inside the citrate compartment. The notch in the wire should be lined up with the end of the tube nearest the glucose compartment.

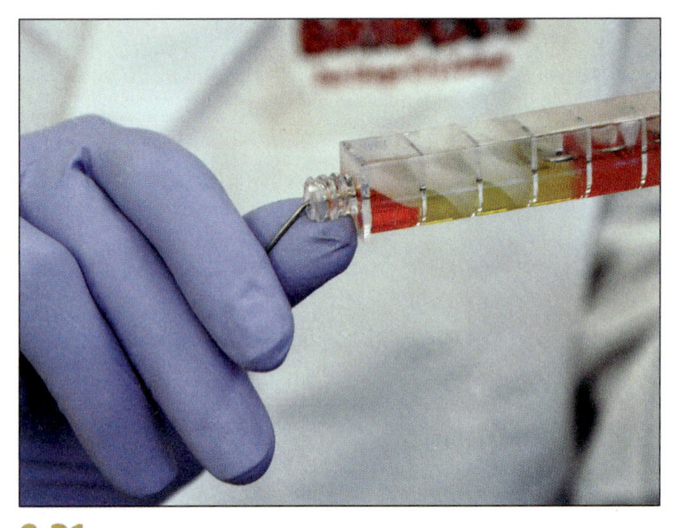

**9.31 Break Wire** ■ Bend the wire until it breaks off at the notch.

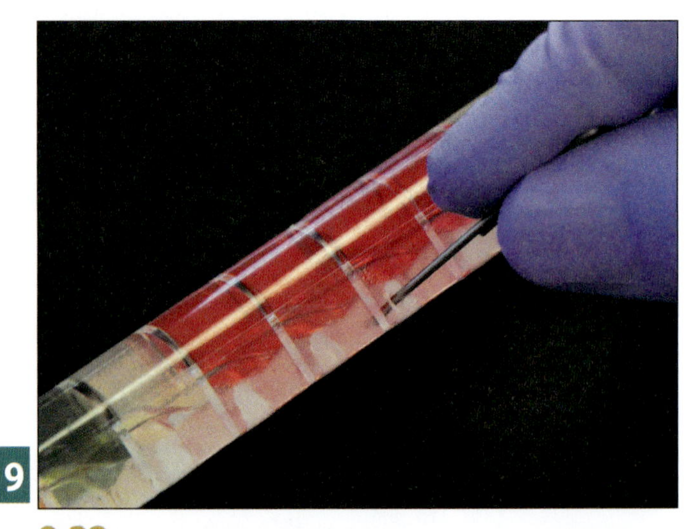

**9.32 Puncture Air Inlets** ■ Using the broken wire, puncture the plastic covering the eight air inlets on the back side of the tube (all but glucose, lysine, ornithine, and indole/$H_2S$ have them). Replace the caps and dispose of the wire in a sharps container.

After incubation, most results can be read as a simple color change (Table 9-3 and Fig. 9.33). The indole and VP tests require addition of reagents, which is done with a syringe (Fig. 9.34) after the first tests are read. All results are recorded on an EnteroPluri-*Test* Data Chart (Fig. 9.35A). As shown in the figure, the combination of positives entered on the data chart results in a five-digit numeric code. This code is used for identification in the EnteroPluri-*Test* Codebook (Fig. 9.35B).

The codebook is a master list of all enterics and their assigned numeric codes. In most cases, the five-digit number applies to a single organism, but when two or more species share the same code, a confirmatory test is performed to further differentiate the organisms.

## ■ Application

The EnteroPluri-*Test* is a multiple-test system used for rapid identification of bacteria from the family *Enterobacteriaceae* and other Gram-negative, oxidase-negative bacteria.

## ■ In This Exercise

You will use a multi-test system to run an entire battery of tests designed to identify a member of the *Enterobacteriaceae*.

## ▼ Materials

### Per Student

- ☐ Lab coat
- ☐ Disposable gloves
- ☐ Chemical eye protection

### Per Student Group

- ☐ EnteroPluri-*Test* System
- ☐ EnteroPluri-*Test* System Data Chart
- ☐ Kovac's reagent
- ☐ VP reagents (KOH and α-naphthol)
- ☐ Needles and syringes or disposable pipettes for addition of reagents
- ☐ Streak plate (numbered to an instructor's key) containing colonies of an unknown enteric (**BSL-2 organisms should not be chosen as unknowns by schools or classes not adequately prepared to handle them.**)
- ☐ Sterile wooden sticks or disposable loops
- ☐ Microscope slides
- ☐ Gram-stain kit
- ☐ Bibulous paper
- ☐ Immersion oil

- □ Lens paper
- □ Compound microscope with oil-immersion lens

## Reagent Recipes

### Indole Test

Kovac's Reagent

| | |
|---|---|
| □ Amyl alcohol | 75.0 mL |
| □ Hydrochloric acid, concentrated | 25.0 mL |
| □ *p*-dimethylaminobenzaldehyde | 5.0 g |

### VP Test

KOH Reagent

| | |
|---|---|
| □ Potassium hydroxide | 20.0 g |
| □ Distilled water to bring volume to | 100.0 mL |

α-Naphthol Reagent

| | |
|---|---|
| □ α-naphthol | 5.0 g |
| □ Absolute ethanol to bring volume to | 100.0 mL |

### Per Class

- □ EnteroPluri-*Test* Code Book

### Lab One

1. Wear a lab coat, gloves, and chemical eye protection when performing this procedure.
2. Obtain an unknown and an EnteroPluri-*Test* Data Chart. Write the number of your unknown on the data chart. **If notified by your instructor that BSL-2 organisms have been distributed as unknowns, follow BSL-2 procedures throughout.**
3. Perform a Gram stain on your isolate to verify that it is a Gram-negative rod.
4. Place a paper towel on the tabletop and soak it with disinfectant.
5. Remove the blue cap and then the white cap from the EnteroPluri tube, being careful not to contaminate the sterile wire tip (from under the white cap). Place the caps open end down on the paper towel.
6. Aseptically remove a large amount of growth from one of the plated colonies (Fig. 9.28). (This presumes your streak plate has only one organism on it. If not, you will need to inoculate with the same colony you Gram stained.) Try not to remove any of the agar with it.
7. Grasp the looped end of the wire and, while turning, gently pull it back through all of the EnteroPluri tube compartments (Fig. 9.29). You do not have to completely remove the wire, but be careful to pull it back far enough to inoculate the last compartment.
8. Using the same turning motion as described above, slide the wire back into the tube until the notch in the wire lines up with the end of the tube (Fig. 9.30).
9. Bend the wire at the notch until it breaks off (Fig. 9.31).
10. Locate the air inlets on the side of the tube opposite the label. Using the removed piece of inoculating wire, puncture the plastic membrane in the adonitol, lactose, arabinose, sorbitol, VP, dulcitol/PA, urea, and citrate compartments (Fig. 9.32). These openings will create the necessary aerobic conditions for growth. Be careful not to perforate the plastic film covering the flat side of the tube.
11. Discard the wire in disinfectant or a sharps container and replace the blue and white caps.
12. Incubate the tube lying on its flat surface or vertically with glucose on top at 35 ± 2°C for 18 to 24 hours. (***Note***: If the tube cannot be read at 24 hours, have someone place it in a refrigerator until it can be read.)
13. Save or dispose of the original culture plate as directed by your instructor.

### Lab Two

1. Examine the tube and record the results of all tests but indole and VP on the data chart (Fig. 9.35). Use Table 9-3 and Figure 9.33 as guides.
2. Next, puncture the plastic membrane of the $H_2S$/Indole compartment and (with a needle and syringe or disposable pipette) add four drops of Kovac's reagent to the compartment (Fig. 9.34). Formation of a red color within 15 seconds is a positive result. Record the result.
3. Puncture the plastic on the bottom of the VP compartment and add the VP-A and VP-B reagents. A red color within 20 minutes is a positive result. Record the result.
4. For each positive result, circle the corresponding number (4, 2, or 1) on the data chart. Add the numbers in each section to obtain a five-digit number. Find the number in the EnteroPluri-*Test* Codebook for identification. Note any "atypical tests" associated with that number. Your instructor will determine if a confirmatory test should be run, or if your identification is satisfactory.
5. Discard the EnteroPluri-*Test* in an appropriate autoclave container.
6. Attach your data chart the data sheet, page 587 and answer the questions.

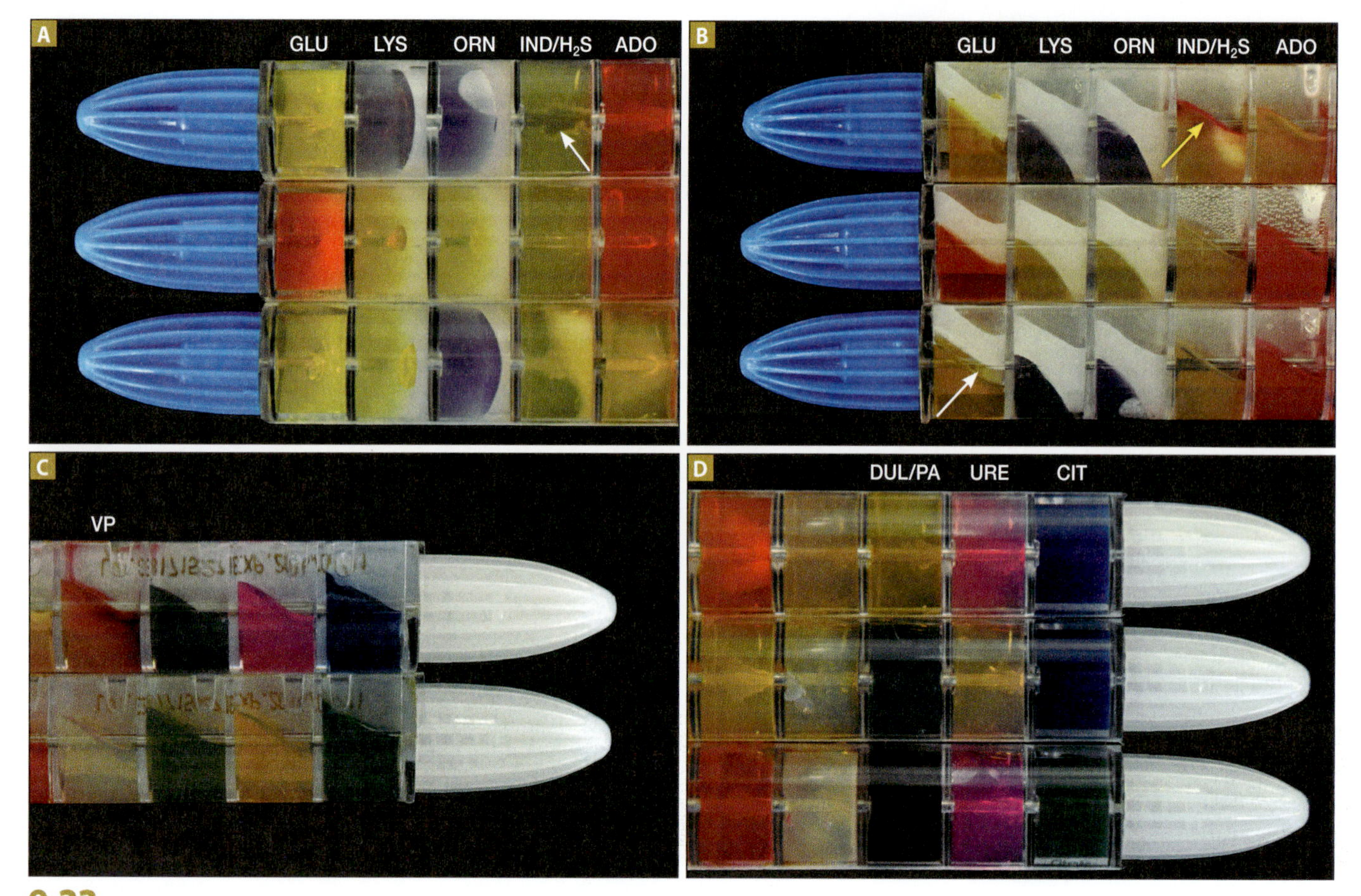


**9.33 EnteroPluri-*Test* Results** ■ (**A**) Compartments viewed from above (L to R): glucose acid (GLU), lysine decarboxylase (LYS), ornithine decarboxylase (ORN), indole/$H_2S$ (IND/$H_2S$), and adontiol acid (ADO). The tube in the center is uninoculated for comparison. The other tubes are both positive for acid production from glucose, as they should be if they are enterics. The tube below shows the difference between a negative (yellow LYS) and a positive (purple ORN) decarboxylase result. The tube on top is positive for both decarboxylases and also has reduced sulfur ($H_2S$-positive). Note the black in the center (arrow). The indole compartment in the bottom tube is cloudy from addition of Kovac's reagent. The last visible compartment is acid from adonitol. The tube below is positive (yellow); the other tubes are negative (red). Acid from lactose, arabinose, and sorbitol are read the same way and are not shown. (**B**) In this photo, the compartments are viewed from the side. The center tube is uninoculated, and the other were inoculated with different organisms than shown in (**A**). This photo is used to illustrate positive glucose-gas and indole results. Note the space between the white wax and the yellow agar in the lower tube, indicating gas production from glucose (white arrow). Compare with the agar/wax interface in the other two tubes. Also notice the pink layer in the fourth (indole) compartment of the upper tube (yellow arrow). This is a positive result for indole production and occurred after addition of Kovac's reagent (which soaked into the agar by the time the photo was taken and produced the white artifact in the medium). (**C**) Viewed from the side, these tubes show a positive VP (above) and a negative VP (below) result. In this specimen, a positive reaction was seen in about 10 minutes after addition of reagents (see Fig. 9.34), but it may take up to 20 minutes to become positive. (**D**) All three of these tubes were inoculated and are viewed from above. Compartments from left to right: dulcitol/phenylalanine deaminase, urea, and citrate. The PA/dulcitol compartment will be dark brown if the organism is phenylalanine deaminase positive (below), yellow if dulcitol positive (above) and green if negative for both (center). Conflicting results for these two tests are rare because organisms for which the EnteroPluri-*Test* is designed have an extremely low probability of being positive for both. A positive urease test is pink (above and below) and yellow if negative (center). A positive citrate test is blue (top two) and green if negative (below).

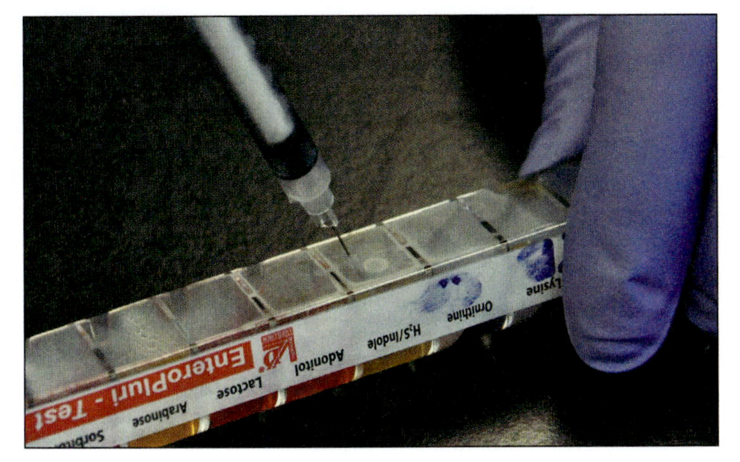


**9.34 Add Reagents after Incubation** ■ After recording results of tests that can be read directly, add reagents to the indole and VP compartments. Use a needle and syringe with Kovac's reagent to penetrate the plastic film on the flat side of the indole test and add 4 drops. A positive indole result will be seen within 15 seconds. Do the same with the VP test. Add 3 drops of VP-A and 2 drops of VP-B. A positive VP result may take up to 20 minutes to develop.

**9.35 EnteroPluri Data Chart and Codebook** ■ (**A**) The compiled results of the inoculated tube in Figure 9.27 are shown on this Data Chart, where the tests are arranged into five groups of three. Positive results are assigned their corresponding number (4, 2, or 1) and the sum of the positive results for each group is recorded. The resulting 5-digit number is then found in the EnteroPluri-*Test* Codebook. (**B**) Shown is a portion of a page from the codebook. The 5-digit code for the unknown scored in (**A**) is 74161, which corresponds to *Serratia liquefaciens*. Because of genetic variability, not all *S. liquefaciens* strains will give identical results for the 15 tests. Note that the codes 74147, 74160, and 74162 (and several others not shown in this figure) also code for *S. liquefaciens*. The codebook indicates which results are atypical for the isolate. For instance, 74160 identifies an isolate as *S. liquefaciens* that is atypically citrate-negative. (Can you see how a citrate-negative result would change the code from 74161 to 74160?) Also notice that in some cases a single code may be produced by two organisms (e.g., 74150). In these cases additional testing would be required to determine the isolate's identity.

Code data from EnteroPluri–*Test* reprinted with permission. ©Liofilchem® S.r.l. – *Microbiology Products*

**A**

**MODULO DATI / *DATA CHART***

| DATA<br>*DATE* | CAMPIONE<br>*SAMPLE* |
|---|---|

| | GROUP 1 | | | GROUP 2 | | | GROUP 3 | | | GROUP 4 | | | GROUP 5 | | |
|---|---|---|---|---|---|---|---|---|---|---|---|---|---|---|---|
| Test | Glucose | Gas | Lysine | Ornithine | $H_2S$ | Indole | Adonitol | Lactose | Arabinose | Sorbitol | VP | Dulcitol | PA | Urea | Citrate |
| Codice di positività<br>*Positivity code* | 4 | 2 | 1 | 4 | 2 | 1 | 4 | 2 | 1 | 4 | 2 | 1 | 4 | 2 | 1 |
| Risultati<br>*Results* | 4 | 2 | 1 | 4 | 0 | 0 | 0 | 0 | 1 | 4 | 2* | 0 | 0 | 0 | 1 |
| Somma dei codici<br>*Code sum* | 7 | | | 4 | | | 1 | | | 6 | | | 1 | | |

CODICE NUMERICO / *NUMERICAL CODE* — MICRORGANISMO / *MICROORGANISM* — *VP not visible from this view.

**B**

***ENTEROBACTERIACEAE***
***ENTEROBACTERIACEAE***

| Codice numerico<br>*Code number* | Microrganismo<br>*Microorganism* | Test atipici<br>*Atypical tests* |
|---|---|---|
| 74147 | *Serratia liquefaciens* | PA, URE |
| 74150 | *Salmonella sp. subsp. choleraesuis* | $H_2S$, CIT |
| | *Escherichia coli* | IND, LAC |
| 74151 | *Salmonella sp. subsp. choleraesuis* | $H_2S$ |
| 74160 | *Serratia liquefaciens* | CIT |
| 74161 | *Serratia liquefaciens* | NONE |
| 74162 | *Serratia liquefaciens* | URE, CIT |

9

TABLE **9-3** **EnteroPluri-*Test* Results and Interpretations**

| Compartment | Test | Result | Interpretation | Symbol |
|---|---|---|---|---|
| 1 | Glucose (acid/gas) | Red | Organism does not ferment glucose | − |
| | | Yellow, wax not lifted | Organism ferments glucose to acid | + |
| | | Yellow, wax lifted | Organism ferments glucose to acid and gas | + |
| 2 | Lysine | Yellow | Organism does not decarboxylate lysine | − |
| | | Purple | Organism decarboxylates lysine | + |
| 3 | Ornithine | Yellow | Organism does not decarboxylate ornithine | − |
| | | Purple | Organism decarboxylates ornithine | + |
| 4 | $H_2S$ | Beige | Organism does not reduce sulfur | − |
| | | Black | Organism reduces sulfur | + |
| | Indole (after addition of Kovac's reagent) | No red in Kovac's reagent (after 15 seconds) | Organism does not produce indole | − |
| | | Red in Kovac's reagent (within 15 seconds) | Organism produces indole | + |
| 5 | Adonitol | Red | Organism does not ferment adonitol | − |
| | | Yellow | Organism ferments adonitol | + |
| 6 | Lactose | Red | Organism does not ferment lactose | − |
| | | Yellow | Organism ferments lactose | + |
| 7 | Arabinose | Red | Organism does not ferment arabinose | − |
| | | Yellow | Organism ferments arabinose | + |
| 8 | Sorbitol | Red | Organism does not ferment sorbitol | − |
| | | Yellow | Organism ferments sorbitol | + |
| 9 | VP (after addition of VP-A and VP-B reagents) | Colorless (after 20 minutes) | Organism does not produce acetoin | − |
| | | Red (within 20 minutes) | Organism produces acetoin | + |
| 10 | Dulcitol | Green | Organism does not ferment dulcitol | − |
| | | Yellow | Organism ferments dulcitol | + |
| | Phenylalanine deaminase | Green | Organism does not deaminate phenylalanine | − |
| | | Black-smoky-gray | Organism deaminates phenylalanine | + |
| 11 | Urea | Beige | Organism does not hydrolyze urea | − |
| | | Red-purple/pink | Organism hydrolyzes urea | + |
| 12 | Citrate | Green | Organism does not utilize citrate | − |
| | | Blue | Organism utilizes citrate | + |

9

## *References*


EnteroPluri–*Test* package insert.

EnteroPluri–*Test* Codebook.

Name ______________________________

Date ______________________________

Lab Section ______________________________

I was present and performed this exercise (initials) ______________

DATA SHEET

9-5

# EnteroPluri-*Test*

**1** Tape your EnteroPluri-*Test* result sheet here.

**2** If your organism produced atypical results, enter those here along with any confirmatory test(s) your instructor assigned to verify or refute the provisional identification.

| Five-Digit Code | Possible Organisms | Atypical Result(s) | Confirmatory Test Result/ Confirmation (Y/N) |
|---|---|---|---|
| | | | |
| | | | |
| | | | |
| | | | |

# QUESTIONS

**1** *Fecal coliforms such as* Escherichia coli, Enterobacter aerogenes, *and* Klebsiella pneumoniae *are enterics that ferment lactose to acid and gas at 35°C within 48 hours. For most strains of these organisms, the table below summarizes reactions in the EnteroPluri-*Test. *Fill in the missing information and, using colored pencils, fill in the appropriate colors for each positive test.*

| | GLU | GAS | LYS | ORN | $H_2S$ | IND | ADO | LAC | ARA | SOR | VP | DUL | PA | URE | CIT |
|---|---|---|---|---|---|---|---|---|---|---|---|---|---|---|---|
| *E. coli* | | | + | – | – | + | – | | + | + | – | – | – | – | – |
| *E. aerogenes* | | | + | + | – | – | + | | + | + | + | – | – | – | + |
| *K. pneumoniae* | | | + | – | – | – | + | | + | + | + | – | – | + | + |

**2** *Based on the information above, enter the five-digit codes for the three organisms.*

E. coli ________________ E. aerogenes ________________ K. pneumoniae ________________

**3** *Examine the results table below. Fill in the color reactions and symbols for each test based on the ID value given.*

| | GLU | GAS | LYS | ORN | $H_2S$ | IND | ADO | LAC | ARA | SOR | VP | DUL | PA | URE | CIT |
|---|---|---|---|---|---|---|---|---|---|---|---|---|---|---|---|
| 45543 | | | | | | | | | | | | | | | |
| 51100 | | | | | | | | | | | | | | | |
| 00303 | | | | | | | | | | | | | | | |
| 60340 | | | | | | | | | | | | | | | |

**4** *Which organism(s) in question 3 is/are not member(s) of* Enterobacteriaceae? *Why?*

______________________________________________

______________________________________________

______________________________________________

______________________________________________

**5** *Which of the four ID values is questionable? Why?*

______________________________________________

______________________________________________

______________________________________________

______________________________________________

APPENDIX A

# Biochemical Pathways

So much of what is done in microbiology relies on an understanding of basic biochemical pathways (Fig. A.1). It is not as important to memorize them (although, with exposure they will become second nature) as it is to understand their importance in metabolism and to interpret diagrams of them when available. The following discussion along with Figure A.1 are provided so you can see how the various biochemical tests presented in this manual fit into the overall scheme of cellular chemistry.

## Oxidation of Glucose: Glycolysis, Entner-Doudoroff, and Pentose-Phosphate Pathways

Most organisms use glycolysis, also known as the Embden-Meyerhof-Parnas pathway (Fig. A.2), in energy metabolism. It performs the stepwise disassembly of glucose into two pyruvates, releasing some of its energy and electrons in the process. The exergonic (energy-releasing) reactions are associated with ATP synthesis by a process called **substrate phosphorylation**. Although a total of four ATPs are produced per glucose in glycolysis, two ATPs are hydrolyzed early in the pathway, leaving a net production of two ATPs per glucose.

In one glycolytic reaction, the loss of an electron pair (oxidation) from a 3-carbon intermediate occurs simultaneously with the reduction of $NAD^+$ to $NADH + H^+$. The $NADH + H^+$ then may be oxidized in an electron transport chain or a fermentation pathway, depending on the organism and the environmental conditions. The former yields ATP (by **oxidative phosphorylation**) and the latter generally does not. In summary, each glucose oxidized in glycolysis yields two pyruvates, $2\ NADH + 2\ H^+$, and a net of 2 ATPs (Table A-1).

Although the intermediates of glycolysis are carbohydrates, many are entry points for amino acid, lipid, and nucleotide catabolism. Many glycolytic intermediates also are a source of carbon skeletons for the synthesis of these other biochemicals. Some of these are shown in Figure A.1.

The Entner-Doudoroff pathway (Fig. A.3) is an alternative means of degrading glucose into two pyruvates. This pathway is found (almost) exclusively among *Bacteria* (e.g., *Pseudomonas* and *E. coli*, as well as other Gram negatives) and certain *Archaea*. Some obligate aerobes use this pathway because they lack the enzymes required to convert glucose to glyceraldehyde 3-phosphate in glycolysis. The pathway also allows utilization of a different category of sugars (aldonic acids) than glycolysis and therefore improves the range of resources available to the organism. It is less efficient than glycolysis because only one ATP is phosphorylated and only one NADH is produced. Table A-2 summarizes this pathway.

The pentose-phosphate pathway (Fig. A.4) is a complex set of cyclic reactions that provides a mechanism for producing 5-carbon sugars (pentoses) from 6-carbon sugars (hexoses). Pentose sugars produced are used in ribonucleotides and deoxyribonucleotides, as well as being precursors to aromatic amino acids. Further, this pathway produces NADPH, which is used as an electron

TABLE **A-1** **Summary of Glycolytic Reactants and Products per Glucose (Maximum Yield)**

| Reactant | Product |
|---|---|
| Glucose ($C_6H_{12}O_6$) | 2 Pyruvates ($C_3H_3O_3$) |
| (2 ATP + 4 ADP) = NET: 2 ADP | (2 ADP + 4 ATP) = NET: 2 ATP |
| 2 $NAD^+$ | 2 $NADH+2\ H^+$ |

TABLE **A-2** **Summary of Entner-Doudoroff Reactants and Products per Glucose (Maximum Yield)**

| Reactant | Product |
|---|---|
| Glucose ($C_6H_{12}O_6$) | 2 Pyruvates ($C_3H_3O_3$) |
| (1 ATP + 2 ADP) = NET: 1 ADP | (1 ADP + 2 ATP) = NET: 1 ATP |
| 1 $NAD^+$ | 1 NADH + 1 $H^+$ |
| 1 $NADP^+$ | 1 NADPH + 1 $H^+$ |

A

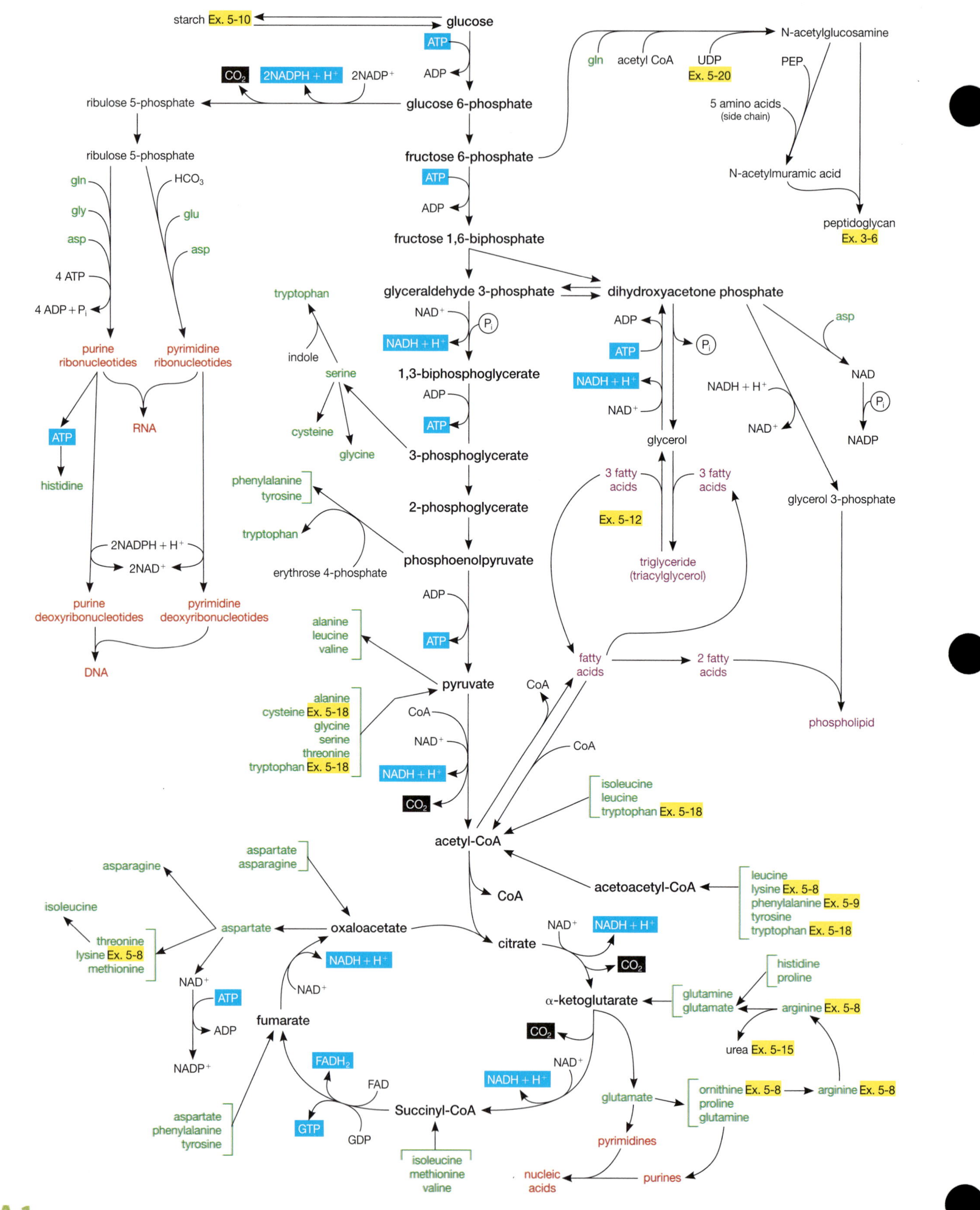


**A.1 Integrated Metabolism** ■ This diagram of metabolism shows basic anabolic and catabolic pathways in bacterial cells along with lab exercises where those reactions are relevant. Remember this diagram is here and check in with it periodically to regain your "forest" perspective if you feel you're getting lost in the "trees." Some details have been left out because of space limitations.

KEY TO **FIGURE A.1**

| **Black:** Carbon intermediates and miscellaneous compounds |
|---|
| **CoA** = coenzyme A |
| **$CO_2$** = carbon dioxide |
| **$P_i$** = inorganic phosphate |
| **PEP** = phosphoenolpyruvate |
| **Blue:** Short term energy carrying molecules |
| **ATP** = adenosine triphosphate |
| **ADP** = adenosine diphosphate |
| **GTP** = guanosine triphosphate |
| **GDP** = guanosine diphosphate |
| **UDP** = uridine diphosphate |
| **$NAD^+$/NADH** = nicotinamide adenine dinucleotide (oxidized and reduced forms) |
| **$NADP^+$/NADPH** = nicotinamide adenine dinucleotide phosphate (oxidized and reduced forms) |
| **FAD/$FADH_2$** = flavin adenine dinucleotide (oxidized and reduced forms) |
| **Green:** Amino acids |
| **asp** = aspartate |
| **glu** = glutamate |
| **gln** = glutamine |
| **gly** = glycine |
| **Red:** Nucleotides and nucleic acids |
| **Violet:** Lipids |

donor in anabolic pathways. Unlike NADH, produced in glycolysis and Entner-Doudoroff, NADPH is not used as an electron donor in an electron transport chain for oxidative phosphorylation of ADP.

The pentose-phosphate reactants and products are listed in Table A-3. To completely oxidize one hexose to $6CO_2$, a total of six hexoses must enter the cycle as glucose 6-phosphate and follow one of three different routes (notice the symmetry of pathways as drawn). Notice in Figure A.4 that each hexose loses a $CO_2$ upon entry into the cycle, but at the end five hexoses are produced. Thus, the net reaction is one hexose being oxidized to $6CO_2$. Notice also the reactions that transfer 2-carbon and 3-carbon fragments between the 5-carbon intermediates. Transketolase catalyzes the 2-carbon transfer, whereas transaldolase catalyzes the 3-carbon transfer. Alternatively, the 5-carbon intermediates can be redirected into pathways for synthesis of aromatic amino acids and nucleotides (not shown).

The decarboxylation of glucose 6-phospate in the pentose phosphate pathway is not as simple as shown in Figure A.4. The preliminary steps include production of 6-phosphogluconate, an intermediate of the Entner-Doudoroff pathway, which allows passage of carbon skeletons between the two pathways. The pentose phosphate pathway also intersects with glycolysis at glyceraldehyde 3-phosphate and fructose 6-phosphate.

TABLE **A-3** **Summary of Pentose-Phosphate Reactants and Products per Glucose-6-phosphate (Maximum Yield)**

| Reactant | Product |
|---|---|
| Glucose-6-phosphate ($C_6$) | $6CO_2 + 1\ P_i$ |
| 12 $NADP^+$ | 12 NADPH + 12 $H^+$ |



## Oxidation of Pyruvate: The Citric Acid Cycle and Fermentation

Pyruvate represents a major crossroads in metabolism, and all three pathways discussed so far have the potential to make it, though not in equal quantities. Some organisms are able to further disassemble the pyruvates produced in glycolysis and the other paths and make more ATP and NADH + $H^+$ in the citric acid cycle. Other organisms simply reduce the pyruvates with electrons from NADH + $H^+$ without further (or minimal) energy production in fermentation.

The citric acid cycle is a major metabolic pathway used in energy production by organisms that respire aerobically or anaerobically (Fig. A.5). Pyruvate produced in glycolysis or other pathways is first converted to acetyl-coenzyme A during the entry step (also known as the intermediate, or gateway, step). Acetyl-CoA enters the citric acid cycle through a condensation reaction with oxaloacetate.

Products for each pyruvate that enters the cycle via the entry step are: 3 $CO_2$, 4 NADH + $H^+$, 1 $FADH_2$, and 1 GTP. (Because two pyruvates are made per glucose, these numbers are doubled in Table A-4, which shows maximum yield per glucose.) The energy released from oxidation of reduced coenzymes (NADH + $H^+$ and $FADH_2$) in an electron transport chain is then used to make ATP. ATP yields are summarized in Table A-5.

Like glycolysis, many of the citric acid cycle's intermediates are entry points for amino acid, nucleotide, and lipid catabolism, as well as a source of carbon skeletons for synthesis of the same compounds. These pathways are shown in Figure A.1. Single arrows may represent several reactions, and other carbon compounds, not illustrated, may be required to complete a given reaction.

Figure A.6 illustrates some major fermentation pathways exhibited by microbes (though no single organism is capable of all of them). Pyruvate (shown in the blue box) is typically the starting point for each. End products of fermentation are shown in red. Fermentation allows some cells living under anaerobic conditions to oxidize reduced coenzymes (such as NADH + $H^+$ and shown in blue) generated during glycolysis or other pathways. Some bacteria (aerotolerant anaerobes) rely solely on fermentation and do not use oxygen even if it is available. Table A-6 summarizes major fermentations and some representative organisms that perform each.

Notice that fermentation end products typically fall into three categories: acid, gas, or an organic solvent (an alcohol or a ketone). The specific fermentation performed is the result of the enzymes present in a species and often is used as a basis of classification.

A

**TABLE A-4 Summary of Reactants and Products per Glucose in the Entry Step and the Citric Acid Cycle (Maximum Yield)**

| Entry Step | | Citric Acid Cycle | |
|---|---|---|---|
| **Reactant** | **Product** | **Reactant** | **Product** |
| 2 Pyruvates + 2 Coenzyme A | 2 Acetyl-CoA + 2 $CO_2$ | 2 Acetyl CoA | 4 $CO_2$ + 2 Coenzyme A |
| 2 $NAD^+$ | 2 NADH + 2$H^+$ | 6 $NAD^+$ | 6 NADH + 6$H^+$ |
| | | 2 GDP + 2 $P_i$ (= 2 ADP + 2 $P_i$) | 2 GTP (= 2 ATP) |
| | | 2 FAD | 2 $FADH_2$ |

**TABLE A-5 ATP Yields from Complete Oxidation of Glucose to $CO_2$ by a Prokaryote Using Glycolysis, Entry Step, and the Citric Acid Cycle with $O_2$ as the Final Electron Acceptor**

| Compound | Number Produced | ATP Value[1] | Total ATPs per Glucose |
|---|---|---|---|
| NADH + $H^+$ | 10 | 3 | 30 |
| $FADH_2$ | 2 | 2 | 4 |
| ATP (by substrate phosphorylation) | 4 | | 4 |

1 Experimental evidence indicates that the ATP values of 3 per NADH and 2 per $FADH_2$ in the aerobic electron transport chain are probably too high. For mitochondria, values of 2.5 and 1.5, respectively, are closer to reality. However, because multiple electron transport chains have been identified within the domain *Bacteria* (and even within single species!) no single, universal number for ATP yields per NADH and $FADH_2$ can be assigned. So, we fall back on the comfortable numbers of 3 and 2, recognizing that they are wrong but still can be used to illustrate how ATP yields could be calculated from the various pathways if we only had the correct information.

**TABLE A-6 Major Fermentations, Their End-Products, and Some Organisms That Perform Them**

| Fermentation | Major End Products | Representative Organisms |
|---|---|---|
| Alcoholic fermentation | Ethanol and $CO_2$ | *Saccharomyces cerevisiae* |
| Homofermentation | Lactate | *Streptococcus* and some *Lactobacillus* |
| Heterofermentation | Lactate, ethanol, and acetate | *Streptococcus*, *Leuconostoc*, and *Lactobacillus* |
| Mixed acid fermentation | Acetate, formate, succinate, $CO_2$, $H_2$, and ethanol | *Escherichia*, *Salmonella*, *Klebsiella*, and *Shigella* |
| 2,3-butanediol fermentation | 2,3-butanediol | *Enterobacter*, *Serratia*, and *Erwinia* |
| Butyrate/butanol fermentation | Butanol, butyrate, acetone, and isopropanol | *Clostridium*, *Butyrivibrio*, and some *Bacillus* |
| Propionic acid fermentation | Propionate, acetate and $CO_2$ | *Propionibacterium*, *Veillonella*, and some *Clostridium* |

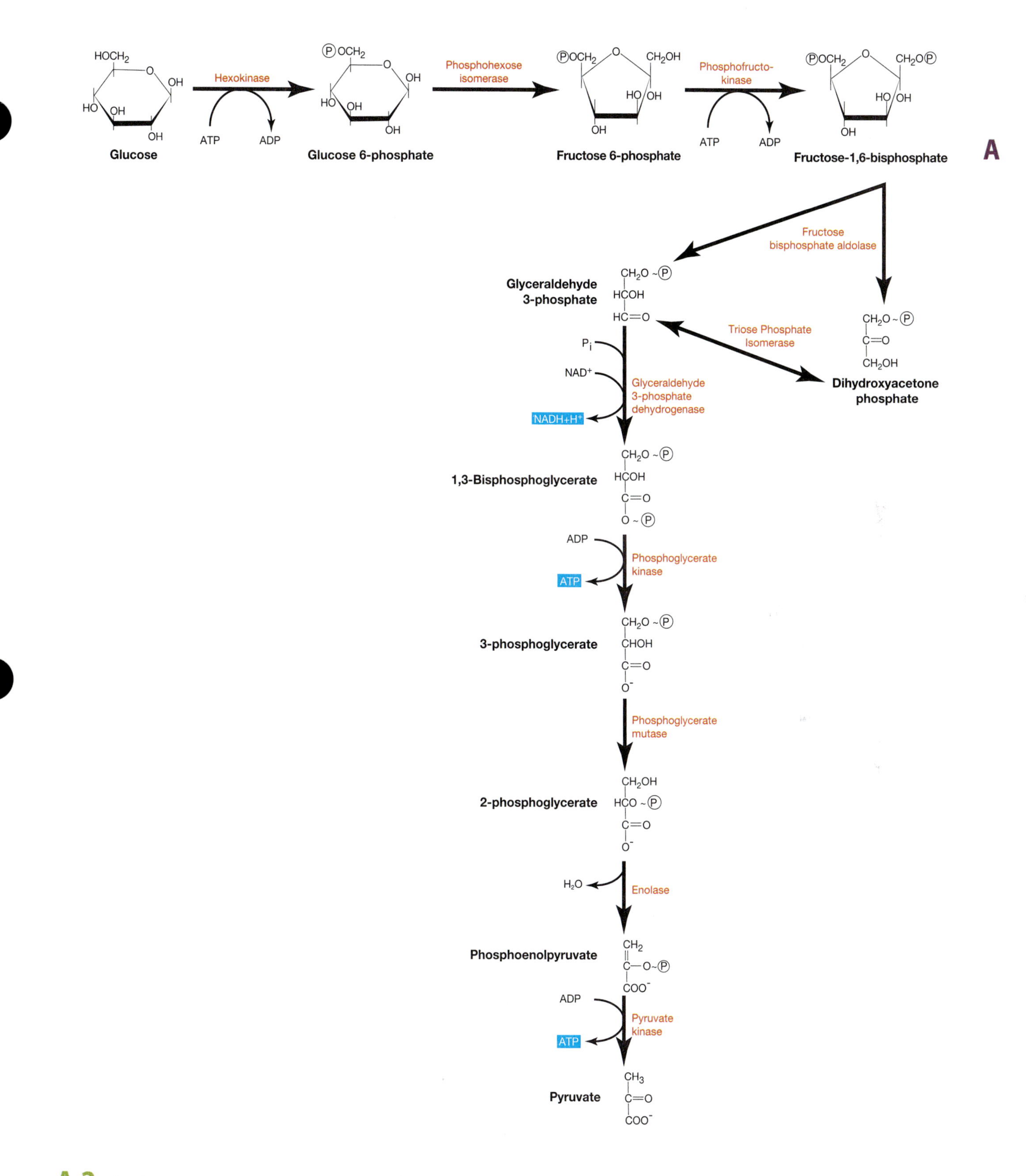


**A.2 Glycolysis** ■ The names of glycolytic intermediates are printed in black ink; the enzyme names are in red. Reducing power (in the form of NADH+$H^+$) and ATP are highlighted in blue. The major key to getting product yields correct is to recognize that both $C_3$ compounds (glyceraldehyde 3-phosphate, GAP, and dihydroxyacetone phosphate, DiHAP) produced from splitting fructose 1,6-bisphosphate can pass through the remainder of the pathway because of the triose phosphate isomerase reaction. The conversion of each into pyruvate results in the formation of 2 ATPs and 1 NADH+$H^+$ (Table A-1).

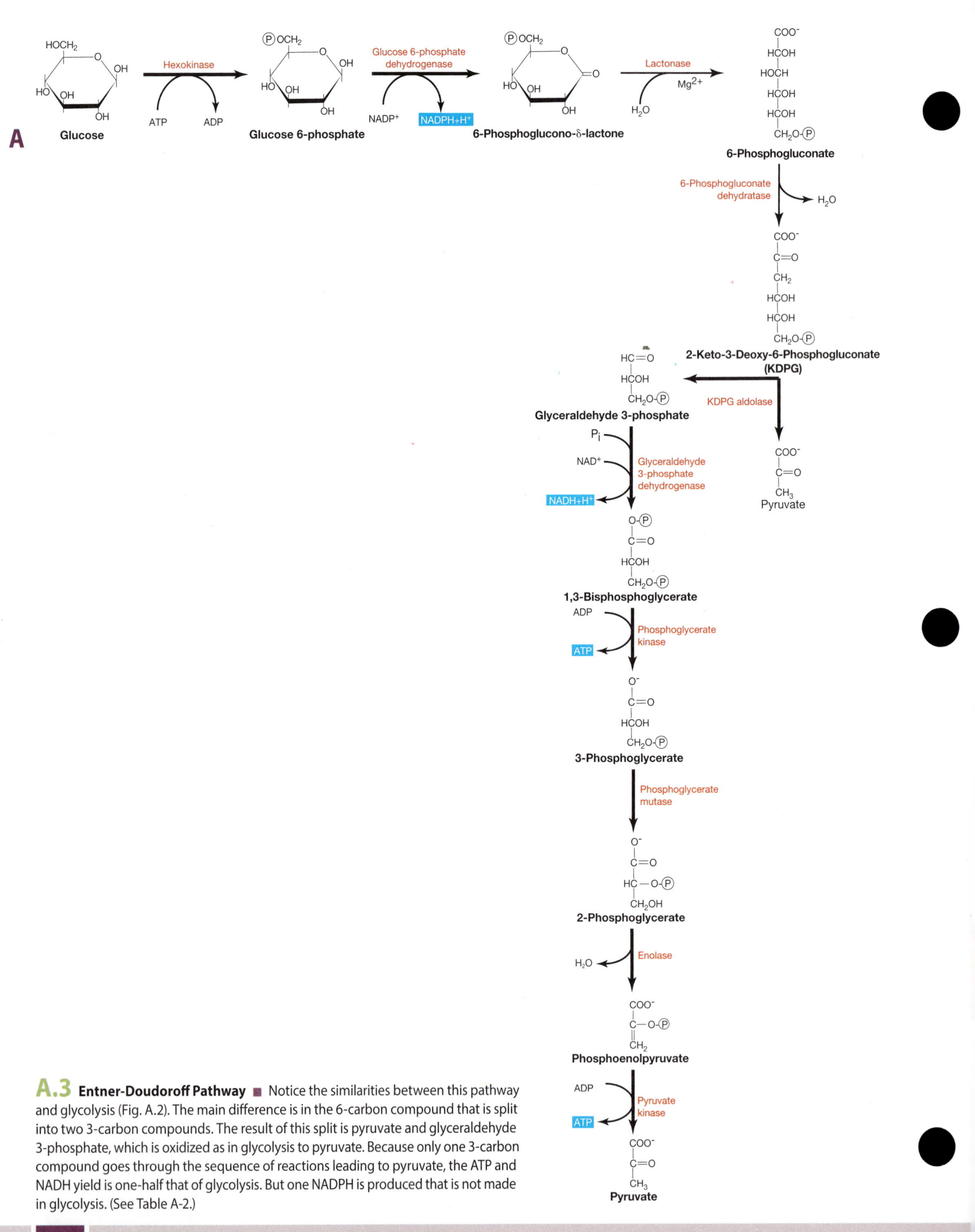


**A.3 Entner-Doudoroff Pathway** ■ Notice the similarities between this pathway and glycolysis (Fig. A.2). The main difference is in the 6-carbon compound that is split into two 3-carbon compounds. The result of this split is pyruvate and glyceraldehyde 3-phosphate, which is oxidized as in glycolysis to pyruvate. Because only one 3-carbon compound goes through the sequence of reactions leading to pyruvate, the ATP and NADH yield is one-half that of glycolysis. But one NADPH is produced that is not made in glycolysis. (See Table A-2.)

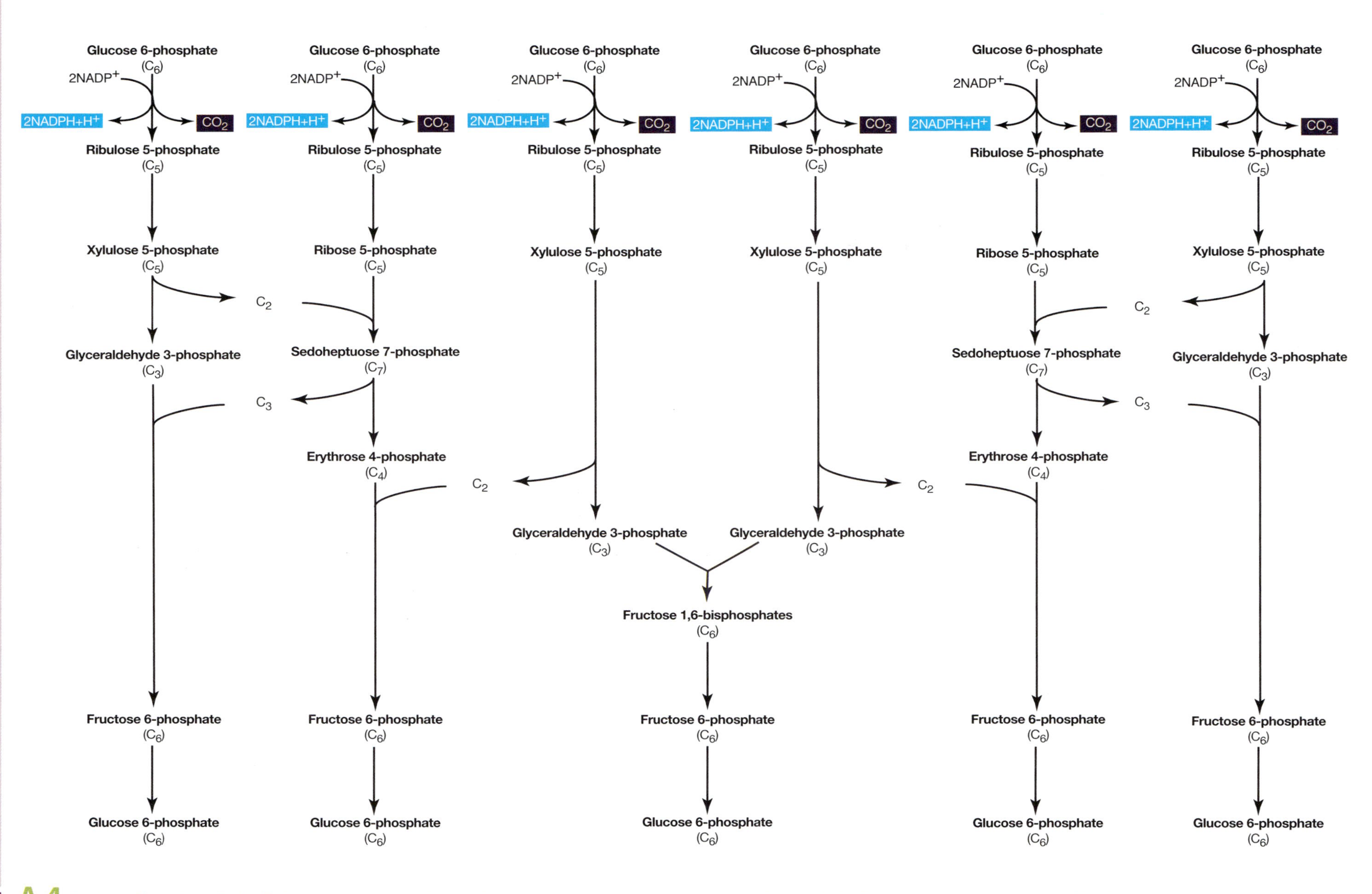


**A.4 Pentose-Phosphate Cycle** ■ For every six glucose-6-phosphates that enter and complete the cycle, $6CO_2$ and 12 NADPH+H$^+$ are produced. Some of the 5-carbon intermediates, however, may be redirected into synthesis of aromatic amino acids and nucleotides. If the cycle is performed as shown, 36 carbons enter as six glucose 6-phosphates ($6 \times C_6 = 36C$). Six $CO_2$ are immediately lost, leaving a total of 30C to get shuffled around by the remaining reactions to form five glucose 6-phosphates ($5 \times C_6 = 30C$). (See Table A-3.)

A

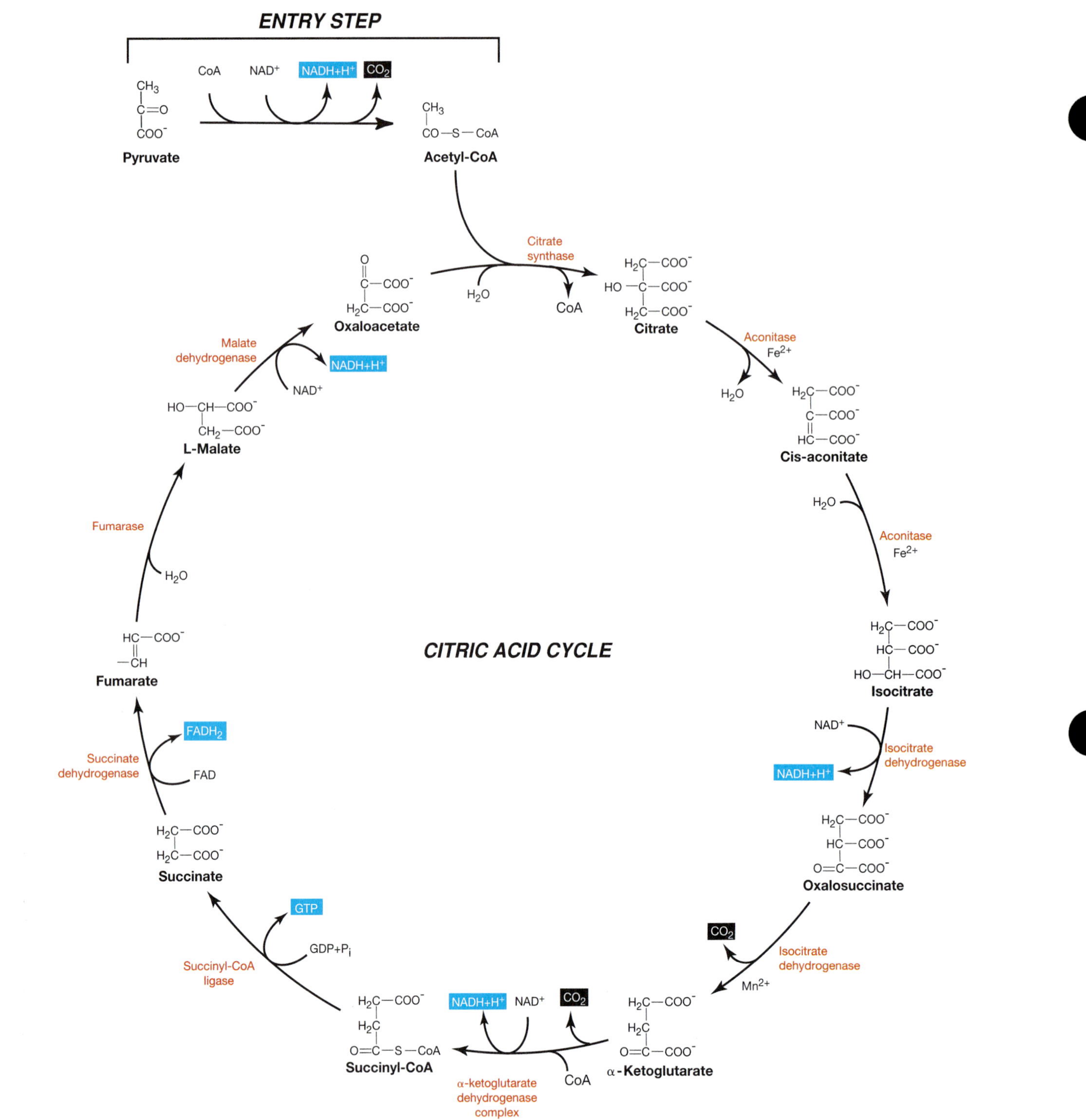


**A.5 Entry Step and Citric Acid Cycle** ■ The names of intermediates are printed in black; enzymes are in red. Reducing power (in the form of NADH+$H^+$ and $FADH_2$) and GTP are highlighted in blue. $CO_2$ produced from the oxidation of carbon is highlighted in black. (See Table A-4.)

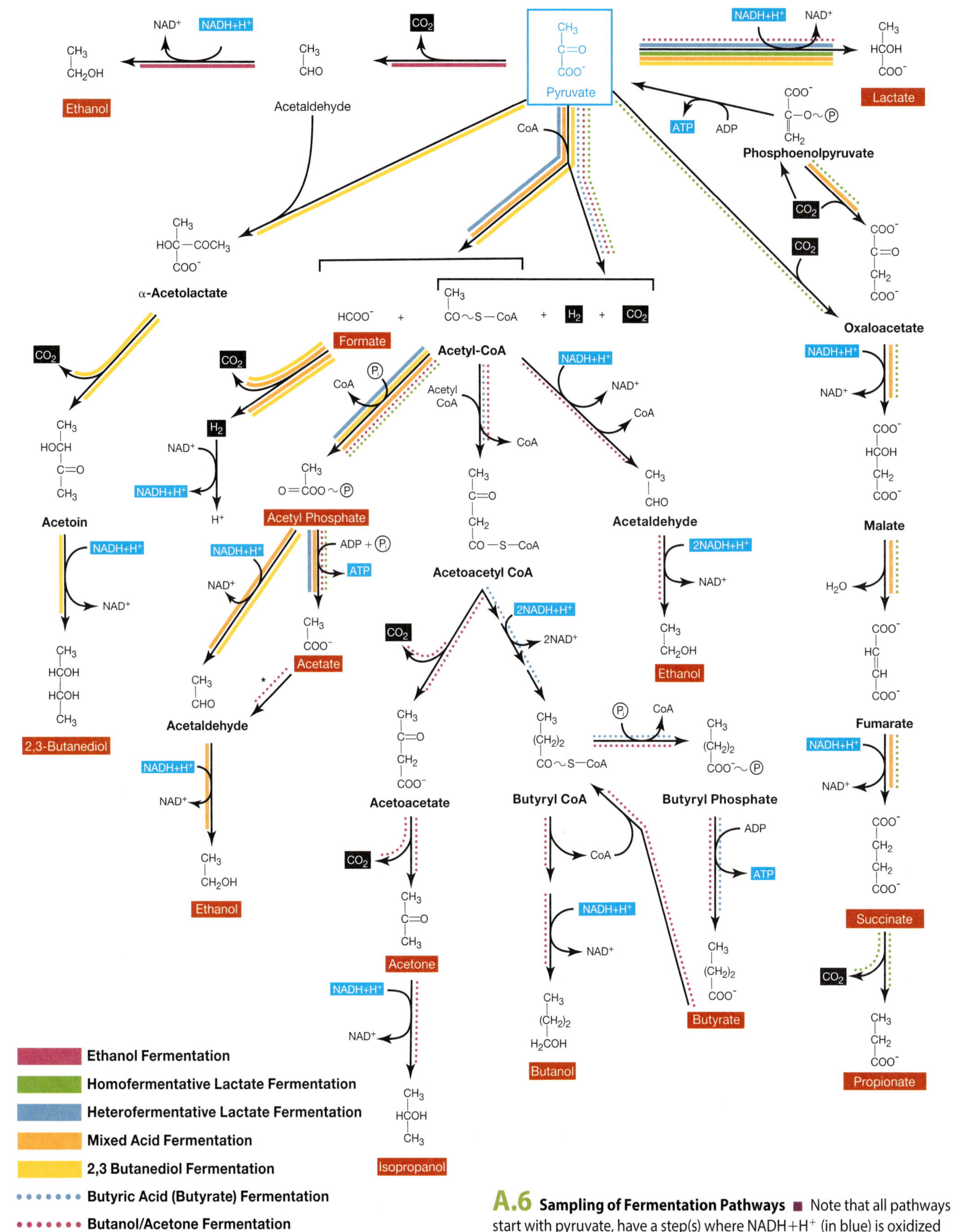


**A.6 Sampling of Fermentation Pathways** ■ Note that all pathways start with pyruvate, have a step(s) where NADH+H$^+$ (in blue) is oxidized to NAD$^+$, and produce end-products falling into one of three categories: acid, gas, or alcohol. Major fermentation pathways are highlighted by different colors and are summarized in Table A-5.

## References

Madigan, Michael T., John M. Martinko, Kelly S. Bender, Daniel H. Buckley, and David A. Stahl. Chap. 3 in *Brock's Biology of Microorganisms,* 14th ed. San Francisco: Pearson/Benjamin Cummings, 2015.

A Moat, Albert G., John W. Foster, and Michael P. Spector. Pages 373–376 in *Microbial Physiology*, 4th ed. New York, NY: Wiley–Liss, 2002.

White, David, James Drummond, and Clay Fuqua. Chap. 5 in *The Physiology and Biochemistry of Prokaryotes*, 4th ed. New York, NY: Oxford University Press, 2012.

APPENDIX B

# Miscellaneous Transfer Methods

Following are instructions for transfer methods that are performed less routinely than those in Exercise 1-4.

## Transfers Using a Sterile Cotton Swab

A sterile swab generally is used to obtain a sample from a primary source—either a patient or an environmental site. Occasionally, swabs are used to transfer pure cultures (e.g., Kirby-Bauer test, Exercise 7-2). Sterile swabs may be dry, or they may be in sterile water, depending on the nature of the sample to be taken (wet or dry). In either case, care must be taken not to contaminate the swab by unintentionally touching other surfaces with it. Your instructor may provide specific instructions on collection of samples from sources other than the ones below.

### Obtaining a Sample from a Patient's Throat with a Swab

1. **Use BSL-2 precautions throughout.** Wear a lab coat, disposable gloves, and eye protection.
2. Use a sterile tongue depressor and swab prepared in sterile water to obtain a sample from the throat. Have the patient open his/her mouth, and then gently press down on the tongue (Fig. B.1).
3. With the swab in your dominant hand, carefully sample the patient's throat with a swirling motion. Touching other parts of the oral cavity is likely to cause contamination. Also, avoid touching the soft palate or it may initiate a gag reflex!
4. Transfer the sample to an appropriate plated medium as quickly as possible. If plating is to be done at a later time, place the swab in an appropriate sterile container (such as a sterile, capped test tube with sterile water deep enough to immerse the cotton). Store in the lab refrigerator if necessary.
5. Use the swab to begin a quadrant streak (Exercise 1-5, Fig. 1.39). Properly dispose of the swab according to your lab's protocol, and then complete the streaking with your loop. Dispose of the loop in the appropriate autoclave container.

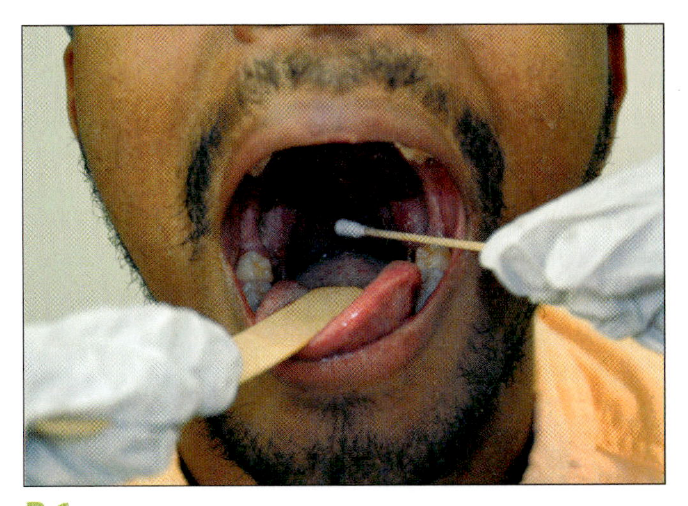

**B.1 Taking a Throat Sample** ■ Use a sterile tongue depressor and swab to obtain a sample from a patient's throat. Be careful not to touch other parts of the oral cavity or the sample will get contaminated. (The communities of microorganisms differ in different parts of the oral cavity and throat!) Transfer the sample to a sterile medium as soon as possible.

6. Label the plate with your name, the sample source, date, and medium.
7. Incubate the plate in an inverted position at the appropriate temperature for the designated time.

### Obtaining an Environmental Sample with a Cotton Swab

1. **Use BSL-2 precautions throughout.** Wear a lab coat, disposable gloves, and eye protection.
2. Use a sterile swab prepared in sterile water to obtain a sample from an environmental source.
3. Rotate the swab to collect from the area to be sampled (Fig. B.2).
4. Transfer the sample to an appropriate plated medium as quickly as possible. If plating is to be done at a later time, place the swab in an appropriate sterile container (such as a sterile, capped test tube with sterile water deep enough to immerse the cotton). Store in the lab refrigerator if necessary.

5. Depending on the anticipated cell density, use the swab to perform a zig-zag streak pattern (Exercise 1-5, Fig. 1.38) or to begin a quadrant streak with a disposable loop (Exercise 1-5, Fig. 1.39). Properly dispose of the swab and loop (if used) in the appropriate autoclave container.
6. Label the plate with your name, the sample source, date, and medium.
7. Incubate the plate in an inverted position at the appropriate temperature for the designated time.

## Stab Inoculation of Agar Tubes Using an Inoculating Needle

**B** Stab inoculations of agar tubes are used for several types of differential media (usually to examine growth under anaerobic conditions or to observe motility). A stab is *not* used to produce a culture of microbes for transfer to another medium. Generally, the culture to be transferred from will be growing on a solid medium, but a turbid broth will also work. You just won't be able to transfer the same amount of organisms.

1. Label the tube with your name, date, medium, and organism or sample source.
2. Aseptically obtain a visible amount of the specimen on the tip of a sterile inoculating needle. If transferring from a broth, just submerge the needle in the mixed broth.
3. Remove the cap of the sterile medium with the little finger of your inoculating needle hand, and hold it there. (Culture caps are never stored on the table top.)
4. Flame the tube by quickly passing it through the Bunsen burner flame a couple of times. Keep your needle hand still.
5. Hold the open tube on an angle to minimize airborne contamination. Keep your needle hand still.
6. Carefully move the agar tube over the needle wire (Fig. B.3). Insert the needle into the agar to about 1 cm from the bottom.
7. Withdraw the tube carefully so the needle leaves along the same path it entered. (When removing the tube, be especially careful not to catch the needle tip on the tube lip. This springing action of the needle creates bacterial aerosols.)
8. Flame the tube lip as before. Keep your needle hand still.
9. Keeping the needle hand still (remember, it has growth on it), move the tube to replace its cap. If using a screw cap tube, simply set it on the tube and screw it on once you've flamed your needle.

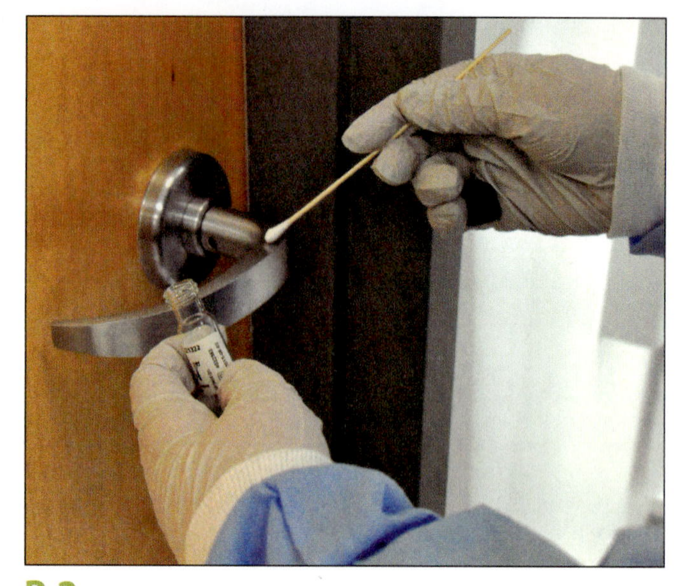

**B.2 Taking an Environmental Sample** ■ Use a spinning motion of a sterile swab to sample inanimate objects in the environment. The swab may be placed in a sterile test tube until it is convenient to transfer the sample to a growth medium. Notice the tube's cap being held in the pinky of the swab hand.

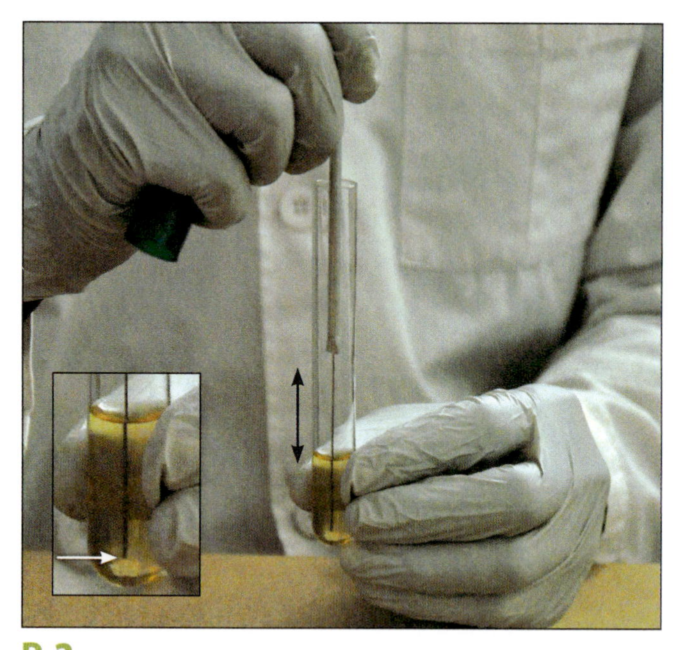

**B.3 Agar Deep Stab** ■ Use the inoculating needle to stab the agar to a depth about 1 cm from the bottom (tip of white arrow, inset). It is generally desirable to remove the needle along the original stab line and not create a new one. Upon completion, sterilize the needle.

10. Incinerate the needle from base to tip to minimize aerosol production. **If transferring a BSL-2 organism, use a bacterial incinerator.**
11. Incubate at the appropriate temperature for the assigned time.
12. Save or dispose of the original culture(s) as directed by your instructor.

## Spot Inoculation of an Agar Plate

Sometimes an agar plate may be used to grow several different specimens at once. This is a typical practice with plated *differential* media (i.e., media designed to differentiate organisms based on growth characteristics).

Prior to beginning the transfer, divide the plate into as many as four sectors using a marking pen. (Some plates already have marks on the base for this purpose.) Then each may be inoculated with a different organism.

Inoculation involves touching the loop to the agar surface once so growth is restricted to a single spot—hence the name "spot inoculation."

1. Wear a lab coat, gloves, and chemical eye protection when performing this procedure.
2. Label the plate's base with your name, date, medium, and organism(s) inoculated. Write these toward the periphery, not across the center, which makes results difficult to read.
3. With a marking pen, divide the base of the plate into the desired number of sectors. Figure B.4 shows a plate divided into four sectors.
4. Aseptically get growth from the source culture on the inoculating loop.
5. Lift the lid of the sterile agar plate and use it as a shield to prevent airborne contamination.
6. Touch the agar surface toward the periphery of the sector (Fig. B.4).
7. Remove the loop and replace the lid.

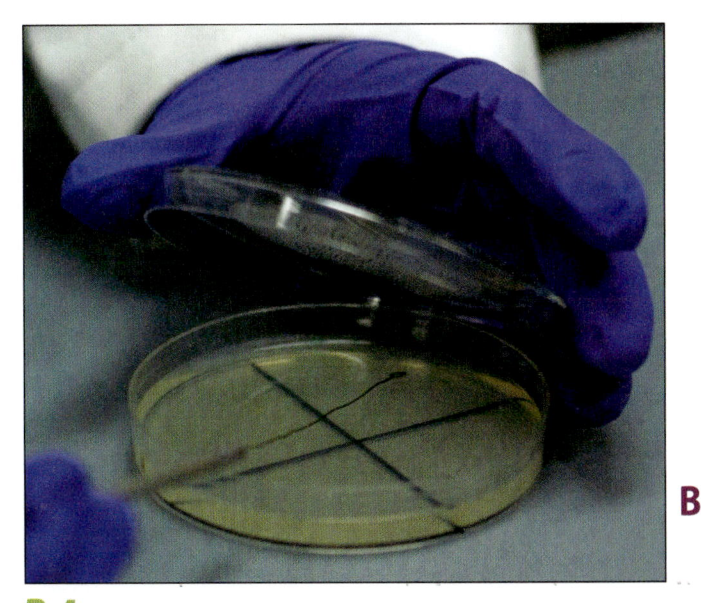

**B.4 Spot Inoculation of a Plate** ■ Each of four sectors is spot inoculated with a different organism by touching the loop to the surface and making a mark about 1 cm in length. Generally, spot inoculations are done toward the edge (rather than the crowded middle) to prevent overlapping growth and/or test results.

8. Incinerate your loop as before. It is especially important to flame it from base to tip now because the loop has bacteria on it.
9. Incubate the plate in an inverted position for the assigned time at the appropriate temperature.
10. Save or dispose of the original culture as directed by your instructor.

# APPENDIX C

# Transfers Using a Glass Pipette

Glass pipettes are used to transfer a known volume of liquid diluent, media, or culture. Originally pipettes were filled by sucking on them like a drinking straw, but mouth pipetting is dangerous and has been replaced by mechanical pipettors. Three examples are shown in Figure C.1, each with its own method of operation. Your instructor will show you how to properly use the type of pipettor available in your lab.

To use a pipette correctly, you must be able to correctly read the calibration. Examine Figure C.2. The numbers at the top indicate the pipette's *total volume* and its *smallest calibrated increments*. This is a 10.0 mL pipette divided into 0.1 mL increments.

When reading volumes, use the base of the meniscus (Fig. C.3). The volume in the center pipette is read as exactly 3.0 mL because the meniscus is resting on the line. The left pipette is read as 2.9 mL and the right pipette is read as 3.1 mL (0 is always at the top of the pipette). Although the difference in volume between these three pipettes may seem negligible (1 part in 70[1] is about a 1.4% error), it introduces avoidable error into your work.

Two pipette styles are used in microbiology (Fig. C.4): the serological pipette and the Mohr pipette. A serological pipette is calibrated to deliver (TD) its volume by completely draining it and blowing out the last drop. The tip of a Mohr pipette is not graduated, so fluid flow must be stopped at a calibration line. Stopping the fluid beyond the last line on a Mohr pipette results in an unknown volume being dispensed. In either case, volumes are read at the bottom of the fluid meniscus.

[1] This number presumes the pipette shown has a total volume of 10 mL. If the meniscus is on the 3, the pipette will dispense 7 mL. If emptied completely, an error of 0.1 mL above (delivering 7.1 mL) or below the 3 (delivering 6.9 mL) would represent 0.1 mL in 7 mL, or 1 part in 70, a 1.4% error that is avoidable.

***Important***

If pipetting a bacterial culture, work over a disinfectant-soaked towel in case any culture drips from the pipette before disposing of it in the autoclave container. This is especially important if transferring a BSL-2 organism. Clean up any spills with additional disinfectant. Also, it is recommended that you use a Mohr pipette to transfer BSL-2 organisms. These are not "blown-out," so aerosol production is reduced.

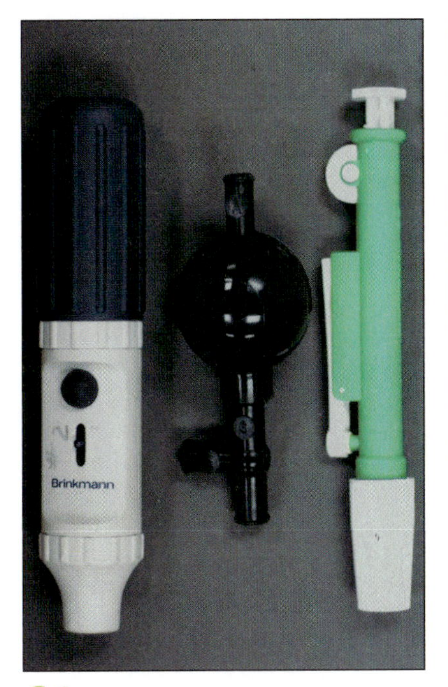


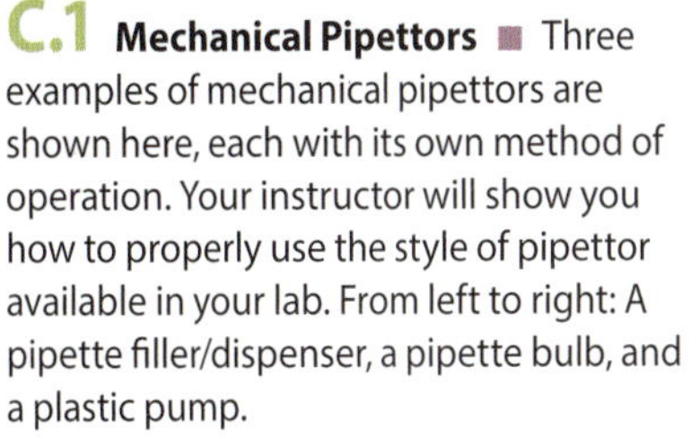

**C.1 Mechanical Pipettors** ■ Three examples of mechanical pipettors are shown here, each with its own method of operation. Your instructor will show you how to properly use the style of pipettor available in your lab. From left to right: A pipette filler/dispenser, a pipette bulb, and a plastic pump.

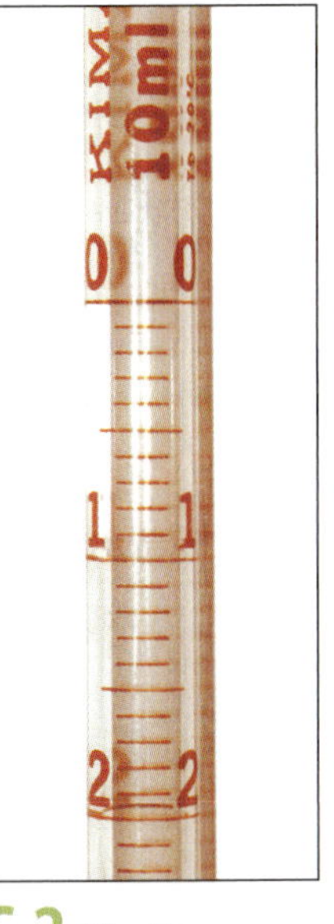


**C.2 Pipette Calibration** ■ Prior to using a pipette, read its calibration. The numbers indicate the pipette's total volume and its smallest calibrated increments. This is a 10.0 mL pipette divided into 0.1 mL increments.

## Filling a Glass Pipette

1. Wear a lab coat, gloves, and chemical eye protection when performing this procedure.
2. Bacteria should be suspended in the broth with a vortex mixer (Fig. 1.18) or by agitating with your fingers (Fig. 1.19). Be careful not to splash the broth into the cap or lose control of the tube. If transferring a sterile liquid, mixing is not necessary.
3. Pipettes are sterilized in metal canisters, individually in sleeves, or as multiples in packages (if disposable). They typically are stored in groups of a single size (Fig. C.5). *Be sure you know what volume your pipette will deliver.* Set the canister at the table edge and remove its lid. (Canisters—even cylindrical ones—should not be stored in an upright position, because they may fall over and break the pipettes or become contaminated.) If using pipettes in a package, open the end *opposite the tips*. Grasp *one pipette only* and remove it.
4. Carefully insert the pipette into the mechanical pipettor (Fig. C.6). Grasp the pipette near the end with your fingertips. This gives you more control and reduces the chance that you will break the pipette and cut your hand or jam the broken pipette into your palm. **Notice how the microbiologist is holding it in Figure C.6.** *Do not touch* any part of the pipette that will contact the specimen or the

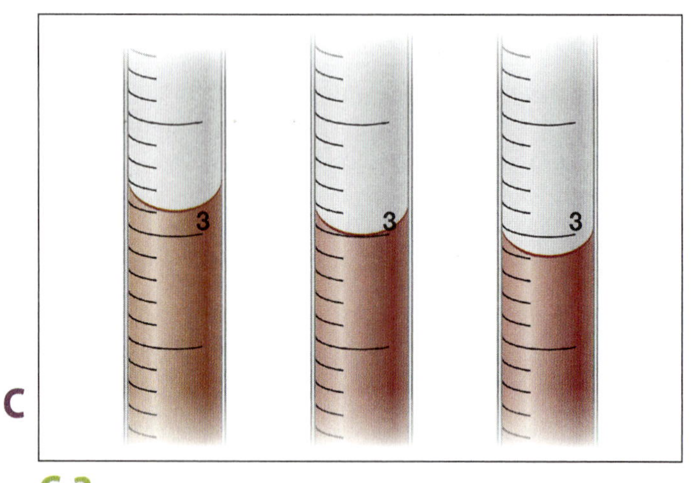


**C.3 Read the Base of the Meniscus** ■ When reading volumes, use the base of the meniscus. The volume in the center pipette is read at exactly 3.0 mL because the meniscus is resting on the line. The left pipette is read as 2.9 mL and the right pipette is read as 3.1 mL (0 is always at the top of the pipette). Note that the number read usually isn't the volume delivered. If these are all 10.0 mL pipettes, then completely emptying the center one would deliver 10.0 mL − 3.0 mL = 7.0 mL. The pipette on the left would deliver 7.1 mL and the one on the right 6.9 mL.

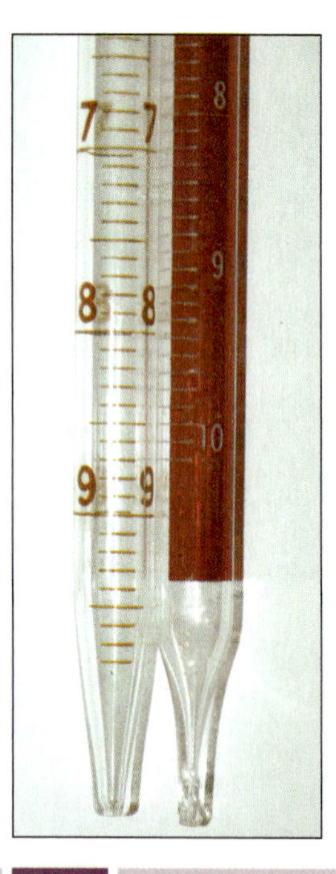

**C.4 Two Types of Pipettes** ■ Two pipette styles—the serological pipette (left) and the Mohr pipette (right)—are used in microbiology. A serological pipette is calibrated to deliver (TD) its volume by completely draining it and blowing out the last drop. The tip of a Mohr pipette is not graduated, so fluid flow must be stopped at a calibration line. Stopping the fluid beyond the last line on a Mohr pipette results in an unknown volume being dispensed.

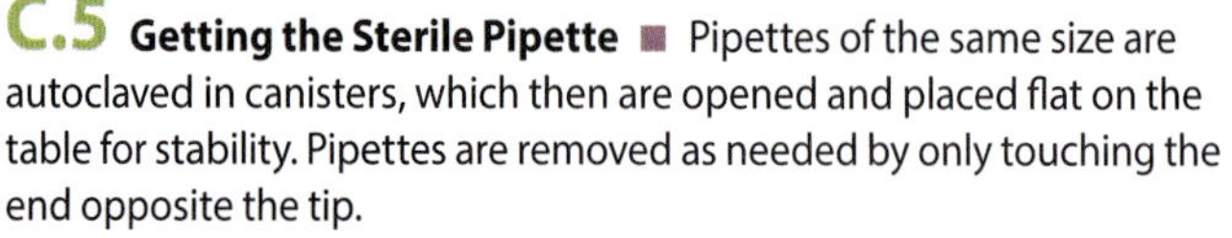

**C.5 Getting the Sterile Pipette** ■ Pipettes of the same size are autoclaved in canisters, which then are opened and placed flat on the table for stability. Pipettes are removed as needed by only touching the end opposite the tip.

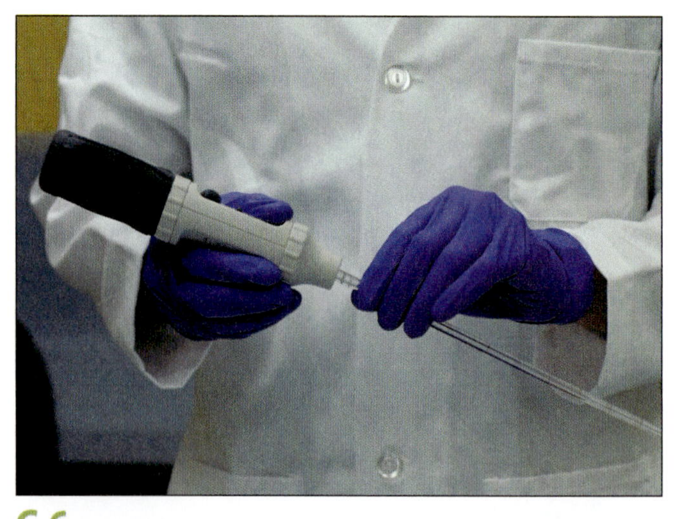

**C.6 Assembling the Pipette** ■ Carefully insert the pipette into a mechanical pipettor. Notice that the pipette is held near the end with the fingertips. For safety, the hand is out of the way in case the pipette breaks.

medium or you risk introducing a contaminant. Also, do not lay the pipette on the tabletop while you continue.

5. While keeping your pipette hand still, bring the culture tube toward it. Use your little finger to remove and hold its cap.
6. Flame the open end of the tube by passing it through a Bunsen burner flame two or three times.
7. Hold the tube at an angle to prevent contamination from above.
8. Insert the pipette and withdraw the appropriate volume (Fig. C.7). Bring the pipette to a vertical position briefly to read the meniscus accurately. (***Remember:*** The volumes in the pipette are correct only if the meniscus of the fluid inside is resting *on* the line, not above or below it.) Then carefully remove the pipette from the tube.
9. Flame the tube lip as before. Keep your pipette hand still.
10. Keeping the pipette hand still (remember, it may contain fluid with microbes in it), move the tube to replace its cap.
11. What you do next depends on the medium to which you are transferring the liquid. Please continue with the appropriate inoculation section.

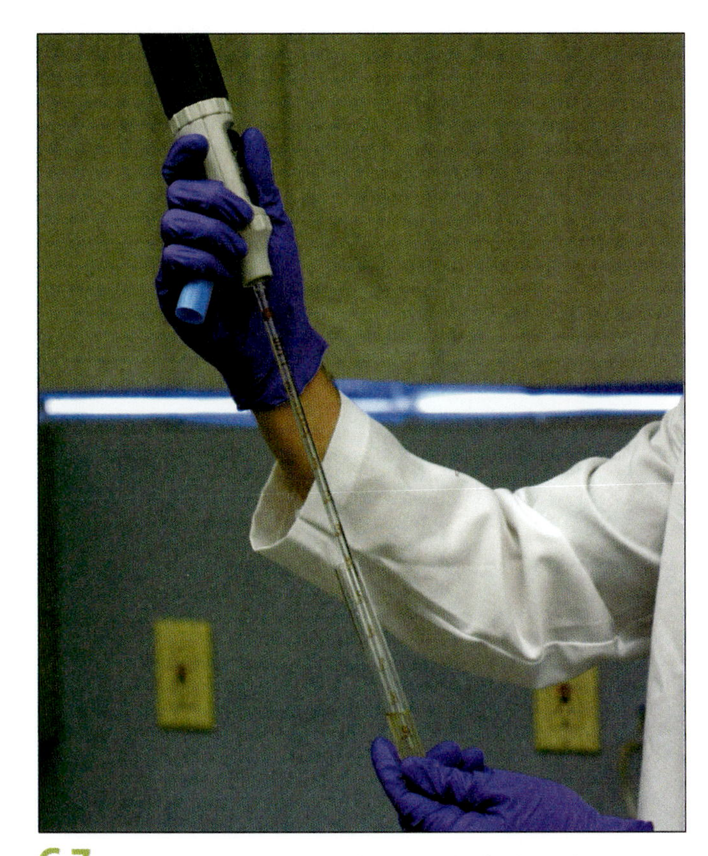

**C.7 Filling the Pipette** ■ Carefully draw the fluid into the pipette. Briefly bring it to a vertical position and read the volume. Notice how the tube's cap is being held by the pinky finger of the pipettor hand.

## Inoculation of Broth Tubes with a Pipette

Pipettes are often used to inoculate a known volume of culture into a tube or bottle containing a known volume of diluent during serial dilutions. They also would have been used to deliver the volume of diluent waiting in the tube or bottle.

1. While keeping the pipette hand still, bring the dilution tube/bottle toward it. Use your little finger to remove and hold its cap.
2. Flame the lip of the tube/bottle by quickly passing it through the Bunsen burner flame two or three times. Keep your pipette hand still.
3. Hold the open tube/bottle on an angle to minimize airborne contamination. Keep your pipette hand still.
4. Insert the pipette tip and gently dispense the correct volume of inoculum. Aerosols can be produced if you dispense too quickly.
5. Withdraw the tube/bottle from over the pipette. If using a serological pipette touch its tip to the glass to remove any excess broth before completely removing it.
6. Completely remove the pipette, but avoid waving it around. This can create aerosols.
7. Flame the tube/bottle lip as before. Keep your pipette hand still.
8. Keeping the pipette hand still, move the tube/bottle to replace its cap.
9. The pipette may be contaminated with microbes and must be disposed of correctly. Each lab has its own specific procedures, and your instructor will advise you what to do. Glass pipettes typically are placed in a pipette disposal receptacle containing a small amount of disinfectant until they are autoclaved and reused (Fig. C.8). Disposable pipettes

C

**C.8 Dispose of the Pipette** ■ The pipette must be disposed of correctly, especially if it is contaminated with microbes. Each lab has its own specific procedures, and your instructor will advise you what to do. Shown here is a canister used for glass pipettes. Disposable pipettes must be placed in an appropriate biohazard container. In either case, be careful when removing the pipette from the mechanical pipettor. There is danger of culture dripping from the pipette or of breaking the glass or plastic.

must be placed in an appropriate biohazard container. In either case, be careful when removing the pipette from the mechanical pipettor. There is danger of culture dripping from the pipette or of breaking the glass or plastic.

## Inoculation of Agar Plates with a Pipette

Pipettes also can be used to dispense a known volume of inoculum to an agar plate.

1 Lift the lid of the plate and use it as a shield to protect it from airborne contamination. It also will help in containing any aerosols you inadvertently produce delivering the inoculum.

2 Hold the pipette over the agar and dispense the correct volume (often 0.1 mL) onto the agar surface between the middle and one edge of the dish (Fig. C.9). From this point, the remainder of steps should be completed within about 15 seconds to prevent the inoculum from soaking into the agar.

3 The pipette is contaminated with microbes and must be disposed of correctly. Each lab has its own specific procedures, and your instructor will advise you what to do. Glass pipettes typically are placed in a pipette disposal receptacle containing a small amount of disinfectant until they are autoclaved and reused (Fig. C.8). Disposable pipettes must be placed in an appropriate biohazard container. In either case, be careful when removing the pipette from the mechanical pipettor. There is danger of culture dripping from the pipette or of breaking the glass or plastic.

4 Continue with the spread plate technique (Exercise 1-6).

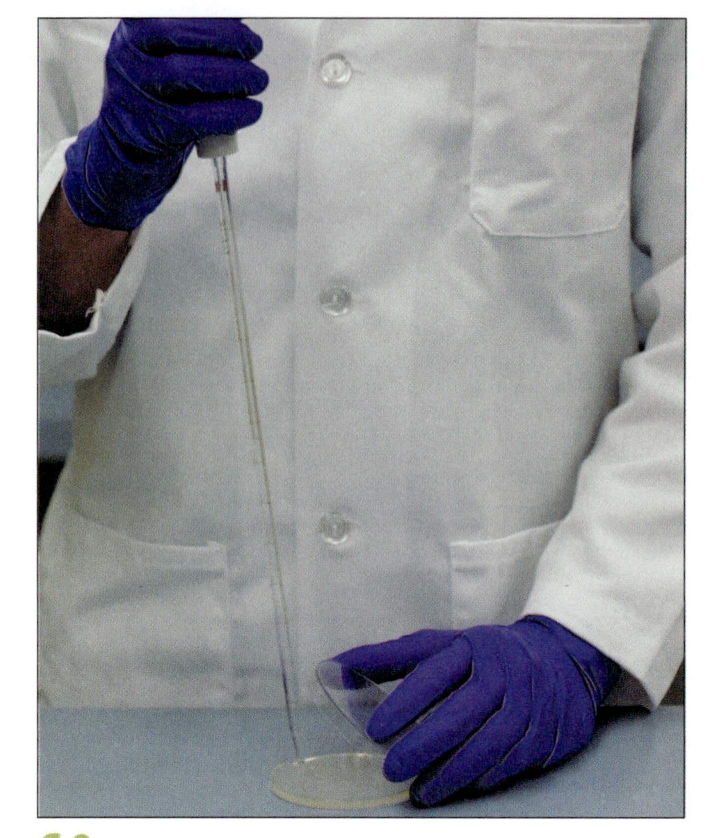

**C.9 Dispensing to an Agar Plate** ■ Open the lid and dispense the inoculum onto the surface of the agar between the middle and edge. This provides open agar on which to cool the spreading rod. Notice how the plate's lid is held to prevent aerial contamination and to contain aerosols. The plate may be placed on a disinfectant-soaked towel in case the pipette drips (not shown).

APPENDIX D

# Transfer from a Broth Culture Using a Digital Pipette

Modern molecular biology procedures often involve transferring extremely small volumes of liquid with great precision and accuracy. This has led to the development of digital pipettors or micropipettors (Fig. D.1) that can be set to dispense microliter volumes through milliliter volumes (recall that 1 mL equals 1,000 μL). Common digital pipettor calibrations include volumes of 1 to 10 μL, 10 to 100 μL, 100 to 1,000 μL, or 1,000 to 5,000 μL.

## Filling a Digital Pipettor

Many manufacturers make digital pipettors, but they all work basically the same way. The following instructions are for Eppendorf Series 2100 models. **Use BSL-2 precautions, if appropriate**. Before you begin, label the destination tube(s) or plate(s) with your name, the date, source of inoculum, and medium.

1. Wear a lab coat, gloves, and chemical eye protection when performing this procedure.
2. Growth may be suspended in the broth with a vortex mixer (Fig. 1.18) or by agitating with your fingers (Fig. 1.19). Be careful not to splash the broth into the cap or lose control of the tube. If transferring a sterile liquid, mixing is not necessary.
3. Determine which digital pipettor should be used to dispense the desired volume.
4. Turn the setting ring to set the desired volume (Fig. D.2). Do not turn the dial past either extreme of the pipettor's volume range. Doing so will damage it.

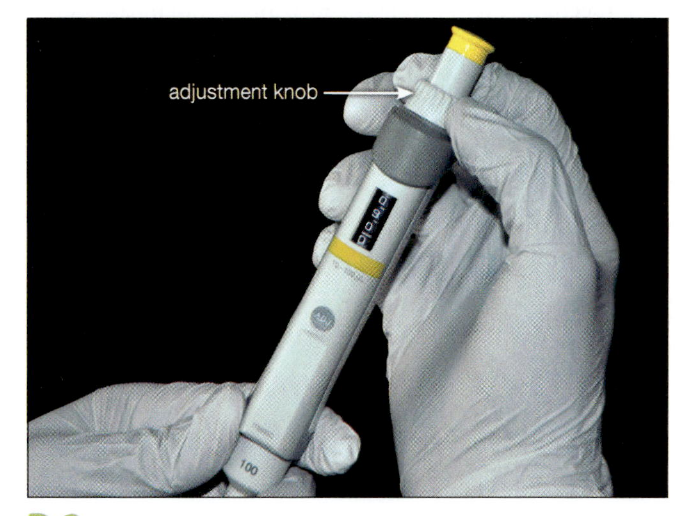


**D.2 Setting the Volume** ■ Rotate the adjustment knob (setting ring) to set the volume on a digital pipettor. This pipettor has been set at 90.0 μL (the line between the last two zeroes is a decimal point). Never rotate the adjustment knob beyond the volume limits of the pipettor as it may become damaged.

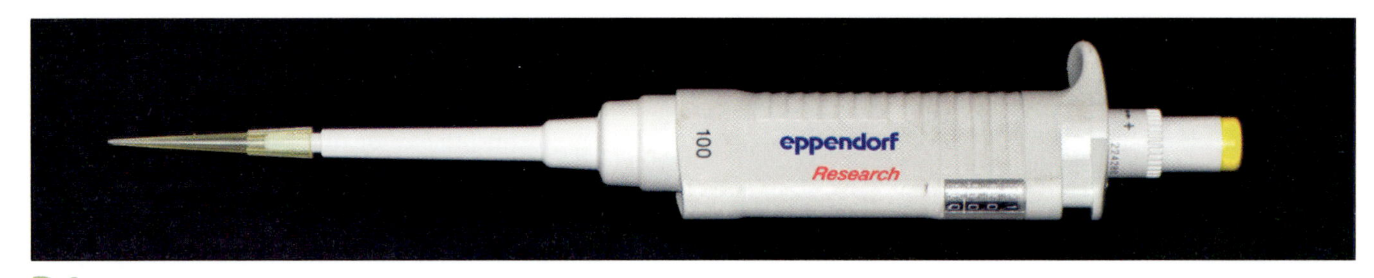


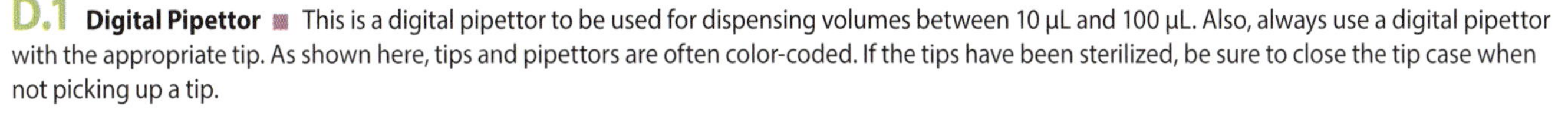

**D.1 Digital Pipettor** ■ This is a digital pipettor to be used for dispensing volumes between 10 μL and 100 μL. Also, always use a digital pipettor with the appropriate tip. As shown here, tips and pipettors are often color-coded. If the tips have been sterilized, be sure to close the tip case when not picking up a tip.

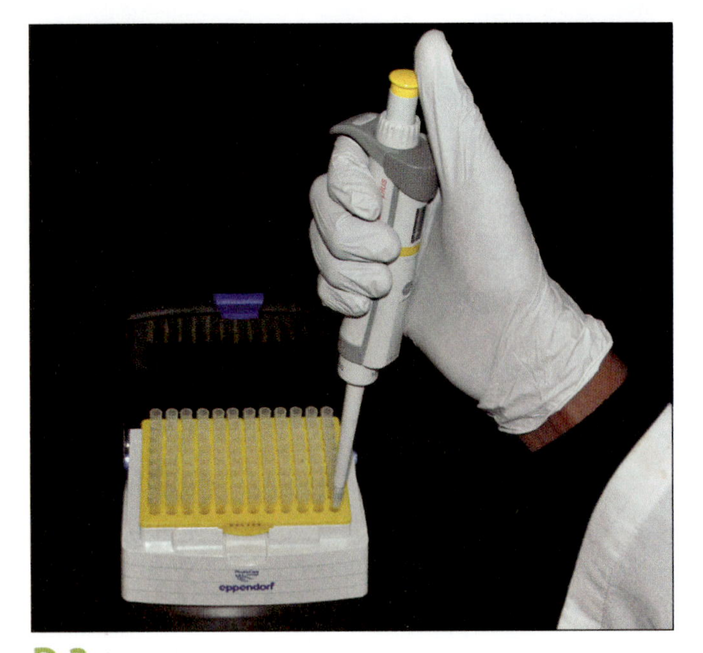

**D.3 Digital Pipettor Tips** ■ Digital pipettors must be fitted with a (sterile) tip of appropriate size (often color-coded to match the pipettor). Open the case and press the pipettor into a tip, then close the case to maintain sterility. Do not touch the pipettor tip.

5 Hold the digital pipettor in your dominant hand.

6 Open the case of appropriate pipette tips for your pipettor (these are often color-coded and match the pipettor) and push the pipettor into a sterile tip (Fig. D.3). Close the case. Do not touch the pipette tip with your hands or leave the case open. *Never use a digital pipettor without a tip.*

7 Remove the cap from the culture tube with the little finger of your dominant hand, and flame the tube. If transferring from a microtube, simply open the cap, but don't flame it!

D

8 Press down the control button with your thumb to the first stop. This is the measuring stroke.

9 Insert the tip into the broth approximately 3 mm while holding it vertically (Fig. D.4). Make sure you put the tip in deep enough—but not too deep—so it doesn't come out as the liquid level drops while filling.

10 *Slowly* release pressure with your thumb to draw fluid into the tip. Be careful not to pull any air into the tip.

11 Remove the pipettor and hold it still as you flame the tube as before or close the microtube's cap (without flaming).

12 Keeping the pipettor hand still, move the tube to replace its cap.

13 What you do next depends on the medium to which you are transferring the growth. Please continue with the appropriate inoculation section.

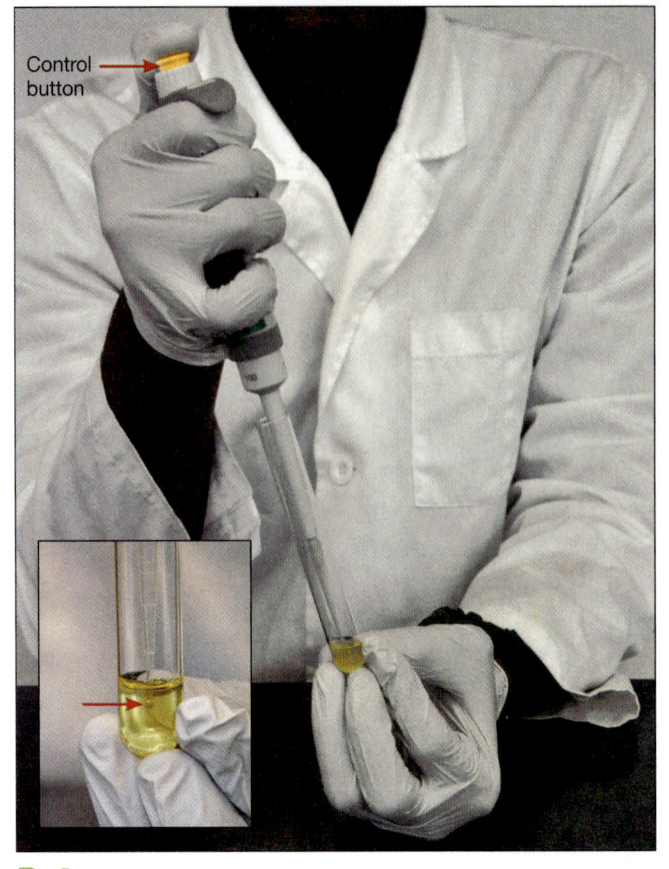


**D.4 Filling the Pipette** ■ Depress the control button with your thumb to the first stop (measuring stroke). Holding the tube and pipettor in a vertical position, insert the tip deep enough into the fluid (about 3 mm, see inset) that the tip won't be exposed as the liquid level drops during filling. Slowly release pressure with your thumb to fill the pipettor.

## Inoculation of Broth Tubes with a Digital Pipettor

Digital pipettes are often used to inoculate a known volume of culture into a tube or bottle containing a known volume of diluent during serial dilutions. They also would have been used to deliver the volume of diluent waiting in the tube or bottle.

1 Remove the cap of the sterile medium with the little finger of your pipettor hand, and hold it there. If using a microtube, simply open the cap and continue with step 3.

2 Flame the tube by quickly passing it through the Bunsen burner flame a couple of times. Keep your pipettor hand still.

3 Insert the pipette tip into the tube. Hold it at an angle against the inside of the tube (Fig. D.5).

4 Depress the control button slowly with your thumb to the first stop and pause when no more liquid is dispensed. Then continue pressing gently to the second stop to deliver the remaining volume. This is the blow-out stroke, but don't do it so forcefully that you splash liquid.

5 While keeping pressure on the control button, carefully remove the pipettor from the tube by sliding it along the inside of the tube. Once it is out, slowly release pressure on the control button.

6 Keeping your pipette hand still, flame the tube lip as before. If using a microtube, close the cap and continue with step 8.

7 Keeping the pipette hand still, move the tube to replace its cap.

8 The pipettor tip is contaminated with microbes, and must be disposed of correctly. Use the ejector button to remove the tip into an appropriate biohazard container (Fig. D.6). Each lab has its own specific procedures and your instructor will advise you what to do.

## Inoculation of Agar Plates with a Pipettor

1 Lift the lid of the plate and use it as a shield to protect from airborne contamination as well as limiting the spread of any aerosols as you deliver the inoculum.

2 Place the pipette tip over, but not touching, the agar surface. Hold the pipettor in a vertical position between the middle and one edge of the dish (Fig. D.7).

3 Depress the control button slowly with your thumb to the first stop and pause when no more liquid is dispensed. Then continue gently pressing to the second stop to deliver the appropriate volume. This is the blow-out stroke. From this point, the remaining steps should be completed within about 15 seconds to prevent the inoculum from soaking into the agar.

4 Because the pipettor tip is contaminated with microbes, you must dispose of it correctly. Use the ejector button to remove the tip into an appropriate biohazard container (Fig. D.6). Each lab has its own specific procedures and your instructor will advise you what to do.

5 Continue with the spread plate technique (Exercise 1-6).

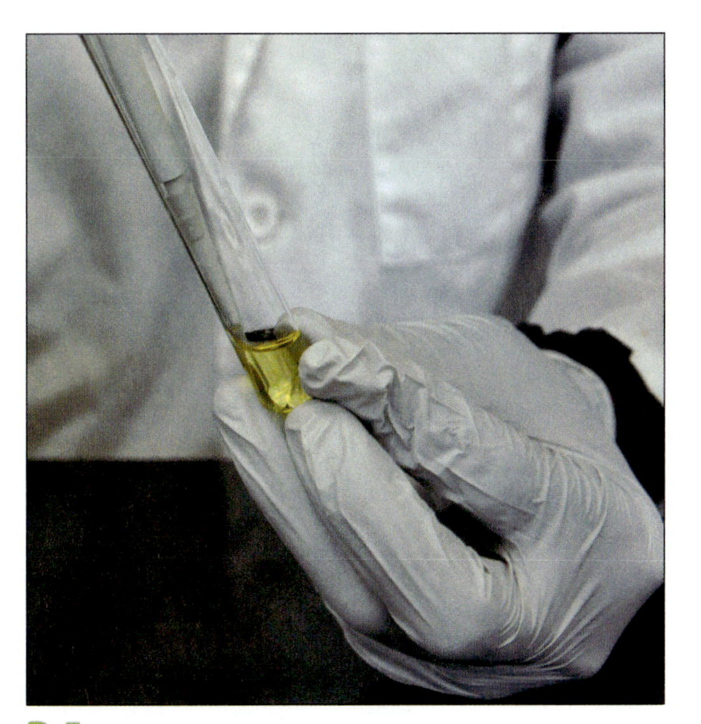

**D.5 Dispensing to a Liquid** ■ Hold the pipettor on a slight angle with the tip immersed in the liquid and held against the glass of the test tube (or plastic of a microtube). Press the control button gently to the first stop (don't shoot the liquid out!). When no more fluid comes out, continue pressing to the second stop to completely empty the pipette. Remove the pipettor along the glass (plastic), then slowly release pressure on the control button and replace the cap on the test tube. Eject the tip into a biohazard container if transferring a culture (Fig. D.6).

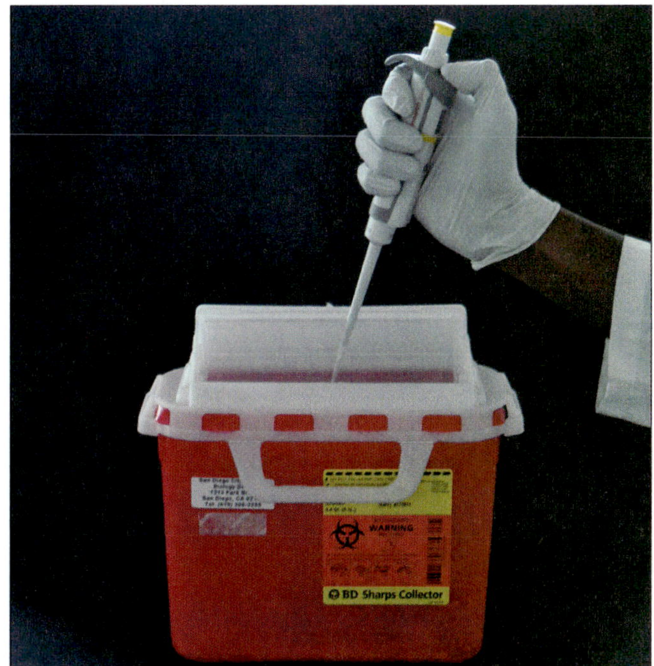


**D.6 Ejecting the Tip** ■ Use the tip ejector button to eject the tip into an appropriate biohazard container. Be sure the tip goes into the container and doesn't hit an edge and bounce back. (If this happens, treat it as a spill and wipe down all areas the tip contacted with disinfectant.) Manually put the tip in the biohazard container while wearing gloves, then properly remove and dispose of them. Your instructor will advise you as to the proper procedure in your laboratory. If transferring nonbiohazardous liquids (e.g., sterile water or saline), tips can be ejected into a cup or other suitable container for disposal.

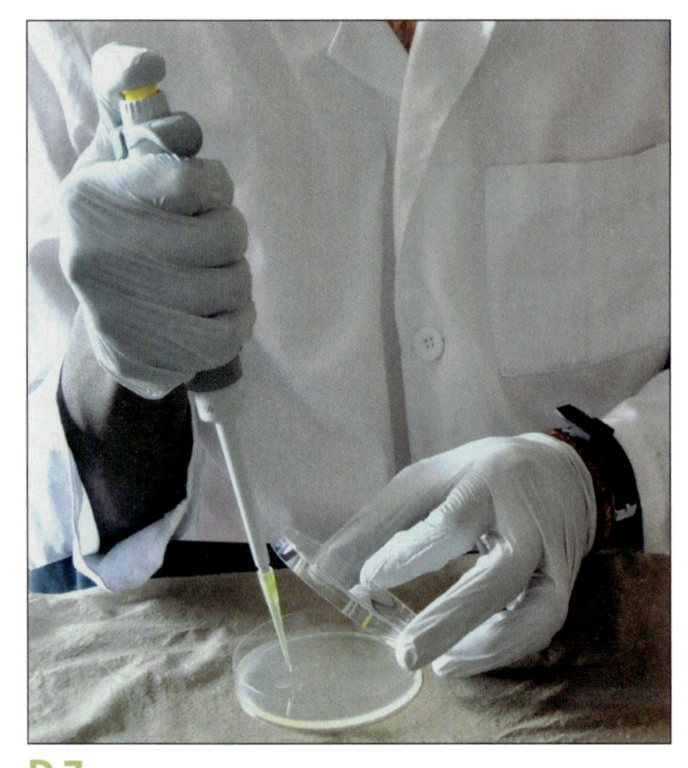

**D.7 Dispensing to a Plate** ■ Notice how the lid is held as a shield to prevent aerial contamination of the plate and to contain aerosols. Hold the pipette in a vertical position and press the control button to the first stop. When no more fluid is released, continue pressing to the second stop. If transferring a BSL-2 organism, try to deliver the inoculum as far under the lid as is practical. Then, slowly release pressure on the control button and eject the tip into a proper biohazard container (Fig. D.6). Notice the work area is covered with a disinfectant soaked towel. This is always a good idea in case of drips, but it is a necessity when transferring a BSL-2 organism.

## Modifications for Transferring BSL-2 Microorganisms

D Several modifications to the instructions above should be made to minimize aerosol production when using a digital pipettor. They may also be employed for other transfers, but are especially important when transferring BSL-2 organisms.

In all cases, transfers should be made over a disinfectant-soaked towel in case inoculum leaks from the pipette tip. Put more disinfectant on any spills. Also, keep the culture and destination media as close together as is safe in order to minimize hand movements.

Following are suggested modifications for transferring BSL-2 organisms with a digital pipettor.

- The pipette tip containing inoculum may be inserted into the broth. Then, gently deliver the inoculum into the broth. Thus, the "blow-out" stroke (pressing to the second stop) occurs within the broth and does not produce aerosols. However, if you use too much force you can cause bubbling of the broth with concurrent aerosol production, so be sure to *gently* dispense the inoculum. Also be sure to keep pressure on the control button as you withdraw the pipette from the broth or you risk changing an intended concentration as you pull back some broth. Use the ejector button to dispose of the tip according to your lab's conventions with a minimum amount of pipettor movement.
- If you are delivering the inoculum to a plate, keep the lid down low and *gently* dispense as far under the lid as possible while still holding the pipettor more-or-less vertically. This will help to contain aerosols. Use the ejector button to dispose of the tip according to your lab's conventions with a minimum amount of pipettor movement.
- Another method involves using the pipettor "in reverse." Instead of pressing to the first stop to fill, filling can be done by pressing to the *second* stop. Then, instead of dispensing by pressing to the second stop, delivery of the inoculum is done by pressing to the *first* stop to avoid the inevitable spraying that occurs when completely emptying the pipette tip. Use the ejector button to dispose of the tip and the remaining inoculum within it according to your lab's conventions with a minimum amount of pipettor movement.

### Reference

Brinkmann Instruments, *Eppendorf Series 2100 Pipette Instruction Manual.* Westbury, New York.

APPENDIX E

# Medium, Reagent, and Stain Recipes

## MEDIA

### General Preparation Methods

Unless otherwise specified, prepare all media as follows.

#### Broths

1. Suspend the dry ingredients in water. Agitate and heat slightly (if necessary) to dissolve completely.
2. Dispense 7.0 mL portions into 16 × 150 mm test tubes and cap loosely.
3. Autoclave for 15 minutes at 121°C to sterilize the medium.

#### Slants and Stabs

1. Suspend the dry ingredients in the water, mix well, and boil until completely dissolved.
2. Dispense 7.0 mL (or 10.0 mL for deep stabs) portions into 16 × 150 mm test tubes, and cap loosely.
3. Autoclave for 15 minutes at 121°C to sterilize the medium.
4. Slant the tubes or leave upright as needed for stabs and cool to room temperature.

#### Plates

1. Suspend the dry ingredients in the water, mix well, and boil until completely dissolved.
2. Cover loosely and sterilize in the autoclave at 15 lbs. pressure (121°C) for 15 minutes.
3. Remove from the autoclave, allow the medium to cool to approximately 50°C, and aseptically pour into sterile 100 mm Petri dishes (20.0 mL/plate).
4. Allow the medium to cool to room temperature. Store upside down.

### Specific Recipes

#### Bile Esculin Agar (Plate or Slant) (Ex. 5-16)

- Pancreatic digest of gelatin 5.0 g
- Beef extract 3.0 g
- Oxgall (ox bile) 20.0 g
- Ferric citrate 0.5 g
- Esculin 1.0 g
- Agar 14.0 g
- Distilled or deionized water 1.0 L

*pH 6.6–7.0 at 25°C*

#### Blood Agar (Ex. 5-20, 5-21, 5-22, and 9-2)

- Infusion from beef heart (solids) 2.0 g
- Pancreatic digest of casein 13.0 g
- Sodium chloride 5.0 g
- Yeast extract 5.0 g
- Agar 15.0 g
- Defibrinated sheep blood 50.0 mL
- Distilled or deionized water 1.0 L

*pH 7.1–7.5 at 25°C*

1. Suspend the dry ingredients in water, mix well, and boil until completely dissolved. This is blood agar base.
2. Cover loosely and sterilize in the autoclave at 15 lbs. pressure (121°C) for 15 minutes.
3. Remove from the autoclave and cool to 45°C.
4. Aseptically add the sterile, room-temperature sheep blood to the blood agar base and mix well.
5. Pour into sterile Petri dishes and allow the medium to cool to room temperature.

#### Brain-Heart Infusion Agar (Plate or Slant)

- Calf brains, infusion from 200 g 7.7 g
- Beef heart, infusion from 250 g 9.8 g
- Proteose peptone 10.0 g
- Dextrose 2.0 g
- Sodium chloride 5.0 g
- Disodium phosphate 2.5 g
- Agar 15.0 g
- Distilled or deionized water 1.0 L

*pH 7.2–7.6 at 25°C*

### Brain-Heart Infusion Broth

| | |
|---|---|
| Calf brains, infusion from 200 g | 7.7 g |
| Beef heart, infusion from 250 g | 9.8 g |
| Proteose peptone | 10.0 g |
| Dextrose | 2.0 g |
| Sodium chloride | 5.0 g |
| Disodium phosphate | 2.5 g |
| Distilled or deionized water | 1.0 L |

*pH 7.2–7.6 at 25°C*

### Brilliant Green Lactose Bile Broth (Ex. 7-6)

| | |
|---|---|
| Peptone | 10.0 g |
| Lactose | 10.0 g |
| Oxgall | 20.0 g |
| Brilliant green dye | 0.0133 g |
| Distilled or deionized water | 1.0 L |

*pH 7.0–7.4 at 25°C*

1. Suspend the dry ingredients in water, mix well, and boil until completely dissolved.
2. Dispense 10.0 mL portions into test tubes.
3. Place an inverted Durham tube in each broth and cap loosely.
4. Sterilize in the autoclave at 15 lbs. pressure (121°C) for 15 minutes.
5. Remove the medium from the autoclave and allow it to cool before inoculating.

### Citrate Agar (Simmons) (Ex. 5-7 and 9-1)

| | |
|---|---|
| Ammonium dihydrogen phosphate | 1.0 g |
| Dipotassium phosphate | 1.0 g |
| Sodium chloride | 5.0 g |
| Sodium citrate | 2.0 g |
| Magnesium sulfate | 0.2 g |
| Agar | 15.0 g |
| Bromothymol blue | 0.08 g |
| Distilled or deionized water | 1.0 L |

*pH 6.7–7.1 at 25°C*

1. Suspend the dry ingredients in water, mix well, and boil until completely dissolved.
2. Dispense 7.0 mL portions into test tubes and cap loosely.
3. Sterilize in the autoclave at 15 lbs. pressure (121°C) for 15 minutes.
4. Remove from the autoclave and position tubes in such a way as to produce a slant nearly to the tube's base.
5. Cool to room temperature.

### Columbia CNA with 5% Sheep Blood Agar (Ex. 4-2 and 9-2)

| | |
|---|---|
| Pancreatic digest of casein | 12.0 g |
| Peptic digest of animal tissue | 5.0 g |
| Yeast extract | 3.0 g |
| Beef extract | 3.0 g |
| Corn starch | 1.0 g |
| Sodium chloride | 5.0 g |
| Colistin | 10.0 mg |
| Nalidixic acid | 10.0 mg |
| Agar | 13.5 g |
| Sheep blood (defibrinated, sterile) | 50.0 mL |
| Distilled or deionized water | 1.0 L |

*pH 7.1–7.5 at 25°C*

1. Suspend the dry ingredients in water, mix well, and boil until completely dissolved.
2. Cover loosely and sterilize in the autoclave at 15 lbs. pressure (121°C) for 15 minutes.
3. Remove from the autoclave and cool to 45 to 50°C.
4. Add the sheep blood to 950 mL of the mixture and aseptically pour into sterile Petri dishes (20 mL/plate).
5. Allow the medium to cool to room temperature. Store upside down.

### Decarboxylase Medium (Møller) (Ex. 5-8 and 9-1)

| | |
|---|---|
| Peptone | 5.0 g |
| Beef extract | 5.0 g |
| Glucose (dextrose) | 0.5 g |
| Bromocresol purple | 0.01 g |
| Cresol red | 0.005 g |
| Pyridoxal | 0.005 g |
| L-lysine, L-ornithine, or L-arginine | 10.0 g |
| Distilled or deionized water | 1.0 L |

*pH 5.8–6.2 at 25°C*

1. Suspend the dry ingredients in water, mix well, and boil until completely dissolved. (Use only one of the listed L-amino acids.)
2. Adjust pH by adding NaOH if necessary.
3. Dispense 7.0 mL volumes into test tubes and cap.
4. Sterilize in the autoclave at 15 lbs. pressure (121°C) for 10 minutes.
5. Remove from the autoclave and cool to room temperature.

### DNase Test Agar with Methyl Green (Ex. 5-11)

- Tryptose 20.0 g
- Deoxyribonucleic acid 2.0 g
- Sodium chloride 5.0 g
- Agar 15.0 g
- Methyl green 0.05 g
- Distilled or deionized water 1.0 L

*pH 7.1–7.5 at 25°C*

### EC Broth (Ex. 7-6)

- Tryptose 20.0 g
- Lactose 5.0 g
- Bile Salts No. 3 1.5 g
- Dipotassium phosphate 4.0 g
- Monopotassium phosphate 1.5 g
- Sodium chloride 5.0 g
- Distilled or deionized water 1.0 L

*pH 6.7–7.1 at 25°C*

1. Suspend the dry ingredients in water, and mix well until completely dissolved.
2. Dispense 10.0 mL portions into test tubes.
3. Place an inverted Durham tube in each broth and cap loosely.
4. Sterilize in the autoclave at 15 lbs. pressure (121°C) for 15 minutes.
5. Remove the media from the autoclave and allow it to cool before inoculating.

### m-Endo Agar LES (Ex. 7-5)

- Yeast extract 1.2 g
- Casitone 3.7 g
- Thiopeptone 3.7 g
- Tryptose 7.5 g
- Lactose 9.4 g
- Dipotassium phosphate 3.3 g
- Monopotassium phosphate 1.0 g
- Sodium chloride 3.7 g
- Sodium desoxycholate 0.1 g
- Sodium lauryl sulfate 0.05 g
- Sodium sulfite 1.6 g
- Basic fuchsin 0.8 g
- Agar 15.0 g
- Distilled or deionized water 1.0 L

*pH 7.3–7.7 at 25°C*

### Enriched TSA (See Tryptic Soy Agar)

### Eosin Methylene Blue Agar (Levine) (Ex. 4-5)

- Pancreatic digest of gelatin 10.0 g
- Lactose 10.0 g[1]
- Dipotassium phosphate 2.0 g
- Agar 15.0 g
- Eosin Y 0.4 g
- Methylene blue 0.065 g
- Distilled or deionized water 1.0 L

*pH 6.9–7.3 at 25°C*

### Glucose Salts Medium (Ex. 2-5)

- Glucose 5.0 g
- NaCl 5.0 g
- $MgSO_4$ 0.2 g
- $(NH_4)H_2PO_4$ 1.0 g
- $K_2HPO_4$ 1.0 g
- Distilled or deionized water 1.0 L

### Modified *Halobacterium* Broth (Ex. 2-10)

- Sodium chloride 0 g, 50 g, 100 g, 150 g, 200 g, or 250 g

(for 0%, 5%, 10%, 15%, 20%, and 25% saline broths)

- Magnesium sulfate, heptahydrate 20.0 g
- Trisodium citrate, dihydrate 3.0 g
- Potassium chloride 2.0 g
- Casamino acids 5.0 g
- Yeast extract 5.0 g
- Deionized water 1.0 L

*Adjust pH to 7.2 using 5 M or concentrated HCl*

### Hektoen Enteric Agar (Ex. 4-6)

- Yeast extract 3.0 g
- Peptic digest of animal tissue 12.0 g
- Lactose 12.0 g
- Sucrose 12.0 g
- Salicin 2.0 g
- Bile salts 9.0 g
- Sodium chloride 5.0 g
- Sodium thiosulfate 5.0 g
- Ferric ammonium citrate 1.5 g
- Bromothymol blue 0.064 g
- Acid fuchsin 0.1 g
- Agar 13.5 g
- Distilled or deionized water 1.0 L

*pH 7.4–7.8 at 25°C*

[1] An alternative recipe replaces the 10.0 g of lactose with 5.0 g of lactose and 5.0 g of sucrose.

E

1. Suspend the dry ingredients in the water, mix well, and boil until completely dissolved.
2. Do not autoclave.
3. When cooled to 50°C, pour into sterile plates.
4. Cool to room temperature with lids slightly open.

### Kligler Iron Agar (Ex. 5-19)

- Pancreatic digest of casein 10.0 g
- Peptic digest of animal tissue 10.0 g
- Lactose 10.0 g
- Dextrose (glucose) 1.0 g
- Ferric ammonium citrate 0.5 g
- Sodium chloride 5.0 g
- Sodium thiosulfate 0.5 g
- Agar 15.0 g
- Phenol red 0.025 g
- Distilled or deionized water 1.0 L

*pH 7.2–7.6 at 25°C*

1. Suspend the dry ingredients in water, mix well, and boil until completely dissolved.
2. Transfer 7.0 mL portions to test tubes and cap loosely.
3. Sterilize in the autoclave at 15 lbs. pressure (121°C) for 15 minutes.
4. Remove from the autoclave and slant in such a way as to form a deep butt.
5. Allow the medium to cool to room temperature.

### Lauryl Tryptose Broth (Ex. 7-6)

- Tryptose 20.0 g
- Lactose 5.0 g
- Dipotassium phosphate 2.75 g
- Monopotassium phosphate 2.75 g
- Sodium chloride 5.0 g
- Sodium lauryl sulfate 0.1 g
- Distilled or deionized water 1.0 L

*pH 6.6–7.0 at 25°C*

1. Suspend the dry ingredients in water until completely dissolved. Heat slightly if necessary.
2. Dispense 10.0 mL portions into test tubes.
3. Place an inverted Durham tube in each broth and cap loosely.
4. Sterilize in the autoclave at 15 lbs. pressure (121°C) for 15 minutes.
5. Remove the medium from the autoclave and allow it to cool before inoculating.

E

### Litmus Milk Medium

- Skim milk 100.0 g
- Azolitmin 0.5 g
- Sodium sulfite 0.5 g
- Distilled or deionized water 1.0 L

*pH 6.3–6.7 at 25°C*

1. Suspend and mix the ingredients in water and heat to approximately 50°C to dissolve completely.
2. Transfer 7.0 mL portions to test tubes and cap loosely.
3. Sterilize in the autoclave at 113°C–115°C for 20 minutes.
4. Remove from the autoclave and allow the medium to cool to room temperature.

### Luria-Bertani Agar (Ex. 8-2)

- Tryptone 10.0 g
- Yeast extract 5.0 g
- NaCl 10.0 g
- Agar 15.0 g
- Distilled or deionized water 1.0 L

*pH 7.4 at 25°C*

### Luria-Bertani Broth (Ex. 8-1)

- Tryptone 10.0 g
- Yeast extract 5.0 g
- NaCl 10.0 g
- Distilled or deionized water 1.0 L

*pH 7.4 at 25°C*

### MacConkey Agar (Ex. 4-4 and 9-1)

- Pancreatic digest of gelatin 17.0 g
- Pancreatic digest of casein 1.5 g
- Peptic digest of animal tissue 1.5 g
- Lactose 10.0 g
- Bile salts 1.5 g
- Sodium chloride 5.0 g
- Neutral red 0.03 g
- Crystal violet 0.001 g
- Agar 13.5 g
- Distilled or deionized water 1.0 L

*pH 6.9–7.3 at 25°C*

### Mannitol Salt Agar (Ex. 4-3)

- Beef extract 1.0 g
- Peptone 10.0 g
- Sodium chloride 75.0 g
- D-Mannitol 10.0 g
- Phenol red 0.025 g
- Agar 15.0 g
- Distilled or deionized water 1.0 L

*pH 7.2–7.6 at 25°C*

### Milk Agar (Ex. 5-13)

- Pancreatic digest of casein 5.0 g
- Yeast extract 2.5 g
- Powdered nonfat milk 100.0 g
- Glucose 1.0 g
- Agar 15.0 g
- Distilled or deionized water 1.1 L

*pH 6.9–7.1 at 25°C*

1. Suspend the powdered milk in 500.0 mL of water in a 1-liter flask, mix well, and cover loosely.
2. Suspend the remainder of the ingredients in 500.0 mL of water in a 1-liter flask, mix well, boil to dissolve completely, and cover loosely.
3. Sterilize in the autoclave at 113–115°C for 20 minutes.
4. Remove from the autoclave and allow the mixtures to cool slightly.
5. Aseptically pour the agar solution into the milk solution. Mix *gently* (to prevent foaming).
6. Aseptically pour into sterile Petri dishes (15 mL/plate).
7. Allow the medium to cool to room temperature.

### Modified *Halobacterium* Broth (see *Halobacterium* Broth)

### Motility Test Medium (Ex. 5-25, 9-1, and 9-3)

- Beef extract 3.0 g
- Pancreatic digest of gelatin 10.0 g
- Sodium chloride 5.0 g
- Agar 4.0 g
- Triphenyltetrazolium chloride (TTC) 0.05 g
- Distilled or deionized water 1.0 L

*pH 7.1–7.5 at 25°C*

### MR-VP Broth (Ex. 5-3 and 9-1)

- Buffered peptone 7.0 g
- Dipotassium phosphate 5.0 g
- Dextrose (glucose) 5.0 g
- Distilled or deionized water 1.0 L

*pH 6.7–7.1 at 25°C*

### Mueller-Hinton II Agar (Ex. 7-2)

- Beef extract 2.0 g
- Acid hydrolysate of casein 17.5 g
- Starch 1.5 g
- Agar 17.0 g
- Distilled or deionized water 1.0 L

*pH 7.2–7.4 at 25°C*

1. Suspend the dry ingredients in water, mix well, and boil until completely dissolved.
2. Cover loosely and sterilize in the autoclave at 121°C (15 lbs.) for 15 minutes.
3. Remove from the autoclave, cool to approximately 50°C.
4. Aseptically pour into sterile Petri dishes to a depth of 4 mm.
5. Allow the medium to cool to room temperature.

### Nitrate Broth (Ex. 5-6)

- Beef extract 3.0 g
- Peptone 5.0 g
- Potassium nitrate 1.0 g
- Distilled or deionized water 1.0 L

*pH 6.8–7.2 at 25°C*

1. Suspend the ingredients in water, mix well, and warm until completely dissolved.
2. Transfer 8.0 mL portions to test tubes, and cap loosely. (Add inverted Durham tubes before capping, if desired.)
3. Sterilize in the autoclave at 15 lbs. pressure (121°C) for 15 minutes.
4. Remove from the autoclave and allow the medium to cool to room temperature.

### Nutrient Agar (Plate or Tube)

- Beef extract 3.0 g
- Peptone 5.0 g
- Agar 15.0 g
- Distilled or deionized water 1.0 L

*pH 6.6–7.0 at 25°C*

#### NA Tubes

1. Suspend the dry ingredients in water, mix well, and boil until completely dissolved.
2. Dispense 7 mL or 10 mL portions into test tubes, and cap loosely.
3. Autoclave for 15 minutes at 121°C to sterilize the medium.
4. Cool to room temperature with the tubes in an upright position (10 mL) for agar deep tubes. Cool with the tubes on an angle (7 mL) for agar slants.

### Nutrient Broth

- Beef extract 3.0 g
- Peptone 5.0 g
- Distilled or deionized water 1.0 L

*pH 6.6–7.0 at 25°C*

### Nutrient Gelatin (Ex. 5-14)

- Beef extract 3.0 g
- Peptone 5.0 g
- Gelatin 120.0 g
- Distilled or deionized water 1.0 L

*pH 6.6–7.0 at 25°C*

1 *Slowly* add the dry ingredients to the water while stirring.
2 Warm to >50°C and maintain temperature until completely dissolved.
3 Dispense 7.0 mL volumes into test tubes and cap loosely.
4 Sterilize in the autoclave at 15 lbs. pressure (121°C) for 15 minutes.
5 Remove from the autoclave immediately and allow the medium to cool to room temperature in the upright position.

## O-F Medium (Hugh and Leifson) (Ex. 5-1)

### Basal Medium

- Pancreatic digest of casein — 2.0 g
- Sodium chloride — 5.0 g
- Dipotassium phosphate — 0.3 g
- Agar — 2.5 g
- Bromothymol blue — 0.03 g
- Distilled or deionized water — 1.0 L

*pH 6.6–7.0 at 25°C*

### O-F Carbohydrate Solution

- Carbohydrate (glucose, lactose, sucrose) — 1.1 g
- Distilled or deionized water to total — 10.0 mL

1 Suspend the dry basal medium ingredients, *without the carbohydrate*, in water, mix well, and boil to dissolve completely. This is basal medium.
2 Divide the medium into 10 aliquots of 100.0 mL each.
3 Cover loosely and sterilize in the autoclave at 121°C for 15 minutes.
4 Prepare carbohydrate solution, cover loosely, and autoclave at 118°C for 10 minutes.
5 Allow both solutions to cool to 50°C.
6 Aseptically add 10.0 mL sterile carbohydrate solution to a basal medium aliquot, and mix well.
7 Aseptically transfer 7.0 mL volumes to sterile test tubes, and allow the medium to cool to room temperature.

E

## Phenol Red (Carbohydrate) Broth (Ex. 5-2, 9-1, 9-2, and 9-3)

- Pancreatic digest of casein — 10.0 g
- Sodium chloride — 5.0 g
- Carbohydrate (use any; e.g., glucose, lactose, sucrose, etc.) — 5.0 g
- Phenol red — 0.018 g
- Distilled or deionized water — 1.0 L

*pH 7.1–7.5 at 25°C*

1 Suspend the dry ingredients in water, mix well, and warm slightly to dissolve completely.
2 Dispense 7.0 mL volumes into test tubes.
3 Insert inverted Durham tubes into the test tubes and cap loosely.
4 Sterilize in the autoclave at 116–118°C for 15 minutes.
5 Remove from the autoclave and allow the medium to cool to room temperature.

## Phenylalanine Deaminase Agar (Ex. 5-9)

- DL-Phenylalanine — 2.0 g
- Yeast extract — 3.0 g
- Sodium chloride — 5.0 g
- Sodium phosphate — 1.0 g
- Agar — 12.0 g
- Distilled or deionized water — 1.0 L

*pH 7.1–7.5 at 25°C*

1 Suspend the dry ingredients in water, mix well, and boil until completely dissolved.
2 Dispense 7.0 mL volumes into test tubes and cap loosely.
3 Sterilize in the autoclave at 15 lbs. pressure (121°C) for 10 minutes.
4 Remove from the autoclave, slant, and allow the medium to cool to room temperature.

## Phenylethyl Alcohol Agar (Ex. 4-1 and 9-2)

- Pancreatic digest of casein — 15.0 g
- Papaic digest of soybean meal — 5.0 g
- Sodium chloride — 5.0 g
- β-Phenylethyl alcohol — 2.5 g
- Agar — 15.0 g
- Distilled or deionized water — 1.0 L

*pH 7.1–7.5 at 25°C*

## Purple Broth (Optional, Ex. 5-2)

- Peptone — 10.0 g
- Beef extract — 1.0 g
- Sodium chloride — 5.0 g
- Bromocresol purple — 0.02 g
- Carbohydrate (use any; e.g., glucose, lactose, sucrose, etc.) — 10.0 g
- Distilled or deionized water — 1.0 L

*pH 6.6–7.0 at 25°C*

1 Suspend the dry ingredients in water, mix well, and warm slightly to dissolve completely.
2 Dispense 9.0 mL volumes into test tubes.

3 Insert inverted Durham tubes into the test tubes and cap loosely.
4 Sterilize in the autoclave at 118°C for 15 minutes.
5 Remove from the autoclave and allow the medium to cool to room temperature.

### Sabouraud Dextrose Agar (Ex. 3-3)

(with antibiotics added to inhibit bacterial growth)

- Peptone 10.0 g
- Dextrose 40.0 g
- Agar 15.0 g
- Penicillin[3] 20,000.0 units
- Streptomycin[3] 0.00004 g
- Distilled or deionized water 1.0 L

*pH 5.2–5.6 at 25°C*

1 Suspend the peptone, dextrose, and agar in water, mix well, and boil until completely dissolved.
2 Autoclave for 15 minutes at 15 lbs. pressure (121°C).
3 Remove the agar mixture from the autoclave and cool to 50°C.
4 Aseptically add antibiotics. Mix and pour into sterile Petri dishes.
5 Allow the medium to cool to room temperature.

### SIM (Sulfur-Indole-Motility) Medium (Ex. 5-18, 9-1, and 9-3)

- Pancreatic digest of casein 20.0 g
- Peptic digest of animal tissue 6.1 g
- Ferrous ammonium sulfate 0.2 g
- Sodium thiosulfate 0.2 g
- Agar 3.5 g
- Distilled or deionized water 1.0 L

*pH 7.1–7.5 at 25°C*

### Skim Milk Agar (See [Skim] Milk Agar)

### Snyder Test Medium (Ex. 7-1)

- Pancreatic digest of casein 13.5 g
- Yeast extract 6.5 g
- Dextrose 20.0 g
- Sodium chloride 5.0 g
- Agar 16.0 g
- Bromocresol green 0.02 g
- Distilled or deionized water 1.0 L

*pH 4.6–5.0 at 25°C*

[3] To obtain the desired proportions of antibiotics in the medium, prepare as follows:
1. Dissolve 100,000 units penicillin in 10 mL sterile distilled or deionized water. Add 2 mL to 1.0 liter of agar medium.
2. Dissolve 1.0 g streptomycin in 10 mL sterile distilled or deionized water. Add 1.0 mL of this mixture to 9.0 mL sterile distilled or deionized water. Add 4 mL of this diluted mixture to 1.0 liter of agar medium.

1 Suspend the dry ingredients in water, mix well, and boil until completely dissolved.
2 Transfer 7.0 mL portions to test tubes and cap loosely.
3 Sterilize in the autoclave at 118°C–121°C for 15 minutes.
4 Remove from the autoclave and place in a hot water bath set at 45°C–50°C. Allow at least 30 minutes for the agar temperature to equilibrate before beginning the exercise.

### Soft Agar (Ex. 6-4)

- Beef extract 3.0 g
- Peptone 5.0 g
- Sodium chloride 5.0 g
- Tryptone 2.5 g
- Yeast extract 2.5 g
- Agar 7.0 g
- Distilled or deionized water 1.0 L

1 Suspend the dry ingredients in water, mix well, and boil until completely dissolved.
2 Transfer 2.5 mL portions to test tubes and cap loosely.
3 Sterilize in the autoclave at 15 lbs. pressure (121°C) for 15 minutes.
4 Remove from the autoclave and place in a hot water bath set at 45°C. Allow 30 minutes for the agar temperature to equilibrate.

### Sporulating Agar (Ex. 3-9 and 3-12)

- Pancreatic digest of gelatin 6.0 g
- Pancreatic digest of casein 4.0 g
- Yeast extract 3.0 g
- Beef extract 1.5 g
- Dextrose 1.0 g
- Agar 15.0 g
- Manganous sulfate 0.3 g
- Distilled or deionized water 1.0 L

*Final pH 6.6±0.2 at 25°C*

### Starch Agar (Ex. 5-10, 9-2, and 9-3)

- Beef extract 3.0 g
- Soluble starch 10.0 g
- Agar 12.0 g
- Distilled or deionized water 1.0 L

*pH 7.3–7.7 at 25°C*

### Thioglycollate Medium (Fluid) (Ex. 2-6 and 9-3)

- Yeast extract 5.0 g
- Casitone 15.0 g
- Dextrose (glucose) 5.5 g
- Sodium chloride 2.5 g
- Sodium thioglycolate 0.5 g

- L-cystine 0.5 g
- Agar 0.75 g
- Resazurin 0.001 g
- Distilled or deionized water 1.0 L

*pH 7.1–7.5 at 25°C*

1. Suspend the dry ingredients in water, mix well, and boil until completely dissolved.
2. Dispense 10.0 mL into sterile screw-cap test tubes.
3. Autoclave for 15 minutes at 15 lbs. pressure (121°C) to sterilize. Allow the medium to cool to room temperature before inoculating.

### Tributyrin Agar (Ex. 5-12)

- Beef extract 3.0 g
- Peptone 5.0 g
- Agar 15.0 g
- Tributyrin oil 10.0 mL
- Distilled or deionized water 1.0 L

*pH 5.8–6.2 at 25°C*

1. Suspend the dry ingredients in water, mix well, and boil until completely dissolved.
2. Cover loosely, and sterilize together with the tube of tributyrin oil in the autoclave at 15 lbs. pressure (121°C) for 15 minutes.
3. Remove from the autoclave, and aseptically pour agar mixture into a sterile glass blender.
4. Aseptically add the tributyrin oil to the agar mixture and blend on "High" for 1 minute.
5. Aseptically pour into sterile Petri dishes (20 mL/plate). Allow the medium to cool to room temperature.

### Triple Sugar Iron Agar (TSIA) (Ex. 5-19)

- Pancreatic digest of casein 10.0 g
- Peptic digest of animal tissue 10.0 g
- Dextrose (glucose) 1.0 g
- Lactose 10.0 g
- Sucrose 10.0 g
- Ferrous ammonium sulfate 0.2 g
- Sodium chloride 5.0 g
- Sodium thiosulfate 0.2 g
- Agar 13.0 g
- Phenol red 0.025 g
- Distilled or deionized water 1.0 L

*pH 7.1–7.5 at 25°C*

1. Suspend the dry ingredients in the water, mix well, and boil until completely dissolved.
2. Transfer 7.0 mL portions to test tubes and cap loosely.
3. Sterilize in the autoclave at 15 lbs. pressure (121°C) for 15 minutes.
4. Remove from the autoclave and slant in such a way as to form a deep butt.
5. Allow the medium to cool to room temperature.

### Tryptic Soy Agar (Plate or Tube)

- Tryptone 15.0 g
- Soytone 5.0 g
- Sodium chloride 5.0 g
- Agar 15.0 g
- Distilled or deionized water 1.0 L

*pH 7.1–7.5 at 25°C*

### Tryptic Soy Broth

- Tryptone 17.0 g
- Soytone 3.0 g
- Sodium chloride 5.0 g
- Dipotassium phosphate 2.5 g
- Distilled or deionized water 1.0 L

*pH 7.1–7.5 at 25°C*

### Tryptic Soy Broth Plus 1% Glucose

- Tryptic Soy Broth (see Tryptic Soy Broth recipe)
- Glucose 10 g

### Urea Broth (Rustigian and Stuart) (Ex. 5-15)

- Yeast extract 0.1 g
- Potassium phosphate, monobasic 9.1 g
- Potassium phosphate, dibasic 9.5 g
- Urea 20.0 g
- Phenol red 0.01 g
- Distilled or deionized water 1.0 L

*pH 6.6–7.0 at 25°C*

1. Suspend the dry ingredients in the water and mix well.
2. Filter sterilize the solution. *Do not autoclave.*
3. Aseptically transfer 1.0 mL volumes to small sterile test tubes and cap loosely.

## REAGENTS

### Kovac's Reagent

- Amyl alcohol 75.0 mL
- Hydrochloric acid, concentrated 25.0 mL
- *p*-dimethylaminobenzaldehyde 5.0 g

E

## MR-VP (Methyl Red–Voges-Proskauer) Test Reagents

### Methyl Red

- Methyl red dye 0.1 g
- Ethanol 300.0 mL
- Distilled water to bring volume to 500.0 mL

1. Dissolve the dye in the ethanol.
2. Add water to bring the total volume up to 500 mL.

### VP Reagent A (Barritt's)

- α-naphthol 5.0 g
- Absolute ethanol to bring volume to 100.0 mL

1. Dissolve the α-naphthol in approximately 95 mL of ethanol.
2. Add ethanol to bring the total volume up to 100 mL.

### VP Reagent B (Barritt's)

- Potassium hydroxide 40.0 g
- Distilled water to bring volume to 100.0 mL

1. Dissolve the potassium hydroxide in approximately 60 mL of water. (*Caution:* This solution is highly concentrated and will become hot as the KOH dissolves. It should be prepared in appropriate glassware on a stirring hot plate.) Allow it to cool to room temperature.
2. Add water to bring the volume up to 100 mL.

## Methyl Red (See MR-VP)

## Methylene Blue Reductase Reagent

- Methylene blue dye 8.8 mg
- Distilled or deionized water 200.0 mL

## Nitrate Test Reagents

### Reagent A

- Sulfanilic acid 0.8 g
- 5N Acetic acid 100.0 mL

### Reagent B

- N,N-Dimethyl-α-naphthylamine 0.6 g
- 5N Acetic acid 100.0 mL

## Oxidase Test Reagent

- Tetramethyl-*p*-phenylenediamine dihydrochloride 1.0 g
- Distilled or deionized water 100.0 mL

## Phenylalanine Deaminase Test Reagent

- Ferric chloride 12.0 g
- Concentrated hydrochloric acid 2.5 mL
- Distilled or deionized water ≅100.0 mL

1. Dissolve the ferric chloride in approximately 87 mL of distilled or deionized water.
2. Carefully add 2.5 mL concentrated HCl.
3. Add water to bring the total volume up to 100 mL.

## Voges-Proskauer Reagents A and B (See MR-VP)

# SOLUTIONS

## McFarland Turbidity Standard (0.5)

- Barium chloride ($BaCl_2 \bullet 2H_2O$) 1.175 g
- Sulfuric acid, concentrated ($H_2SO_4$) 1.0 mL
- Distilled or deionized water ≅200.0 mL

1. Pour approximately 90 mL of water into a small Erlenmeyer flask.
2. Add the $BaCl_2$ and mix well.
3. Remeasure and add water to bring the total volume up to 100 mL.
4. Add the $H_2SO_4$ to approximately 90 mL of water.
5. Remeasure and add water to bring the total volume up to 100 mL.
6. Add 0.5 mL of the $BaCl_2$ solution to 99.5 mL of $H_2SO_4$ and mix well.
7. While keeping the solution well mixed (the barium sulfate will precipitate and settle out) distribute 7 mL to 10 mL volumes into very clean screw-cap test tubes.

## 10× Phosphate Buffered Saline (PBS) Stock Solution

- $Na_2HPO_4$, anhydrous, reagent grade 12.36 g
- $NaH_2PO_4 \times H_2O$, reagent grade 1.80 g
- NaCl, reagent grade 85.00 g

Dissolve ingredients in distilled water to a final volume of 1.0 L.

## 1× PBS Working solution (0.01 M phosphate, pH 7.6)

- 10× PBS Stock solution 100 mL
- $dH_2O$ 900 mL

## 10× Tris-Acetate-EDTA (TAE) buffer

- Trisma base 48.4 g
- Glacial acetic acid 11.42 mL
- 0.5 M EDTA, pH 8.0 20.0 mL
- Deionized water 1.0 L

## 1× Tris-Acetate-EDTA (TAE) buffer

- 10× TAE buffer 100 mL
- Deionized water 900 mL

## 0.1× Tris-Acetate-EDTA (TAE) buffer

- 1× TAE buffer 100 mL
- Deionized water 900 mL

E

## STAINS

### Acid-Fast Stain, Modified Kinyoun Procedure (Ex. 3-7, 3-12, and 9-3)

#### Carbolfuchsin (Primary Stain)

- Basic fuchsin 1.5 g
- Phenol 4.5 g
- Ethanol (95%) 5.0 mL
- Isopropanol 20.0 mL
- Distilled or deionized water 75.0 mL

1. Dissolve the basic fuchsin in the ethanol and add the isopropanol.
2. Mix the phenol in the water.
3. Mix the solutions together and let stand for several days.
4. Filter before use.

#### Decolorizer

- $H_2SO_4$ 1.0 mL
- Ethanol (95%) 70.0 mL
- Distilled or deionized water 29.0 mL

#### Brilliant Green (Counterstain)

- Brilliant green dye 1.0 g
- Sodium azide 0.01 g
- Distilled or deionized water 100.0 mL

### Acid-Fast Stain, Ziehl-Neelson Procedure (Ex. 3-7, 3-12, and 9-3)

#### Carbolfuchsin (Primary Stain)

- Basic fuchsin 0.3 g
- Ethanol 10.0 mL
- Distilled or deionized water 95.0 mL
- Phenol 5.0 mL

1. Dissolve the basic fuchsin in the ethanol.
2. Dissolve the phenol in the water.
3. Combine the solutions and let stand for a few days.
4. Filter before use.

#### Decolorizer

- Ethanol 97.0 mL
- HCl (concentrated) 3.0 mL

E

#### Methylene Blue (Counterstain)

- Methylene blue chloride 0.3 g
- Distilled or deionized water 100.0 mL

### Basic Fuchsin Stain (0.5%)

- Basic fuchsin 0.5 g
- $dH_2O$ 99.5 mL

### Capsule Stain Procedure (Ex. 3-8 and 3-12)

#### Congo Red Stain (Primary Stain)

- Congo red dye 5.0 g
- Distilled or deionized water 100.0 mL

#### Maneval's Stain (Counterstain)

- Phenol (5% aqueous solution) 30.0 mL
- Acetic acid, glacial (20% aqueous solution) 10.0 mL
- Ferric chloride (30% aqueous solution) 4.0 mL
- Basic fuchsin (1% aqueous solution) 2.0 mL

### Carbolfuchsin (See Acid Fast, Kinyoun Procedure)

### Congo Red Stain (See Capsule Stain Procedure)

### Crystal Violet (See Gram Stain Procedure)

### Gram Stain Procedure (Ex. 3-6, 3-12, 9-1, 9-2, and 9-3)

### Gram Crystal Violet (Modified Hucker's) (Primary Stain)

#### Solution A

- Crystal violet dye (90%) 2.0 g
- Ethanol (95%) 20.0 mL

#### Solution B

- Ammonium oxalate 0.8 g
- Distilled or deionized water 80.0 mL

1. Combine solutions A and B. Store for 24 hours.
2. Filter before use.

#### Gram Decolorizer

- Ethanol (95%)
- Can use 75% ethanol and 25% acetone, but decolorization time must be reduced

#### Gram Iodine (Mordant)

- Potassium iodide 2.0 g
- Iodine crystals 1.0 g
- Distilled or deionized water 300.0 mL

1. Dissolve the potassium iodide in the water *first.*
2. Dissolve the iodine crystals in the solution.
3. Store in an amber bottle.

#### Gram Safranin (Counterstain)

- Safranin O 0.25 g
- Ethanol (95%) 10.0 mL
- Distilled or deionized water 100.0 mL

1. Dissolve the safranin O in the ethanol.
2. Add the water.

### Methylene Blue Stain (See Acid Fast, Z-N Procedure)

## Nigrosin Stain (Ex. 3-5)

- Nigrosin 10.0 g
- Distilled or deionized water 100.0 mL

## Safranin Stain (See Gram Stain)

## Spore Stain Procedure

### Malachite Green (Primary Stain)

- Malachite green dye 5.0 g
- Distilled or deionized water 100.0 mL

## References


Baron, Ellen Jo, Lance R. Peterson, and Sydney M. Finegold. *Bailey & Scott's Diagnostic Microbiology,* 9th ed. St. Louis, MO: Mosby Year Book, 1994.

Eisenstadt, Bruce, C. Carlton, and Barbara J. Brown. *Methods for General and Molecular Bacteriology*. Philipp Gerhardt, R. G. E. Murray, Willis A. Wood, and Noel R. Krieg, eds. Washington, DC: American Society for Microbiology, 1994.

Farmer, J. J. III. Chap. 41 in *Manual of Clinical Microbiology,* 8th ed. Patrick R. Murray, Ellen Jo Baron, James H. Jorgensen, Michael A. Pfaller, and Robert H. Yolken, eds. Washington, DC: ASM Press, 2003.

Forbes, Betty A., Daniel F. Sahm, and Alice S. Weissfeld. Chap. 19 in *Bailey & Scott's Diagnostic Microbiology,* 11th ed. St. Louis, MO: Mosby, 2002.

Koneman, Elmer W., et al. *Color Atlas and Textbook of Diagnostic Microbiology,* 5th ed. Philadelphia: Lippincott-Raven Publishers, 1997.

MacFaddin, Jean F. *Biochemical Tests for Identification of Medical Bacteria*, 2nd ed. Baltimore: Lippincott Williams & Wilkins, 1980.

Zimbro, Mary Jo and David A. Power, eds. *Difco*™ and BBL™ Manual—Manual of Microbiological Culture Media. Sparks, MD: Becton Dickinson and Co., 2003.

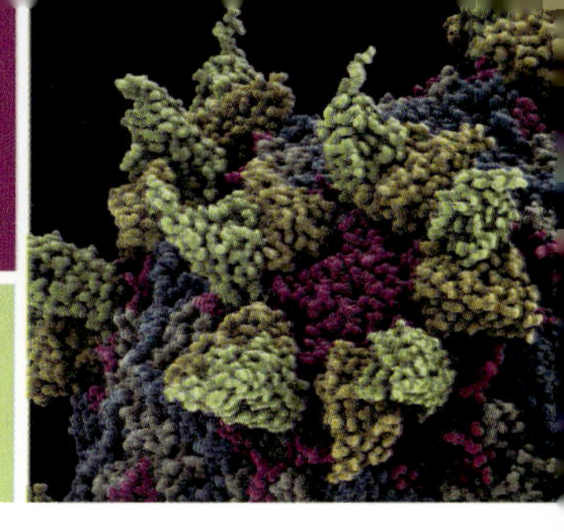

# GLOSSARY

## A

**absorbance** A measurement, using a spectrophotometer, of how much light entering a substance is *not* transmitted.

**acetoin** A four-carbon intermediate in the conversion of pyruvic acid to 2,3-Butanediol; the compound detected in the Voges-Proskauer test.

**acetyl-CoA** Compound that enters the Krebs cycle by combining with oxaloacetate to form citrate; may be produced from pyruvate in carbohydrate metabolism or from oxidation of fatty acids.

**acid clot** Pink clot formed in litmus milk from the precipitation of casein under acidic conditions; an indication of lactose fermentation.

**acidic stain** Staining solution with a negatively charged chromophore.

**acidophil** *See* eosinophil.

**acidophile** Microorganism adapted to a habitat below pH 5.5.

**acid reaction** Pink color reaction in litmus milk from acidic conditions produced by lactose fermentation.

**adenosine triphosphate (ATP)** In hydrolysis of ATP to adenosine diphosphate (ADP), supplies the energy necessary to perform most work in the cell.

**aerobe** Microorganism that requires oxygen for growth.

**aerobic respiration** A type of energy releasing metabolism in which oxygen is the final electron acceptor in the pathway. It also has the highest potential ATP yield of all catabolic and exergonic reactions.

**aerosol** Droplet nuclei that remain suspended in the air for a long time.

**aerotolerance** Designation of an organism based on its ability to grow in the presence of oxygen.

**aerotolerant anaerobe** Anaerobe that grows in the presence of oxygen but does not use it metabolically.

**agar overlay** Used in plaque assays, the soft agar containing a mixture of bacteriophage and host poured over nutrient agar as an inoculum.

**agglutination** Visible clumping produced when antibodies and particulate antigens react.

**agranulocyte** Category of white blood cells characterized by the absence of prominent cytoplasmic granules; includes monocytes and lymphocytes.

**akinete** A thick-walled resting spore produced by some cyanobacteria. It is derived from a vegetative cell.

**alkaliphile** Microorganism adapted to a habitat above pH 8.5.

**alpha ($\alpha$) hemolysis** The greening around a colony on blood agar as a result of partial destruction of red blood cells; typical of certain members of *Streptococcus*.

**ammonification** The metabolic production of ammonia. *See* oxidative deamination.

**amphitrichous** Describes a flagellar arrangement with flagella at both ends of an elongated cell.

**amylase** Family of enzymes that hydrolyze starch.

**anaerobe** Microorganism that cannot tolerate oxygen.

**anaerobic respiration** A type of energy releasing metabolism in which an inorganic compound other than oxygen is the final electron acceptor. These include $NO_3$, $NO_2$, $SO_4$, and $CO_2$. The ATP yield is higher than fermentation, but not as high as aerobic respiration.

**annealing** The attachment of a primer to a complimentary sequence of DNA. Annealing is done on each strand of the denatured DNA in positions flanking the desired sequence (template) to be replicated. Annealing is the second of three phases in a PCR cycle, following denaturation and prior to extension.

**antibiotic** Refers to an antimicrobial substance produced by a microorganism such as a bacterium or fungus.

**antibody** A glycoprotein produced by plasma cells, in response to an antigen, that reacts with the antigen specifically.

**antigen** A molecule of high molecular weight and complex three-dimensional shape that stimulates the production of antibodies and reacts specifically with them.

**antimicrobic(al)** Describes any substance that kills microorganisms, natural or synthetic.

**antiporter** A system that simultaneously transports substances in opposite directions across the cytoplasmic membrane.

**antiserum** General term applied to a serum that contains antibodies of a specific type.

**Ascomycetes** A division of fungi that produces sexual spores in a saclike structure called an ascus.

**aseptic** Describes the condition of being without contamination.

**assimilation (nitrogen)** The incorporation of $NH_4$ or $NO_3$ into organic form (amino acids) by plants.

**assimilatory sulfate reduction** The conversion of sulfate ($SO_4^{2-}$) to a sulfhydryl group (–SH) in the biosynthesis of amino acids cysteine and methionine.

**autoinducer** Substance secreted by bioluminescent bacteria that, upon reaching sufficient concentration, triggers the bioluminescent reaction; believed to help conserve energy by synchronizing individual cellular reactions.

**autotroph** An organism that is able to make all of its organic molecules using $CO_2$ as the source of carbon. *See also* heterotroph.

**auxotroph** A nutritional mutant; a cell incapable of synthesizing the nutrient that is acting as the marker in a specific genetic experiment; will grow only on a complete medium. *See also* prototroph.

## B

**bacillus** Rod-shaped cell.

**back mutation** Mutation in a gene that reverses the effect of an original mutation.

**bacteriophage** Virus that attacks bacteria; usually host-specific.

**basic stain** Staining solution with a positively charged chromophore.

**Basidiomycetes** Division of fungi characterized by club-like appendages (basidia) that produce haploid spores; include common mushrooms and rusts.

**basophil** Category of white blood cells; one of three types of granulocyte (along with eosinophils and neutrophils) characterized by a cytoplasm with dark-staining granules and a lobed nucleus (often difficult to see because of the dark cytoplasmic granules).

**β-hemolysis** The clearing around a colony on blood agar resulting from complete destruction of red blood cells; typical of certain members of *Streptococcus*.

**β-lactamase** Enzyme found in some bacteria that breaks a bond in the β-lactam ring of β-lactam antibiotics, rendering the antibiotic ineffective; the source of resistance against these antibiotics.

**β-lactam antibiotic** A group of structurally related antimicrobial chemicals that interfere with cross-linking of peptidoglycan subunits, which results in a defective wall structure and cell lysis; examples are penicillins and cephalosporins.

**biofilm** A layer of bacterial cells adhering to and reproducing on a surface; used commercially for acetic acid production and water purification.

**bioluminescence** Process by which a living organism emits light.

**biosafety level (BSL)** One of four sets of minimum standards for laboratory practices, facilities, and equipment to be used when handling organisms at each level; BSL-1 requires the least care, BSL-4 the most.

**bound coagulase** An enzyme (also called clumping factor) bound to the bacterial cell wall, responsible for causing the precipitation of fibrinogen and coagulation of bacterial cells.

## C

**capnophile** Microaerophile that requires elevated $CO_2$ levels.

**capsule** Insoluble, mucoid, extracellular material surrounding some bacteria.

**carbohydrate** One of four families of biochemicals; characterized by containing carbon, hydrogen, and oxygen in the ratio 1:2:1.

**carcinogen** Any substance that causes cancer.

**cardinal temperatures** Minimum, optimum, and maximum temperatures for an organism.

**casease** A family of hydrolytic enzymes that break down casein, often extracellularly.

**casein** Milk protein that gives milk its white color.

**catalase** Enzyme produced by some bacteria that catalyzes the breakdown of metabolic $H_2O_2$.

**cestode** The class of parasitic worms commonly called tapeworms; characterized by having a head (scolex) with suckers and hooks and numerous segments (proglottids) that are little more than sex organs.

**chemoheterotroph** Any organism that uses chemicals for energy and organic compounds as a carbon source. Usually the same organic compounds serve both purposes.

**chemolithotroph** Any organism that gets its energy from chemicals and its electrons from an inorganic source. No eukaryotes are chemolithotrophic, only some prokaryotes such as sulfur and ammonia oxidizing bacteria.

**chromogenic reducing agent** A substance that produces color when it gives up electrons (becomes oxidized).

**chromophore** The charged region of a dye molecule that gives it its color.

**coagulase** *See* bound coagulase; free coagulase.

**coagulase-reacting factor** Plasma component that reacts with free coagulase to trigger the clotting mechanism.

**coccus** Spherical cell.

**coenzyme** Nonprotein portion of an enzyme that aids the catalytic reaction (usually by accepting or donating electrons).

**coliform** Member of *Enterobacteriaceae* that ferments lactose with production of gas within 48 hours at 37°C.

**colony** Visible mass of cells produced on culture media from a single cell or single CFU (cell forming unit); a pure culture.

**colony forming unit** Term used to define the cell or group of cells (e.g., staphylococci, streptococci) that produces a colony when transferred to plated media.

**commensal** Describes a synergistic relationship between two organisms in which one benefits from the relationship and the other is affected neither negatively nor positively.

**common source epidemic** An epidemic in which the disease is transmitted from a single source (such as water supply) and is not transmitted person-to-person.

**compatible solutes** Compounds such as amino acids that function both as metabolic constituents and solutes necessary to maintain osmotic balance between the internal and external environments.

**competent cell** Cell capable of picking up DNA from the environment; some cells are naturally competent, in other cases the cells are made to be competent artificially.

**competitive inhibition** Process whereby a substance attaches to an enzyme's active site, thereby preventing attachment of the normal substrate.

**complete medium** A medium used in genetic experiments that supplies all nutrients for growth required by prototrophs and auxotrophs.

**complex medium** A medium in which at least one nutritional ingredient is of unknown composition or amount. Media with any plant or animal extract are always complex.

**confirmatory test** One last test after identification of an unknown, to further support the identification.

**conjugate acid** A compound (one of a conjugate pair) that donates a proton in solution.

**conjugate base** A compound (one of a conjugate pair) that accepts a proton in solution.

**conjugate pair** A compound that alternates between acid and base forms by losing or gaining one proton. Example: $HCl/Cl^-$.

**constitutive** Term used to describe an enzyme that is produced continuously, as opposed to an enzyme produced only when its substrate is present. *See also* induction.

**cos site (cohesive end sites)** In Lambda phage (linear) DNA, the single stranded tails at the 5' ends of both strands are complimentary to each other and can form a circular molecule. These "sticky" ends can be a source of error in DNA fragment size determinations if not properly heated before running gel electrophoresis.

**counterstain** Stain applied after decolorization to provide contrast between cells that were decolorized and those that weren't.

**culture** A liquid or solid medium with microorganisms growing in or on it.

**cyst** "Resting stage" in life cycle of certain protozoans. *See also* trophozoite.

**cytochrome** A class of iron-containing enzymes in electron transport chains; characterized by a porphyrin ring structure.

**cytochrome oxidase** An electron carrier in the electron transport chain of aerobes, facultative anaerobes, and microaerophiles that makes the final transfer of electrons to oxygen.

**cytoplasm** The semifluid component of cells inside the cellular membrane in which many chemical reactions take place.

## D

**deaminase** Enzyme that catalyzes the removal of the amine group ($NH_2$) from an amino acid.

**death phase** Closed system microbial growth phase immediately following stationary phase; characterized by population decline usually resulting from nutrient deficiencies or accumulated toxins.

**decarboxylase** Enzyme that catalyzes the removal of the carboxyl group (COOH) from an amino acid.

**decimal reduction value** Amount of time at a specific temperature to reduce a microbial population by 90% (one log cycle).

**defined medium** A growth medium in which the exact amount and chemical formula (and thus identity) of each ingredient are known.

**denaturation** The first of three phases in a PCR cycle, followed by annealing and extension. The separation of the strands in double stranded DNA.

**denitrification** The bacterial process in which nitrate is reduced to various forms of nitrogen in its role as final electron acceptor of anaerobic respiration.

**deoxyribonuclease** A family of hydrolytic enzymes that depolymerize DNA into polynucleotides. The precise location of hydrolysis within the sugar-phosphate backbone depends on which DNase is acting. These are usually secreted enzymes.

**deuteromycetes** An unnatural grouping of fungi in which the sexual stages are either unknown or are not used in classification.

**differential medium** Growth medium that contains an indicator (usually color) to detect the presence or absence of a specific metabolic activity.

**differential stain** Staining procedure that allows distinction between cell types or parts of cells; often involves more than one stain, but not necessarily.

**dilution blank** A test tube containing a measured volume of sterile diluent (water, saline, or buffer) used to serially dilute a concentrated solution or broth.

**dilution factor** Proportion of original sample present in a new mixture after it has been diluted; calculated by dividing the volume of the original sample by the total volume of the new mixture. (Subsequent dilutions in a serial dilution must be multiplied by the dilution factor of the previous dilution.)

**diploid** Defines a cell that has two complete sets of genetic information; one stage in the life cycle of all eukaryotic microorganisms (which also have a haploid [one set] stage).

**direct agglutination** Serological test in which the combination of antibodies and *naturally* particulate antigens, if positive, form a visible aggregate. *See also* indirect agglutination.

**disaccharide** A sugar made of two monosaccharide subunits; e.g., the monosaccharides glucose and galactose can combine to make the disaccharide lactose.

**disk diffusion test (Kirby-Bauer Test)** This is one way to check susceptibility and resistance of a microbe to an antibiotic. A paper disk containing an antibiotic is placed on a plate inoculated to produce confluent growth of a bacterium. Susceptibility is determined by the size of the zone of inhibition of growth around the disk.

**disproportionation reaction** A redox reaction of some sulfur reducers whereby a substrate is split into two molecules, one that is more reduced and one that is more oxidized than the original compound. Example: $S_2O_3^{2-} \longrightarrow SO_4^{2-} + H_2S$.

**dissimilatory sulfate reduction** A respiratory reaction in which sulfate is reduced exclusively for the purpose of gaining energy.

**DNA polymerase** A group of enzymes that catalyze the addition of deoxyribonucleotides to the 3' end of an existing polynucleotide chain.

**DNase** *See* deoxyribonuclease.

**Durham tube** A small, inverted test tube used in some liquid media to trap gas bubbles and indicate gas production.

## E

**electromagnetic energy** The energy that exists in the form of waves (including x-rays, ultraviolet, visible light, and radio waves).

**electron donor** A compound that can be oxidized to transfer electrons to another compound, which in turn becomes reduced.

**electron transport chain** A series of membrane-bound electron carriers that participate in oxidation-reduction reactions whereby electrons are transferred from one carrier to another until given to the final electron acceptor (oxygen in aerobes, other inorganic substances in anaerobes) in respiration.

**endergonic** Any metabolic reaction in which the products have more potential energy than the reactants. These are usually associated with anabolic (synthesis) reactions.

**endospore** Dormant, highly resistant form of bacterium; produced only by species of *Bacillus, Clostridium*, and a few others.

**enteric bacteria** Informal name given to bacteria that occupy the intestinal tract.

***Enterobacteriaceae*** A group of Gram-negative rods that also are oxidase-negative, ferment glucose to acid, have polar flagella if motile, and usually are catalase-positive and reduce nitrate to nitrite.

**enzyme** A protein that catalyzes a metabolic reaction by interacting with the reactant(s) specifically; each metabolic reaction has its own enzyme.

**eosinophil** A category of white blood cells; one of three types of granulocyte (along with basophils and neutrophils) characterized by a cytoplasm with red (in typical stains) granules and lobed nucleus.

**epitope** That portion of an antigen that stimulates the immune system and reacts with antibodies.

**eukaryote** Type of cell with membranous organelles, including a nucleus, and having 80S ribosomes and many linear molecules of DNA.

**exergonic** Any metabolic reaction in which the reactants have more potential energy than the products. These are usually associated with catabolic (degradative) reactions.

**exoenzyme** Enzyme that operates in the external environment after being secreted from the cells.

**exponential phase** Closed-system microbial growth phase immediately following the lag phase; characterized by constant maximal growth during which population size increases logarithmically.

**extension** The third of three phases in PCR, following denaturation and annealing, whereby a polymerase (usually *Taq*I) catalyzes the replication of a DNA template.

**extracellular** Pertains to the region outside a cell.

**extreme halophile** Organism that grows best at 15% or higher salinity.

**extreme thermophile** Organism that grows best at temperatures above 80°C.

## F

**facultative anaerobe** Microorganism capable of both fermentation and respiration; grows in the presence or absence of oxygen.

**facultative thermophile** Microorganism that prefers temperatures above 40°C but will grow in lower temperatures.

**false negative** Test result that is negative when the sample is actually positive; usually a result of lack of sensitivity in the test system.

**false positive** Test result that is positive when the sample is actually negative; usually a result of lack of specificity of the test system.

**fastidious (microorganism)** Describes a microorganism with strict physiological requirements; difficult or impossible to grow unless specific conditions are provided.

**fatty acid** A long chain, organic molecule with a carboxylic acid at one end with the rest of the carbons being bonded to hydrogens.

**fermentation** Metabolic process in which an organic molecule acts as an electron donor and one or more of its organic products act as the final electron acceptor, marking the end of the metabolic sequence (differs from respiration in that an inorganic substance is not needed to act as final electron acceptor).

**final electron acceptor (FEA)** The molecule receiving electrons (it becomes reduced) at the end of a metabolic sequence of oxidation/reduction reactions.

**flavin adenine dinucleotide (FAD)** Coenzyme used in oxidation/reduction reactions that acts as an electron acceptor or donor, respectively.

**flavoprotein** A flavin-containing protein in the electron transport chain, capable of receiving and transferring electrons as well as entire hydrogen atoms; sometimes bypasses normal route of transfer and reduces oxygen directly, forming hydrogen peroxide and superoxide radicals.

**free coagulase** An enzyme produced and secreted by some microorganisms that initiates the clotting mechanism in plasma; seen as a solid mass in the coagulase tube test or clumps in the slide test.

**free energy (of formation)** The energy released by a molecule in a chemical reaction that is able to do work (e.g., produce ATP). It is equivalent to the energy required to form the molecule from its individual elements.

**free living** A term used to describe nonparasitic organisms.

## G

**gametangium** A structure that produces gametes; used in describing fungi.

**gamete** Reproductive cell that must undergo fusion with another gamete to continue the life cycle; usually are haploid. *See also* spore.

**gelatinase** Enzyme secreted by microorganisms, which catalyzes the hydrolysis of gelatin.

**generation time** The time required for a population to produce offspring (i.e., in bacteria that reproduce by binary fission, the time needed for the population to double); inverse of mean growth rate.

**germ theory** Theory holding that diseases and infections are caused by microorganisms; first hypothesized in the 16th century by Girolamo Fracastoro of Verona.

**glycolysis** Metabolic process by which a glucose molecule is split into two 3-carbon pyruvic acid molecules, producing two ATP (net) and two $NADH_2$ molecules.

**gradient bacteria** Microaerophiles with very narrow nutritional requirements living in habitats characterized by nutrients diffusing upward from sediments below and oxygen diffusing downward from air above.

**granulocyte** A category of white blood cells characterized by prominent cytoplasmic granules; includes neutrophils, eosinophils, and basophils.

**group A streptococci** α-hemolytic members of the genus *Streptococcus*, characterized by possessing the Lancefield group A antigen; belong to the species *S. pyogenes*, an important human pathogen.

**group B streptococci** β-hemolytic members of the genus *Streptococcus*, characterized by possessing the Lancefield group B antigen; belong to the species *S. agalactiae*.

**group D streptococci** Streptococci possessing the Lancefield group D antigen; include *S. bovis* and species of the genus *Enterococcus*.

## H

**halophile** Microorganism that grows best at 3% or higher salinity.

**haploid** Describes a cell that has one complete set of genetic information; all eukaryotic microorganisms have a haploid stage and a diploid (two sets) stage in their life cycle; all prokaryotes are haploid.

**hemagglutination** General term applied to any agglutination test in which clumping of red blood cells indicates a positive reaction.

**hemoglobin** Iron-containing protein of red blood cells that is responsible for binding oxygen.

**hemolysin** A class of chemicals produced by some bacteria that break down hemoglobin and produce hemolysis.

**heterocyst** This is a hollow-appearing cell found in some cyanobacteria specialized to perform nitrogen fixation.

**heterotroph** An organism that requires carbon in the form of organic molecules. *See also* autotroph.

**host** An organism that serves as a habitat for another organism such as a parasite or a commensal.

**hydrolysis** The metabolic process of splitting a molecule into two parts, adding a hydrogen ion to one part and a hydroxyl ion to the other. (One water molecule is used in the process.)

**hyperosmotic** A term used to describe extracellular solute concentration relative to the cell; a solution that contains a higher concentration of solutes than the cell, such that water tends to move down its concentration gradient and diffuse out of the cell.

**hypha** (*pl.* hyphae) A filament of fungal cells.

**hyposmotic** A term used to describe extracellular solute concentration relative to the cell; a solution that contains a lower concentration of solutes than the cell, such that water tends to move down its concentration gradient and diffuse into the cell.

## I

**IMViC** Acronym representing the four tests used in identification of *Enterobacteriaceae*: Indole, Methyl Red, Voges-Proskauer, and Citrate.

**incomplete medium** *See* minimal medium.

**incubation** The process of growing a culture by supplying it with the necessary environmental conditions.

**index case** First occurrence of an infection or disease that results in an epidemic.

**indirect agglutination** Serological test in which artificially produced particulate antigens or antibodies are used to form a visible aggregate if positive. *See also* direct agglutination.

**induction** Process by which a substrate (inducer) causes the transcription of the genes used in its digestion.

**infectious disease** A transmissible illness or infection.

**inoculum** The organisms used to start a new culture or transferred to a new place.

**inorganic molecule** A molecule that does *not* contain carbon and hydrogen.

**intracellular** Within the cell; as an intracellular enzyme catalyzing reactions inside the cell.

**iron-sulfur protein** Iron-containing electron carrier in electron transport chains. Differentiated from cytochromes by the lack of porphyrin ring structure.

**isolate** (*v.*) The process of separating individual cell types from a mixed culture; (*n.*) the group of cells resulting from isolation.

**isosmotic** Describe extracellular solute concentration relative to the cell; a solution that contains the same concentration of solutes as the cell, such that water tends to move equally into and out of the cell.

## K

**karyogamy** The process of nuclear fusion that occurs after fertilization (plasmogamy) in sexual life cycles.

**Krebs cycle** A cyclic metabolic pathway found in organisms that respire aerobically or anaerobically.

## L

**lag phase** Closed-system microbial growth phase immediately preceding the exponential phase; characterized by a period of adjustment in which no growth takes place.

**limit of resolution** The closest two points can be together for the microscope lens to make them appear separate; two points closer than the limit of resolution will blur together.

**ligate** To join together.

**lipase** A family of hydrolytic enzymes that break down fats into their component parts: glycerol and up to three fatty acids.

**lipid** A fat.

**lophotrichous** Describes a flagellar arrangement with a group of flagella at one end of an elongated cell.

**lymphocyte** A category of white blood cells; one of two types of agranulocyte (along with monocytes); characterized by a large nucleus and little visible cytoplasm; involved in specific acquired immunity as T-cells and B-cells.

**lysozyme** A naturally occurring bactericidal enzyme in saliva, tears, urine, and other body fluids; functions by breaking peptidoglycan bonds.

**lytic cycle** The viral life cycle from attachment to lysis of the host cell.

## M

**mean growth rate constant** The number of generations produced per unit time; the inverse of generation time.

**medium** A substance used for growing microbes; may be liquid (usually a broth) or solid (usually agar).

**meiosis** The process in which the nucleus of a diploid eukaryotic cell nucleus divides to make four haploid nuclei.

**mesophile** A microorganism that grows best at temperatures between 15°C and 45°C.

**methylase** An enzyme that functions to add methyl groups ($-CH_3$) to adenine or cytosine bases within the DNA recognition sequence, which is thus modified and protected from the endonuclease.

**methylation** A protective enzymatic mechanism in many organisms whereby methyl groups ($-CH_3$) attach to DNA, thus blocking attachment and destruction by restriction endonucleases.

**microaerophile** A microorganism that requires oxygen, but can't tolerate atmospheric levels of it.

**microbial growth curve** Graphic representation of microbial growth in a closed system, consisting of lag phase, exponential (log) phase, stationary phase, and death phase.

**minimal medium** A medium used in genetic experiments, supplying all nutrients for growth *except* the one required by auxotrophs, and thus supporting growth of only prototrophs.

**minimum inhibitory concentration** The lowest concentration of an antimicrobial substance required to inhibit growth of all microbial cells it contacts; on an agar plate, typically the outer edge of the zone of inhibition where the substance has diffused to the degree that it no longer inhibits growth.

**mitosis** The process in which a nucleus divides to produce two identical nuclei.

**mixed acid fermentation** Vigorous fermentation producing many acid products including lactic acid, acetic acid, succinic acid, and formic acid, and subsequently lowering the pH of the medium to pH 4.4 or below.

**mold** Informal grouping of filamentous fungi. *See also* yeast.

**monocyte** A category of white blood cells; one of two types of agranulocyte (along with lymphocytes); characterized by large size and lack of cytoplasmic granules; the blood form of macrophages.

**monotrichous** Describes a flagellar arrangement consisting of a single flagellum.

**morbidity** Epidemiological measurement of incidence of a disease; typically accompanied by "incidence rate," referring to the incidence of a disease over time.

**morphology** The shape of an organism.

**mortality** Epidemiological measurement of death caused by a disease; typically accompanied by "incidence rate," referring to the incidence of death from a disease over time.

**mutagen** A substance that causes mutation in DNA; most mutagens are carcinogens.

**mutation** Alteration in a cell's DNA.

**mutualistic** Describes a synergistic relationship between two organisms in which both benefit from the interaction.

**mycelium** A mass of fungal filaments (hyphae).

## N

**nematode** A class of roundworms; environmentally abundant in some parasitic species.

**neutrophil** Category of white blood cells; one of three types of granulocyte (along with eosinophils and basophils) characterized by a granular cytoplasm and lobed nucleus; also known as "polymorphonuclear granulocytes" or "PMNs."

**neutrophile** Microorganism adapted to a habitat between pH 5.5 and 8.5.

**nicotinamide adenine dinucleotide (NAD)** A coenzyme used in oxidation/reduction reactions that acts as an electron acceptor or donor, respectively.

**nisin** Antibiotic produced by *Lactococcus lactis*.

**nitrate** A highly oxidized form of nitrogen; $NO_3$.

**nitrate reductase** An enzyme produced by all members of *Enterobacteriaceae* (and others) that catalyzes the reduction of nitrate ($NO_3$) to nitrite ($NO_2$).

**nitrification** The chemical process in which nitrogen compounds, such as ammonia and nitrite are oxidized to form nitrate. Typically, different bacteria perform these two steps in which ammonia is first oxidized to nitrite, which is subsequently oxidized to nitrate.

**nitrite** $NO_2$, an oxidized form of nitrogen.

**nitrogen fixation** The bacterial process in which gaseous nitrogen ($N_2$) becomes reduced to $NH_4$. This is ecologically important because $N_2$, while abundant in the atmosphere, is not in a form usable by most organisms, whereas $NH_4$ is usable by plants and can thus enter food chains.

**nitrogenase** The enzyme involved in nitrogen fixation, that is, converting gaseous $N_2$ to $NH_4$. This is performed by free-living bacteria (e.g., *Azotobacter*), symbiotic bacteria (e.g., *Rhizobium*), and many cyanobacteria.

**noninfectious diseases** Conditions not caused by microorganisms; examples are stroke, heart disease, and emphysema.

## O

**objective lens** The microscope lens that first produces magnification of the specimen in a compound microscope.

**obligate (strict) aerobe** Microorganism that requires oxygen to survive and grow.

**obligate (strict) anaerobe** Microorganism for which oxygen is lethal; requires the complete absence of oxygen.

**obligate thermophile** Microorganism that grows only at temperatures above 40°C.

**ocular lens** The lens the microscopist looks through; produces the virtual image by magnifying the real image.

**ocular micrometer** A uniformly graduated linear scale placed in the microscope ocular used for measuring microscopic specimens.

**oligonucleotide** A short nucleic acid molecule.

**operon** A prokaryotic structural and functional genetic unit consisting of two or more structural genes that code for enzymes in the same pathway and that are regulated together.

**opportunistic pathogen** A microorganism not ordinarily thought of as pathogenic (i.e., most enterics) that will cause infection when out of its normal habitat.

**organic molecule** A molecule made of reduced carbon—that is, containing at least carbon and hydrogen.

**osmosis** Diffusion of water across a semipermeable membrane.

**osmotolerant** Microorganism that will grow outside of its preferred salinity range.

**oxidation/reduction** Chemical reaction in which electrons are transferred. (The molecule losing the electrons becomes oxidized; the molecule gaining the electrons becomes reduced.)

**oxidative deamination** The metabolic process in which an amino acid has its amine removed, producing ammonia and an organic acid.

**oxidative phosphorylation** The process by which an electron transport chain is used to add a phosphate to ADP to make ATP.

**oxidizing agent** A substance that removes electrons from (oxidizes) another. *See also* reducing agent.

## P

**palindrome** A word, phrase, number, or other sequence of characters that reads the same backward or forward.

**parasite** An organism that lives symbiotically with another, but to the detriment of the other organism (called a host).

**parthenogenesis** A process in which females produce offspring from an unfertilized egg.

**peptidoglycan** The insoluble, porous, cross-linked polymer comprising bacterial cell walls; generally thick in Gram-positive organisms and thin in Gram-negative organisms.

**peptone** A digest of protein used in formulating some bacteriological media.

**peritrichous** Describes a flagellar arrangement in which flagella arise from the entire surface of the cell.

**pH** The measure of a solution's alkalinity or acidity; the negative logarithm of the hydrogen ion concentration.

**phage** *See* bacteriophage.

**phage host** Bacteria attacked by a virus.

**phototaxis** Phototaxis is a response to light stimuli. Positive phototaxis is movement toward a light stimulus, whereas negative phototaxis is movement away from light.

**plaque** The clearing produced in a bacterial lawn as a result of cell lysis by a bacteriophage; used to calculate phage titer (PFU/mL) when accompanied by a serial dilution.

**plaque forming unit (PFU)** Term that replaces "viral particle" or "single virus" when referring to phage titer; accounts for multiple particle arrangements in which more than one virus is responsible for initiating a plaque.

**plasma** The noncellular (fluid) portion of blood; consists of serum (including serum proteins) and clotting proteins.

**plasmid** Small, circular, extrachromosomal piece of DNA found in prokaryotic cells; often carries genes for antibiotic resistance.

**plasmogamy** The cytoplasmic fusion of gametes at the time of fertilization. *See also* karyogamy.

**plasmolysis** Shrinking of cell membrane (pulling away from the rigid cell wall) because of loss of water to the environment and reduced turgor pressure.

**polar flagellum** A single flagellum at one end of an elongated cell.

**pour plate technique** Method of plating bacteria in which the inoculum is added to the molten agar prior to pouring the plate.

**preadsorption period** When performing a plaque assay, the time given to allow a bacteriophage to attach to the host before adding the mixture to the molten agar being plated.

**precipitate** An insoluble material that comes out of a solution during a precipitation reaction.

**presumptive identification** Tentative identification of an isolate based on one or more key test results.

**primary stain** The first stain applied in many differential staining techniques; usually subjected to a decolorization step that forms the basis for the differential stain.

**primer** A short nucleotide sequence (typically 10–20 nucleotides) used in PCR to attach to DNA for the purpose of replicating a desired sequence of nucleotides. Typically, two primers attach to sites on opposite strands, flanking the area (template) to be replicated.

**proglottid** Tapeworm segments posterior to the scolex, used for absorption of nutrients and containing reproductive organs.

**prokaryote** Type of cell lacking internal compartmentalization (membranous organelles, including a nucleus) and having 70S ribosomes and a circular molecule of DNA; more primitive than eukaryotes.

**promoter site** The patch of DNA upstream from the structural gene(s), which binds RNA polymerase to begin transcription.

**propagated transmission** Conveying a disease person-to-person.

**proteolytic** Refers to catabolism of protein; e.g., a proteolytic enzyme.

**prototroph** A strain that is capable of synthesizing the nutrient that is acting as the marker in a particular genetic experiment; prototrophs will grow on complete and minimal medium. *See also* auxotroph.

**pseudohypha** A chain of fungal cells produced by budding (rather than by cytokinesis that produces two equally sized cells) and which is characterized by constrictions at cell junctions.

**psychrophile** A microorganism that grows only at temperatures below 20°C, with an optimum around 15°C.

**psychrotroph** A microorganism that grows optimally at temperatures between 20 and 30°C but will grow at temperatures as low as 0°C and as high as 35°C.

**pure culture** Microbial culture containing only a single species.

**purine nucleotide** Purines are made of a sugar (ribose or deoxyribose), phosphate, and a nitrogen-containing base composed of two joined rings. Examples are adenine (A) and guanine (G). These are found in DNA and RNA.

**putrefaction** The process of digesting dead organic material; decay.

**pyrimidine nucleotide** Pyrimidines are made of a sugar (ribose or deoxyribose), phosphate, and a nitrogen-containing base composed of a single ring. Examples are cytosine (C) in DNA and RNA, thymine (T) in DNA, and uracil (U) in RNA.

**pyruvic acid (pyruvate)** A three-carbon compound produced at the end of glycolysis that may enter a respiration or a fermentation pathway; also serves as a starting point for synthesis of certain amino acids and an entry point for their digestion.

## Q

**quinoidal (compound)** A color-producing compound containing quinone as its central structure.

**quorum sensing** The phenomenon in bioluminescing bacteria whereby the light-emitting reaction of all cells takes place simultaneously when a threshold concentration of secreted autoinducer is reached.

## R

**real image** Magnified image of a specimen produced by the objective lens of a microscope; the real image is magnified again by the ocular lens to produce the virtual image.

**recognition site** A short DNA sequence where a restriction enzyme attaches. Each enzyme recognizes and attaches to its own specific sequence.

**reducing agent** A substance that donates electrons to (reduces) another. *See also* oxidizing agent.

**reductase** An enzyme that catalyzes the transfer of electrons from donor molecule to acceptor molecule, thereby reducing the acceptor.

**reduction** *See* oxidation/reduction.

**refraction** The bending of light as it passes from a medium with one refractive index into another medium with a different refractive index.

**reservoir** A nonhuman host or other site in nature serving as a perpetual source of pathogenic organisms.

**resolution** The clarity of an image produced by a lens; the ability of a lens to distinguish between two points in a specimen; high resolution in a microscope is desirable.

**resolving power** *See* limit of resolution.

**respiration** Metabolic process by which an organic molecule acts as an electron donor and an inorganic substance—such as oxygen, sulfur, or nitrate—acts as the final electron acceptor in an electron transport chain, marking the end of the metabolic sequence (differs from fermentation, which uses one of its own organic products as the final electron acceptor).

**restriction enzyme** Enzymes involved in cutting DNA at specific recognition sites, unique to the enzyme. Involved in removing damaged DNA and destroying foreign (e.g., phage) DNA.

**reticuloendothelial system** Combination of macrophages and associated cells located in the liver, spleen, bone marrow, and lymph nodes.

**reverse citric acid cycle** An energy consuming process used by green sulfur bacteria whereby molecules ordinarily associated with the oxidative respiratory reactions of the Krebs cycle perform reactions in the opposite direction to reduce carbon dioxide and form pyruvate.

**reverse electron flow** An energy consuming process used by purple sulfur bacteria whereby electrons transferred down the ETC of the photosystem are boosted upward against their thermodynamic gradient to reduce $NADP^+$ needed by the Calvin cycle.

**reversion** In carbohydrate fermentation tests, the phenomenon of a microorganism fermentively depleting the carbohydrate and reverting to amino acid metabolism, thereby neutralizing acid products with alkaline products; produces a false negative.

**rhizoid** A root-like structure used for attachment of some fungi to the substrate.

**RNA polymerase** A group of enzymes that catalyze the addition of ribonucleotides to the 3' end of an existing polynucleotide chain.

## S

**saltern** Salterns are low pools of saltwater. Evaporation of the water leaves salt, which can then be harvested.

**saprophyte** A heterotroph that digests dead organic matter; a decomposer.

**scolex** The "head" of a tapeworm, often with suckers and hooks for attachment.

**selective medium** Growth medium that favors growth of one group of microorganisms and inhibits or prevents growth of others.

**sensitivity** This is a measure of a test's ability to detect small amounts of the item being tested for. The better the sensitivity, the fewer the false negatives (due to smaller amounts triggering a positive reaction), and the more useful the test is.

**serial dilution** Series of dilutions used to reduce the concentration of a culture and thereby produce between 30 and 300 colonies when plated, providing a means of calculating the original concentration.

**serology** A discipline that utilizes a serum containing antibodies (antiserum) to detect the presence of antigens in a sample; also refers to identification of antibodies in a patient's serum.

**serum** Fluid portion of blood minus the clotting factors.

**soft agar** A semisolid growth medium containing a reduced concentration of agar; used in plaque assay to allow diffusion of bacteriophage while arresting movement of the bacteriophage host.

**solute** The dissolved substance in a solution.

**solution** The mixture of dissolved substance (solute) and solvent (liquid).

**solvent** The liquid portion of a solution in which solute is dissolved.

**specificity** This is a measure of a test's ability to produce a positive response only when reacting with the particular item being tested for. The better the test's specificity, the fewer the false positives (due to better discrimination), and the more useful it is.

**spirillum** Spiral-shaped cell.

**sporangium** Structure that produces spores.

**spore** In bacteria, a dormant form of a microbe protected by specialized coatings produced under conditions of, and resistant to, adverse conditions; also known as an endospore; in fungi and plants, spores are specialized reproductive cells; frequently a means of dissemination.

**spread plate technique** Method of plating bacteria in which the inoculum is transferred to an agar plate and spread with a sterile bent-glass rod or other spreading device.

**stage micrometer** A microscope ruler used to calibrate an ocular micrometer.

**standard curve** A graph constructed from data obtained using samples of known value for the independent variable; once made, can be used to experimentally determine the value of the independent variable when the dependent variable is measured on an unknown.

**stationary phase** Closed system microbial growth phase immediately following exponential phase; characterized by steady, level growth during which death rate equals reproductive rate.

**stolons** Surface hyphae of some molds (e.g., *Rhizopus*) that attach to the substrate with rhizoids.

**stormy fermentation** Vigorous fermentation produced in litmus milk by some species that produce an acid clot but subsequently break it up because of heavy gas production (members of *Clostridium*).

**streptolysin** Hemolysin (blood hemolyzing exotoxin) produced and secreted by members of *Streptococcus*.

**superoxide dismutase** Enzyme produced by some bacteria that catalyzes the conversion of superoxide radicals to hydrogen peroxide.

**syntrophy** A situation where two or more organisms, having differing metabolic capabilities, derive mutual benefit by providing and/or receiving essential nutrients otherwise not available to them.

## T

**template DNA** A particular sequence of DNA to be replicated in the polymerase chain reaction process.

**thermal death time** The amount of time required to kill a population of a specific size at a specific temperature.

**thermophile** A microorganism that grows best at temperatures above 40°C.

**titer** A measurement of concentration of a substance or particle in a solution; used in measurements of phage concentration.

**transformant cells** Cells that have undergone transformation by picking up foreign DNA in a genetic engineering experiment.

**transformation** A form of genetic recombination performed by some bacteria in which DNA is picked up from the environment and incorporated into its genome.

**trematode** A class of parasitic flatworms; also known as "flukes."

**trend line** A line drawn on a graph to show the general relationship between X and Y variables; also known as a regression line.

**triacylglycerol** *See* triglyceride.

**tricarboxylic acid cycle** *See* Krebs cycle.

**trichome** A filament of cyanobacterial cells. It does not include the sheath, if present.

**triglyceride** A molecule composed of glycerol and three long chain fatty acids.

**trophozoite** "Feeding" stage in the life cycle of certain protozoans. *See also* cyst.

**tryptophan** An amino acid.

**tryptophanase** Enzyme that catalyzes the hydrolysis of tryptophan into indole and pyruvic acid; detected in the indole test.

**turgor pressure** The pressure inside a cell that is required to maintain its shape, tonicity, and necessary biochemical functions.

**2,3-butanediol fermentation** The end-product of a metabolic pathway leading from pyruvate through acetoin with the associated oxidation of NADH; detected in the Voges-Proskauer test.

## U

**undefined medium** *See* complex medium.

**urease** Enzyme that catalyzes the hydrolysis of urea into two ammonias and one carbon dioxide.

**utilization medium** A differential medium that detects the ability or inability of an organism to metabolize a specific ingredient.

## V

**variable** A factor in a scientific experiment that is changed; the control and experimental groups in a good experiment differ in only one variable, the one being tested.

**vector** In genetic engineering, a means, often a plasmid or a virus, of introducing DNA into a new host.

**vegetative cell** An actively metabolizing cell.

**virtual image** The image produced when the ocular lens of a microscope magnifies the real image; appears within or below the microscope.

**Voges-Proskauer test** A differential test used to identify organisms that are capable of performing a 2,3-butanediol fermentation.

## W

**wavelength** Measurement of a wave from crest to crest, usually in nanometers—as in electromagnetic energy.

**whey** Watery portion of milk as seen upon coagulation of casein in the production of a curd.

## X

**X-Y scatter plot** Graph presenting the relationship between two variables.

## Y

**yeast** An informal grouping of unicellular fungi. *See also* mold.

## Z

**zone of inhibition** On an agar plate, the area of nongrowth surrounding a paper disk containing an antimicrobial substance. (The zone typically ends at the point where the diffusing antimicrobial substance has reached its minimum inhibitory concentration, beyond which it is ineffective.)

**zygospore** The product of fertilization and the site of meiosis in some molds.

**zygote** The product of gamete fusion (plasmogamy) and nuclear fusion (karyogamy); a fertilized egg.

# INDEX